Titanium Powder Metallurgy

Science, Technology and Applications

Titanium Powder Metallurgy

Science, Technology and Applications

Ma Qian

RMIT University, School of Aerospace, Mechanical and Manufacturing Engineering, Centre for Additive Manufacture, Melbourne, Victoria, Australia

Francis H. (Sam) Froes

Consultant to the Titanium Industry, Tacoma, WA, USA

ELSEVIER

AMSTERDAM • BOSTON • HEIDELBERG • LONDON
NEW YORK • OXFORD • PARIS • SAN DIEGO
SAN FRANCISCO • SINGAPORE • SYDNEY • TOKYO
Butterworth-Heinemann is an imprint of Elsevier

B H

Butterworth-Heinemann is an imprint of Elsevier
225 Wyman Street, Waltham, MA 02451, USA
The Boulevard, Langford Lane, Kidlington, Oxford OX5 1GB, UK

Notices

Knowledge and best practice in this field are constantly changing. As new research and experience broaden our understanding, changes in research methods, professional practices, or medical treatment may become necessary.

Practitioners and researchers must always rely on their own experience and knowledge in evaluating and using any information, methods, compounds, or experiments described herein. In using such information or methods they should be mindful of their own safety and the safety of others, including parties for whom they have a professional responsibility.

To the fullest extent of the law, neither the Publisher nor the authors, contributors, or editors, assume any liability for any injury and/or damage to persons or property as a matter of products liability, negligence or otherwise, or from any use or operation of any methods, products, instructions, or ideas contained in the material herein.

Library of Congress Cataloging-in-Publication Data

Qian, Ma, author.
Titanium powder metallurgy / by Ma Qian, Francis H. Froes. – First edition.
pages cm
Includes bibliographical references and index.
ISBN 978-0-12-800054-0
1. Powder metallurgy. 2. Titanium alloys. I. Froes, F. H., author. II. Title.
TN697.T5Q27 2015
669'.7322–dc23

2014034817

British Library Cataloguing-in-Publication Data

A catalogue record for this book is available from the British Library

For information on all Butterworth-Heinemann publications visit our website at http://store.elsevier.com/

This book has been manufactured using Print On Demand technology. Each copy is produced to order and is limited to black ink. The online version of this book will show color figures where appropriate.

ISBN: 978-0-12-800054-0

Contents

List of contributors

Stanley Abkowitz Dynamet Technology, Inc. (now RTI Advanced Powder Materials a unit of RTI International)

Susan Abkowitz Dynamet Technology, Inc. (now RTI Advanced Powder Materials a unit of RTI International)

Kamal Akhtar Director of Technology and Quality, Cristal Metals Inc., Lockport, IL, USA

Kerem Araci Process Development Engineer, Technology and Quality, Cristal Metals Inc., Lockport, IL, USA

Daniel P. Barbis AMETEK Specialty Metal Powders

Christopher C. Berndt Industrial Research Institute Swinburne, Faculty of Science, Engineering and Technology, Swinburne University of Technology, Hawthorn, Australia; Department of Materials Science and Engineering, Stony Brook University, NY, USA

R.R. Boyer Retired Boeing Technical Fellow, Seattle, WA, USA

G.M.D. Cantin CSIRO Process Science and Engineering, Clayton South MDC, Victoria, Australia

Joseph A. Capone AMETEK Specialty Metal Powders

Richard R. Chromik Department of Mining and Materials Engineering, McGill University, Montréal, Québec, Canada

L.P. Clark Retired Boeing, Phoenix, AZ, USA

Melchiorre Conti Metalysis Limited, Wath-upon-Dearne, Rotherham, United Kingdom

James Deane Metalysis Limited, Wath-upon-Dearne, Rotherham, United Kingdom

Greg Doughty Metalysis Limited, Wath-upon-Dearne, Rotherham, United Kingdom

B. Dutta DM3D Technology, Auburn Hills, MI, USA

Z. Zak Fang Department of Metallurgical Engineering, University of Utah, Salt Lake City, UT, USA

Thomas Ebel Helmholtz-Zentrum Geesthacht, Zentrum für Material- und Küstenforschung GmbH, Geesthacht, Germany

Sami M. El-Soudani Associate Technical Fellow, The Boeing Company, Huntington Beach, CA, USA

Harvey Fisher Dynamet Technology, Inc. (now RTI Advanced Powder Materials a unit of RTI International)

V. Friederici Fraunhofer-Institut für Fertigungstechnik und Angewandte Materialforschung IFAM, Bremen, Germany

Francis H. (Sam) Froes Consultant to the Titanium Industry, Tacoma, WA, USA

Jo Ann Gan Industrial Research Institute Swinburne, Faculty of Science, Engineering and Technology, Swinburne University of Technology, Hawthorn, Australia; Research Services, La Trobe University, Melbourne, Australia

Robert M. Gasior AMETEK Specialty Metal Powders

Randall M. German Professor, Mechanical Engineering, San Diego State University, San Diego, CA, USA

M.A. Gibson CSIRO Process Science and Engineering, Clayton South MDC, Victoria, Australia

Dina Goldbaum Department of Mining and Materials Engineering, McGill University, Montréal, Québec, Canada

Lucy Grainger Metalysis Limited, Wath-upon-Dearne, Rotherham, United Kingdom

T. Hartwig Fraunhofer-Institut für Fertigungstechnik und Angewandte Materialforschung IFAM, Bremen, Germany

M. Ashraf Imam Materials Science and Technology Division, Naval Research Laboratory, Washington DC, USA

P. Imgrund Fraunhofer-Institut für Fertigungstechnik und Angewandte Materialforschung IFAM, Bremen, Germany

Eric Irissou National Research Council Canada, Boucherville, Québec, Canada

Orest Ivasishin Institute for Metal Physics, Kiev, Ukraine

Mingtu Jia Waikato Centre for Advanced Materials, School of Engineering, The University of Waikato, Hamilton, New Zealand

Katsuyoshi Kondoh Osaka University, Joining and Welding Research Institute (JWRI), Ibaragi, Osaka, Japan

Jean-Gabriel Legoux National Research Council Canada, Boucherville, Québec, Canada

Bin Liu State Key Lab of Powder Metallurgy, Central South University, Changsha, P.R. China

Yong Liu State Key Lab of Powder Metallurgy, Central South University, Changsha, P.R. China

Shudong D. Luo The University of Queensland, School of Mechanical and Mining Engineering, Brisbane, Australia

Damien Mangabhai Quality Superintendent, Technology and Quality, Cristal Metals Inc., Ottawa, IL, USA

Ian Mellor Metalysis Limited, Wath-upon-Dearne, Rotherham, United Kingdom

Vladimir Moxson ADMA Products, Inc., Hudson, OH, USA

James D. Paramore Department of Metallurgical Engineering, University of Utah, Salt Lake City, UT, USA

Ma Qian RMIT University, School of Aerospace, Mechanical and Manufacturing Engineering, Centre for Additive Manufacture, Melbourne, Victoria, Australia

Kartik Rao Metalysis Limited, Wath-upon-Dearne, Rotherham, United Kingdom

V. Samarov President, LNT PM Inc. (Laboratory of New Technologies)

Teddi S. Schaeffer AMETEK Specialty Metal Powders

D. Seliverstov President, LNT PM Inc. (Laboratory of New Technologies)

Pei Sun Department of Metallurgical Engineering, University of Utah, Salt Lake City, UT, USA

H.P. Tang State Key Laboratory of Porous Metal Materials, Northwest Institute for Nonferrous Metal Research, Xi'an, China

David S. van Vuuren The CSIR, PO Box 395, Pretoria, South Africa

Dion Vaughan Metalysis Limited, Wath-upon-Dearne, Rotherham, United Kingdom

Phuong Vo National Research Council Canada, Boucherville, Québec, Canada

Graham P. Walker AMETEK Specialty Metal Powders

J. Wang State Key Laboratory of Porous Metal Materials, Northwest Institute for Nonferrous Metal Research, Xi'an, China

David Whittaker DW Associates 231, Coalway Road, Merryhill Wolverhampton, United Kingdom

J.C. Williams Professor Emeritus, The Ohio State University, Columbus, OH, USA

James C. Withers Materials & Electrochemical Research (MER) Corporation, Tucson, AZ, USA

Wilson Wong Department of Mining and Materials Engineering, McGill University, Montréal, Québec, Canada

X. Wu ARC Centre for Design in Light Metals, Monash University, Melbourne, Australia

Ming Yan RMIT University, School of Aerospace, Mechanical and Manufacturing Engineering, Centre for Additive Manufacture, Melbourne, Victoria, Australia

Ya F. Yang RMIT University, School of Aerospace, Mechanical and Manufacturing Engineering, Centre for Additive Manufacturing, Melbourne, Victoria, Australia

C.F. Yolton CTO, Summit Materials LLC, McDonald, PA, USA

Stephen Yue Department of Mining and Materials Engineering, McGill University, Montréal, Québec, Canada

Deliang Zhang State Key Laboratory of Metal Matrix Composites, School of Materials Science and Engineering, Shanghai Jiao Tong University, Shanghai, China

About the editors

Dr. Ma Qian is Professor and Deputy Director of the Centre for Additive Manufacturing of RMIT University (Royal Melbourne Institute of Technology), Australia. He received his BSc (1984), MSc (1987), and PhD (1991) all from the University of Science and Technology Beijing. He then worked as a postdoctoral research fellow and lecturer with Tsinghua University Beijing from 1991 to 1994. Before joining RMIT University in 2013, he was a Reader in Materials Engineering (2008–2013) and leader of the Powder Metallurgy Group at the University of Queensland, Australia. Prior to that, he worked as a researcher or academic with several other institutions in Japan, Singapore, Australia, and the United Kingdom. He has more than 200 peer-reviewed publications, with about half focused on titanium powder metallurgy (PM) and additive manufacturing of titanium alloys. Recent publications include understanding the effect of oxygen on the ductility of as-sintered Ti-6Al-4V (*Acta Mater.* 68 (2014) 196–206), Additive manufacturing of strong and ductile Ti-6Al-4V (*Acta Mater.* 85 (2015) 74–84) and understanding the impacts of trace carbon on the microstructure of as-sintered biomedical Ti-15Mo (*Acta Biomater.* 10 (2014) 1014–1023). He initiated the Titanium PM conference in 2011, cosponsored by Materials Australia, the Minerals, Metals & Materials Society (TMS), Japan Society of Powder and Powder Metallurgy (JSPM), Titanium Industry Development Association (TiDA), and the PM Branch of China Society for Metals (CSM). He was leader organizer for the TMS Symposium Novel Synthesis and Consolidation of Powder Materials (2013 and 2015). Currently he serves as a board member of the Asian Powder Metallurgy Association (APMA) and an editorial/review/advisory board member of a number of journals including *Metallurgical and Materials Transactions A*, *JOM*, *Powder Metallurgy*, *International Journal of Powder Metallurgy*, *Korean Journal of Powder Metallurgy*, *Acta Metallurgica Sinica (English Letters)*, and *Powder Metallurgy Technology*. He is also advisory editor to Elsevier on powder materials science and engineering.

Dr. Francis H. (Sam) Froes has been involved in the Titanium field with emphasis on powder metallurgy (PM) for more than 40 years. After receiving a BSc (Liverpool University) and an MSc and PhD (from Sheffield University) he was employed by a primary titanium producer, Crucible Steel Company, where he was leader of the titanium group and led a major effort on PM titanium under US Air Force (USAF) funding. He then spent time at the USAF Materials Lab, where he was supervisor of the Light Metals group (which included titanium) and again involved an emphasis on PM. While at the USAF Laboratory, he coorganized a landmark TMS-sponsored Conference on Titanium PM (1980) and presented the keynote speech at the first International Titanium Association Conference in 1984. This was followed by 17 years at the University of Idaho, where he was a director and department head of the

Materials Science and Engineering Department, again leading a number of programs on titanium PM. During this tenure, he was the Chairman of the World Titanium Conference held in San Diego in 1992. He has more than 800 publications, in excess of 60 patents, and has edited almost 30 books – the majority on various aspects of titanium. Recent publications include a comprehensive review of titanium PM and an article on titanium additive manufacturing. Since the early 1980s, he has taught the ASM International course on "Titanium and Its Alloys." He has organized more than 10 symposia on various aspects of Titanium Science and Technology, including in recent years cosponsored four TMS Symposia on Cost Effective Titanium (which included a large number of papers on titanium PM). He is a Fellow of ASM and a member of the Russian Academy of Science, and he was awarded the Service to Powder Metallurgy by the Metal Powder Association.

Preface

The first known research effort of titanium powder metallurgy (PM) was made by Dr Kroll (W. Kroll, Verformbare Legierungen des Titans, *Z Metallkunde,* 29 (1937) 189–192). In the work published, Dr Kroll compacted and sintered 14 different binary titanium alloys (Mo, W, Ni, Fe, Co, Be, Si, Mn, Cr, Cu, Al, Zr, V, and Ta with one addition for each element in the range of 2–9 wt%) in argon soon after he was able to produce about 0.5 kg batches of sponge fines in 1937. Thanks to the persistent development of the Kroll process since 1937 (Dr Kroll switched to the magnesium approach from the calcium in late 1937), titanium as a metal of industrial stature was established in 1948 in the United States, marked by world's first high-quality sponge production by DuPont (3 metric tons of >99% pure sponge in 1948) and the 1st Titanium Symposium in Washington, DC, held also in 1948. Although more than seven decades have passed, the high cost of titanium components still limits its usage compared to the lower-cost structural material options such as steel and aluminum alloys. A major proportion of this high cost is associated with the machining of wrought products to final configurations, suggesting that fabrication of near-net shaped titanium products could lead to dramatically increased use. In that regard, PM techniques remain to be an attractive solution to the production of cost-effective near-net shaped titanium components. In addition, it offers the potential for rapid turnaround prototype parts, manufacture of complex parts without having to make dies or molds, and a method for repairing mismachined parts or worn parts. Another important advantage is that Powder Metallurgy (PM) offers the potential of producing alloys that could not be produced via ingot metallurgy due to segregation problems.

The purpose of this book is to review the developments of titanium PM technologies to date. The subjects covered include titanium powder production methods, including both well-established and developing potential lower-cost approaches, and various near-net-shape-forming PM techniques, including the blended elemental approach, the prealloyed plus hot isostatic pressing method, additive manufacturing, metal injection molding, and spray forming. The last two chapters discuss the current and future markets for Ti PM and the editors' perspectives on the future of Ti PM. In total, the field has been covered in a comprehensive manner and we hope that this effort could help in bringing titanium components made from powder into widespread use in the future.

Ma Qian
RMIT University, School of Aerospace, Mechanical and Manufacturing Engineering, Centre for Additive Manufacture, Melbourne, Victoria, Australia

Francis H. (Sam) Froes
Consultant to the Titanium Industry, Tacoma, WA, USA

A historical perspective of titanium powder metallurgy

1

Francis H. (Sam) Froes
Consultant to the Titanium Industry, Tacoma, WA, USA

1.1 Introduction

Powder metallurgy (PM) is the production, processing, and consolidation of fine particles to make a solid metal. A primary advantage of the PM approach over other methods lies in the more efficient use of material. Other advantages include greater shape flexibility and reduced processing steps.

For many years a large, well-established PM industry involving materials such as iron, copper, and nickel-based alloys has been in existence. The main driver here is cost reduction; mechanical properties play only a secondary role. During the 1980s, PM of titanium alloys was established. Cost and material savings have been the major goal, with two PM approaches, the prealloyed (PA) technique and the blended elemental (BE) method, mainly on Ti-6Al-4V, the workhorse alloy of the titanium industry.

The PA approach is typically designed to produce demanding aerospace components in which mechanical property levels (particularly fatigue behavior) need to be equivalent to those of cast and wrought ingot metallurgy. These requirements have been met with the establishment of clean powder production and handling procedures.

In contrast, titanium compacts produced by the BE method have generally not achieved the fatigue levels required for critical aerospace components because of inherent salt and porosity. These products are aimed at applications not requiring high dynamic (fatigue) properties. In the late 1980s, low-chloride start-stock became available that led to improved density and enhanced fatigue behavior. Production costs for the BE technique are lower than those for the PA technique.

This chapter reviews the history of titanium PM from the early days of the titanium industry (late 1940s) to the present day (late 2013). The early unsuccessful attempts at producing a solid article was followed by a number of years before the technique was commercialized. A landmark TMS Conference was held in 1980 followed by a gap of almost 20 years before commercialization of titanium PM techniques began to occur. At the present time, a number of approaches are poised for widespread commercialization, including metal injection molding (MIM, for structural and nonstructural components), BE (a high point being the qualification of parts for Boeing Commercial aircraft), PA/hot isostatic press (HIP), and additive manufactured technologies.

Titanium Powder Metallurgy. http://dx.doi.org/10.1016/B978-0-12-800054-0.00001-0

1.2 The early years (late 1940s to early 1950s)

The conversion of titanium sponge to a solid article presented very difficult engineering challenges during the early development of the titanium industry in the late 1940s. There were no facilities available at that time for melting the reactive metals without severe contamination. Liquid titanium reacts rapidly with, or dissolves, all solids, liquids, and gases except the inert gases, including argon and helium.

The first serious efforts to consolidate titanium were made by William J. Kroll, the inventor of the magnesium-reduction process bearing his name. Two basic methods were studied by Kroll, the US Bureau of Mines, and private industries. One method employed PM techniques for cold compaction and sintering. These steps were followed, in some cases, by hot or cold working to produce small amounts of metal products suitable for inspection and testing. Such products, however, were found to have serious shortcomings that could not be corrected at that time despite major efforts. Residual magnesium chloride ($MgCl_2$) salts entrapped within the metal particles impaired the mechanical properties and weldability. High hydrogen levels also lowered property values.

The other method explored by Kroll was an electric-arc furnace invented by Werner von Bolton for melting tantalum. This small furnace had three particularly important features: a water-cooled copper hearth, a consumable electrode, and a vacuum system. Kroll modified the furnace to use a nonconsumable thoriated-tungsten electrode. He also replaced the hearth with a small, water-cooled crucible. These initial efforts, with numerous improvements by others, eventually led to the consumable-electrode, vacuum-arc-remelt electric furnaces. The PM approach was discarded at this time.

1.3 The 1980 TMS Conference

The landmark 1980 TMS Conference organized by Froes and Smugeresky [1] provided a well-defined state of the art of the work conducted to that date on titanium PM (the proceedings were so popular that they were even translated and published in the Soviet Union). The conference covered powder production (both the rotating electrode process [REP], and the hydride–dehydride technique, the former using a tungsten electrode), properties of HIP PA powders, BE powder cold pressed and sintered compacts, welding of REP compacts, and the PA/HIP Ti-6Al-4V approach for the production of large and complex components for demanding applications.

At the conference, the capability of the BE technique was demonstrated for very complex shapes, such as the impeller shown in Figure 1.1 produced by cold isostatic pressing (CIP) using elastomeric molds. However, part size was limited to 600 mm (2 ft) diameter because of the lack of availability of CIP equipment. Dimensional tolerance was about ±0.5 mm (±0.02 in.). The press consolidation technique was capable of producing parts up to 30 m (170 ft.), but the shape-making capability was probably not as good as in CIP. The connector link arm shown in Figure 1.2 is an example of a part made by the press consolidation method.

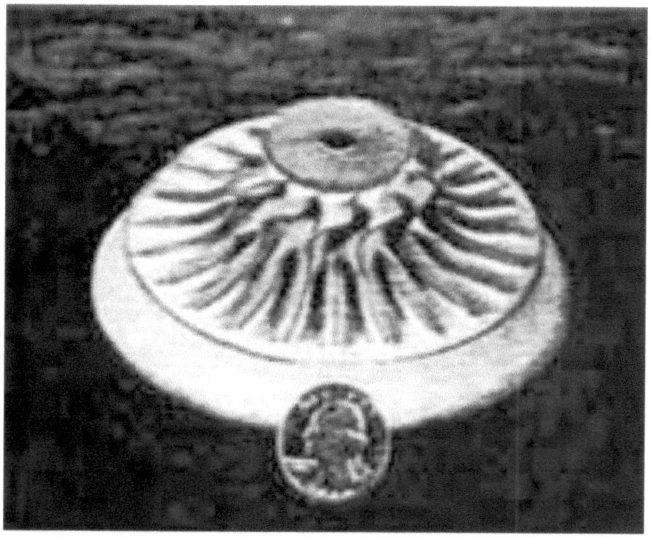

Figure 1.1 CIP-blended elemental Ti-6Al-4V impeller produced using an elastomeric mold.

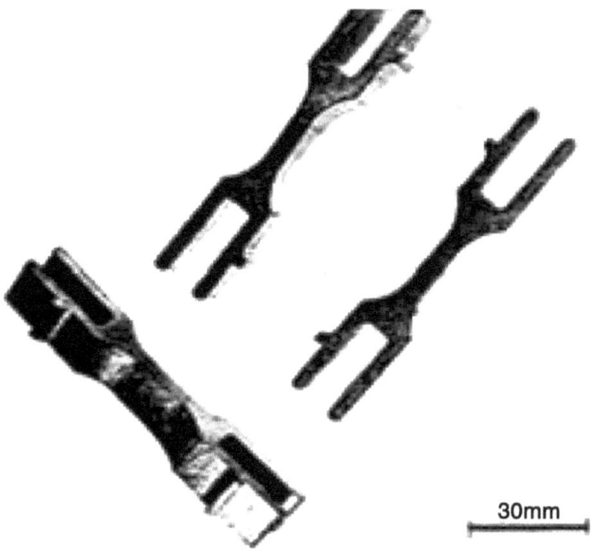

Figure 1.2 Press and sintered blended elemental Ti-6Al-4V connector link arm for the Pratt & Whitney F100 engine.

It was pointed out that larger BE components were difficult to fabricate by welding smaller parts because the BE material did not weld well as a result of the inherent porosity and chloride content. However, it was noted that recent work showed how below 150 ppm chlorine, the weldability can be improved.

Table 1.1 Typical tensile properties of blended elemental Ti-6Al-4V compacts compared to mill-annealed wrought products

Material	0.2% Yield strength, MPa (ksi)	Ultimate tensile strength, MPa (ksi)	Elongation, %	Reduction in area, %
Cold isostatic press and HIP (CHIP)	827 (120)	917 (133)	13	26
Press and sinter (no HIP)	868 (126)	945 (137)	15	25
Wrought mill anneal	923 (134)	978 (142)	16	44
Typical minimum properties (MIL-T-9047)	827 (120)	896 (130)	10	25

The typical tensile properties of BE compacts produced by a number of routes were compared with IM properties in Table 1.1. In critical components, such as rotating parts, fatigue behavior is very important. The most recent fatigue results for BE and PA powders are compared to IM data in Figure 1.3, where it can be seen that the remnant salt from the sponge-making process and associated voids cause early crack initiation in the BE material. Lower chloride contents and control of the microstructure by subsequent processing or heat treatment can enhance the fatigue behavior. One method of producing powder with a very low chloride content was pointed out to be comminution of titanium ingot stock.

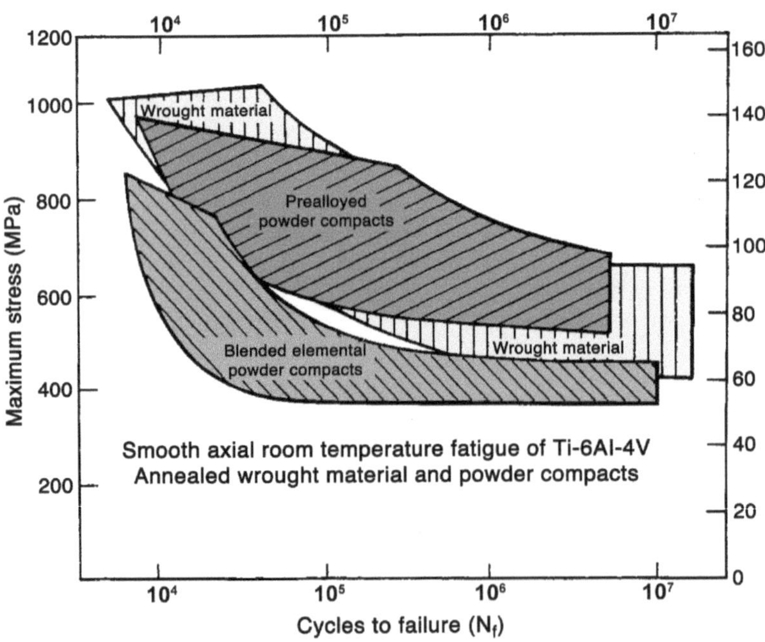

Figure 1.3 Comparison of fatigue behavior of annealed blended elemental and prealloyed Ti-6Al-4V PM compacts with ingot metallurgy material.

Other mechanical properties of BE material such as fracture toughness and fatigue crack growth rate appeared to be at the same levels as expected from IM material of the same chemistry and microstructure, when the density is at least 98% of full density.

In concluding remarks, it was noted that the economics of the elemental approach are attractive; sponge fines blended with master alloy additions, costing only one-fifth of PA powder.

In discussions of the PA approach, it was pointed out that production of complex shapes can be accomplished by one of four competing techniques: metal can, ceramic mold, fluid die or electroplated nickel can. The metal can is shaped by forming methods such as brake bending, press forming, spinning, and superplastic forming. Of these possibilities, the ceramic mold process, as practiced by Crucible Research Center, has received the most attention, particularly under US Air Force funding.

The ceramic mold process (Figure 1.4) [2] relies basically on the technology developed by the investment casting industry, in that molds are prepared by the lost-wax process. By combining this process and PM HIP, low-cost complex-configuration near-net shapes can be produced. The fluid die process also allows production of near-net shapes of complex configurations. Typical components produced by the ceramic mold process are shown in Figure 1.5. It was pointed out that since the cost of titanium component fabrication lies mainly in forging and machining, with material cost also a consideration, the selection of appropriate parts can be made only after evaluation of these factors. Generally, the PA/HIP approach is most attractive for large, complex parts with a high buy-to-fly ratio when fabricated by conventional means. The working volume of the autoclave available (1.2 m diameter by 2.4 m height [47 in. by 96 in.]) limits the size, unless subsequent welding or similar techniques are used to form larger components. Table 1.2 presents Crucible Research Center data, which listed the forging weight, PM product weight, final part weight, and the anticipated potential cost savings for the various parts that have been produced by the PA technique. This information was based on a large multi million-dollar US Air Force/Navy contract, which had as its goal production of cost-effective high-integrity large titanium components using the PA/HIP approach. These estimates suggest that cost savings with the PA route over forged pans could range between 20% and 50%, depending on the size and complexity of the part and quantity of parts produced. Higher-volume runs result in higher savings.

The mechanical properties of parts produced by the PA ceramic mold process (Figure 1.3 and Table 1.3) indicated that these components exhibited properties at the same level as conventional cast and wrought material.

1.4 Developments 1980–present

For the first decade following the 1980 Conference, there was little development in any segment of titanium PM. However, in the last 20 years or so, a number of titanium PM techniques have been initiated and grown, including powder production techniques, the PA/HIP technique, the BE approach, and both the MIM technology and the additive manufacturing method.

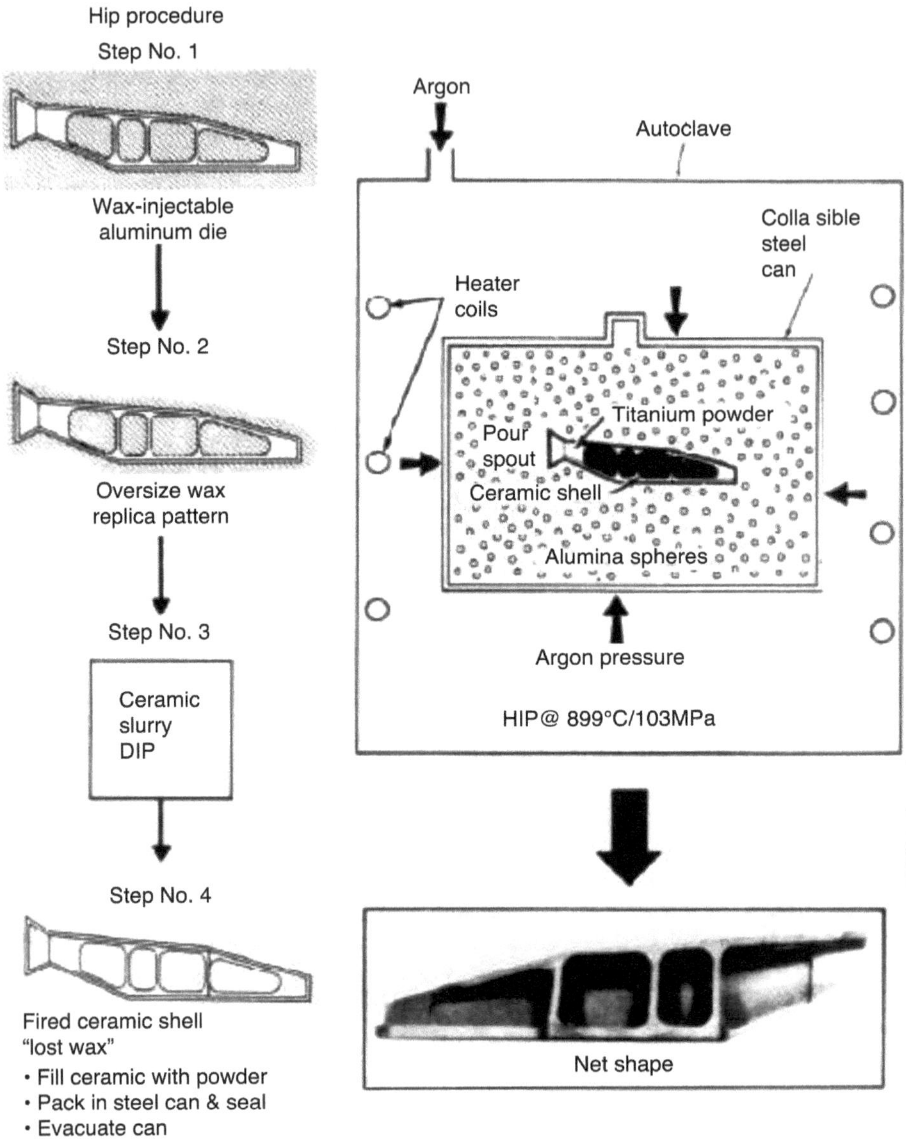

Figure 1.4 The sequence of operations to produce a part using the ceramic mold process [2].

1.5 Developments in the PA/HIP technology

Crucible Research Center abandoned the ceramic mold process when aerospace engine and airframe manufacturers were reluctant to use PA/HIP components because of the possibility of ceramic shell pieces becoming trapped in fabricated components (with a deleterious effect on fatigue properties in particular). The deleterious effect of second-phase particles was confirmed in a research study [3] (which is discussed in

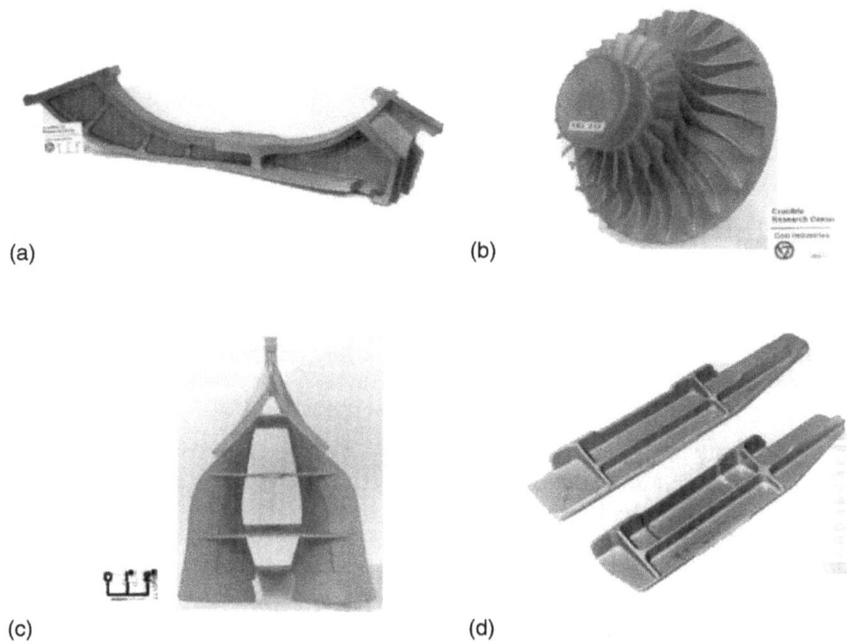

Figure 1.5 Components produced from prealloyed titanium powder, using HIP and the ceramic mold process: (a) a nacelle frame for F14A, Ti-6Al-6V-2Sn; (b) radial impeller for F107 cruise missile engine, Ti-6Al-4V; (c) a complex airframe component for the stealth bomber, Ti-6Al-4V; and (d) engine mount support, Ti-6Al-4V. (Courtesy Crucible Materials Corporation.)

Table 1.2 Weight and cost savings of Ti-6Al-4V prealloyed PM parts*

Compaction technique	US Air Force Part	Forged billet weight, kg (lb)	PM part weight, kg (lb)	Final part weight, cost savings kg (lb)	Potential, %
Ceramic mold	F-14 fuselage brace	2.8 (1.3)	1.1 (0.5)	0.8 (0.4)	50
Ceramic mold	F-18 engine mount support	7.7 (3.5)	2.5 (1.1)	0.5 (0.2)	20
Ceramic mold	F-18 arrestor hook support fitting	80.0 (36.4)	25.0 (11.4)	13.0 (5.91)	25
Ceramic mold	F-107 radial compressor impeller	15.0 (6.82)	2.8 (1.3)	1.7 (0.77)	40
Fluid die	AH64 radial compressor impeller	10.0 (4.55)	2.3 (1.0)	1.1 (0.5)	35
Ceramic mold	F-14 nacelle frame mold	144.0 (65.45)	82.0 (37.3)	24.0 (10.9)	50

*Data developed by Crucible Research Center.

Table 1.3 Properties of Ti-6Al-4V PA Powder Compacts

0.2% Yield strength, MPa (ksi)	Ultimate tensile strength, MPa (ksi)	Elongation, %	Reduction in area, %	Fracture toughness (K_{Ic}), MPa/m (ksi/in.)
930 (135)	992 (144)	15	33	77 (70)

more detail in the chapter on PA/HIP). In the last 10 years, Synertech PM Inc. has developed a PA/HIP metal can approach to production of high-integrity near-net-shape titanium components, based on prior work at VILS (the All Russian Institute for Light Alloys, Moscow, Russia). Further details can be found in the chapter on PA/HIP.

During the late 1990s and the early twenty-first century, a number of new sources of titanium PA powder have become available, including gas atomization (with Crucible Research leading the way with the introduction of a 50-lb-capacity unit in 1988). Further details can be found in the chapter on conventional powder production methods. There is also the possibility that some of the developing powder production methods may be capable of producing a spherical or near spherical powder (with good flow and a high packing density) for use in the PA/HIP technique (see the appropriate chapters in this book).

1.6 The BE method

In the BE arena, two companies, ADMA Products and Dynamet Technologies, have spearheaded this technology.

ADMA Products was founded in 1985 [4] and started to make Ti parts the same year. The parts were hex nuts (electroplating applications and desalination plants), artificial limbs (prosthesis), spacer rings, porous titanium plates (submarines, hydrogen, and oxygen generation systems), etc. Sales reached $10 million by 1995, but both RMI and Deeside closed their Na-reduced titanium sponge production facilities, and raw material sources were lost. Hydrogenated Ti powder was originally produced by a joint venture called Polema (with a company called POLAD, hence Polyma/ADMA) in Tula, Russia. At this time, ADMA Products were using calcium hydride reduction and Ti powder had 0.5% H_2. When the price of calcium escalated, ADMA Products started to work with Orest Ivasishin of Ukraine using hydrogenated sponge. (For further details, see the two chapters on hydrogenated powder production and hydrogenated compacts.) Currently, ADMA Products are working with major aircraft companies, fabricating qualification test samples for them. The samples are being made from TiH_2 powders by CIP/sinter + high-temperature postprocessing (forging, extrusion, rolling, etc.) and by die-pressing (small parts) plus sintering with and without postprocessing (HIP, coining/sizing, forging). Unfortunately, this work is being done under a nondisclosure agreement with their customers and the companies involved do not allow

Figure 1.6 PM CP Ti Spacer rings for desalination plants. (Courtesy ADMA Products Inc.)

ADMA Products to disclose the parts and the test data, but the information should be available soon. Examples of titanium PM parts, produced in the millions, are shown in Figures 1.6–1.11. The first five components (Figures 1.6–1.10) were produced from conventional titanium sponge fines, and the extrusions in Figure 1.11 are from TiH_2. Current titanium PM parts, as noted above, are made exclusively from TiH_2.

The mechanical properties of the compacts fabricated using the TiH_2 powder, including S-N fatigue, meet or exceed cast and wrought (ingot metallurgy) levels. No commercialization can be achieved without a TiH_2 powder supply. In this regard, ADMA Products has purchased two manufacturing buildings in Twinsburg, Ohio, installed a pilot scale unit of 250,000-lb annual capacity, and will produce the first TiH_2

Figure 1.7 PM CP Ti hex nuts of more than 50 various sizes for electroplating industry. (Courtesy of ADMA Products Inc.)

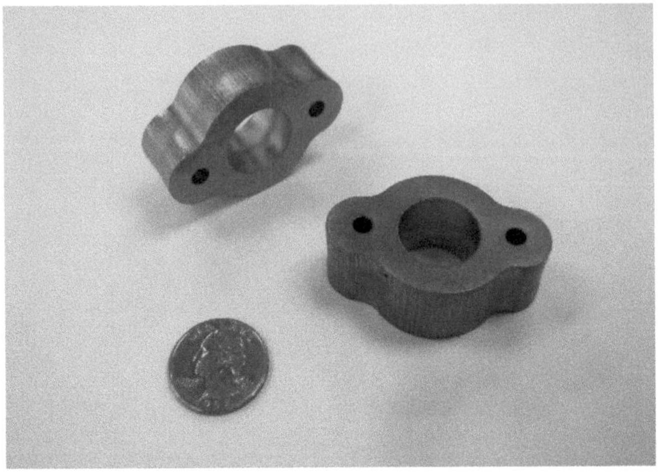

Figure 1.8 PM CP Ti C collars for electroplating industry. (Courtesy ADMA Products Inc.)

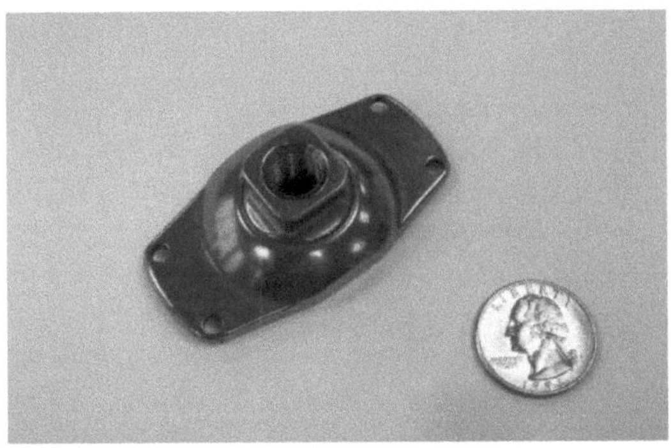

Figure 1.9 PM Ti-6Al-4V foot adapter for artificial limb. (Courtesy ADMA Products Inc.)

Figure 1.10 Porous titanium parts for different application in corrosion environment, produced by different PM processes: (a) CIP/sinter porous filters, (b) direct powder rolled porous plates. (Courtesy ADMA Products, Inc.)

Figure 1.11 Ti-6Al-4V extruded shapes of various cross sections extruded by RTI, TIMET, Plymouth for Boeing. (Courtesy ADMA Products Inc.)

powder lot late December 2013. A production plant in Sheffield, Ohio, is currently planned with a 5000-lb-per-run powder production unit.

Dynamet Corporation, a Massachusetts company, was founded in 1968 by Stanley Abkowitz [5] and investors. In May 1972, it was reorganized as Dynamet Technology, Inc. (not to be confused with Dynamet, a Division of Cartech of Pennsylvania, a titanium mill product converter). Dynamet Technology has pioneered the application of its PM technology with continuing improvement over the years in manufacturing and processes and innovative alloy developments. Its first products were metal matrix composites (Al/B4C) thermal neutron shielding shapes for Navy nuclear propulsion applications. Its first titanium product was the Ti-6Al-4V near-net-shape powder metal dome housing preform for Raytheon's Sidewinder missile (using a CIP approach). This was followed by production of the Warhead Casing from PM Ti-6Al-6V-2Sn alloy for Stinger missile applications. Missile applications started in the late seventies and continue to the present day with well over 100,000 near-net-shape parts supplied to specifications ranging from 95% minimum density to 100% density produced in the press/sinter and also in the CHIP (CIP followed by HIP) condition.

The development of CermeTi® metal matrix composites was commercialized in the mid-1980s. This material offers enhanced elevated temperature strength properties and greater wear resistance than any commercial titanium alloy. This is a PM titanium alloy matrix material reinforced with ceramic particulate (TiC or TiB_2). Current production applications include medical (orthopedic implants) and industrial components such as shot sleeve liners for Al alloy die casting tooling. Early titanium die pressing studies resulted in successful titanium automotive valve retainers and commercial production of its CermeTi face inserts for golf club drivers. At the 1992 World Conference in San Diego, die pressed titanium key chains were given to all attendees.

More recent evaluation and testing of Dynamet Technology material at Boeing Commercial Airplane has resulted in the development of applications at Boeing for aircraft components for advanced 737 aircraft and the 787 Dreamliner. This development was described in the 2013 ITA Applications Award received by Dynamet Technology at the October 2013 ITA Convention at Caesars Palace, Las Vegas. The Boeing specifications were issued in January 2012. Dynamet is currently working with Boeing to incorporate selected high buy/fly ratio components on commercial aircraft. While static properties (tensile data, etc.) meet or exceed cast and wrought levels, the sintered material (CIP) density is 98% minimum density, and the CHIP is ≥99% minimum density, and both conditions exhibit S-N fatigue behavior approximately 75% of wrought levels [5]. This means that first applications will be in static parts, with fatigue-critical uses coming later as S-N fatigue properties are enhanced (see Chapter 17 for more recent data). In view of these recent developments, Dynamet is planning for plant expansion and significant growth. Another indication that Boeing is serious about titanium PM is the memorandum of understanding that it has recently signed with the Council for Scientific and Industrial Research (CSIR), South Africa [6].

Although Boeing parts are not displayed, Figure 1.12 represents a wide variety of diverse component configurations enabled by the Dynamet elemental blend sinter (EBS) technology.

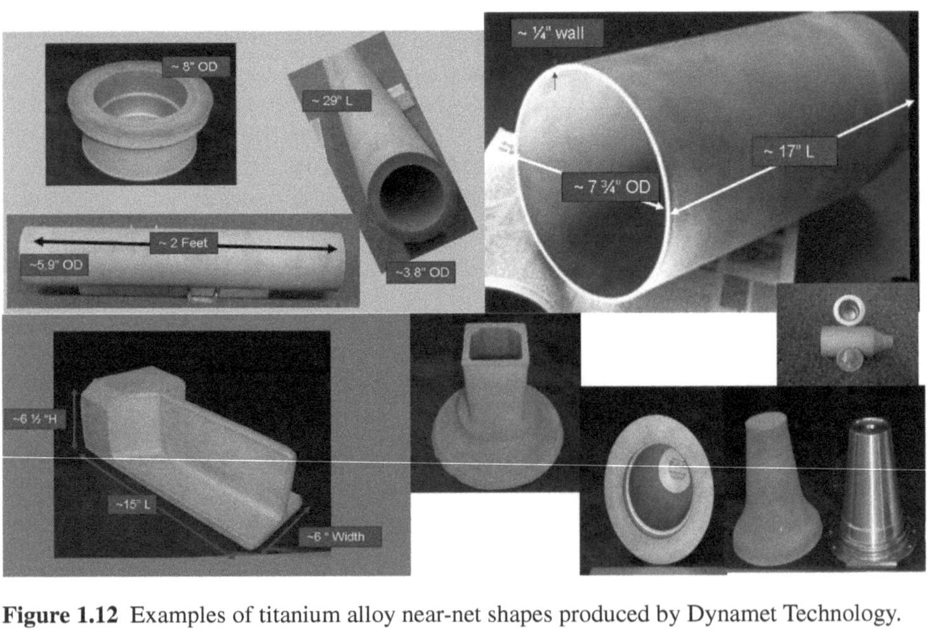

Figure 1.12 Examples of titanium alloy near-net shapes produced by Dynamet Technology.

1.7 Metal injection molding

Metal powder injection molding is a hybrid between PM and plastic molding (see the chapters on this topic in this book). Titanium component production by the metal powder injection molding process (termed Ti-MIM) has advanced considerably based on several reports detailing the powders, binders, debinding, sintering, and post-sintering details (for more details, see the two chapters on MIM). A distinction must be made between (a) decorative items where mechanical and other properties are not so demanding, such as in watch cases, and (b) mechanical components where mechanical and corrosion properties must exceed specified levels (often in competition with stainless steel), such as in surgical instruments. Decorative items, for example, have been fabricated for 20 years, including watch cases, sunglass frames, and cell phone and firearm components. Production of mechanical components with acceptable tensile properties (although S-N fatigue in as-produced MIM is significantly below cast and wrought material, subsequent HIP material is close to ingot metallurgy levels) has increased much more recently with the use of appropriate powders and debinding/sinter cycles. The same binders as initially used for steel and stainless steel (BASF feedstock based on polyacetal, PolyMIM binder based on PEG, and Ryer binder based on wax-polymer), with nothing significantly different proved to be successful with titanium-MIM. A study funded by Witec, Remington, and Bausch & Lomb in the early 1990s showed success, with a wax-polymer binder system first disclosed by Ray Wiech in 1980 [7]. So it was the debinding details that made a difference, and not the binder itself. These acceptable binders do not result in excessive oxygen pickup during the debinder heat-treatment cycle. The worldwide interest in titanium MIM is demonstrated in a recently published article on work in this area in China, including a study on superior implant components featuring a dense structure in the interior and porous outer structure (which facilitates cell adhesion) [8]. The commercial production of titanium MIM parts in Europe is documented in an article by Schlieper in which Element 22 GmbH is credited with spearheading ASTM standards for titanium MIM in surgical implants, and the article further reported on the use of this technology in aerospace applications (bolts, rivets, fasteners, and other small components) and high-end consumer products [9]. See the chapter on titanium MIM authored by German for details on the market for components manufactured by this method, estimated at 10,000 kg/year with sales at the $10-million mark.

1.8 Additive manufacturing

Additive manufacturing (AM) of metals such as stainless steel has been around for a number of years [10], but it is only within the past 20 years that AM of titanium with acceptable mechanical properties has been achieved [11]. The predominant usage of titanium AM is in the aerospace and medical industries. Both these industries are very slow to adapt to new processes because of the huge risks involved with any new technology. Primarily, this is the reason for the slow development of titanium AM. While the reactivity of titanium poses some practical challenges, these have been overcome through inert chamber processing.

All the AM technologies for titanium are based on the principle of slicing a solid model in multiple layers and building the part up layer by layer following the sliced model data. Following ASTM classification, additive manufacturing technologies for metals can be broadly classified into two categories: directed energy deposition and powder bed fusion. There are several technologies under each category as branded by different manufacturers. While the powder bed fusion technologies enable building of complex features, hollow cooling passages, and high-precision parts, these are limited by build envelope, single material per build, and horizontal layer-building ability. In comparison, the directed energy deposition technologies offer larger build envelope and higher deposition rate, while their ability to build hollow cooling passages and finer geometry is limited. Direct metal deposition [11] and laser-engineered net shapes [11] technology also offer the ability to deposit multiple materials in a single build and the ability to add metal on existing parts. For further details, see the chapter on AM of titanium.

The vast majority of Ti AM components have been fabricated using high-cost spherical GA/PREP powder. However, recent work has demonstrated that titanium AM parts can be successfully produced, in one using much lower-cost angular powders [12,13]. In one program [12], titanium sponge was blended with Al and V powder or Al/V master alloy to produce the Ti-6Al-4V composition. After processing this combination through a plasma transferred arc (PTA) (Figure 1.13), the as-fabricated tensile properties were at cast and wrought levels – 142 ksi ultimate tensile strength (UTS), 128 ksi yield strength (YS) and 10% elongation – and S-N Fatigue was at ingot metallurgy levels. A connecting rod fabricated by this AM technique is shown in Figure 1.14. Later work with ADMA Products using TiH_2 was equally successful. In the other program [13], angular Metalysis (FFC/Cambridge) was first converted to a spherical morphology (by melting) and then used to produce quite acceptable titanium AM products using laser heating and the powder bed fusion approach (see the chapter on additive manufacturing). Renishaw's new laser melting additive manufacturing process produces fully dense metal parts from 3D computer-aided design (CAD) data using a high-powered fiber laser. Components such as the turbocharger for an auto engine shown in Figure 1.15 have been fabricated. The process is digitally driven, direct from sliced 3D CAD data, in layer thicknesses ranging from 20 to 100 μm that form a 2D cross section. The process then builds the part by distributing an even layer of metallic powder using a recoater, then fusing each layer in turn under a tightly controlled inert atmosphere. Once complete, the part is removed from the powder bed and undergoes heat treatment and finishing depending on the application. Metalysis plans to fabricate parts using angular as-produced powder. It is interesting that both organizations choose to demonstrate AM with inexpensive titanium powders to fabricate auto parts, low cost always being a primary concern with the auto industry.

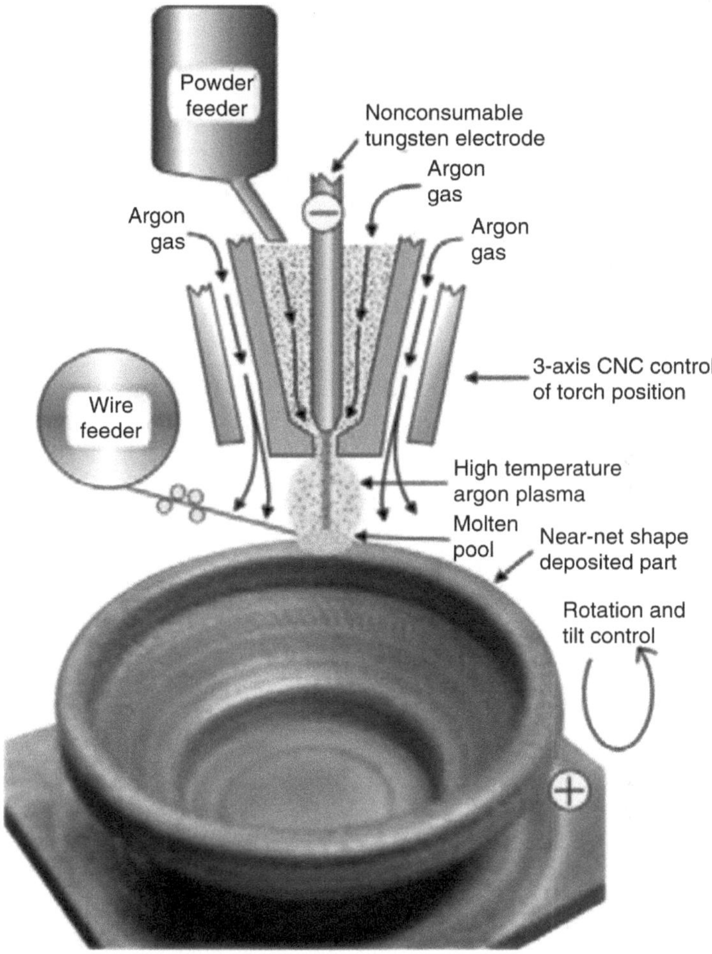

Figure 1.13 Schematic of PTA rapid manufacturing process [12] used by MER Corporation.

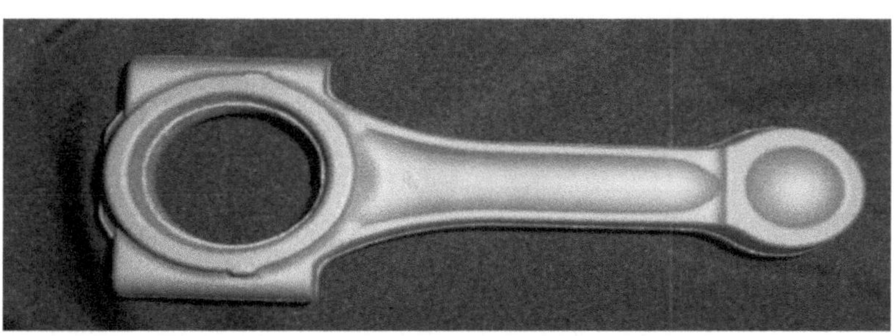

Figure 1.14 A connecting rod produced by forging a preform [12].

Figure 1.15 Additive manufactured turbo charger for auto engine fabricated from spheroidized metalysis powder [13].

1.9 Other developments

In recent years, there have been a number of other developments pertaining to titanium PM. This includes large sums of dollars being injected by the US Defence Advanced Research Projects Agency (DARPA) into the development of alternate, lower cost, processes for production of titanium (both pure and alloy), as substitutes to the conventional Kroll process. Most of the processes developed produced a powder product that could be used in PM approaches to fabrication of titanium components. These processes included processes featuring reduction of titanium oxide and direct reduction of titanium tetrachloride. Details of the majority of these processes are contained in other chapters of this book, in publications by Ed Kraft (Table 1.4) [14,15], and a recent review of titanium PM [16].

There have also been a series of four TMS-sponsored conferences on low-cost titanium that have contained a number of articles on titanium PM [17–20]. Chapter 8 of the ASM International Course on "Titanium and Its Alloys" (which is offered approximately once per year at Materials Park, Ohio) contains a comprehensive review of the status of all aspects of titanium PM.

1.10 Research-based processes

A number of titanium PM processes, developed in the last 20 years, are discussed in a chapter contained in this book. These include rapid solidification, mechanical alloying, vapor deposition, thermohydrogen processing, and porous structures. The first three of these are considered to be "Far from Equilibrium Processes." These processes, with the exception of THP [4], are not currently being scaled up to commercialization.

Table 1.4 **Titanium Extraction Processes**

Techniques	Comments
FFC	Oxide, electrolytic molten $CaCl_2$
MER	Oxide, electrolytic
SRI	Fluidized bed H_2 reduction of $TiCl_4$
BHP (Billiton, Australia)	Oxide electrolytic, prepilot plant
Idaho Ti	Plasma quench, chloride
Ginatta, Italy	Electrolytic, chloride
OS (Ono, Japan)	Electrolytic/calciothermic oxide
MIR, Germany	Iodide reduction
CSIR, South Africa	Electrolysis of oxide
Okabe-1, Tokyo, Japan	Oxide, reduction by Ca
Okabe-11, Tokyo, Japan	Oxide, Ca vapor reduction
Vartech, Idaho	Oxide, Ca vapor reduction
Northwest Institute for Non-Ferrous Metals	Innovative hydride–dehydride
CSIRO, Australia	Chloride, fluidized bed, Na
Armstrong/1TP	Chloride, continuous reduction with Na
DMR	Aluminothermic rutile feedstock
MIT	Oxide, electrolysis
QIT/Rio Tinto	Slag, electrolysis
Tresis	Argon plasma, chloride
Dynamet Technology	Low-cost feedstock

1.11 The 2011 conference on titanium PM

In late 2011, an International Conference on titanium PM was held in Brisbane, Australia, organized by Ma Qian [21]. This conference covered a wide range of topics, including powder production, consolidation, MIM, biomedical applications (including shape memory alloys), alloy development, additive manufacturing, modeling of diffusion, effect of impurities on behavior, titanium composites, forging and equal-channel pressing, fabrication of novel porous titanium structures, laser cladding, welding, slip casting, and other processing developments. It was pointed out that the high cost of titanium powder has limited commercialization but that a number of developing processes show promise (see the various chapters on powder production in this book). An interesting paper by Z. Zak Fang and Pei Sun [22], which builds on the work by ADMA Products and the Institute for Metal Physics, Ukraine [4], and Froes et al. [23] discussed the advantages of using hydrogen as a temporary alloying element to enhance the processing and mechanical properties of titanium alloys such as Ti-6Al-4V (i.e., the thermohydrogen processing technique [23]). This approach was partially patented by Swatch for the fabrication of watch cases about 20 years earlier [24]. The work by Fang and coworkers is currently receiving sizable US Government funding.

A further conference on titanium PM was held in Hamilton, New Zealand, in December 2013; however, the proceedings of this conference did not appear in time to be included in this chapter.

1.12 Thoughts for the future

At this time the future for titanium PM looks bright, especially if lower-cost powders become available for use in the fabrication methods noted below. The PA/HIP approach is commercially available and ready for expansion; the BE method has now been qualified for use on Boeing commercial airplanes, which will stimulate growth, and both MIM (for small parts) and additive manufacturing technology (for larger parts and repair of damaged components) are ready for commercialization and should see growth. The future of titanium PM is discussed in a chapter in this book.

Acknowledgments

The author would like to acknowledge useful discussions and supply of information from the following: Stan and Susan Abkowitz, Bhaskar Dutta, Dave Bourell, Rand German, Vladimir Moxson, Jim Withers, and Fred Yolton.

References

[1] F.H. Froes, J. Smugeresky (Eds.), Proceedings of the Conference on "Powder Metallurgy of Titanium Alloys," TMS, Warrendale, PA, 1980.
[2] R.W. Witt, W.T. Highberger, "Hot Isostatic Pressing of Near Net Titanium Structural Parts" in Reference #1, 255.
[3] S.W. Schwenker, D. Eylon, F.H. Froes, Influence of foreign particles on fatigue behavior of Ti-6Al-4V prealloyed powder compacts, Met. Trans. A 17A (1986) 271.
[4] V. Moxson, ADMA Products, Private Communication, December 13, 2013.
[5] S. Abkowitz, Dynamet Technology, Private Communication, December 23, 2013.
[6] Anon, CSIR signs titanium powder research MOU with Boeing, PM Mold. Int. 7 (4) (2013) 21.
[7] R. German, San Diego State University, Private Communication, December 22, 2013.
[8] Li Yimin, He Hao, Metal molding in China: an overview of current status, opportunities and challenges, Powder Inj. Mold. Int. 7 (4) (December 2013) 31.
[9] G. Schlieper, "Element 22 GmbH: pushing the boundaries of titanium MIM in the medical and aerospace sectors," Ibid, 39.
[10] D. Bourell, "History of laser-based additive manufacturing," presented at the workshop on Laser Additive Manufacturing, University of Texas, February 16–17, 2011.
[11] B. Dutta, F.H. (Sam) Froes, "The additive manufacture (AM) of titanium alloys" to be published in Advanced Materials and Processes, 2014 and a chapter in this book.
[12] J.C. Withers, V. Shapovalov, R. Storm, R.O. Loutfy, "There is Low Cost Titanium Componentry Today" in Reference #20, 11.

[13] P. Whittaker, "'Metalysis' titanium powder used to 3D print automobile parts," http://www.ipmd.net/news/002519.html, January 24, 2014.
[14] E.H. Kraft, "Opportunities for low cost titanium in reduced fuel consumption, improved emissions, and enhanced durability heavy-duty vehicles," July 2002, Oak Ridge National Laboratory.
[15] E.H. Kraft "Summary of emerging titanium cost reduction technologies," December 2003, Oak Ridge National Laboratory.
[16] F.H. (Sam) Froes, Titanium powder metallurgy: developments and opportunities in a sector poised for growth, Powder Metal. Rev. 2 (4) (2013) 29.
[17] F.H.(Sam) Froes, M.A. Imam, D. Fray (Eds.), Cost Affordable Titanium, TMS, Warrendale, PA, 2004.
[18] M.N. Gungor, M.A. Imam, F.H. Froes (Eds.), Innovations in Titanium Technology, TMS, Warrendale, PA, 2007.
[19] M.A. Imam, F.H. Froes, K.F. Dring (Eds.), Cost-Affordable Titanium III, Trans Tech Publications, Zurich-Durnten, Switzerland, 2010.
[20] M.A. Imam, F.H.(Sam) Froes, G.R. Reddy (Eds.), Cost Affordable Titanium IV, Trans Tech Publications, Zurich-Durnten, Switzerland, 2013.
[21] M. Qian (Ed.), Proceedings of the Conference on "Powder Metallurgy of Titanium – Powder Processing, Consolidation and Metallurgy of Titanium," Trans Tech Publications, Durnten-Zurich, Switzerland, 2012.
[22] Z. Zak Fang, P. Sun, "Pathways to optimize performance/cost ratio of powder metallurgy titanium – a perspective" in Reference #21, 15.
[23] F.H. Froes, O.N. Senkov, J.I. Qazi, Hydrogen as a temporary alloying element in titanium alloys: thermohydrogen processing, Inter. Mat. Rev. 49 (2004) 227.
[24] J. Greenspan, F.J. Rizzitano, E. Scala, Titanium powder metallurgy by decomposition sintering, in: R.I. Jaffee (Ed.), Titanium Science and Technology (Proceedings of the Second International Conference), Plenum Press, New York, NY, 1973, pp. 365–379.

Conventional titanium powder production

2

C.F. Yolton, Francis H. (Sam) Froes***
*CTO, Summit Materials LLC, McDonald, PA, USA
**Consultant to the Titanium Industry, Tacoma, WA, USA

2.1 Introduction

Conventional titanium powder production methods, for the purpose of this chapter, are defined as processes that are well established, such as the plasma rotating electrode process (PREP), gas atomization (GA), electrode induction melting–gas atomization (EIGA), plasma atomization (PA), and induction plasma spheroidization (IPS). It does not cover sponge fines, hydride–dehydride (HDH), hydrogenated sponge fines, or developing extraction/powder production techniques, all of which are covered in other chapters of this book. The processes covered are all processes which produce prealloyed spherical powders which are used in (a) the prealloyed/hot isostatic pressing method of fabricating net/near net shapes, (b) metal injection molding, where a size of less than about 40 μm is preferred, and (c) additive manufacturing, each of which is covered in other chapters of this book.

2.2 Prealloyed spherical powder (conventional titanium powder production)

There are a number of processes that produce prealloyed spherical titanium powder. A listing of the various spherical powders that are commercially available is given in Table 2.1. Spherical powders are hot consolidated generally by hot isostatic pressing. Initial complex shape-making work was carried out using ceramic molds. However, partly because of concerns that ceramic or other particles could get into the titanium parts fabricated using the ceramic mold process with a degradation in mechanical properties (particularly S-N fatigue), this process has been discontinued (see Chapter 1 for more details). However, parts produced using a shaped metal can and removable mild steel inserts (removed by chemical dissolution) are commercially available. See Chapter 18.

Titanium Powder Metallurgy. http://dx.doi.org/10.1016/B978-0-12-800054-0.00002-2

Table 2.1 Spherical titanium powder producers

Type	Producer
Gas atomized	ATI Powder (USA), formerly Crucible Research, division of Crucible Materials Corp
Gas atomized	Iowa Powder Atomization Technologies (USA) developmental process
Gas atomization/PREP	Affinity International (China)
Gas atomized (EIGA)	TLS Technik (Germany)
Gas atomized (EIGA)	Puris LLC (USA)
Gas atomized (TILOP, EIGA type)	Osaka Titanium Technologies (Japan)
PREP	Timet (USA), formerly Advanced Specialty Metals
PREP	Phelly Materials (China)
PREP	Baoji Orchid Ti (China)
Plasma atomized	ARCAM AB (Sweden), formerly Advanced Powders and Coatings, division of Raymor Industries
Induction plasma spheroidization	Tekna Plasma Systems (Canada)

2.3 Gas atomization

A gas atomization process for titanium, designated TGA, was developed by Crucible Research Division of Crucible Materials Corporation (now ATI Powder) in 1988. In this process, the starting charge is induction skull melted in a 45-kg (100 lb) water-cooled copper crucible under vacuum or an inert gas. The molten charge is then bottom poured into an induction heated nozzle and the resultant metal stream is atomized with high-pressure argon gas. A schematic of this atomizer is shown in Figure 2.1 [1].

Starting stock for this process can be elemental raw materials or prealloyed materials such as ingots, bars, or revert material. As the starting charge is melted in the crucible, a thin skull forms on the crucible wall so that the molten titanium is always contained in a solid skull of the same composition. After the charge is melted, it can be held in the molten state for an extended time to ensure complete homogenization of the melt.

The pour is initiated in this process by a second smaller induction coil that causes localized melting at the center of the bottom of the skull. The molten titanium is then bottom poured through a refractory metal nozzle and free-falls into an atomizing die, which contains a ring of high-pressure jets. The force of the atomization gas disintegrates the stream into tiny droplets that rapidly solidify into powder particles as they fall through the cooling tower. The atomization gas carries the powder particles into a cyclone collector where they fall into a collection canister. The TGA powder tends to have more satellite particles than other gas atomization processes.

The powder produced by this process (Figure 2.2) is spherical and free flowing. Typical size distributions of titanium alloys and gamma titanium aluminide alloys are shown in Figure 2.3. Typical packing densities for titanium alloys range from

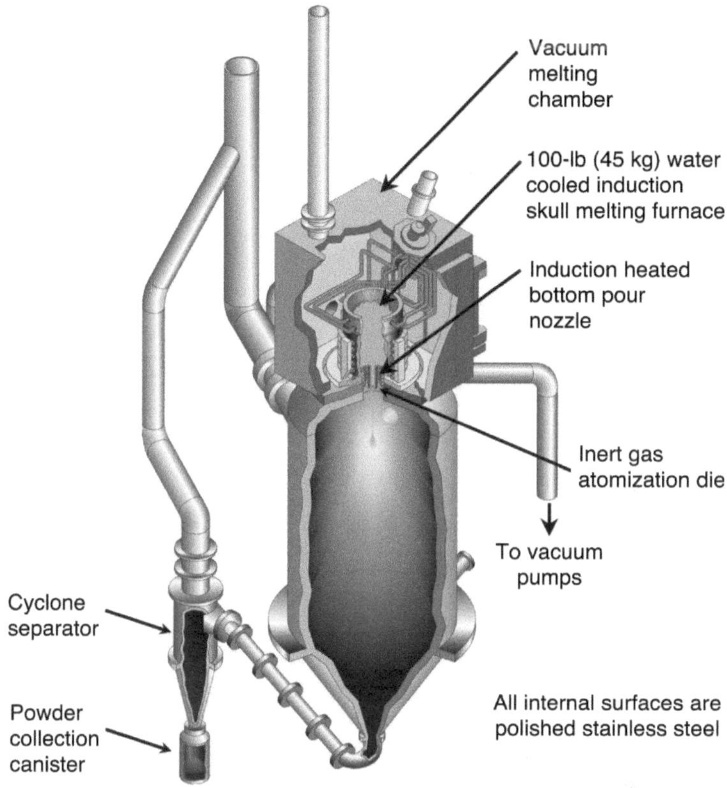

Figure 2.1 ATI Powder 45 kg (100 lb) titanium gas atomizer.

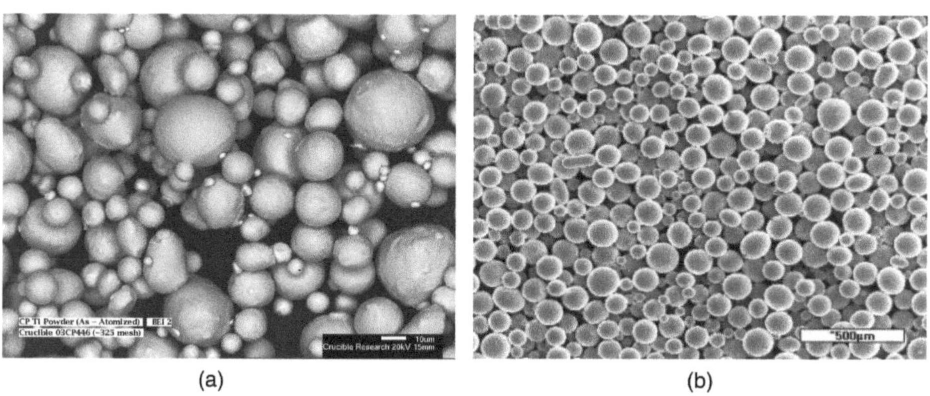

Figure 2.2 SEM micrographs of gas atomized powder: (a) commercially pure Ti powder produced by ATI Powder, (b) Ti-6Al-4V powder produced by Affinity International.

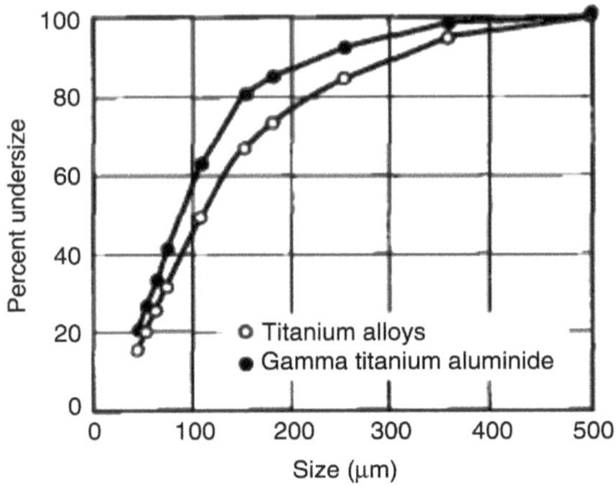

Figure 2.3 Typical size distributions of gas atomized titanium and gamma titanium aluminide alloys.

60% to 70% of solid density. Flow rates are in the range of 25–35 s as measured by ASTM B213. The interstitial content of the powder, specifically oxygen, is dependent on particle size as shown in Figures 2.4 and 2.5. The elements carbon, nitrogen, and hydrogen are much less dependent on particle size. The increase in oxygen content from starting melt stock to 500-μm powder is typically 200–300 wppm.

A titanium gas atomization process is under development at Iowa Powder Atomization Technologies Inc. which utilizes close coupled atomization to produce gas atomized powder [2,3]. This process is directed toward producing a large-size fraction of a small-diameter product (less than 40 μm).

2.4 Plasma rotating electrode process

The PREP is a centrifugal atomization process for making titanium prealloyed powder developed by Nuclear Metals/Starmet (now part of TIMET). In this process, a helium plasma is used to melt the end of a centerless ground rapidly rotating bar [4]. Molten

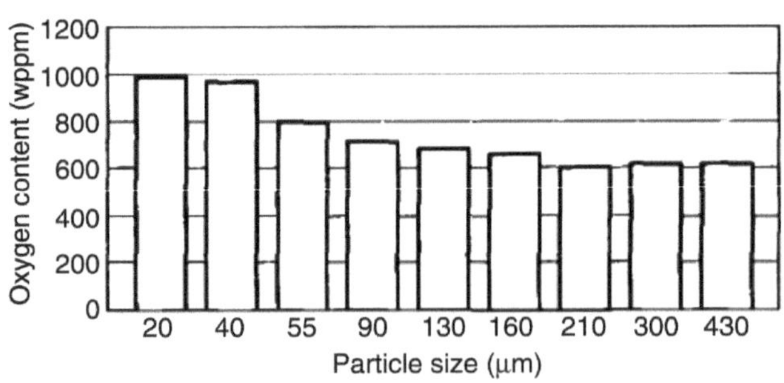

Figure 2.4 Effect of particle size on oxygen content of gas atomized Ti-6Al-4V powder.

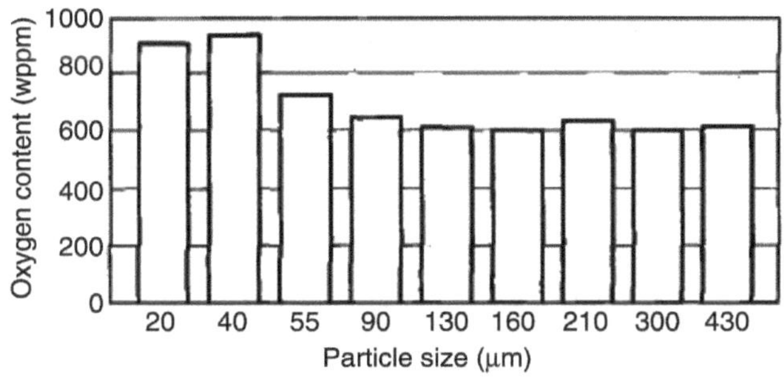

Figure 2.5 Effect of particle size on oxygen content of gas atomized gamma titanium aluminide powder.

droplets are spun off and solidify in flight in a helium atmosphere. A schematic of this process is shown in Figure 2.6. Melting and atomization take place in a stainless steel chamber 2.4 m (96 in.) in diameter maintained at a positive pressure of helium. The starting electrodes are prealloyed bars nominally 64 mm (2.5 in.) in diameter and are rotated at speeds up to 15,000 rpm.

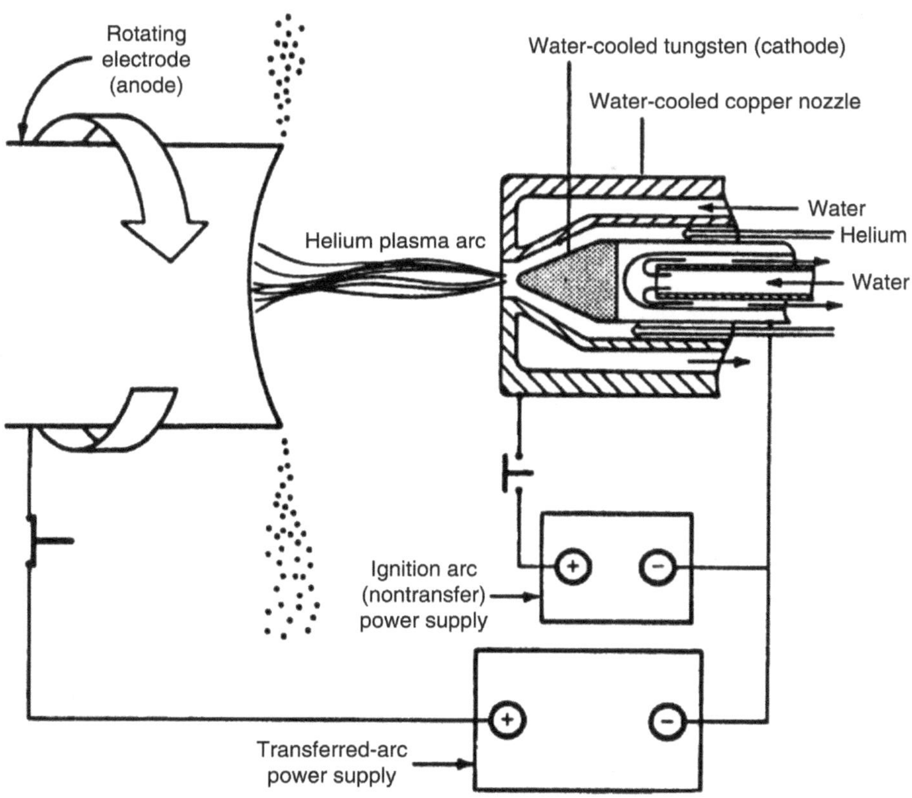

Figure 2.6 Schematic diagram of PREP.

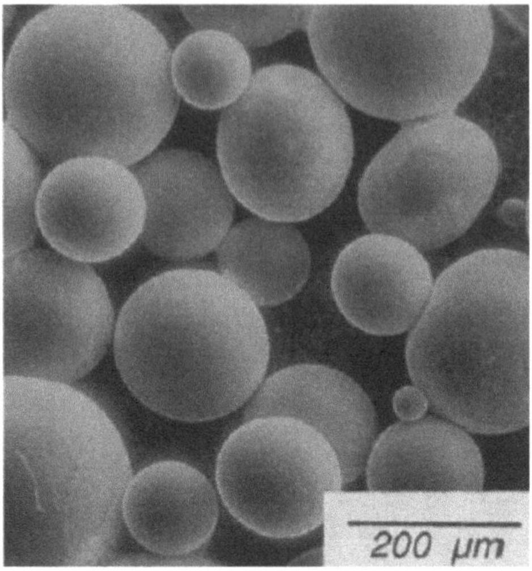

Figure 2.7 Ti-6Al-4V powder made by the PREP.

The PREP powder (Figure 2.7) is spherical and has good flow and packing characteristics. The size distribution of the powder depends on the alloy, electrode diameter, and rotational speed. Typical size for Ti-6Al-4V powder is between 100 and 300 μm with a d_{50} of about 175 μm [5]. This process produces 1% or less powder finer than 50 μm.

2.5 Electrode induction–melting gas atomization

Like the PREP process the EIGA process is a completely crucible-free melting and atomization process developed by ALD Vacuum Technologies. In this process, a slowly rotating prealloyed bar is fed into a conical induction coil. The end of the bar is inductively heated and molten titanium falls into an atomization die where the liquid titanium is atomized with a high-pressure inert gas [6]. A schematic of the EIGA process is shown in Figure 2.8. The starting bar diameter is 25–70 mm (1.0–2.75 in.) and the feed rate is 40–60 mm/min (1.6–2.4 in./min).

The size distribution of EIGA test runs on 45 mm (1.77 in.) and 60 mm (2.36 in.) bar is shown in Figure 2.9. The size distribution is compared with stainless steel and a niobium alloy powder. Atomization of the larger-diameter 60 mm (2.36 in.) titanium bar results in a small shift of the powder d_{50} from 60 to 65 μm. The EIGA powder has good flow and packing characteristics. Experience to date indicates that the EIGA process has the ability to produce finer-mesh powder than both the PREP and TGA processes. Osaka Titanium Technologies Co. Ltd. (formerly Sumitomo Sitix) has developed a similar process for producing titanium powder [7].

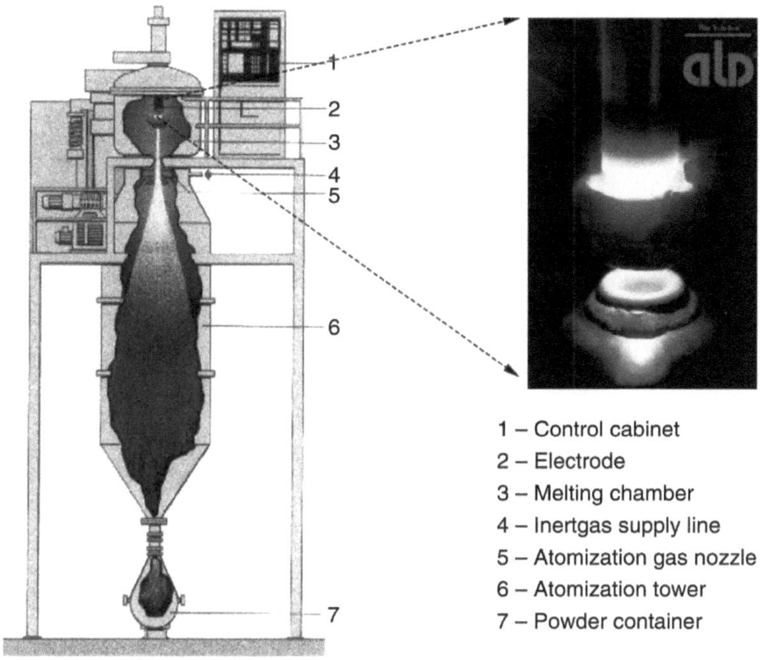

Figure 2.8 Schematic of the EIGA process.

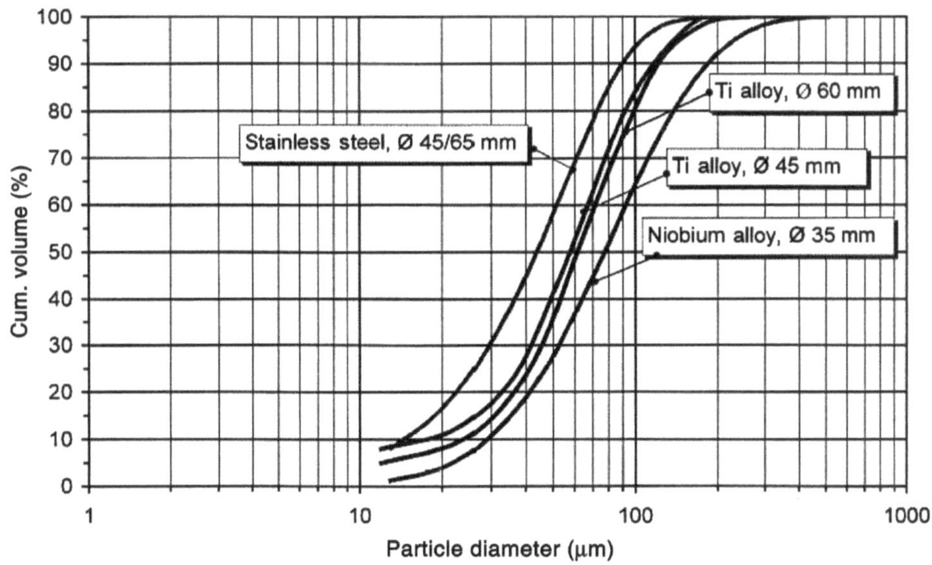

Figure 2.9 Size distribution of EIGA titanium alloy powder compared with stainless steel and niobium alloy powder.

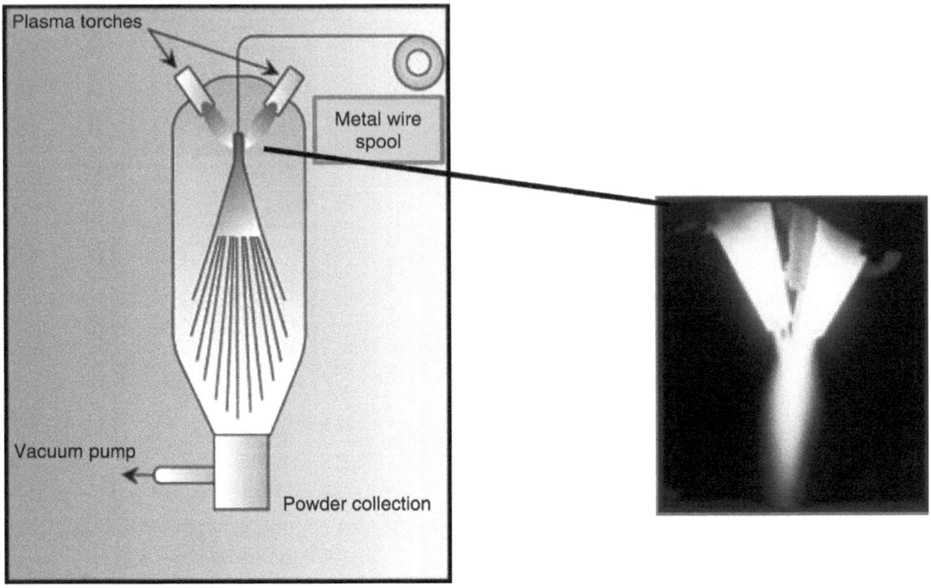

Figure 2.10 Schematic of the plasma atomization process and photograph of atomization in progress.

2.6 Plasma atomization

A schematic of the PA process is shown in Figure 2.10. In this atomization process, powder is atomized by feeding titanium wire into nontransferred arc plasma torches. The high-velocity plasma melts the wire and breaks the liquid into fine droplets that solidify in flight [8]. The powders produced by this process are spherical and satellite free. The powder has a narrow size distribution, with an average particle size of 40 μm. A typical scanning electron microscope (SEM) micrograph of titanium powder produced by PA is shown in Figure 2.11.

2.7 Induction plasma spheroidization

The induction plasma spheroidization process converts irregular powders such as HDH powder to spherical powder. A schematic of this process, developed by Tekna Plasma Systems Inc, is shown in Figure 2.12. In this process, the starting nonspherical powder is fed into an induction plasma, where it is melted and immediately solidified into spherical powder particles [9]. Figure 2.13 shows an example of titanium powder produced by this process.

As explained in the process descriptions, atomized powders are generally prealloyed and spherical. All of the processes addressed involve going through the molten state followed by a rapid cooling to ambient temperatures; thus, a

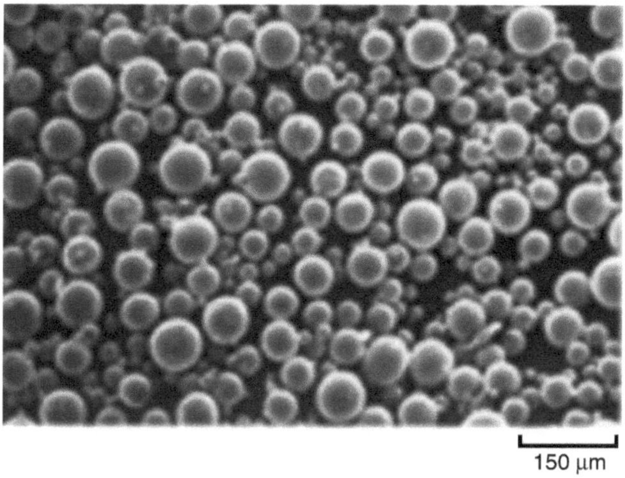

Figure 2.11 SEM micrograph titanium powder made by the plasma atomization process.

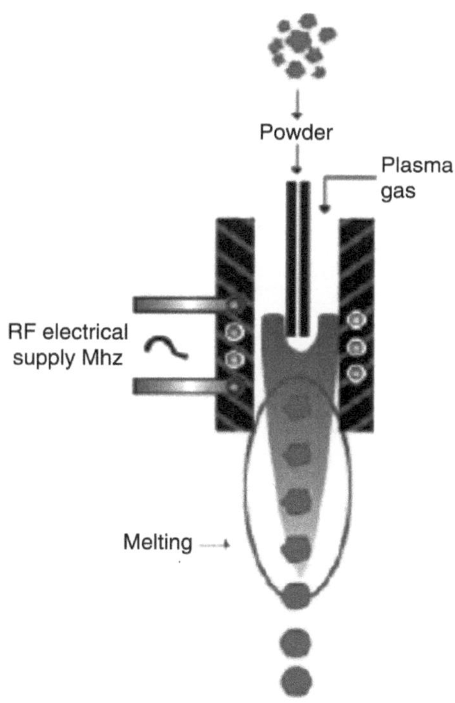

Figure 2.12 Schematic of the induction plasma spheroidization process.

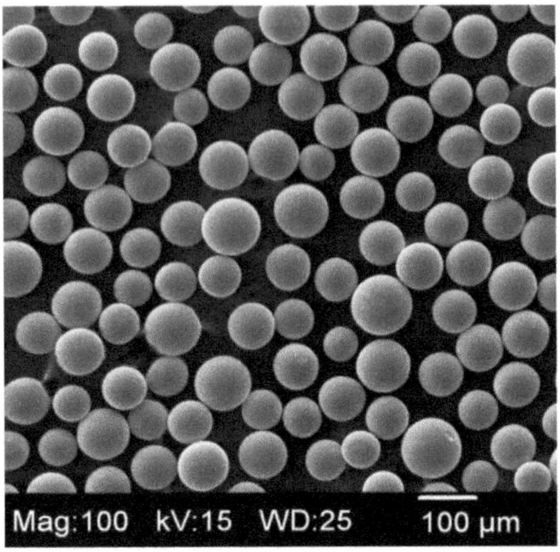

Figure 2.13 SEM micrograph titanium powder made by induction plasma spheroidization.

martensitic microstructure results in alpha-beta alloys such as Ti-6Al-4V, while in richer alloys such as Beta III, equiaxed beta grains are present. On subsequent hot compaction by sub-beta transus hot isostatic pressing for both types of alloys alpha phase precipitates during the elevated temperature excursion. An example of a gas atomized powder martensitic microstructure in the as-atomized condition is shown in Figure 2.14. Figure 2.15 shows the microstructure of Ti-6Al-4V gas atomized powder consolidated by HIP at 1750°F, which produces an alpha plus beta microstructure.

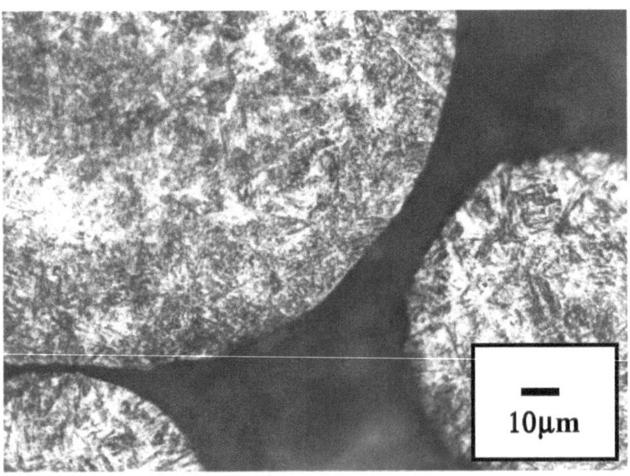

Figure 2.14 Microstructure of gas atomized Ti-6Al-4V powder (original magnification 500×).

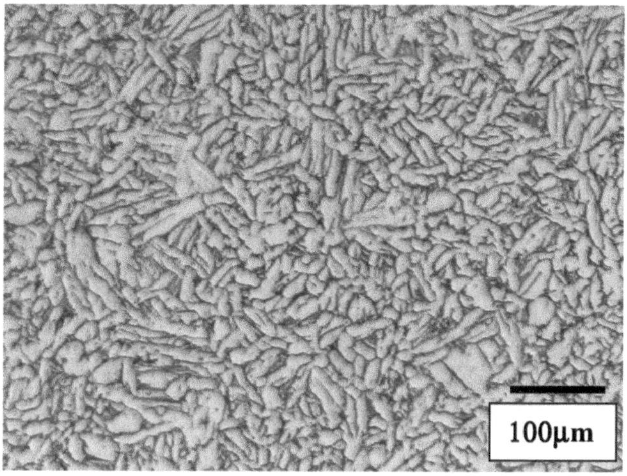

Figure 2.15 Microstructure of gas atomized Ti-6Al-4V powder hot isostatically pressed at 1750°F.

2.8 Conclusions

There are a number of commercial processes for production of conventional titanium powder (with a prealloyed spherical morphology), all of which result in very acceptable mechanical properties in compacted material (near net shapes via the prealloyed/hot isostatic allied pressed method, titanium metal injection molded parts, and additive manufactured components). Current 2013–2014 prices for spherical titanium powder range from $40 to $200 per pound depending on alloy, mesh size, and quantity. The major challenge is to reduce the price of these powders, and for the injection-molded approach, to produce a large size fraction below about 40 μm.

References

[1] C.F. Yolton, Gas atomized titanium and titanium aluminide alloys, in: F.H. Froes (Ed.), P/M in Aerospace and Defense Technologies, vol. 1, MPIF, Princeton, NJ, 1990, pp. 123–131.

[2] F.H. Froes, Titanium powder metallurgy: developments and opportunities in a sector poised for growth, powder metallurgy review, Winter 2013, Inovar Communications Ltd. 2013 pp. 27–41.

[3] J.R. Rieken, A.J. Heidloff, I.E. Anderson, Improved fine powder production of titanium alloys using close-coupled gas atomization, Proceedings of MS&T 2013, Montreal, QC, Oct 27–31, 2013.

[4] P. Lowenstien, Specialty metal powders by the rotating electrode process, Prog. Powder. Metall. 37 (1982).

[5] W.T. Nachtrab, P.R. Roberts, H.A. Newborn, Powder metallurgy of advanced titanium alloys, Key Eng. Mater. 77–78 (1993) 115–140.

[6] S. Pleir, W. Goy, B. Schaub, M. Hohmann, M. Mede, R. Schumann, EIGA-innovative production method for metal powder from reactive and refractory alloys, PM2TEC, MPIF, Princeton, NJ, 2004.
[7] H. Shiraishi, N. Arimoto, K. Yamasaki, S. Mori, Manufacturing technique of titanium powders by gas atomizing process, Materia Japan 34 (6) (1995) 792–794.
[8] M. Entezarian, F. Allaire, P. Tsantrizos, R.A.L. Drew, Plasma atomization: a new process for the production of fine, Spherical Powders, JOM (June 1996) 53–55.
[9] M.I. Boulos, New frontiers in thermal plasmas from space to nanomaterials, Nucl. Eng. Technol. 44 (1) (February 2012) 1–8.

Production of titanium powder by an electrolytic method and compaction of the powder

3

James C. Withers
Materials & Electrochemical Research (MER) Corporation, Tucson, AZ, USA

3.1 Introduction

Since the Kroll route [1,2] titanium became the dominant process to produce titanium, which is a large irregular particulate morphology more than a half-century ago, there have been significant efforts from a number of reduction technologies to develop the process to produce titanium particulate in a morphology suitable for direct use in powder metallurgy processing. The morphology of sponge particulate from the Kroll process limits its direct use in powder metallurgy processing due to its high cost. Kroll sponge is subsequently melt refined and wrought processed into mill forms which are then reprocessed into powder suitable for direct powder metallurgy processing. Consequently, high-quality titanium powder results in 15–30 times the original value of sponge. In addition, the ingot melt metallurgy alloy processing limits the possible alloy combinations because of density and melting point wide differences with titanium. Expanding possible alloy compositions and producing powders with a morphology for direct use in powder metallurgy processing at lower energy and processing cost than Kroll processing, provides the impetus to broaden the use of titanium while making it affordable for broad industrial use.

The first, if not one of the first, electrolytic isolations of titanium was by Hupperty [3] in 1905. In spite of Kroll's success of using electrolytically produced magnesium to metallothermically reduce titanium tetrachloride ($TiCl_4$), which is referred to as "tickle," that is, the Kroll process, Kroll suggested that the most preferred process to produce metallic titanium would be electrolytic. There have been extensive electrolytic investigations over the past half-century to produce titanium [4–53] and develop electrolytic processing to a commercial scale [54–64]. The morphology of electrolytically produced titanium ranges from nanoparticles, to flakes, to dendrites that relates to the fused salt electrolyte and operating parameters as well as the metallic titanium ion concentration in the electrolyte. Because of the potential of titanium for electromotive decomposition relative to hydrogen in the electromotive thermodynamic series, titanium can only be electrolytically produced from electrolytes that do not contain any ionizable hydrogen. This means fused salts in which all water or ionizable hydrogen is eliminated and maintained in isolation from the atmosphere to avoid picking up any water vapor as well as interstitials of oxygen, nitrogen, or carbon, to which titanium is quite sensitive. Almost

Titanium Powder Metallurgy. http://dx.doi.org/10.1016/B978-0-12-800054-0.00003-4

without exception, the alkali and alkaline earth halides have been used as electrolytes whose eutectic compositions can melt at as low a temperature as 350°C.

Electrolytic processes have primarily utilized a titanium halide or a complex of the halide as the feed to electrolytically produce titanium with most emphasis in $TiCl_4$. The electrolysis of titanium is an ionizable process whereas $TiCl_4$ is a covalent bonded compound that has very low solubility in ionized fused halide salts. This has led to many architectural configurations of electrolytic processes in order to achieve sufficient solubility of a titanium ion in solution to support electrolysis at a voltage potential less than that needed for decomposing the alkali or alkaline earth metal at the cathode. The disproportionation reversibility between the variable valences of titanium ions in fused salt electrolytes complicates processing to electrolytically produce titanium. In case of covalent compound feed of $TiCl_4$, possible reactions include the following:

$$TiCl_4 + Ti = 2TiCl_2$$

$$TiCl_4 + TiCl_2 = 2TiCl_3$$

$$2TiCl_3 + Ti = 3TiCl_2$$

In the ionized state, the reversible reactions include the following:

$$2Ti^{3+} \leftrightarrow Ti^{4+} + Ti^{2+}$$

$$4Ti^{3+} \leftrightarrow 3Ti^{4+} + Ti$$

$$2Ti^{2+} \leftrightarrow Ti^{4+} + Ti$$

$$3Ti^{2+} \leftrightarrow 2Ti^{3+} + Ti$$

When a potential is applied, possible reactions include the following:

$$Ti^{4+} + e^- \rightarrow Ti^{3+}$$

$$Ti^{3+} + e^- \rightarrow Ti^{2+}$$

$$Ti^{2+} + 2e^- \rightarrow Ti$$

To minimize disproportionation reversibility of titanium ions in an electrolytic reactor that consumes electrical power without producing titanium metal, a separator diffusion barrier between the anode and the cathode has been utilized [63]. Many materials of construction and configurations of separator diffusion barrier have been investigated

[65]. The most reported successful separators diffusion barriers were those described by US Bureau of Mines [63,65] and the author [66–68].

Although success to achieve Faradaic efficiencies in the late 1980s up to the mid-1990s was reported using $TiCl_4$ as a feed in electrolytic processes [23–33], large-scale commercialization was impeded by the batch nature of the electrolytic process similar to the batch operation of the Kroll process. As in all electrolytic processes where the electrolytic deposition is in the solid state below the melting point of the metal being deposited (i.e., in contrast to magnesium and aluminum), deposition is carried out on large flat cathode surfaces analogous to electrolyte copper. However, in the case of titanium, the electrolytic cell must be sealed and protected from the atmosphere, which complicates harvesting titanium electrolytically deposited on a large sheet cathode and replacing a new cathode for additional deposition. Elaborate chambers were used to exchange large flat cathodes and/or cylinder cathodes without exposing the electrolytic cell to the atmosphere. Another disadvantage in the electrolytic process was drag out of salt electrolyte containing valuable titanium ion on the very high surface area of the electrowon titanium on the cathode surface. After exchange of the cathodes from the airlocks and exposure to the atmosphere, the titanium ion in the salt oxidizes, leaving a titanium oxygen composition on the electrowon titanium as well as loss of this valuable titanium ion material. The titanium cathodes were then washed to remove the salt electrolyte, which produced an environmentally hazardous salt-water stream that had to be dealt with by discarding or recycling, which could not contain any residual titanium oxygenated material.

The cathode harvesting while maintaining the electrolytic cell free of atmosphere contamination and dealing with freeing the cathode of salt with water washing without exacerbating oxidation passivation of the electron titanium as well as the environmental issues of a water-salt stream, and the batch nature of the process all contributed to the electrolytic process not replacing the Kroll process even with the proven large commercial pilot plants of Dow-Howmet and Timet that were believed to be more economical than the Kroll process.

3.2 New and advanced processing

The Defense Advanced Research Project Agency (DARPA) initiated a program in 2003 to produce titanium more economically and different from the Kroll process, with the oxygen content of the titanium not to exceed 500 ppm. A process based on Withers' patents [69,70] consisting of a composite anode of a metal oxide, and carbon was selected by DARPA for investigation of electrolytically producing titanium powder. The proposed process eliminated using large plate cathodes and thus the expensive, cumbersome, and batch processing of previous electrolytic approaches. Additional patents were granted to Withers [71–74] for utilizing an electrolytic cell wherein the cathode electrowon titanium powder was washed or flushed from the cathode by pumping high-flow-rate electrolyte over the cathode surface into a separate closed cell where the electrowon titanium powder was harvested by filtering and centrifugal processing to separate all the electrolyte except that held on the titanium powder by surface tension. The separated electrolyte containing the valuable titanium flows back to the electrolytic cell without any exposure to contamination. This process provides for the electrolytic cell to operate continuously without ever having to be opened for contamination exposure.

The DARPA program demonstrated that the composite anode could be a stoichiometric mixture of titanium dioxide (TiO_2) and carbon or a reduced form consisting of TiO_xC_y with the ideal stoichiometry of Ti_2OC. The x and y ratio is preferred to be such that in electrolysis CO is emitted from the anode without leaving an excess of O or C. The reduction of TiO_2 with carbon in the absence of air can be heat treated with a hydrocarbon gas or with solid carbon. TiO_2 reduction begins at approximately 1150–1200°C, with the reduction rate increasing with increasing temperature. At lower temperatures, the reduction to a specific stoichiometry takes more time, whereas it is complete in a matter of minutes at approximately 1800°C. There is extensive literature related to carbothermic reduction of TiO_2 [75–112].

The TiO_2 for reduction can be natural or synthetic rutile, or an ore. In the DARPA program [67,68] and other government agency–sponsored programs, Withers [66,75] investigated, in addition to natural and synthetic rutile, off-specification very fine to nano TiO_2 and the ores ilmenite and perovskite. Processing demonstrated that a contamination-free TiO_xC_y that was typically Ti_2OC could be produced. The TiO_xC_y could be utilized as an anode to electrolytically produce titanium in a fused salt electrolyte consisting of alkali or alkaline earth halides. At the anode, the Ti_2OC is ionized to Ti^{2+} ions, which are then reduced to titanium metal at the cathode.

$$Ti_2OC = 2Ti^{+2} + CO \text{ anode reaction}$$

$$Ti^{+2} = 2e + Ti \text{ cathode reaction}$$

While the electrochemistry is desirably straightforward, the mechanics of operating commercial-size cells are challenging. The electrical conductivity is desirably high of Ti_2OC at under 1000 μΩ-cm. The Ti_2OC is a ceramic that can be processed into solid blocks to serve as the anode by standard ceramic processing techniques. Anode lead rodding to the Ti_2OC anode blocks becomes challenging. Graphite is the only anode connector rod that prevents anodic dissolution that would contaminate the electrolytic cell. It is thus necessary to rod a dense ceramic Ti_2OC with a graphite rod that has minimum resistance within the joint that will not overheat under the power of electrolysis.

Since electrolysis is typically carried out above approximately 500°C to maintain solubility and avoid precipitation of the alkali titanium halide complex such as K_2TiCl_6 or $KTiCl_4$, the dense Ti_2OC anode is subjected to cracking when pieces of anode dislodge, which can short across the small spacing of 1.25–3 cm between the anode and the cathode that can shut the cell down, requiring opening for repair. Further, as the anode electrochemically dissolves, the micrometer- to nanometer-size Ti_2OC particles can dislodge and be swept into the electrowon titanium cathode powder, resulting in contamination.

An alternative to large solid Ti_2OC anodes and the rodding connection issues is to use pieces of Ti_2OC in a graphite fabric basket that serves as the anode in which the graphite basket is made anodic as well as graphite rods within the basket. The graphite fabric basket must be porous enough to pass ions and prevent high resistance yet prevent small particulate Ti_2OC from passing through to contaminate the electrowon titanium powder. Such an anode configuration has higher resistance than solid Ti_2OC, which means higher consumption of power to produce titanium powder.

The inherency of Ti_2OC to ionize into Ti ions at low energy suggests there is a reactivity bias that can be advantageously used to produce $TiCl_4$ [113]. Ti_2OC begins to chlorinate at approximately 200°C and can achieve a self-sustaining reaction at 350–400°C, which is confirmed by the chlorination thermodynamics [113].

In the DARPA program, Withers [67,68] demonstrated that titanium-bearing ores could be converted to Ti_2OC, which could be chlorinated at ±400°C or to produce $TiCl_4$. This low-energy, low-cost $TiCl_4$ can then be used for Kroll processing or as a feed to electrolytic cells that calculates to produce titanium at substantially less energy than the Kroll route. Comparatively the energy of the Kroll process to produce sponge is approximately \$9/kg, which consumes approximately 100 kWh/kg of energy [114–116]. The estimated electrolytic cost is \$3.50/kg for pure Ti and to codeposit aluminum and vanadium to produce the alloy is \$7/kg, both at approximately 50% energy efficiency. Also there is considerably less CO_2 emission for the electrolytic process versus the Kroll route.

3.3 Electrolytic production of titanium powder

In the DARPA program, large pilot commercial demonstration cells that operated continuously were fabricated and operated to demonstrate producing titanium powder. An illustration of such a system is shown in Figure 3.1, with the actual cell shown in Figure 3.2.

It was confirmed that a titanium ion concentration of approximately 2% was required to operate at high cathode current density and produce particulate generally greater than approximately 10–25 μm and large particle sizes up to approximately 500 μm. Low cathode current densities under approximately 0.25 amp/cm^2 tended to result in fine powder regardless of titanium ion concentrations. Particle sizes generally maximized in the cathode current density of approximately 1.0–1.25 amp/cm^2. Particle size is also controlled by the periodicity of flushing the depositing particles from the cathode surface by pumping the electrolyte at high velocity across the cathode surface to sweep off the deposited powder. Frequent flushing does not give the nucleated particles on the cathode sufficient time to grow to a prescribed size. Thus, cathode current density and flushing periodicity can be regulated to produce select particle sizes. Some electrowon titanium particles are shown in Figures 3.3–3.5.

Since the electrolytic cell remains closed, it becomes very pure from interstitial contaminates with continued electrolysis and is capable of producing titanium particles with oxygen contents less than 100 ppm.

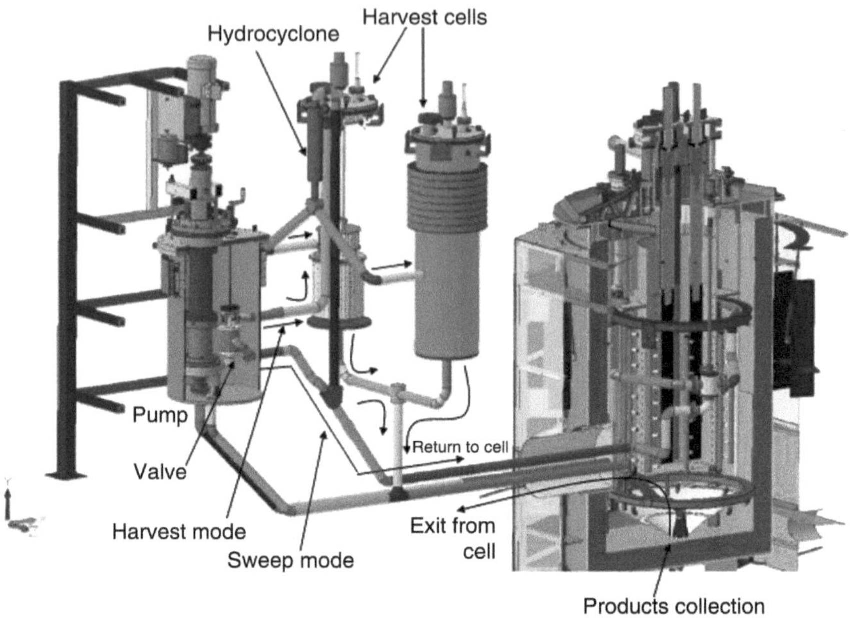

Figure 3.1 Illustration of a production cell to continuously produce titanium.

The residual oxygen content of electrowon titanium particles will contain passivation oxygen relative to particle size as well as the processing used to separate the residual electrolyte from the particles. In the continuous electrolysis system, pumping of the electrolyte at high velocity over the cathode surface transports the particles to metal screen filters that concentrate the particles, which are then transferred to a centrifuge operating above the melting point of the salt that removes all the salt electrolyte except that adhering from surface tension. That residual salt is then removed by one of three techniques consisting of vacuum vaporizing at temperatures just below the sintering temperature of the powder, which is approximately 900°C. Using the eutectic salt of KCl–LiCl, residual chlorine can be reduced to below detection at 10 ppm. If lower vapor pressure salts such as NaCl, $CaCl_2$, etc. are used, higher temperatures can be used in a fluid bed, which generally prevents the titanium powder from sintering. Ball-milling the salt-laden titanium powder at the ambient or a lower temperature will crack the brittle salt off the titanium particulate.

3.4 Titanium alloy powder

Currently, titanium alloy powder is produced from alloyed Kroll sponge that is formed into a wrought product that is then reduced to a powder by a melt process, resulting in a high-cost powder in the range of approximately 15–30 times the cost of sponge. Armstrong/ITP/Crystal has experimentally produced Ti-6Al-4V

Figure 3.2 Electrolytic cell assembly continuously producing titanium powder.

Figure 3.3 Electrolytically produced from halide salt electrolytes containing Ti^{2+} ion.

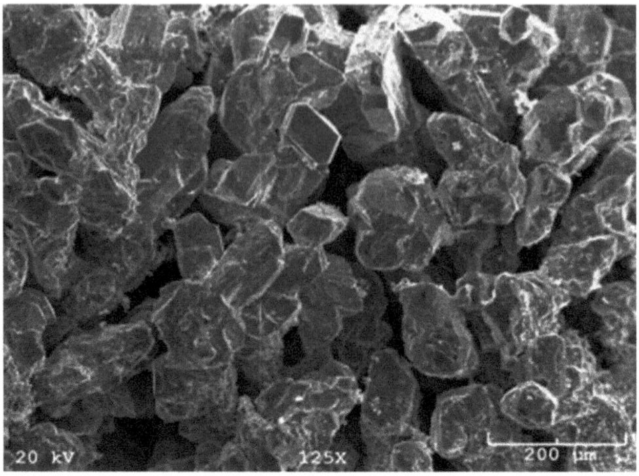

Figure 3.4 Median size electrolytically produced titanium particulate.

powder by the coreduction of the precursor chlorides $TiCl_4$–$AlCl_3$–VCl_4 with sodium (Hunter processing), which results in high residual chlorine because of the difficulty in fully removing the by-product salt NaCl and unreacted sodium; in addition, the Hunter process is known to produce a small particle size of approximately 10 μm. In the DARPA [67,68] and Air Force programs [112], it was demonstrated that the Ti-6Al-4V alloy could be electrolytically produced if $AlCl_3$ and VCl_3 or VCl_2 was added to the electrolytic processing described above. Control of the concentration of the alloying compound relative to the activity coefficient of the alloying ion and the voltage to achieve codeposition provides a window to electrolytically produce Ti-6Al-4V alloy in desirable particle sizes for direct

Figure 3.5 Good uniform crystalline electrolytically produced titanium particulate.

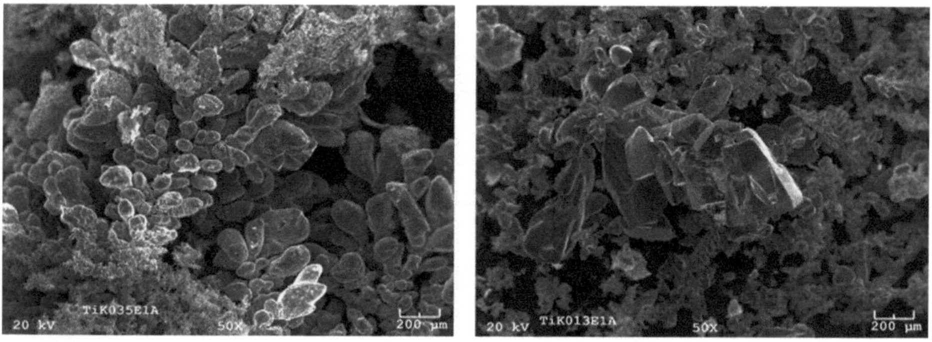

Figure 3.6 Alloy powder produced electrolytically from fused salt containing $TiCl_2$–$AlCl_3$–VCl_3.

use in powder metallurgy processing. The decomposition voltages of the alloying elements are sufficiently close to that of titanium to provide a sufficiently wide operating window for commercial production of alloy powder. Since the cost of electrolytically producing titanium powder is projected to be less than the Kroll sponge cost, it can be estimated to electrolytically produce Ti-6Al-4V powder for the approximate cost of sponge that is significantly lower than that for current alloy powder. An example of electrolytically produced alloy powder is shown in Figure 3.6.

3.5 Compaction of electrolytically produced titanium powder

Electrolytic titanium powder was cold pressed in steel dies and/or cold isostatically pressed followed by sintering at 1000°C for 4 h. Some of the electrolytic powder was also melted and along with the sintered material rolled into sheet for mechanical property characterization. The properties achieved by an outside titanium company are shown in Table 3.1.

The electrolytically produced powder was also used as a feed to a plasma transferred arc (PTA) layer-by-layer rapid additive manufacturing (AM) system that consolidated the powder into billets that were hot and cold rolled down to 1 mm thickness. In the PTA system shown in Figure 3.7, aluminum and vanadium powder as well as prealloyed aluminum and vanadium powder were co-fed to produce the alloy Ti-6Al-4V.

A variety of other alloy compositions have been produced that are not possible by standard ingot metallurgy processing. The alloying elements coadded with titanium singularly and in combination include iron, chromium, manganese, and TiB_2 to produce an alloy with higher modulus than standard Ti-6Al-4V. Titanium (Kroll) sponge has also been used as a titanium feed to the PTA system. A variety of net shapes produced by the PTA AM system is shown in Figure 3.8.

Table 3.1 Properties of electrolytic powder and parts made from electrolytic powder

Chemistry of CP Electrolytic Ti Powder

	ASTM Grade 1	Direct Rolled Electrolytic Compact	Direct Rolled Electrolytic Compact
	CP Ti	CP Ti	CP Ti
		2A	**2B**
Carbon	0.08 max	0.011	0.011
Oxygen	0.07 max	0.022	0.031
Nitrogen	0.03 max	0.002	0.006
Hydrogen	0.015 max	0.015	0.016
Iron	0.20 max	0.001	0.004
Chlorine	0.10 max	0.03	0.01

	ASTM Grade 1 – CP			**ASTM Grade 5 Ti-6Al-4V**		
Properties	**ASTM Gr 1 Specs**	**2A Direct Roll**	**2B Melt / Roll**	**ASTM Gr 5 Specs**	**3A Melt and Roll**	**3B Melt and Roll**
Oxygen ppm	1800	200	300	2000	1500	1500
UTS ksi	35	34	40	130	143	140
0.2% YS ksi	25	22	26	120	123	116
Elongation %	24%	49%	41%	10%	13%	13%

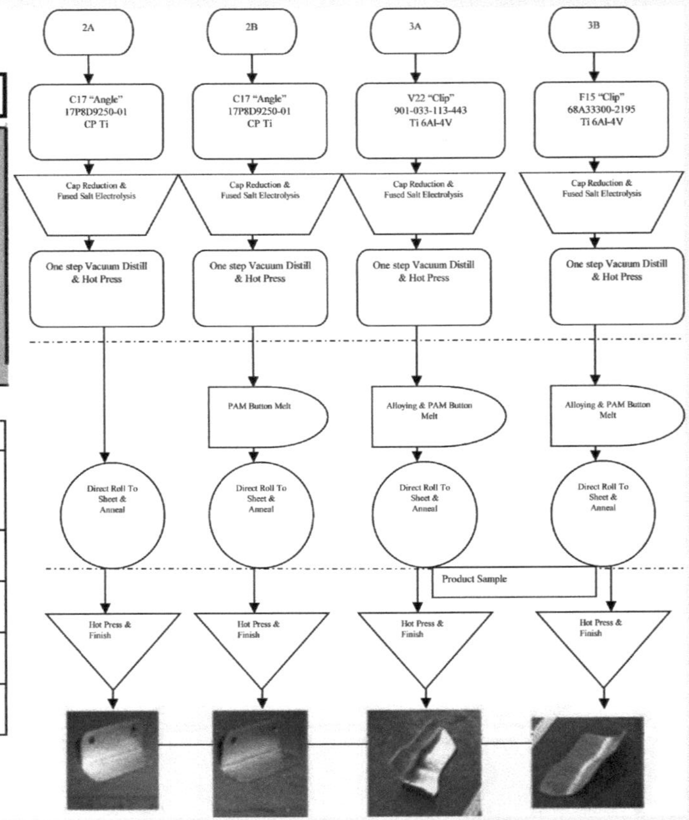

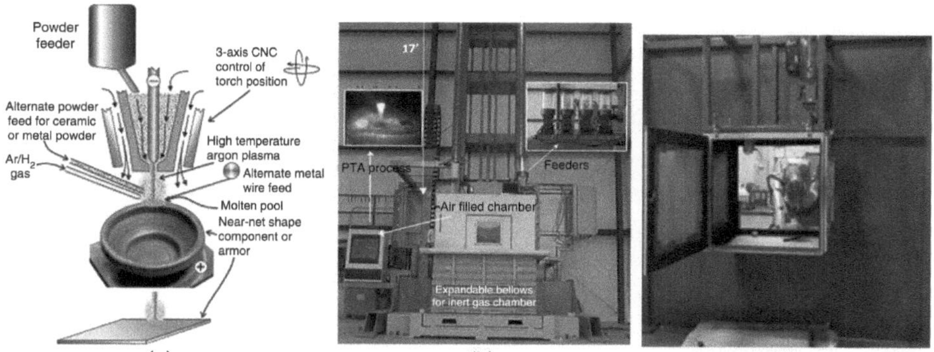

Figure 3.7 (a) Schematic of PTA rapid additive manufacturing system, (b) operational MER unit, and (c) robotic controlled system.

Some examples of properties of PTA-produced alloys are shown in Table 3.2.

The reason the PTA AM-produced titanium alloys exhibit higher strength than standard commercially produced alloy is that the grain size is much smaller, as shown in Figures 3.9 and 3.10. The PTA AM-produced alloy is rapidly cooled from the minimelt pool, also contributing to smaller grain formation.

In addition to the PTA system producing fully dense material to near-net shape, processing parameters can be used to produce melt-formed foams. An example is shown in Figure 3.11 that provides a cross section and an actual component. The porosity can be varied from approximately 20% to 70% with varying thickness skins as

Figure 3.8 Varying net-shape 3-dimensional geometries of PTA produced low-cost titanium from low-cost sponge powder.

Table 3.2 Tensile/modulus properties of PTA-produced titanium alloys

Alloy composition	Tensile strength (MPa)	Modulus (GPa)	Elongation (%)
Ti-6Al-4V literature	896	147	10–14
Ti-6Al-4V PTA	1470	148	9–14
Ti-$_x$Al-$_y$Fe	1470	145	8–12
Ti-$_x$Al-$_y$Mn	1700	145	8–10
Ti-$_x$Al-$_y$Cr	1525	149	7–10
Ti-$_x$ + $_y$	1560	175	4–8

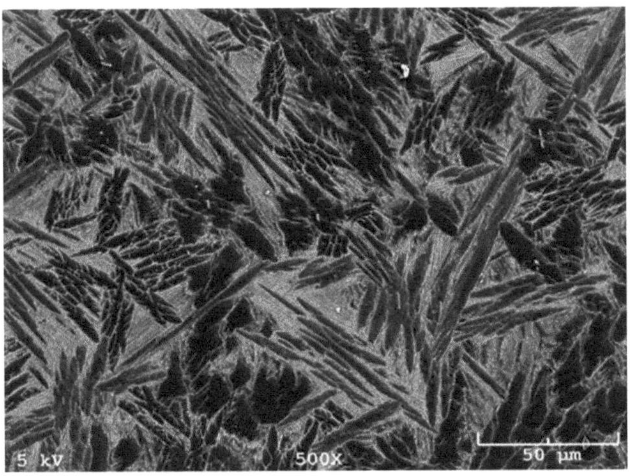

Figure 3.9 Micrograph of Ti-6Al-4V from standard commercial plate.

Figure 3.10 Micrograph of PTA-produced Ti-6Al-4V from electrolytic powder.

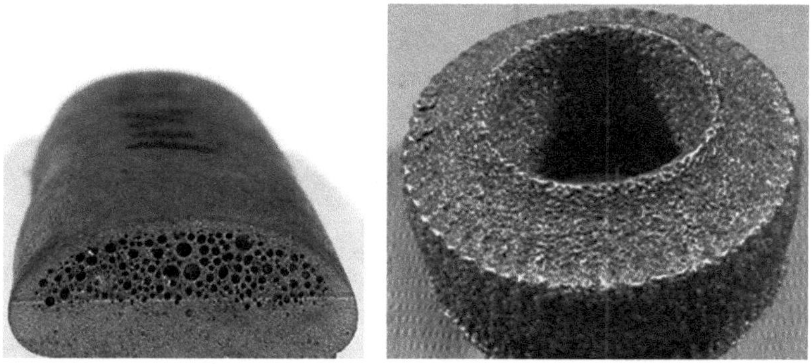

Figure 3.11 Porous foam titanium produced by PTA processing using electrolytic powder feed or sponge feed.

well as solid nonporous areas within the foam. Since the foam is melt formed, its properties are on the order of two times those of powder metallurgy–produced porous materials.

References

[1] W. Kroll, Method for manufacturing titanium and alloys thereof. German patent application July 10, 1937, U.S. patent application July 6, 1938, U.S. Patent 2, 205, 854 (1937).

[2] W. Kroll, Tr. Electrochem. Soc. 78 (1940) 35–47.

[3] E. Hupperty, Electrochem. Met. Ind. 3 (1905) 35.

[4] R.P. Lee. Corrosion and Materials Engineering Data for $TiCl_4$ and Ti Metal Manufacture, National Lead Company, Titanium Division, Development and Engineering Department, 1952.

[5] H.R. Palmer, O.W. Moles, J.A. Hamilton, Construction of a Twelve Pound Per Day Electrolytic Titanium Cell for Use with a Chlorine Atmosphere, National Lead Company, Titanium Division, 1955.

[6] O.W. Moles, R.C. Gardner, L.W. Gendvil, J.A. Hamilton, H.R. Palmer, Operation of a Basket Cathode Titanium Cell Utilizing a Chlorine Protective Atmosphere, National Lead Company, Titanium Division, 1956.

[7] H.R. Palmer, C. Mac Dye, Electrolytic production of titanium, TMCA, 1957.

[8] L.E. Snyder, Operation of the research electrolytic cell, Project Report, TMCA, 1958, pp. 47–43.

[9] A. Boozenny, Mechanism for electrolytic reduction of $TiCl_2$, TMCA, 1958.

[10] F.P. Haver, D.H. Baker, Development of a 10,000-ampere cell for electrorefining titanium, Bureau of Mines, 1960.

[11] A. Boozenny, Na_0-$TiCl_2$-$TiCl_3$ Equilibrium in NaCl melts, TMCA, 1961.

[12] A. Boozenny, E.P. Rheinfeld, W.F. Plott, J.J. Henderson, The effects of alternating current density and atmospheric contamination on titanium electrorefining cell operation, TMCA, 1961.

[13] L.R. Lyons, E.P. Rheinfeld, J.J. Henderson, Electrodeposition of titanium at low temperatures, TMCA, 1962.

[14] L.R. Lyons, E.P. Rheinfeld, J.J. Henderson, Electrodeposition of titanium in NaCl-KCl electrolyte at 700°C, TMCA, 1962.

[15] L.R. Lyons, E.P. Rheinfeld, J.J. Henderson, Electrowinning of titanium at 700°C, TMCA, 1963.
[16] L.R. Lyons, E.P. Rheinfeld, Electrorefining titanium from basket anodes at 700°C, TMCA, 1963.
[17] J.C. Priscu, L.E. Snyder, E.R. Poulsen, Electrolytic production of titanium, Project Report, TMCA 1965, pp. 46–56.
[18] L.C. Covington, E.H. Groshan, Electrorefining of titanium in the molten salt system NaCl-KCl-$MgCl_2$-$TiCl_2$, TMCA, 1966.
[19] J.C. Priscu, E.R. Poulsen, Alteration of reactor No. 8 for a 46" × 46" × 56" cathode, reactor 8B progress report no. 10, Project Report, TMCA, 1967, pp. 44–56.
[20] L.E. Anderson, L. Vegas, R.E. Adams, G.D. Weese, both of Henderson, all of Nev. Pump for molten salts and metals, U.S. Patent, 3 (776), (1973), 660.
[21] R.E. Adams, E.R. Poulsen, Evaluation of production melting of electrolytic sponge, TMCA, 1974.
[22] E.R. Poulsen, Reactor no. 12 prototype electrowinning cell construction & operation, TMCA, 1977.
[23] A.M. Martinez, Y. Castrillejo, E. Barrado, G.M. Haarberg, G. Picard, A chemical and electrochemical study of titanium ions in the molten equimolar $CaCl_2$ + NaCl mixture at 500°C, J. Electroanal. Chem. 449 (1998) 67–80.
[24] J.A. Menzies, D.L. Hill, G.J. Hill, J. Young, J. Bockris, J. Electroanal. Chem. 1 (1959)161.
[25] W.E. Reid, J. Electrochem. Soc (1957) 108, 393.
[26] F. Ouemper, D. Deroo, N. Rigard, J. Electrochem. Soc. 119 (1972) 1353.
[27] E. Chassaing, F. Basile, G. Lorthiair, J. Less-Com. Metals 68 (1979) 153.
[28] S. Tokumoto, E. Tanaka, O. Ogisu, J. Metall. 27 (1974) 175.
[29] S. Mori, T. Kuroda, K. Kawamura, Denki Kagaku 42 (1974) 175.
[30] M.B. Airpert, F.J. Shultz, W.F. Sullivan, J. Electrochem. Soc. 104 (1957) 555.
[31] M.B. Airpert, J.A. Hamilton, F.J. Shultz, W.F. Sullivan, J. Electrochem. Soc. 106 (2) (1959) 142.
[32] K. Komarek, P. Herasynenko, J. Electrochem. Soc. 105 (4) (1958) 210.
[33] M.V. Smirnov, O.V. Skiba, Electrochemistry of Molten and Solid Electrolytes vol. 2 Consultants Bureau, New York, (1964) p. 21.
[34] S.N. Flengas Ann NY Acad. Sci. 79 (1960) 853.
[35] M.J. Rand, L.J. Reimert, J. Electrochem. Soc. 111 (1964) 429.
[36] E. Chassaing, F. Basile, G. Lorthoir, J. Appl. Electrochem. 11 (1981) 187.
[37] D.M. Ferry, Thèse de Doctorate de l'Université Paris 6 (1985).
[38] D.M. Ferry, G.S. Picard, B. Trémillon, Trans. Inst. Min. Metall. C. Min. Proc. Ext. Metall. 97 (1988) C21.
[39] E. Chassaing, F. Basile, G. Lorthoir, Study of Ti (III) solutions in various molten alkali chlorides, J. Appl. Electrochem. 11 (1981) 187–191.
[40] F. Lantelme, K. Kuroda, A. Barhoun, Electrochemical and thermodynamic properties of titanium chloride solution in various alkali chloride mixtures, Electrochim. Acta 44 (1998) 421–431.
[41] V. Vasilesci, N. Ene, The electrochemistry of some titanium and zirconium compounds in molten salts, Rom. Chem. Q. Rev. 2 (1) (1991) 29–49.
[42] N. Ene, S. Zuca, M. Anghel, G. Busila. Influence of current density on the cathode deposition of Ti powder by electrorefining in molten alkali halide mixtures, Revue Roumaine Chim. 36 (2) (1993) 133–138.
[43] S. Zuca, N. Ene, Electrorefining of titanium in molten alkali halide mixtures, Mater. Sci. Forum 73-75 (1991) 643–652.

[44] N. Ene, S. Zuca, Role of free F^- anions in the electrorefining of titanium in molten alkali halide mixtures, J. Appl. Electrochem. 25 (1995) 671–676.
[45] F. Lantelme, A. Salmi, J. Electrochem. Soc. 142 (1995) 3451.
[46] D. Wei, M. Okido, T. Oki, J. Appl. Electrochem. 24 (1994) 923.
[47] H. Numata, H. Takamura, I. Ohno, Molten salt chemistry and technology, in: H. Wendt (Ed.), Proceedings of Fifth International Symposium on Molten Salt Chemistry and Technology, Molten Salt Forum, Dresden, Germany, 5–6 (1998) 311.
[48] F.R. Clayton, G. Mamantov, D.L. Manning, J. Electrochem. Soc. 120 (1973) 1193.
[49] G.R. Stafford, T.P. Moffat, Electrochemistry of titanium in molten $2AlCl_3$-NaCl, J. Electrochem. Soc. 142 (10) (1995) (October).
[50] G.R. Stafford, The electrodeposition of Al_3 Ti from chloroaluminate electrolytes, J. Electrochem. Soc. 141 (4) (1994) (April).
[51] K.W. Fung, G. Mamantov, Electrochemistry of titanium (11) in $AlCl_3$ + NaCl melts, J. Electroanal. Chem. 35 (1972).
[52] V. Ananth, et al. Single step electrolytic production of titanium trans, Indian Inst. Met. 51 (5) (1998) pp. 399–403 (October).
[53] H.Y. Hsu, D.L Chen, H.W. Tsaur, C.C.Yang The Electrodeposition of Ti from the Low Temperature Molten Electrolyte, Electrochemical Society Proceedings, 99–41 (1991) 585–596.
[54] L. Reimert, M. Rand, J. Electrochem. Soc 111 (April) (1964) 4.
[55] A. Myhren. J. Met. May (1968).
[56] R. MacMullin, J. Electrochem. Soc. (November) (1976) pp. 359C–368C.
[57] M. Alpert, J. Electrochem. Soc. 104 (September) (1957) 9.
[58] M. Alpert, J. Electrochem. Soc. 106 (February) (1959) 2.
[59] J.C. Priscu, Titanium Electrowinning Cell. Symposium on Electrometallurgy, Proceedings AIME Extractive Metallurgy Div., Cleveland, Ohio, December, 1968, pp. 83–91.
[60] E. Poulsen, J. Hall, J. Met.(June) (1983).
[61] G. Cobel, (1980). Proceedings of the Fourth International Conference on Titanium, Kyoto, Japan pp. 1969–1976.
[62] G. May Cobel, J. Fisher, L. Snyder, Electrowinning of Titanium from Titanium Tetrachloride, Proceedings of the Forth International Conference on Titanium, Kyoto, Japan, 1980.
[63] O.Q. Leone, H. Knudsen, D. Couch, High purity titanium electrowon from titanium tetrachloride, J. Metall. (1967) pp18–23 March.
[64] E. DiMaria, RMI gets license to make a new type of titanium, Metal Working News, 1988, February.
[65] O.Q. Leone, D.E. Couch, Use of composite diaphragms in the electrowinning of titanium, Bureau of Mines report of investigations, 7648, June 5, 2003.
[66] J.C. Withers, Low cost alternatives to titanium plate production, Contract W911QX-04-C-0009.
[67] J.C. Withers, Contract DAAD17-03-C-0048.
[68] J.C. Withers, Contract HR0011-06-0007.
[69] U.S. Patent 4,342,637.
[70] U.S. Patent 4,409,083.
[71] U.S. Patent 7,410,562.
[72] U.S. Patent 7,794,580.
[73] U.S. Patent 7,914,600.
[74] U.S. Patent 7,985,326.
[75] R.L. Bickerdike, G. Hughes, An examination of part of the titanium-carbon system, J. Alloy. Compd. 2 (1959) 42–49.

[76] C.C. Chou, C.I. Lin, Reaction between titanium dioxide and carbon in flowing helium stream, Brit. Ceram. T. 100 (6) (2001) 197–202.
[77] K.S. Coley, B.S. Terry, P. Grieveson, Simultaneous reduction and carburization of ilmenite, Metall. Mater. Trans. B 26B (1995) 485–494.
[78] K.S. Coley, B.S. Terry, P. Grieveson, Carburisation-reduction of ilmenite ores, pp. 151–175.
[79] M.A.R. Dewan, G. Zhang, O. Ostrovski, Carbothermal reduction of titanium in different gas atmospheres, Metall. Mater. Trans. B 40B (2009) 62–69.
[80] M.A.R. Dewan, G. Zhang, O. Ostrovski, Carbothermal reduction of a primary ilmenite concentrate in different gas atmospheres, Metall. Mater. Trans. B 41B (2010) 182–192.
[81] A. Fernandes, P. Carvalho, F. Vaz, N.M.G. Parreira, P. Goudeau, E. Le Bourhis, et al. Correlation between processing and properties of titanium oxycarbide. TiC_xO_y thin films, Wiley InterScience, (2008).
[82] T. Hashishin, T. Yamamoto, M. Ohyanagi, Z. Munir, Simultaneous synthesis and densification of titanium oxycarbide, Ti (C,O), through gas-solid combustion, J. Am. Ceram. Soc. 86 (12) (2003) pp2067–2073.
[83] R. Koc, J.S. Folmer, Carbothermal synthesis of titanium carbide using ultrafine titanium powders, J. Mater. Sci. 32 (1997) 3101–3111.
[84] R. Koc, J.S. Folmer, Synthesis of submicrometer titanium carbide powders, J. Am. Ceram. Soc. 80 (4) (1997) 952–956.
[85] K.L. Komarek, A. Coucoulas, N. Klinger, Reactions between refractory oxides and graphite, Dept. Metall. Mater. Sci. Coll. Eng. 110 (7) (2009) 783–791.
[86] G.G. Lee, B.K. Kim, Effect of raw material characteristics on the carbothermal reduction of titanium dioxide, Mater. Trans. 44 (10) (2003) 2145–2150.
[87] P. Lefort, A. Maitre, P. Tristant, Influence of the grain size on the reactivity of TiO_2/C mixtures, J. Alloys Compd. 302 (2000) 287–298.
[88] A. Maitre, P. Cathalifaud, P. Lefort, Thermodynamics of titanium carbide and the oxycarbide Ti_2OC, High Temp. Mater. Process. 1 (1997) 393–408.
[89] A. Maitre, P. Lefort, Carbon oxidation at high temperature during carbothermal reduction of titanium dioxide, Phys. Chem. Chem. Phys. 1 (1999) 2311–2318.
[90] J.L. Murray, The C-Ti (carbon-titanium) system, (1987) 47–51.
[91] J.L. Murray, H.A. Wriedt, The O-Ti (oxygen-titanium) system, Bull. Alloy Phase Diag. 8 (2) (1987) 148–165.
[92] N. Setoudeh, A. Saidi, N.J. Welham, Effect of elemental iron on the carbothermic reduction of the anatase and rutile forms of titanium dioxide, J. Alloys Compd. 395 (2005) 141–148.
[93] K. Sun, W.K. Lu, Mathematical modeling of the kinetics of carbothermic reduction of iron oxides in ore-coal composite pellets, Metall. Mater. Trans. B 40B (2009) 91–103.
[94] G.A. Swift, R. Koc, Formation studies of TiC from carbon coated TiO_2, J. Mater. Sci. 34 (1999) 3080–3093.
[95] Y. Wang, Y. Zhangfu, Reductive kinetics of the reaction between a natural ilmenite and carbon, Int. J. Miner. Process. 81 (2006) 133–140.
[96] N.J. Welham, J.S. Williams, Carbothermic reduction of ilmenite ($FeTiO_3$) and rutile (TiO_2), Metall. Mater. Trans. B 30B (1999) 1075–1081.
[97] G. Zhang, O. Ostrovski, Reduction of titanium by methane-hydrogen-argon gas mixture, Metall. Mater. Trans. B 31B (2000) 129–139.
[98] G. Zhang, O. Ostrovski, Reduction of ilmenite concentrates by methane-containing gas: part I. Effects of ilmenite composition, temperature and gas composition, Can. Metall. Quarter. 40 (3) (2001) 317–326.

[99] G. Zhang, O. Ostrovski, Reduction of ilmenite concentrates by methane-containing gas: part II. Effects of preoxidation and sintering, Can. Metall. Quarter. 40 (4) (2001) 489–497.
[100] G. Zhang, O. Ostrovski, Effect of preoxidation and sintering on properties of ilmenite concentrates, Int. J. Miner. Process. 64 (2002) 201–218.
[101] A. Afir, M. Achour, N. Saoula, X-ray diffraction study of Ti-O-C system at high temperature and in a continuous vacuum, J. Alloys Compd. 288 (1999) 124–140.
[102] P.V. Ananthapadmanabhan, P.R. Taylor, W. Zhu, Synthesis of titanium nitride in a thermal plasma reactor, J. Alloys Compd 287 (1999) 126–129.
[103] L.M. Berger, P. Ettmayer, B. Schultrich, Influencing factors on the carbothermal reduction of titanium dioxide without and with simultaneous nitridation, Int. J. Refract. Metals Hard Mater. 12 (1994) 161–172.
[104] W.-Y. Li, F.L. Riley, The production of titanium nitride by the carbothermal nitridation of titanium dioxide powder, J. Eur. Ceram. Soc. 8 (1991) 345–354.
[105] T. Licko, V. Figusch, J. Puchyova, Carbothermal reduction and nitriding of TiO_2, J. Eur. Ceram. Soc. 5 (1989) 257–265.
[106] S.A. Rezan, G. Zhang, O. Ostrovski, Carbothermal reduction and nitridation of titanium dioxide in a H_2-N_2 gas mixture, J. Am. Ceram. Soc. 94 (11) (2011) 3804–3811.
[107] S.A. Rezan, G. Zhang, O. Ostrovski, Effect of gas atmosphere on carbothermal reduction an nitridation of titanium dioxide, Metall. Mater. Trans. B 43B (2012) 73–81.
[108] X. Chen, Y. Li, L. Yawei, J. Zhu, S. Jin, L. Zhao, Z. Lei, X. Hong, Carbothermic reduction synthesis of Ti(C,N) powder in the presence of molten salt, Ceram. Int. 34 (2008) 1253–1259.
[109] H.G. Branstatter, Recovery of titanium values, U.S. Patent 4 521 (1985) 385.
[110] S. Jiao, H. Zhu, Electrolysis of Ti_2CO solid solution prepared by TiC and TiO_2. J. Alloys Compd. (2006) 1–4.
[111] S. Jiao, H. Zhu, Novel metallurgical process for titanium production, Mater. Res. Soc. 21 (9) (2006) 2172–2175.
[112] J.C. Withers, Contract FA8650-09-C-5229.
[113] B.K. Chadwick, Y.K. Rao, Carbon and energy requirements of the chlorination of titanium oxides: an equilibrium analysis. TMS Technical Paper A86-25.
[114] A. Choragudi, M.A. Kuttolamadom, J.J. Jones, M.L. Mears, T. Kurfess, Investigation of the machining of titanium components in lightweight vehicles. SAE International Congress, 2010.
[115] S. Froes, Titanium powder metallurgy, a review part 1, 2012.
[116] J. Onillon, Aperam, Goldman Sachs Global Metals. Mining & Steel Conference, 2012.

Titanium powder production via the Metalysis process

4

Ian Mellor, Lucy Grainger, Kartik Rao, James Deane, Melchiorre Conti, Greg Doughty, Dion Vaughan
Metalysis Limited, Wath-upon-Dearne, Rotherham, United Kingdom

4.1 Introduction

Metalysis is a startup company spun out from the University of Cambridge, established to commercialize the FFC® process. It was originally founded in 2000 [1] as FFC Limited, subsequently adopting its current and more recognized name in 2003. The headquarters relocated to its present location near Sheffield, South Yorkshire, in 2004, prior to which it was based in Cambridge. The company was created to exploit the research of Fray, Farthing, and Chen [2], reporting the possibility of directly reducing solid TiO_2 to pure titanium in a molten salt electrolyte. In the region of 60 individual elements, and an indeterminate number of alloy combinations therein, are feasible via the FFC® process, with the extent of the former being depicted in Figure 4.1 below.

The basis of the selection criteria for each separate metal is that the voltage required to dissociate its corresponding oxide lies within the potential window of the solvent, generally calcium chloride containing a small amount of dissolved calcium oxide. The opportunity to produce a range of product types other than titanium illustrates a key advantage of FFC® versus its competing titanium powder manufacturing processes, which are somewhat limited in this regard.

Furthermore, as electrolysis is conducted solely in the solid state, it is possible to generate alloys in a single stage, which are either currently impossible, or challenging to synthesize via conventional means [1]. Titanium–tungsten [3–5] offers a prime example of this, as boiling of the former (3560 K or 3285°C) occurs below the onset of melting for the latter (3695 K or 3422°C). Moreover, numerous treatment steps would be necessary to produce this using more traditional methods, to overcome segregation resulting from mismatched densities (4,505 vs. 19,250 kg/m^3), and achieve full homogenization. The ability to accomplish this provides additional evidence of the vast potential and wide-ranging applicability of the FFC® process.

At the outset, Metalysis was granted a license for all metals bar titanium and its alloys above a threshold of 40 wt% [1], and only when an agreement was later reached with QinetiQ (formerly the Defence Evaluation and Research Agency [DERA]) did it secure the rights for the entire range of elements. As part of this arrangement, Metalysis acquired the intellectual property and associated knowledge developed by QinetiQ, which was marketed under the acronym EDO® (electro-de-oxidation) process. During this period, BHP Billiton were also conducting research on their own variant of FFC, termed the Polar® Process. Following negotiation, a strategic alliance

Titanium Powder Metallurgy. http://dx.doi.org/10.1016/B978-0-12-800054-0.00004-6

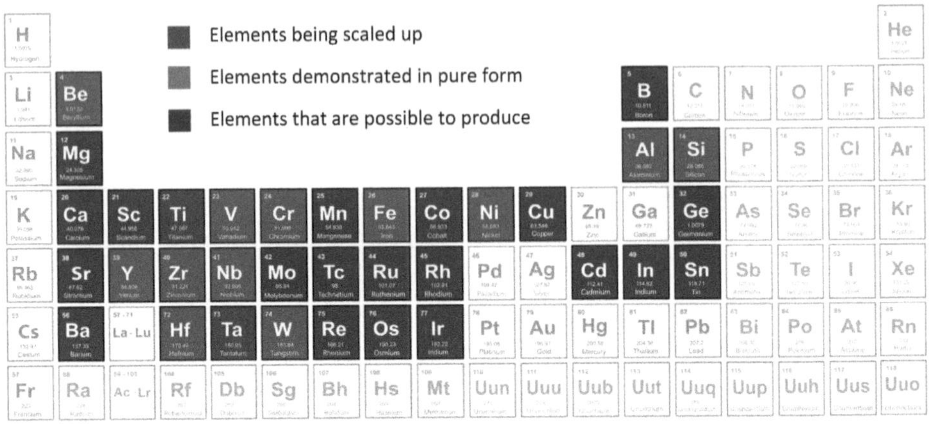

Figure 4.1 Periodic table highlighting the range of elements possible and those demonstrated via the FFC® process.

was struck between the two parties, where the generated intellectual property and technical know-how were transferred to Metalysis, in return for equity in the company. Not until all legal matters relating to this period were resolved in late 2007 was it possible to commence titanium activities in earnest. Currently, Metalysis intellectual property portfolio consists of more than 25 live patent families, granted in more than 90 countries worldwide, created via a combination of consolidating the inventions filed of partnering organizations and capturing new learning conceived internally. The detail of these traverse a multitude of areas specific to raw materials and feedstock, FFC® electrolysis fundamentals, plant design and operation, downstream processing of metallic powders, and product.

4.2 FFC® process overview

The elegance of the FFC® process lies in its simplicity. The metal oxide (or mixture in the case of an alloy) acts as the cathode in an electrochemical cell, where on application of a nominal voltage between this and the counter electrode, ionization occurs, releasing the liberated oxygen ion (O^{2-}) into the electrolyte. This is illustrated via the following reaction mechanism:

$$M_xO_y + 2ye^- \longrightarrow xM + yO^{2-} \tag{4.1}$$

Migration of O^{2-} to the anode ensues, where it subsequently combines with the electron passed via the external circuit and is evolved as either CO and CO_2 in the case of graphite (carbon), or O_2 when using an inert (nonconsuming) anode. These pathways can be described as

$$C + 2O^{2-} \longrightarrow CO_2 + 4e^- \tag{4.2}$$

$$C + O^{2-} \longrightarrow CO + 2e^- \tag{4.3}$$

$$O^{2-} \longrightarrow \tfrac{1}{2}O_2 + 2e^- \tag{4.4}$$

The net result is that the converted metal product is revealed at the cathode, where it is consequently harvested.

Reduction is performed in the range 1073–1273 K (800–1000°C) in a molten halide salt, generally calcium chloride. A distinct feature of this is its wide operating potential window, allowing a range of elements and alloys to be manufactured via the FFC® process (cf. Figure 4.1). Other benefits are its high ionic (O^{2-}) but poor electronic conductivity, attractive cost, and low toxicity, where one of its alternative applications is to deice roads in winter.

On completing the reaction, the titanium is removed from the cell in the form of a loosely bound mass. This is encapsulated in solidified calcium chloride, acting as a preventive barrier to re-oxidation, while also providing a degree of temporary structural integrity to the extracted cathode bed. Washing in water dissolves the present calcium chloride layer, and occasionally a dilute mineral acid is employed to eliminate any residual calcium oxide present. Light pressure can be applied where necessary, to yield the finished powder.

A schematic of the overall Metalysis process to synthesize powder is shown in Figure 4.2, in parallel to that for Kroll, by means of a comparison. One of the clear advantages of FFC® is its low environmental impact, as CO and CO_2 are its only

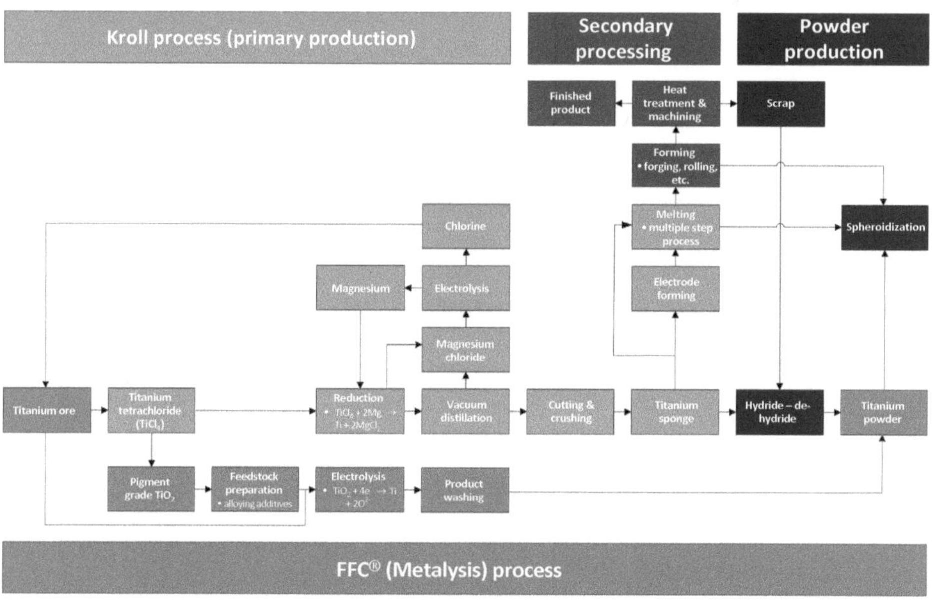

Figure 4.2 Schematic illustrating the steps required to produce titanium powder via FFC® compared to the more traditional Kroll process.

by-products (Equations 4.2 and 4.3), and under certain conditions these can be eliminated by selecting an inert anode, thus generating pure oxygen (Equation 4.4). Furthermore in stark contrast to Kroll, no highly reactive and hazardous chemicals are either utilized as the feedstock (titanium tetrachloride) or produced *in situ* (chlorine), and the most economically available reductant, the electron, is employed directly. This opinion has been independently corroborated by van Vuuren [6], and interestingly, Kroll himself predicted his method would eventually be replaced by an electrolytic process [1]. Conversely magnesium, which is used to manufacture conventional titanium sponge, involves a two-stage process. This initially comprises $MgCl_2$ electrolysis to generate the base metal, followed by subsequent reaction with $TiCl_4$.

In terms of feedstock selection, FFC® has traditionally consumed high-purity TiO_2, and although it commands a higher price than $TiCl_4$ (typically double), this is offset by the fact that only 1.66 kg of TiO_2 is required per kilogram of titanium produced, versus 4 kg of $TiCl_4$. However, from a historical perspective, early pioneers of the process transformed the titania feedstock into a sintered body prior to commencing reduction, thus creating an avoidable expense. Therefore, a key feature of Metalysis strategy has been to eradicate this, forming the basis of the succeeding section. In a more recent development, it has been demonstrated that it is possible to synthesize powder directly from either synthetic or naturally occurring rutile ore (beach sand) [7], which is considered to be the holy grail in terms of titanium production [8]. Negating the requirement to chlorinate the raw material input has a significant impact on the cost base, signaling a paradigm shift in the production of low-cost titanium powder. Alternative mineral bodies such as ilmenite [9–13], partially upgraded titania-rich slags (ilmenite reacted with coal) [9,14,15], and intermediates/by-products of the chloride/sulfate routes to manufacture pigment-grade TiO_2 such as calciner discharge/dust [9,15] and metatitanic acid ($TiOSO_4$) [15], have been assessed as potential raw material inputs for the FFC process, both by Metalysis and numerous research institutions.

4.3 Preforms: evolution to elimination

A number of preliminary investigations into the FFC® process were dedicated to understanding how reduction proceeds, leading to the significance of the three-phase interface between the feedstock, electrolyte, and conduction path for electrons (either via the cathode current collector, or metallic front advancing through the oxide) being emphasized [15–19]. As a result of this, it was assumed that a porous but sintered mass providing a network of interconnects allowing current flow and facilitating conversion was crucial. Consequently, all titanium-related scientific literature published on this subject advocates the electrolysis of pellets [2,10,15,17–22], or a consolidated component [12–14], and this approach was adopted by Metalysis during initial experimentation, in the guise of a honeycomb [23].

While preforms are appropriate for conducting fundamental academic research and mechanistic studies [21,22], they present a considerable challenge in terms of scale-up, both operationally and in relation to electrode design. Each individual ceramic element is painstakingly loaded, and subsequently removed from supports, forming part of a cathode hanger. Two further concerns associated with pellets are that they represent an engineered product, adding unnecessary cost to the feedstock, and the ensuing titanium offers no functional benefit to the customer, unless used in the form of a master alloy addition, that is, Ti-6Al-4V, for example, as it is essentially a substitute for conventional Kroll sponge.

With particular reference to the honeycomb, it exhibits a low packing density and therefore needs to be fragmented prior to subsequent refining. Furthermore, in terms of its suitability as a precursor to powder, milling is not appropriate as titanium smears rather than crushes because of its enhanced ductility, and it provides no competitive advantage to sponge or wrought scrap for hydride–dehydride (HDH) processing. It should be noted that a proportion of the FFC literature is somewhat misleading, when inferring reduced pellets directly lead to powder during washing [1,18,19]. Since electrolysis is conducted at temperatures approaching two thirds the melting point of titanium (1933 K or 1660°C), localized sintering occurs between bound particles forming part of the internal structure, resulting in a near-net-shape product [10].

Unsurprisingly, an alternative mind set was necessary, and with an initial focus on improved operability, significance was placed on random-packed beds, where inspiration for feedstock design was taken from distillation (fractionating) columns, in the form of raschig rings and berl saddles. This approach challenges the conventional wisdom that each individual preform requires intimate contact with the current collector, to provide an electrically conductive path throughout the cathode layer, promoting full conversion. The success of this demonstrated that loose packing still enables electron flow within the bed, heralding a significant step forward in the commercialization of the FFC® process for titanium at Metalysis. However, as with the honeycomb, a limitation of both raschig rings and berl saddles is that they still represent a solid engineered part, albeit a cheaper one, and again the resulting titanium product is only a variant of traditional sponge. Therefore, further innovation was required, to reach the ultimate objective of producing powder directly from the cell. The remaining obstacle to surmount demanded provision of a technical solution, in combination with overcoming indoctrinated perception, that a random-packed bed of powdered feedstock could not be reduced, because of insufficient electrolyte flow. Further nuances associated with this were to ensure that adequate connectivity was incorporated into the cathode bed to pass current, but not to a degree where this results in sintering of the titanium.

This concept was demonstrated by initially synthesizing coarse (1–2 mm diameter) [24], followed by more conventional (sub-200 μm) powder from a TiO_2 pigment grade precursor, to produce traditional commercially pure ASTM grades 1–4. More recently, it has been revealed that it is possible to reduce granules of naturally occurring rutile ore (beach sand) [7], totally eliminating the requirement for a preforming step, signaling a further accomplishment in the quest to lower cost. Principally, the titanium-bearing ore is extracted from source, subjected to minimal beneficiation, predominantly to remove zircon ($ZrSiO_4$), and literally dispensed straight onto the cathode of the electrochemical cell. An overview of the progress achieved at Metalysis in relation to the direct manufacture of titanium powder is summarized in Figure 4.3.

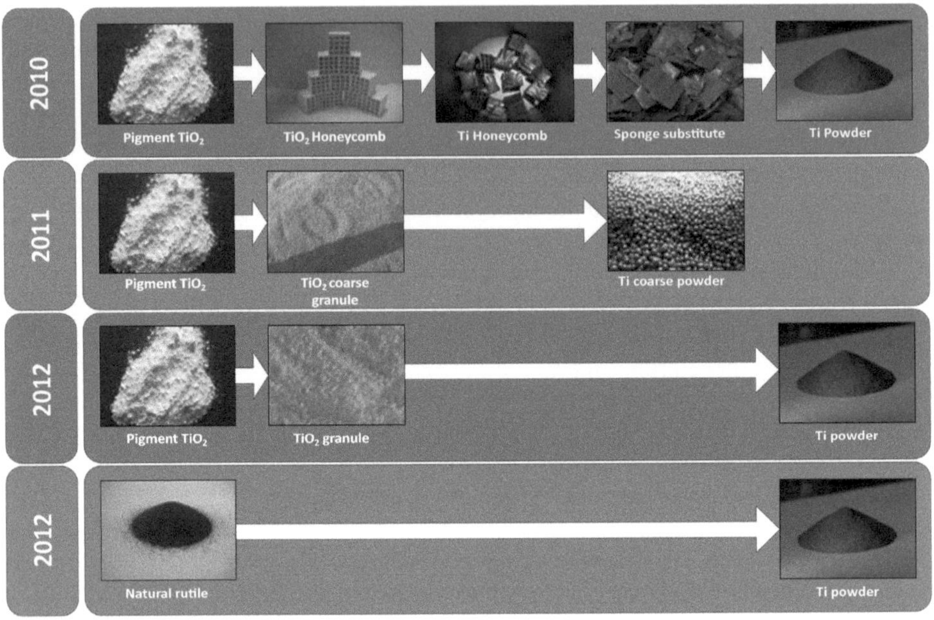

Figure 4.3 Preform evolution to elimination in the quest for titanium powder via the FFC® process.

4.4 Titanium alloys via the FFC® process

A distinct advantage of the FFC® process lies in the fact that it is not solely limited to commercially pure grades of titanium (ASTM 1–4); it can also manufacture a wide range of alloys. Several examples have been reported both in the scientific literature, such as Ti-W [3–5], Ni-Ti [25–29], Ni-Ti-Hf [30], Ti-Mo [31], Ti-Si [32], Ti-Nb [33], Ti-Al [34], and Ti-10V-2Fe-3Al [5], and demonstrated internally at Metalysis, the latter providing the focus of attention here. Early studies were dedicated to synthesizing Ti-6Al-4V, the work horse titanium alloy, accounting for ca. 70% of current global titanium output. A scanning electron microscope (SEM) micrograph of the observed microstructure obtained directly from the cell is given in Figure 4.4, alongside energy-dispersive X-ray spectroscopy (EDX) mapping, quantifying the distribution of critical elements throughout the thickness of a particle. Interpretation of the data implies aluminum and vanadium concentrations of 5.8 and 4 wt%, respectively.

While Ti-6Al-4V is an obvious and sensible choice to benchmark FFC®, it does not showcase the full capabilities of the process; therefore, it has been further exploited to fabricate more challenging beta (β) alloys, such as metastable Ti-8V-5Fe-1Al and Ti-10Mo. The former was first introduced by RMI Titanium, a subsidiary of RTI International Metals, aimed toward high ultimate tensile and shear strength applications, with a specific reference to aerospace fasteners. This is difficult to form via conventional means, because of a tendency for iron segregation during melting [35]. The

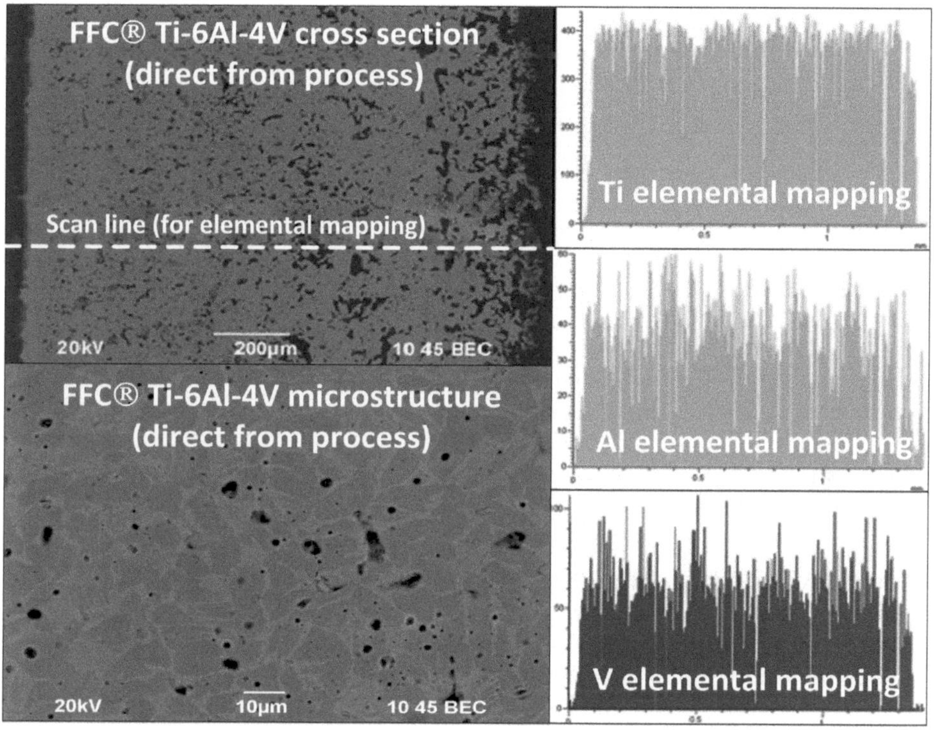

Figure 4.4 SEM micrographs of Ti-6Al-4V generated via the FFC® process.

cross-sectional image of FFC® material is provided in Figure 4.5, where again it should be noted that the microstructure observed, has been obtained directly from the process.

Similarly, a range of Ti-*x*Mo alloys (where $x \leq 40$ wt%) were originally developed in the 1950s by Rem Cru Titanium Corporation, primarily to utilize their promising corrosion resistance. More recently, because of the long-term health concerns associated with elements such as aluminum, vanadium, and nickel, a Ti-10Mo variant is being targeted toward selected biomedical markets. Further benefits associated with a pure beta (β) alloy, is the reduced elastic modulus compared to more common titanium

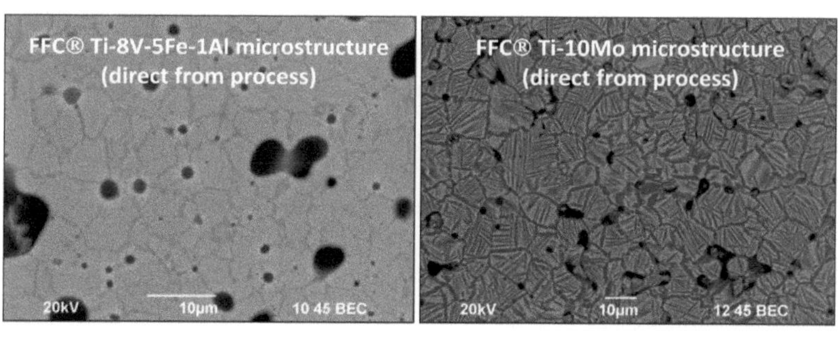

Figure 4.5 SEM micrographs of Ti-8V-5Fe-1Al and Ti-10Mo generated via the FFC® process.

Figure 4.6 SEM micrographs of Metalysis pigment TiO_2 and natural rutile-derived titanium powders.

compositions (i.e., Ti-6Al-4V and Ti-6Al-7Nb), being closer to that of natural bone. An example of this is also included in Figure 4.5, where again the microstructure presented has been obtained without application of any additional treatment steps, for a product where achieving the required compositional homogeneity is demanding [36].

4.5 Metalysis titanium powder characterization

SEM micrographs of both Metalysis unalloyed pigment- and natural rutile-derived titanium powders, and a combination of their key chemical and physical properties are summarized in Figure 4.6 and Table 4.1, respectively, in conjunction with commercially available HDH product [37] by means of an analogy.

Both pigment- and rutile-derived titanium powders currently manufactured directly from the Metalysis process compare favorably with commercially available HDH materials, synthesized from both wrought and sponge precursors [37]. A distinct advantage associated with the Metalysis route is that the tendency to increase impurity levels, both in terms of interstitial gases and materials of construction [38], is eliminated, as no further treatment step other than the application of light pressure to separate

Table 4.1 **Chemical and physical characteristics of Metalysis titanium powders compared to commercially available hydride–dehydride (HDH) powders**

Property	Coarse powder	Pigment powder	Rutile powder	HDH from wrought[16]	HDH from sponge[16]
Oxygen (wt%)	0.15–0.40	0.20–0.40	0.15–0.40	0.22	0.13
Apparent density (g/cm^3)	n/a	1.56	1.86	1.58	1.62
Tap density (g/cm^3)	n/a	1.93	2.24	n/a	n/a
Hall flow (s/50 g)	n/a	88	47	33	34
PSD d_{10} (μm)	1000–2000	101	94	45–150	45–150
PSD d_{50} (μm)		144	141		
PSD d_{90} (μm)		213	217		

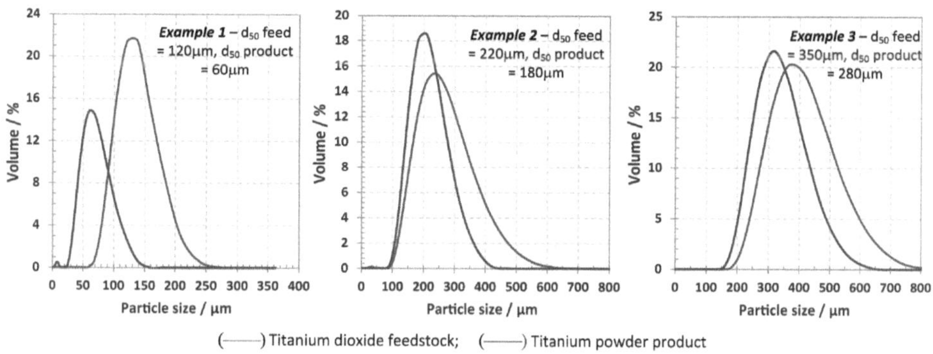

Figure 4.7 Plots demonstrating the ability to tailor the average particle diameter and corresponding size distribution of titanium powder based on selection of the feed.

the reduced cathode bed to produce powder is necessary (i.e., no milling/crushing is needed). Additionally, the price of hydrogen, both in terms of the volume required, and the additional capital investment necessary to construct a plant, can detract from HDH as a preferred method [39]. Finally this technique has a propensity to create an appreciable volume of fines [39], essentially deemed as scrap because of their high residual oxygen concentrations. Regarding this particular point, Metalysis is able to control both the average particle diameter and size distribution of the product, based on appropriate selection of the feedstock. To illustrate this fact, Figure 4.7 describes how the average particle size (d_{50}) is affected following the transition from oxide to metal powder, for an array of titania granules exhibiting varying starting diameters.

The consequence of this is that it is possible to tailor the titanium product toward targeted PM techniques (i.e., metal injection molding, selective laser melting, electron beam melting, cold isostatic pressing, hot isostatic pressing [HIP], etc.), hence minimizing the generation of waste – in terms of both nonconforming powder and energy consumption.

As stated earlier in the text, it is possible to fabricate near-net-shape articles via the Metalysis process [10], also being demonstrated internally with honeycombs [23], raschig rings, berl saddles, and coarse powder [24]. Therefore, in addition to manipulating particle size, future activity will focus on developing bespoke particle morphologies to suit particular PM applications.

Moreover, Metalysis unalloyed pigment- and rutile-derived titanium can be spheroidized, directly converting the powder generated in the cell by utilization of a plasma-based process. Implementing this leads to an improvement in both flow and apparent density, while having minimal or no effect on bulk oxygen and nitrogen levels. The salient properties and micrographs of the ensuing products are portrayed in Figure 4.8 and Table 4.2.

The attractive features pertaining to the average particle size and its accompanying narrow distribution are retained, albeit with a slight downward shift in relation to the feed, which can be compensated for accordingly. Concerns linked to gas atomized powders such as the tendency to form satellites [40], or hollow spheres [41] containing

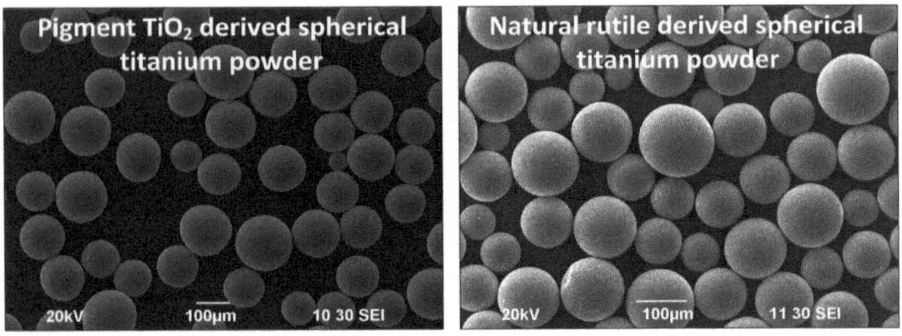

Figure 4.8 SEM micrographs of Metalysis pigment TiO_2 and natural rutile-derived spherical titanium powders.

entrapped carrier gas (e.g., argon), which can cause issues during downstream processing, are not discernible via this approach. Recovery is near 100%, and overall costs are favorable in comparison to using an HDH precursor, for example, based on the yield considerations outlined above (i.e., the undersize is discarded because of its high interstitial gas content).

4.6 Additive manufacturing (AM)

Because of their excellent intrinsic physical properties, that is, flowability, packing density, particle diameter, and its narrow size distribution, etc., Metalysis spherical titanium powders are an ideal feedstock for additive manufacturing (AM). It is also reported that the prohibitive raw material price, which constitutes in the region of 50% of the final component [42], currently presents an obstacle in the widespread deployment of AM for metal-based applications. As stated earlier in the chapter, Metalysis cost base is favorable in comparison to other competing technologies (cf. Figure 4.2),

Table 4.2 **Physical properties of Metalysis spherical titanium powders compared to those obtained directly from the FFC® process**

Property	Pigment from cell	Pigment spherical	Rutile from cell	Rutile spherical
Apparent density (g/cm^3)	1.56	2.71	1.86	2.76
Tap density (g/cm^3)	1.93	2.92	2.24	2.95
Hall flow (s/50 g)	88	23	47	22
PSD d_{10} (μm)	101	88	94	77
PSD d_{50} (μm)	144	117	141	105
PSD d_{90} (μm)	213	166	217	151

Figure 4.9 Additive manufactured automotive turbocharger part fabricated via selective laser melting (SLM) using Metalysis spherical titanium powders.

perfectly positioning the company to shape the landscape of the pending 3D printing revolution.

A range of parts were fabricated from both unalloyed pigment- and rutile-derived spherical product, using a Renishaw SLM 125 (selective laser melting) machine. Figure 4.9 shows an example of an automotive turbocharger component, in conjunction with a variety of trial cubes, generated to ascertain the preferred build parameters (i.e., deposition rate, speed of movement across the bed, and laser power output) required to accomplish this.

A guide vane, forming part of an aerospace gas turbine, is depicted in Figure 4.10. Also included are computed tomography (CT) X-ray scans, indicating that near total consolidation has been achieved, and an SEM image illustrating the cross-sectional microstructure.

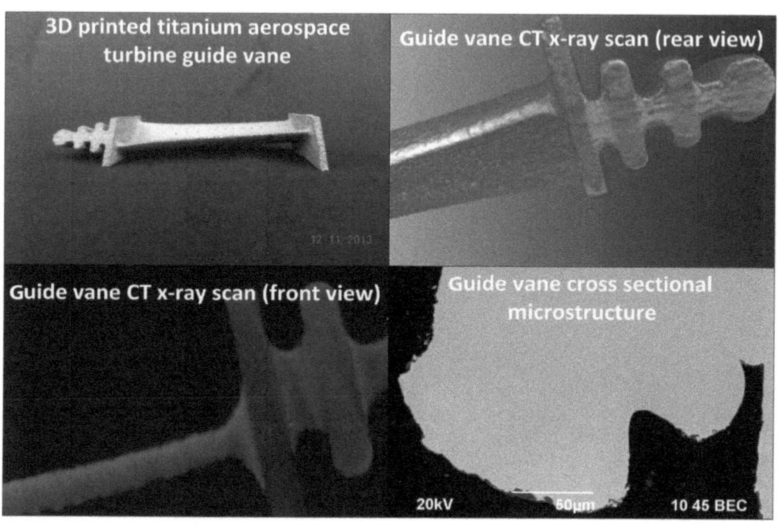

Figure 4.10 Additive manufactured aerospace turbine guide vane part fabricated via selective laser melting (SLM) using Metalysis spherical titanium powders.

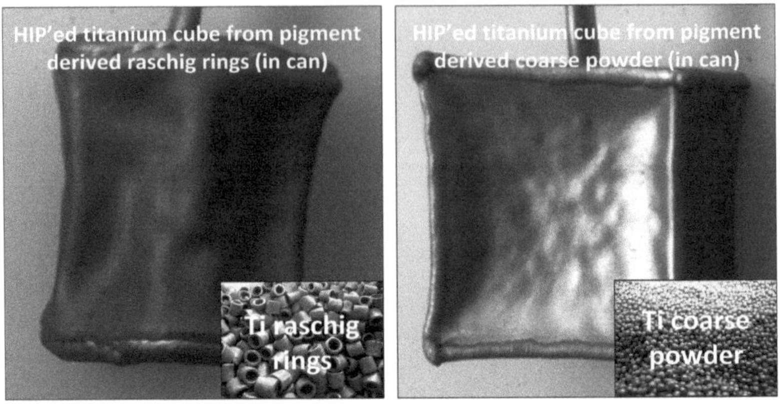

Figure 4.11 Hot isostatically pressed (HIP) cans of cubes using raschig rings and coarse titanium powder.

4.7 Hot isostatic pressing

Preliminary HIP trials were performed on unalloyed titanium raschig rings and coarse powder, both synthesized from a pigment-grade precursor. Consolidation parameters consisting of an applied pressure of 103 MPa under an argon atmosphere at 1193 K (920°C) for 2 h were applied, following an outgassing cycle at room temperature. Photographs of the cans post treatment are depicted in Figure 4.11.

Close inspection of these suggest a greater degree and an inconsistent deformation, when employing the raschig rings, attributed to their limited packing density and high voidage. Crushing of the feedstock prior to charging in an attempt to reduce porosity had minimal effect. Shrinkage of the coarse powder was considerably less and more uniform, where the resulting block is shown in Figure 4.12, on removal of the can. A relative density of 95% was calculated in relation to theoretical maximum

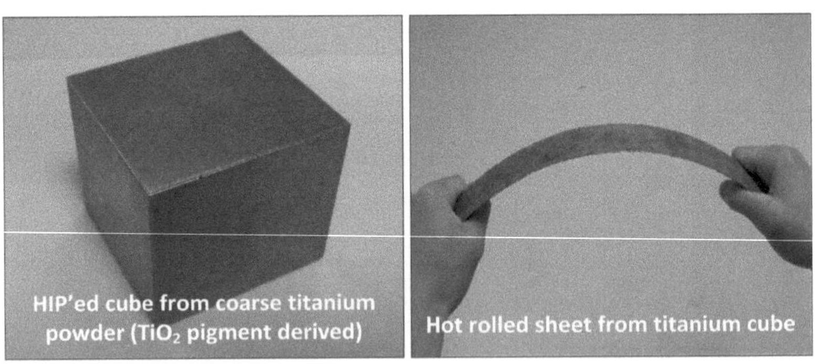

Figure 4.12 Hot isostatically pressed (HIP) cube from coarse titanium powder and subsequently hot rolled sheet.

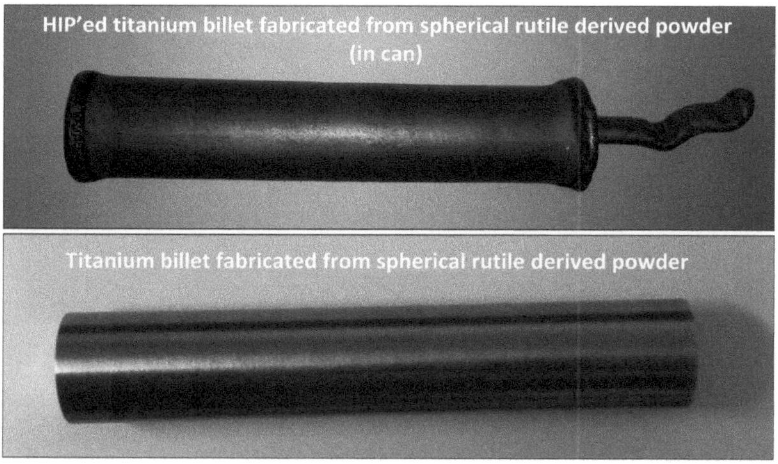

Figure 4.13 Hot isostatically pressed (HIP) billet from spherical rutile-derived titanium powder.

(4.51 g/cm^3), using a combination of the dimensions and weight of the cube. Slicing thin sections and attempting hot rolling yields titanium strip, which is also represented in Figure 4.12, illustrating the inherent ductility of the material.

A billet was produced from spherical rutile-derived powder under the same conditions as previously, where images of this both pre- and post encapsulation are given in Figure 4.13. In stark contrast to the earlier HIP investigation using a combination of raschig rings and coarse powder, the extent of contraction during processing was minimal, conveying the excellent physical properties of the material, as previously indicated in Table 4.2. This is further portrayed in the relative density of the part, determined to be in excess of 95%.

A number of specimens were machined from the billet to conduct monotonic tensile testing, as per method BS EN 2002-2:2005. The results of this are outlined in Figure 4.14. Close inspection of the plot indicates the ultimate tensile strength (UTS) exceeds that of ASTM titanium grades 1–4 [43], and is approaching that of Ti-6Al-4V (grade 5 [43]). Furthermore, this value is equivalent to the Department of Defense MIL-DTL-46077G requirement, for weldable titanium alloy armor plate [44]. Also included are the respective specifications for a range of duplex (2205 and 2507) and stainless (303, 316, 409, and 430) steels, which because of the cost benefits of Metalysis titanium, are vulnerable to displacement in targeted applications.

Elongation and reduction of area were lower than expected, measured at 2% and 5%, respectively. On closer inspection, this behavior can be attributed to the presence of discrete impurity particles, introduced to the powder during processing. These are causing the test piece to shear rather than deform, and the result therefore does not provide a true representation of the material's inherent ductility. Current effort is focused on elucidating the source and, subsequently, eradicating these.

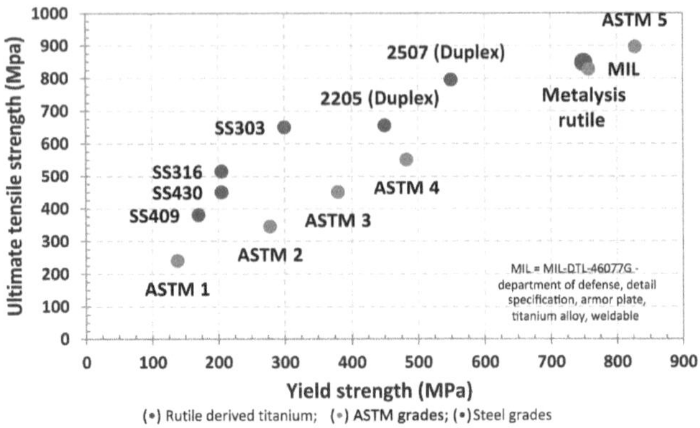

Figure 4.14 Plot illustrating the ultimate tensile strength (UTS) of hot isostatically pressed (HIP) spherical rutile-derived titanium powder compared to traditional ASTM grades and a selection of steels.

4.8 Spark plasma sintering (SPS) and hot rolling

Spark plasma sintering (SPS), a method combining pressing and rapid heat treatment in a single step, has been explored as a potential route for producing sheet, by hot rolling compacts prepared via this technique. Discs of 80 mm diameter and 14 mm thickness were generated by subjecting a preform of spherical rutile-derived titanium, to a pressure of 21 MPa at 1473 K (1200°C). The result of this is illustrated in Figure 4.15, in conjunction with a cross-sectional bright-field optical image, indicating the presence of negligible residual porosity – any remaining is characterized by dark specks.

Titanium sheet of gauge ca. 1.8 mm was subsequently fabricated by hot rolling the material at 1223 K (950°C) in a flow of argon, being shown in Figure 4.16. The majority of the voidage has been eradicated; however, where it still exists, this has been elongated and compressed during processing, as indicated by the corresponding cross-sectional bright field optical image.

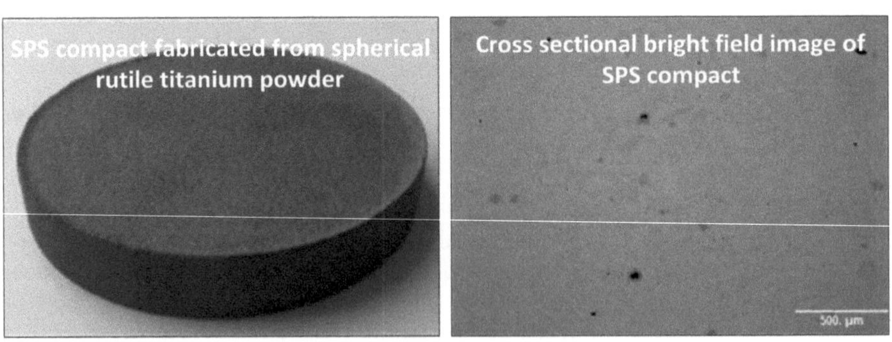

Figure 4.15 Spark plasma sintered (SPS) compact from spherical rutile-derived titanium powder.

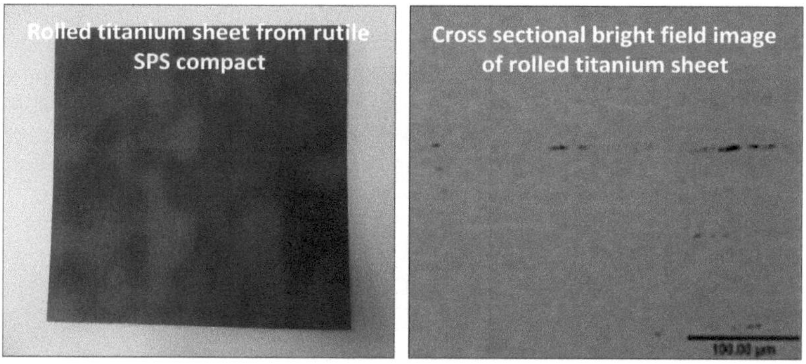

Figure 4.16 Rolled titanium sheet from a spark plasma sintered (SPS) spherical rutile-derived compact.

4.9 Summary

Over the recent years, Metalysis has demonstrated that it is possible to eradicate the need for engineered preforms in the FFC® process, where titanium powder can now be generated directly from the cell. Furthermore, in a more recent development, both synthetic and naturally occurring rutile (beach sand) have been revealed as suitable feedstocks, as a cheaper alternative to pigment-grade titania. The ability to generate products tailored toward targeted powder metallurgy (PM) techniques, by controlling the average particle diameter and its associated size distribution, in combination with their amenability to create a spherical morphology by exploiting a plasma-based route, allows Metalysis to simultaneously traverse two current global revolutions being (1) shaping the economics and complexity of titanium production by collapsing steps in conventional metallurgy and (2) delivering low-cost consumables bespoke to AM.

Acknowledgments

The authors would like to thank Prof. Iain Todd, Dr. Fatos Derguti, and James Hunt of the Mercury Centre, and Dr. Martin Jackson and Nicholas Weston in the Department of Materials Science and Engineering, all at the University of Sheffield, for their support and assistance in performing the downstream consolidation and associated mechanical measurements.

References

[1] A.J. Fenn, G. Cooley, D. Fray, L. Smith, Exploiting the FFC Cambridge process, Adv. Mater. Process. 162 (2004) 51–53.

[2] G.Z. Chen, D.J. Fray, T.W. Farthing, Direct electrochemical reduction of titanium dioxide to titanium in molten calcium chloride, Nature 407 (2000) 361–364.

[3] K. Dring, R. Bhagat, M. Jackson, R. Dashwood, D. Inman, Direct electrochemical production of Ti-10W alloys from mixed alloy precursors, J. Alloys Compd. 419 (2006) 103–109.

[4] R. Bhagat, M. Jackson, D. Inman, R. Dashwood, Production of Ti-W alloys from mixed oxide precursors via the FFC Cambridge process, J. Electrochem. Soc. 156 (2009) E1–E7.
[5] R. Dashwood, M. Jackson, K. Dring, K. Rao, R. Bhagat, D. Inman, Development of the FFC Cambridge process for the production of titanium and its alloys, in: Innovations in Titanium Technology, TMS (The Minerals, Metals & Materials Society), 2010, pp. 49–58.
[6] D.S. van Vuuren, A critical evaluation of processes to produce primary titanium, J. South Afr. Inst. Mining Metall. 109 (2009) 455–461.
[7] K. Rao, J. Deane, L. Grainger, J. Clifford, M. Conti, J. Collins, Electrolytic production of powder. World Patent WO 2013/050772 A2, April 11, 2013.
[8] R. Hill, Titanium: 21st century metal. PROCESS – CSIRO research in mineral processing and metal production. February 1–2, 2002.
[9] Metalysis Ltd. Unpublished results.
[10] S.J. Oosthuizen, In search of low cost titanium: the Fray Farthing Chen (FFC) Cambridge process, J. South Afr. Inst. Mining Metall. 111 (2011) 1–6.
[11] M. Ma, D.H. Wang, X.H. Hu, X.B. Jin, G.Z. Chen, A direct electrochemical route from ilmenite to hydrogen storage ferrotitanium alloys, Chem.A Eur. J. 12 (2006) 5075–5081.
[12] M. Hu, C. Bai, X. Liu, X.I. Lv, J. Du, Deoxidation mechanism of preparation FeTi alloy using ilmenite concentrate, J. Mining Metall. B Metall. 47 (2011) 193–198.
[13] R. Shi, C. Bai, M. Hu, X. Liu, J. Du, Experimental investigation on the formation mechanism of the TiFe alloy by the molten-salt electrolytic titanium concentrate, J. Mining Metall. B Metall. 47 (2011) 99–104.
[14] J. Mohanty, Electrolytic reduction of titania slag in molten calcium chloride bath, J. Mater. 64 (2012) 582–584.
[15] M. Ma, D. Wang, W. Wang, X. Hu, G.Z. Chen, Extraction of titanium from different titania precursors by the FFC Cambridge process, J. Alloys Compd. 420 (2006) 37–45.
[16] T. Nohira, K. Yasuda, Y. Ito, Pinpoint and bulk electrochemical reduction of insulating silicon dioxide to silicon, Nat. Mater. 2 (2003) 397–401.
[17] D.J. Fray, G.Z. Chen, Extraction of titanium from solid titanium dioxide in molten salts, in: Cost-Affordable Titanium: Symposium Dedicated to Harvey Flower, TMS (The Minerals, Metals & Materials Society), 2004, pp. 9–17.
[18] K.S. Mohandas, D.J. Fray, FFC Cambridge process and removal of oxygen from metal-oxygen systems by molten salt electrolysis: an overview, Trans. Indian Inst. Metals 57 (2004) 579–592.
[19] G.Z. Chen, D.J. Fray, A morphological study of the FFC chromium and titanium powders, Trans. Inst. Min. Metall. C 115 (2006) 49–54.
[20] D.J. Fray, Novel methods for the production of titanium, Int. Mater. Rev. 53 (2008) 317–325.
[21] C. Schwandt, D.J. Fray, Determination of the kinetic pathway in the electrochemical reduction of titanium dioxide in molten calcium chloride, Electrochim. Acta 51 (2005) 66–76.
[22] D.T.L. Alexander, C. Schwandt, D.J. Fray, Microstructural kinetics of phase transformations during electrochemical reduction of titanium dioxide in molten calcium chloride, Acta Mater. 54 (2006) 2933–2944.
[23] M. Bertolini, L. Shaw, L. England, K. Rao, J. Deane, J. Collins, The FFC Cambridge process for production of low cost titanium and titanium powders, Key Eng. Mater. 436 (2010) 75–83.
[24] K. Rao, M. Bertolini, I. Mellor, J. Collins, J. Deane, L. Grainger, M. Conti, Development of a new generation FFC pilot plant for production of low cost titanium and titanium alloys, Proceedings of the 12th World Conference on Titanium, Beijing, China 1 (19–24 June 2011) pp. 181–184
[25] B. Jackson, M. Jackson, D. Dye, D. Inman, R. Dashwood, Production of NiTi via the FFC Cambridge process, J. Electrochem. Soc. 155 (2008) E171–E177.

[26] B.K. Jackson, D. Dye, D. Inman, R. Bhagat, R.J. Talling, S.L. Raghunathan, M. Jackson, R.J. Dashwood, Characterisation of the FFC Cambridge process for NiTi production using in situ X-ray synchrotron diffraction, J. Electrochem. Soc. 157 (2010) E57–E63.
[27] B.K. Jackson, D. Inman, M. Jackson, D. Dye, R.J. Dashwood, NiTi production via the FFC Cambridge process: refinement of process parameters, J. Electrochem. Soc. 157 (2010) E36–E43.
[28] S. Jiao, L. Zhang, H. Zhu, D.J. Fray, Production of NiTi shape memory alloys via electro-deoxidation utilizing an inert anode, Electrochim. Acta 55 (2010) 7016–7020.
[29] H. Zhu, M. Ma, D. Wang, K. Jiang, X. Hu, X. Jin, G.Z. Chen, Electrolytic reduction of mixed solid oxides in molten salts for energy efficient production of the TiNi alloy, Chinese Sci. Bull. 51 (2006) 2535–2540.
[30] B.X. Wang, R. Bhagat, X.I. Lan, R.J. Dashwood, Production of Ni-35Ti-15Hf alloy via the FFC Cambridge process, J. Electrochem. Soc. 158 (2011) D595–D602.
[31] R. Bhagat, M. Jackson, D. Inman, R. Dashwood, The production of Ti-Mo alloys from mixed oxide precursors via the FFC Cambridge process, J. Electrochem. Soc. 155 (2008) E63–E69.
[32] X. Zou, X. Lu, Z. Zhou, C. Li, W. Ding, Direct selective extraction of titanium silicide Ti_5Si_3 from multi-component Ti-bearing compounds in molten salt by an electrochemical process, Electrochim. Acta 56 (2011) 8430–8437.
[33] X.Y. Yan, D.J. Fray, Electrosynthesis of NbTi and Nb_3Sn superconductors from oxide precursors in $CaCl_2$-based melts, Adv. Funct. Mater. 15 (2005) 1757–1761.
[34] P.K. Tripathy, D.J. Fray, On the preparation of TiAl alloy by direct reduction of the oxide mixtures in calcium chloride melt. Fray International Symposium, Cancun, Mexico, 27 November–1 December, http://www.osti.gov/scitech/servlets/purl/1042346, 2011.
[35] G. Welsch, R. Boyer, E.W. Collings, Materials Properties Handbook Titanium Alloys, ASM International, Materials Park, Ohio, (1994).
[36] Z. Gao, H. Luo, Q. Li, Y. Wan, Preparation and characterisation of Ti-10Mo alloy by mechanical alloying, Metall. Microstr. Anal. 1 (2012) 282–289.
[37] T.M. Zwitter, P. Nash, X. Xu, C. Johnson, Energy efficient press and sinter of titanium powder for low-cost components in vehicle applications. Final Scientific Report – DOE NETL – NT 01913, (2011).
[38] X. Goso, A. Kale, Production of titanium metal powder by the HDH process, J. South Afr. Inst. Mining Metall. 111 (2011) 203–210.
[39] W.H. Peter, Y. Yamamoto, Near net shape manufacturing of new, low cost titanium powders for industry. DOE Award Number: CPS Agreement No. 17881, 2013.
[40] I.-H. Oh, H. Segawa, N. Nomura, S. Hanada, Microstructures and mechanical properties of porosity-graded pure titanium compacts, Mater. Trans. 44 (2003) 657–660.
[41] C. Leyens, M. Peters, Titanium and Titanium Alloys: Fundamentals and Applications, Wiley-VCH GmbH & Co. KGaA, Weinheim, Germany, (2003).
[42] F.H. Froes, Titanium powder metallurgy: a review – part 2, Adv. Mater. Process. 170 (2012) 26–29.
[43] ASTM B265-13ae1. Standard Specification for Titanium and Titanium Alloy Strip, Sheet and Plate, (2013). DOI: 10.1520/B0265.
[44] MIL-DTL-46077G. Detail Specification – Armor Plate, Titanium Alloy, Weldable, (2006).

Direct titanium powder production by metallothermic processes

5

David S. van Vuuren
The CSIR, PO Box 395, Pretoria, South Africa

5.1 Introduction

Many different approaches have been proposed and tried by a multitude of different people and organizations over almost six decades for the direct and cost-effective production of titanium powder. To assess and compare the different approaches, it is useful to consider both their commonalities and differences.

All the approaches, including non-metallothermic processes and processes to produce primary titanium in nonpowder form, require the use of a precursor (or feed material containing titanium), a reducing agent, and a separation stage where the titanium product is separated from the medium in which the reaction occurs and the by-products of the reaction (if applicable). The differences between the approaches arise, inter alia, from different choices with respect to the following process factors:

- Precursor (e.g., ilmenite, TiO_2, TiO/TiC, TiN, $TiCl_4$, TiF_4, $TiBr_4$, and TiI_4).
- Reducing agent or reductant (e.g., electricity, H_2, Al, Mg, Na, Ca/CaH_2, and Li).
- Reactor type (e.g., static, stirred molten salt tank, electrolytic, agitated solid salt, fluidized bed, molten metal, mechanical mill, spray, gas flow, plasma, and organic ionic liquid reactors).
- Separation principle (e.g., leaching, distillation, extrusion, deposition, Ti extraction, and Ti melting).
- Titanium product form (e.g., sponge, powder, ingot, and preform).

In addition, many studies have been undertaken with the aim of rendering the process continuous, in contrast to the industry standard batch-operated Kroll and Hunter processes. Furthermore, developments have been directed toward making Ti alloy powders, TiH_2, or hydride alloy powders directly.

The implications of specific choices are discussed below, followed by analyses of recent developments to produce titanium powder directly via metallothermic reduction of the relevant precursors.

5.2 Precursors

Precursors (or feedstock) considered for the production of titanium metal in order of increasing cost include

- Ilmenite (±\$0.93/kg contained Ti @ \$300/t ore).
- Metallurgical-grade TiO_2 (±\$2.45/kg contained Ti @ \$1400/ton synthetic rutile).
- $TiCl_4$ (±\$4/kg contained Ti @ \$1000/t $TiCl_4$).
- Pigment-grade TiO_2 (±\$5/kg contained Ti @ \$3000/t pigment).

Titanium Powder Metallurgy. http://dx.doi.org/10.1016/B978-0-12-800054-0.00005-8

- TiF_4 or fluotitanate salts (similar to pigment-grade TiO_2).
- TiO/TiC (no information available).
- TiN (no information available).
- $TiBr_4$ (no information available).
- TiI_4 (no information available).

(Note: The prices given are estimates in 2013. Actual costs or prices fluctuate as market conditions change, and negotiated prices for large contracts are not in the public domain.)

Ilmenite is the cheapest commercial feedstock but it is not suitable for use in metallothermic processes for producing commercially pure titanium. However, there have been attempts to use it in electrochemical processes to produce titanium ingot directly [1,2]. The principle is basically that the iron in the ilmenite is first electrowon at a lower potential than the decomposition potential of dissolved titanium oxide species and tapped from the cell before the cell potential is increased to recover titanium. The envisaged process conditions to electrowin molten titanium are very demanding and it remains to be seen whether such processes can be developed successfully.

Another approach proposed is to reduce ilmenite aluminothermically to produce an unrefined body of titanium metal followed by electrorefining to produce titanium of higher purity [3].

Metallurgical-grade TiO_2 is also being considered in processes to electrowin titanium powder. The key in such developments is to reduce the titanium feedstock and electrorefine the titanium product in the same process and not in subsequent processes. These developments are alluded to in Chapter 3.

$TiCl_4$ is used as precursor in the commercially proven Kroll and Hunter processes. The main advantages of using $TiCl_4$ are as follows:

- It is relatively easy to remove all undesired impurities from $TiCl_4$ by distillation.
- $TiCl_4$ is a liquid at room temperature, and hence it is simple to feed it at an accurately controlled feed rate into a reactor.

Disadvantages of using $TiCl_4$ include:

- It is relatively expensive, for example, ±$4/kg contained Ti.
- It is highly hazardous and hence undesirable and expensive to transport or transfer over long distances.
- When exposed to the atmosphere, it reacts with moisture to form HCl and TiO_2. HCl is highly corrosive, and hence although $TiCl_4$ is compatible with ordinary steels, the slightest leak causes undesirable damage to most metal structures and electrical equipment in the vicinity of such a leak.
- It is very volatile and has a boiling point of only 136.4°C (277.5°F), causing process complications when reacting at elevated temperatures (normally well above 600°C [1112°F]) with metals to form titanium and molten or solid salts.
- Although it is theoretically possible to make alloys such as Ti-6Al-4V directly by blending or cofeeding chlorides of aluminum and vanadium with $TiCl_4$ and then reducing the mixture at the same time, the chlorides of alloying metals (e.g., $AlCl_3$ and VCl_4) are not the usual form in which feedstocks of these elements are traded.
- On reduction, by-product Cl_2 or a chloride salt is formed, which must be either recycled to a plant producing the $TiCl_4$, sold as a by-product, or disposed of in a safe and environmentally responsible manner.

Because of some of these disadvantages or combinations of them, many alternatives to the use of $TiCl_4$ to produce titanium have been considered in the past, and some are still being pursued.

Although high-purity *pigment-grade TiO_2* is more expensive than $TiCl_4$ (perhaps less than shown by the indicative prices above), its use as precursor is attractive for the following reasons:

- It is safe to work with.
- It is readily transported and hence it is less essential to have a titanium metal plant and a titanium precursor production plant in proximity to each other.
- It can alternatively be produced via sulfates instead of via $TiCl_4$, which may be better in some situations.
- Most of the other elements used in titanium alloys are commercially available as oxides that can be blended with TiO_2 and then be reduced together with the TiO_2 to make titanium alloys directly.

Technical disadvantages of using TiO_2 include the following:

- It is more difficult to reduce the oxygen content in the final product to commercially accepted levels than to reduce chlorine levels, because in the case of oxygen, it is not only oxides that have to be reduced but also the oxygen that is dissolved in the titanium metal.
- The oxides that are produced as by-products in metallothermic reduction processes have much higher melting points than the chlorides produced when reducing $TiCl_4$, making the design of a suitable continuous-flow reactor for the process more complex.
- Separation of the oxide by-products from the titanium is normally done by acid leaching. This introduces another main chemical component to the overall process, leading to increased complexities and cost.
- When it is used as precursor in electrolytic processes, the anode becomes a consumable, in contrast to graphite anodes, which are essentially inert in alkali and alkaline earth electrowinning processes.

A number of studies were undertaken into using TiO_2 as feedstock in both metallothermic and electrolytic processes to make powder [4–6], preforms [7,8], TiH_2 [9], alloys [10], or ingot [1,2,11]. Electrochemical processes to reduce TiO_2 are discussed in Chapters 3 and 4, and various metallothermic routes are discussed below.

Although *fluotitanates* are not produced on a large scale, a number of different fluotitanates have been considered as feedstock in quite a number of developments [12–16]. The single biggest advantage of using fluotitanates as precursor for the production of titanium is that pure fluotitanate salts can be precipitated from aqueous solutions containing virtually no hydrates or oxides. The difficulty of ensuring that the oxygen content of the final titanium product meets commercial specifications is therefore considerably less than when using TiO_2.

Other technical advantages include the following:

- Similarly to TiO_2 and $TiCl_4$, titanium fluorides can be prepared in very pure forms, making them suitable as precursors for the production of high-quality titanium.
- Titanium fluorides are much more stable than the bromides and chlorides and are therefore easier to handle. For example, TiF_3 has a melting point of 1200°C (2192°F) and is stable in dry air at room temperature.

- Fluotitanates, for example, K_2TiF_6, are quite soluble in chloride salts, and they are therefore a logical choice as feedstock for electrowinning of titanium.
- A titanium metal plant using fluotitanates as feedstock does not have to be in close proximity to a $TiCl_4$-producing plant.
- Effluents (if any) can be easily treated to acceptable environmental standards with lime, forming highly stable and sparingly soluble CaF_2.
- In contrast to the oxides, the halides do not form a range of intermediate phases when being reduced. In typical electrowinning processes, the trihalide is converted directly to metal [16].
- Halide gases are significantly less soluble in titanium than oxygen, so that contamination of the product by dissolved gases is significantly reduced.

Disadvantages include the following:

- The use and handling of hydrogen fluoride to produce the fluotitanates are extremely dangerous.
- Separation of the titanium product and the fluoride by-product is problematic.
- The fate of the by-product fluorides is also problematic. (Recycling is not conventional because the reducing metals are normally not produced from their fluoride salts, and disposal is likely to be too expensive, making selling off of the by-product the only practical option.)

Titanium oxycarbide, TiO and/or TiC are produced by the carbothermic reduction of TiO_2 at elevated temperature (>1400°C [2522°F]). Whereas further capital and operating costs will always be incurred to convert titanium dioxide to titanium oxycarbide, it is more expensive than TiO_2. However, it is possible that relatively pure titanium oxycarbide produced from metal-grade TiO_2 is a bit cheaper than pigment-grade TiO_2.

Titanium oxycarbide is not suitable for use as precursor for titanium production via metallothermic reduction because the affinity for carbon of the suitable reductants used to reduce titanium oxides is insufficient to remove the carbon from the titanium carbide that will form in the process. However, TiO and/or titanium oxycarbide was [17] and is again [18,19] being considered as a possible precursor for electrowinning of titanium.

It is conceptually possible to use *$TiBr_4$* and *TiI_4* as precursors to produce titanium. However, this is more of academic interest since to date no process that seems to be commercially viable to produce large quantities of titanium for the metallurgical industry has been reported.

One advantage of $TiBr_4$ versus $TiCl_4$ is that $TiBr_4$ has a higher boiling point than $TiCl_4$ (230 vs. 136.4°C [446 vs. 277.5°F]) and it is thus possible to partially reduce $TiBr_4$ to $TiBr_3$ with hydrogen at 500°C (932°F) and 40 bar under $TiBr_4$ refluxing conditions [20]. Following partial reduction of the tetrahalide, titanium can be produced from $TiBr_3$ or $TiCl_3$ by disproportionation reactions [21,22]. A process for doing this with $TiCl_3$ was patented in 1959 [23] and more recently a concept for achieving this using the bromides was developed and tested by the Council for Scientific and Industrial Research (CSIR) in South Africa [20]. One of the reasons why the CSIR abandoned this approach is that the achievable conversion for reducing the tetrahalide to the trihalide is low and consequently a very large and hence expensive system is required for separating the unconverted reactants and by-product hydrogen halide from each other before recycling the unconverted reactants.

TiI_4 is used as an intermediate in the purification of titanium to produce electronic-grade Ti using the Van Arkel and De Boer process [24]. More recently, a process was described [25] for reacting TiO_2 with I_2 and CO to form TiI_4 and CO_2. Apart from the expected high cost of TiI_4, it is unlikely that it will be a practical feedstock for producing commercially pure titanium on a large scale. The reasons for this include the long processing times required and the high mass of iodine required, that is, the molecular mass of TiI_4 is 555.5 compared with that of $TiCl_4$, which is 189.7. The large molecular mass of TiI_4 implies that for an equivalent scale of production, approximately two-and-a-half times the mass of material has to be moved through the process.

In conclusion, it seems that the only precursors that can be considered practical for metallothermic reduction processes to produce titanium are TiO_2, $TiCl_4$, and fluotitanates.

5.3 Reducing agents

Reducing agents that can be used to reduce appropriate precursors of titanium include H_2, electricity (electrons), Al, Mg, Na, Ca/CaH_2, and Li. In order to convey a sense of the relative costs of these reductants, recent prices published on the internet or prices quoted for relatively large orders of particular metals are shown in Table 5.1. The stoichiometric amounts required to produce a kilogram of titanium are also given, as well as the cost per kilogram of titanium.

As can be seen, there would potentially be a significant cost advantage if titanium could be produced by non-metallothermic means (i.e., using reduction with H_2 or electrowinning). The problem is just that none of numerous attempts [26–29] to do so over the last 50 years has been commercially successful.

The main advantage if *hydrogen* can be used as reducing agent is its low cost relative to that of other reducing agents, whereas the main disadvantage is the low conversions, resulting in the need for large hydrogen and unconverted precursor recycle loops.

Table 5.1 Approximate prices of reducing agents

Reductant	Price	Units	Use (unit/kg Ti)	Cost ($/kg Ti)
H_2	2	$/kg	0.08 (kg)	0.17
Electricity*	8.5	c/kWh	4.03 (kWh)	0.34
Al	1.9	$/kg	0.75 (kg)	1.43
Mg	2.6	$/kg	1.01 (kg)	2.64
Na	2.0	$/kg	1.92 (kg)	3.84
Ca	3.0	$/kg	1.67 (kg)	5.02
Li	60	$/kg	0.58 (kg)	34.78

* For electricity, a current efficiency of 100% was assumed to electrowin Ti from a tetravalent precursor in an electrochemical cell operating at 1.8 V.

Direct use of *electricity* to electrowin titanium from virtually any of the mentioned precursors is attractive because of the low cost of electricity relative to that of metallic reducing agents. Another advantage is that it might be possible to save on the overall capital cost of a titanium production plant because separation of the titanium product from the reducing agent containing by-products is not required.

Major disadvantages include the following:

- The technology of electrowinning the relevant metallic reducing agents has been used commercially and improved over many decades, but suitable electrolytic cells to electrowin titanium have still not been scaled up, despite much effort to do so by numerous entities.
- The risk of contaminating the titanium product with carbon originating from graphite anodes requires additional measures that are not necessary when electrowinning aluminum or alkali and alkaline earth metals.
- Removing titanium powder from an electrolytic cell is much more complicated than simply draining or siphoning a molten metal from such a cell as is done with the metallic reducing agents.

Aluminum has, inter alia, been used in different developments to reduce $TiCl_4$ [30], fluotitanates [13,15], and TiO_2 [3,31]. Apart from it being significantly cheaper than other reducing metals, advantages of using aluminum include the following:

- It is a commodity that is generally available and it is traded and transported all over the world.
- It is much safer to handle than Mg, Na, Ca, and Li.

Disadvantages include the following:

- It forms alloys with titanium and it is therefore difficult to prevent contamination of the Ti product when it is desired not to have Al in the product.
- It is not sufficiently reducing to reduce the level of dissolved oxygen in aluminothermically reduced TiO_2 to that required for commercially pure Ti grades.
- It is impractical to use the salt by-product from the reaction in the molten state as a medium of reaction because at atmospheric pressure both $AlCl_3$ and AlF_3 sublime before they melt, and the melting point of Al_2O_3 is very high (ca. 2015°C [3659°F]).
- Separation and recycling of by-product $AlCl_3$, AlF_3, or Al_2O_3 from titanium production is more difficult than separation of alkali or alkaline earth metal halides or oxides from titanium. ($AlCl_3$ can be leached, but the solubility of AlF_3 in water is low. The sublimation point of $AlCl_3$ is 178°C [352°F] and that of AlF_3 is 1291°C [2356°F]. Since $AlCl_3$ and AlF_3 do not form molten phases at atmospheric pressure, they cannot be separated from Ti by draining it as in the Kroll process.)

Instead of recycling the aluminum-containing by-product, it would probably be more economical to sell it. $AlCl_3$ is, for example, used for water treatment and AlF_3 is used in the aluminum smelting industry in cryolite, which is the electrolyte used in electrolytic cells. Selling as an extra product can generate extra revenue, but it has the drawback that the output of the titanium production plant is then set by the volume of by-product that can be sold.

Magnesium is currently the preferred choice of reducing agent in the titanium industry as used in the Kroll process. Reasons for this include the following:

- It is cheaper than Na, Ca, and Li, and requires less energy to produce.
- It is significantly less reactive than Na, Ca, or Li and it can be handled safely in the solid state without special (expensive) precautions.

- The technology to produce Mg is mature.
- In the Kroll process the bulk of the $MgCl_2$ by-product is separated in the anhydrous form by simply draining and tapping it from the sponge product.
- The vapor pressure of hot $MgCl_2$ is high enough for the remainder of the $MgCl_2$ by-product to be distilled from the Kroll reactors and recovered in the anhydrous form.
- Anhydrous $MgCl_2$ can be electrolyzed directly to produce Cl_2, which can be recycled to produce $TiCl_4$ and Mg for reducing the $TiCl_4$. Little waste is therefore generated in an integrated plant.
- Even if waste is generated, $MgCl_2$ is not particularly hazardous.
- The vapor pressure of Mg metal at the reactor conditions is not very high (e.g., at 800°C [1472°F] it is approximately 2 mbar), which makes the Kroll process much safer than the Hunter process, where the vapor pressure of Na at reaction conditions is approximately 0.5 bar.
- Mg is virtually insoluble in Ti.

Apart from its advantages in the Kroll process, there are also advantages in using Mg in some other processes, for example:

- The boiling point of Mg is 1107°C (2025°F), making it easier to volatilize and feed into a reactor in the gaseous state than Ca and Li, which have boiling points of 1487 (2709) and 1317°C (2403°F), respectively.

When Ti powder is being made, using Mg has a number of disadvantages. These include the following:

- Separation of the by-product $MgCl_2$ from the Ti powder to recover anhydrous $MgCl_2$ by simply draining it from Ti sponge as in the case of the Kroll process is not efficient.
- Separation of $MgCl_2$ from Ti powder by leaching followed by subsequent crystallization gives magnesium hydrate crystals. When magnesium hydrate is calcined, it decomposes to form MgO and HCl. MgO is not suitable for electrolytic recovery of Mg using conventional cells. Recovering anhydrous $MgCl_2$ that is suitable for electrolytic recovery is possible, but it is relatively expensive and energy intensive [32].
- Separation of $MgCl_2$ from Ti powder in a continuous manner by distillation requires nonconventional process equipment [33] and is energy intensive.
- Mg is not sufficiently reducing to reduce the level of dissolved oxygen in magnesiothermically reduced TiO_2 to that required for commercially pure Ti grades. However, it might be suitable for partially reducing TiO_2 to make an intermediate product that would have to be further deoxidized [34], or to make TiH_2 in the presence of hydrogen [35].
- The solubility of Mg in $MgCl_2$ is very low. This has negative process implications [36] when reducing $TiCl_4$ continuously in a medium of molten $MgCl_2$.

Sodium has been used commercially in the Hunter and similar processes for a number of reasons, including the following:

- It has been produced on a large scale commercially as precursor for the production of tetraethyl lead, and the technology to produce it is mature.
- It has a low melting point (98°C [208°F]) and is therefore relatively easy to handle as a liquid.
- NaCl can easily be leached with water and recovered as anhydrous NaCl for direct recycling to electrolytic cells to produce Na and Cl_2.

- It has a relatively low boiling point (892°C [1638°F]) and can therefore be fed relatively easily as a gas into processes that use reactants in the gas phase.
- Little waste is produced and, if any, it is quite harmless.
- It is virtually insoluble in Ti.
- It has a fairly high solubility in molten NaCl and is therefore suitable for use in a process where it is predissolved in NaCl before reacting with $TiCl_4$ [36,37].

Disadvantages include the following:

- It has a high vapor pressure at the reaction conditions of the Hunter process (ca. 0.5 bar) and can be highly dangerous.
- The vapor pressure of NaCl is about four times less than that of $MgCl_2$ at 900°C (1652°F) and it is therefore more difficult to distil from Ti than $MgCl_2$.
- It is not sufficiently reducing to reduce the level of dissolved oxygen in sodiothermically reduced TiO_2 to that required for commercially pure Ti grades.
- The melting point of NaCl is the highest of all the suitable alkali and alkaline earth metal chlorides, making the operating temperature of processes that work with molten chloride by-products the highest in the case of sodium (MP_{LiCl} = 614°C [1137°F], MP_{MgCl2} = 712°C [1314°F], MP_{CaCl2} = 772°C [1422°F], MP_{NaCl} = 801°C [1474°F]).

Calcium and CaH_2 have a number of properties that give them particular advantages for use in some process concepts. These include the following:

- Ca and CaH_2 are sufficiently reducing to reduce the level of dissolved oxygen in calciothermically reduced TiO_2 to that required for commercially pure Ti grades [38].
- CaO has a relatively high solubility in molten $CaCl_2$, which is important in processes where cathodes made of TiO_2 are reduced electrolytically [39].
- Ca has a relatively high solubility in molten $CaCl_2$, which is important in processes where Ca is regenerated electrolytically from $CaCl_2$ *in situ* in the reduction reactor system [40].
- The high solubility of both Ca and CaO in molten $CaCl_2$ is important in processes where TiO_2 is reduced calciothermically and Ca regenerated electrolytically from CaO dissolved in $CaCl_2$ *in situ* in the reducing reactor system [38].
- $CaCl_2$ can be easily leached with water and CaO with dilute hydrochloric acid, and both $CaCl_2$ and CaO are nontoxic.
- The boiling point of $CaCl_2$ (±1936°C [3517°F]) is higher than the melting point of Ti (1668°C [3034°F]), which makes it possible to melt and extrude a titanium ingot directly out of $CaCl_2$ [41].
- The vapor pressure of $CaCl_2$ is very low and that of Ca relatively low at typical operating temperatures of Ti production processes.
- CaH_2 is brittle and can readily be crushed and intimately mixed with TiO_2 before initiating a reaction between the two components.
- Ca is virtually insoluble in Ti.

Apart from being nearly twice as expensive as Mg, other disadvantages of using calcium are as follows:

- The melting point of Ca is relatively high (842°C [1548°F]), making it difficult to work with molten Ca.
- Because of the relatively high melting points of $CaCl_2$ and Ca, the operating temperatures of processes that work with these substances are relatively high (>800°C [1472°F]).

Lithium is by far the most expensive reducing agent to use and little work on the use of Li to produce Ti is reported in the literature [42,43]. Advantages of using Li as a reducing agent are as follows:

- LiCl (by-product from lithiothermic reduction of $TiCl_4$) has the lowest melting point of all the relevant alkali and alkaline earth metal chlorides, and hence the operating temperature of processes that involve the handling of molten chlorides is the lowest when using Li.
- The vapor pressures of both LiCl and Li at suitable conditions to operate a process to reduce $TiCl_4$ in molten LiCl are relatively low (e.g., about 4.2 and 16 Pa).
- LiCl is very soluble in water and anhydrous LiCl for recycling to a Li electrowinning step can be readily recovered from an aqueous solution of LiCl.
- Li is virtually insoluble in Ti.
- The melting point of Li is relatively low (179°C [354°F]), making it relatively easy to handle molten Li.
- The operating temperature of commercial electrolytic cells to recover Li is much lower than that of Na and Mg cells (e.g., ±450°C [842°F] vs. ±580°C [1076°F] and ±680°C [1256°F]).

Apart from cost, disadvantages of Li include the following:

- It is highly reactive and dangerous.
- Molten Li is aggressive and care must be taken in selecting suitable materials of construction to handle molten Li.
- There are only a few companies in the world that produce Li metal and there has not been the same incentive to improve Li electrowinning technology as there have been for Na and Mg electrowinning.
- Li is not sufficiently reducing to reduce the level of dissolved oxygen in lithiothermically reduced TiO_2 to that required for commercially pure Ti grades.

5.4 Reactor type

Reactor types considered in different developments include static, stirred molten salt, electrolytic, agitated solid salt, fluidized bed, molten metal, solid state reduction, spray, gas flow, plasma, and organic ionic liquid reactors.

The most basic type of reactor used to produce Ti is a batch, *static* reactor as used for the Kroll process. Its advantages include the following:

- Simplicity of mechanical construction (no moving parts in the reactor).
- Simplicity of control system.
- Simple scaling up by replicating the reactors.

Disadvantages include the following:

- No economy of scale.
- Long residence times (±7 days), resulting in low space-time yields.
- Labor intensiveness of operation.
- Heat integration is more complex in batch compared to continuous processes.
- Part of the product being excessively contaminated by iron at the reactor wall.
- Relatively short life of the reactor shell.
- Unsuitability for direct powder production.

A variant of the Kroll process is employed by the ADMA Corporation to produce TiH_2 *in situ* in the reactor [44]. Details of the process and TiH_2 powder metallurgy are given in Chapters 8 and 9.

Stirred molten salt reactors and combinations of stirred reactors have been used in a number of developments [37,45–51]. Potential advantages include the following:

- Economy of scale resulting from scaling up of the process.
- Continuous operation.
- Relatively low residence time (a couple of hours), resulting in a high throughput and space-time yield.
- Control of reaction conditions affecting Ti powder morphology.
- High heat transfer coefficients to reactor walls and internals.
- Heat integration.

Problems associated with these processes include the following:

- Prevention of the formation of Ti sponge adhering to the reactor internals, eventually causing a forced shutdown of the process.
- Measurement and control of the composition of the reactor contents.
- Mechanical problems related to the handling of hot suspensions of Ti powder and lumps in molten salt in the presence of $TiCl_4$ and/or highly reducing molten alkali or alkaline earth metals.
- High inventory of molten salt in the reactor system, so that means must be provided to handle the salt in the event of plant stoppages.

Recent research to develop a process to produce titanium powder by metallothermic reduction of $TiCl_4$ in molten salt reactors is discussed in Section 5.6.

Electrolytic processes to recover titanium are discussed in detail in Chapters 3 and 4. Inherent advantages of electrolytic cells per se include the following:

- Improved energy efficiency.
- Simple scaling up by replicating electrolytic cells.

Disadvantages include the following:

- Space time yields in metal electrowinning cells are relatively low.
- The operating temperature of typical electrolytic cells to electrowin Ti is between 800 (1472) and 900°C (1652°F), whereas the operating temperatures of Li, Na, and Mg electrowinning cells are typically about 450 (842), 580 (1076), and 680°C (1256°F), respectively. This makes the mechanical design of Ti electrowinning cells much more demanding.
- The use of carbon for or in anodes can cause contamination of Ti in Ti electrowinning cells, whereas this problem has been solved in commercial alkali and alkaline earth metal electrowinning cells.

To overcome the problem of Ti sponge formation in molten salt reactors causing a blockage of the feed lines, Imperial Chemical Industries [52] proposed a process to feed molten Na onto an *agitated particulate bed* of NaCl and then to add $TiCl_4$ vapor to the bed to react with Na. However, drawbacks of this process include the following:

- Mechanical complexity and erosion.
- Dust suppression.
- Low heat transfer rates.

The process was never commercialized.

A simpler method to achieve agitation of a bed of particles is to use a *fluidized bed* instead of a mechanically agitated bed in which $TiCl_4$ vapor is reacted with a reducing metal or hydrogen [28,53–56]. Advantages of fluidized bed reactors include the following:

- Avoidance of the problem of Ti lump formation caused by long-range electronically mediated reactions [57] as in molten salt beds.
- High mixing rates preventing hot spots in the reactor.
- Relatively high heat transfer coefficients that facilitate heat recovery and integration.
- No moving mechanical parts in the hot zone of the reactor.

Potential problems include the following:

- Handling of dust.
- Handling of salt and reducing metal vapors caused by partial sublimation in the hot reactor zone.
- High recirculation rates of inert fluidizing gas and cost of associated equipment.

The development of a fluidized bed process to reduce $TiCl_4$ magnesiothermically is currently being piloted by the CSIRO in Australia. This is discussed further in Section 5.6.

In a number of developments, a *molten metal* was used as reaction medium in order to avoid inter alia the problems associated with using molten salts as reaction medium [13,58–61]. Advantages of using molten metal as reaction medium include the following:

- High heat transfer rates.
- Avoidance of Ti sponge formation.

Disadvantages include the following:

- The separation of Ti becomes more complex because instead of two components that have to be separated, that is, Ti and by-product salt, the molten metal used as medium for the reaction has to be recovered and recycled as well.
- There is virtually no control over Ti particle morphology when using molten alkali or alkaline earth metals as medium.
- Handling large volumes of molten metal adds mechanical complexities.

The ITP/Armstrong process has been under development for a number of years. More details about the process are given in Section 5.6.

In a number of developments, the reactants to produce titanium are intimately mixed (*milled*) in the solid state and, if needed, heated to initiate the reaction between the precursor and the reductant [10,62]. These processes tend to be batch operations. Advantages of this approach include the following:

- It is possible to produce Ti alloy powders directly.
- There is some control over the particle morphology.
- Scaling up is done by replicating the reactor system.

Disadvantages include the following:

- The processes tend to be batch operations, making them capital and labor intensive.
- There is limited economy of scale.
- Opportunities for energy integration are limited.

The University of Waikato and Titanox Ltd. is currently developing a process (the TiPro process) using this approach and more details are given in Section 5.6.

The use of a *spray* reactor in which $TiCl_4$ is reacted with a spray of molten reducing metal has been investigated in a number of studies [63–66]. In this type of reactor, the use of Na might be the most convenient because pumping molten Na is easier than pumping the other metals that are suitable reducing agents as its melting point is the lowest. Advantages of this approach include the following:

- The process is suitable for continuous operation.
- The formation of lumps of Ti sponge is avoided.
- The inventory of reactants and products in the reactor is low and losses during shutdowns or stoppages are therefore low.

Disadvantages include the following:

- The conversion of $TiCl_4$ is incomplete. Kametani and Kurihara [65] reported a Ti yield of 90% in the reactor and further losses of 20% in the leaching operation.
- The formation of a mist of salt and Ti particles causes deposition problems.
- There is no control over the particle morphology.

Another flow reactor type considered in some studies is to use gas-phase reactions between $TiCl_4$ and gaseous reducing metals or H_2 [67–70]. The advantages of a *gas flow* reactor system include the following:

- Reactions are faster and hence more complete than in a spray reactor.
- The process is continuous.
- The operating temperature and pressure can be adjusted so that the by-product salt is in the gas phase and the Ti product either liquid or solid and separated as such from the salt vapor.
- The inventory of reactants and products in the reactor system is low and turnaround time between stoppages is short.

Disadvantages include the following:

- Extra equipment and energy are required to evaporate the reducing metal.
- There is virtually no way to control the particle morphology.
- Efficient collection of fine Ti particles from the gas phase poses challenges.
- Very fine Ti particles are produced that may result in excessive surface contamination when the product is exposed to the atmosphere.

On the upper end of the temperature range considered for reactors to produce titanium, the use of *plasma* reactor have been considered [27,71,72]. Plasma reactors differ from gas-phase reactors in that the reaction temperature is further increased by the use of a plasma torch. This reactor type holds promise for making molten Ti or ingot directly. However, it is unlikely that it is suitable for Ti powder production.

Lastly, at low temperatures, the use of *organic ionic liquids* as reaction media in a high-shear reactor is being considered in recent developments [73]. An example of a suitable ionic liquid is a compound of *N*-methyl,*N*-propyl-piperidinium cations and bis(trifluoromethane sulfonyl)imide anions. The concept is basically to form a dispersion of reducing metal droplets or particles in the reactor that reacts with liquid $TiCl_4$ that is also dispersed in the reactor at temperatures in the range of about 20 to about

100°C (68 to 212°F). The by-product salt from the reaction dissolves in the ionic liquid and is readily separated from the particulate titanium product. More details are also given in Section 5.6.

Advantages of this approach include the following:

- The lowest operating temperatures of all the reactor types, resulting in simplicity of construction and use of low-cost materials of construction.
- Low corrosion rates.
- Use of conventional filtration equipment to separate the titanium product from the reaction medium and dissolved salt.
- The possibility to produce titanium powder and alloys directly.

Disadvantages include the following:

- Cost of the ionic liquids.
- Health and safety hazards resulting from the use of such liquids.
- Noncommercial technology required to separate the solution of by-product salt from the ionic liquid (direct electroplating of the reducing metal used from the ionic salt was proposed).
- No possibility to recover useful energy from the heat of reaction.

5.5 Separation principle

In the Kroll process, the bulk of the by-product $MgCl_2$ is *drained* from the reactor and the remainder is either distilled off or leached out of the product. Draining is the simplest and least expensive form of separation, but is not practical in most of the processes that do not form Ti sponge. Because of the costs originating from the separation technology used, many process concepts have been conceived with the primary aim of simplifying the separation step. Separation technologies used include leaching, distillation, extrusion, deposition, Ti extraction, and Ti melting.

Leaching of by-products with either water or dilute hydrochloric acid is simple. However, in many countries, it is not acceptable to dispose of the leachate into the environment. Recovering and recycling of the salt is preferred. In order to do so, the leachate has to be concentrated and the salt crystallized and dried. These steps are energy intensive, and in the case of recovering $MgCl_2$, extra costs are incurred to ensure that anhydrous $MgCl_2$ is recovered and not MgO.

An alternative is to distil the by-product salt from the Ti. With *distillation*, anhydrous salt suitable for use in an electrolytic cell is obtained directly. Unfortunately, the vapor pressures of the by-product salts typically produced are very low, and high temperatures and vacuum have to be used to achieve the distillation. A further problem is that when the vaporized salt is condensed, it deposits as a solid that has to be removed and extracted mechanically. The result is that salt distillation processes are costly and difficult to operate.

A number of developments have been aimed at concentrating the titanium powder in the reaction mixture and *extruding* it from the reaction vessel [74,75]. The mechanical designs proposed are in principle simple but becomes complex when used to operate at the harsh process conditions in Ti production reactors. These concepts have not been proven on a large scale.

In gas-phase and plasma-phase reactors, the temperature of the process can be adjusted so that the by-product salt is in the gas phase and the Ti powder is either solid or molten. If the Ti droplets or powder are *deposited* on a suitable surface, they are separated from the by-product salt before the salt condenses. In order to apply this separation technique properly, the temperature profile in the outlet ducting from the reaction zone of the process must be controlled adequately to ensure that the bulk of the Ti is recovered before the salt condenses. A disadvantage of the process is that mechanical devices operating at high temperature are required to remove the titanium and the salt if the process is to be operated continuously.

A number of developments have aimed at *extracting* the Ti from the reaction bath with molten Zn [13,52]. The titanium in the Zn is free of salt, and once the solution of Ti in Zn has been removed from the reactor, the Zn is distilled from the Ti and recycled. The main disadvantage of this technique is that continuous distillation of Zn from Ti is also energy intensive and rather complex. None of the developments has been commercialized.

Concepts to produce Ti ingot directly involve *melting* the produced Ti while separating $MgCl_2$ from it by evaporation [46,73], or molten $CaCl_2$ by floating it on top of the molten Ti [41]. The idea is to extrude the ingot while maintaining a molten pool of Ti on the solidified section. This method of separation is obviously not suitable for producing Ti powder.

5.6 Recent developments

Recent studies on producing titanium powder by metallothermic reduction of a titanium-bearing precursor on which sufficient information has been disclosed in the public domain to deduce the overall process intent include research in Japan to produce Ti by metallothermic reduction of $TiCl_4$; the CSIR-Ti process; the ITP/Armstrong process; the TiRO™ process of the CSIRO; the ADMA process; the Ono Suzuki (OS) process; the TiPro process; the Peruke process and the Ionic Liquid process of GM Global. Apart from the ADMA process, which is discussed in Chapter 8, brief summaries of the processes are given below. In addition a short summary of the Chinuka process, which is not a metallothermic but rather an electrolytic process, is also given since it is not discussed elsewhere in the book.

The aim of the *Japanese studies* [48,50,51,76,77] on the metallothermic reduction of $TiCl_4$ in a medium of molten salt was to develop a reactor system that could be operated continuously. The research included fundamental studies on the solubility of different chlorides in molten $MgCl_2$, considerations of the reaction mechanism leading to sponge and powder formation, and of the effect of agitation on the morphology of the titanium powder product.

Problems were experienced when feeding the $TiCl_4$ into a steel reactor through a steel lance [77] because of the adherence and growth of Ti particles on the steel walls and Ti sponge formation on the tip of the lance, eventually choking the lance. The cause of these problems was ascribed to electrochemical reactions taking place via the steel, which is an electrical conductor. When using reactor crucibles and a lance made of ceramic (either Al_2O_3 or MgO), only powder and no sponge was formed.

Problems were also experienced with mechanical agitation [48], and a special agitator that generated a dispersion of fine Mg droplets in the reactor was developed and found to work well.

Eventually, it was proposed to do the reduction of $TiCl_4$ in two steps, first to reduce it to $TiCl_2$ (or $TiCl_3$), which has a relatively high solubility in molten $MgCl_2$, and then to reduce the subchloride to Ti. Unfortunately, it seems that the work was not taken any further than laboratory-scale tests.

Final block flow diagrams for the processes proposed by the different Japanese research groups referred to were not given. However, considering the process considerations discussed by Tanaka et al. [48] and the two-stage reactor proposed by Suzuki et al. [50], Figure 5.1 is a diagram illustrating the key process steps. In the process, $TiCl_4$ is firstly reduced in a prereduction reactor with titanium powder that is produced in and recycled from a separate $TiCl_2$ reduction reactor to form a solution of $TiCl_2$ in molten $MgCl_2$. The medium in which both the reactions in the prereduction and $TiCl_2$ reduction reactor occurs is molten $MgCl_2$, and both reactors operate at a temperature of between about 780 and 900°C (1436 and 1652°F).

The solution of $TiCl_2$ in molten $MgCl_2$ produced in the prereduction reactor is transferred to the $TiCl_2$ reduction reactor, where it is reduced with molten magnesium that is dispersed into the molten $MgCl_2$ using a specially designed agitator. The titanium powder product is formed as a suspension in the molten $MgCl_2$ medium, and some of it is recycled to the $TiCl_4$ prereduction step and some of it is withdrawn from the reactor. The withdrawn product stream of titanium powder suspended in molten $MgCl_2$ is first passed to a sedimentation step to separate the bulk of the $MgCl_2$ from the titanium powder, whereafter the remainder is removed by distillation.

The *CSIR* in South Africa is developing a continuous process similar to that proposed by White and Oden [37] to reduce $TiCl_4$ metallothermically in a medium

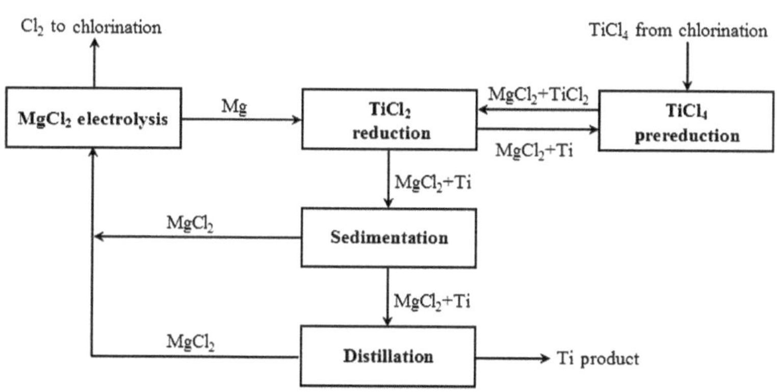

Figure 5.1 Process block flow diagram for processes researched in Japan.

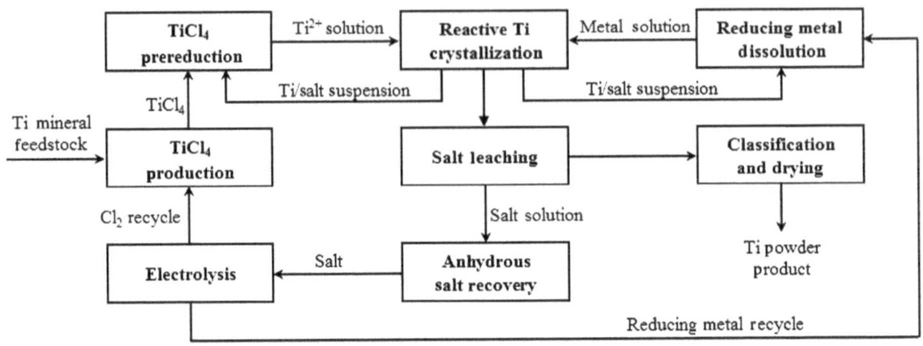

Figure 5.2 CSIR-Ti process block flow diagram.

of molten salt formed as a by-product of the reaction. In the process illustrated in Figure 5.2, $TiCl_4$ is first reduced by Ti powder suspended in molten salt to form a solution of $TiCl_2$ in molten salt. Before reducing metal is added to reduce the $TiCl_2$, it is dissolved in a separate step in molten salt, and the solutions of reducing metal and $TiCl_2$ are then mixed in a well-stirred reactor to produce more Ti powder suspended in the molten salt. A part of the suspension of Ti powder is recycled to the $TiCl_4$ partial reduction reactor and another part to the reducing metal dissolution step. The remainder overflows into a cooling stage where the salt is solidified and cooled. The cold mixture of Ti powder in salt is then passed to a separation stage to recover the Ti powder product and anhydrous salt. The salt is recycled to an electrolysis step to recover the reducing metal and Cl_2.

In the CSIR's titanium pilot plant, an aqueous salt-extraction system is used because that is deemed to be the simplest way to separate the by-product salt from the Ti powder.

Conceptually the process is simple, but many mechanical problems are being experienced with nonstandard process equipment and custom-designed instrumentation and control systems. Key challenges are to demonstrate that the process can operate continuously without forming lumps of Ti powder that adhere to or accumulate in the reactors of the process and proving that titanium crystals with a large enough size can be produced to limit oxygen pick-up by surface oxidation.

In the *ITP/Armstrong process* [61,78,79], illustrated in Figure 5.3 gaseous $TiCl_4$ is injected into a stream of excess molten Na flowing through a pipe reactor. The Na is a good conductor of heat and apart from serving as a reactant and reaction medium, it also helps to absorb and transfer the heat generated by the highly exothermic reaction. The outlet from the reactor is cooled down to allow filtration of Na still in the liquid state from solid NaCl and Ti particles formed by the reaction. The remaining Na in the filter cake is distilled from the cake, the Ti powder is passivated, and the NaCl is leached with water from the Ti powder. The Ti powder is dried and packed, and the NaCl can be crystallized out of the aqueous solution and recycled to electrolytic cells to regenerate Na and Cl_2.

The engineering of the system is not trivial. It is envisaged that many of the items of equipment needed are non-standard and require custom design and development.

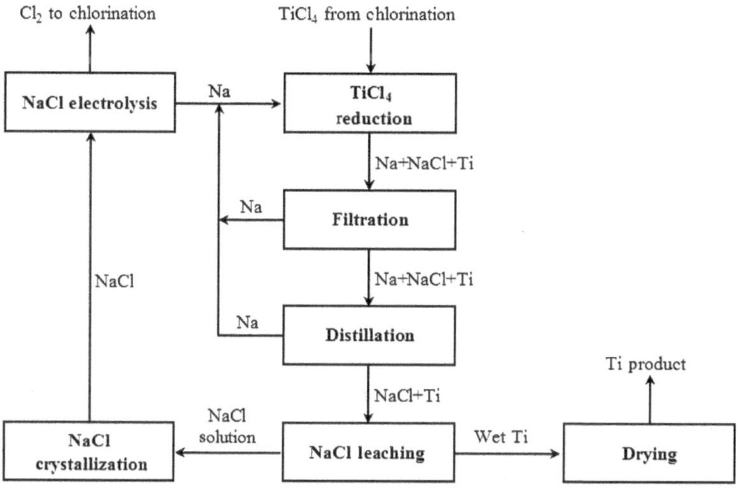

Figure 5.3 Armstrong process block flow diagram.

Examples of the potential problems include the choice of materials of construction to handle molten Na in a process with both hot and cold zones; choice of filter; choice of process instrumentation, such as flow and slurry density measurement sensors; sealing of the process at the inlets and outlets of the Na distillation section, which has to operate under vacuum to ensure that the NaCl does not melt in the unit; heat transfer to and from the processed materials; and design of the system for startup, continuous operation, and shutdown conditions. These problems are not insurmountable, but time is required to solve them and to demonstrate that the solutions actually work as intended.

In the *TiRO process* [33,56] illustrated in Figure 5.4 gaseous $TiCl_4$ and Mg powder are fed into a fluidized bed of $MgCl_2$ and Ti particles operating in the temperature range above the melting point of Mg (651°C [1204°F]) and below the melting point of $MgCl_2$ (708°C [1306°F]). The Mg forms a molten coating on the particles in the bed and reacts with $TiCl_4$ vapor to form more Ti and $MgCl_2$. The bed is fluidized with Ar, which is recirculated via dust collectors and heat exchangers. Composite Ti and

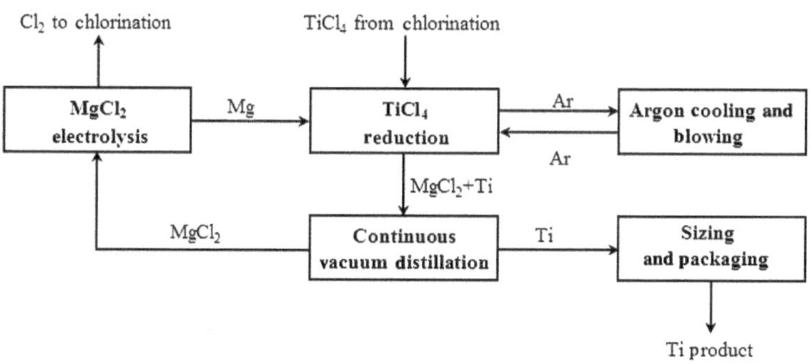

Figure 5.4 TiRO process block flow diagram.

$MgCl_2$ powder is extracted from the fluidized bed and collected from the dust collectors and then fed into a continuous vacuum distillation unit operating below the melting point of $MgCl_2$ to distil the $MgCl_2$ from the Ti powder, forming a "biscuit" of loosely sintered Ti powder and anhydrous $MgCl_2$. Finally, the anhydrous $MgCl_2$ can be recycled directly to electrolytic cells to produce Mg and Cl_2.

A 2-kg/h pilot plant was built, has recently been commissioned, and some test runs have already been completed.

As with the ITP/Armstrong process and for that matter all the processes to produce Ti powder directly in a continuous manner, the basic chemistry of the process is simple, but the engineering problems involved in designing and operating the process are complex. For example, fluidized bed reactors are well known, but operating a unit that has one component in the molten state (i.e., Mg) can result in excessive particle agglomeration; handling of vapors that can deposit in downstream equipment can be problematic; feeding Mg powder or a spray of Mg droplets into the fluidized bed may lead to entrainment of fine Mg droplets, which may cause downstream handling problems; and the design and operation of the novel vacuum distillation unit is difficult.

These problems are also not insurmountable, but it takes time to encounter unforeseen problems in a first-of-a-kind process and then to develop satisfactory solutions.

In the *OS (Ono Suzuki) process* [38,80], TiO_2 powder is metallothermically reduced with Ca dissolved in molten $CaCl_2$ to produce Ti and CaO dissolved in the molten $CaCl_2$. The Ti powder is separated from the salt and the CaO-containing salt is directly electrolyzed to again produce dissolved Ca and a mixture of CO_2 and CO.

The reported work was limited to theoretical and laboratory-scale studies and not further developed to a pilot-scale level. Thermodynamic studies explained why only Ca and Y are suitable to produce Ti containing less than 1000 ppm oxygen and that Mg and Al are not sufficiently reducing to achieve the same result. It was shown that the ratio of the activities of CaO to Ca in the $CaCl_2$ must be less than one. Also reported were the experimental conditions to show at what concentration levels and residence time sufficient reduction and oxygen removal can be achieved.

In the *TiPro or Titanox process* [10], TiO_2 is first annealed at an elevated temperature to homogenize the crystal structure of the powder and increase the TiO_2 powder grain size. It is then crushed, screened, and mixed with CaH_2 and metal or metal oxide powders of alloying elements. The mixture is heated under vacuum or in an Ar atmosphere to a temperature below the melting point of the alloy. Upon heating, the CaH_2 decomposes to form Ca and H_2, and the Ca reacts with the blended mixture of oxides to reduce it to corresponding elements. The different alloying elements diffuse into the resulting Ti grains to form Ti alloy powder. After the reaction mixture has cooled down, it is crushed and the CaO produced by the reaction is leached from the metal powder with dilute acid.

The grain size of the Ti alloy powder is determined by the grain size of the TiO_2 powder from the screening step that follows TiO_2 annealing.

The process does not lend itself to continuous operation, requires an additional high-temperature annealing step, and uses rather expensive CaH_2 as reducing agent, but it is probably the most convenient approach to producing different types of Ti alloy powders directly.

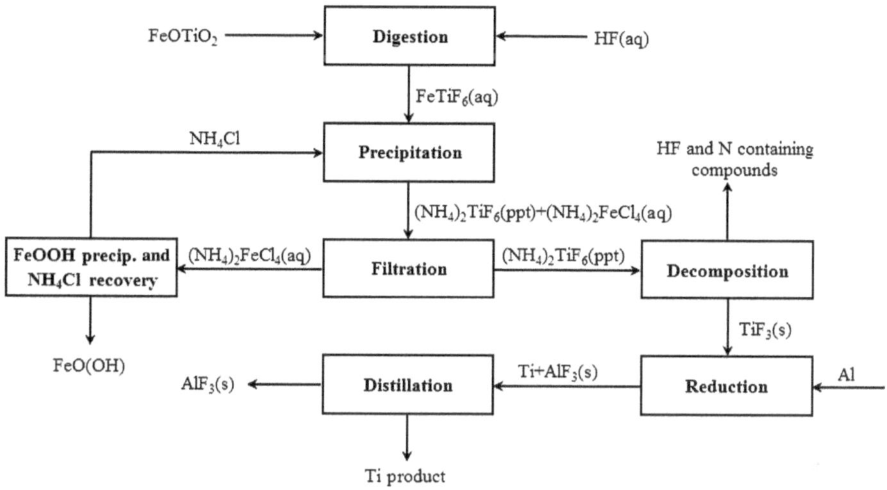

Figure 5.5 Peruke process block flow diagram.

In the *Peruke process* [15], TiO_2-bearing raw feedstock is digested with an aqueous solution of hydrogen fluoride (HF) to form an aqueous solution of fluotitanate salts. A solution of an ammonium or alkali metal salt is then added to the fluotitanate solution to precipitate selectively M_2TiF_6 where M is ammonium or an alkali metal cation. When an ammonium salt is used, $(NH_4)_2TiF_6$ is precipitated, which can conveniently be decomposed to form particulate TiF_3 at the reaction conditions. The TiF_3 is reacted with Al to produce Ti and AlF_3, which are separated by distilling the AlF_3 from the Ti, leaving Ti powder. Figure 5.5 is a block flow diagram of one variation of the process.

A pilot plant was built to scale up the process, but no results from the pilot plant have been reported in the open literature. The project has apparently been abandoned.

Advantageous features of the process are that high-purity fluotitanate salts can be extracted from the cheaper feed materials using relatively conventional hydrometallurgy process steps and a subfluoride with no cations other than titanium; that is, TiF_3 can also be produced readily. A particular disadvantage is the difficulty in separating the formed AlF_3 from the product while ensuring that titanium aluminides are not formed as well.

In the *ionic liquid process* [73] of GM Global illustrated in Figure 5.6 an organic ionic liquid is used as a medium in which $TiCl_4$ is reduced with an alkali or alkaline earth metal. The reaction proceeds at temperatures in the range of 20 to 100°C (68 to 212°F). Key attributes of the selected ionic liquids are that they are stable against oxidation by Cl_2 and reduction by the reducing metal. Furthermore, it is desirable that the by-product salt produced by the reaction between $TiCl_4$ and the reducing metal has a high solubility in the ionic liquid.

In the Reduction reactor, the reducing metal is dispersed with a high-shear mixer as droplets when using metals with a low melting point such as mixtures of metallic sodium and potassium. Alternatively, the reducing metal is dispersed as solid particles. $TiCl_4$ is also fed into the reactor and is also dispersed as fine droplets in the ionic

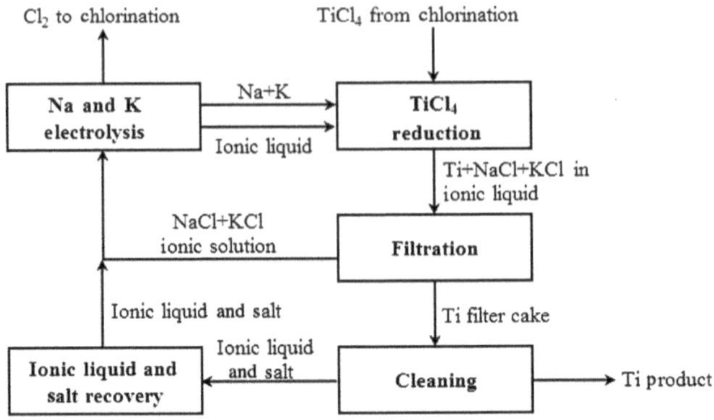

Figure 5.6 Ionic liquid process block flow diagram.

liquid. The $TiCl_4$ and reducing metal react to form titanium powder and the chloride salt of the reducing metal. The salt is soluble in the ionic liquid and as such most of the separation between the titanium product and the by-product salt is achieved already in the reactor.

The suspension of titanium powder in the solution of by-product salt and ionic liquid is then filtered. Following filtration, the filter cake of titanium powder is washed, whereafter it is dried before packaging.

It is proposed to feed the filtrate of dissolved salt in ionic liquid directly to the electrolytic cells, where it is electrolyzed to electroplate the reducing metal on the cathodes of the cells and to recover chlorine at the anodes.

The advantageous features of this process are a direct result of the low temperature of operation; for example, standard materials of construction can be used, the separation of titanium from by-product salt is seemingly very simple, and materials-handling problems experienced with the alternative high-temperature processes are significantly simplified.

Uncertainties regarding the process are the cost of ionic liquids, the financial impact caused by losses of the liquid in the various process steps, washing of the filter cake to recover both the titanium product and all the ionic liquid, and the development required to not only scale the reaction step up but also the novel electrowinning technology.

The *Chinuka process* [19] is similar to the electrolytic process of Materials and Electrochemical Research Corporation discussed in Chapter 3. Figure 5.7 is a block flow diagram of the process. The titanium feedstock is rutile (natural or synthetic), which is partially reduced in the first step of the process to produce titanium oxycarbide. The reaction is thermodynamically not favorable and is therefore done at a high temperature (>1500°C [2732°F]) in vacuum. After reduction, the titanium oxycarbide is formed into the shape of suitable anodes and then sintered at about 1700°C (3092°F). The anodes are then inserted into novel electrolytic cells to electrorefine and recover titanium nodules and crystallites at the cathodes of the cell. The operating temperature of the electrolysis cells depends on the electrolyte used. When using a eutectic mixture of NaCl and KCl, experimental conditions were varied between 700

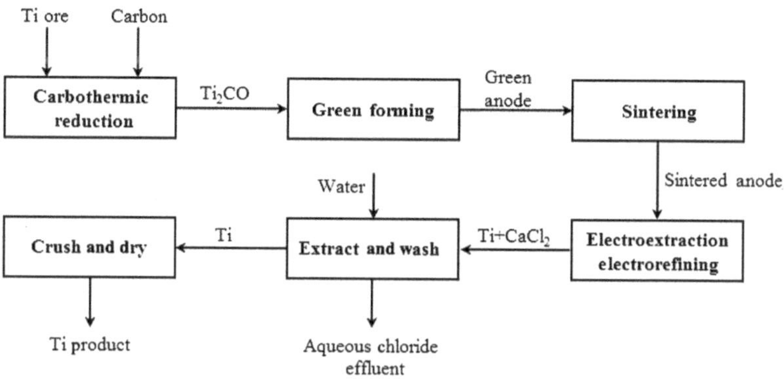

Figure 5.7 Chinuka process block flow diagram.

and 900°C (1292 and 1652°F). Following the electrolysis step, the titanium product is extracted from the cells and then washed with water to remove electrolyte from the product. Finally the product is dried.

The main advantages of the process compared to the standard Kroll process is that much cheaper feedstock, namely, natural or synthetic rutile, is used rather than $TiCl_4$. In addition, electricity is used directly to extract and reduce the titanium product instead of indirectly as in metallothermic processes, and the process might therefore be more energy efficient.

On the other hand, the process would be difficult to perform in a continuous manner since the carbothermic reduction step, the sintering step, and the electrolysis steps are all very difficult to do in a truly continuous manner. Furthermore, the process requires a lot of mechanical handling of intermediate products, which is either labor intensive or requires automation. In addition, although the effluent from the washing step of the process is not particularly hazardous, it results in a loss of electrolyte that has to be recovered or replaced. Finally, whereas the process might consume less energy and more specifically energy in the form of electricity, it remains to be seen if in practice it would indeed consume less since both the carbothermic reduction and sintering steps are also energy intensive and the current efficiency of the process is expected to be less than that of conventional alkali and alkaline earth electrowinning processes.

5.7 Concluding remarks

Many different approaches to produce Ti powder continuously by metallothermic reduction have been proposed and tested over a period spanning about six decades. To date, none of these has been commercialized successfully. However, the work has led to a much better understanding of the problems underlying the technologies and a number of promising approaches are currently being piloted. It is expected that cost-effective Ti powder will be produced and become commercially available to the titanium industry in the near future.

References

[1] H. Lu, W. Jia, Method of preparing metallic titanium with high temperature fused salt electrolysis of titanium dioxide, Chinese Patent CN101343756 (2008).
[2] F. Cardarelli, Method for electrowinning of titanium metal or alloy from titanium oxide containing compound in the liquid state, US Patent 7,504,017 (2009).
[3] J.R. Cox, C.L. De Alwis, B.A. Kohler, M.G. Lewis, System and method for extraction and refining of titanium. US Patent Application Publication US 20,130,164,167 (2013).
[4] Peter Spence & Sons Ltd. A process for preparing titanium metal. British Patent GB713,446 (1954).
[5] H.L. Slatin, Process for the electrolytic production of metals, US Patent 3,003,934 (1961).
[6] Y. Hashimoto, Influences of fluoride salt baths on fused salt electrodeposition of titanium metals from TiO_2, Denki Kagaku 39 (12) (1971) 938–943.
[7] G.Z. Chen, D.J. Fray, T.W. Farthing, Direct electrochemical reduction of titanium dioxide to titanium in molten calcium chloride, Nature 407 (2000) 361–364.
[8] X. Nie, L. Dong, C. Bai, D. Chen, G. Qiu, Preparation of Ti by direct electrochemical reduction of solid TiO_2 and its reaction mechanism, Trans. Nonferrous Met. Soc. China 16 (2006) 723–727.
[9] P.P. Alexander, Production of titanium hydride. US Patent 2,427,338 (1947).
[10] J. Liang, Production of titanium alloys in particulate form via solid state reduction process. International Patent Publication WO2010036131 (2010).
[11] T. Takenaka, T. Suzuki, M. Ishikawa, E. Fukusawa, M. Kawakami, The new concept for electrowinning process of liquid titanium, Electrochemistry 67 (6) (1999) 661–668.
[12] Horizons Titanium Corporation, Electrolytic production of titanium and zirconium. British Patent GB771,679 (1957).
[13] J.A. Megy, Process for making zero valent titanium from an alkali metal fluotitanate. US Patent 4,668,286 (1987).
[14] T.A. O'Donnell, J. Besida, T.K.H. Pong, D.G. Wood, Process for the production of metallic titanium and intermediates useful in the processing of ilmenite and related minerals, US Patent 5 397 (1995) 375.
[15] G. Pretorius, A method to produce titanium. International Patent Publication WO2006079887 (2006).
[16] J.G. Wurm, L. Gravel, R.J.A. Potvin, The mechanism of titanium production by electrolysis of fused halide baths containing titanium salts, J. Electrochem. Soc. 104 (1957) 301–308.
[17] Horizons Titanium Corporation, Préparation du titane métallique par electrolyse. French Patent FR1051539 (1954).
[18] J.C. Withers, R.O. Loutfy, Thermal and electrochemical process for metal production. International Patent Publication WO2005/019501 (2005).
[19] D. Jewell, M. Kurtanjek, S. Jiao, D. Fray, Titanium extraction – production of titanium by electrorefining of titanium oxycarbide. http://www.wmtcorp.com/es/Chinuka.pdf, 2013 (downloaded on 16.11.2013).
[20] D.S. Van Vuuren, In search of low-cost titanium. PhD Thesis, University of Pretoria, South Africa (2010).
[21] R.C. Young, W.C. Schumb, The anhydrous lower bromides of titanium, J. Am. Chem. Soc. 52 (1930) 4233–4239.
[22] H. Hartmann, G. Rinck, Das disproportionierungsgleichgewicht des titanchlorides, Z. Phys. Chem. Neue Folge 11 (1957) 213–233.
[23] R.H. Singleton, Method of producing titanium, US Patent 2 889 (1959) 221.

[24] P.C. Turner, A. Hartman, J.S. Hansen, S.J. Gerdemann, Low-cost titanium – myth or reality? Report DOE/ARC-2001-086, EPD Congress held at the 2001 TMS Annual Meeting, New Orleans, LA, Feb. 11–15 (2001).
[25] R. Ottensmeyer, P.J. Plath, A new process for production of titanium. Paper Presented at the 10th World Conference on Titanium, Hamburg, Germany, July 14 (2003).
[26] J.J. Casey, J.W. Berhman, Method of producing titanium, US Patent 3 123 (1964) 464.
[27] B.A. Detering, A.D. Donaldson, J.R. Fincke, P.C. Kong, R.A. Berry, Fast quench reactor method, US Patent 5 935 (1999) 293.
[28] K. Lau, D. Hildenbrand, E. Thiers, G. Krishnan, E. Alvarez, D. Shockey, L. Dubois, A. Sanjurjo, Direct production of titanium and titanium alloys. Paper Presented at the 19th Annual Conference of the International Titanium Association, Monterey, California, October 13–15 (2003).
[29] A. Klevtsov, A. Nikishin, J. Shuvalov, V. Moxson, V. Duz, Continuous and semi-continuous process of manufacturing titanium hydride using titanium chlorides of different valency, US Patent 8 388 (2013) 727.
[30] J. Haidar, Method and apparatus for forming titanium-aluminium based alloys. International Patent Publication WO2009129570 (2009).
[31] D. Zhang, D. Ying, Z. Li, Z. Cai, J. Liang, Titanium-based composites and coatings and method of production, US Patent 6 692 (2004) 839.
[32] H. Eklund, P.B. Engseth, B. Langseth, T. Mellerud, O. Wallewik, An improved process for the production of magnesium, in: H.I. Kaplan (Ed.), Magnesium Technology, TMS (The Minerals, Metals & Materials Society), Warrendale, PA, 2002, pp. 9–12.
[33] C. Doblin, D. Freeman, M. Richards, The TiRO™ process for the continuous direct production of titanium powder, Key Eng. Mater. 551 (2013) 37–43.
[34] R. Bolívar, B. Friedrich, Synthesis of titanium via magnesiothermic reduction of TiO_2 (pigment). Proceedings of EMC 2009, European Metallurgical Conference, Innsbruck, Austria, 2009.
[35] X. Song, N. Chen, C. Zhao, Method for preparing metal Ti by using hydrogen to induce Mg to reduce TiO_2. Chinese Patent CN102528067 (2012).
[36] D.S. Van Vuuren, Fundamental reactor design configurations for reducing $TiCl_4$ metallothermically to produce Ti powder, Key Eng. Mater. 520 (2012) 101–110.
[37] J.C. White, L.L. Oden, Continuous production of granular or powder of Ti, Zr or Hf or their alloy products, US Patent 5 259 (1993) 862.
[38] R.O. Suzuki, K. Ono, OS process – a new calciothermic reduction of TiO_2 in the molten state. Paper Presented at the 18th Annual Conference of the International Titanium Association, Orlando, Florida, October 7 (2002).
[39] D.J. Fray, Emerging molten salt technologies for metals production, J. Metall. (Oct. 2001) 26–31.
[40] T. Ogasawara, M. Yamaguchi, T. Yenishi, M. Hori, K. Takemura, K. Dakeshita, Process for producing Ti and apparatus there for. US Patent Application Publication US20,100,089,204 (2010).
[41] T. Ogasawara, M. Yamaguchi, K. Dakeshita, M. Hori, Method of high-melting-point metal separation and recovery. US Patent Application Publication US20,080,250,901 (2008).
[42] M.P. Nelpert, R.D. Blue, Production of titanium metal, US Patent 2 913 (1959) 332.
[43] F. Seon, P. Nataf, Production of metals by metallothermia, US Patent 4 725 (1988) 312.
[44] V.A. Drozdenko, A. Petrunko, A.E. Andreev, O.P. Yatsenko, O.M. Ivasishin, D.G. Savvakin, V.S. Moxson, F.H. Froes, Manufacture of cost-effective titanium powder from magnesium reduced sponge, US Patent 6 638 (2003) 336.
[45] C.M. Olson, Method of producing titanium metal, US Patent 2 839 (1958) 385.

[46] Y. Okura, Titanium sponge production technology, in: P.A. Blenkinsop, W.J. Evans, H.M. Flower (Eds.), Titanium'95: Science and Technology, Proceedings of the Eighth World Conference on Titanium, Birmingham, UK, 1995, pp. 1427–1437.
[47] S.J. Gerdemann, L.L. Oden, J.C. White, Continuous production of titanium powder, Proc. Materials Week'97: Titanium Extraction and Processing, Indianapolis, Indiana, 1997, pp. 49–54.
[48] J. Tanaka, T.H. Okabe, N. Sakai, T. Fujitani, K. Takahashi, N. Michishita, Y. Umetsu, K. Nikami, New titanium production process with molten salt mediator, J. Jpn. Inst. Metals 65 (8) (2001) 659–667.
[49] G.R.B. Elliott, The continuous production of titanium powder using circulating molten salt, J. Metall. (Sept. 1998) 48–49.
[50] R.O. Suzuki, T.N. Harada, T. Matsunaga, T.N. Deura, K. Ono, Titanium powder prepared by magnesiothermic reduction of Ti^{2+} in molten salt, Metall. Mater. Trans. B 30B (Jun 1999) 403–410.
[51] A. Fuwa, S. Takaya, Producing titanium by reducing $TiCl_2$-$MgCl_2$ mixed salt with magnesium in the molten state, J. Metall. (Oct. 2005) 56–60.
[52] Imperial Chemical Industries, Improvements in or relating to the manufacture of titanium, British Patent GB828 (1960) 374.
[53] S. Okudaira, T. Irie, H. Uehida, E. Fukazawa, K. Kobayashi, M. Yamaguchi, Method for producing metal from its halide, US Patent 4 877 (1989) 445.
[54] K. Hyodo, J. Izeki, A. Moriya, K. Maehara, S. Anpo, H. Watanabe, H. Ito, T. Mitsuya, N. Sakata, N. Sato, Manufacture of metallic Ti, Japanese Patent Publication 03 (150327) (1991).
[55] L. Zhou, F.E.L. Schneider, R.J. Daniels, T. Messer, J.P.R. Peeling, Process for the production of elemental material and alloys, US Patent 6 955 (2005) 703.
[56] G.A. Wellwood, C. Doblin, Low-temperature industrial process, US Patent 7 837 (2010) 759.
[57] T.H. Okabe, D.R. Sadoway, Metallothermic reduction as an electronically mediated reaction, J. Mater. Res. 13 (12) (1998) 3372–3377.
[58] Union Carbide Corporation, Process for producing metals, British Patent GB820 119 (1959).
[59] C.M. Olson, Method of producing titanium, US Patent 2 753 (1956) 256.
[60] R.W. Bartlett, R.J. McClincy, K.D. Bowles, Titanium powder by magnesium reduction of $TiCl_4$ in liquid zinc alloy. Light Metals 1994 Advances in Aluminium Production, TMS, 1994, pp. 1181–1186.
[61] D.R. Armstrong, S.S. Borys, P. Anderson, Method of making metals and other elements, US Patent 5 779 (1998) 761.
[62] F.H. Froes, B.G. Eranezhuth, O.N. Senkov, Reduction of metal oxides through mechanochemical processing, US Patent 6 152 (2000) 982.
[63] K.F. Griffiths, Method for producing the refractory metals hafnium, titanium, vanadium, silicon, zirconium, thorium columbium and chromium, US Patent 3 085 (1963) 871.
[64] R.B. Worthington, Production of metal powder, US Patent 4 445 (1984) 931.
[65] H. Kametani, Y. Kurihara, Developmental studies of the spray-reduction process for the production of titanium powder, in: P.A. Blenkinsop, W.J. Evans, H.M. Flower (Eds.), Titanium '95: Science and Technology, Proceedings of the Eighth World Conference on Titanium, Birmingham, UK, 1995, pp. 2602–2609.
[66] J.D. Leland, Aerosol reduction process for metal halides, US Patent 5 460 (1995) 642.
[67] D.G. Tisdale, J.M. Toguri, W. Curlook, Vapour phase titanium production, CIM Bull. 90 (1008) (1997) 159–163.

[68] H.Y. Sohn, Ti and TiAl powders by the flash reduction of chloride vapors, J. Metall. (Sept. 1998) 50–51.
[69] R. Ma, S. Li, W. Wang, S. Jiao, K. Huang, Preparation method of high-purity titanium. Chinese Patent Publication, CN101984101 (2010).
[70] G. Han, T. Shoji, S. Uesaka, M. Fukumaru, M.I. Boulos, J. Guo, J. Jurewics, Metal titanium production device and metal titanium production method. US Patent Application Publication, US20,130,095,243 (2013).
[71] S.G. Hutchison, C.M. Wai, J. Dong, R.J. Kearney, Titanium production in a plasma reactor: a feasibility investigation, Plasma Chem. Plasma Proc. 15 (2) (1995) 353–357.
[72] A. Joseph, Low-cost high-speed titanium and its alloy production, US Patent 6 824 (2004) 585.
[73] M.P. Balogh, I.C. Halalay, Titanium metal powder produced from titanium tetrachloride using an ionic liquid and high-shear mixing. US Patent Application Publication US20,140,069,233 (2014).
[74] N. Morash, Process and apparatus for producing titanium metal continuously, US Patent 2 826 (1958) 492.
[75] C.L. Schmidt, C.K. Stoddard, Continuous process for the production of titanium metal, Canadian Patent 598 (1960) 473.
[76] T.N. Deura, T. Matsunaga, R.O. Suzuki, K. Ono, Production of titanium powder by reduction of titanium chloride in molten salt, Molten Salts High Temp. Chem. 41 (1) (1998) 7–16.
[77] T.N. Deura, M. Wakino, T. Matsunaga, R.O. Suzuki, K. Ono, Titanium powder production by $TiCl_4$ gas injection into magnesium through molten salts, Metall. Mater. Trans. B 29B (1998) 1167–1174.
[78] R. Anderson, D. Armstrong, L. Jacobsen, Screw device for transfer of Ti-containing reaction slurry into a vacuum vessel. International Patent Publication WO2004022798 (2004).
[79] R.P. Anderson, L. Jacobsen, Separation system of metal powder from slurry and process, US Patent 7 501 (2009) 007.
[80] K. Ono, R.O. Suzuki, A new concept for producing Ti sponge: calciothermic reduction, J. Metall. (Feb. 2002) 59–61.

Research-based titanium powder metallurgy processes

6

Francis H. (Sam) Froes
Consultant to the Titanium Industry, Tacoma, WA, USA

6.1 Introduction

A number of titanium powder metallurgy processes that have not yet reached commercialization are discussed in this paper. These include rapid solidification (RS), mechanical alloying (MA), vapor deposition (VD), thermohydrogen processing, and porous structures. The first three of these are considered to be "far from equilibrium processes."

6.2 Rapid solidification, mechanical alloying, and vapor deposition

RS, MA, and VD all fall in the category of "far from equilibrium processes" [1]. Novel constitutional (such as extension of solubility levels) and microstructural (in particular microstructural refinement and production of very stable dispersions of second-phase particles) effects can be obtained by all three processes; however, commercial processes are not on the near horizon.

An example of the fine dispersion of second-phase particles that can be obtained by RS is shown in Figure 6.1 including nanograined material produced by MA [2,3]. These nanograins show surprisingly good stability on exposure to elevated temperatures, especially when yttria particles are dispersed throughout the matrix.

The VD approach can be used to alloy together normally virtually immiscible Mg with Ti to create low-density alloys akin to Al-Li alloys, and fabricate layered structures at the nano level (Figure 6.2) [4].

6.3 Thermohydrogen processing (THP)

Also, currently in the research base is thermohydrogen processing (THP) [1,5], although other papers in this book discuss use of hydrogen in producing hydrogenated titanium sponge and work which builds on the present work which could lead to commercialization of THP. By intentionally adding hydrogen to a titanium alloy such as Ti-6Al-4V with a normal PM microstructure, the microstructure can be refined in the dehydrogenated conditions (Figure 6.3), with an enhancement in mechanical properties [5].

Titanium Powder Metallurgy. http://dx.doi.org/10.1016/B978-0-12-800054-0.00006-X

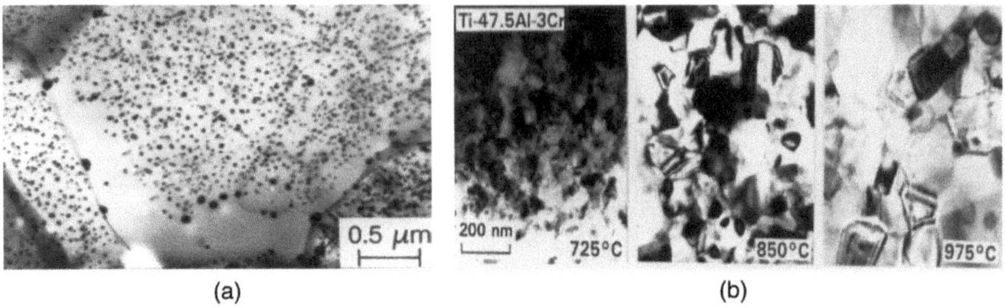

Figure 6.1 Titanium aluminide intermetallic alloys exhibiting (a) a fine dispersion of second-phase erbia particles (Ti_3Al-based alloy) and (b) nanograins after hot isostatic pressing at the temperatures indicated (TiAl-based alloy).

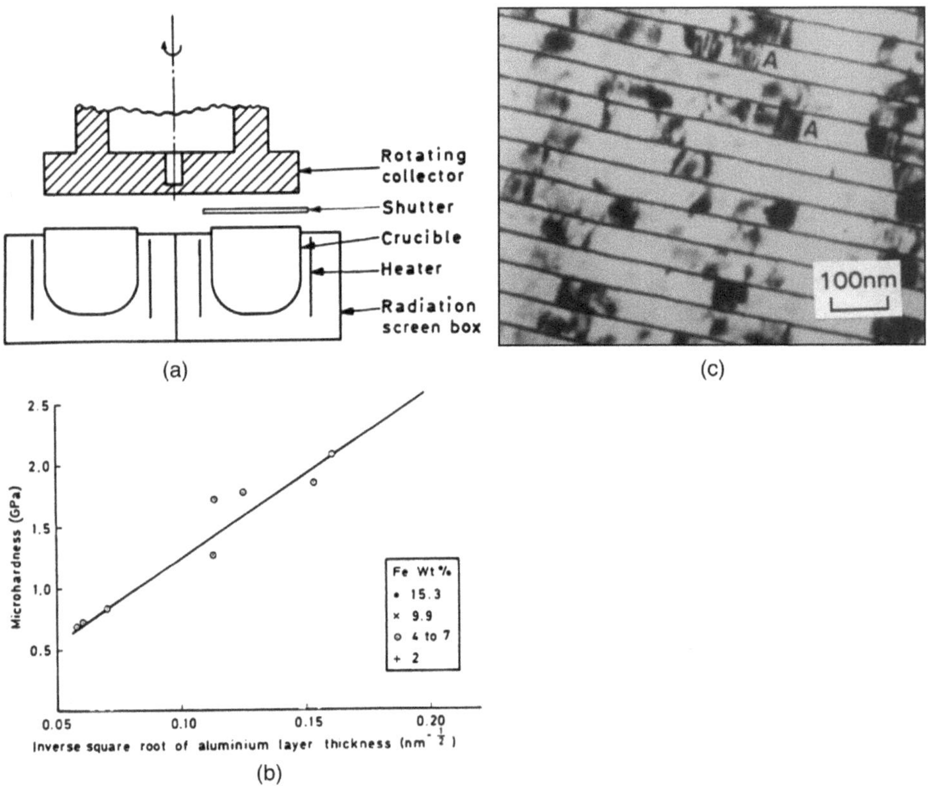

Figure 6.2 Schematic cross-section of rotating collector used in vapor deposition of layered nanostructured materials (a). Increase in hardness with decreasing layer spacing (b), and layered nanostructured consisting of layers of Al and Fe (c).

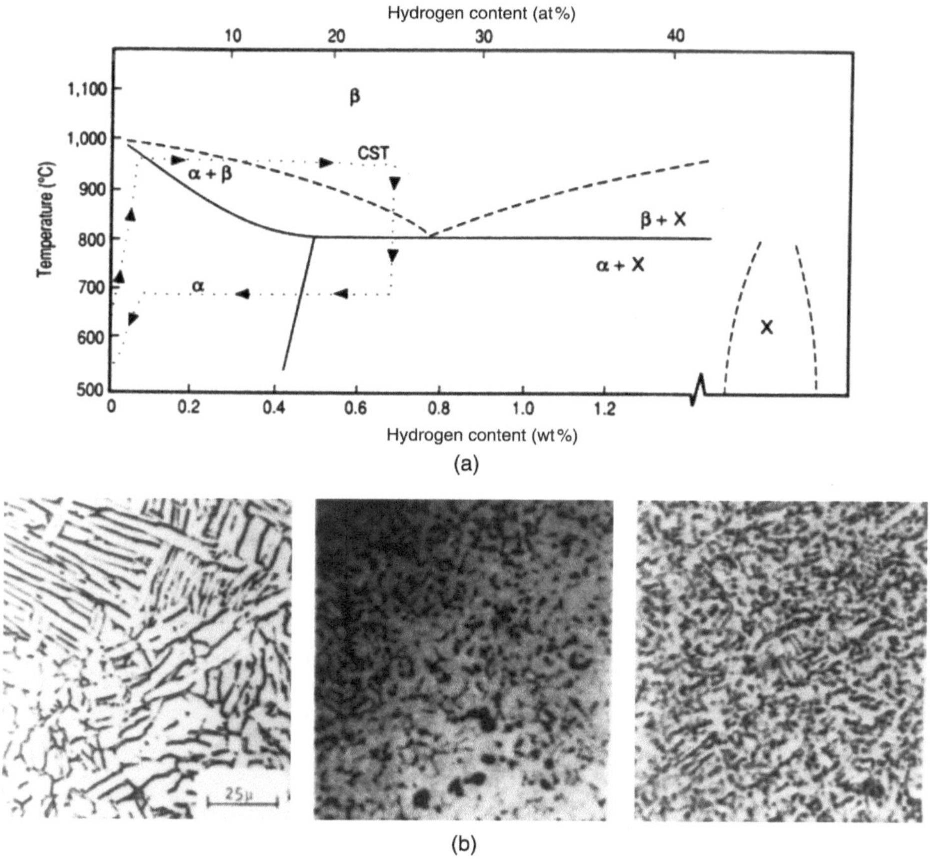

Figure 6.3 (a) Pseudo binary-phase diagram for Ti-6Al-4V. X represents the hydride phase and CST (constitutional solution treatment) is one possible thermohydrogen processing treatment, and (b) refinement of the microstructure of Ti-6Al-4V powder compact using the thermohydrogen processing technique, (left) as hot isostatic pressed coarse alpha laths, (center) hydrogenated then compacted, (right) hydrogenated in compacted state. The latter two conditions are after dehydrogenation, both showing a refined alpha microstructure, (center) equiaxed grains, (right) fine alpha laths.

6.4 Porous structures

A novel type of porous low-density titanium alloy can be produced by HIP consolidation of alloy powder in the presence of an inert gas such as argon (Figure 6.4) [1]. The tensile strength decreases in a linear manner as the porosity level increases, following the "rule of mixtures" relationship at least up to the 30% porosity level. This material exhibited excellent damping characteristics, suggesting a generic area of application. There may also be applications in body implants with the foam integrated in various

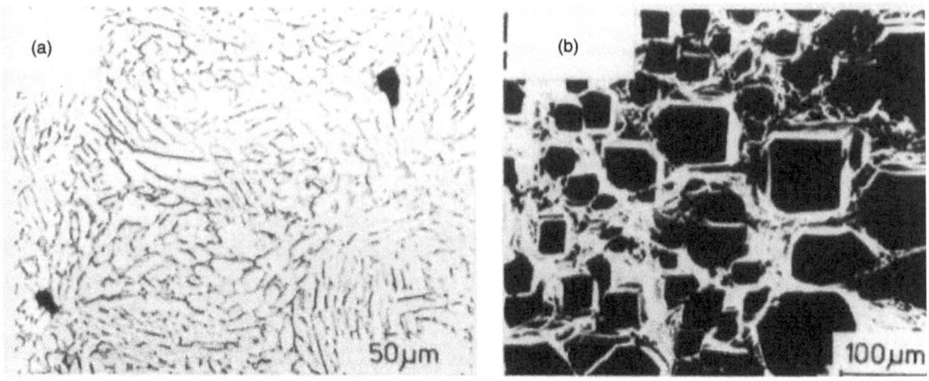

Figure 6.4 (a) Optical micrograph of pores in hot isostatic pressed Ti-6Al-4V containing argon after annealing at 700°C (1290°F), and (b) SEM of sample containing up to 40% porosity after an anneal in excess of 1000°C (1830°F).

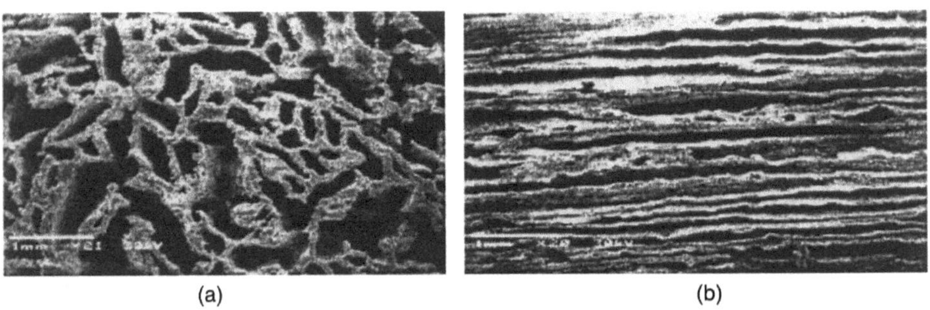

Figure 6.5 Micrographs of foams produced by die compaction (a) and extrusion (b). a: Cross section is perpendicular to the compaction direction; b: cross section is parallel to the extrusion direction.

locations to facilitate growth of bone/flesh into the porous regions promoting a stronger joint. Tensile testing of the bond between foam and dense material indicated a bond strength in excess of 85 MPa well above the FDA requirement of 22 MPa for porous coatings on orthopedic implants.

Porous structures with potential use in honeycomb structures or in sound attenuating or firewall applications are now possible with very precisely controlled porosity levels and architecture using a novel blended metal-plastic approach. (Figure 6.5) [6].

Further discussion of porous structures can be found in another chapter in this book.

Acknowledgments

The author gratefully acknowledges useful input from R. Loutfy, C. F. Yolton, V. Moxson, and Z. Fang.

References

[1] F.H. (Sam) Froes, C. Suryanarayana, Powder processing of titanium alloys, in: A. Bose, R.M. German, A, Lawley (Eds.), Particulate Materials, MPIF 1, Princeton, NJ, 1993, p. 233.
[2] N. Srisukhumbowornchai, O.N. Senkov, F.H. Froes, M.L. Öveçoglu, J. Hebeisen, Stability of nanocrystalline structures in a Ti-47.5Al-3Cr (at %) alloy produced by mechanical alloying and hot isostatic pressing, in: C.M. ward-Close, F.H. Froes, D.J. Chellman, S.S. Cho (Eds.), Synthesis/Processing of Lightweight Metallic Materials – II, TMS, Warrendale, Pennsylvania, 1997, p. 243.
[3] P.B. Trivedi, E.G. Baburaj, A. Geng, M.L. Ovecoglu, S. Patankar, F.H. Froes, Grain size control in Ti-48Al-2Cr-2Nb with Yttrium additions, J. Alloys Compd. (2003) 100–106.
[4] Ward-Close Malcolm, Private Communication, Nov. 11, 2003.
[5] F.H. Froes, O.N. Senkov, J.I. Qazi, Hydrogen as a temporary alloying element in titanium alloys: thermohydrogen processing, Int. Mater. Rev. 49 (3–4) (2004) 227–245.
[6] R. Loutfy, MER Corp., Tucson, AZ, unpublished work, 2000.

Titanium powders from the hydride–dehydride process

7

Daniel P. Barbis, Robert M. Gasior, Graham P. Walker, Joseph A. Capone, Teddi S. Schaeffer
AMETEK Specialty Metal Powders

7.1 Introduction

The hydride–dehydride process (HDH) for the production of titanium powders began around 1957 at the Titanium Metals Corporation, leading to a pair of US Patents [1,2]. The TIMET process was developed to enable easier sizing of mill scrap for incorporation back into electrodes for remelting. Basic jaw crushing produced output of –2-in. pieces for easier electrode fabrication.

In more recent years, the desire was to create near-net-shape parts from both prealloyed and blended elemental powders with wrought mechanical properties [3–6]. Other applications included plasma-spray powder coatings on orthopedic implants and highly porous titanium structures that mimic the modulus of elasticity of human bone. As a result of these high-technology applications, the production of titanium powders [7–10] evolved into specialized businesses developing new and innovative products. Commercial production of high-quality titanium powders with low oxygen and nitrogen began.

The HDH process can produce titanium powders from, among others, both the Kroll and Hunter process sponges, from commercially pure titanium (CP-Ti) and from the workhorse alloy, Ti-6Al-4V. Titanium sponge and CP-Ti powders are typically produced to Grades 1–4 of ASTM B-348, and Ti-6Al-4V to either Grade 5 or Grade 23. For medical applications, ASTM F67 and ASTM F1580 are the two main specifications referenced.

7.2 HDH titanium feedstock

Nearly any source of titanium can be converted to powder through the HDH process. All sources must be clean and below 5 cm (2 in.) in thickness to yield powder. Any thicker sections may not become fully hydrided, leaving a solid core behind that would require a second hydride operation. Feedstock with heavy surface oxidation or with a nitride coating, for example, may be unsuitable as these can act as a barrier to hydrogen penetration without proper cleaning.

Hunter and Kroll process sponges, produced via sodium and magnesium metal reduction [11,12] of titanium tetrachloride ($TiCl_4$), respectively, are normally acquired

Titanium Powder Metallurgy. http://dx.doi.org/10.1016/B978-0-12-800054-0.00007-1

as granules less than 1 in. in size. As the "sponge" name implies, these sources yield porous powders with relatively low apparent densities (1.4–1.8 g/cm^3 [87–110 lb/ft^3]).

Ingot produced by vacuum arc remelting (VAR) or by electron beam (EB) melting can be used as starting feedstock for titanium and titanium alloy powders. Mill products like sheet, plate, and turnings are also typically good, high-quality, sources for powder manufacturing. Angular, blocky, and denser powders (2.0–2.4 g/cm^3 [120–150 lb/ft^3]) are recovered from these ingot and mill products. Often, the aspect ratio of these powders can be correlated to the grain structures produced by these melting, forging, rolling, and extrusion processes.

Melting operations greatly reduce residual $MgCl_2$ and NaCl salt remnants from the sponge reduction process, from upwards of 1400 ppm to often well less than 20 ppm. These residual salts are problematic in the HDH process, as they will evaporate and later deposit inside the vacuum chamber or in cooler areas, such as vacuum lines. These salts accumulate over time and can restrict pumping capabilities if allowed to build up. These crystalline deposits may also deflagrate if exposed to atmospheric conditions in an uncontrolled manner. Cold wall furnaces require frequent hot zone maintenance to remove these formations. Regular vacuum pump maintenance, especially oil changes, is required for reproducible furnace processing.

7.3 The HDH process

7.3.1 Process background

The HDH process, as the name implies, is a reversible two-step process where hydrogen is used as a temporary alloying element. Other transition elements such as tantalum, zirconium, niobium, nickel, and iron may also be processed using this technique. The basic process is shown below. Hydriding the raw material allows it to be crushed, milled, and screened to a desired size target. Hydride may only be powdered while it is embrittled. After dehydriding, the mechanical properties of a ductile titanium powder are regained and further attrition is not possible.

HYD → Size/Screen → DEH → Size/Screen

Dehydriding the powder results in some sintering, the amount being a function of the particle size distribution of the hydride. Agglomerates created during the dehydride cycle must be again crushed, milled, and screened to recover the original particle size distribution of the hydride. With careful management of the two lengthy vacuum processes, time spent at temperatures that encourage sintering may be minimized.

High-quality powders can be successfully produced only if carefully inerted during all crushing, milling, and screening operations. The term *high quality* typically refers to levels of the interstitial elements oxygen and nitrogen. This is true of all reactive metals such as titanium, tantalum, niobium, etc. In some cases, a further deoxidation step is used to control oxygen.

A refractory-lined or water-jacketed batch vacuum furnace capable of holding as much as 2500 kg (5500 lb) or more of product and achieving temperatures approaching 1000°C (1800°F) is typical. Furnace pumping capacity must be matched with load capacity for maximum cycle efficiency and adequate control of furnace pressure, particularly while dehydriding. Water-jacketed vacuum lines are also typically used to cool the hot hydrogen gas while dehydriding and protect vacuum isolation valves. Similarly, forced cooling with argon through a heat exchanger is normally employed for maximum throughput.

The basis for the HDH process can be examined in the Ti-H phase diagram as shown in Figure 7.1. The brittle intermetallic δ hydride phase (TiH_{1-x}) has a wide compositional range with a minimum of 2.16 wt% hydrogen. Most often, however, hydrogen is absorbed very aggressively, yielding contents approaching 5 wt% in pure titanium.

The hydride phase forms exothermically and is generally a self-sustaining reaction if begun above 700–725°C (1292–1337°F). This aggressive reaction drives the embrittlement deep into the targeted material. The distorted face-centered tetragonal hydride phase, when formed, yields a volumetric expansion of roughly 3–5% and in the case of pure titanium can cause spalling of the hydrided surface layers, similar to peeling an onion. This can facilitate the embrittlement of thicker sections above 4 cm (~1.5 in.).

In Ti-6Al-4V, embrittlement can be less obvious, though it will normally still be easy to crush and mill. Hydrogen contents in this alpha-beta alloy is typically noticeably lower than in alpha alloys, with maximum levels approaching 4 wt%. The depth to which any form of titanium can be hydrided may be controlled by thermal management, but in some cases, the process may need to be repeated after removing the outer shell to expose the unembrittled core.

7.4 The hydriding process

7.4.1 *Furnace seals and leaks*

The presence of moisture or volatile organic compounds within the furnace load can compromise the accuracy of leak checks. The same may occur if the environment contains excessive moisture (humidity for example). At these times, a vacuum dwell period can be employed to evacuate any volatiles that may temporarily increase the apparent leak rate. Furnace components (refractory or insulation) may also become laden with moisture as a result of high humidity, so it is advantageous to minimize time when the furnace is open to atmosphere.

In cases where the furnace leak rate is above recommended levels, a helium leak detector may be used to locate the leak. The helium detection unit may be attached to a flange on a vacuum pump or onto the main vacuum line. The vessel is then evacuated for investigation of the source. A handheld wand discreetly discharges helium across potential leak sources, such as O-rings, thermocouple wells, or any other flange or valve. Monatomic helium gas, being one of the smallest elements on the periodic

Ti-H phase diagram

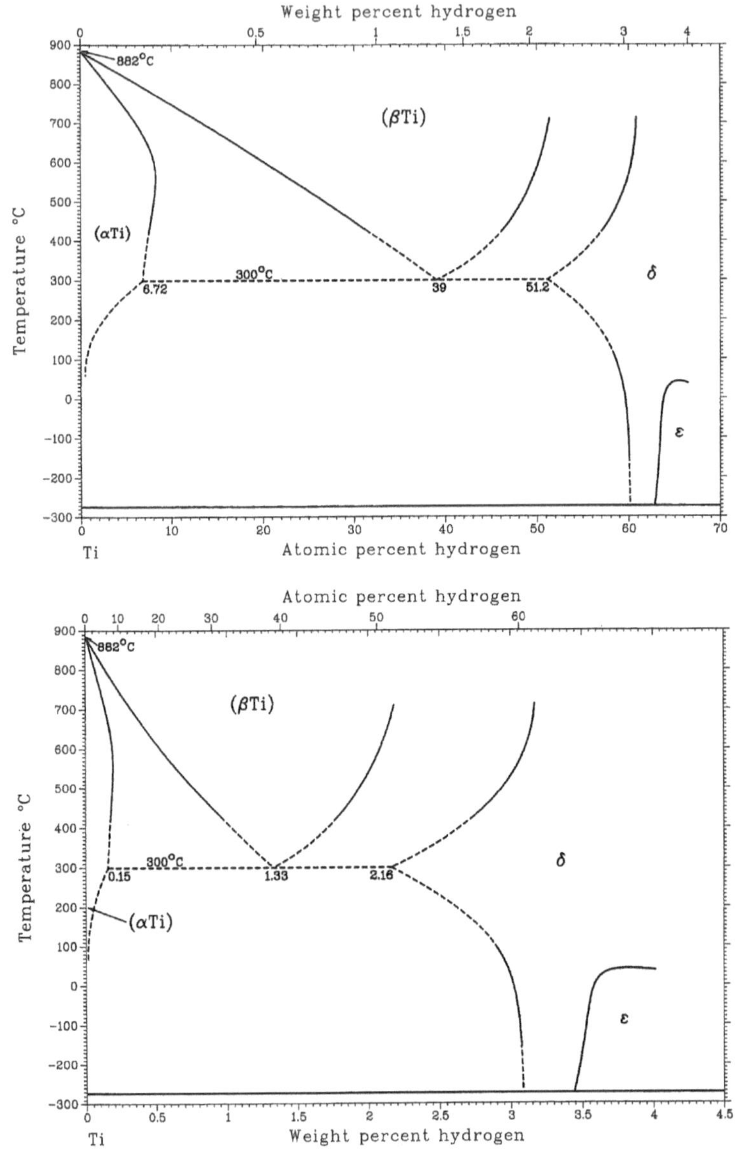

Obtained from a *P-T-X* surface. A. San-Martin and F.D. Manchester, 1987.

Figure 7.1 Hydrogen-titanium phase diagram.

table, is most effectively pulled into small leaks. The helium detector rates the magnitude of the vacuum leak, and any major leak may be addressed first.

Vacuum leaks can not only dramatically impact interstitial levels in the hydride but can also lead to serious safety risks by creating a potentially explosive atmosphere. Heating elements such as graphite, molybdenum or nickel-based alloys may also unexpectedly fail if exposed to a high-temperature oxidizing environment.

Explosive atmospheres with hydrogen and air occur above the lower explosive limit (LEL) and below the upper explosive limit (UEL). These values are roughly 5% and 95%, respectively, according to the *Handbook of Chemistry and Physics*. Outside this range, the atmosphere is less dangerous, but far from sufficient to produce high-quality product. Great efforts are taken to maintain the purest of environments for both hydriding and dehydriding. Refer to process safety specialists for complete procedures.

7.4.2 Hydriding

Convective heating in a vacuum is poor, so a partial or full argon atmosphere can be more effective for more rapid heating, provided gas pressures can be balanced as the temperature increases. Diminished vacuum levels may also neutralize or minimize any potential leaks into the furnace. Like most furnace heat treatments, an initial thermal soak period is normally employed to homogenize the entire furnace load. This will help facilitate rapid hydriding and lead to more uniform hydrogen uptake, particularly in thicker cross sections. This soak will cause the incorporation of the oxide film into the bulk of the raw material and provide a cleaner surface for hydrogen penetration. After the load is sufficiently heated, hydrogen may be admitted.

At the onset of hydrogen gas flow above roughly 700°C (1292°F), the exothermic hydride reaction begins (Equation 7.1), and the furnace temperature may rise for an extended period, dependent on load weight, before slow cooling is begun. At the onset, where hydrogen is most reactive, the embrittlement is the greatest. Hydride formed at lower temperatures tends to be less friable. The hydride phase of CP-Ti and the Ti-sponges typically absorb up to 4–5% hydrogen by weight, while Ti-6Al-4V absorbs approximately 3–4%.

$$\mathrm{Ti(s)} + \mathrm{H_2(g)} \rightarrow \mathrm{TiH_2} + \mathrm{heat} \tag{7.1}$$

After substantially cooling the hydrided load, the hydrogen atmosphere is replaced by argon to finish the cooling cycle. On reaching ambient temperature, the argon may be totally or partially removed and systematically replaced with air to allow a slow passivation to occur. This can lead to some reheating of the load while the passive oxide film stabilizes. Passivation must be done precisely, especially in the case of higher surface area loads packed into deep beds, such as may be done with sponge fines. Uncontrolled passivation may lead to a runaway exothermic reaction and loss of the entire load. A hydride cycle typically runs 7–10 days, dependent upon many factors, including load size and composition, hydrogen target levels, and furnace capabilities.

7.4.3 Preparing to size hydride

The delta phase on the Ti-H phase diagram has a wide range of hydrogen solubility, from 2.16 wt% to above 5 wt%. The friability of the hydride is a strong function of the hydrogen content. Below 3.5 wt% hydrogen, the best control of hydride sizing can be achieved, with minimization of dust (fines) generation. Residual toughness of lower hydrogen content powder can lead to several challenges; however, including higher crushing and grinding forces and a greater likelihood of contamination through abrasive wear of the tooling.

The propensity to create dust clouds while milling powders with hydrogen contents above 4 wt% must be minimized or avoided. Titanium hydride (sponge, CP-Ti, and Ti-6Al-4V) is classified as a flammable solid. While the hydride powder can be physically difficult to ignite in bulk containers, a dust cloud will readily generate a bright white flash without significant sound if uncontained. Deflagration forces generated by ignition of a dust cloud in a highly confined milling environment can be devastating, potentially leading to secondary explosions.

Any hazardous powder should be classified by an accredited laboratory to understand the maximum explosion pressure (P_{max}), the maximum rate of pressure rise $(dP/dt)_{max}$, and the explosion class (or *Kst*) before attempting these operations. These laboratory services can also offer guidance for safest practices.

Many factors must be taken into account when designing milling and screening circuits. Potential ignition sources can include sparks from metal–metal contact or static discharge, excessive heat in the powder from the grinding operation, or from hot surfaces or motors. Control over these sources must be employed at all steps, especially while milling and screening and during powder transfer operations.

7.4.4 Sizing of hydride

Once the hydrided raw materials are unloaded and at room temperature, the materials are reduced in size to below approximately 5 cm (2 in.) in either a jaw crusher or roll crusher. Argon gas is fed into the crushing circuit and dust collection is used to minimize airborne fines that may be generated. After crushing, an attrition mill circuit is employed to continue the size reduction to the size target.

Control over the generated particle size distribution of hydrided titanium alloys is particularly difficult. Carefully defined attrition practices are a leading factor in producing an optimal yield of sized powder. Wider distributions than desired usually occur, often producing a large percentage of fines. While desirable for some product applications, these fines will contribute a disproportionate amount to overall powder oxygen and nitrogen content, because of a higher surface area. These fines will also contribute significantly to sintering during dehydriding and lead to agglomerate formation with their higher surface area.

Sizing of hydride to powder in this inert environment can take 2–4 days per load, dependent on the raw material form, the alloy, and the desired particle size distribution output. All powder is kept under an inert environment while the chemistry and screen distribution is tested. After these are deemed acceptable, the powder is loaded into trays for dehydriding.

7.5 The dehydriding process

After sizing and screening the hydride powder to obtain the desired particle size distribution, hydrogen is removed via a high-temperature vacuum dehydride process [13]. This restores the mechanical properties to the previously tough and ductile material. The dehydride process (Equation 7.2) is the reverse of the hydride process and is endothermic, requiring substantial heat input to break the intermetallic hydride bonds and liberate hydrogen gas out of the system through the vacuum pumps.

$$TiH_2 + \text{heat} \rightarrow Ti + H_2 \tag{7.2}$$

Deliberate application of heat is required to avoid furnace overpressure during dehydriding. Thermal damage to furnace seals, valves, and vacuum pump(s) can result if the dehydriding is not properly controlled and adequately cooled while evacuating the hot hydrogen. Rapid liberation of large quantities of hydrogen gas can fluidize the dehydriding powder if the furnace furniture is not designed properly. This may cause furnace contamination and potentially damage vacuum pumps.

Dehydriding begins near 350°C (662°F) and ends near 700°C (1292°F). After soaking at temperature sufficiently, the furnace pressure recovers to well below 20×10^{-3} torr (0.02 mbar). This is an indication that dehydriding is complete and the heat may be turned off. A heat exchanger may be used to accelerate the cool-down phase and minimize time at temperature, which contributes heavily to sintering of the dehydride powder.

The degree to which titanium powder sinters during dehydriding is a function of oxygen content, with higher purity dehydride powders tending to sinter harder and quicker than powders with high interstitial contents. On top of this, finer powders also sinter quicker because of high surface area. This makes the production of 45-μm, low-oxygen powders particularly challenging.

A typical dehydride cycle runs 3–14 days, depending on many factors, including load size and composition, the particle size distribution being dehydrided, and most importantly, furnace capabilities.

7.6 Dehydride recovery

Once the process of hydrogen removal is complete [14], breaking apart the dehydrided product can begin. Unlike “sizing” hydride, dehydride cannot be “sized.” The task here is simply to reduce the powder “cake” back to the distribution that was loaded into the dehydride process. Attempting to size the titanium beyond this will be both futile and potentially dangerous.

Also, unlike hydride, the dehydride powder can easily ignite without a dust cloud. Table salt is the most common agent used to subdue a fire by starving the fire of oxygen by melting of the salt. Met-L-X extinguishers may also be used, but can potentially make the problem worse by spreading the fire on discharge of the extinguisher. Refer to the NFPA standards for guidelines to safe handling of titanium.

Deagglomeration of titanium powder can be particularly abrasive to milling equipment. This can be managed by careful selection of contacting materials and the milling technology. Any contamination may be effectively counteracted by subsequent magnetic separation and other aqueous techniques.

After dehydride recovery operations are complete, an optional deoxidation step may be employed to lower the oxygen content of the powders. This process requires specialized equipment and is not suitable to all products or final applications. Recovery of a single load of dehydride back to powder can take 1–3 days or more, depending mainly on the particle size distribution of the hydride and the amount of sintering that occurred.

7.7 Magnetic separation and acid washing

Titanium powders can be very abrasive and cause deterioration of any wear surface, such as attrition mill blades. Selecting magnetic materials as the wear surfaces allows for magnetic separation of any debris generated during powder recovery operations after each process operation. In combination with the use of rare earth magnetics, the powder products are often also washed in dilute muriatic acid to largely eliminate ferromagnetic contamination.

As titanium is paramagnetic, magnetic separation is not straightforward. Good product may act as a ferromagnetic contaminant and can be removed, causing yield loss and partial blinding of magnets. Analysis of the iron content on separated product can be done to better understand the progress of this step and minimize loss of good product. As a consequence, specifications requiring complete absence of ferromagnetic particulate require careful evaluation.

7.8 Interstitial contents

Table 7.1 shows typical gas contents for CP-Ti, Hunter-process sponge, and Ti-6Al-4V powders without deoxidation as a function of particle size distribution. These are only general tendencies and depend on many factors. Batch-to-batch variation as well as starting material quality are always important considerations. No attempt has been made here to show statistical trend data across a large number of batches, largely because of the wide number of product specifications and proprietary data.

Ti-6Al-4V tends to show a rapid increase in gas contents as particle size decreases significantly below the 200-mesh. This increase becomes nearly exponential near to and below 325-mesh. This exponential tendency also occurs for CP-Ti but at particle sizes closer to 500-mesh. Ti-sponge has a tendency between that of Ti-6Al-4V and CP-Ti, depending on its porosity/surface area. Examples of these general tendencies are shown in Figure 7.2a and b for oxygen and nitrogen, respectively.

Finer distributions contribute heavily to higher interstitial contents. Not all distributions may be offered in all grades. It is often found that sponge and CP-Ti grades are

Table 7.1 Typical oxygen/nitrogen contents for titanium powders by screen distribution

	60/80 mesh		120/200 mesh		200/325 mesh		325/500 mesh		-500 mesh	
	O	N	O	N	O	N	O	N	O	N
CP-Ti	0.12–0.20	0.004–0.008	0.18–0.26	0.006–0.011	0.30–0.35	0.007–0.012	0.31–0.40	0.011–0.017	0.47–0.50	0.020–0.026
Ti-Na	0.17–0.24	0.004–0.009	0.18–0.27	0.006–0.012	0.20–0.32	0.008–0.016	0.33–0.35	0.012–0.032	0.38–0.44	0.020–0.039
Ti-6-4	0.30–0.36*	0.007–0.018	0.35–0.44*	0.018–0.022	0.45–0.52*	0.025–0.038	0.48–0.54	0.035–0.047	0.80–1.0	0.08–0.1

* Indicated items that may be deoxidized.

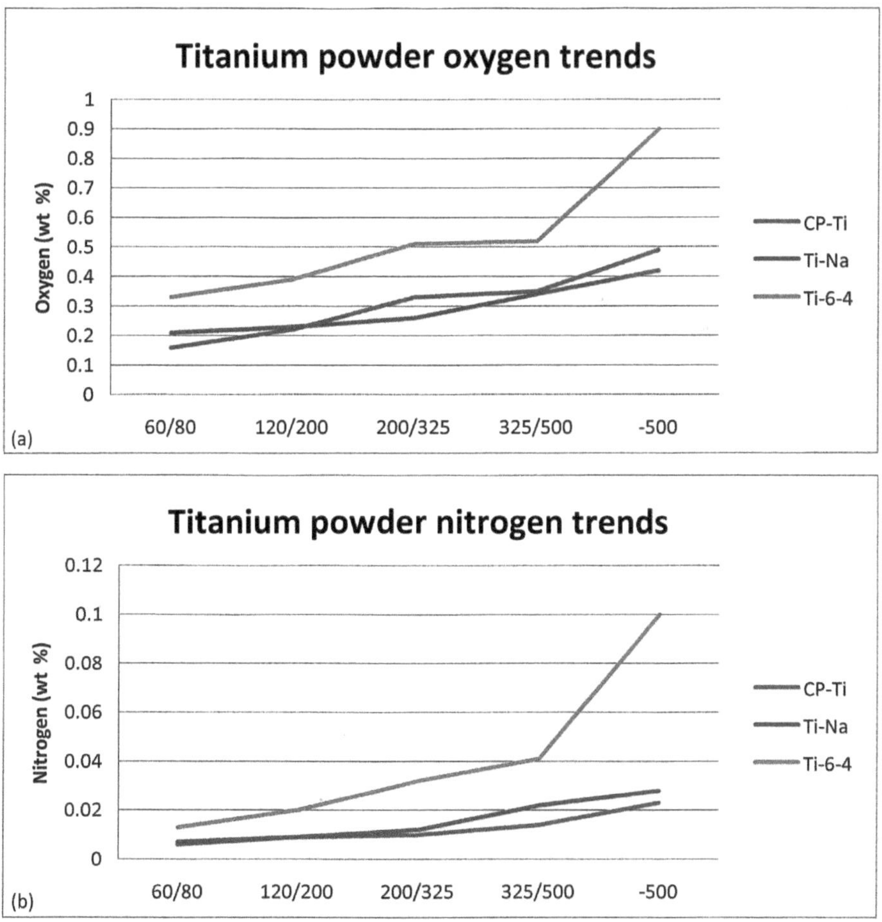

Figure 7.2 (a) Titanium oxygen trends by screen distribution. (b) Titanium nitrogen trends by screen distribution.

more easily produced in finer distributions than Ti-6Al-4V. Oxygen and nitrogen are always the limiting factor and deoxidation cannot diminish increasing nitrogen.

7.9 Screening and screen specifications

The HDH operation produces a wide powder distribution. Narrow distributions extracted from HDH powders are typically low yielding (<10%) and at worst inconsistently recovered. The powder's morphology plays a significant factor in the screening operation. The sharpness with which a screen fraction may be extracted is limited, particularly when compared to spherical powders.

Titanium powders are a challenge to screen without ultrasonic deblinding systems, as are available from several manufacturers. Generally, to avoid blinding the mesh, these systems need to be employed below 60-mesh. Other forms of deblinding, such as ball decks or similar, are a risk for contamination and avoided. However, extreme care must be taken to employ *electrically* powered ultrasonic transducers when screening these hazardous powders, particularly with distribution pans between screen decks. The power cord for the transducer is subject to abrasion if not sufficiently supported. This may lead to a significant fire, even if measures have been taken to inert the screening environment.

Also relevant to the screening process are the qualities of the screen mesh itself. Tolerances on "average opening size" and "maximum allowable opening" make screening itself a very statistical process. Refer to ASTM E-11-95 for additional information on standardized quality criteria of woven wire screen. A significant trade-off exists between the use of longer-lasting market-grade mesh with its heavier-gauge wire and tensile bolting mesh with its lighter-gauge wire and significantly higher "open area" for improved throughput. The life span and stability of these meshes are a significant financial consideration, particularly with large-diameter ultrasonic frames below 200-mesh.

Many factors are important when screening titanium to any distribution. Wider distributions with tolerance for a minimum of 5% oversize and undersize are generally achievable. Narrow distributions with no mesh sizes available between the top and bottom specification are exceptionally costly and very low yielding. In high-volume production operations, it is standard practice to inset one sieve size from the top and bottom screens to achieve 5% oversize and undersize limits. For example, on a requirement of 100/325-mesh, it may be far more productive to use 120/270 screens to achieve 5% oversize and undersize on 100 and 325 screens. This may not be necessary if 10% tolerances are permissible.

Specifications pinning the percentages on all intermediate screens are particularly difficult to achieve. These types of specifications normally require blending of multiple distributions to achieve the desired output. Repairs to an intermediate screen fraction requires significant time and often leads to multiple screen changes.

Particularly challenging are the sponge powders. These powders are fragile and difficult to sieve without breaking down their porous structures. It is typically better to make additions to these powders to achieve the desired distribution than it is to attempt removal of a particular component. Sponge powders also produce the most fines and normally require insetting screens to achieve 10% oversize and undersize.

7.10 Laser specifications

When producing angular HDH powders to laser specifications, a correlation between screened fractions and the laser results received from it must be made. This correlation is straightforward once it is understood that HDH powders, because of their shape, screen differently than spherical powders. Whereas the spherical powder can only go through a mesh opening greater than the diameter of the sphere, an angular particle may pass through while standing on end. This angular particle, once converted to a sphere in the laser analyzer, yields a diameter larger than should have passed through the sieve.

For example, an HDH powder screened to 325-mesh will not yield a laser distribution's 90th percentile, or d(90) of 45 μm. The d(90) of an HDH powder screened to –325 mesh can approach 60–65 μm. This 15–20-μm increase in the d(90) is only a guideline, but fairly accurate across most mesh sizes.

As a consequence of this morphology, a laser specification requiring a d(90) of 45 μm may require screening at 400- or 500-μm. The impact of this change will often cause difficulties with conformance interstitial requirements. For example, meeting ASTM B-348 Grade 4 chemistry (0.40 oxygen) *and* a d(90) of 45-μm can be far more challenging than meeting the oxygen requirement and achieving 5% oversize on a 325-mesh screen.

To demonstrate this, Figure 7.3 shows two Ti-6Al-4V powders sieved to 230/270 mesh. One powder is an HDH powder (in red), and the other a plasma spheroidized HDH powder (in green). Both were tested via a wet laser particle size analysis technique. The spherical powder yields a smaller d(90) than the HDH powder—a result of how higher aspect ratio particles may stand on end and pass through a screen it otherwise could not if it were spherical.

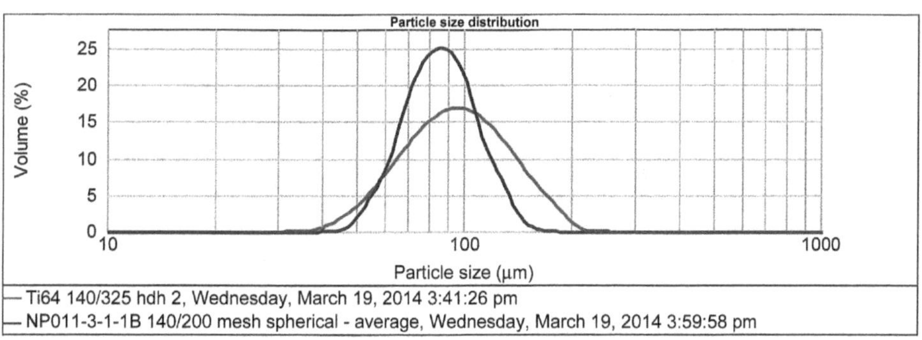

Figure 7.3 Comparative laser PSD analysis for two 230/270-mesh Ti-6Al-4V powders.

Many of these blocky, dense, and angular powders may be spheroidized to improve flow characteristics and make them usable for additive manufacturing processes. The plasma spheroidization process can convert an unagglomerated CP-Ti or Ti-6Al-4V distribution to a 99% spherical product with no internal porosity. This can be done while still maintaining compliance with ASTM B-348 chemistries. Fine distributions of Ti-6Al-4V Grade 23 powders may also be generated, but low nitrogen contents can be particularly difficult to achieve as the particle size distribution decreases significantly below 200-mesh.

7.11 Powder morphologies

Morphologies created by the HDH process vary according to the starting raw materials used. Hunter Process sponge has the highest surface area as shown in Figure 7.4a. This structure is also the most brittle and its particle size distribution may become skewed toward the finer side simply by the action of screening (Ro-Tap). It is often necessary to specify 10% or more undersize on this type of product to facilitate manufacturing capability. Hunter Process sponge, if not processed properly, may also contain highly acicular or needle-like particles (Figure 7.4b). Fully hydriding this sponge will prevent this type of generally undesirable morphology.

Kroll process sponge shows somewhat less porosity, as shown in Figure 7.5a, and may sometimes be difficult to discern from agglomerated forms of CP-Ti, shown in Figure 7.5b. Hybrid morphologies can be created as blends of agglomerated and unagglomerated powders, which may often provide target structural goals.

A typical CP-Ti powder morphology for a 100/325-mesh distribution is shown in Figure 7.6a, while a much finer CP-Ti powder with a laser distribution of below 90 μm is shown in Figure 7.6b. Ti-6Al-4V morphologies are typically slightly blockier than CP-Ti, as shown in Figure 7.7a and b, respectively for 50/100-mesh and 80/200-mesh.

Figure 7.4 (a, b) Typical Hunter process sponge powder morphologies.

Figure 7.5 (a, b) Typical Kroll process sponge powder morphologies.

Figure 7.6 (a, b) Typical CP-Ti powder morphologies.

Figure 7.7 (a, b) Typical Ti-6Al-4V powder morphologies.

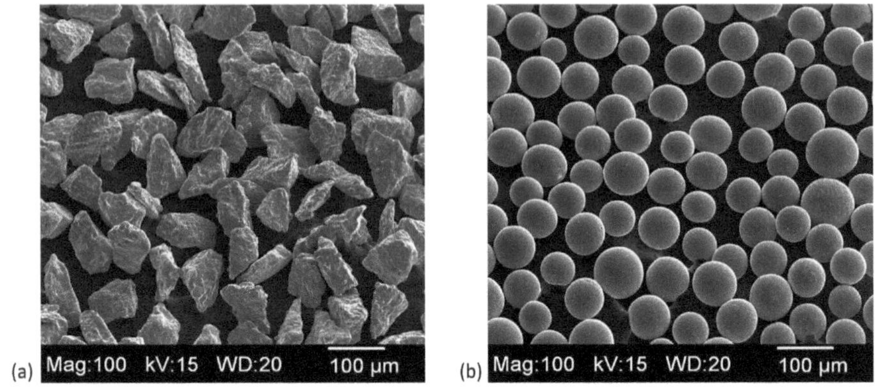

Figure 7.8 (a, b) Ti-6Al-4V powder (140/200-mesh) before and after plasma spheroidization.

7.12 Spherical powders

CP-Ti and Ti-6Al-4V powders are angular, blocky, and fully dense (no porosity). For some applications, this is desirable. For applications such as PM or additive manufacturing, good flowability and consistent shrinkage are required. HDH titanium does not reliably flow below 200-mesh and yields inconsistent die compaction and nonuniform shrinkage during sintering.

Plasma spheroidization yields a consistently flowable, virtually spherical powder, out of the HDH product. This process has exceptional yield down to 325-mesh and is porosity-free, potentially suitable for additive manufacturing. A specific example of powder morphology before and after spheroidization for a 140/200-mesh Ti-6Al-4V is shown in Figure 7.8a and b, respectively.

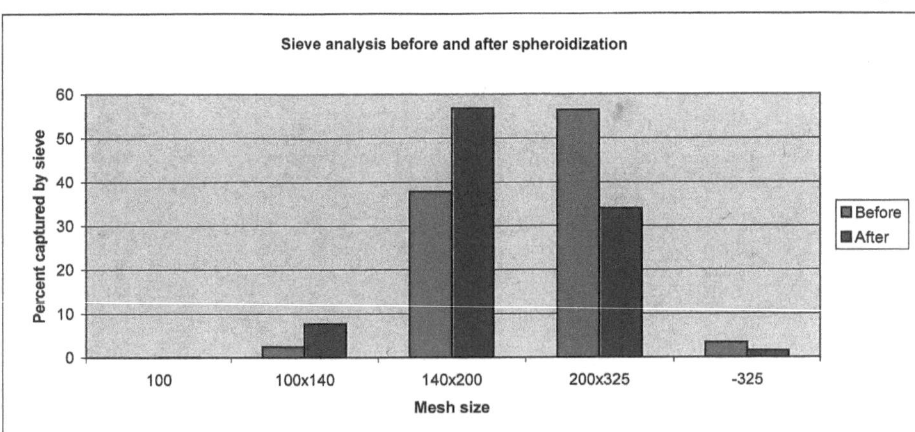

Figure 7.9 Change in screen distribution before and after plasma spheroidization.

The morphology change associated with plasma spheroidization also creates a change in the sieved particle size distribution before and after the process. The before (in blue) and after (in red) particle size distribution, as tested on test sieves, is shown in Figure 7.9. This shows a coarsened distribution after spheroidization. This can be explained very similarly as was done with the two laser distributions above in Figure 7.3. A spheroidized angular particle that may once have passed a 325-mesh sieve can no longer pass while standing on end. It has effectively coarsened, although its volume remains unchanged.

7.13 Summary

The manufacturing process to convert titanium and titanium alloy ingots and mill products into powders is complex and time consuming. Careful methods and safety procedures allow for successful production of high quality, low interstitial content, powders. These powders show a wide range of morphologies, from the highly porous and low-density powders produced from sponges, to blocky, angular, and porosity-free powders of CP-Ti and Ti-6Al-4V. Careful tailoring and blending of the different powders can create a wide range of products. Others powders may be further processed into highly flowable and spherical powders suitable for additive manufacturing.

References

[1] L.C. Tao, Producing brittle titanium metal, Titanium Metals Corporation, US Patent #3005698 (1961).
[2] R.L. Powell, Reclaiming scrap titanium, Titanium Metals Corporation, US Patent #2992094 (1961).
[3] H. Wang, M. Lefler, Z.Z. Fang, T. Lei, S. Fang, J. Zhang, Q. Zhao, Titanium and titanium alloy via sintering of TiH_2, Key Eng. Mater. 436 (2010) 157–163.
[4] O.N. Senkov, F.H. Froes, Hydrogen as a temporary alloying element in titanium alloys, in: V.A. Goltsov (Ed.), Progress in Hydrogen Treatment of Materials, 2001, pp. 255–279.
[5] S. Zhang, Hydrogenation behavior, microstructure and hydrogen treatment for titanium alloys, in: V.A. Goltsov (Ed.), Progress in Hydrogen Treatment of Materials, 2001, pp. 282–298.
[6] D. Eylon, F.H. Froes, Method for refining microstructures of prealloyed powder metallurgy titanium articles, US Patent #4534808 (1985).
[7] F.H. Froes, D. Eylon, Production of titanium powder, in: Metals Handbook, Powder Metallurgy, ninth ed., vol. 7, 1984, pp. 164–168.
[8] C. McCracken, C. Motchenbacher, D.P. Barbis, Review of titanium-powder-production methods, Int. J. Powder Metall. 46 (5) (2010).
[9] C. McCracken, Manufacture of hydride-dehydride low oxygen Ti-6Al-4V (Ti-6Al-4V) powder incorporating a novel powder deoxidation step, Euro PM Conference, 2009.
[10] C. McCracken, D. Barbis, Production of fine titanium powders via the hydride-dehydride (HDH) process, Powder Inj. Mould. Int. 2 (2) (2008) 55–57.

[11] F.H. Froes, Titanium powder metallurgy: a review – part 1, in: Advanced Materials and Processes, 2012, pp. 16–22.
[12] S.J. Gerdemann, Titanium process technologies, in: Advanced Materials and Processes, 2001, pp. 41–43.
[13] J.T. Fraval, M.T. Godfrey, Method of producing titanium powder, US Patent #6,475,428 (2002).
[14] V. Bhosle, E.J. Baburah, M. Miranova, K. Salama, Dehydrogenation of TiH_2, Mater. Eng. A356 (2003) 190–199.

Low-cost titanium hydride powder metallurgy

8

Orest Ivasishin, Vladimir Moxson***
*Institute for Metal Physics, Kiev, Ukraine
**ADMA Products, Inc., Hudson, OH, USA

8.1 Introduction

Titanium and titanium-based materials are widely used for many applications, including in the aerospace industry, because of the unique combination of excellent mechanical properties, low density, and corrosion resistance in most aggressive environments. However, they are expensive to produce and fabricate. PM offers an effective means to reduce the cost of titanium parts as it produces near-net shapes, thus minimizing material waste typical to ingot metallurgy (IM) processing.

Especially attractive from an economic viewpoint is the blended elemental powder metallurgy (BEPM) approach. In this process, the titanium matrix powder is blended with other elemental or master alloy additives to achieve the required alloy composition and then cold pressed into shapes, followed by sintering at time-temperature conditions sufficient to increase the density and chemically homogenize the material. BEPM can also be used as an intermediate step to produce ingots or preforms for hot working processing.

It is well recognized that the low cost of BEPM-processed parts is often sacrificed by their lower performance because the mechanical properties of the titanium PM products strongly depend on the sintered density, impurity content, and microstructure, which, in turn, depend on the quality of the starting materials (titanium and alloying additives) and specific processing parameters. Hence, BEPM is not often used for producing near-fully dense high-performance components. Elsewhere in this book, there are many examples of how substantial research efforts have aimed at improving the titanium PM process, from powder production to post sintering treatments. The main goal of these efforts is always unchanged, that is, to produce components including those for critical applications at a *low cost* without sacrificing their *performance* ability. As for powder production, this means to develop technologies resulting in the *low-cost* but *high-quality* particles that could be easily adjusted to compaction techniques. With regards to the consolidation and sintering stages developed for such specific powder, they should be capable to produce the desired density and microstructure while minimizing gas contamination of the product. Post sintering hot-pressing operations should be eliminated to avoid increased cost of production.

This chapter presents a potential solution to the *cost/performance* dilemma of the BEPM, referred to as "Low Cost Titanium Hydride Powder Metallurgy" since its key feature is the use of titanium hydride (TiH_2) powder instead of titanium powder as the starting material. However, the main advantages of this approach are not just reduced

Titanium Powder Metallurgy. http://dx.doi.org/10.1016/B978-0-12-800054-0.00008-3

to a simple substitution of the starting material. The use of TiH_2 powder or, at wider consideration, of hydrogenated titanium powder leads to several specific features in the compaction and sintering stages that eventually result in properties equivalent to or better than those produced through conventional IM processing.

Using of TiH_2 powder in PM was mentioned much before we started to use it in BEPM method of Ti PM. However, earlier trials with TiH_2 powder were just aimed to use it, fully or partially, in place of corresponding amounts of titanium metal powder [1,2], either because of economic reasons (lower cost) or some specific features of TiH_2, for example, its lower flammability and oxidation ability versus Ti metal powder whose pyrophoric characteristics require delicate conditions when handling and working under contact with air. In some cases, TiH_2 powder has been used either alone or blended with Ti metal powder as a source of hydrogen, which was believed to be useful for the desired microstructural evolution of the compact [3,4]. Such processes were based on the fact that the hydrogen released from TiH_2 can hydrogenate the titanium, thus making it more ductile because of partial transformation of *hcp* α-phase to *bcc* β-phase. Hydrogen-induced ductility was considered to be useful in hot compaction of particulate blends widely used in the pre-alloyed (PA) powder methods. Intensive hydrogen release from TiH_2 can result in high pressure inside the material, and this feature has been used to make foam materials [5] and even to disintegrate bulk electrode into disperse particles to produce titanium powder [6].

Historically, use of TiH_2 powder as the only titanium constituent in the BEPM process was never tried because of the lack of ductility of TiH_2 powder, which seemed to make it impossible to compact the green part to the strength sufficient for further handling. It was thought that a binding agent was needed to keep the TiH_2 powder particles connected in the desired shapes [7]. However, the binder generally consists of elements considered as impurities aggressively reacting with titanium and contaminating final products even just in a small quantity.

In the early 1990s, TiH_2 powder was employed as a starting material in its blend with Al for reactive sintering of titanium aluminides [8]. In those experiments, the blends were consolidated because of ductile aluminum in which brittle TiH_2 particles were refined on ball milling of the blend. Certainly, such consolidation mechanism is impossible if TiH_2 powder is the only or major component of the compact.

The experiments on reactive sintering of titanium aluminides led to a very important finding; namely, an activated diffusivity of the titanium was observed after partial or full dehydrogenation. It resulted in solid-state reactions between titanium and aluminum and therefore made it possible to process dense titanium aluminides without post-sintering hot deformation operations. In contrast, when conventional titanium metal powder is used, such reactions proceeded only after melting of the aluminum, which led to a tremendous swelling of the compact. It was also suggested [8] and confirmed by thermodynamic calculations that diffusion activation could be further promoted by reduction of the Al_2O_3 surface scale by the atomic hydrogen released from the TiH_2 particles.

In the same way, titanium aluminide–based composites strengthened with SiC fibers were produced [9]. Formation of nearly dense aluminide matrix at reasonable time–temperature conditions ensured its strong adhesion with SiC fibers and reduced fiber–matrix interface reaction.

The success achieved with using TiH_2 powder in reactive sintering of titanium aluminides led to its use as a main constituent in BEPM processing of CP Ti and Ti-based alloys. Although the first publication appeared in 2000 [10], experiments started several years earlier at the Institute for Metal Physics (Ukraine) in cooperation with USA partners (University of Idaho and ADMA Products, Inc.). It was shown with Ti-6Al-4V as an example that compacts based on Ti or TiH_2 powder behaved in a very different way. In the former case, because of the melting of aluminum and the subsequent intensive exothermic reaction between liquid aluminum and titanium particles, the sintered density was low, even lower (!) than the green density, while in the case of TiH_2-based compact, it exceeded 98% of the theoretical density (TD). It was also found that TiH_2-based blends were cold compacted to strengths sufficient for further handling without any binder. Although the first explanation was that it was due to the other elemental powders (aluminum and/or vanadium), further experiments showed that TiH_2 powder itself exhibited sufficient cold compactibility.

Even more striking was that the sintered densities of TiH_2-based compacts exhibited only negligible, if any, dependence on compaction pressure, contrary to the sintering of Ti-based compacts. And finally, because of the specific mechanism of compaction, TiH_2-based blends exhibited much lower contact interaction with die walls, thereby eliminating seizing and galling on the die walls, typical of ductile titanium powders.

Use of TiH_2 or hydrogenated titanium powder was always accompanied by concerns of whether the final hydrogen content can be kept low enough to meet respective specifications. Fears were disseminated after carefully controlled measurements. As expected from the long time experience of using thermohydrogen treatment [11], hydrogen could be used as a temporary alloying element in the BEPM processing since its content could be minimized to the safe level by heating in vacuum. In the PM processing, the release of hydrogen from the compact is even easier than from bulk materials since it proceeds mostly in the condition of open porosity.

It was thus concluded from the first experiments on the BEPM processing based on TiH_2 powder that this innovative approach is worthwhile of extensive research to develop fundamental aspects of the Titanium Hydride PM and to estimate its possible commercial perspectives. Results of such efforts are presented in this chapter.

8.2 Titanium hydride: physical and mechanical properties and phase transformations upon heating

Binary Ti-H phase equilibrium is presented by a diagram of eutectoid type [12]. The solubility of hydrogen is only 1500 ppm at the eutectoid temperature 300 °C, below which excessive hydrogen content results in the formation of brittle titanium hydride precipitates in the α-phase. At hydrogen content above 3.0 wt% *fct* δ-hydride becomes the only phase, which transforms to *fct* when the hydrogen content exceeds 3.5 wt%. In the *fct* hydride, hydrogen atoms are located mostly in tetrahedral interstitial positions. The nature of titanium–hydrogen atomic bonds in titanium hydride is disputable. Since the hydride has some properties inherent to metals, for example, metallic conductivity, it is often considered as a titanium-hydrogen alloy in which hydrogen electrons occupy the *d*-level of titanium atoms, thus transforming hydrogen atoms into protons H^+. In fact, the structure of titanium hydride structure seems more complicated than that and can include hydrogen as anions H^- or atoms that form covalent bonds with titanium as well [13]. Since hydrogen can saturate titanium to 4 wt% or 66 at%, titanium hydride is often designated as TiH_2, although it exists over a wide hydrogen content range of 3–4 wt%. Density of titanium decreases with increasing hydrogen content from 4.51 g/cm^3 for pure Ti to 3.75 g/cm^3 at stoichiometric TiH_2 composition. Interaction of titanium with hydrogen is reversible and accompanied with significant thermal effect:

$$Ti + H_2 \leftrightarrow TiH_2 + Q(15.15\,\text{kcal/mol}) \tag{8.1}$$

Titanium hydride is a brittle, low-strength (150–250 MPa) material. This defines its specific compaction behavior upon pressing. Its brittleness combined with its low strength has proved to be amenable to cold compaction. Brittleness of hydrogenated titanium is a property that has been used for years for the production of HDH Ti powder. The higher the hydrogen content, the more easily the bulk material can be crushed. TiH_2 powder particles behave in the same way under compaction in the die. The particles, irregularly shaped, break into smaller sizes on compaction at pressures exceeding 150–250 MPa.

Figure 8.1 shows the particle size distribution before and after compaction. Irregular shapes favor crushing, promoting formation of smaller fragments to fill the free space between coarser particles. Reciprocal displacements of the neighboring powder particles and their fragmentation form specific interlocking to keep the green part shaped after the pressure is relieved. If powder particles were strong, only particle displacements would take place, which would be inadequate for green shape formation. Irregular particle shapes are utterly important. This is confirmed by experiments with spherical TiH_2 particles that have proved to be not suitable for cold compaction (Figure 8.2). They exhibit some ductility followed by cracking at higher pressures.

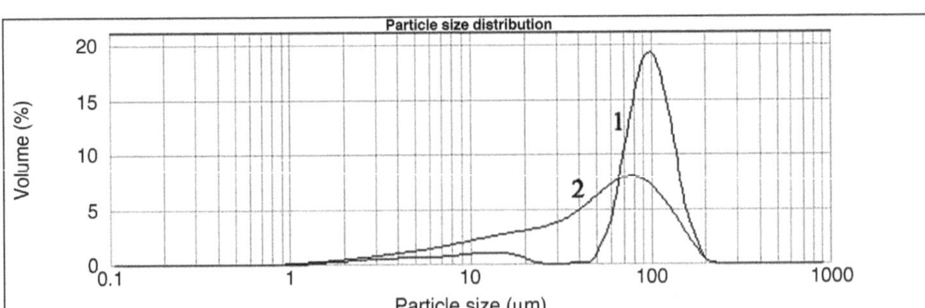

Figure 8.1 Size distribution of TiH_2 powder particles before (1) and after (2) compaction.

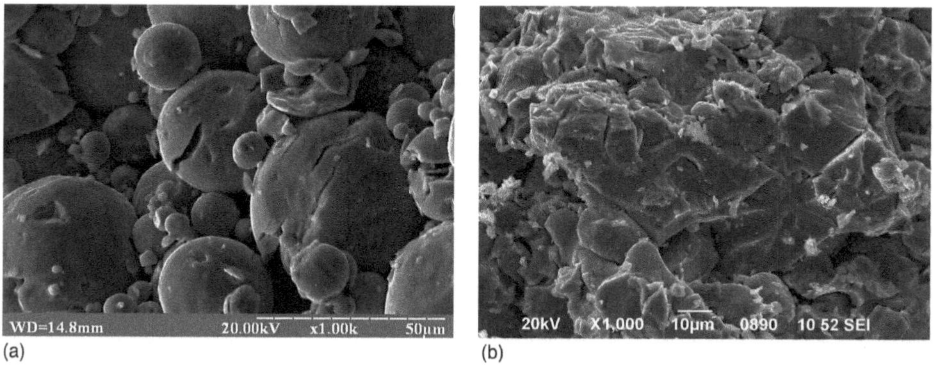

Figure 8.2 Spherical-shape TiH_2 powder particles exposed to compaction pressure of 640 MPa (a). For comparison, fragmentation of conventional, irregular-shaped TiH_2 powder particles at the same compaction pressure is shown (b).

The limited ductility is insufficient for keeping the compact consolidated, while its cracking tendency is inadequate to realize desired fragmentation.

TiH_2 powder can be compacted to densities of 2.7–3.2 g/cm^3 depending on molding pressure (320–960 MPa range, Figure 8.3), while densities of compacted Ti metal powder of the same size and morphology are noticeably higher [14], 2.9–3.8 g/cm^3. The difference comes from the lower density of bulk titanium hydride as compared to titanium. Relative densities of TiH_2 compacts are equal to those of Ti compacts at high compaction pressure or even higher at low pressure (see numbers on Figure 8.3); that is, volume fractions of voids in both compacts are comparable despite different compaction mechanisms for ductile titanium and brittle TiH_2 powders.

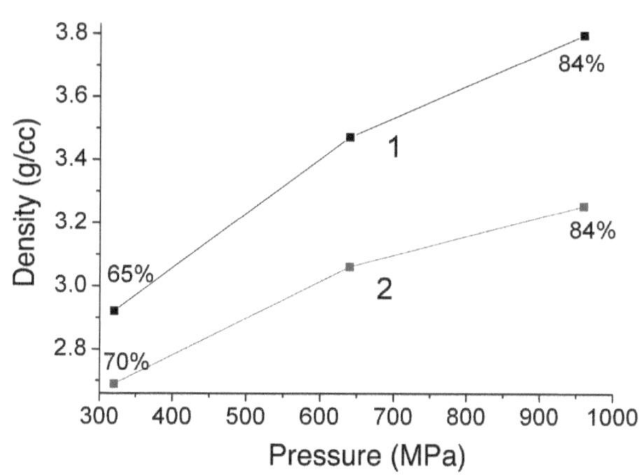

Figure 8.3 Green densities of compacts produced from Ti (1) and TiH_2 (2) powders of the same size. The densities related to Ti and TiH_2 correspondingly are noted. (Source: Data from Ref. [14].)

Titanium hydride is stable at room temperature; however, above 300–400°C, it decomposes and releases hydrogen. Literature data [15,16] on dehydrogenation and corresponding phase transformations varies because of the dependence on experimental techniques employed, sample size, impurity content, surface condition, heating rate, etc. This makes the use of these data for the analysis of the BEPM process difficult.

Detailed studies of phase transformations on continuous heating of powder compacts identical to those used in the BEPM process were performed [17]. The compacts were consolidated from the specific TiH_2 powder (hydrogen content: 3.5 wt%, size: 100 μm) under a pressure of 640 MPa. The heating rate used was 7°C/min, close to that conventionally used for sintering. Experimental techniques included *in situ* high-temperature X-ray analysis, dilatometry combined with mass spectrometry [18] to study the gases evolved from the material, and thermogravimetric analysis (TGA) coupled with differential scanning calorimetry (DSC).

Intensive hydrogen desorption was found to begin at 300–320°C (Figure 8.4, curve 1) and end at around 700°C. Small portions of hydrogen still emit at higher temperatures. The intensity of hydrogen emission exhibits two maxima whose positions depend on partial hydrogen pressure in the heating chamber and correlate with phase transformations proceeding in titanium hydride. Although hydrogen desorption begins at 320°C, the single-phase TiH_2 (δ) state exists up to 450°C, then $TiH_2(\delta) \rightarrow \beta$ transformation occurs. The single β phase state is observed above 500°C; displacement of the β phase diffraction peaks upon heating implies an intensive decrease of hydrogen content in the β-phase to the level at which $\beta \rightarrow \alpha$ transformation takes place. The single α-phase state is eventually realized at 650°C.

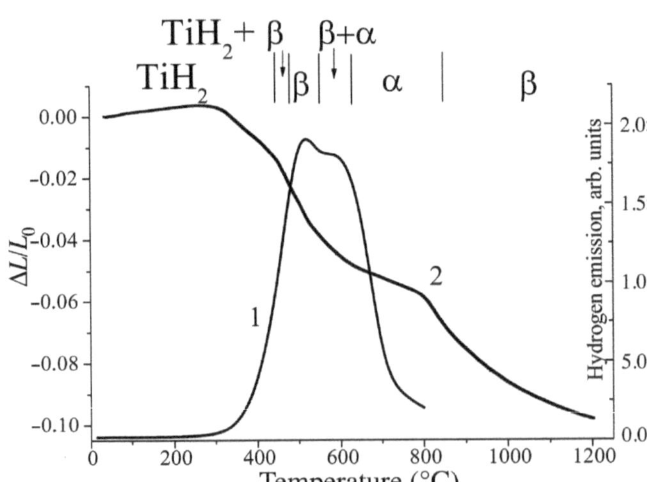

Figure 8.4 Hydrogen emission (1) and shrinkage behavior (2) upon heating of TiH_2 compact (compaction pressure 640 MPa). Temperature intervals of phase fields under dehydrogenation process, determined by *in situ* X-ray analysis, are indicated at the top.

According to the TGA/DSC data [17], mass loss due to dehydrogenation, accompanied by endothermic effects, starts at 440°C and ends at 830°C. The intensity of mass loss varies with temperature, thus confirming the influence of phase transformations noted above on dehydrogenation. The value of total mass loss (3.53%) corresponds to the starting hydrogen content. The TGA/DSC experiment is carried out under an inert atmosphere which shifts dehydrogenation to higher temperatures compared with X-ray and mass-spectrometry experiments performed in vacuum. Two endothermic effects observed are explained by hydrogen emission from the δ-hydride and β-phase, respectively. Mass loss is most intensive at temperatures corresponding to the second endothermic effect.

Dehydrogenation is accompanied with significant volume changes (shrinkage) due to difference in densities of TiH_2 and Ti. Shrinkage on heating of the compact results from the sintering of powder particles as well. Total volume shrinkage in temperature ranges up to 1250°C is 27%, estimated from dilatometric measurements (Figure 8.4, curve 2) based on the assumption that the volume changes are isotropic. First diffusion contacts between powder particles are observed at 700–710°C; at lower temperatures, shrinkage results mostly from dehydrogenation, while after dehydrogenation is completed, shrinkage is due to sintering. It is accelerated in the 820–880°C temperature interval, which can be attributed to the activated diffusivity of titanium at $\alpha \rightarrow \beta$ phase transformation [19].

Expected bulk material shrinkage due to $TiH_2 \rightarrow Ti$ transformation is 6.7%. In fact, powder compacts exhibited length changes close to the same percentage (see Figure 8.4), confirming that compacted TiH_2 particles and bulk TiH_2 behave similarly. This is an important result as it confirms that TiH_2 powder is suited to PM.

Temperature interval and kinetics of dehydrogenation depend not only on the diffusivity of hydrogen in the crystal lattices of the TiH_2, β-phase, and α-phase but also on the size of samples, which determines the diffusion path to the surface; the presence and thickness of oxide scale on particle surface, a barrier for hydrogen atoms; and the dynamic partial pressure of hydrogen in the heating chamber. Generally, dehydrogenation of powder compacts is faster than that of the bulk form because of small particle sizes and open porosity between them. At equal particle size, dehydrogenation depends on compaction pressure, which determines which part of the particle surface remains free. Lower vacuum slows down the dehydrogenation. Influence of the environment of the chamber is also pronounced: mass loss under inert atmosphere starts and ends later than in vacuum.

The phase transformations of $TiH_2 \rightarrow \beta \rightarrow \alpha$ observed are caused by the decrease in hydrogen content, the rate of which, in turn, depends on hydrogen diffusivity in various phases. Hydrogen diffusivity in titanium hydride is low (10^{-14} m^2/s) at room temperature [20,21]. Because of this and also the surface oxide scale, the hydride phase is stable at room temperature even in high vacuum. Increase in diffusivity to 10^{-12}–10^{-11} m^2/s on heating results in the initiation of dehydrogenation, which then accelerates in the temperature interval of $\delta \rightarrow \beta$ phase transformation since the diffusivity of hydrogen in *bcc* β titanium is 2–3 orders of magnitude higher than that in hydride [22]. At this stage, hydrogen concentration in the β-titanium is rapidly reduced to the level that allows $\beta \rightarrow \alpha$ transformation, but the formation of the *hcp* α-phase in the final stage of dehydrogenation again slows down the rate of dehydrogenation because of the lower hydrogen diffusivity [23].

8.3 Surface contamination of titanium hydride powder

Surface condition affects the dehydrogenation of TiH_2 powder particles. Not only oxygen as oxide but other impurities are generally located on powder particle surfaces. Surface contamination is considered to be a main source of the impurities in sintered PM products since they either dissolve into the bulk upon sintering, resulting in embrittlement of the PM products, or reside in the residual pores, impeding pore closure – like chlorine, which was found to be a reason for excessive porosity, deteriorating fatigue properties, and decreased weldability [24].

The surface contamination of TiH_2 powder and HDH Ti powder is similar [25], determined using the XPS technique. Oxygen is the main surface impurity for both powders. Besides, impurities originated from the Kroll process such as Mg, Cl, and Ca are identified in the XPS spectra. Unexpectedly, impurity peaks are weaker for titanium powder.

The O 1s XPS spectra (Figure 8.5a) can be fitted with at least two curves that correspond to different types of oxygen states on the particle surface. The one curve with a peak at the binding energy 531 eV has been assigned to TiO_2, while the other one with a maximum at binding energy 532.2 eV has been assigned to water and hydroxyl groups.

Positions and shapes of the dominating peaks in the Ti 2p spectra (Figure 8.5b) taken from the very surface and referenced to respective spectra of pure TiO_2 reveal 4+ oxidation state of Ti. This reveals that the only titanium form on the surface is oxide film. Small peaks or shoulders on the low binding energy side from Ti $2p_{3/2}$ peak could be due to the presence of reduced Ti cations (oxidation state Ti^{x+}, $x < 4$) as well as due to Ti bonding with chlorine and carbon.

After sputtering by 2 keV Ar+ ion beam, which allows analyzing deeper layers, the spectrum changed as compared to that obtained for the starting conditions. No peaks corresponding to chlorine and calcium were observed; the intensity of Mg peak noticeably decreased. The contribution from Ti in the XPS spectra became more pronounced; the contribution from water-related species declined.

These XPS results show that TiH_2 as well as Ti powder particles have similar core-shell structures. Particles are enveloped with TiO_2 scale and water absorbed on the surface. The thickness of TiO_2 scales and the amount of surface-absorbed water are lower for Ti powder. In subsurface layers, oxygen is bounded with titanium in TiO_y ($y < 2$) compounds and Ti-O solid solution. Other impurities are also preferentially located on the surface.

Despite similarity in surface condition between TiH_2 and Ti powders, their behavior becomes conceptually different on heating in vacuum. Water vapor emission is observed upon heating TiH_2 compact (Figure 8.6); the intensity of emission varies, with two distinct peaks at around 100°C and at 300–400°C. The second peak matches exactly with the emission of hydrogen; that is, water emission takes place within the dehydrogenation interval. Upon heating of Ti compact, only one peak at around 100°C is observed.

It can be concluded that water desorbs from the surface of both powders at low temperatures; desorption is more intensive from the TiH_2 powder. Water emission from

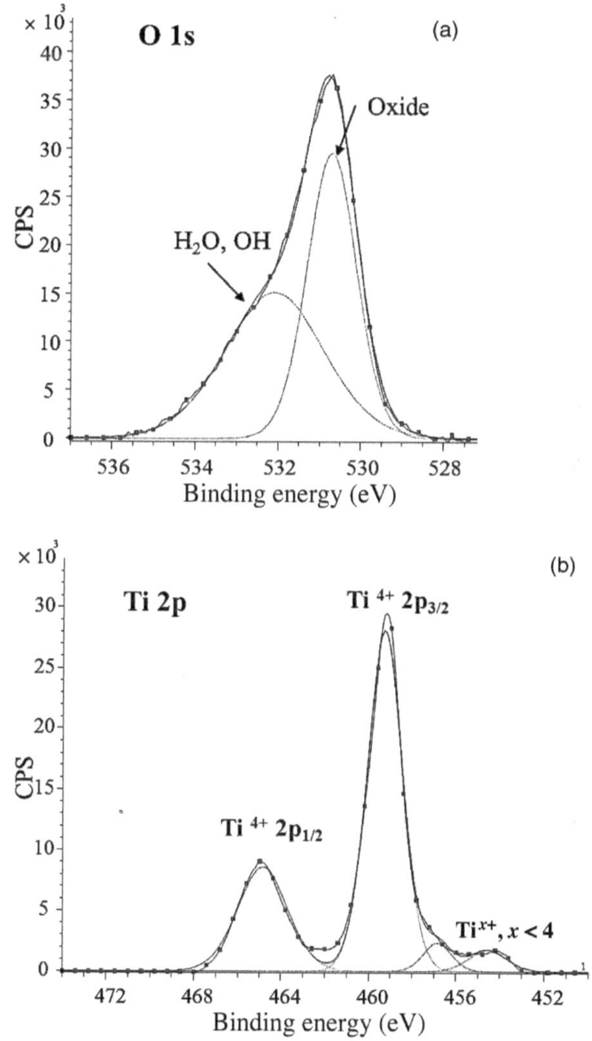

Figure 8.5 O 1s (a) and Ti 2p (b) electron binding energy spectra for TiH_2 powder surface, obtained with Al X-ray source. (Source: From Ref. [25], reprinted with permission.)

TiH_2 powder coupled with the dehydrogenation gives direct experimental evidence of the reduction of surface TiO_2 scale by hydrogen. Such possibility is predicted by thermodynamic calculations performed [26]. Two reactions were analyzed:

$$TiO_2 + 2H_2 = 2H_2O + Ti \tag{8.2}$$

$$TiO_2 + 4H = 2H_2O + Ti \tag{8.3}$$

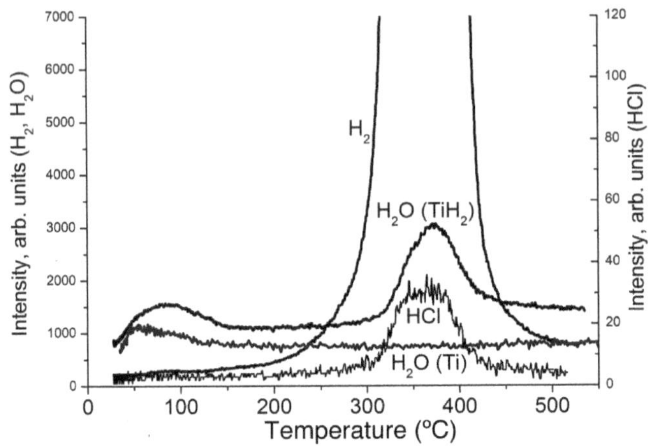

Figure 8.6 Temperature dependencies of H_2, H_2O, and HCl emission upon heating of TiH_2 powder. For comparison, H_2O emission upon heating of Ti metal powder is shown. (Source: Data from Ref. [25].)

The second reaction is justified because of the atomic state of hydrogen in hydrogenated titanium. It remains in such a state while passing through the particle surface and transforms into molecular state only after desorption from metal surface.

Respective equations are available for the change of Gibbs energy $\Delta G(T)$:

$$\Delta G(T) = 101600 - 27.95T + 3.46T \lg T + 2RT\left(\ln\frac{P_{H_2O}}{P_0} - \ln\frac{P_{H_2}}{P_0}\right) \tag{8.4}$$

$$\Delta G(T) = -110800 + 27.95T + 3.46T \lg T + 2RT\left(\ln\frac{P_{H_2O}}{P_0} - 2\ln\frac{P_{H_2}}{P_0}\right), \tag{8.5}$$

where P is the gas pressure, P_0 is the atmospheric pressure (10^5 Pa), and T is the temperature in Kelvin.

Reactions become possible at negative $\Delta G(T)$ values only, but calculations show that this is not the case for the first reaction at any reasonable combination of P_{H_2O} and P_{H_2}; that is, reduction of TiO_2 with molecular hydrogen is impossible. Contrary to that, calculations for the second reaction predict that oxide reduction is possible for the majority of gas pressures and temperatures combinations (Figure 8.7). It was found that only deep vacuum and high temperatures can prevent the reducing of oxide scale. Probability of the reaction increases with increasing hydrogen pressure, which is observed at dehydrogenation.

Reduction of surface oxide by atomic hydrogen has two important consequences. First, it promotes faster and better bonding of adjacent particles and activates mass

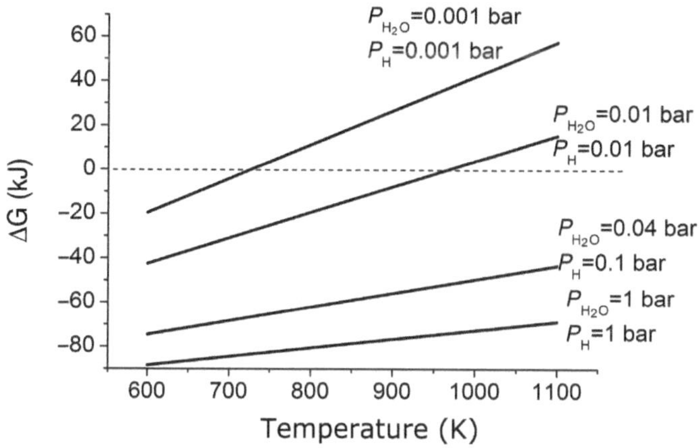

Figure 8.7 Change of Gibbs energy for TiO_2 reduction by atomic hydrogen at various hydrogen and H_2O pressures. (Source: Data from Ref. [26].)

transfer through the interface. Second, oxygen from the oxide scale can be removed from the compact, thus decreasing its content in the sintered product. For that, water vapor should have a possibility to come out from the compact through the open porosity. Clearly, reduction of oxide scale must proceed before the scale dissolves. Since on vacuum heating, oxide scales dissolve at around 700°C [27] and hydrogenation is generally completed by this temperature, prerequisites for cleaning of compact from oxygen indeed exist.

The TiH_2 powder compacts are cleaned by hydrogen not only for oxygen. Mass spectrometry experiments show that another volatile product, HCl, emits in the dehydrogenation temperature range (see Figure 8.6) indicating that at least part of the chlorine leaves the compact. Chlorine is an inevitable impurity in titanium from the Kroll process used today for titanium sponge manufacturing. It exists as hydrated $MgCl_2$ crystals located on powder surface or inside the closed pores. Trapped residual chlorides do not allow full densification of the sintered products [24] during Ti BEPM processing. Using TiH_2 powder instead of Ti metal powder provides a unique opportunity to decrease chlorine content in BEPM products. As brittle TiH_2 particles are crushed during compaction, thus increasing fraction of chlorides on the surface of fragments, atomic hydrogen can bound chlorine in volatile HCl through the reaction:

$$MgCl_2 + 2H = Mg + 2HCl \qquad (8.6)$$

The reaction is confirmed by thermodynamic calculations [25] and experiments with "dirty" TiH_2 powder prepared from so-called underseparated sponge containing chlorine around 1000 ppm. It was reduced to 150 ppm by proper heating in vacuum. This opens a possibility for using starting powders excessively contaminated with chlorine in TiH_2-based BEPM, especially for applications not strictly controlled by weldability.

More examples of "cleaning action" of hydrogen are given in Refs [14,25]. It is now clear that the lower contamination of HDH Ti metal powder compared to TiH_2 powder is due to such "cleaning action" upon dehydrogenation stage. In this regard, fresh HDH powder can be considered as a low-impurity product unless it is "spoiled" by further handling.

8.4 PM processing of CP Ti

TiH_2 and Ti powders compacted at 640 MPa and sintered at 1350°C for 4 h achieved the sintered densities of 98.5%TD and 95%TD, respectively. To understand the positive influence of hydrogen, dilatometric curves were recalculated to show the temperature dependence of density (Figure 8.8). The density of TiH_2 compact, much lower at the beginning, increases very fast on dehydrogenation and becomes nearly equal to that of the Ti compact by the end of dehydrogenation. Moreover, densification of the TiH_2 compact proceeds more intensively on further heating, thus giving advantage to the TiH_2 compact over Ti compact already at the heating stage. The reasons for such activated sintering are related to the reduction of surface TiO_2 scale, promoting faster interface diffusion, and also to phase transformations on dehydrogenation. By the end of heating, relative densities reached 96.5% TD and 93% TD for TiH_2 and Ti compacts, respectively. The difference in density remains after sintering.

The advantage of TiH_2 over Ti in the BEPM press-and-sinter processing results in more balanced mechanical properties [14,25,26]. Typical properties of CP Ti processed from TiH_2 powders are yield stress 470–480 MPa, ultimate tensile stress 560–570 MPa, elongation 25–26%, reduction in area 35–38%, and fatigue endurance limit 250 MPa. As-sintered materials show typical microstructures for CP Ti. Such balanced mechanical properties reflect the cooperative influence of the density, residual

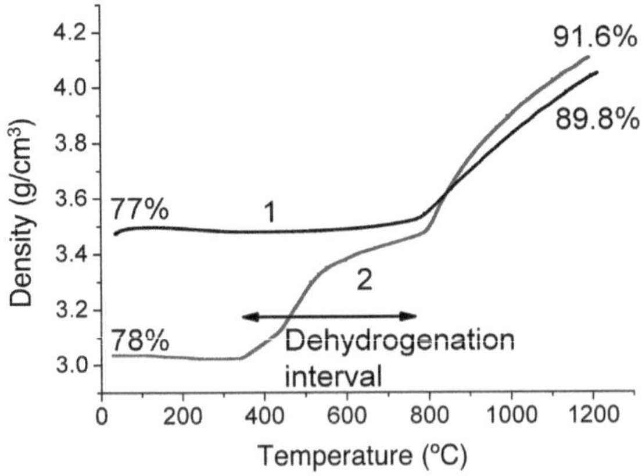

Figure 8.8 Change of densities for Ti (1) and TiH_2 (2) compacts (compaction pressure 640 MPa) calculated from corresponding dilatometric curves. (Source: Data from Ref. [14].)

oxygen content, and grain size and possibly variations of processing parameters and sizing of the starting materials. For example, coarser hydride particles result in a slight decrease in sintered density, especially at low compaction pressure, but with lower oxygen content. Both characteristics decrease the strength but can leave the ductility unchanged. It is important that at equal processing parameters, the tensile properties of CP Ti processed from TiH_2 powder are always better and show smaller variations than those processed from Ti metal powder.

Residual porosity is undesirable, especially for fatigue-controlled applications. That is why, it is important to analyze a possibility to achieve a pore-free material. Transformation of a powder compact into a bulk material by sintering includes two interrelated processes: transformation of a pore system and formation and evolution of the grain boundary network. Reduction in the size of a pore up to its full healing occurs via vacancy migration to any boundaries between either the particles or the grains. The pores can effectively inhibit the migration of the boundaries and corresponding grain growth. As the volume fraction and size of pores decrease, their role as a barrier to grain growth reduces, leading to increased average grain size. As a result, the distance from pores to the nearest boundaries increases; vacancies are no longer absorbed by the boundaries but migrate between the pores, causing their coalescence by Ostwald ripening mechanism, by which the average grain size increases with time as $t^{1/3}$ while keeping the volume fraction constant [28]. Therefore, sintered CP Ti often preserves some pores, whose volume fraction could not be decreased by further exposure. Nevertheless, residual porosity keeps the grain size within tolerable limits (about 200 μm for CP Ti) even after 4 h of sintering at 1250°C. For comparison, in cast CP Ti, the grain size would increase to several millimeters after such a heat treatment.

Clearly, for effective densification, it is necessary to create the conditions under which the pore system evolves, mainly by healing of pores rather than by coalescence. This can be achieved by activated diffusion at lower temperatures when grain growth is slow. The use of TiH_2 powder favors such activation because of dehydrogenation prior to sintering. Also, increasing compaction pressure reduces the initial pore fraction and the size of pores, and that will make their healing easier. Diffusion acceleration due to the influence of impurities within the specifications of different CP Ti grades should be taken into account as well. It was shown [25] that increasing the oxygen content to 0.3% increases the sintered density to 99% TD. Pore-free materials were obtained by sintering of TiH_2 nanopowders, for which high oxygen content is inherent [29]. TiH_2 powder obtained from three grades of sponge is processed to CP Ti of various grades. The observation is that higher impurity content corresponds to higher density and higher strength but much lower ductility (Table 8.1). Oxygen changes the correlation between pore healing and grain growth processes because it, like other interstitials, increases the temperature of polymorphous $\alpha \rightarrow \beta$ transformation and in this way affects the temperature dependence of grain growth. It is also known that iron is abnormally mobile in titanium and even in small quantities accelerates by several orders of magnitude the diffusion of other elements, including self-diffusion of titanium [30]. Iron, however, contrary to α-stabilizers, decreases the β-transus and, therefore, can accelerate the grain growth. On the other hand, excessive impurity embrittles PM Ti products and should be strictly controlled. To get high density through increasing

Table 8.1 Properties of CP Ti PM processed from titanium hydride powder of various contamination

Density g/cm^3 (% of theoretical)	Impurities (%)	UTS (MPa)	Elongation (%)
4.46 (98.8)	O: 0.19; N: 0.02	605	20
4.48 (99.3)	O: 0.24; N: 0.037	678	9.7
4.49 (99.6)	O: 0.27; N: 0.042	712	3.7

impurity content is not desired in most cases, although for some applications limits on their content in PM products could be less critical.

8.5 BEPM processing of titanium alloys

Mechanical properties of specific titanium PM products depend on the chemical composition, density, final microstructure, and contamination level. BEPM processing should homogenize the chemistry, provide particle bonding leaving minimum possible porosity, and eliminate grain growth since the sintering temperature is generally well above the β transus where grains grow fast. Uptake of gases should be minimized to avoid contaminating the compact. These requirements are often contradicting to each other. For example, chemical homogenization and densification would require the highest possible temperature that, however, favors unwanted microstructural evolution (grain growth and respective coarsening of intragrain microstructure). Therefore, real processing is always a compromise to reach optimized properties of the BEPM alloy.

8.5.1 *Ti-6Al-4V*

To date, the most common titanium alloys is Ti-6Al-4V. Therefore, the majority of research efforts in titanium hydride PM were done for this alloy [10,31–35]. The advantages of this approach can be seen from experiments based on Ti and TiH_2 powders of similar sizing <100 μm at equivalent cold compaction and sintering parameters [31,32]. HDH Ti powders were produced from titanium sponge of TG130 grade, and the TiH_2 powder (hydrogen content 3.5 wt%, density 3.9 g/cm^3) was extracted before final dehydrogenation. Three options of alloying additives were used for each base powder, namely, EP: relatively coarse (<100 μm) Al and V elemental powders; EPD: fine-size Al (<20 μm) and V (<40 μm) elemental powders; and MAP: master alloy powders 25Al-75V (<100 μm) and 65Ti-35Al (<40 μm). Blends were die-pressed into green preforms at room temperature, with pressure in the range of 320–960 MPa.

Respective green densities are presented in Figure 8.9 as a function of compaction pressure. Despite the difference in compaction mechanisms, both TiH_2- and Ti-based blends exhibit similar green density dependence on compaction pressure in their relative scales, although the real density values differ significantly. Finer particle size of the EPD powders slightly increases the compaction ability. Blends in which MAP particles (harder than elemental powders) are used exhibit the lowest green density.

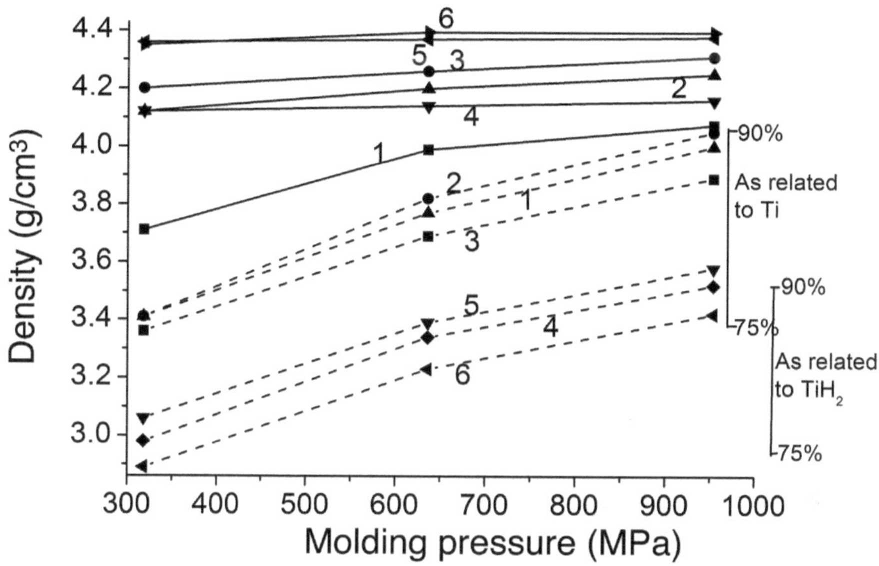

Figure 8.9 Green (dashed lines) and sintered (solid lines) densities of Ti-6Al-4V compacts based on Ti (1, 2, 3) and TiH_2 (4, 5, 6) powders with various alloying additives: 1, 4, elemental powders EP; 2, 5, fine-size elemental powders EPD; 3, 6, master alloy powders MA. Densities related to Ti and TiH_2 ones are indicated in right. (Source: Data from Refs [31,32].)

Alloying powders introduce some specific features into the BEPM processing of Ti-6Al-4V as compared to CP Ti, since the necessity to homogenize the mixture of titanium and alloying additives becomes a main requirement for successful sintering. Time-temperature conditions for homogenization depend not only on diffusivity of Al and V but also on their interference.

Introduction of Al and V as two elemental powders results in stabilization of the corresponding phases in areas around the respective alloying powder particles. Because of difficult penetration of Al into the β-phase and vice versa, V into the α-phase the overall homogenization is delayed. Faster homogenization can be reached by an increase in temperature to provide α→β transformation even in aluminum-rich areas, or long exposures sufficient for aluminum to gradually penetrate into the β-phase in which its solubility, although lower than in the α-phase, is not as low as that of vanadium in α-phase, or finer alloying powders. Neither of these options is optimal since they lead to either coarsening of the grain structure or increase of impurity content, or make processing more expensive.

If alloying elements are added as Al-V MA particles, another difficulty arises. Al, whose solubility in the α-phase is much higher than that of V, will penetrate into areas around the MA particles before the β-transus, thereby stabilizing the α-phase in them. In turn, such α-areas act as barriers for V, preserving its high concentration at the locations of the MA particles. With this, formation of homogeneous alloy is limited by slow V redistribution and again needs long time exposure or increased temperature to attain chemical homogeneity. Similar phenomenon is observed upon BEPM processing of Ti-5Al-2.5Fe when Al blocks iron in Fe-Al MA particles [36,37].

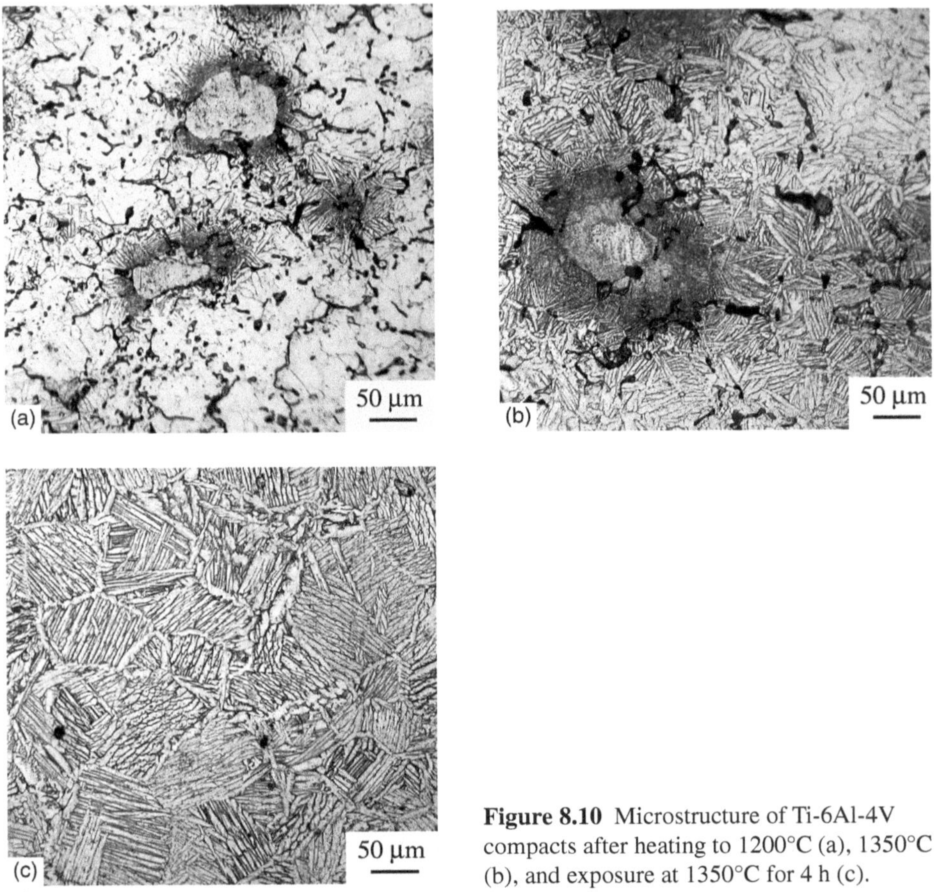

Figure 8.10 Microstructure of Ti-6Al-4V compacts after heating to 1200°C (a), 1350°C (b), and exposure at 1350°C for 4 h (c).

So, independent of the alloying approach, homogenization of Ti-6Al-4V is controlled by the redistribution of V, which begins only at 1200°C (Figure 8.10a). Markedly heterogeneous microstructure due to uneven vanadium concentration is seen even after heating to 1350°C (Figure 8.10b). Modeling with DICTRA™ software [37] shows that 60Al-40V MA composition is not optimal and lower aluminum content in MA would provide faster homogenization. Such finding can be considered as a recommendation for commercial BEPM processing of Ti-6Al-4V if supported by MA producers.

To reach chemical homogeneity (Figure 8.10c), the above compacts were exposed at 1350°C for 4 h. The sintered densities are shown in Figure 8.9. They vary considerably depending on the matrix and alloying additives employed. However, TiH_2-based compacts always exhibit higher sintered densities. In both cases (TiH_2 and Ti), the lowest densities correspond to compacts with coarse EP and the highest ones correspond to compacts with MA additives. For TiH_2 based compacts, EPD and MA alloying options result in nearly the same densities.

Generally, low sintered densities of EP alloyed compacts are understandable because melting of Al followed by its exothermic reaction with Ti hinders consolidation [10,31–33]. To avoid this, Al should react with Ti before it melts.

However, even after 15 minutes' exposure at 600°C, solid-state reaction is still incomplete. Liquid Al penetrates between Ti powder particles, causing swelling of the compact and leaving coarse voids at the former places of Al particles. The sintered density of the Ti-EP compact is as low as 83–91% TD, depending on compaction pressure. Use of TiH_2 powder matrix can only partially eliminate the swelling as solid-state reaction, although activated by dehydrogenation, is still incomplete and the sintered density levels at around 93% TD. Finer alloying additives result in a noticeable increase in density. It reached 96.5% TD and 99% TD in the Ti-based and TiH_2-based compacts, respectively. The highest densities are attained in the MAP compacts, since liquid phase is not formed on heating. A distinctive feature of the TiH_2-based compacts is that they exhibit a negligible dependence on compaction pressure from 320 to 640 MPa and no dependence at the higher pressures.

The TiH_2-based approach thus leads to higher sintered densities. It can also reduce the final oxygen content. Regarding alloying options, MAP additives show the highest densification ability (98.5% TD). TiH_2 based compacts with EPD additives also are densified to 98.5% but excessive contamination due to high specific surface of the EPD particles does not allow the final oxygen content to reach an acceptable level. Because of low porosity, fine-grained uniform microstructure and acceptable final oxygen content, the BEPM-processed Ti-6Al-4V alloy from the TiH_2 powder blended with MAP exhibits mechanical properties attractive for practical applications: ultimate tensile strength of 950–1050 MPa and elongations of 8–13%. The strength and ductility meet the ASTM specification for this alloy, with the ductility value occasionally below the required minimum of 10%.

Further optimization of the BEPM process of Ti-6Al-4V is based on a better understanding of the influence of specific processing variables that can affect the residual porosity, microstructure, and chemistry and, through these characteristics, mechanical properties. Such variables include chemistry and nominal size of starting powders, compaction pressure, and sintering temperature and time. Qualitatively, their possible influence on the characteristics of the powder-based Ti-6Al-4V is presented on Figure 8.11 [34], which shows that each material characteristic is affected by several variables. Sometimes, this influence is only conditional (dashed lines); that is, it appears only at some specific values of the variables. Some comments on Figure 8.11 are given below.

Variation in compaction pressure from 320 to 960 MPa does not affect the sintered density (see Figure 8.9) and microstructure. This unique feature is attributed to the specific compaction mechanism of the TiH_2 matrix, which leaves only small-size pores between crushed fragments of the base powder. Although MA particles are not crushed upon compaction, they maintain reliable contact with the base powder.

Variation in the base powder size results in a negligible change in sintered density. Once again, the explanation is related to the compaction mechanism, and the pore network in the green compacts is similar for all starting powder sizes (use of very fine hydride powders slightly reduces the porosity). The base powder size does not affect the microstructure. However, a decrease in powder size generally leads to increased oxygen absorption during presintering processing. The MA powder size is also important for the final impurity content. Besides, this variable is critical in attaining chemical and microstructural homogeneity. Smaller MA particle size is preferred since it allows for lower sintering temperature and shorter exposure, but care

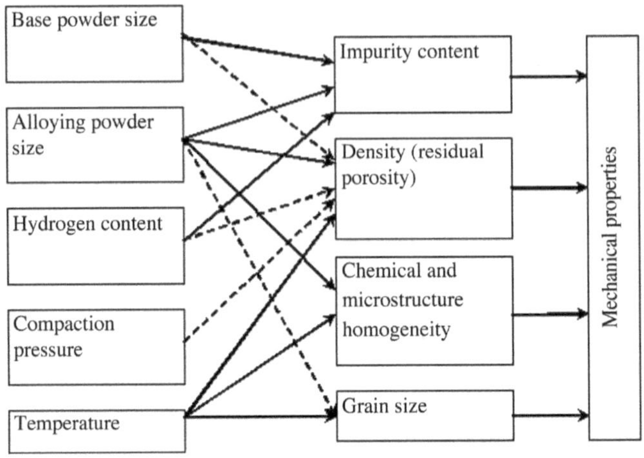

Figure 8.11 Schematic influence of process variables on characteristics of synthesized alloys. Dashed arrows denote either conditional influence under specific combination of variables or minor influence. (Source: From Ref. [34], with permission.)

should be taken to avoid excess contamination. Decrease in hydrogen content from 3.9 to 1.1% and then to 0.2% results in some decrease in sintered density. At low hydrogen content and low pressure, the density was reduced to 96.5% TD, which deteriorates the fatigue endurance [35]. On the other hand, lower hydrogen content could be advantageous from the viewpoint of lower shrinkage upon sintering.

Decrease in sintering temperature from 1350°C to 1260°C and then to 1200°C at the same exposure time (4 h) results in a decrease in the average beta-grain size from 110 to 77 μm, but with coarser MA particles, there is an indication of uneven microstructure along with a drop in sintered density of about 1% after 1200°C sintering. On the other hand, with finer MA powder, uniform low-porous fine-grained microstructures can be obtained even at a short (2-h) exposure at 1100°C. Tensile elongation of optimized BEPM processed Ti-6Al-4V alloy never drops below 10%, most often being in the range of 13 to 15%.

8.5.2 *High-strength alloys*

High-strength metastable beta alloys, such as Ti-5Al-5V-5Mo-3Cr (Ti-5553) and Ti-10V-2Fe-3Al (Ti-1023), are being increasingly used because of their excellent combination of properties. BEPM processing of these alloys presents a real challenge [38], since they contain up to 18 wt% of alloying elements, both α and β stabilizers, whose diffusivity in titanium lies in a wide range above and below the self-diffusivity of titanium. Ti-5553 and Ti-1023 alloys fabricated using TiH_2 powder and sintered at 1250°C for up to 4 h achieved only 96–97% TD, much lower than that of Ti-6Al-4V sintered under the same conditions. High residual porosity is noticeable in the microstructure (Figure 8.12a). Increasing isothermal exposure to 6 h and/or sintering temperature to 1350°C failed to noticeably increase the sintered density. Instead, the

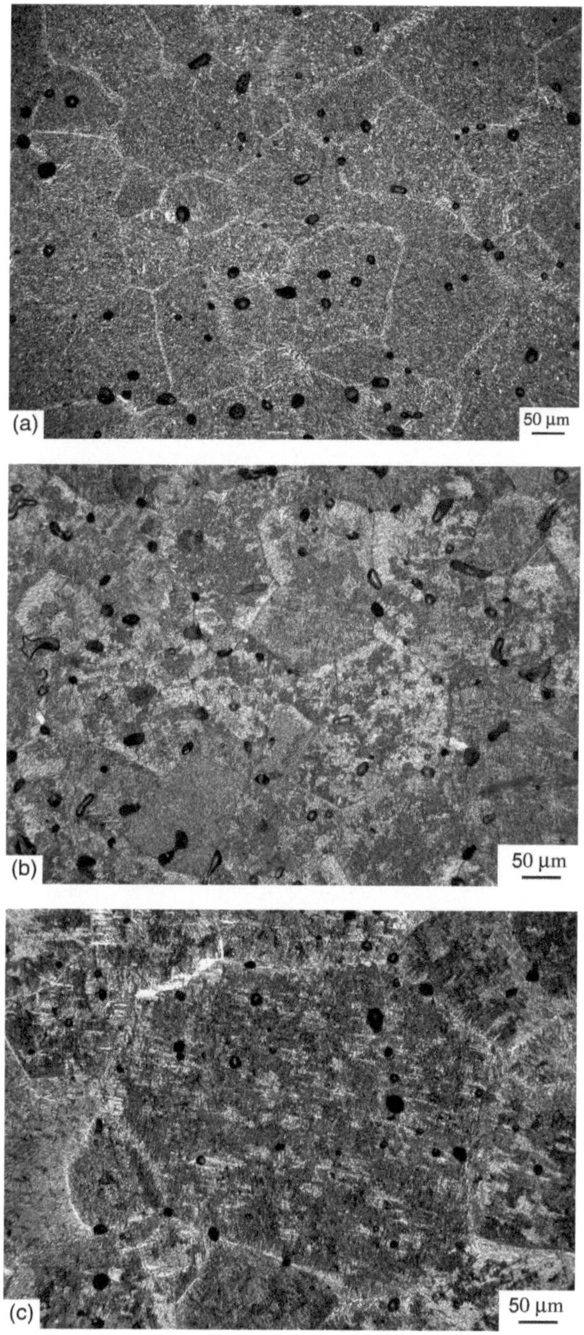

Figure 8.12 Microstructure and residual porosity in Ti-5553 (a) and Ti-1023 (b) alloys synthesized at 1250°C for 4 h; anomalously large grain formed in Ti-1023 alloy synthesized at 1350°C, 4 h (c). (Source: From Ref. [38], with permission.)

use of longer times or higher temperatures caused remarkable coarsening of the beta grains (Figure 8.12c). Grain growth is more pronounced in the Ti-1023 than in Ti-5553.

As-sintered mechanical properties of the above alloys with β-transformed α+β intragrain microstructure formed by slow cooling from the sintering temperature (with isothermal exposure at 750°C, 1 h, for Ti-5553) are presented in Table 8.2. They are typical of these alloys processed through IM and better than those of Ti-1023 BEPM processed using Ti metal powder [39]. The ductility is slightly higher for alloys sintered at 1350°C because of lower porosity. At such a strength level, the ductility does not deteriorate noticeably by excess porosity. The grain size is not a critical parameter for β-transformed microstructures. Rather, the colony size is more important, which depends on the cooling rate. Hence, Ti-5553 and Ti-1023 alloys can be fabricated from the TiH_2 base powder mixed with MA powders and achieve properties inherent to the annealed condition. However, the high residual porosity and relatively coarse-grained microstructure leads to unacceptably low ductility in high-strength STA condition. In particular, poor ductility was found to be in the alloys sintered at 1350°C because of the formation of anomalously large grains.

The poor densification of these alloys can be understood from their shrinkage behavior upon heating compared to that of the TiH_2 powder (Figure 8.13). The shrinkage of Ti-1023 is distinctly different. Since dehydrogenation of TiH_2 and Ti-1023 occurs over the same temperature range and the resulting mass losses are similar, when normalized against the amount of TiH_2, it can be found that the Ti-1023 compact shrinks less than expected according to the volume changes related to the transformation of

Table 8.2 **Structure and properties of Ti-5553 and Ti-1023 alloys**

Alloy	Sintering temperature (°C)	Condition	Density (%)	Grain size (μm)	YS (MPa)	UTS (MPa)	Elongation (%)	RA (%)
Ti-5553	1250	BEPM	97.3	75	966	1067	10.1	16.9
	1350	BEPM	97.8	129	933	1031	11.3	14.9
	1250	STA	97.3	75	1172	1268	2.6	4.3
	1350	STA	97.8	129	1258	1350	0.76	1.4
	1200	STA*	98.5	100	1200	1304	6.8	10.3
	BMS 7-360 Specification. High strength condition				1170	1240	5	–
Ti-1023	1250	BEPM	96.0	105	944	1033	8.0	13.5
	1350	BEPM	97.2	159	939	1033	12.0	19.5
	1350	STA	97.2	159	1166	1205	0.5	4.2
	1200	STA*	98.0	82	1115	1250	5.2	11.0
	AMS 4984 Specification. High strength condition				1103	1193	4	–

BEPM, as-sintered condition; STA, solution treated and aged after BEPM condition; STA*, solution treated and aged after optimized BEPM condition.

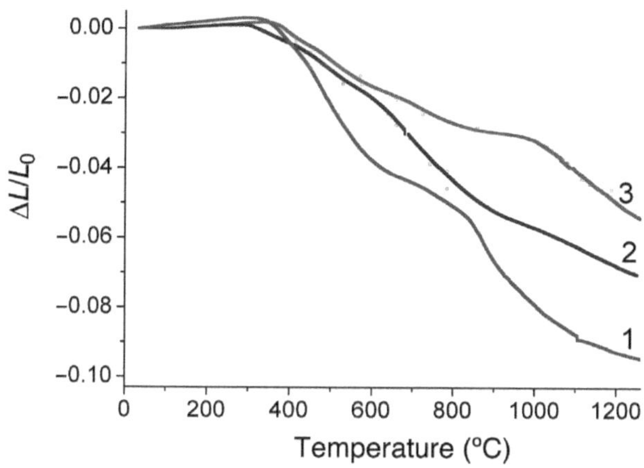

Figure 8.13 Dilatometric curves for TiH_2 (1), Ti-5553 (2), and Ti-1023 (3) compacts on its base. Heating rate 7°C/min. (Source: From Ref. [38], with permission.)

$TiH_2 \rightarrow Ti$. This leads to increased porosity. This issue is most likely related to the matrix/MA interfaces at which formation of additional porosity is observed [37,38].

Another distinctive feature of the Ti-1023 is the much delayed accelerating of densification related to the $\alpha \rightarrow \beta$ transformation due to the penetration of Al from the MA particles into matrix and the corresponding increase of the β-transus temperature. Fe remains locked in the MA particles until the $\alpha \rightarrow \beta$ transformation completes. After the phase barrier disappears, Fe spreads fast over the matrix, leaving vanadium behind because of its much lower diffusivity. Fast diffusion of Fe is a main reason for fast sintering in the final stage of heating, but this does not compensate the initial delay.

As with Ti-1023, homogenization and densification in Ti-5553 are determined by interference of Al and Cr. Therefore, the relative densities of both alloys at the end of heating are markedly lower than that of CP-Ti or Ti-6Al-4V. On isothermal exposure, their densities first increased but then stabilized at 96–97% TD. They do not approach their pore-free density, most probably because of activation of the beta grain growth promoting a transition from pore healing to pore coalescence. The Ti-1023, in which grain growth is more pronounced because of fast diffusion of iron, has higher residual porosity.

It can be concluded that although the addition of fast diffusive elements like Fe or Cr improves titanium sinterability [40], complex combinations of alloying elements and their interference lead to poor sinterability compared with CP-Ti and Ti-6Al-4V, and eventually poor ductility in the solution-treated and aged (STA) condition (see Table 8.2).

It should be noted that the use of Ti metal powder and various combinations of alloying additives in numerous experimental trials led to even worse results [41]. In Ref. [39], the as-sintered microstructure of the Ti-1023 was modified by hot isostatic pressing (HIP) before the STA treatment. To avoid this additional costly operation, it was proposed [38,41] to modify the sintering regime such that the phase transformation of the TiH_2 skips over the multistep route of $TiH_2(\delta) \rightarrow \beta \rightarrow \alpha \rightarrow \beta$ and goes directly to the β-phase, omitting the α-phase field. This can be achieved either by fast heating or by keeping hydrogen partial pressure sufficient to avoid β→α+β transformation. As a result, β stabilizers are not shielded by Al-enriched zones and can penetrate into matrix at lower temperatures. This way, Fe in Ti-1023 becomes a leading element that penetrates into titanium before it causes drastic changes in beta grain structures. Additionally, use of finer alloying particles uniformly distributed in the matrix has proved to be effective for sintering modification. Homogeneous distribution of finer MA particles in the matrix favors finer initial porous structure and more uniform diffusion flows, thus accelerating the sintering.

Modifications of BEPM processing as described above reduce residual porosity in both alloys under the same sintering regimes. Moreover, sintering can be performed at lower temperatures (e.g., 1200°C), which minimizes grain growth. Lower porosity and finer grain structure makes balance of strength and ductility in high-strength STA condition sufficient to meet requirements of corresponding specifications (see Table 8.2).

8.6 Production of hydrogenated titanium powder

Commercial success of titanium hydride PM depends on the availability of low-cost but high-quality titanium hydride powder. For laboratory experiments, it can be prepared in small amounts via hydrogenation of titanium sponge. For large-scale use, several methods have been developed for commercial production of hydrogenated titanium.

Conventionally, titanium hydride is being produced by hydrogenation of low-grade titanium sponge, turnings, and other titanium materials. Use of titanium scrap is encouraged for cost reasons [5]. The titanium charge is first heated to temperatures above 600°C in vacuum for dissolution of surface oxides, which are barriers for hydrogen penetration into the material. Then, the vacuumized chamber is filled with hydrogen and the titanium charge is exposed under a hydrogen atmosphere of prescribed pressure at temperatures of 400–600°C. This method is relatively simple; however, it needs high vacuum condition (10^{-2}–10^{-3} Pa) at the first stage and long exposure to get uniform distribution of hydrogen through the final product, especially in the case of using bulky charge.

To speed up hydrogenation, self-propagation high-temperature synthesis (SHS) process was developed [42,43]. The essence of the SHS process is use of heat released from absorption of hydrogen by titanium for sustaining sufficiently high temperature necessary to complete the reaction. SHS hydrogenation is initiated locally by external source. Heat of the reaction spreads fast over the whole material and supports the temperature sufficient for its hydrogenation. The SHS is most suitable for hydrogenation of materials with a high surface area. Low energy consumption and short time provides high cost-efficiency of the SHS hydrogenation. On the other hand, SHS hydrogenation needs a precise control of the hydrogen pressure and temperature during heating and cooling to provide uniform hydrogen distribution. Another drawback of the SHS method is the risk of excessive contamination of hydrogenated titanium.

Production of hydrogenated titanium in the above methods is based on hydrogenation of already produced titanium materials; therefore, its cost cannot be lower than that of the titanium. However, in order to meet the requirements of low-cost BEPM, the hydride powder should be made at least no more expensive than the primary titanium product.

Process based on calcium hydride reduction of titanium dioxide was a first attempt to produce hydrogenated titanium powder directly from the raw material, omitting a stage of primary titanium production [5,44]:

$$TiO_2\,(\text{solid}) + 2CaH_2\,(\text{gas}) = TiH_2\,(\text{solid}) + 2CaO\,(\text{solid}) + H_2\,(\text{gas}) \tag{8.7}$$

Although the above process was originally developed for the production of titanium metal powder, it was found that by change of temperature and hydrogen pressure, powders hydrogenated to various hydrogen content can be obtained. Since the calcium hydride process does not include titanium tetrachloride, the powders contain low levels of chlorine. Typical composition of Ti powder produced by calcium hydride reduction process is presented in Table 8.3.

Calcium hydride reduced titanium powder was widely used in the late 1980s for manufacturing CP Ti and Ti alloy parts with the BEPM press-and-sinter approach [44]. However, the increased cost of the starting materials (calcium and pure titanium oxide) resulted in the high cost of this powder and, consequently, did not allow to achieve any commercial success.

A revolutionary new approach for manufacturing hydrogenated titanium powder was developed in Ukraine in collaboration with US partners PNNL and ADMA Products, Inc. [45,46]. Its key feature is integration of titanium production and its hydrogenation in one continuous cycle, including reduction of $TiCl_4$, vacuum distillation, and hydrogenation of the sponge block in the same vessel upon cooling from the distillation temperature. This differs from normal operations of removing the titanium sponge block from the vessel at the end of the reduction/distillation cycle, crushing of the block, and, finally, performing hydrogenation of crushed sponge in another vessel. Such innovations certainly need modification of the magnesium-thermic process, including redesign of the equipment and adjusted correlations between sponge production and hydrogenation regimes. Regarding the later requirement, the magnesium–thermic reduction process was modified in a way that promotes formation of a sponge block having an axial-type cavity and excess porosity for accelerating vacuum distillation from excessive Mg and $MgCl_2$ at 800–850°C, and for uniform penetration of hydrogen into the sponge.

Table 8.3 Chemical composition of calcium hydride reduced CP titanium powder (wt%)

Ti	Fe	Ni	C	N	O	H	Ca	Si	Cl
Base	0.11	0.07	0.03	0.06	0.19	0.34	0.04	0.05	0.004

With these modifications, the time of high-temperature distillation stage becomes shorter, providing labor and saving energy. Besides, cooling of the sponge block in a hydrogen-containing medium cleans the sponge from chlorine-containing remnants. Porosity and brittleness of hydrogenated titanium block facilitates the subsequent crushing and grinding of the block into powder. Shortening of distillation stage and respective savings compensates expenditures related to hydrogenation, thus making the cost of the hydrogenated titanium powder at least equivalent to that of pure titanium sponge.

Further developments for manufacturing the titanium hydride powder was done by ADMA Products, Inc. [47]. Magnesium–hydrogen reduction process was proposed as the most cost-effective approach to producing low-cost, high-quality hydrogenated titanium powder. In this process, magnesium is partially replaced with hydrogen as a reducing agent. Hydrogen is charged into the retort before the $TiCl_4$ and solutes in molten magnesium. In the first stage, $TiCl_4$ is reduced with magnesium and hydrogen:

$$TiCl_4 + (1\text{-}2)H_2 + (1\text{-}2)Mg = TiH_2 + (1\text{-}2)MgCl_2 + (0\text{-}2)HCl \tag{8.8}$$

Use of hydrogen shortens reduction time and increases the utilization of magnesium. Moreover, using of hydrogen on the reduction stage significantly shortens successive vacuum distillation stage and allows lower distillation temperature. Distillation is done in such a way that hydriding/dehydriding steps follow each other several times, resulting in a highly developed porosity and cracking due to an 18% difference in titanium and titanium hydride densities. Sponge comminution becomes simpler, hydrogenated titanium can be mechanically ground to the desired particle size even inside the retort and then extracted and compacted in an inert environment for subsequent compaction steps. Furthermore, if necessary, the TiH_2 powder after grinding can be dehydrided in the same retort and used like any other titanium metal powder. Composition of titanium hydride powder averaged from 25 lots produced by the ADMA process is presented in Table 8.4.

A pilot-scale unit (Figure 8.14) with annual capacity of 250,000 lb of TiH_2 powder was installed in 2013 at ADMA facilities. At least 50% energy savings and 20% cost reduction relative to existing commercial Kroll's process are expected at full commercialization of the ADMA process.

Recently, new processes for manufacturing titanium hydride powder directly from titanium slag or synthetic rutile have been developed by ADMA Products, Inc. [48].

Table 8.4 Chemical composition of ADMA titanium hydride powder, wt%

Material	Fe	N	C	O	H	Ti
ADMA TiH_2 powder	0.03–0.16	0.030	0.010	0.063	3.80–3.85	Bal.
ASTM B348 Grade 2	0.300	0.030	0.080	0.250	0.015	Bal.

Figure 8.14 Pilot scale unit for manufacturing TiH_2 powder at ADMA Products, Inc., Twinsburg facilities.

8.7 Scaling up titanium hydride powder metallurgy

Significant efforts have been made to scale up Titanium Hydride PM from laboratory experiments to commercial applications. This is being done in three directions, which include die pressing, cold isostatic pressing (CIP), and direct powder rolling (DPR) as methods of cold compaction. They all provide high productivity, sufficient strength of green parts, and capability to produce different sizes and shapes.

Die pressing is used mostly for relatively small-sized parts and does not need any specific measures to be taken compared to the conventional titanium PM, except for accounting much higher shrinkage at tooling design. Press-and-sinter TiH_2-based BEPM processing [49–52] is being used by ADMA Products, Inc., for manufacturing various parts with controlled shape, uniform density, and mechanical properties meeting requirements of respective specifications, from CP Ti, Ti-64, Ti-3-2.5, Ti-10-2-3, and Ti-5553 for aerospace, automotive, and medical applications [53,54]. Oxygen content control is being of special attention. Careful experiments have been performed to quantify the oxygen content upon various stages of the processing. It has been found that oxygen is uptaken upon milling/ blending/pressing stages but its content decreases upon vacuum sintering of consolidated parts, thus confirming on a commercial scale the possibility to reduce oxides by atomic hydrogen. Exposure of the green parts in open air before sintering has to be strictly eliminated in order to keep a low oxygen content in the final product.

Adjusting the results obtained from laboratory experiments to commercial die press-and-sinter production of specific parts is generally based on Figure 8.11, in which the influence of the BEPM processing variables on the mechanical properties is qualitatively presented. A possibility to use a cooperative action of several variables, such as the content of hydrogen in the base powder, alloying powder size, and sintering temperature adds flexibility to the processing windows and allows for good dimensional stability of the parts.

A distinctive feature of the BEPM processed CP Ti and titanium alloys is that their tensile properties are not very sensitive to sintered density variations over the range of 98–99% TD. Below that a drop in the ductility is often observed, especially if the residual pores are prone to merge under loading. However, even if process variables are not critical for the tensile properties, other properties such as fatigue endurance could suffer more significantly. Therefore, components for fatigue-controlled applications should be more carefully processed.

CIP followed by vacuum sintering is a favorable processing option to form "chunky" components of relatively simple shapes and, more importantly, to produce preforms as semifinished products for further hot deformation (rolling, forging, extrusion, etc.). This additional processing step somewhat increases the cost of final components, but has a positive impact on their mechanical properties through microstructure refinement and density increase up to 100% TD. Moreover, the hot deformation step allows formation of components with relatively complex shapes that may not be achieved through other powder compaction methods.

Titanium materials produced by the CIP powder consolidation and sintering technology are an emerging field with substantially reduced cost, which could replace the current labor-intensive and costly methods of IM technology [55–58]. With the development of direct consolidation of powder-based titanium alloys, substantial cost reductions can be achieved through bypassing numerous laborious operations in ingot melting and thermomechanical processing. Besides, with direct powder consolidation, titanium product makers are able to use much closer to near-net thickness preforms and/or billets (i.e., reduced thickness, diameters, etc.) for extrusion, rolling and/or forging with just enough thermomechanical processing to meet the material specification requirements, thus reducing the overall cost of the required processing steps. With this in mind, a low-cost titanium-powder technology for canless extrusion processing was developed [55,56] for replacing the present IM-based processing. TiH_2 powder has been selected as a material well suited for implementation of these cost-reduction measures in a big program performed jointly by ADMA Products, Inc. (fabrication of Ti-6Al-4V powder-based billets), RTI International Metals, Inc., Plymouth Engineered Shapes, Inc. (fabrication of extrusions), and Boeing (material microstructure and processing-microstructure-property characterization). The titanium hydride material used for this program was supplied to ADMA Products Inc. by Zaporozhje Titanium-Magnesium Co. (Ukraine). ADMA's vacuum sintering furnace is used to sinter large-scale billets (Figure 8.15), but first experiments on manufacturing the billets has revealed necessity to modify the furnace since its vacuum pumping system

Figure 8.15 Vacuum sintering furnace at ADMA Products, Inc., Hudson facilities.

was not efficient enough to cope with the huge amount of hydrogen released from the large-sized TiH_2-based billets. Modification of the furnace pumping and cooling systems eliminated the possibility of the billets acting as oxygen getters in the case of a rather long sintering cycle, with even normally expected air entry into the vacuum chamber. This, along with careful shielding measures in the out-of-vacuum-chamber steps preceding the sintering (the use of inert gaseous environment during milling and blending steps and especially during packing blended powder into the tooling in preparation for CIP), allowed to keep the interstitial element concentrations within the respective specifications. It is important to stress that not only oxygen was low but chlorine content as well in the as-sintered BEPM billets was under 100 ppm. This ensures that issues related to excessive chlorine content, for example, weldability, will not exist for these materials.

Results of careful examination of the extrusions produced from TiH_2 powder-based billets are described in details in Ref. [55]. It is concluded that the overall mechanical property of the powder-based product is equivalent to that of the double-arc-remelted ingot-based extrusions. The property equivalence is achieved not only in terms of static mechanical properties (tensile and monolithic fracture toughness K_{IC} [K_Q] values), but also in terms of dynamic fatigue properties (combined S/N plus *da/dN* properties), as well as stress-corrosion resistance measured in terms of K_{ISCC} threshold values.

A similar program has been carried out for flat products [57,58]. Ti-6Al-4V sintered billets were manufactured from TiH_2 powder by ADMA Products, Inc.; rolling was performed by RTI [57] and TIMET [58]. Powder-based rolled mill products processed with optimized pathways show static tensile and fatigue S/N and fracture properties matching similarly processed ingot-based hot-rolled products (Table 8.5). They meet the AMS Specification mechanical property requirements for aerospace applications. For fatigue-driven structures, full densification and microstructure refinement are necessary in rolled mill products, which are achieved with rolling reduction no less than 3:1.

Table 8.5 Properties of Ti-6Al-4V products manufactured in various processing routes

Processing history	Product form	Number of tests used in average values	UTS (ksi)	YS (ksi)	Elongation (%)	K_{IC} or K_C (ksi.in.$^{1/2}$)
BEPM CIP/ sintered, hot rolled, mill annealed	Plate	10	144	130	15	67
	Sheet	34	152	135	9	131
Double-arc remelted, hot rolled, mill annealed standard grade	Plate	2	138	126	13	79
	Sheet	8	132	121	10	147

The largest powder-based billet produced by ADMA using "in-house" facilities has dimensions of 8.50 in. × 12.0 in. × 52.0 in. and weighs 725 lb. Plates produced by hot rolling of such billets have been tested for many critical applications. One of the successful examples is the plate used for Commander's Hatch of Bradley Fighting vehicle [54]. Extruded or hot-rolled powder-based bars have been successfully tested for the Applique Armor Attachment program [59] and recently also for the fasteners and springs intended for automotive applications [60]. Of special interest are PNNL-headed efforts to introduce powder-based high-strength Ti-185 bar feedstock into the automotive industry. Static and cyclic S/N fatigue properties of this BEPM processed alloy were assessed, which provide the opportunity to develop an attractive material for fastening applications. The conventional ingot-processing approach for this alloy was discounted because of the significant segregation issue of iron (beta flecks).

Another development is that using hollow cylinder billets piping systems of titanium CP Grade 2 were developed for naval applications from low-cost titanium hydride powder [54,61].

The DPR process is a low-cost method to produce flat products (foils, sheets, plates) by the continuous compaction of titanium powders. The process has been successfully used by ADMA Products, Inc., for manufacturing of CP Ti, Ti-6Al-4V, TiAl, and other Ti alloy strips of various thicknesses (0.05–0.25 in.) and densities (50% to ≥99%). The green density/strength of DPR compacts made from pure TiH_2 powder was found to be insufficient for further handling but better results were achieved when it was mixed with titanium metal powder. The sintered flat products generally are too contaminated by oxygen and therefore did not pass requirements of aerospace use products [57,58]. However, the DPR of titanium hydride–based powder blends has been successfully used for manufacturing porous flat products.

References

[1] E. Gregory, Fabrication of niobium superconductor alloys, U.S. Patent 3,472,705 (1967, 1969).

[2] K. Obara, Y. Nishino, S. Matsumoto, Process to producing a sintered article of a titanium alloy. U.S. Patent 3,950,166 (1974, 1976).

[3] C. Yolton, F. Froes, Method for producing powder metallurgy articles. U.S. Patent 4,219,357 (1978, 1980).

[4] A.P. Pankevich, A.F. Chertovich, G.A. Libenson, Production of titanium specimens from a titanium hydride powder by hot pressing, Sov. Powder Metall. Metal Ceram. 25 (2) (1986) 89–92 (278).

[5] V.S. Ustinov, Yu.G. Olesov, L.N. Antipin, V.A. Drozdenko, Titanium Powder Metallurgy (in Russian), Metallurgiya publisher, Moscow, 1973, p. 248.

[6] E.J. Dulis, V.K. Chandhok, F.H. Froes, L.P. Clark, Manufacturing procedure for the production of large titanium PM shapes: current status, SAMPE 10 (1978) 316–329.

[7] T. Gladden. Process for the manufacture by sintering of a titanium part and a decorative article made using a process of this type. U.S. Patent 5,441,695 (1994, 1995).

[8] O.M. Ivasishin, A.N. Demidik, D.G. Sawakin, Phase transformations on synthesis of titanium aluminides from TiH_2 and Al powders, in: B.A. Blenkinsop, W.J. Evans, H.M. Flower (Eds.), Titanium-95 Science and Technology: Proceedings of eight World Conference on Titanium, The Institute of Materials, London, UK, 1995, pp. 440–447, 1996.

[9] O.M. Ivasishin, A.N. Demidik, V.M. Prozorov, D.G. Sawakin, Synthesis of Ti*Al-SiC composites by reactive sintering using TiH_2 powder. in: R.V. Lang (Ed.), High Tech in Saltzburg, Proceedings of 16th International SAMPE Europe Conference, SAMPE, pp. 281–292.

[10] O.M. Ivasishin, V.M. Anokhin, A.N. Demidik, D.G. Savvakin, Cost-effective blended elemental powder metallurgy of titanium alloys for transportation application, Key Eng. Mater. 188 (2000) 55–62.

[11] A.A. Illin, B.A. Kolachev, V.K. Nosov, A.M. Mamonov, Hydrogen Technology of Titanium Alloys, Moskow Institute for Steels and Alloys (MISIS) publisher, Moscow, 2002, p. 392 (in Russian).

[12] A. San-Martin, F.D. Manchester, The Ti-H (titanium – hydrogen) system, Bull. Alloy Phase Diagrams 8 (1) (1987) 30–42.

[13] P.V. Held, P.A. Ryabov, L.P. Mokhracheva, Hydrogen and Physical Properties of Metals and Alloys (in Russian), Metallurgy publisher, Moscow, 1985, p. 232.

[14] D.H. Savvakin, M.M. Humenyak, M.V. Matviichuk, O.H. Molyar, Role of hydrogen in the process of sintering of titanium powders, Mater. Sci. 47 (5) (2012) P651–P661.

[15] H. Liu, P. He, J.C. Feng, J. Cao, Kinetic study on nonisothermal dehydrogenation of TiH_2 powders, Int. J. Hydrogen Energy 34 (2009) 3018–3025.

[16] V. Bhosle, E.G. Baburaj, M. Miranova, K. Salama, Dehydrogenation of TiH_2, Mater. Sci. Eng. A356 (2003) 190–199.

[17] O.M. Ivasishin, D.G. Savvakin, M.M. Gumenyak, Dehydrogenation of titanium hydride powder and role of this process in a sintering activation, Metal Phys. Novel Technol. 33 (7) (2011) 899–918 (in Russian).

[18] O.M. Ivasishin, V.T. Cherepin, V.N. Kolesnik, M.M. Gumenyak, An automated dilatometric system, Instrum. Exp. Techniques 53 (3) (2010) 457–460.

[19] V.B. Brik, Diffusion and Phase Transformations in Metals and Alloys (in Russian), Naukova Dumka publisher, Kiev, 1985, p. 232.

[20] K. Mizuno, Y. Furuya, K. Hirano, H. Okamoto, Hydrogen diffusion in titanium-hydride observed by the diffraction-enhanced X-ray imaging method, Phys. Stat. Sol. (a) 204 (2007) 2734.

[21] H. Wipf, B. Kappesser, R. Werner, Hydrogen diffusion in titanium and zirconium hydrides, J. Alloys Compd. 310 (2000) 190–195.
[22] K.W. Kehr, Theory of the diffusion of hydrogen in metals, in: G. Alefeld, J. Volke (Eds.), Hydrogen in Metals I, Springer, Berlin, 1978, p. 197.
[23] G. Lütjering, J.C. Williams, Titanium, Springer-Verlag, Berlin, 2003.
[24] F.H. Froes, D. Eylon, Powder metallurgy of titanium alloys, Inter. Mater. Rev 35 (3) (1990) 162–182.
[25] O.M. Ivasishin, D.G. Savvakin, O.B. Bondarchuk, M.M. Gumenyak, Role of surface contamination in titanium PM, Key Eng. Mater. 520 (2012) 121–132.
[26] D.G. Savvakin, Physical background of solid state synthesis of titanium alloys. The thesis for Dr. Sci. Degree, G.V. Kurdyumov Institute for Metal Physics NAS of Ukraine, Kiev, 2013.
[27] W. Mo, G.Z. Deng, F.C. Luo, Titanium Metallurgy, second ed., Metallurgical Industry Press, Beijing, 2007, pp. 11–20, 48, 49
[28] I.M. Lifshits, V.V. Slesov, On kinetic of diffusion decomposition of supersaturated solid solutions, J. Exp. Theor. Phys. 35 (1959) 479–492 (in Russian).
[29] H. Wang, M. Lefler, Z.Z. Fang, et al., Titanium and titanium alloy via sintering of TiH_2, Key Eng. Mater. 436 (2010) 157–163.
[30] C. Herzig, S. Divinski, Y. Mishin, Bulk and interface boundary diffusion in group IV hexagonal close-packed metals and alloys, Metall. Mater. Trans. A 33 (3) (2002) 765–775.
[31] O.M. Ivasishin, D.G. Savvakin, V.S. Moxson, K.A. Bondareva, F.H. Froes, Synthesis of alloy Ti-6Al-4V with low residual porosity by a powder metallurgy method, Powder Metall. Metal Ceram. 41 (7/8) (2002) 382–390.
[32] O.M. Ivasishin, D.G. Savvakin, V.S. Moxson, K.A. Bondareva, F.H. Froes, Titanium powder metallurgy for automotive components, Mater. Technol. Adv. Performance Mater. 17 (1) (2002) 20–25.
[33] O.M. Ivasishin, D.G. Savvakin, X.O. Bondareva, O.I. Dekhtyar, Synthesis of PM titanium alloys using titanium hydride powder: mechanism of densification, in: G. Lutjering, J. Albrecht (Eds.), Ti-2003 Science and Technology: Proceedings of 10th World Conference on Titanium, vol. 1, WILEY-VCH Verlag, Weinheim, 2003, pp. 495–502, 2004.
[34] O.M. Ivasishin, D.G. Savvakin, V.S. Moxson, V.A. Duz, C. Lavender, Production of titanium components from hydrogenated titanium powder: optimization of parameters, in: M. Niinomi, S. Akiyama, et al. (Eds.), Ti-2007 Science and Technology: Proceedings of 11th World Conference on Ti, Japan Institute of Metals, 2007, pp. 757–760 .
[35] O.M. Ivasishin, D.G. Savvakin, I.S. Bielov, V.S. Moxson, V.A. Duz, R. Davies, C. Lavender, BEPM synthesis of Ti-6Al-4V alloy using hydrogenated titanium, in: Proceedings of EURO PM2005 Conference (Prague, Oct. 2–5, 2005), Printed by EPMA, vol. 1, pp. 115–120.
[36] O.M. Ivasishin, D.G. Savvakin, K.A. Bondareva, et al., Synthesis of Ti-Fe and Ti-Al-Fe alloys using elemental powder blends, Metal Phys. Novel Technol. 26 (7) (2004) 963–980 (in Russian).
[37] O.M. Ivasishin, D. Eylon, V.I. Bondarchuk, D.G. Savvakin, Diffusion during powder metallurgy synthesis of titanium alloys, Defect Diffus. Forum 277 (2008) 177–185.
[38] O.M. Ivasishin, D.G. Savvakin, The impact of diffusion on synthesis of high-strength titanium alloys from elemental powder blends, Key Eng. Mater. 436 (2010) 113–121.
[39] R.R. Boyer, D. Eylon, C. Yolton, F.H. Froes, Powder metallurgy of Ti-10V-2Fe-3Al, in: F.H. Froes, D. Eylon (Eds.), Titanium Net-Shape Technologies, The Metallurgical Society of AIME, 1984, pp. 63–78 .
[40] Y. Liu, L.F. Chen, H.P. Tang, et al., Design of powder metallurgy titanium alloys and composites, Mater. Sci. Eng. A 418 (2006) 25–35.

[41] M.V. Matviychuk, Structure and properties of high-strength titanium alloys synthesized using hydrogenated titanium powder. The thesis for Ph.D. Degree, G.V. Kurdyumov Institute for Metal Physics NAS of Ukraine, Kiev, 2010.
[42] S.K. Dolukhanyan, V.D. Nersesyan, A.D. Nalbandyan, I.P. Borovinskaya, A.G. Merzhanov, Combustion of transition metals in hydrogen, Dokl. Akad. Nauk. SSSR 231 (3) (1976) 675–678.
[43] V.I. Ratnikov, I.P. Borovinskaya, V.K. Prokudina, SHS hydrogenation and subsequent dehydrogenation of titanium sponge, Int. J. Self-Prop. High-Temp. Synth. 15 (2) (2006) 193–202.
[44] V. Moxson, O.N. Senkov, F.H. (Sam) Froes, Production and characterization of titanium powder products for environmental, medical, and other applications, in: Advanced Particulate Materials and Processes – 97. MPIF, Princeton, New Jersey, 1997, pp. 387–394.
[45] V.A. Drozdenko, A.M. Petrunko, O.M. Ivasishin, et al., Method of titanium powder production. Ukrainian Patent No. 65654 (2001, 2004).
[46] V.A. Drozdenko, A.M. Petrunko, A.E. Andreev, et al., Manufacture of cost-effective titanium powder from magnesium reduced sponge. U.S. Patent 6,638,336 B1 (2002, 2003).
[47] S. Kasparov, A. Klevtsov, A. Cheprasov, et al., Semi-Continuous Magnesium-Hydrogen Reduction Process for Manufacturing of Hydrogenated, Purified Titanium Powder. U.S. Patent 8,007,562 (2008, 2011).
[48] A. Klevtsov, A. Nikishin, Yu. Shuvalov, et al., Continuous and Semi-Continuous Process of Manufacturing Titanium Hydride Using Titanium Chlorides of Different Valency. U.S. Patent 8,388,727 B2 (2010, 2013).
[49] O. Ivasishin, D. Savvakin, V. Drozdenko, et al., The method for titanium alloy component production. Ukraine Patent 70366 (2001, 2004).
[50] V.A. Duz, O.M. Ivasishin, V.S. Moxson, et al., Cost-effective titanium alloy powder compositions and method for manufacturing flat or shaped articles from these powders. U.S. Patent 7,993,577 (2007, 2011).
[51] O. Ivasishin, V. Duz, V. Moxson, et al., The method for titanium alloy component production. Ukraine Patent 92714 (2009, 2010).
[52] V. Duz, O. Ivasishin, V. Moxson, et al., The method for manufacturing of titanium alloy components. Eurasian Patent 018035 (2009, 2013).
[53] G. Abakumov, V.A. Duz, et al., High performance titanium powder metallurgy components produced from hydrogenated titanium powder by low cost blended elemental approach, in: Lian Zhou, Hui Chang, Yafeng Lu, Dongsheng Xu (Eds.), Ti-2011: Proceedings of the 12th World Conference on Titanium, Science Press, Beijing, 2011, pp. 1639–1643, 2012.
[54] G. Abakumov, V.S. Moxson, V.A. Duz, et al., Powder metallurgy titanium and titanium alloy components manufactured from hydrogenated titanium powders. Presented at ITA-2012 Conference (Oct. 7–10, 2012, Atlanta, USA), 2012.
[55] S.M. El-Soudani, K.-O. Yu, E.M. Crist, et al., Optimization of blended-elemental powder-based titanium alloy extrusions for aerospace applications, Metall. Mater. Trans. A 44 (2) (2013) 899–910.
[56] S.M. El-Soudani, M. Campbell, J. Phillips, T. Esposito, V. Moxson, V. Duz, Canless extrusion process development for blended-elemental powder-based titanium Ti-6AL-4V alloy, AeroMat 2008, June 23–26, 2008 Austin, Texas, 2008.
[57] S. El-Soudani, K.-O. Yu, E.M. Crist, et al., Development and optimization of rolled product forms using blended-elemental powder-based Ti-6AL-4V alloy. Presented at AeroMat 2012, June 18–21, 2012, Charlotte, NC, USA, 2012.

[58] S. El-Soudani, J. Fanning, M. Harper, et al., Rolled product form development and optimization using blended-elemental powder-based billets of Ti-6AL-4V alloy. Presented at AeroMat 2012, June 18–21, 2012, Charlotte, NC, USA, 2012.

[59] J.W. Adams, E. Chin, W. Roy, V.S. Moxson, V. Duz, J. Deters, C. Niese, C. Suminski, Powder metallurgy titanium trims cost of extrusion billets for making applique armor attachments, Presented at 26th Army Science Conference, Orlando, Florida, 2008.

[60] V.V. Joshi, C. Lavender, V.S. Moxson, et al., Development of Ti-6Al-4V and Ti-1Al-8V-5Fe alloys using low-cost TiH_2 powder feedstock, J. Mater. Eng. Perform. 22 (4) (2013) 995–1003.

[61] C. Lavender, V.V. Joshi, E. Stephens, et al., Development of low cost seamless CP-Ti tubes. Presented at AeroMat 2012, June 18–21, 2012, Charlotte, NC, USA.

Production of titanium by the Armstrong Process®

9

Kerem Araci, Damien Mangabhai**, Kamal Akhtar†*
*Process Development Engineer, Technology and Quality,
Cristal Metals Inc., Lockport, IL, USA
**Quality Superintendent, Technology and Quality, Cristal Metals Inc.,
Ottawa, IL, USA
†Director of Technology and Quality,
Cristal Metals Inc., Lockport, IL, USA

9.1 Process overview

The Armstrong Process® is metal halide reduction process for titanium powder production in which a gaseous stream of titanium tetrachloride is injected into a stream of liquid sodium metal to produce titanium metal and sodium chloride, according to the reaction

$$TiCl_4(g) + 4Na(l) \rightarrow Ti(s) + 4NaCl(s)$$

The resulting mixture is washed to remove the sodium chloride and is subsequently dried.

In addition to titanium tetrachloride, other metal chlorides such as aluminum trichloride and vanadium tetrachloride can be introduced into the reaction stream to produce a homogeneous, pre-alloyed Ti-6Al-4V. The process yields a powder with irregular, "coral-like" particle morphology. Figure 9.1 shows the typical morphology for Armstrong Process–produced powder. The material has a high surface area to volume ratio, which offers some unique advantages to the compressibility and sinterability, but also some challenges in overall bulk density and flow. These benefits and challenges will be discussed in greater detail in subsequent sections.

9.2 Powder characteristics

Different titanium powder applications in today's world require a variety of powder particle characteristics such as bulk density, particle size distributions, and morphology. All these characteristics have significant effects on powder flow and compaction behavior. Therefore, consolidation processes and parameter selections are determined as a function of these characteristics [1]. The Armstrong Process powders appear with dendritic "coral-like" morphology by their nature. This unique morphology leads

Titanium Powder Metallurgy. http://dx.doi.org/10.1016/B978-0-12-800054-0.00009-5

Figure 9.1 SEM images of the Armstrong Process Ti-6Al-4V powder representing unique, coral-like morphology.

to their high compressibility and low bulk density. The typical morphology of the Armstrong Process powders is shown in Figure 9.1.

Postprocessing for powder modifications plays an important role in obtaining the desired tap densities and particle size distributions for different consolidation methods. The intent in any sort of post–powder production activities is to minimize or control contamination levels while achieving desired powder properties.

Over the years, studies have shown that the ball milling of as-produced Armstrong Process powders results in significant increase in the tap density and improvement in the size distribution without going through hydride–dehydride (HDH) post processing [1–3]. Two main controllable process parameters, processing time and powder-to-media weight ratio, have significant effects on final product properties.

Figure 9.2 shows the effect of the increasing milling time on the particle size distribution (PSD) of the Armstrong Process CP-Ti powder. As the processing time

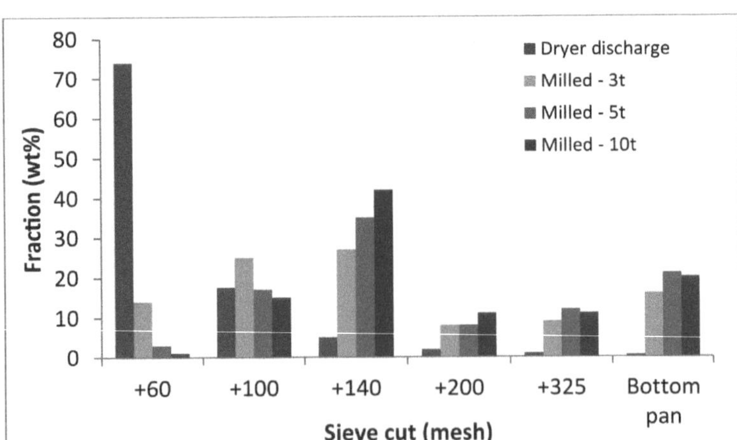

Figure 9.2 Particle size distribution change of the Armstrong Process CP-Ti powder as a function of milling time.

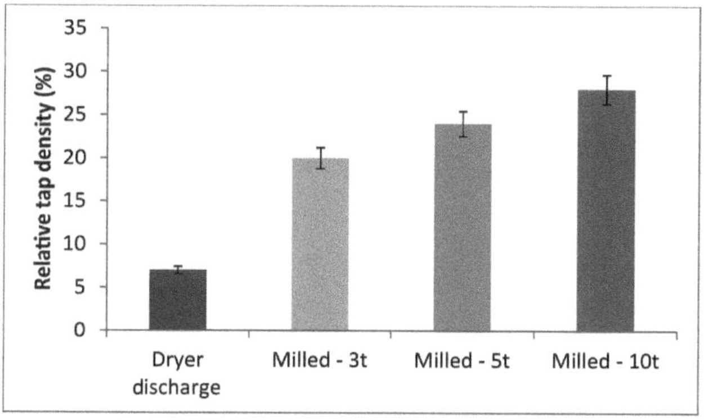

Figure 9.3 Tap density change of the Armstrong Process CP-Ti powder as a function of milling time.

increases, coarser end of the PSD breaks down and shifts to the finer end. This behavior results in a finer final mean particle, thereby increasing the bulk density, as shown in Figure 9.3 [3].

Increasing the grinding media weight in the milling container is another controllable parameter that promotes more aggressive milling behavior of the powders. In theory, increasing the grinder media weight increases the milling efficiency as a result of greater interaction between powder and the grinder media. Figure 9.4 shows the effect of increasing media weight on the Armstrong Process Ti-6Al-4V powder PSD. It is important to note that the increase in the amount of media used is limited to the design specification of the milling equipment. Powder particles tend to break up easier with increasing media presence, especially for the finer end of the PSD, in which the increase from 1/40 to 1/60 powder-to-media ratio results in more significant changes [3].

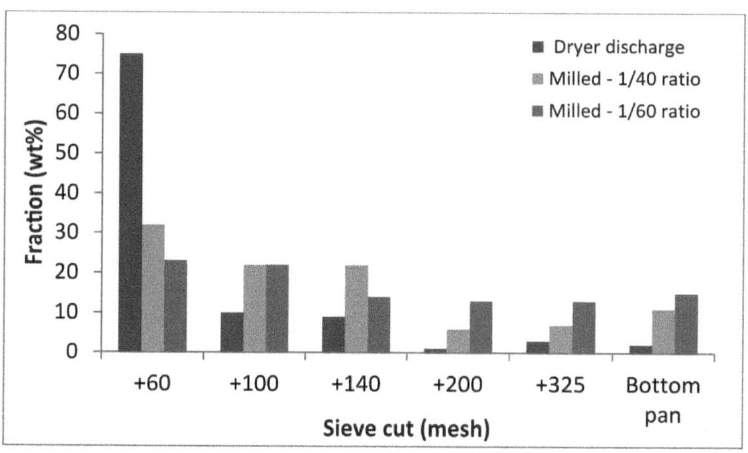

Figure 9.4 Particle size distribution change of the Armstrong Process Ti-6Al-4V powder as a function of powder-to-media weight ratio.

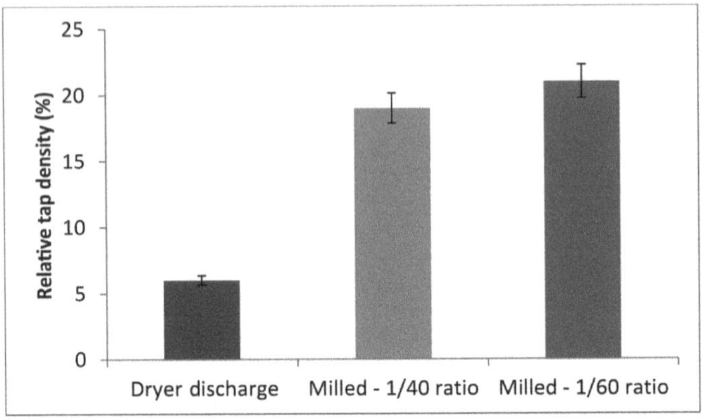

Figure 9.5 Tap density change of the Armstrong Process Ti-6Al-4V powder as a function of powder-to-media weight ratio.

Tap density is slightly affected by the media weight in the milling container for the same amount of powder. A small increase of 3% is shown in Figure 9.5.

Safety is one of the most important aspects of titanium particle size modification via dry milling. Titanium metal is highly reactive and has an extremely high affinity for oxygen. During dry milling, newly created titanium surfaces are created by the breaking down of powder particles and have a tendency to oxidize instantaneously. This behavior raises the safety concern of the autoignition of the powders via oxidation during milling or when exposed to air for the first time from an inert atmosphere [4].

Wet milling is one of the safer methods for powder particle size modification of titanium. Having a liquid component in the chamber provides a cooling mechanism for the temperature rise observed due to particle–media and interparticle friction during milling. Since titanium has a lower affinity for oxygen at low temperatures (close to room temperature), the presence of a liquid helps fresh surface areas to undergo passivation in a more controlled fashion during milling [5].

As it is for dry milling, processing time has a significant effect on particle size reduction for wet milling. Figure 9.6 shows the decreasing trend of mean particle size for two powders with different starting mean particle size as a function of increasing processing time.

Tap densities achieved via wet milling are significantly higher than those achieved by the dry milling method as presented in Figure 9.3. Figure 9.7 shows an increasing trend for the tap density as a function of the processing time.

Different types of liquids are available for wet milling processing, yielding different effects on the final product. As shown in Figure 9.8, Armstrong Process Ti-6Al-4V powders produce different mean particle size depending on whether they are milled in deionized (DI) water or heptane. Powder is broken down more aggressively in DI water compared to that in heptane. Furthermore, DI water has additional advantages over heptane in terms of safety, handling, and long-term toxicity [5].

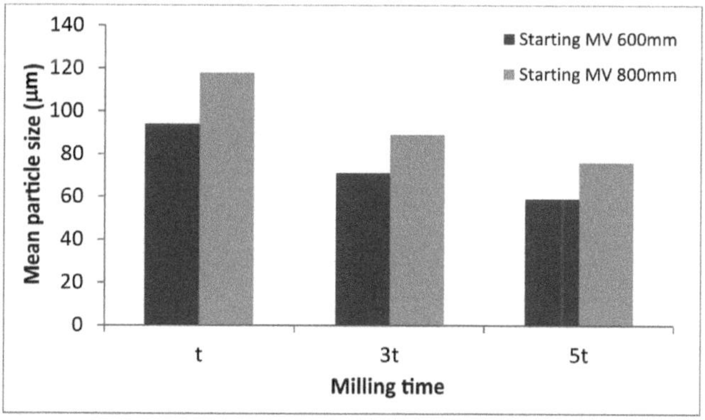

Figure 9.6 Mean particle size change of the Armstrong Process Ti-6Al-4V powder as a function of wet milling processing time.

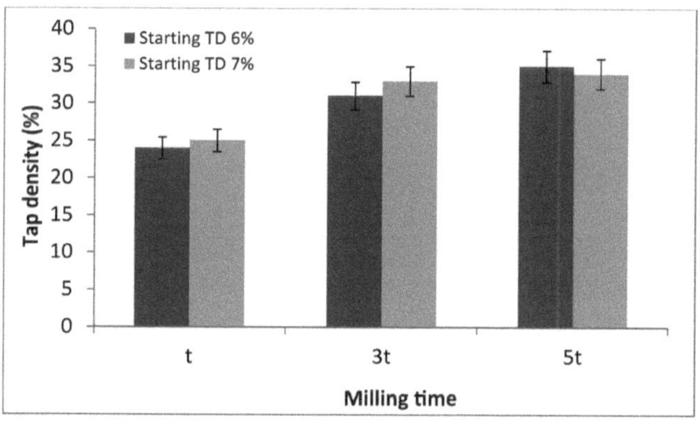

Figure 9.7 Tap density change of the Armstrong Process Ti-6Al-4V powder as a function of wet milling processing time.

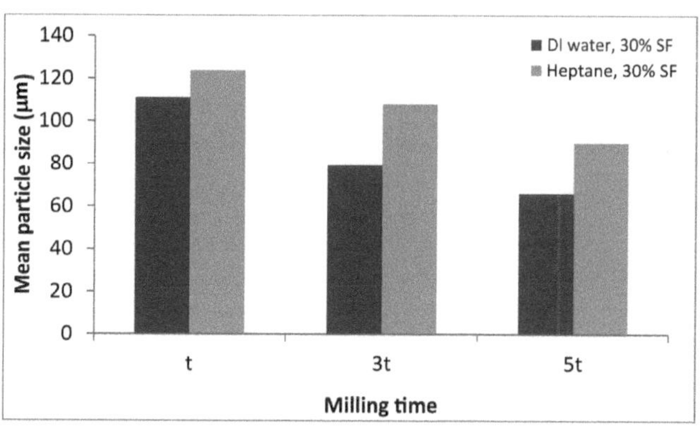

Figure 9.8 Processing liquid effect on the mean particle size of the Armstrong Process Ti-6Al-4V powder as a function of wet milling processing time.

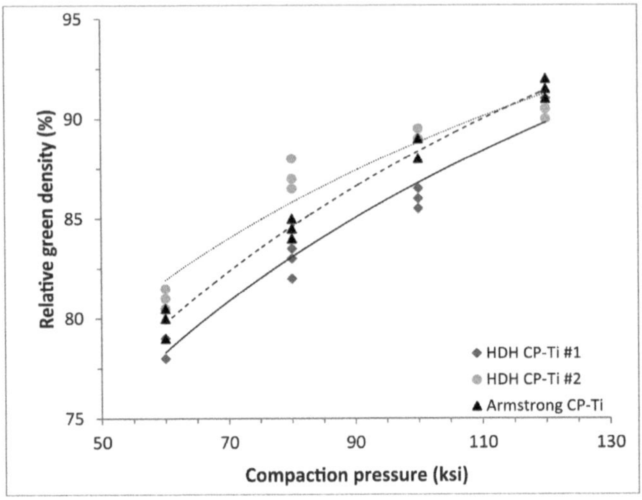

Figure 9.9 Compaction curves of CP-Ti powders.

9.3 Compaction

The unique dendritic morphology of the Armstrong Process powders in its as-produced condition presents many challenges for component fabrication using traditional powder metallurgy techniques. In some cases, however, the "coral-like" structure offers advantages over other commercially available titanium powders [5].

Cold compaction performance of titanium powders is usually evaluated by the density and strength of the green compact. There are several parameters affecting cold compaction behavior and green compact properties such as lubrication method, compaction pressure, and polyurethane bag strength in the case of cold isostatic pressing (CIP) [3,5].

Uniaxial cold compaction is the most common method for pressing metal powders into simple geometrical configurations. Compaction curves for uniaxial pressed samples are shown in Figure 9.9 for the Armstrong Process CP-Ti powders with comparison to two hydride–dehydride (HDH) powders from different sources. At the compaction pressure of 50 TSI, CP-Ti compacts achieve 84–89% relative green density [5].

Even though the green density limits are similar for different type of powders, the morphology plays an important role to achieve required green strength of the compacts. The Armstrong Process product does not go through any high temperature processing (melting or HDH) in its production. Therefore, lower residual stresses result in higher plastic deformation during cold compaction. Furthermore, the coral-like morphology supports this behavior structurally and exhibits much better interlocking behavior under mechanical forces. This provides higher green strength for the Armstrong Process green compacts at the same compaction pressure and similar density compared to HDH compacts [3,5].

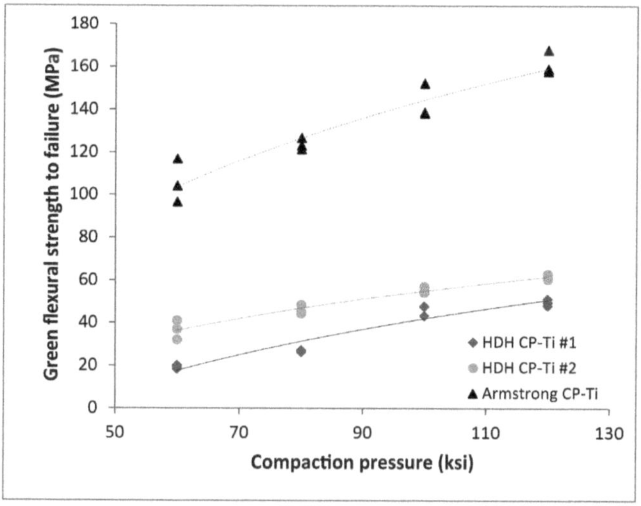

Figure 9.10 Green strength of the CP-Ti compacts as a function of compaction pressure.

Green strength comparison shown in Figure 9.10 present that the Armstrong Process powder compacts achieves more than two and a half times higher green strengths at same compaction pressures.

Green strength of the PM compacts is measured by three-point flexural test on transverse rupture strength (TRS) samples. Significantly noticeable, unconsolidated powder particles and structural voids can be seen via scanning electron microscopic (SEM) examination of the fracture surfaces of broken TRS bars produced from HDH. This is shown in Figure 9.11. However, the fracture surface of the Armstrong Process green compact is free of any definite particle boundaries, with less topographic features, serving as a proof of high-level interlocking during mechanical bonding under the applied forces. Discrete powder particles are no longer distinguishable after compaction.

Isostatic powder pressing is another mass-conserving compaction method that is broadly used for titanium powders, especially for relatively larger size and complex geometries. Compaction pressure, CIP bag strength, and feedstock powder properties are several important process parameters affecting final compact density and surface quality [5].

Compaction pressure for isostatic pressing is usually much lower than that used in uniaxial pressing of titanium powders. Studies have shown that the Armstrong Process green compacts reach approximately 85% theoretical density for CP-Ti and between 70% and 75% for Ti-6Al-4V at the optimum compaction pressures. To achieve these density levels, the CIP method requires half the compaction pressure required in uniaxial pressing. This behavior is shown in Figure 9.12 via compaction curves when compaction pressure is in kilo pounds per square inch (ksi) [3,5].

CIP bag strength is another critical parameter affecting final compact size, shape and surface finish. Commonly used CIP bags are made of very flexible polyurethane material and tend to wear out after certain number of process cycles. There is a

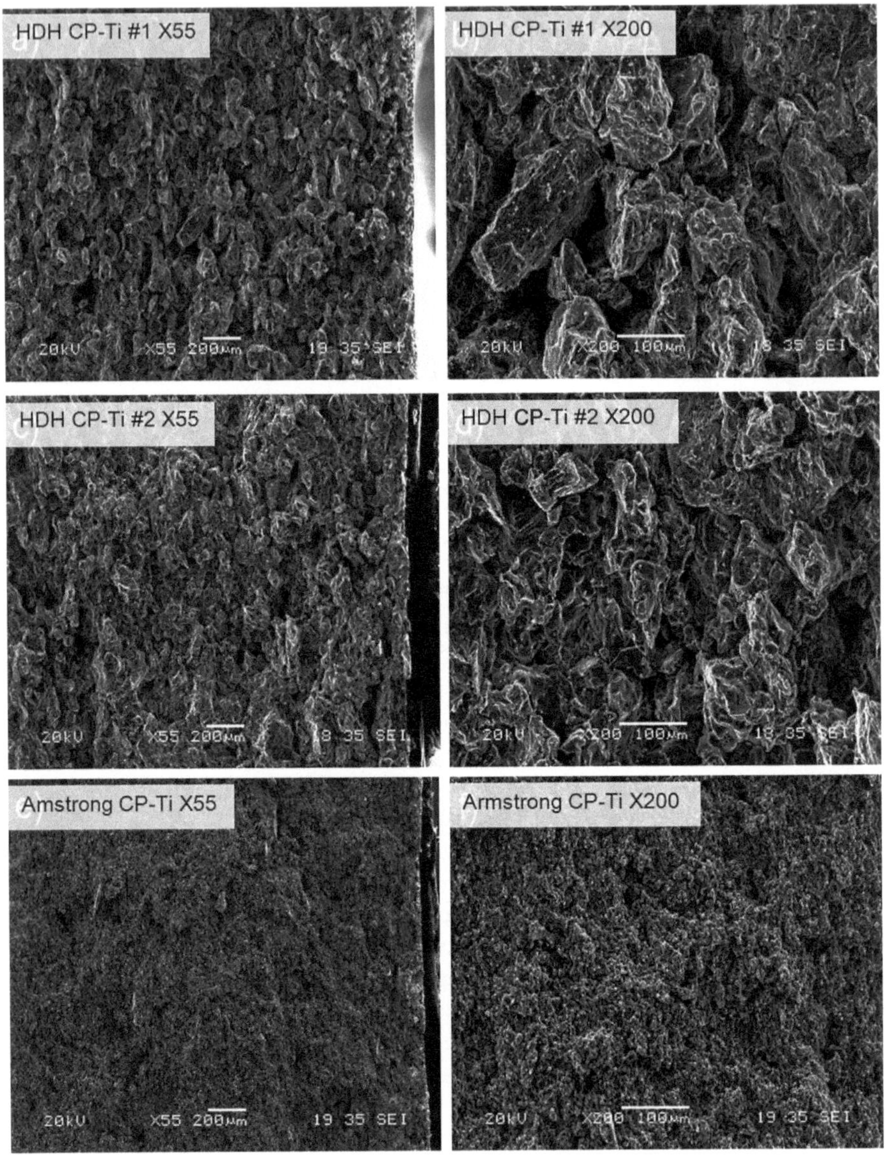

Figure 9.11 Fractography of green TRS bars.

sensitive interaction between the compaction pressure and the bag strength in terms of protecting the bag and achieving optimum product properties. Delamination and distortion of the bag is very common, which affects the cycle life significantly. These behaviors also provoke undesired shapes, rough surface finish and low levels of green density. As it is shown in Figure 9.13, it was observed that using more rigid CIP bags result in higher density and better shape and size control [3].

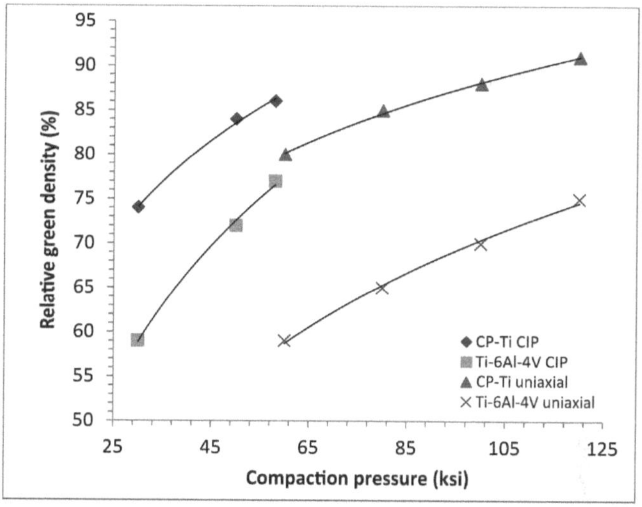

Figure 9.12 Compaction curves for the Armstrong Process green compacts: CIP versus uniaxial pressing.

In addition to the bag strength and compaction pressure, starting feedstock powder condition has also a significant effect on the green compact quality. Modified and finer particle size feedstock material tends to settle better in the CIP bag with its higher bulk density. Milled powders deliver higher-density green bars with better shape and size control under the same process conditions compared to powder in its as-produced condition. This behavior is shown in Figure 9.14.

The Armstrong Process powders can be roll compacted into a strip shape through a direct powder rolling (DPR) method. Key parameters include roll diameter, roll gap setting, roll surface coefficient of friction, rolling speeds, rolling loads, and the powder feeding system, which must provide uniform and consistent volume fill conditions. As the green sheet does not contain any organic binders, sheet strength is a critical issue, especially when operating under continuous conditions. Green strips made out of the Armstrong Process powders at the Commonwealth Scientific and Industrial Research Organization (CSIRO) provide strip lengths up to 5 m with excessive interlocking

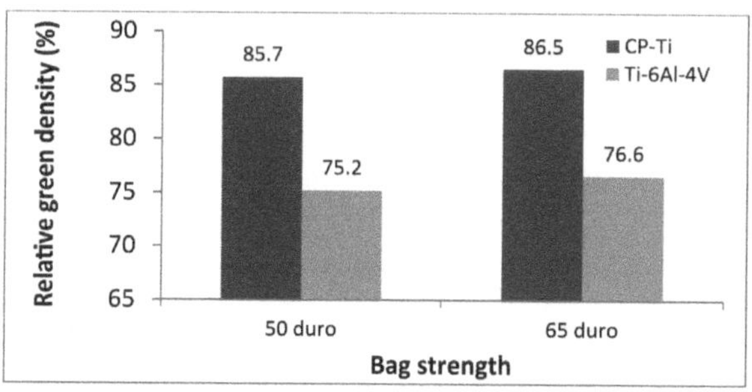

Figure 9.13 Green densities as a function of the CIP bag strength.

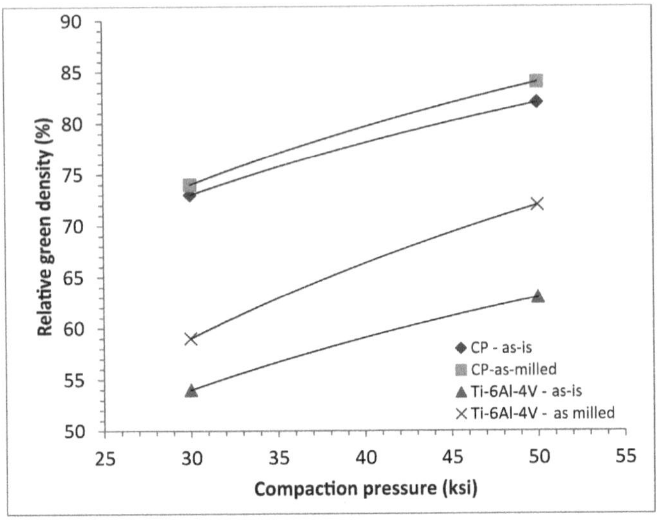

Figure 9.14 Powder condition effect on the green strength as a function of compaction pressure.

behavior and green strength. This creates critical advantages for production in terms of product quality as well as handling or transportation. The Armstrong Process green strips produced through the DPR process are shown in Figure 9.15.

In most cases, the green density increases with decreasing strip thickness in DPR process as it is seen in Figure 9.16. All green strip thicknesses range between 1 and 1.7 mm. Light-colored data points are the later trials, with optimized conditions achieving 75–80% relative green density with the average strip thickness of 1.5 mm.

9.4 Densification

Sintering is a heat treatment process applied to a powder compact in order to impart strength and integrity by densification. Sintering parameters such as temperature and dwell time at sintering temperature are very critical for the final characteristics of the near-full dense part.

Diffusion is the main mechanism that triggers the densification during sintering. Because of the dendritic "coral-like" shape, the Armstrong Process powders have a larger surface area than that of HDH powder. With lower sintering activation energy because of the larger surface area, the sintering characteristics of the Armstrong Process powders are significantly different from that of HDH powders [2,3].

As shown in Figure 9.17, the Armstrong Process Ti-6Al-4V compact starts densifying via diffusion at much earlier stages of sintering process compared to those of HDH blended elemental (BE) or pre-alloyed (PA) powder compacts. This encourages less aggressive process conditions for sintering such as lower sintering temperatures or shorter dwell times for the Armstrong Process compacts. Furthermore, higher shrinkage rate promotes higher final density at more than 99.5% [2,6].

Figure 9.15 The Armstrong Process CP-Ti green strips.

Densification of the Armstrong Process compacts is shown via compaction curves in Figure 9.18. Because of the harder nature of Ti-6Al-4V powders, lower plastic deformation during cold compaction results in lower green density compared to those of CP-Ti compacts. However, the higher diffusion rate in β-Ti plays an important role in increasing the final density of Ti-6Al-4V compacts significantly greater than that of CP-Ti through sintering.

Increasing density during sintering elevates the structural stability of the part as shown in Figure 9.19 with strength curves at different compaction pressures. At the

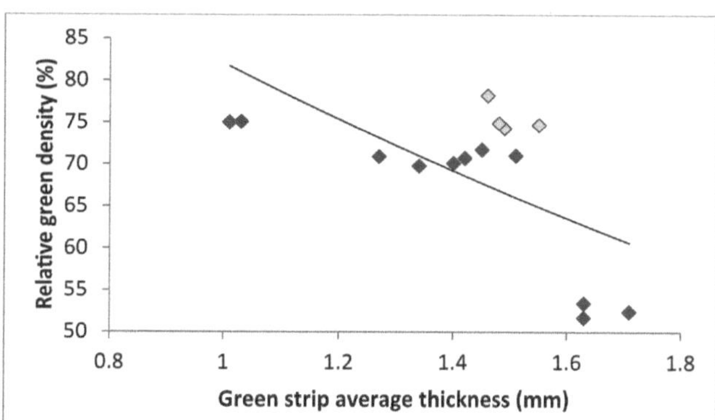

Figure 9.16 Green density–thickness relationship for the Armstrong Process CP-Ti strips.

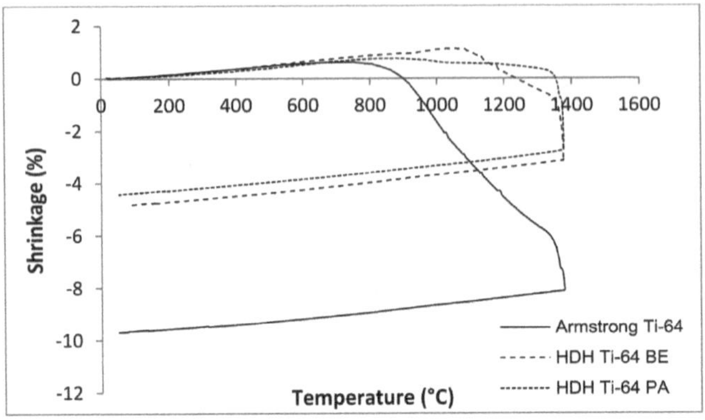

Figure 9.17 Dilatometry curves showing densification behavior during sintering.

optimum compaction pressure, the strength achieves more than 2000 MPa for the Armstrong Process Ti-6Al-4V parts in sintered state with 97% theoretical density [6,7].

9.5 Spheroidization

Powder densification via spheroidization is widely employed to transform the morphology of either agglomerated or angular powders produced by conventional crushing methods into spherical powder. This technique is primarily used to improve the flow properties of the powders. Spherical powders are ideal for applications such as powder injection molding, thermal spraying of coatings, three-dimensional printing (additive manufacturing), or manufacturing of near-net-shape parts.

The Armstrong Process powders can be spheroidized by induction plasma technology to modify the morphology of powders to achieve better bulk density and flow

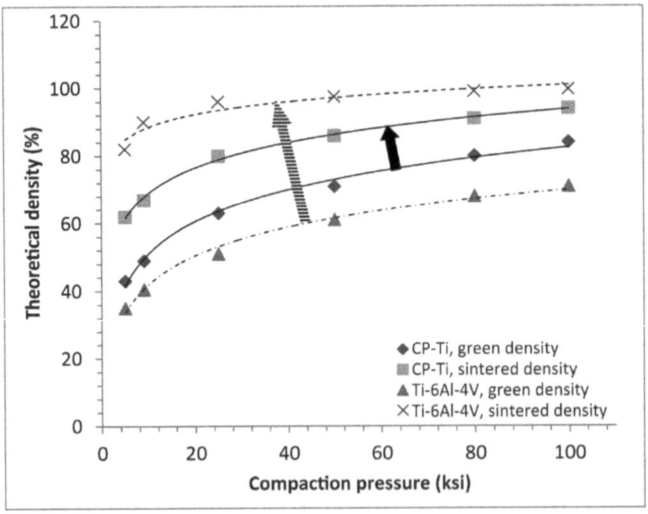

Figure 9.18 Compaction and densification curves of the Armstrong Process powders.

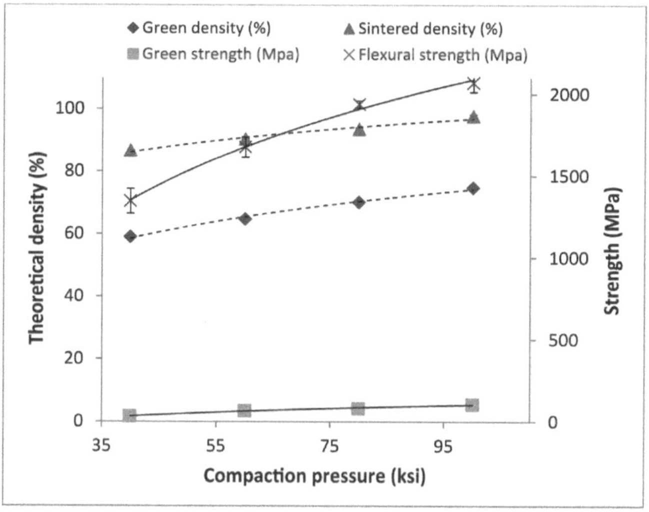

Figure 9.19 Density and strength improvement of the Armstrong Process Ti-6Al-4V from green to near-full dense state.

capabilities. It is feasible to achieve considerable improvement on the Armstrong Process powders properties in multiple aspects such as powder flow, decreased porosity, increased powder density, reduced powder friability and in some cases improved powder chemistry.

Powders with different-size fractions can be spheroidized through the process. Feedstock particle size characteristics are transferred to the final spherical product without any significant changes. Feedstock and spherical product morphologies are shown in Figure 9.20 for the Armstrong Process CP-Ti with a particle size range of 45–105 μm [5].

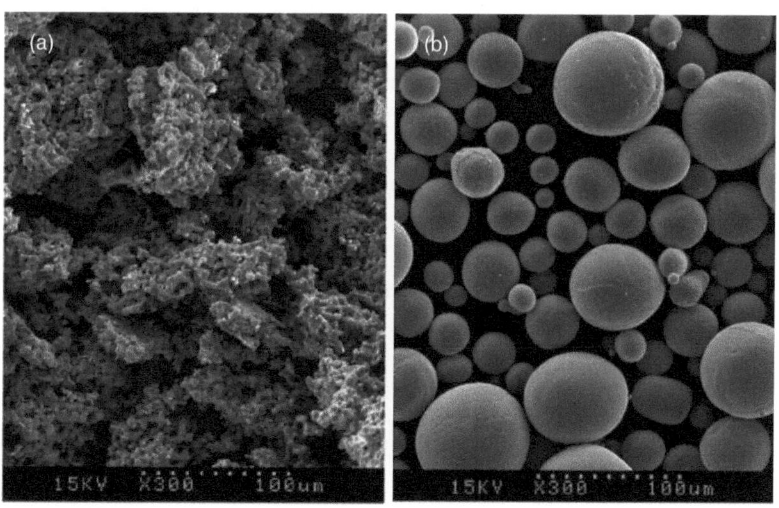

Figure 9.20 (a) SEM images of the feedstock material, (b) SEM images of the final product – induction plasma spheroidization.

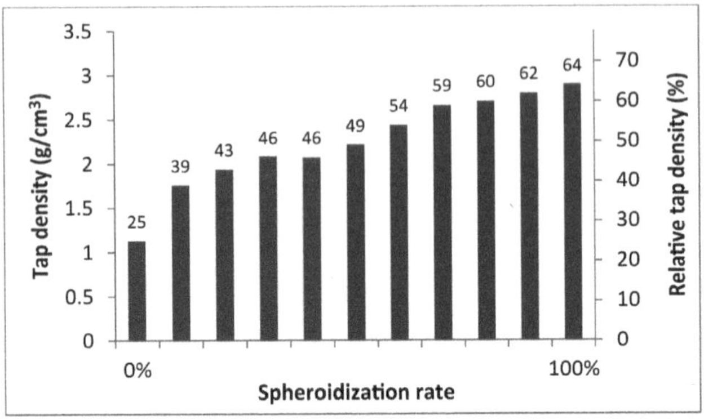

Figure 9.21 Improving density as a function of spheroidization rate.

Spherical powder density can be controlled by the spheroidization rate via process parameters. In Figure 9.21, it is shown that an increased spheroidization rate provokes a higher number of fully spheroidized particles, and therefore higher final density. Feedstock material starts from 25% theoretical tap density and reaches more than 60% for fully spheroidized products [5].

References

[1] M. Qian, Cold compaction and sintering of titanium and its alloys for near-net-shape or preform fabrication, Int. J. Powder Metall. 46 (5) (2010) 29–44.

[2] X. Xu, K. Araci, P. Nash, T. Zwitter, Cost effective press and sinter of titanium and Ti-6Al-4V powder for low cost components in vehicle applications, PowderMet, June 27–30, 2010, Hollywood, FL.

[3] K. Araci, K. Akhtar, D. Mangabhai, From powder to low-cost parts, The 12th World Conference on Titanium (Ti-2011), June 19–24, 2011, Beijing, China.

[4] D. Mangabhai, K. Araci, K. Akhtar, N. Stone, D. Cantin, Y. Yamamoto, T. Muth, Processing of titanium into consolidated parts and sheet, TMS, March 3–7, 2013, San Antonio, TX.

[5] K. Araci, K. Akhtar, D. Mangabhai, Cost affordable titanium powder metallurgy components, PowderMet, June 27–30, 2012, Hollywood, FL.

[6] X. Xu, P. Nash, T. Zwitter, K. Araci, K. Akhtar, Single press and sinter of Armstrong process Ti powder, The 12th World Conference on Titanium, June 19–24, 2011, Beijing, China.

[7] X. Xu, P. Nash, Y. Yan, K. Araci, D. Mangabhai, T. Zwitter, Single press and sinter of Ti and Ti6Al4V powders, PowderMet, June 24–27, 2013, Chicago, IL.

Hydrogen sintering of titanium and its alloys

10

James D. Paramore, Z. Zak Fang, Pei Sun
Department of Metallurgical Engineering, University of Utah, Salt Lake City, UT, USA

10.1 Introduction

Hydrogen sintering and phase transformation (HSPT) is a blended elemental (BE) press and sinter process for the production of titanium alloys [1,2]. During this process, compacts of hydrogenated titanium, metallic titanium, and/or master alloy powders are sintered under a dynamically controlled hydrogen partial pressure. The use of hydrogenated powder and the presence of hydrogen in solution during sintering has been shown in several studies to significantly improve purity and densification of PM titanium alloys [3–8]. However, HSPT differs from other PM processes in that a dynamically controlled hydrogen partial pressure is used to control the hydrogen content within the alloy during sintering. This is done as a means to refine the microstructure during sintering via phase transformations in the Ti–H alloy systems. The goal of HSPT is to produce engineered microstructures and application-tailored mechanical properties in the as-sintered state, thereby eliminating thermomechanical work compulsory to traditional titanium-processing routes.

Figure 10.1 is a theoretical flow sheet of processing steps necessary to produce titanium alloys with high strength and ductility via melt-wrought processing, traditional PM, and HSPT. PM has long been sought as a means to decrease the cost of titanium alloys [9]. As a convention, titanium PM can be grouped into two generalized processes: BE and prealloyed (PA). PA generally produces better mechanical properties, but the PA powders tend to be relatively expensive. On the other hand, BE titanium alloys often have relatively poor mechanical properties in the as-sintered state because of high interstitial concentrations, residual porosity, and undesirable microstructures [9–11]. These properties can be improved through the implementation of special powder handling techniques to minimize oxygen content, pressure-assisted sintering such as hot isostatic pressing (HIP) to improve densification, and thermomechanical work after sintering to close porosity and refine the microstructure through recrystallization [10]. However, these additional processing steps tend to be energy-intensive and increase cost. Therefore, the goal of HSPT is to maximize the performance-to-cost ratio of titanium by eliminating the need for these types of processing steps in a BE process [12]. Finally, whereas in other titanium PM processes vacuum is the preferred sintering atmosphere [13], HSPT can be conducted at atmospheric pressure. Because of this, the possibility exists for HSPT to be commercialized as a continuous process.

HSPT should not be confused with hydrogen sintering processes used commercially for other alloy systems such as steels, tungsten alloys, etc. During sintering of

Titanium Powder Metallurgy. http://dx.doi.org/10.1016/B978-0-12-800054-0.00010-1

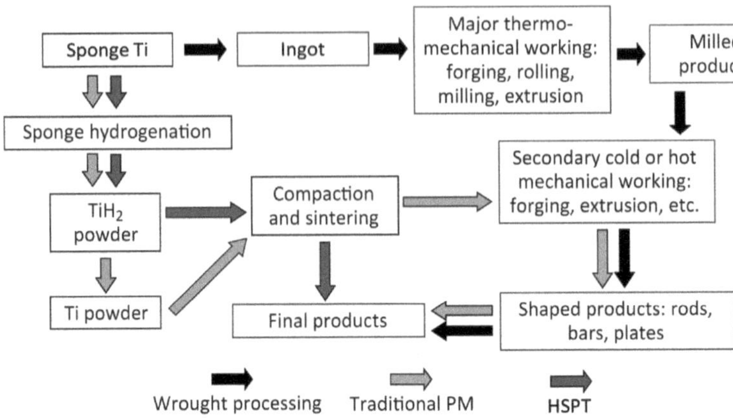

Figure 10.1 Flow sheet of a titanium melt-wrought process, a traditional PM process, and the HSPT process.

these alloys, hydrogen is chosen for its reducing character [14]. However, titanium lies below hydrogen on the Ellingham diagram [15]. As such, hydrogen is not a viable reducing agent for titanium and plays a very different role in HSPT. As mentioned above, the hydrogen partial pressure is used to control the hydrogen content within the titanium alloy during sintering. Because the phase equilibria and transformation kinetics of titanium alloys varies as a function of hydrogen content, the hydrogen partial pressure is employed as an additional parameter to control microstructural evolution during sintering and heat treatments. The mechanisms and implications of these phenomena are discussed in detail later in this chapter.

10.2 Background and history

Hydrogen embrittles titanium and is, therefore, generally an undesirable component in titanium alloys. However, because of rapid diffusion and the reversible nature of the hydrogenation reaction, hydrogen can be easily removed by annealing in vacuum or inert gas atmospheres. Additionally, the presence of hydrogen has several benefits for a PM titanium process. As such, there is a long history of using hydrogenated titanium in a PM process.

10.2.1 History of using hydrogenated titanium in PM

The hydride–dehydride (HDH) process is a well-known, commercial process for titanium powder production. HDH takes advantage of the brittleness of titanium hydride to improve comminution behavior in powder production [16]. However, hydrogenated titanium is also employed in sintering processes. The use of hydrogenated titanium versus metallic titanium results in better control of oxygen content in the sintered sample [17,18]. Additionally, titanium hydride exhibits exceptional compaction behavior, resulting in excellent green density and strength [7,19]. Finally, densification during sintering is significantly improved when hydrogenated titanium powder is used [20]. For these reasons, there is a long history of hydrogenated titanium being used as the feedstock material for a sintering process.

The first patent to use hydrogenated titanium in a sintering process, filed by Gregory, dates back nearly 50 years to 1967 [21]. This patent describes a method in which titanium hydride powder is used in the production of niobium-based superconductor alloys. Shortly afterwards, a process to produce titanium alloys and metal matrix composites using titanium hydride powder was developed at the United States Army Materials and Mechanics Research Center. This process was called "decomposition powder metallurgy" and was first reported in 1970 by Greenspan et al. as a means to produce titanium alloys and metal matrix composites via hot pressing [22,23]. The first press and sinter process to produce titanium alloys from titanium hydride was patented by the Japanese team of Obara et al. in 1974 [24]. In the Japanese process, sintering was achieved under vacuum or inert gas atmospheres. Following these pioneering publications and patents, a wide variety of processes have been published and/or patented that use hydrogenated titanium powder. These processes include both pressure-assisted (i.e., hot pressing or HIP) as well as press and sinter methods, and have been used to produce a range of products, including titanium alloys [19,25–31], sputtering targets [32], decorative titanium products [33], and shape-memory alloys [34]. Since 2000, a significant amount of work regarding vacuum sintering of hydrogenated titanium powder has been reported by Ivasishin et al. [3,4,8,17,20,35–38]. This work has led to a series of processes developed and commercialized by an Ohio-based company called The Advanced Material Group (ADMA). ADMA has patented several processes for the production of hydrogenated titanium powder as well as vacuum sintering of these powders to produce titanium alloys and metal matrix composites [18,39–43].

10.2.2 History of thermo-hydrogen processing

Utilizing the phase transformations in the Ti–H system as a means to refine microstructure and improve mechanical properties of titanium alloys has a history in the literature as well. Thermo-hydrogen processing (THP) and thermochemical processing are the names given to a group of processes during which hydrogen is used as a temporary alloying element. In some processes, hydrogen is used to improve the workability of titanium. In others, THP utilizes the phase transformations in the Ti–H system to refine the microstructure of titanium alloys. During THP, bulk titanium alloys are hydrogenated, subjected to heat treatments and/or hot working, and subsequently dehydrogenated.

The use of hydrogen as a temporary alloying element was first patented in 1959 by Zwicker et al. as a means to improve the workability and deformability of titanium during hot working [44]. The first studies to investigate THP as a means to refine microstructure via phase transformations and the effect on mechanical properties were conducted by Kerr et al. in the late 1970s and early 1980s [45,46]. Since then, many investigations of THP have been published [47–68], including a review paper by

Froes et al. published in 2004 [69]. THP has been used on both wrought material and sintered PM material as a means to refine the microstructure and improve mechanical properties. Across these studies, THP consistently increased strength in both wrought and PM titanium. In some studies, the increase in strength was accompanied by a decrease in ductility [47,48,50,65,67,68]. However, other studies reported an increase in strength as well as ductility [46,56,61].

10.2.3 Ti–H phase diagram

When studying the Ti–H phase diagram (Figure 10.2) [70], one realizes it bears a resemblance to the Fe–C system, particularly the famous eutectoid reaction that is responsible for the vast array of steel microstructures available from that system. As mentioned in the previous section, these phase transformations were the motivation for THP research. A hypothesis and concept was thus proposed that the effects of hydrogen on phase transformations could be incorporated in sintering of titanium to control microstructural evolution during sintering.

As mentioned above, the Ti–H system contains a eutectoid transformation. Therefore, when cooling the material below the eutectoid temperature at a controlled rate and with a controlled concentration of hydrogen in the system, a eutectoid decomposition of the β-phase will occur. If this is done at a sufficient rate, a high degree of undercooling will provide significant driving force to precipitate α-Ti and δ-TiH_2, but

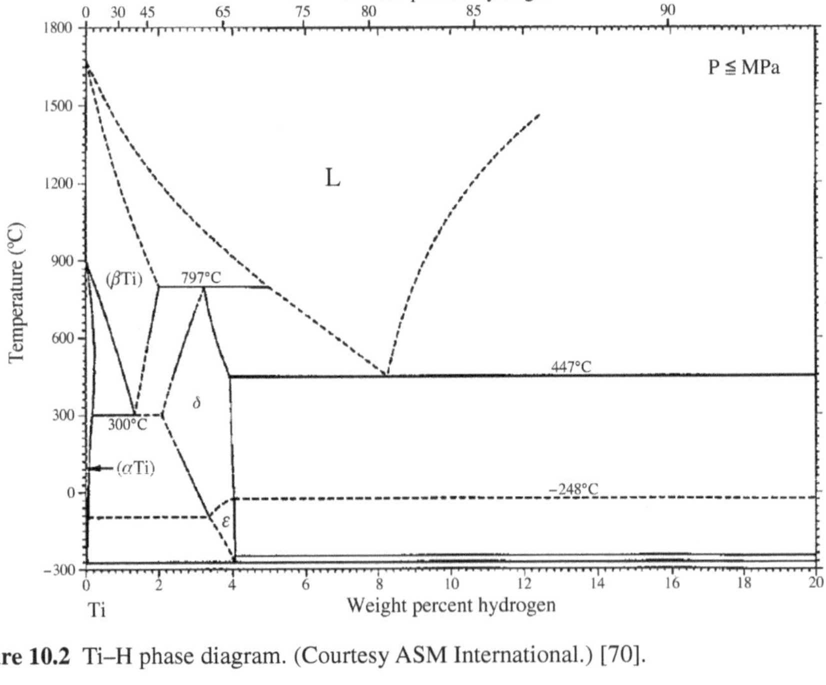

Figure 10.2 Ti–H phase diagram. (Courtesy ASM International.) [70].

limit grain growth in the new phases. Other phase transformations are also possible, such as the formation of pro-eutectoid α and pro-eutectoid δ-TiH_2 for hypoeutectoid and hypereutectoid compositions, respectively. Additionally, α_2-Ti_3Al will precipitate if the temperature is held below the solvus temperature [71], or if α becomes supersaturated with aluminum because of large volume fractions of β and δ-TiH_2. Hydrogen is a β stabilizer. Therefore, the volume fractions of β and δ-TiH_2 are directly related to hydrogen concentration. Additionally, the β-transus temperature is also dependent on the hydrogen concentration. Therefore, all of these phase transformations may be controlled by adjusting the hydrogen concentration as a function of temperature and hydrogen partial pressure during sintering and subsequent cooling.

10.2.4 Hydrogen concentration versus hydrogen partial pressure

The hydrogen concentration within titanium changes as a function of both temperature and hydrogen partial pressure. As an example, the equilibrium hydrogen partial pressure of β-Ti(H), β phase with dissolved hydrogen, is given in Equation 10.1:

$$\text{In}(p_{\text{H}_2}) = 15.84 + 0.7C + 2\text{In}\left(\frac{C}{2-C}\right) - \frac{16.0\times10^3}{T} \quad (10.1)$$

where p_{H_2} is the equilibrium partial pressure of hydrogen, C is the H/Ti ratio, and T is absolute temperature [72].

Because hydrogen is a volatile alloying element, using the phase diagram to design heat treatments is not as straightforward for Ti–H as it is with other eutectoid systems, such as Fe–C. To subject a steel specimen to a purely eutectoid reaction, one would simply prepare an alloy of the eutectoid composition, austenitize the material by holding above the eutectoid temperature, and then cool the material below the eutectoid temperature. However, since the composition of the Ti–H system is dependent on hydrogen partial pressure, the phase equilibria and transformation kinetics change as a function of hydrogen partial pressure as well as temperature. Therefore, controlling phase transformations in the Ti–H system requires dynamic control of the atmosphere in addition to temperature.

Example: Assume a titanium specimen was held above the eutectoid temperature in the β-Ti(H) phase field under a hydrogen partial pressure and temperature that corresponded to the eutectoid composition. Upon cooling, the hydrogen concentration within the specimen would increase if the atmosphere was unchanged. Therefore, a cooling curve superimposed on the phase diagram would fall to the right of the eutectoid point, and pro-eutectoid δ-TiH_2 would form. If it was desired that the cooling curve passed through the eutectoid point to produce a fully eutectoid reaction, the hydrogen partial pressure must be simultaneously decreased with temperature to maintain the eutectoid composition during cooling.

This additional process parameter, coupled with the rapid diffusion rates of hydrogen, is a unique advantage to working with the Ti–H system. While this certainly makes designing an HSPT process more complicated, it also offers an additional degree of freedom to control the microstructural evolution during heat treatments. Gas carburization of steel is an analogous phenomenon [73]. However, the diffusion kinetics of carbon in steel limit that process to surface treatments and the carbon is not a temporary alloying element. Because the diffusion of hydrogen in titanium is rapid, HSPT is capable of engineering bulk properties of titanium alloys via atmospheric control.

10.2.5 Pseudo-binary phase diagram

The phase diagram shown in Figure 10.2 is only valid as a reference for unalloyed titanium. In order to fully understand and predict the phase transformations of an alloy during HSPT, such as Ti-6Al-4V, one needs a pseudo-binary phase diagram. In such a diagram, the abscissa is the hydrogen concentration within the alloy and the origin represents the pure alloy free of hydrogen. It is important to realize that a pseudo-binary phase diagram in actuality represents a system with more than two components. Therefore, by Gibb's Phase Rule, the number of phases possible in each field is increased by the number of additional components in the alloy beyond titanium and hydrogen. As an example, the (Ti-6Al-4V)–H diagram can have phase fields with four phases in equilibrium.

Multiple studies have endeavored to generate the (Ti-6Al-4V)–H pseudo-binary phase diagram as a reference tool for THP studies [45,52,60,74]. The results of these studies have been superimposed in Figure 10.3 to show the general form of the phase diagram. Generating such a phase diagram is difficult, resulting in contradictory findings between the publications. As mentioned earlier, the equilibrium concentration of hydrogen within the alloy varies as a function of temperature and pressure, meaning care must be taken to determine the instantaneous concentration of hydrogen within the alloy at a given temperature. One study endeavored to

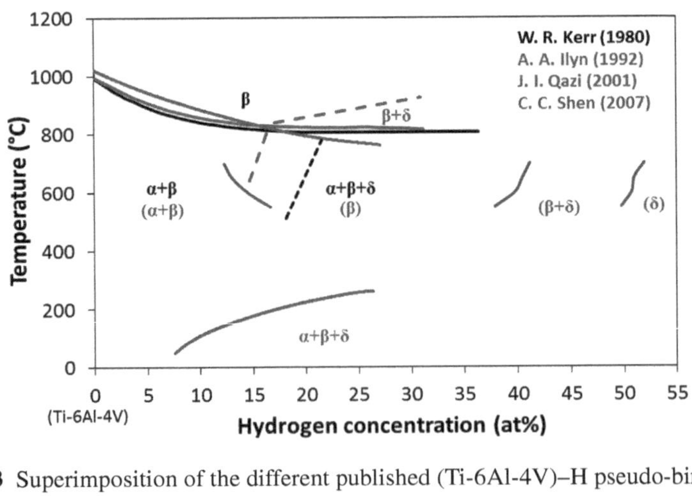

Figure 10.3 Superimposition of the different published (Ti-6Al-4V)–H pseudo-binary phase diagrams [45,52,60,74].

stabilize the hydrogen concentration by oxidizing the sample surface after hydrogenation to create a hydrogen diffusion barrier [60]. However, the rapid diffusion of hydrogen in titanium would likely result in an appreciable degree of dehydrogenation during the oxidation step. Additionally, when small samples are studied, oxygen, which is an α stabilizer, would diffuse into the sample at high temperatures and affect the phase transformations. Furthermore, quenching is not a viable solution to this problem because of the rapid diffusion. Lastly, rapid phase changes during cooling make observation of the high-temperature phases very difficult without *in situ* methods.

To address the difficulty in obtaining a reliable phase diagram for the (Ti-6Al-4V)–H system, extensive experimentation has been performed by Sun et al. to observe the phase evolution of this system in real-time as a function of temperature and hydrogen concentration [75]. These experiments have been performed using *in situ* X-ray diffraction (XRD). A custom-designed tube furnace with atmospheric control was installed in the beamline of the synchrotron at the Advanced Photon Source (Beamline 11-ID-C) located at the Argonne National Lab. The increased resolution of the synchrotron was necessary to effectively differentiate between α-Ti and α_2-Ti_3Al peaks, which lie nearly on top of one another on the diffractogram. Complimentary experiments were also conducted using a Panalytical X-ray diffractometer with an installed hot-stage. Additionally, extensive experimentation was performed using thermogravimetric analysis (TGA) and differential scanning calorimetry (DSC). By combining the real-time XRD phase data with the equilibrium hydrogen concentration data from TGA, an updated pseudo-binary phase diagram has been generated. The results of this study are currently being prepared for publication.

10.2.6 Phase transformation kinetics

Another important consideration regarding microstructural evolution during HSPT is the kinetics of the phase transformations. It has been reported that the kinetics of the phase transformations in (Ti-6Al-4V)–H change as a function of hydrogen concentration. Qazi et al. reported that as the hydrogen concentration is increased, the nose temperature is decreased and the nose time is increased for the time–temperature–transformation (TTT) curve of the eutectoid reaction [60]. Therefore, as hydrogen concentration is increased, the kinetics slow and the degree of undercooling required to maximize kinetics is increased.

The dependence of hydrogen concentration on temperature and atmosphere adds an additional degree of complication to the problem of kinetics as well. If the hydrogen partial pressure is held constant, according to Equation 10.1, the hydrogen concentration in titanium is increased as the temperature is decreased. However, as the hydrogen concentration is increased, the nose temperature of the TTT curve is simultaneously decreased. Therefore, when trying to identify the temperature at which transformation kinetics are maximized, one is chasing a moving target. To overcome this complication and determine the temperature at which transformation kinetics are maximized, a curve for the hydrogen concentration as a function of temperature at a constant hydrogen partial pressure (Figure 10.4a) and a curve for the nose temperature

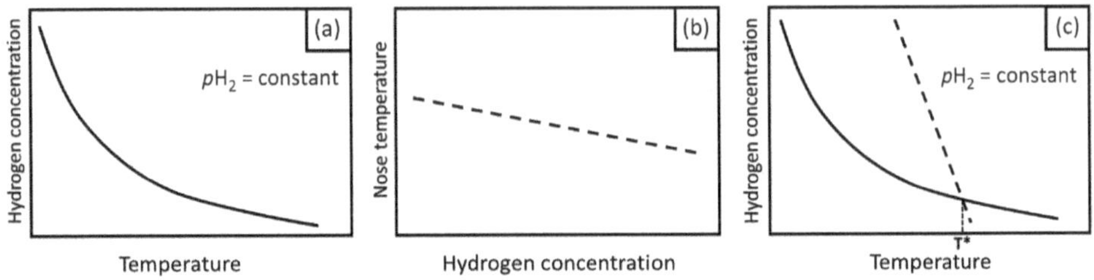

Figure 10.4 Schematic plots of (a) equilibrium hydrogen concentration as a function of temperature for a given hydrogen partial pressure, (b) nose temperature of the eutectoid reaction as a function of hydrogen concentration, and (c) superimposition of a and b (axes flipped) to determine temperature at which transformation kinetics are maximized.

as a function of hydrogen concentration (Figure 10.4b) are needed. The intersection of these two curves would indicate the temperature at which transformation kinetics are maximized for a given hydrogen partial pressure (Figure 10.4c).

10.3 HSPT process description

Typically, HSPT is a BE PM process. Of course, PA powders could also be used with HSPT. However, PA powders are generally more expensive and are unnecessary based on current results from the BE method. The most common HSPT practice is to blend a commercially pure titanium hydride (CP-TiH_2) powder with a master alloy powder via mechanical mixing to produce the desired titanium alloy. In the case of Ti-6Al-4V, CP-TiH_2 is blended with a 9.6 wt% 60Al-40V master alloy. Therefore, after dehydrogenation, the alloy will have the appropriate composition (90 wt% Ti, 6 wt% Al, and 4 wt% V). The CP-TiH_2 can come from traditional and/or low-cost sources including virgin metal from Kroll or an alternative extractive process, recycled scrap or turnings, and/or sponge fines. It is also possible to use a low-cost, PA source, such as Ti-6Al-4V machine turnings that have been hydrogenated [76]. The excellent milling behavior of hydrogenated titanium make it easy to produce a powder of the required size from almost any source. The only important concern when sourcing a powder would be the initial cleanliness of the feedstock.

10.3.1 Sintering

During HSPT, the partial pressure of hydrogen is dynamically controlled throughout the thermal cycle. The desired hydrogen partial pressure is produced either by mixing hydrogen gas with an inert gas at atmospheric pressure or by creating a partial vacuum on a pure hydrogen atmosphere. The thermal profile for an HSPT process consists of three general steps, detailed below:

Step 1. Sintering: During this step, sintering is achieved at a temperature above the β-transus (1000–1300°C) and a relatively high hydrogen partial pressure, generally 1–100 kPa (0.1–1 atm). Higher hydrogen partial pressures, up to 1 MPa (10 atm), can be used to increase hydrogen content at the sintering temperature, but this is generally not required. At these temperatures and pressures, the material is entirely β-Ti(H), β phase with dissolved hydrogen. The high temperature and presence of hydrogen in solution during this step facilitates near full densification, resulting in less than 1 vol% porosity after sintering.

Step 2. Homogenization and Phase Transformation: The second step in HSPT refines the microstructure by taking advantage of phase transformations present in Ti–H phase diagrams. During this step, the hydrogen partial pressure is controlled to accommodate a higher or lower hydrogen concentration. Because the thermodynamics and kinetics of the phase transformations are a function of hydrogen concentration, the resulting microstructure can be controlled via the hydrogen partial pressure as well as the temperature profile. At the sintering temperature, titanium has very low hydrogen solubility. Therefore, depending on the alloy and desired microstructure, a homogenization step may be included to increase the overall hydrogen content by homogenizing at a temperature below the sintering temperature but above the phase transformation temperature. The material is then cooled to a multiphase field to allow β-Ti (*bcc*) to decompose into low-temperature phases such as α-Ti (*hcp*) and δ-TiH_2 (*fcc*). Under certain circumstances, α_2-Ti_3Al (ordered *hcp*) can also form, if desired. Nucleation of the low-temperature phases refines the microstructure. Additionally, δ-TiH_2 ($\rho_{TiH_2} = 3.75\,g/cm^3$) is significantly less dense than titanium ($\rho_{Ti} = 4.51\,g/cm^3$). Therefore, the formation of a significant volume fraction of δ-TiH_2 produces internal strain that can lead to subsequent grain refinement due to recrystallization during dehydrogenation or other subsequent heat treatments.

Step 3. Dehydrogenation: In the final step of the process, the hydrogen is removed at a temperature below the β-transus under a hydrogen-free atmosphere. This is achieved either by vacuum or by flowing an inert gas essentially free of hydrogen. This step may be performed either in a separate dehydrogenation furnace or in the same furnace immediately following Step 2 by adjusting the atmosphere and temperature. During dehydrogenation, further grain refinement is facilitated by the decomposition of δ-TiH_2 to α and β, as well as recrystallization due to internal strain caused during the phase transformation step. After dehydrogenation, hydrogen levels well below the ASTM standard for Grade 5 T-6Al-4V (150 ppm) [77] are easily achieved using inert gas or low vacuum ($>10^{-1}$ Pa). Hydrogen levels below 10 ppm have been achieved by using high vacuum ($<10^{-2}$ Pa).

The microstructure of the as-sintered material can be further modified by incorporating simple heat treatments following dehydrogenation. Heat treatments of titanium are widely used commercially [71]. Therefore, a large body of information exists that could be incorporated into an HSPT process. Preliminary studies have observed the formation of equiaxed α grains during heat treatments following dehydrogenation, resulting in bimodal microstructures. However, heat treating results of HSPT-produced titanium are currently limited.

10.4 Typical results

For all experimental results discussed in this section, compacts were prepared by BE methods to produce the most widely used titanium alloy, Ti-6Al-4V. For Ti-6Al-4V sintered from TiH_2, the blend was 90.4 wt% CP-TiH_2 blended with 9.6 wt% 60Al-40V master alloy. When Ti powder was used, the blend was 90 wt% CP-Ti blended with 10 wt% 60Al-40V master alloy. All powder (CP-TiH_2, CP-Ti, and master alloy) was sieved below 44 μm (–325 mesh) and was supplied by Reading Alloys (AMETEK).

10.4.1 Densification

Figure 10.5 shows typical dilatometry curves for Ti-6Al-4V sintered from blended CP-TiH_2 and master alloy. The sample represented by the blue curve was sintered under a 40.5 kPa (0.4 atm) hydrogen partial pressure, while the sample represented by the red curve was sintered under a 20.3 kPa (0.2 atm) hydrogen partial pressure. During both experiments, the hydrogen partial pressure was controlled by mixing hydrogen and argon gases with programmable mass flow controllers at atmospheric pressure. The hydrogen mass flow controller was turned off before the final cooling segment to prevent the formation of excess δ-TiH_2, which would lead to cracking of the sample.

Both samples expand during the initial stages of the heat-up due to thermal expansion. As the temperature reaches approximately 450–500°C, the hydride begins to rapidly decompose, causing the samples to begin to shrink. It should be noted that

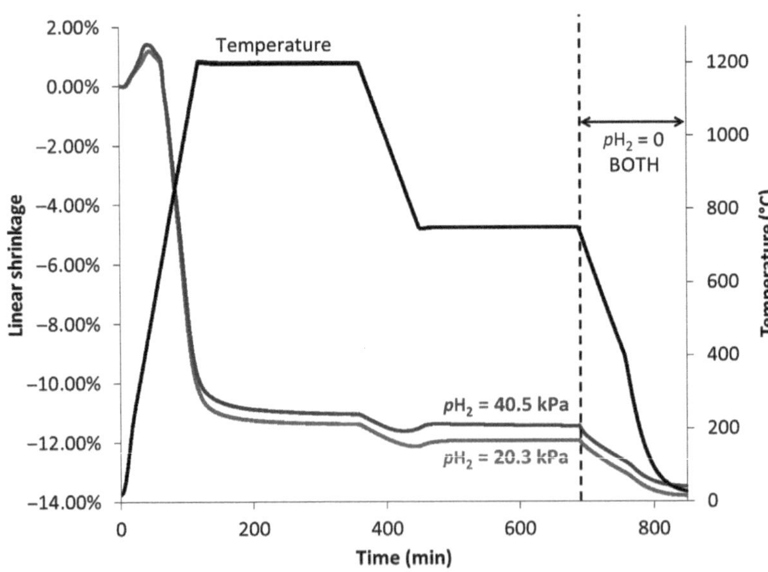

Figure 10.5 Typical dilatometry curves for Ti-6Al-4V sintered via HSPT under different hydrogen partial pressures.

during sintering of hydrogenated powder, shrinkage will occur as a result of hydrogen evolution as well as sintering; decoupling these mechanisms to determine shrinkage due to densification alone can be challenging. As seen, most of the densification occurs during heat-up, before the sintering temperature is reached. During the high temperature hold, a small amount of additional shrinkage is observed. The samples are then cooled and held at the phase transformation temperature to allow formation of the lower temperature phases (α- and δ-TiH_2). As the material is cooled, the hydrogen concentration increases. During the ramp-down segment, shrinkage due to thermal expansion is greater than expansion due to hydrogen absorption, resulting in net shrinkage. However, as the hold temperature is reached, measurable expansion due to hydrogen absorption is visible in the data.

Before dehydrogenation, the densities of samples sintered under higher hydrogen partial pressures are slightly less due to a slightly greater hydrogen concentration. However, after dehydrogenation, the relative density is generally greater than 99% regardless of hydrogen partial pressure during sintering. Sintering of CP-TiH_2 without master alloy in hydrogen-containing atmospheres also consistently produces greater than 99% density, though the homogenization and phase transformation temperatures must be adjusted to accommodate the different phase diagram. While limited research has been conducted on alloys other than Ti-6Al-4V and CP-Ti, it is likely that sintering under hydrogen-containing atmospheres should produce similar results for other alloys.

10.4.2 Micrographs of HSPT samples

The micrographs shown in Figure 10.6 are from three samples produced using three different BE methods: vacuum sintering of Ti, vacuum sintering of TiH_2, and HSPT. All three samples are Ti-6Al-4V and were sintered using the same sintering temperature and hold time. Additionally, all three samples were prepared from powder sieved below 44 μm (−325 mesh) and were all compacted using a cold isostatic press (CIP) at 350 MPa (50.8 ksi).

Figure 10.6a was prepared from CP-Ti powder blended with 10 wt% master alloy and sintered under high vacuum at 1200°C for 4 h. As is evidenced by the micrograph, densification was relatively poor, resulting in ~96% relative density. Additionally, the high temperatures and long hold times necessary for densification resulted in a coarse Widmanstätten lamellar α plate colony structure typical of PM Ti-6Al-4V. The microstructure and properties of this sample could be improved by incorporating pressure-assisted sintering to increase densification and thermomechanical work to remove porosity and refine the microstructure. However, these additional steps would increase energy consumption and cost in a commercial process.

Figure 10.6b was prepared from CP-TiH_2 blended with 9.6 wt% master alloy and sintered under high vacuum at 1200°C for 4 h. Despite having an identical thermal profile and atmosphere, densification is significantly improved versus Figure 10.6a because of the presence of hydrogen in solution during sintering. As such, this sample has a relative density of ~99%. However, the sintering conditions still result in the

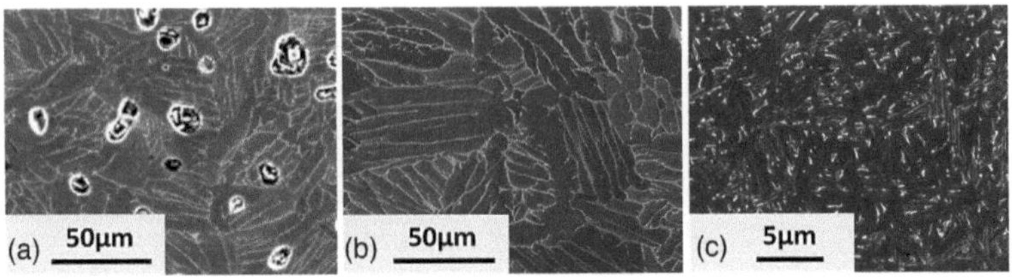

Figure 10.6 Micrographs of Ti-6Al-4V produced by BE PM via (a) vacuum sintering of metallic powder, (b) vacuum sintering of hydrogenated powder, and (c) HSPT of hydrogenated powder.

same coarse lamellar microstructure, which is undesirable for mechanical properties and would require post processing to refine.

Figure 10.6c was prepared from CP-TiH_2 blended with 9.6 wt% master alloy and sintered using HSPT. The sintering conditions were similar to those shown in Figure 10.5. After sintering, the sample was dehydrogenated under high vacuum at 750°C for 8 h. The relative density of this sample was also ~99%. Residual porosity, though none is visible on the shown micrograph, is on the order of 10–20 µm in diameter. The microstructure is still fully lamellar. However, α plates (dark-colored phase) are much finer; α plate widths are submicron and α plate lengths are on the order of 1–5 µm. Additionally, the β phase (light-colored phase) is discontinuous and dispersed, as opposed to the continuous β phase visible in the other two micrographs. The discontinuous β phase is also advantageous with regard to mechanical properties.

10.4.3 Quasi-static mechanical properties

Table 10.1 gives typical as-sintered mechanical properties of Ti-6Al-4V produced by HSPT as well as the ASTM B348 standard for Grade 5 titanium (Ti-6Al-4V) [77]. As seen in the data, HSPT samples consistently exceed ASTM standards for strength and ductility of wrought-processed Ti-6Al-4V. All data presented in this section are from samples in the as-sintered state. Each sample was machined into a round tensile bar after dehydrogenation and tensile tested following the ASTM E8 standard [78]. The microstructures of the samples presented in this section are similar to Figure 10.6c.

Figure 10.7 shows fractographs at low magnification (Figure 10.7a) and high magnification (Figure 10.7b) of a Ti-6Al-4V sample produced via HSPT. This sample experienced 15% elongation at failure. As expected from the elongation data, the

Table 10.1 Typical mechanical properties of Ti-6Al-4V produced via HSPT versus ASTM B348 standard for Grade 5 Ti-6Al-4V [77]

	Tensile strength (MPa)	0.2% Yield strength (MPa)	El (%)	O (wt%)	H (wt%)
ASTM B348	895	828	10	0.20	0.015
HSPT	970 ± 30	890 ± 30	14 ± 2	0.2 ± 0.02	<0.005

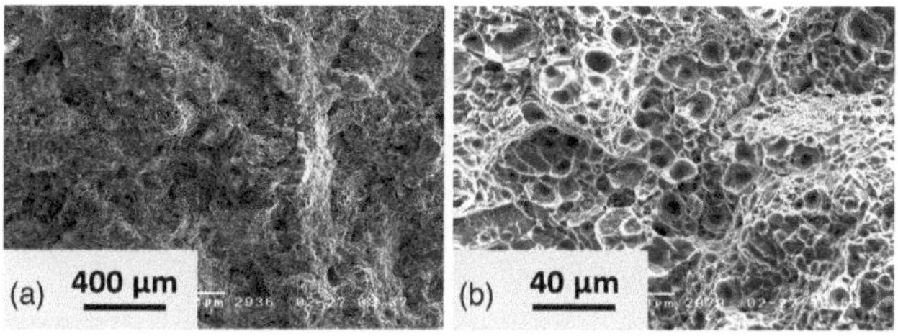

Figure 10.7 Fractographs of Ti-6Al-4V produced via HSPT at (a) 50× and (b) 500× magnification.

fractographs are indicative of ductile fracture. Each fractograph shows dimpling that is characteristic of microvoid formation during uniaxial loading.

Figure 10.8 shows a stress–strain curve from an E8 tensile test of Ti-6Al-4V produced via HSPT. This sample also experienced 15% elongation at failure and had an ultimate tensile strength of 972 MPa. This strength and ductility is typical for the sintering profile shown in Figure 10.5; the typical range of values are given in Table 10.1.

It should be noted that all data presented in this section are for specimens with ~0.2 wt% oxygen. However, Ti-6Al-4V samples have also been produced from CP-TiH_2 powder with higher initial oxygen content, producing interesting results. In these circumstances, the tensile strength of the material is increased by approximately 75 MPa for each 0.1 wt% of additional oxygen. However, among these samples, the increased strength is often achieved without a loss in ductility versus samples with lower oxygen levels.

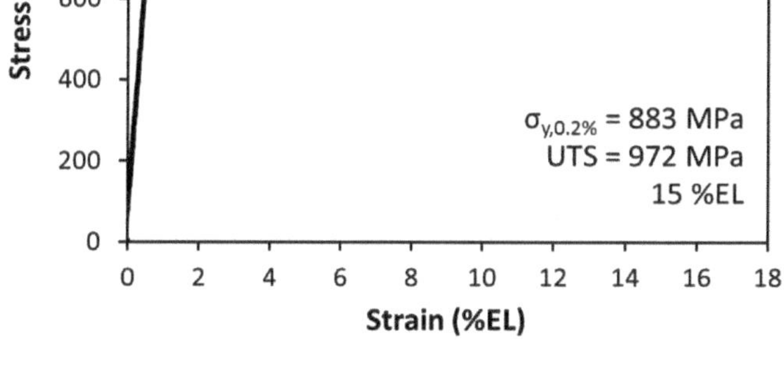

Figure 10.8 Typical tensile stress–strain curve of Ti-6Al-4V produced via HSPT.

10.4.4 Fatigue

Fatigue data for titanium alloys produced by HSPT are currently limited. However, an active study is currently progressing, and preliminary results seem promising. Based on current fatigue results, endurance limits similar to those seen in wrought titanium alloys are expected. However, a definitive correlation between fatigue properties and microstructures produced via HSPT is currently unavailable.

For a PM process, studying fatigue is much more challenging than quasi-static mechanical properties. A single large pore or "rogue pore" will significantly decrease fatigue strength. Therefore, avoiding pressing defects and identifying pore structure of green compacts and sintered samples is imperative to understanding fatigue performance. An extensive study is currently underway to determine the fatigue performance of samples produced via HSPT. In addition, X-ray tomography is being used to map the structure and distribution of pores in green compacts and sintered specimens. Chemical analysis techniques, such as inert gas fusion for oxygen and nitrogen and combustion for carbon, are also being employed to determine interstitial concentrations of fatigue specimens. The data from these studies should give a comprehensive understanding of the fatigue performance of HSPT and the relative effect of each contributing parameter.

10.5 Cost and energy savings

To estimate the cost and energy savings of HSPT versus wrought processing, an energy model has been developed by Paramore et al. and is currently being prepared for publication [79]. In order to quantitatively compare HSPT to traditional wrought processing, a specific component must be identified. Therefore, the energy consumed during the production of round bar stock with a 2-in. diameter via a theoretical HSPT process versus a typical wrought process was calculated.

For HSPT, the energy required per ton of bar stock produced was calculated using the following three equations:

$$E_{\text{HSPT}} = E_{\text{compaction}} + E_{\text{sinter}} \tag{10.2}$$

$$E_{\text{compaction}} = \frac{W_{\text{compaction}}}{\eta_{\text{hydraulic}}} \tag{10.3}$$

$$E_{\text{sinter}} = E_{\text{ramp}} + E_{\text{dehydro}} + E_{\text{gas}} + E_{\text{insulation}} \tag{10.4}$$

where $W_{\text{compaction}}$ is the theoretical work required to compact the powder, $\eta_{\text{hydraulic}}$ is the hydraulic efficiency of the press, E_{ramp} is the energy required to heat the powder and furnace hardware to the sintering temperature, E_{dehydro} is the energy required to overcome the endothermic dehydrogenation reaction, E_{gas} is the energy required to heat the gas flowing into the sintering furnace, and $E_{\text{insulation}}$ is the energy lost through

Table 10.2 Calculated energy consumed per ton to produce round bar stock with a 2-in. diameter via HSPT

	Electricity (kWh/ton)	Fuel (kWh/ton)
Compaction (kWh/ton)	3.93	0
Compaction yield	100%	100%
Sintering (kWh/ton)	0	1667.27
Sintering yield	100%	100%
Total energy	3.93	1667.27
Coal equiv. (ton_{Coal}/ton_{Ti})	0.00197	0.27
Total equiv. coal (ton_{Coal}/ton_{Ti})	**0.27**	

the furnace insulation. The calculated fuel and electricity consumption was then converted to equivalent tons of coal using typical values published by the US Energy Information Agency [80]. Values for each of the energy terms given in Equations 10.3 and 10.4 were calculated using thermodynamics and adjusted using published efficiencies of typical equipment. A description of how these values were calculated and the assumptions that were made will be discussed in detail in the forthcoming publication [79]. The results of these calculations are given in Table 10.2.

Unfortunately, data for the energy consumed during wrought processing of specific titanium components and alloys is not available in the literature. Therefore, for this energy model, the energy required to produce 2-in. round bar stock via wrought processing was also calculated from basic thermodynamic equations and adjusted using published efficiencies of typical equipment. From these calculations, it was determined that wrought processing required approximately 5.5 times the equivalent tons of coal per ton of titanium produced.

From this cost model, it was determined that HSPT offers about an 80% energy savings versus wrought processing for the production of 2-in. round bar stock. For the sake of being conservative and to simplify the calculations, a typical mill product was chosen as the geometry used in this model. However, it should be noted that for any PM process, the cost and energy savings are maximized for a complex geometry because of the ability to produce near-net-shape (NNS) components.

10.6 Conclusions

The impetus of HSPT research is the desire to maximize the performance-to-cost ratio of PM titanium alloys by eliminating thermomechanical work that is compulsory to other processes. During HSPT, the hydrogen content of titanium alloys during sintering is controlled as a function of temperature and hydrogen partial pressure. Because the phase equilibria and transformation kinetics of the Ti–H alloy systems vary as a function of hydrogen content, these parameters can be used to control microstructural evolution during sintering. Therefore, HSPT is capable of producing titanium alloys with engineered microstructures and application-tailored mechanical properties in the as-sintered state. To date, HSPT has consistently produced titanium alloys with mechanical properties that exceed ASTM standards for wrought alloys. A preliminary energy model has predicted about an 80% energy savings versus wrought processing for the production of 2-in. diameter bar stock, a number that could increase for the production of complex geometries that benefit from NNS processing.

Predicting microstructural evolution during HSPT has generated a need to further the fundamental understanding of phase transformations in hydrogen-containing titanium alloy systems. However, the volatile nature and rapid diffusion of hydrogen in titanium creates unique challenges when studying these processes. As such, HSPT research has endeavored to develop new methods to produce reliable data such as phase diagrams, TTT curves, and relationships between microstructure and mechanical properties of these systems.

Current studies are aimed at furthering the capabilities and fundamental understanding of HSPT. The goals of these efforts include producing a wider array of microstructures by incorporating simple heat treatments after sintering, eliminating the currently less than 1% residual porosity after sintering through novel powder processing and sintering methods, and understanding the relationship between dynamic mechanical properties and microstructure of HSPT-produced titanium alloys. Additionally, the feasibility of scaling up this process and incorporating novel production technologies such as green machining and additive manufacturing are also being investigated.

References

[1] Z.Z. Fang, P. Sun, H. Wang, Hydrogen sintering of titanium to produce high density fine grain titanium alloys, Adv. Eng. Mater. 14 (2012) 383–387.

[2] P. Sun, Z.Z. Fang, M. Koopman, A comparison of hydrogen sintering and phase transformation (HSPT) processing with vacuum sintering of CP-Ti, Adv. Eng. Mater. 15 (2013) 1007–1013.

[3] O.M. Ivasishin, V.M. Anokhin, A.N. Demidik, et al. Cost-effective blended elemental powder metallurgy of titanium alloys for transportation application, Key Eng. Mater. 188 (2000) 55–62.

[4] O.M. Ivasishin, D.G. Savvakin, V.S. Moxson, et al. Titanium powder metallurgy for automotive components, Mater. Technol. 17 (2002) 20–25.

[5] O. M. Ivasishin, D.G. Savvakin, V.S. Moxson, et al., Low-cost PM titanium materials for automotive applications, in: TMS-2005 134th Annual Meeting & Exhibition, San Francisco, CA, 2005.

[6] V.A. Duz, O.M. Ivasishin, C. Lavender, et al. Innovative powder metallurgy process for producing low cost titanium, in: Titanium 2008 24th Annual ITA Conference & Exhibition, 2008.

[7] H. Wang, M. Lefler, Z.Z. Fang, et al. Titanium and titanium alloy via sintering of TiH_2, Key Eng. Mater. 436 (2010) 157–163.

[8] D.H. Savvakin, M.M. Humenyak, M.V. Matviichuk, et al. Role of hydrogen in the process of sintering of titanium powders, Mater. Sci. 47 (2012) 651–661.

[9] D. Eylon, F.H.(Sam) Froes, S. Abkowitz, Titanium powder metallurgy alloys and composites, in: ASM Handbook, Powder Metal Technologies and Applications, vol. 7, ASM, 1998, pp. 874–886.

[10] H. Wang, Z.Z. Fang, P. Sun, A critical review of mechanical properties of powder metallurgy titanium, Int. J. Powder Metall. 46 (2010) 45–57.
[11] F.H.(Sam) Froes, D. Eylon, Powder metallurgy of titanium alloys, Int. Mater. Rev. 35 (1990) 162–182.
[12] Z.Z. Fang, P. Sun, Pathways to optimize performance/cost ratio of powder metallurgy titanium – a perspective, Key Eng. Mater. 520 (2012) 15–23.
[13] D.F. Heaney, R.M. German, Advances in the sintering of titanium powders, in: Powder Metallurgy World Congress and Exhibition (PM2004), Vienna, Austria, 2004.
[14] R.M. German, Powder Metallurgy Science, MPIF, Princeton, New Jersey, 1994.
[15] H.J.T. Ellingham, Reducibility of oxides and sulphides in metallurgical processes, J. Soc. Chem. Ind. 63 (1944) 125–133.
[16] C.G. McCracken, C. Motchenbacher, D.P. Barbis, Review of titanium-powder-production methods, Int. J. Powder Metall. 46 (2010) 19–26.
[17] O.M. Ivasishin, D.G. Savvakin, M.M. Gumenyak, et al. Role of surface contamination in titanium PM, Key Eng. Mater. 520 (2012) 121–132.
[18] V.A. Duz, O.M. Ivasishin, V.S. Moxson, et al., Cost-effective titanium alloy powder compositions and method for manufacturing flat or shaped articles from these powders, U.S. Patent 7,993,577 (2011).
[19] I.M. Robertson, G.B. Schaffer, Comparison of sintering of titanium and titanium hydride powders, Powder Metall. 53 (2010) 12–19.
[20] O.M. Ivasishin, D.G. Savvakin, F. Froes, et al. Synthesis of alloy Ti-6Al-4V with low residual porosity by a powder metallurgy method, Powder Metall. Met. Ceram. 41 (2002) 382–390.
[21] E. Gregory, Fabrication of niobium superconductor alloys, U.S. Patent 3,472,705 (1969).
[22] J. Greenspan, F. Rizzitano, E. Scala, Metal Matrix Composites by Decomposition Sintering of Titanium Hydride, Army Materials and Mechanics Research Center, Watertown, MA, 1970.
[23] J. Greenspan, F. Rizzitano, E. Scala, Titanium powder metallurgy by decomposition sintering of the hydride, in: Titanium, Science and Technology: Proceedings of the Second International Conference, 1973, pp. 365–379.
[24] K. Obara, Y. Nishino, S. Matsumoto, Process for producing a sintered article of a titanium alloy, U.S. Patent 3,950,166 (1976).
[25] C.F. Yolton, F.H. Froes, Method for producing powder metallurgy articles, U.S. Patent 4,219,357 (1980).
[26] A.P. Pankevich, A.F. Chertovich, G.A. Libenson, Production of titanium specimens from a titanium hydride powder by hot pressing, Sov. Powder Metall. Met. Ceram. 25 (1986) 89–92.
[27] Y. Li, X.M. Chou, L. Yu, Dehydrogenation debinding process of MIM titanium alloys by TiH_2 powder, Powder Metall. 49 (2006) 236–239.
[28] E. Taddei, V. Henriques, Densification and microstructural behaviour on the sintering of blended elemental Ti-35Nb-7Zr-5Ta alloy, Mater. Sci. Forum 530-531 (2006) 341–346.
[29] E. Carreño-Morelli, W. Krstev, Powder injection moulding of titanium from TiH_2 powders, Euro PM2009 Proc, 2, 2009, pp. 1–6.
[30] I.M. Robertson, G.B. Schaffer, Swelling during sintering of titanium alloys based on titanium hydride powder, Powder Metall. 53 (2010) 27–33.
[31] J.M. Zhang, J.H. Yi, G.Y. Gan, et al. Research on dehydrogenation and sintering process of titanium hydride for manufacture titanium and titanium alloy, Adv. Mater. Res. 616 (2013) 1823–1829.
[32] J. Dunlop, H. Rensing, Method for making tungsten-titanium sputtering targets and product, U.S. Patent 4,838,935 (1989).

[33] T. Gladden, Process for the manufacture by sintering of a titanium part and a decorative article made using a process of this type, U.S. Patent 5,441,695 (1995).
[34] B. Bertheville, M. Neudenberger, J.-E. Bidaux, Powder sintering and shape-memory behaviour of NiTi compacts synthesized from Ni and TiH_2, Mater. Sci. Eng. A 384 (2004) 143–150.
[35] O.M. Ivasishin, K.A. Bondareva, V.I. Bondarchuk, et al. Fatigue resistance of powder metallurgy Ti–6Al–4V alloy, Strength Mater. 36 (2004) 225–230.
[36] O.M. Ivasishin, D.G. Savvakin, I.S. Bielov, et al. Microstructure and properties of titanium alloys synthesized from hydrogenated titanium powders, Mater. Sci. Technol. 2005 Proc. (2005) 151–158.
[37] O.M. Ivasishin, D. Eylon, V.I. Bondarchuk, et al. Diffusion during powder metallurgy synthesis of titanium alloys, Defect Diffus. Forum 277 (2008) 177–185.
[38] O.M. Ivasishin, D.G. Savvakin, The impact of diffusion on synthesis of high-strength titanium alloys from elemental powder blends, Key Eng. Mater. 436 (2010) 113–121.
[39] A. Klevtsov, A. Nikishin, J. Shuvalov, Continuous and semi-continuous process of manufacturing titanium hydride using titanium chlorides of different valency, U.S. Patent 8,388,727 (2013).
[40] S. Kasparov, A. Klevtsov, A. Cheprasov, Semi-continuous magnesium-hydrogen reduction process for manufacturing of hydrogenated, purified titanium powder, U.S. Patent 8,007,562 (2011).
[41] V. Moxson, E. Ivanov, Method for manufacturing fully dense metal sheets and layered composites from reactive alloy powders, U.S. Patent 7,566,415 (2009).
[42] V. Moxson, V. Duz, Process of direct powder rolling of blended titanium alloys, titanium matrix composites, and titanium aluminides, U.S. Patent 7,311,873 (2007).
[43] V. Drozdenko, A. Petrunko, A. Adreev, Manufacture of cost-effective titanium powder from magnesium reduced sponge, U.S. Patent 6,638,336 (2003).
[44] U. Zwicker, H.W. Schleicher, Process for improving the workability of titanium alloys, U.S. Patent 2,892,742 (1959).
[45] W.R. Kerr, P.R. Smith, M.E. Rosenblum, et al. Hydrogen as an alloying element in titanium (Hydrovac), in: Titanium '80, Science and Technology; Proceedings of the Fourth International Conference on Titanium, Kyoto, Japan, 1980, pp. 2477–2486.
[46] W. Kerr, The effect of hydrogen as a temporary alloying element on the microstructure and tensile properties of Ti-6Al-4V, Metall. Trans. A 16A (1985) 1077–1087.
[47] W.H. Kao, D. Eylon, C.F. Yolton, et al. Effect of temporary alloying by hydrogen (Hydrovac) on the vacuum hot pressing and microstructure of titanium alloy powder compacts, Prog. Powder Metall. 37 (1981) 289–301.
[48] R. Smickley, L. Dardi, Microstructural Refinement of Cast Titanium, U.S. Patent 4,505,764 (1985).
[49] D. Kohn, P. Ducheyne, Microstructural refinement of β-sintered and Ti-6Al-4V porous-coated by temporary alloying with hydrogen, J. Mater. Sci. 26 (1991) 534–544.
[50] D. Kohn, P. Ducheyne, Tensile and fatigue strength of hydrogen-treated Ti-6Al-4V alloy, J. Mater. Sci. 26 (1991) 328–334.
[51] C. Zhang, W. Bian, Z. Lai, et al. Structure of hydrides in highly hydrogenated Ti-6Al-4V alloy, Acta Metall. Sin. 5 (1992) 362–368.
[52] A.A. Ilyn, B.A. Kolachev, A.M. Mamonov, Phase and structure transformations in titanium alloys under thermohydrogen treatment, in: Titanium '92, Science and Technology; Proceedings, 1992.
[53] B. Gong, N. Mitsuo, T. Kobayashi, et al. Strength and toughness of microstructurally controlled α+β type titanium alloys by thermomechanical processing with hydrogen, J. Japan Inst. Light Met. 42 (1992) 638–643.

[54] K. Yang, Z. Guo, D. Edmonds, Processing of titanium matrix composites with hydrogen as a temporary alloying element, Scr. Metall. Mater. 27 (1992) 1695–1700.
[55] H. Yoshimura, K. Kimura, M. Hayashi, et al. Ultra-fine equiaxed grain refinement and improvement of mechanical properties of $\alpha+\beta$ type titanium alloys by hydrogenation, hot working heat treatment, Mater. Trans. 35 (1994) 266–272.
[56] M. Niinomi, B. Gong, T. Kobayashi, Fracture characteristics of Ti-6Al-4V and Ti-5Al-2.5 Fe with refined microstructure using hydrogen, Metall. Mater. Trans. A 26A (1995) 1141–1151.
[57] Y. Zhang, S.Q. Zhang, Hydrogenation characteristics of Tt-6Al-4V cast alloy and its microstructural modification by hydrogen treatment, Int. J. Hydrogen Energy 22 (1997) 161–168.
[58] H. Yoshimura, Mezzoscopic grain refinement and improved mechanical properties of titanium materials by hydrogen treatments, Int. J. Hydrogen Energy 22 (1997) 145–150.
[59] T. Fang, W. Wang, Microstructural features of thermochemical processing in a Ti-6Al-4V alloy, Mater. Chem. Phys. 56 (1998) 35–47.
[60] J.I. Qazi, O.N. Senkov, J. Rahim, et al. Phase transformations in Ti-6Al-4V-xH alloys, Metall. Mater. Trans. A 32A (2001) 2453–2463.
[61] J. Qazi, Thermohydrogen processing (THP) of Ti-6Al-4V and TiAl alloys, PhD Dissertation, University of Idaho, 2002.
[62] J.I. Qazi, O.N. Senkov, J. Rahim, et al. Kinetics of martensite decomposition in Ti–6Al–4V–xH alloys, Mater. Sci. Eng. A A359 (2003) 137–149.
[63] C. Yu, C. Shen, T. Perng, Microstructure of Ti-6Al-4V processed by hydrogenation, Scr. Mater. 55 (2006) 1023–1026.
[64] A.A. Il'in, S.V. Skvortsova, A.M. Mamonov, Control of the structure of titanium alloys by the method of thermohydrogen treatment, Mater. Sci. 44 (2008) 336–341.
[65] Z. Sun, W. Zhou, H. Hou, Strengthening of Ti-6Al-4V alloys by thermohydrogen processing, Int. J. Hydrogen Energy 34 (2009) 1971–1976.
[66] T. Zhu, M. Li, Effect of 0.770 wt% H addition on the microstructure of Ti–6Al–4V alloy and mechanism of δ hydride formation, J. Alloys Compd. 481 (2009) 480–485.
[67] A.A. Ilyin, I.S. Polkin, A.M. Manonov, et al., Thermohydrogen treatment – the base of hydrogen technology of titanium alloys, in: Titanium '95: Science and Technology: Proceedings of the Eighth World Conference on Titanium, Birmingham, UK, 1995, pp. 2462–2469.
[68] A. Guitar, G. Vigna, M.I. Luppo, Microstructure and tensile properties after thermohydrogen processing of Ti-6 Al-4V, J. Mech. Behav. Biomed. Mater. 2 (2009) 156–163.
[69] F.H. Froes, O.N. Senkov, J.I. Qazi, Hydrogen as a temporary alloying element in titanium alloys: thermohydrogen processing, Int. Mater. Rev. 49 (2004) 227–245.
[70] H. Okamoto, H-Ti, in: ASM Handbook, vol. 3, Alloy Phase Diagrams, ASM, 1992, p. 2.238.
[71] G. Lütjering, J.C. Williams, Titanium, 2nd ed., Springer, Berlin, Heidelberg, New York, 2003.
[72] W.-E. Wang, Thermodynamic evaluation of the titanium-hydrogen system, J. Alloys Compd. 238 (1996) 6–12.
[73] C. Stickels, Gas carburizing, in: ASM Handbook, vol. 4, Heat Treating, ASM, 1991, pp. 312–324.
[74] C.-C. Shen, T.-P. Perng, Pressure–composition isotherms and reversible hydrogen-induced phase transformations in Ti-6Al-4V, Acta Mater. 55 (2007) 1053–1058.
[75] P. Sun, Z.Z. Fang, M. Koopman, et al. An experimental study of the (Ti-6Al-4V)-xH phase diagram using in situ synchrotron XRD and TGA/DSC techniques. Unpublished.

[76] V.S. Moxson, J. Qazi, S.N. Patankar, et al. Low-cost CP-titanium and Ti-6Al-4V alloys, Key Eng. Mater. 230-232 (2002) 339–343.
[77] ASTM Standard B348-10, Standard Specification for Titanium and Titanium Alloy Bars and Billets, ASTM, West Conshohocken, PA, 2010.
[78] ASTM Standard E8/E8M-11, Standard Test Methods for Tension Testing of Metallic Materials, ASTM, West Conshohocken, PA, 2011.
[79] J. Paramore, Z.Z. Fang, Energy model of titanium production via hydrogen sintering and phase transformation. Unpublished.
[80] U.S. Energy Information Administration, Annual Energy Review, 2013.

Warm compaction of titanium and titanium alloy powders

11

Mingtu Jia, Deliang Zhang***
*Waikato Centre for Advanced Materials, School of Engineering, The University of Waikato, Hamilton, New Zealand
**State Key Laboratory of Metal Matrix Composites, School of Materials Science and Engineering, Shanghai Jiao Tong University, Shanghai, China

11.1 Introduction

Warm compaction patented by Höganäs, Sweden, in the mid-1990s is a cost-effective way to produce ferrous and nonferrous powder metallurgy parts with high performance, with the cost of this process being 25% higher than cold compaction, but 40% lower than forging [1]. It has been used commercially to improve the green density of iron powder compacts by 0.1–0.25 g/cm via a single-press process. The common warm compaction temperature used is in the range of 130–150°C, and the typical warm compaction temperature and pressure are 150°C and 700 MPa, respectively [2]. When iron powder is heated up to 150°C, its yield strength is reduced by 30% from that at room temperature (RT) [3], so a higher relative density of iron powder compact could be achieved by using the same compaction pressure.

In titanium powder metallurgy, achieving high relative density and uniform density distribution of a powder compact is an essential goal of powder compaction, since these two characteristics of a powder compact determine the green strength, sintering rate, and the level of shape distortion during pressureless sintering. As shown in Figure 11.1, the green strength of Ti powder compact increases linearly with its relative density [4]. The amount of shrinkage of a Ti powder compact during vacuum sintering changes with its relative density, and this applies to not only the whole body of the powder compact but also each of its regions. If the density distribution of a powder compact is nonuniform, a higher amount of shrinkage will occur in a lower density region, and the shape of the powder compact would be distorted by the different amounts of shrinkage from different regions. In order to improve the relative density of a green powder compact produced by uniaxial pressing and to reduce the nonuniformity of its density distribution, double pressing, triaxial pressing [5], warm compaction [6], cold isostatic pressing (CIP) [7], and equal channel angle pressing (ECAP) [8] have been used to produce green powder compacts. Among these, warm compaction can produce high density and fairly uniform density distribution with low cost, long die working lifetime, and high productivity.

Warm compaction of titanium patented by Kondoh et al. [9] has been used to compact commercially pure titanium (CP Ti), Ti-6Al-4V, and Ti-6Al-4V-6TiB_2 powders at 150°C and with a compaction pressure ranging from 392 to 1568 MPa. After warm

Titanium Powder Metallurgy. http://dx.doi.org/10.1016/B978-0-12-800054-0.00011-3

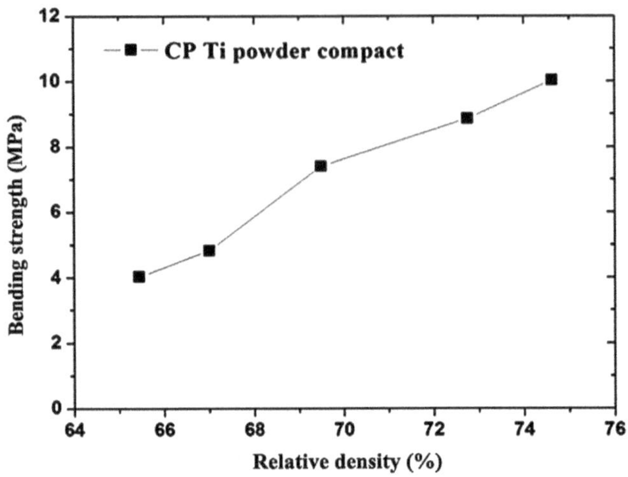

Figure 11.1 Relationship between the relative density and green strength of a Ti powder compact [4].

compaction with a pressure of 588 MPa, the relative density of CP Ti powder compact ranges from 82.5% to 97.5% of the theoretical density (TD). For CP Ti powder compacts, in order to achieve a relative density of ≥99% TD by solid-state sintering, the minimum green density required is 95.5% TD [9,10]. For Ti-6Al-4V and Ti-6Al-4V-6TiB_2 powder compacts, to achieve a sintered density greater than 99% TD, the green densities need to be 91.2% and 94.9% TD, respectively, which can be achieved using a compaction pressure in the range of 784–1568 MPa [9]. Luo et al. [10] applied the warm die compaction to HDH Ti powder at 200°C and achieve a green density increment of 5.0–9.4% TD under the compaction pressure from 200 to 1000 MPa. They found that warm die compaction had no or little effect on prealloyed Ti-10V-2Fe-3Al and Ti-6Al-4V powders. With a lower die compaction temperature in the range of 80–180°C and a compaction pressure over the range from 400 to 600 MPa, the green density increased by 4.7–4.9% TD for HDH Ti powder [11]. The relative density values of titanium and titanium alloy powder compacts produced by warm compaction with die wall lubricant are listed in Table 11.1. At a high compaction pressure of 1568 MPa, the relative density of HDH Ti, mixing elemental (ME) Ti-6Al-4V, and ME Ti-6Al-4V-6TiB_2 powder compacts achieved higher than 95% TD. However, the relative density of PA Ti-6Al-4V, PA Ti-10V-2Fe-3Al, and PA Ti-6.8Mo-4.5Al-1.5Fe powder compacts produced by warm compaction reached only about 80% TD under the same compaction conditions [12].

In titanium powder compaction process, the use of lubricant can help reduce internal friction between powder particles (referred to as admixed lubricant) and particle-die wall friction (referred to as die wall lubricant). The former is admixed with powder before compaction and assists in the powder rearrangement during powder pressing, while the latter assists in the reduction of ejection force and avoiding galling [9]. Unlike warm compaction of iron and iron alloy powders used in industry, hydrogen, nitrogen, carbon, and oxygen in admixed lubricants are dissolved in titanium during the dewaxing and

Table 11.1 **Warm compaction of Ti and Ti alloy powders**

Powder	Compaction temperature (°C)	Compaction pressure (MPa)	Die wall lubricant	Relative density (%)	References
HDH Ti	150	392–1568	High fatty acid-based lubricant	82.5–97.5	[9]
ME Ti-6Al-4V	150	783–1568	High fatty acid-based lubricant	91.1–97.8	[9]
ME Ti-6Al-4V-6TiB$_2$	150	783–1568	High fatty acid-based lubricant	90.2–96.5	[9]
HDH Ti	200	200–1000	Stearic acid	73.5–96.9	[10]
ME Ti-10V-2Fe-3Al	200	400–800	Stearic acid	80.3–89.5	[10]
PA Ti-10V-2Fe-3Al	200	1000	Stearic acid	79.8	[10]
PA Ti-6Al-4V	200	1000	Stearic acid	86.3	[10]
HDH Ti	140	400–600	Macromolecule polymer	80.9–87.8	[11]
PA Ti-6.8Mo-4.5Al-1.5Fe	140	500	Not provided	81.6	[12]

HDH, hydrogenation dehydrogenation; ME, mixing elemental; PA, prealloyed.

sintering. The dewaxing step before sintering also increases the cost of production, so the use of admixed lubricants is less favorable for titanium powder warm compaction [9–11]. In contrast, die wall lubricants (e.g., high fatty acid–based lubricant [9], stearic acid [10], and macromolecule polymer [11]) can be used for titanium powder warm compaction.

The temperature of titanium powder used in warm compaction is typically in the range of 80–200°C. In principle, warm compaction is defined as powder compaction performed at a temperature above the RT but below the recrystallization temperature of the metal, which is normally in the range of 0.6–0.7T_m (T_m is the melting point or solidus temperature of the metal or alloy in Kelvin scale), so the range of compaction temperature could be expanded to be higher than 200°C. However, when compacted at 250°C or higher, a protective gas atmosphere (Ar or He) may need to be used to prevent oxidation and nitrification of the raw powders.

11.2 Warm compaction process

In general, the equipment used for cold compaction of metal and alloy powders can be used for warm compaction. Issues that need to be taken into account include protective atmosphere, die materials suitable for working at the warm compaction temperature, and

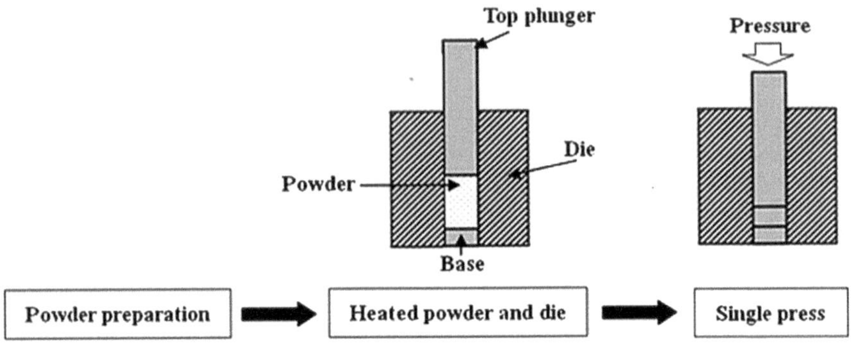

Figure 11.2 Schematic diagram of the warm compaction process.

powder and die heating systems. A typical single-press warm compaction process for titanium and titanium alloy powders, as illustrated in Figure 11.2, has the following steps:

1. *Powder preparation*: If titanium elemental powder is used to make titanium alloy parts, the powder needs to be admixed with other elemental powders and/or master alloy powders to make a powder blend of the desired composition. Lubricant may be added to assist in the powder mixing process. When a prealloyed titanium alloy powder is used, it still needs to be mixed for improved particle size distribution and lubricant distribution (if lubricant is added).
2. *Heated powder and die assembly*: Powders are heated up either prior to filling into the heated die or with the die together, and though powder preheating can maintain the consistency of warm compaction, it can also increase the overall cost [11]. The die assembly can be heated by various heating devices including band heater, cartridge heat elements and resistance heater. High-speed steel with TiN coating is selected as the die assembly material at compaction temperatures up to 200°C [9,10]; for higher compaction temperatures up to 550°C, hardened H13 steel is chosen.
3. *Atmosphere*: At RT, titanium is passive to oxygen and nitrogen because of the dense, stable, and passive surface oxide film that is formed. The passive oxide film crystallized into anatase at about 276°C [13], and loses its protection function, so if the compaction temperature is more than 276°C, vacuum or inert (Ar, He) protective atmosphere is recommended to prevent severe reaction of the powder with O_2 and N_2 in air. Below this temperature (such as 80–150°C [9,11] and 200°C [10]), warm compaction of titanium and titanium alloy powders can be performed in air without serious oxidation or nitridation.
4. *Single press*: Load is applied on the top plunger to compact the powder by a press, including hydraulic and mechanical presses.
5. *Ejection*: In titanium powder cold compaction, high pressure is often used to achieve a high relative density of powder compact, and this results in a higher force needing to be used to eject the powder compact from the die. When the ejection force increases to a certain value (e.g., 24 MPa for pure titanium [9]), galling may occur. Warm compaction can significantly reduce the ejection force for titanium and titanium alloy powder compacts, and as shown in Figure 11.3, for Ti, Ti-6Al-4V, and Ti-6Al-4V-6TiB_2 powders, the increase of compaction pressure beyond a certain value causes little variation of the ejection force [9].

Lubricants are used to reduce the friction between powder particles and also between the die wall and powder particles. Lubricants are divided into two categories: admixed and die wall.

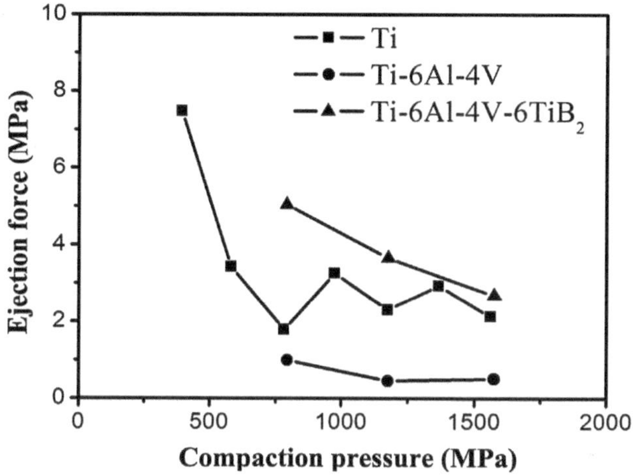

Figure 11.3 Relationship of ejection force and compaction pressure of CP Ti, Ti-6Al-4V and Ti-6Al-4V-6TiB_2 powder compacts made by warm compaction. Source: From Ref. [9], reproduced with permission of Toyota Central Research & Development Laboratories, Inc.

- *Admixed lubricant*: In order to reduce the internal friction between powder particles and improve the density and density distribution of titanium powder compact, lubricants including stearates, stearic acid, polytetrafluoroethylene (PTFE), polyvinyl fluoride, and waxes are mixed with titanium powders before warm compaction. The working temperature of various lubricants needs to be considered when investigating the optimized parameters of warm compaction. The green powder compacts need to be dewaxed before sintering. Because it is difficult to completely remove the lubricant, titanium could react with oxygen, nitrogen, hydrogen, and carbon from the residual lubricant at high temperatures. Consequently, the use of admixed lubricant should be avoided as much as possible.
- *Die wall lubricant*: Die wall lubricant, for instance colloidal graphite, high fatty acid-based lubricant [9], stearic acid [10], and macromolecule polymer [11], coated on the surface of the die wall by spraying guns or electrostatic guns, can reduce the particle-die wall friction. As a result, it can improve the density and density distribution of the titanium and titanium alloy powder compacts without contaminating the powder compact. It can also reduce the ejection force and improve the lifetime of the compaction die. Based on these considerations, warm compaction combined with the use of die wall lubricant is a preferred approach to producing titanium and titanium alloy green powder compacts. The advantages of this approach include easy handling, free of contamination and long die working life.

11.3 Compaction pressure

Compared to cold compaction, warm compaction increases green density via reducing the yield strength of metal powders. At a given compaction temperature, the correlation of the relative density and compaction pressure for warm compaction of

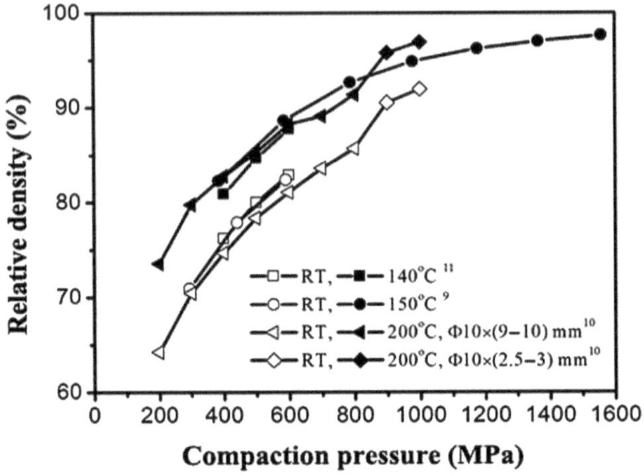

Figure 11.4 Effect of compaction pressure on relative density of HDH Ti powder compacts made by warm and cold compaction. Source: From Ref. [9], reproduced with permission of Toyota Central Research & Development Laboratories, Inc. and from Ref. [10], with permission from Elsevier.

HDH Ti powder is similar to that for cold compaction (Figure 11.4). For instance, at 140°C, the relative density increases from 80.9% to 87.8% TD with increasing compaction pressure from 400 to 600 MPa [11]. At the combination of 1568 MPa and 150°C, the relative density increased dramatically to 97.5% TD [9]. The relative density of the powder compact increases with increasing compaction pressure according to an exponential function, reaching a maximum value of 96.9% at the combination of 200°C and 1000 MPa [10]. Overall, for a given compaction pressure, the relative density of HDH Ti powder compacts made by warm compaction at 200°C is 5.0–9.4% TD higher than that of the HDH Ti powder compacts made by cold compaction.

The effect of compaction pressure on the relative density is similar for the warm compaction of blended elemental titanium alloy (Ti-6Al-4V [9], Ti-6Al-4V-6TiB_2 [9], and Ti-10V-2Fe-3Al [10]) powder and pure titanium powder (Figure 11.5) because Ti powder is the major constituent in the powder blend. The addition of high strength master alloy powder reduces the relative density of the powder compact, and this effect becomes more significant with adding ceramic powder (e.g., TiB_2 powder) to the powder blend [10].

11.4 Compaction temperature

With the increase of temperature, the yield stress of titanium and titanium alloy decrease significantly. For example, the yield stress of Grade 4 CP Ti decreases from 550 to 165 MPa from RT to 300°C, and then changes slightly up to 550°C (Figure 11.6).

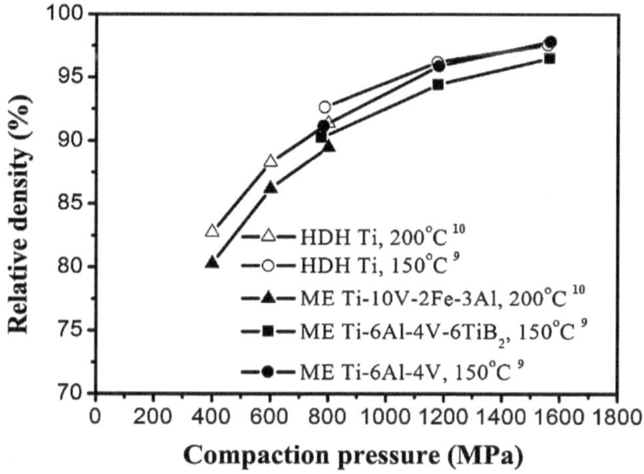

Figure 11.5 Effect of compaction pressure on the relative density of Ti alloy (Ti-6Al-4V, Ti-10V-2Fe-3Al, and Ti-6Al-4V-6TiB_2) powder compacts made by warm compaction. Source: From Ref. [9], reproduced with permission of Toyota Central Research & Development Laboratories, Inc. and from Ref. [10], with permission from Elsevier.

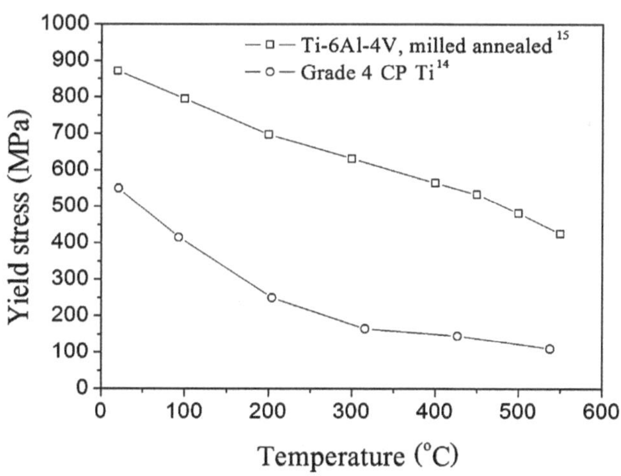

Figure 11.6 Yield stress versus temperature curves of Grade 4 CP Ti and milled annealed Ti-6Al-4V alloy. Source: From Refs [14,15], reproduced with permission from ASM International.

The yield stress of annealed Ti-6Al-4V (wt%) alloy decreases gradually from 871 to 425 MPa with increasing temperature from RT to 550°C.

Even though the yield strength of titanium and titanium alloys decreases with temperature, it is still much higher than that of iron, aluminum, and copper. Figure 11.7 compares the changes of relative density with increasing temperature from RT to 140°C for Ti and Fe powders with admixed lubricant of lithium stearate and for

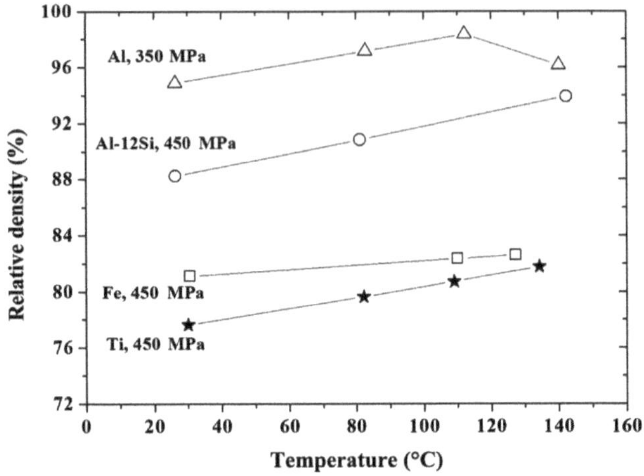

Figure 11.7 Effect of temperature on the green compact density of various metal powders at constant compaction pressure (350, 450, and 600 MPa). Source: From Ref. [16], reproduced with permission from Maney Publishing.

Al-12Si (wt%) and Al powders with admixed lubricant of wax. With the same compaction temperature and pressure, the relative density of the Ti powder compact is the lowest, which increased from 77.6% to 81.8% TD with increasing temperature from RT to 140°C [16].

Compacts of HDH Ti, ME Ti-6Al-4V, and gas atomized (GA) and hydride–dehydride (HDH) prealloyed Ti-6Al-4V powders can all be made by warm compaction with die wall lubricant. As shown in Figure 11.8, the relative density of powder

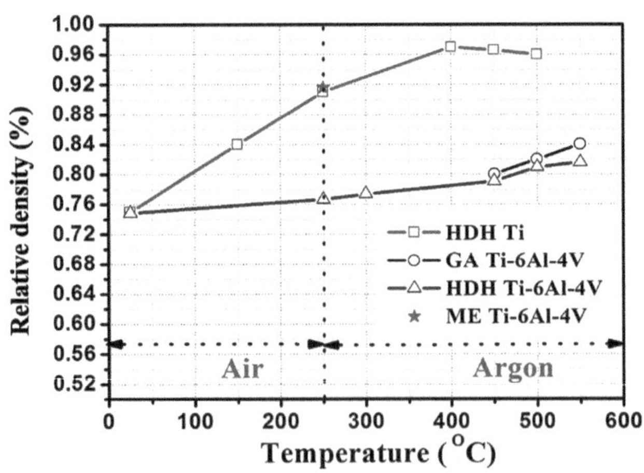

Figure 11.8 Relative density of HDH Ti and HDH and GA Ti-6Al-4V powder compacts as functions of temperature under the constant pressure 544 MPa applied on HDH Ti powder and 726 MPa applied on HDH and GA Ti-6Al-4V powders.

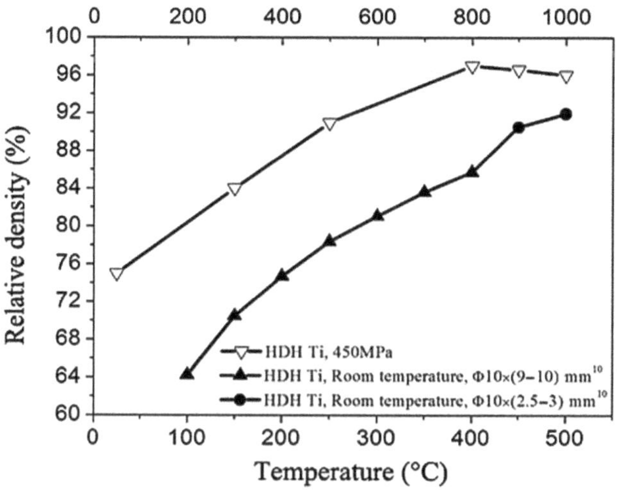

Figure 11.9 Relative density of Ti powder compacts as functions of temperature with a compaction pressure of 544 MPa and pressure with a compaction temperature fixed at room temperature, respectively. Source: From Ref. [10], reproduced with permission from Elsevier.

compacts increases with compaction temperature. The influence of compaction temperature on the relative density of HDH Ti powder compact is more significant than that of HDH and GA Ti-6Al-4V powder compacts. In order to reduce the cost of powder compaction associated with the use of argon and produce relative densities greater than 90% TD, ME powders are compacted at 250°C under 544 MPa, and the relative density of powder compacts achieved is close to that of the CP Ti powder compacts (Figure 11.8). For HDH Ti powder compacts, further increasing compaction temperature beyond 400°C does not improve the relative density. However, it is difficult to compact GA Ti-6Al-4V powder at temperatures lower than 450°C.

Figure 11.9 shows that when compacted at 250°C or higher, HDH Ti powder can be compacted to 92% TD or higher at 544 MPa with die wall lubricant, while at RT a minimum pressure of 900 MPa is needed to produce a similar density [10]. The reduced compaction pressure can extend the working life of the dies.

Increasing compaction temperature is more effective than increasing compaction pressure for achieving a higher density of the powder compacts (Figure 11.10). In addition, it is easier to compact ME Ti-6Al-4V powders consisting of Ti powder and master alloy powder than pre-alloyed Ti-6Al-4V powders.

Various models have been proposed to correlate the relative density of a powder compact and compaction pressure. Heckel's model [18] is given by

$$\mathrm{In}\left(\frac{1}{1-D}\right) = KP + \mathrm{In}\left(\frac{1}{1-D_0}\right) + B, \tag{11.1}$$

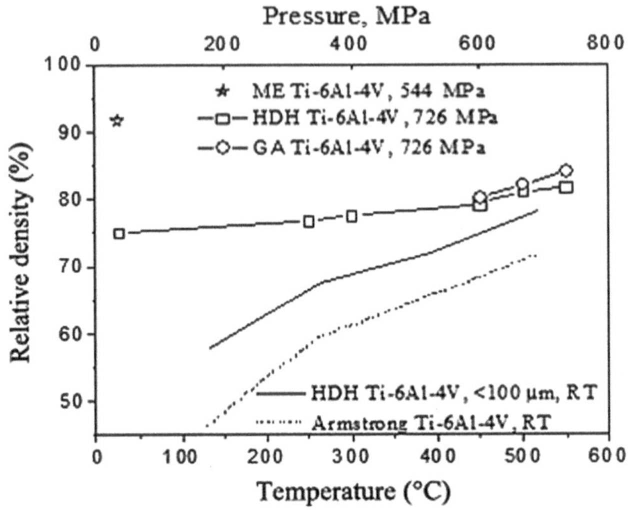

Figure 11.10 Comparison of the relative density of various Ti-6Al-4V powder compacts as functions of temperature with different compaction pressures of 544 and 726 MPa, and as functions of pressure with the compaction temperature being room temperature [17]. (Courtesy of Oak Ridge National Laboratory, US Dept. of Energy.)

where D and D_0 are the relative density and initial density of powder compact, respectively, P is compaction pressure, B is a constant, and K is a coefficient, which is a function of the yield strength of the material.

Heckel's equation has been expressed in another form by Secondi [19]:

$$D = D_0 + (D_\infty - D_0)(1 - e^{-KP}), \tag{11.2}$$

where D_∞ is the maximum achievable density.

Under a constant pressure, P and K are the only variables in Equations 11.1 and 11.2, respectively. Heckel estimated that the value of K is $\sim\sigma/3$, where σ is the yield strength of the powder material. The effect of compaction temperature on the relative density of Ti and Ti-6Al-4V powder compacts can be explained using the relationship between yield stress and temperature of the powder material shown in Figure 11.4.

11.5 Particle shape effects on Ti powder warm compaction

The HDH Ti and Ti-6Al-4V powders are much easier to be compacted than GA Ti-6Al-4V powder (Figure 11.8). This is especially true at temperatures ranging from RT to 300°C, because of their rough particles surfaces and irregular particle shapes. There are two stages of powder compaction with increasing compaction pressure, namely, particle rearrangement and sliding, and particle plastic deformation. Owing to their

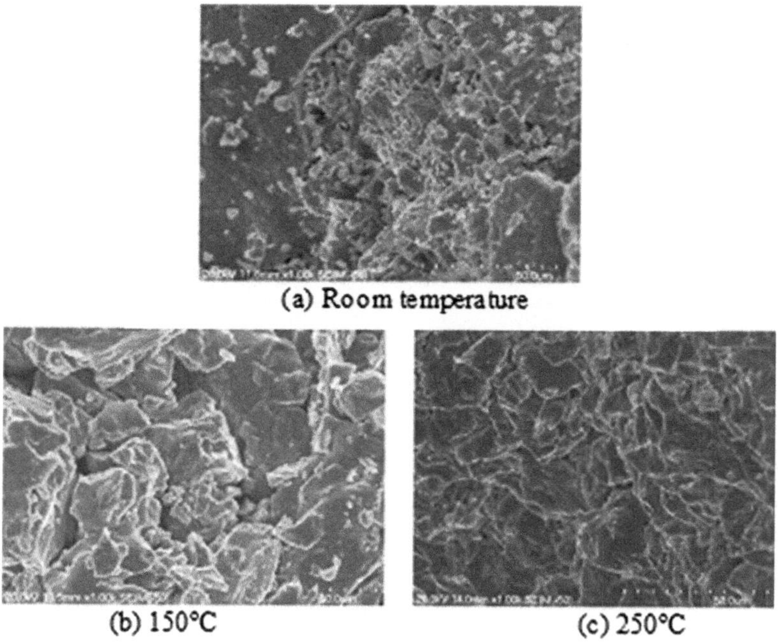

Figure 11.11 Fracture surfaces of HDH Ti powder compacts made by compaction at different temperatures, respectively.

spherical shape and low surface roughness, GA powder particles slide and rearrange themselves more easily than HDH powder particles with irregular shapes.

Cold welding and interlocking are two mechanisms of powder compaction. With increasing compaction temperature, the contact surfaces between the powder particles (Figure 11.11) become smoother, and the powder particles are deformed more severely and warm welded by the friction between adjacent powder particles. This means that the friction between powder particles is beneficial to bond particles and shape the powder into a compact with certain strength. Meanwhile, interlocking of HDH Ti powder particles occurs easily because of their rough surfaces. Based on the particle size distributions of raw powders shown in Figure 11.12, the average particle size of the 100 mesh GA powder is larger than that of the 200 mesh HDH powder, while the range of particle sizes of the HDH powder is much wider than that of the GA powder. Small powder particles are located among the large powder particles as bridges. Hence during powder compaction, small powder particles can be deformed more easily than large powder particles under the same external pressure.

The morphologies of GA Ti-6Al-4V powder particle surfaces after compaction at different temperature are shown in Figure 11.13. Because of good flowability and spherical shape of GA powder particles, particle rearrangement and sliding can occur easily. This makes it difficult to produce GA powder compacts, especially at low temperatures. Warm welding between two adjacent GA powder particles is also difficult to accomplish because of their low surface roughness. Flattened surfaces caused by compaction

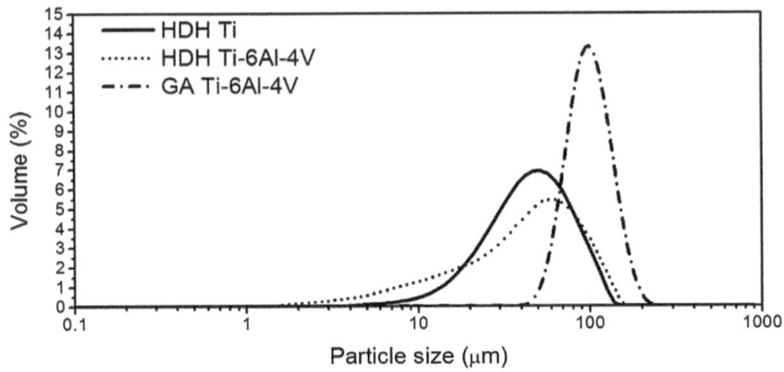

Figure 11.12 Particle size distribution of HDH Ti (200 mesh), HDH Ti-6Al-4V (200 mesh), and GA Ti-6Al-4V (100 mesh) powders.

are readily visible on GA powder particles compared to the compaction of HDH powder particles. With increasing compaction temperature, the areas of flattened surfaces on GA powder particles become larger because of a larger amount of deformation. As shown in Figure 11.13, the area of warm welding between two adjacent powder particles caused by plastic deformation during compaction at 500°C is larger than that during compaction at 450°C. A large amount of shear deformation of the powder particles

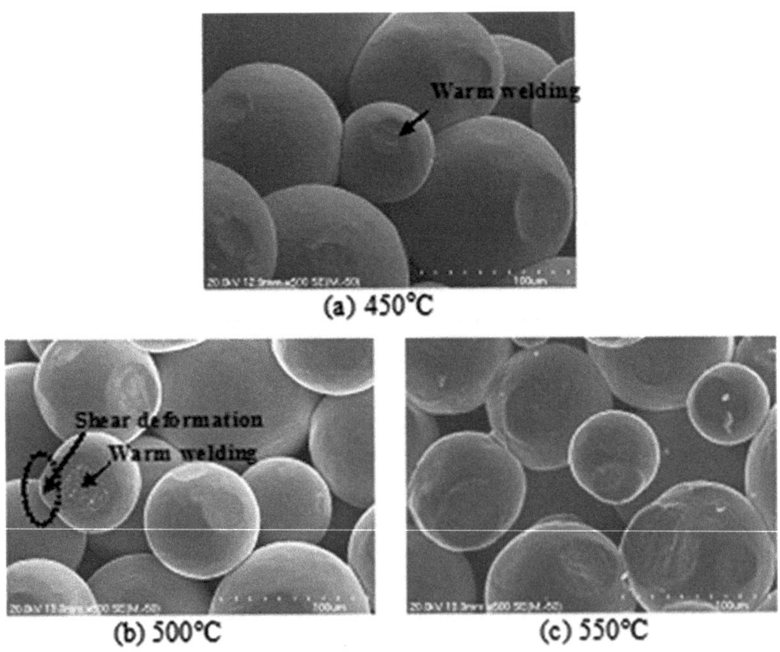

Figure 11.13 GA Ti-6Al-4V powder particle morphologies after compaction at different temperatures with a pressure of 726 MPa, respectively.

occur in the welding area, and this means that welding between GA powder particles is mainly formed by shear deformation of two adjacent powder particles. So the main mechanism of powder warm compaction of GA Ti-6Al-4V powder is warm welding.

11.6 Mechanical properties of sintered titanium and titanium alloy powder compacts produced by warm compaction

11.6.1 Fast initial sintering of Ti and Ti-6Al-4V powder compacts

Green compacts of HDH Ti, GA, and HDH Ti-6Al-4V powders made with die wall lubricant at 250°C and 300°C in air and at 550°C under argon were selected to be induction sintered. The powder compacts were heated up to 1350°C in 3 min, then cooled in a flow argon atmosphere. When the powder compacts are heated to 1350°C, necking between two neighboring particles in the powder compacts was observed (Figure 11.14). The necking between neighboring GA spherical powder particles is more obvious (Figure 11.14b) than that between neighboring HDH powder particles with irregular shapes (Figure 11.14a and c).

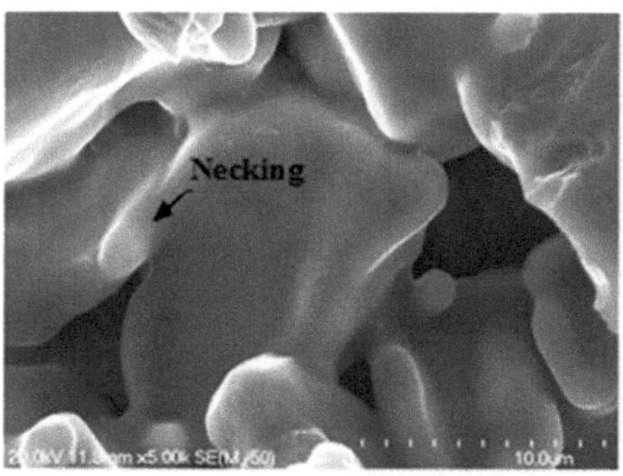

(a) As-sintered HDH Ti powder compact

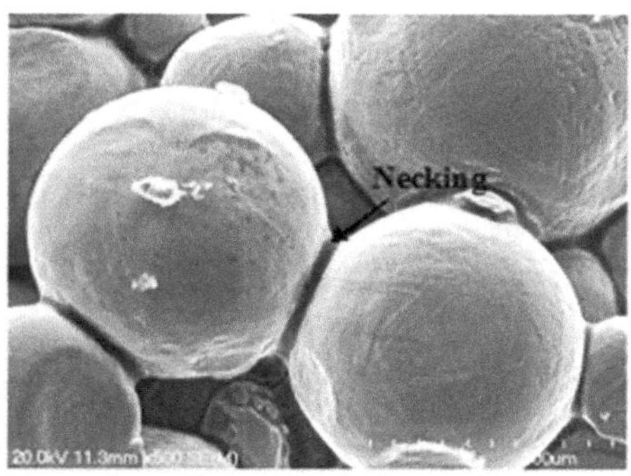

(b) As-sintered GA Ti-6Al-4V powder compact

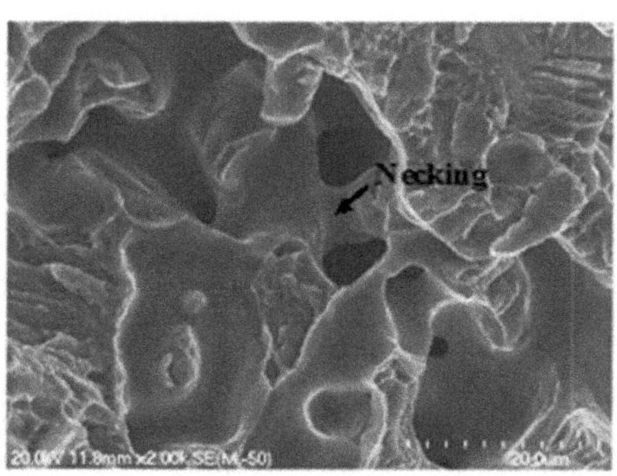

(c) As-sintered HDH Ti-6Al-4V powder compact

Figure 11.14 Powder morphologies of as-sintered powder compacts.

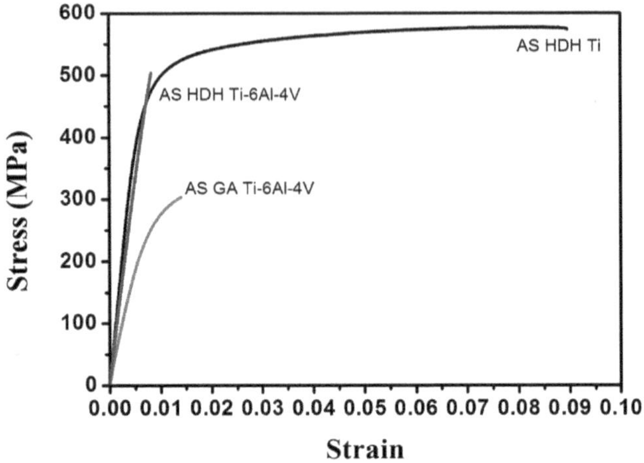

Figure 11.15 Tensile engineering stress-strain curves of specimens cut from the as-sintered HGH Ti, HDH Ti-6Al-4V, and GA Ti-6Al-4V powder compacts. AS, as-sintered.

The tensile engineering stress–strain curves of specimens cut from the as-sintered powder compacts (Figure 11.15) show that their average ultimate tensile strength (UTS) can reach 575 MPa, and their average yield strength is 470 MPa, with the average elongation to fracture being 7.5%. Large isolated pores play an important role as fracture origin and cause reduction in elongation to fracture. As shown in Figure 11.15, the tensile engineering stress–strain curves of specimens cut from as-sintered HDH and GA Ti-6Al-4V powder compacts show very small elongation to fracture, and their fracture stress are both lower than 500 MPa. The brittleness and relative low fracture stress of the as-sintered HDH and GA Ti-6Al-4V powder compacts show that the level of necking and bonding between the powder particles in them is much lower than that in the HDH Ti powder compacts. This can be attributed to the low relative density of about 80% TD of the HDH and GA Ti-6Al-4V powder compacts.

The fracture surfaces of the as-sintered powder compacts show that the fracture of the as-sintered part is mainly the fracture of the necks formed between powder particles, as reflected by the concave and convex features of the fracture surfaces, especially those of the fracture surfaces of as-sintered GA Ti-6Al-4V powder compact shown in Figure 11.16b. At high magnifications (Figure 11.16a and c), the SEM images show that the fracture morphology of the necks between powder particles in as-sintered HDH Ti and GA Ti-6Al-4V powder compacts have dimples indicating that the fracture is ductile. However, the cleavage morphology of the fracture surfaces of the necks between the powder particles in the as-sintered HDH Ti-6Al-4V powder compact (Figure 11.16c) shows that the fracture is brittle.

11.6.2 The effect of green density on sintering

The increased green density of powder compacts by warm compaction prior to sintering is beneficial for dimensional consistency. Under the same compaction pressure,

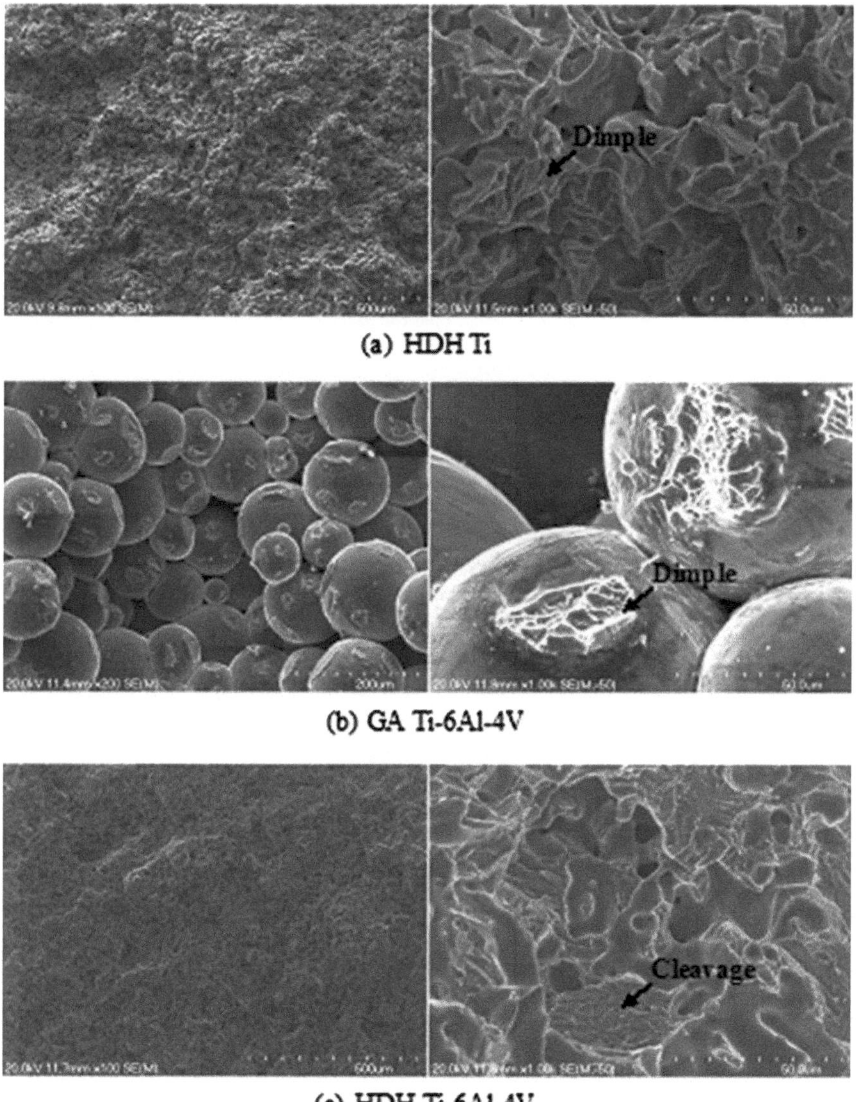

Figure 11.16 SEM images with different magnifications of fracture morphology of as-sintered HDH Ti and GA and HDH Ti-6Al-4V powder compacts after tensile testing.

warm compaction can produce higher green density than cold compaction, and also higher sintered density [10]. The correlation between the green density and sintered density of HDH Ti powder compacts is linear for both warm compaction and cold compaction (Figure 11.17). A minimum relative density of 95.5% TD by warm compaction and 96.6% TD by cold compaction is required to achieve a sintered density greater than 99% TD [9,10]. The similar slopes shown in Figure 11.17 suggest that the

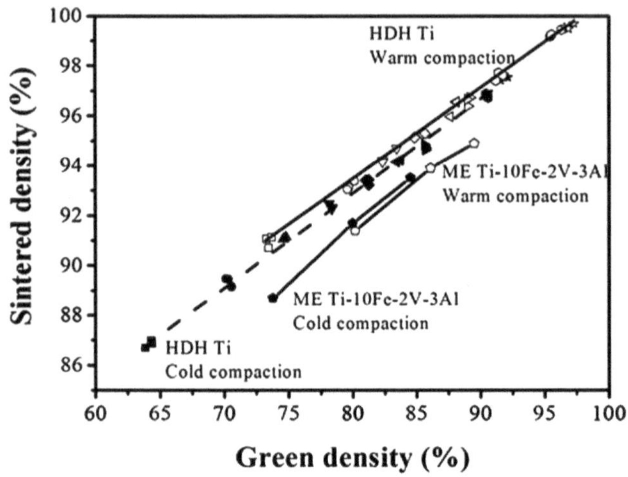

Figure 11.17 Dependence of sintered density on green density of HDH Ti and ME Ti-10Fe-2V-3Al powder compacts made by warm compaction at 200°C and cold compaction at room temperature, respectively. The green compacts were sintered at 1300°C for 120 min [10]. Source: From Ref. [10], reproduced with permission from Elsevier.

densification processes during sintering of the two types of powder compacts (cold and warm) are very similar [10].

As shown in Figure 11.17, for a given compaction pressure, the increase in relative green density of titanium powder compacts by warm compaction at 200°C is 5.0–9.4% TD, which translated to an increase of 2.0–4.4% TD in sintered density [10]. For ME Ti-10V-2Fe-3V alloy powder compacts, the increase in relative green density of the powder compacts is 5.0–6.4% TD while the increase in sintered density is 1.4–2.7% TD.

11.7 Applications

Powder compact preforms of CP Ti, Ti-6Al-4V, Ti-6Al-4V-6TiB$_2$, Ti-10Fe-2V-3Al, Ti-6.8Mo-4.5Al-1.5Fe [12], and Ti-13V-11Cr-3Al have been produced by warm compaction with die wall lubricant. As shown in Figure 11.18, these powder compact performs are made to manufacture near-net-shaped components for automobile and marine industry by powder compact forging.

Warm powder compaction of titanium and titanium alloys can also be used to produce preforms with high density and strength for making parts to be used in automobile, aerospace, marine, and other industries. It is possible to manufacture fully dense near-net-shaped titanium and titanium alloy parts via warm powder compaction and rapid sintering, including induction sintering and sparking plasma sintering in the future.

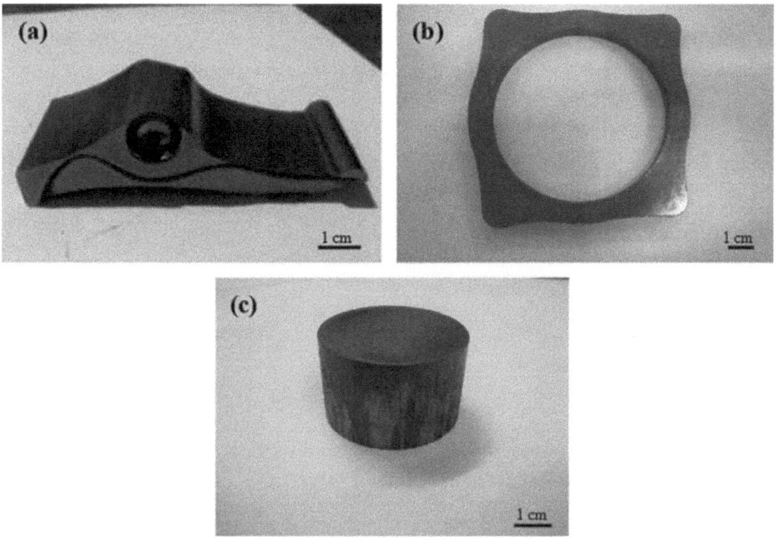

Figure 11.18 Titanium and titanium alloy preforms made by warm compaction: (a) CP Ti rocker arm preform; (b) CP Ti top cover of diving helmet preform; (c) Ti-13V-11Cr-3Al cylinder preform.

References

[1] F.G. Hanejko, Warm compaction, ASM Handbook 7 (1998) 376–381.
[2] R.M. German, Powder Metallurgy of Iron and Steel, Wiley, New York, 1998.
[3] I. Chang, Y. Zhao, Advances in Powder Metallurgy: Properties, Processing and Applications, Elsevier, 2013.
[4] A. Laptev, O. Vyal, M. Bram, H. Buchkremer, D. Stöver, Green strength of powder compacts provided for production of highly porous titanium parts, Powder Metall. 48 (2005) 358–364.
[5] O. Coube, H. Riedel, Numerical simulation of metal powder die compaction with special consideration of cracking, Powder Metall. 43 (2000) 123–131.
[6] C. Sonsino, R. Ratzi, Warm powder compaction substitutes conventionally double pressed and double sintered synchroniser hubs, Powder Metall. 47 (2004) 352–357.
[7] K. Kim, H. Lee, Effect of friction between powder and a mandrel on densification of iron powder during cold isostatic pressing, Int. J. Mech. Sci. 40 (1998) 507–519.
[8] R. Lapovok, D. Tomus, B. Muddle, Low-temperature compaction of Ti–6Al–4V powder using equal channel angular extrusion with back pressure, Mater. Sci. Eng. A 490 (2008) 171–180.
[9] M. Kondoh, T. Saito, H. Takamiya, Green compact and process for compaction the same, metallic sintered body and process for producing the same, worked component part and method of working, Google Patents, 2003.
[10] S. Luo, Y. Yang, G. Schaffer, M. Qian, Warm die compaction and sintering of titanium and titanium alloy powders, J. Mater. Process. Tech. 214 (2014) 660–666.
[11] S. He, Warm compaction behavior of pure titanium powders, Adv. Mater. Res. 189 (2011) 2775–2779.

[12] S. He, Y. Ou, L. Yong, H. Tang, Y. Huang, Study of warm compaction behavior for titanium alloy powder, http://dspace.xmu.edu.cn:8080/dspace/handle/2288/60975, 2005.
[13] E. Gemelli, N. Camargo, Oxidation kinetics of commercially pure titanium, Revista Matéria 12 (2007) 525–531.
[14] Alloy Digest, Datasheet on Crucible A-55, ASM International, 1971.
[15] D. Benjamin, C.W. Kirkpatrick, Properties and Selection: Stainless Steels, Tool Materials and Special Purpose Metals, ASM International, Ohio, 1980.
[16] A. Simchi, G. Veltl, Behaviour of metal powders during cold and warm compaction, Powder Metall. 49 (2006) 281–287.
[17] W.H. Peter, Y. Yamamoto, W. Chen, R.R. Dehoff, S.D. Nunn, A.S. Sabau, J. Kiggans, T.R. Muth, G. Daehn, C. Tallman, Near Net Shape Manufacturing of New, Low Cost Titanium Powders for Industry, Oak Ridge National Laboratory (ORNL); Shared Research Equipment Collaborative Research Center, 2013.
[18] S.J. Gerdemann, P.D. Jablonski, Compaction of titanium powders, Metall. Mat. Trans. A 42 (2011) 1325–1333.
[19] J. Secondi, Modelling powder compaction: from a pressure-density law to continuum mechanics, Powder Metall. 45 (2002) 213–217.

Pressureless sintering of titanium and titanium alloys: sintering densification and solute homogenization

12

Ma Qian, Ya F. Yang*, Shudong D. Luo**, H.P. Tang*[†]

[*]RMIT University, School of Aerospace, Mechanical and Manufacturing Engineering, Centre for Additive Manufacturing, Melbourne, Victoria, Australia
[**]The University of Queensland, School of Mechanical and Mining Engineering, Brisbane, Australia
[†]State Key Laboratory of Porous Metal Materials, Northwest Institute for Nonferrous Metal Research, Xi'an, China

12.1 Introduction

The press-and-sinter approach is technically the simplest and economically the most attractive powder metallurgy (PM) approach for near-net-shape or preform fabrication. The process is typically applied to blended elemental powders or mixed elemental and master alloy powders (MAP), as prealloyed (PA) powders are difficult to press at room temperature because of their high strength. In addition, most PA powders are spherical, which makes the powder particles difficult to hold together. For instance, in order to press 100 mesh PA Ti-6Al-4V powder to 80% of its pore-free or theoretical density (TD), a compaction pressure of 965 MPa is necessary, which is essentially the yield strength of Ti-6Al-4V [1]. In contrast, it requires only 413 MPa to press blended elemental powders of Ti-6Al-4V to the same green density [1]. However, coating PA powders with a thin film of binder by a simple approach can enable easy cold die compaction [2]. In this regard, the press-and-sinter approach is applicable to all types of metal powder materials.

Central to the process is sintering, which produces the final product shapes and dimensions, microstructures, and mechanical properties. Compared to hot isostatic pressing (HIP), hot pressing (HP), and spark plasma sintering (SPS), sintering in vacuum or in a controlled gas atmosphere under atmospheric pressure may be referred to as *pressureless sintering*. To control interstitial contamination and ensure adequate tensile ductility, sintering of titanium and titanium alloys is usually carried out in a low oxygen (O) and nitrogen (N) environment, including good vacuum, high-purity argon, or a purposely designed low-oxygen medium (e.g., the OXYNON furnace, continuous operation using a carbon fiber-reinforced belt). In principle, titanium can also be sintered while submerged in molten calcium or other alkaline earth metals or

Titanium Powder Metallurgy. http://dx.doi.org/10.1016/B978-0-12-800054-0.00012-5

in the atmosphere of such a metal vapor, with or without the assistance of an inert gas [3]. In addition to oxygen and nitrogen, it is also important to avoid carbon contamination during the press-and-sinter process, particularly for the fabrication of titanium alloys containing a high level (e.g., 10 wt%) of molybdenum (Mo) and niobium (Nb) [4–7]. The solubility of carbon in such titanium alloys can be less than 100 ppm by weight. As a result, it is easy to form grain boundary titanium carbides, leading to much reduced ductility or brittleness even though the as-sintered titanium alloy may contain only 200–300 ppm carbon [4–7].

As first approximation in PM, it is important to sinter a green compact to its pore-free or near-pore-free density with a small maximum pore size, particularly for applications requiring good dynamic properties (e.g., for titanium alloys sintered to ≥99% TD, the maximum pore size should be limited to 10 μm for good fatigue properties [8]). Accordingly, sintering of titanium and titanium alloys has been largely focused on densification. This approach works for the sintering of commercially pure titanium (CP-Ti) and titanium alloys that contain no slow diffusers (in relation to the self-diffusion of titanium). However, it has proved to be more challenging to sinter titanium alloys that contain slow diffusers. Ivasishin and coworkers [9–10] discussed various diffusional phenomena during the sintering of several titanium alloys, particularly during heating around the $\alpha \rightarrow \beta$ transformation. Yang et al. [11] discussed the phenomenon of decoupled sintering densification and solute homogenization during the sintering of Ti-10V-2Fe-3Al compacted from elemental powder and MAP mixes. These studies revealed the complexity of the fabrication of titanium alloys by pressureless sintering. In general, high-performance PM Ti alloys are expected to show good chemical homogeneity and uniform microstructures (fine beta grains after isothermal sintering or small colonies) as well as high sintered densities and low interstitial impurities (O, N, and C).

The general aspects of the press-and-sinter approach have been discussed previously [12–14]. This chapter focuses on sintering densification and solute homogenization during the sintering fabrication of titanium and titanium alloys including CP-Ti, Ti-10V-2Fe-3Al, and Ti-6Al-4V. It is shown that achieving high sintering densification is only the first step in the sintering of titanium alloys containing slow diffusers because of the decoupled sintering densification (controlled by titanium self-diffusion) and solute homogenization (controlled by slow diffusers). It is essential to achieve both for desired mechanical properties.

12.2 Stability of the surface titanium oxide film

The surface oxide film on titanium powder is not persistent at elevated temperatures because of the significant solubility of oxygen (O) in both α-Ti and β-Ti [14]. For instance, β-Ti can dissolve 1.75–2.25 wt% O over the typical sintering temperance range of 1200–1350°C while the solidity limit of O in α-Ti is up to 14.25 wt% at 600°C [14,15]. The binary Ti–O phase diagram indicates that under equilibrium heating conditions, titanium oxide will disappear at 600°C [15], corresponding to the maximum solubility limit of O in Ti. For nonequilibrium heating, early work suggested that the oxide film on titanium powder disappears around 550°C (α phase region) [15]. A subsequent study reported that it took about 60 min for the surface oxide film to disappear on loose Ti powder at 1000°C (β phase region) [16]. Mo et al. [17] suggested that significant dissolution of the surface oxide film in Ti began at about 700°C (α phase region). Based on the solubility, the surface oxide film can disappear in either the α or β phase region, depending on heating rate.

The sintering densification of titanium powder can be assumed to start when noticeable dissolution of the surface oxide film begins. In that regard, the dilation curve of CP-Ti powder is informative for a reasonable estimate of the onset temperature. Figure 12.1 shows the dilation curve of 100 mesh hydride-dehydride (HDH)) CP-Ti powder (0.20 wt% O from Kimet China), compacted at 400 MPa, heated to 1300°C at 10°C/min in high-purity argon. The parameters selected are typical of the press-and-sinter approach for PM Ti. Sintering shrinkage started from 674°C (point A), determined using Proteus® Software (Version 6.1, NETZSCH) from the intersection of the two tangents around the approximate onset temperature. This onset temperature depends on powder characteristics and compaction pressure. The lowest temperature measured was 615°C for 100 mesh Sumitomo HDH Ti powder compacted at 200 MPa under the same heating conditions and the highest was 674°C out of six experiments with different powder characteristics. This measured range of onset temperatures (615–674°C) is generally consistent with the phase diagram temperature (600°C) mentioned previously. Considering that the temperature measured by a thermal couple always lags behind the actual temperature, it is reasonable to assume that noticeable dissolution of the surface oxide film on titanium powder could have

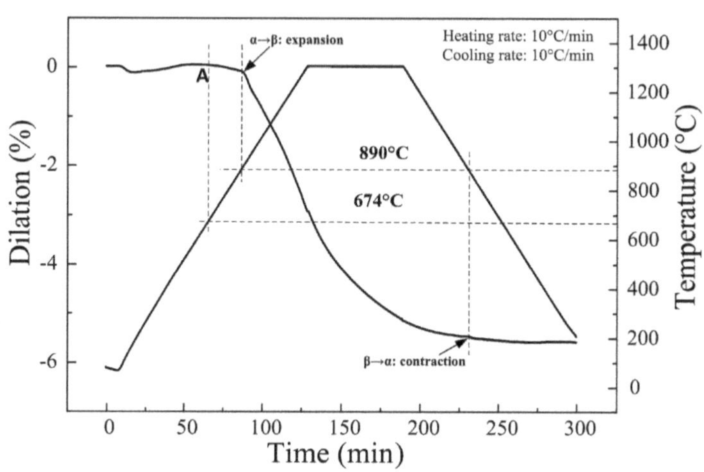

Figure 12.1 Sintering response of CP-Ti powder (100 mesh HDH Ti powder) compacted at 400 MPa, heated at 10°C/min 1300°C for 120 min in high purity argon.

started from a slightly lower temperature range (e.g., 600–650°C) during the sintering of HDH Ti powder under typical heating rates (10°C/min).

12.3 Sintering of CP-Ti

For effective high sintering densification, it is necessary to sinter CP-Ti green compacts in the temperance range of 1200–1350°C. The use of a lower sintering temperature is practical for fine titanium powder but, sintering at temperatures above 1350°C may trigger significant beta grain growth unless grain growth inhibitors are used [18]. As pointed out earlier, CP-Ti green compacts (100 mesh powder) start to shrink from 615–674°C and the shrinkage rate picks up in the beta region following the $\alpha \rightarrow \beta$ transformation due to the accelerated self-diffusion of titanium. Note that the atomic packing factors are 74% for α-Ti and 68% for β-Ti. Hence, there is slight expansion from α to β during heating and slight contraction from β to α during cooling (see Figure 12.1), but the net volume change is negligible.

The apparent activation energy (Q, kJ/mol) for sintering is indicative of the potential densification mechanism and densification during solid-state sintering occurs through lattice and/or grain boundary diffusion. The basic principle for determining the Q value is based on the Arrhenius equation [19]:

$$\ln(r) - n\ln(c) - \frac{nQ}{RT} + n\ln(t) \tag{12.1}$$

where r is the shrinkage rate (%/min), R the universal gas constant, T temperature (K), t time (minutes), n a time exponent independent of temperature, and C a constant. The value of n can be determined from the slope of the linear relationship between $\ln(r)$ and $\ln(t)$ and then the value of Q from the slope of the linear relationship between $\ln(r)$ and $1/T$ [11].

Figure 12.2 plots the experimental data on $\ln(r)$ and $1/T$ obtained over the range of 650–1250°C for the sintering of HDH Ti powder (<25 μm) [20]. A clear discontinuity occurred at about 900°C, resulting in two separate linear relationships and therefore two Q values: 184 ± 12 kJ/mol over the α-Ti range and 113 ± 4 kJ/mol over the beta range [20]. The activation energy value is reported to be in the range of 169–192 kJ/mol for self-diffusion of titanium over the α-Ti range [21–25] and 92.5–158 kJ/mol over the β-Ti range from 900°C to 1250°C [22,26–28]. This suggests that the densification of CP-Ti powder is controlled by the self-diffusion of titanium either in the α-Ti range or the β-Ti range.

It is useful to note that contrary to expectations, as-sintered CP-Ti is usually not a single α-Ti phase material because of the influence of the impurity iron [29]. It has been found that when the impurity iron is greater than about 1200 ppm by weight, iron-enriched grain boundary (GB) β phases can readily develop during cooling of the sintered CP-Ti because of the fast diffusion of Fe toward the GB region. As a result, the as-sintered CP-Ti is essentially an α–β material [29].

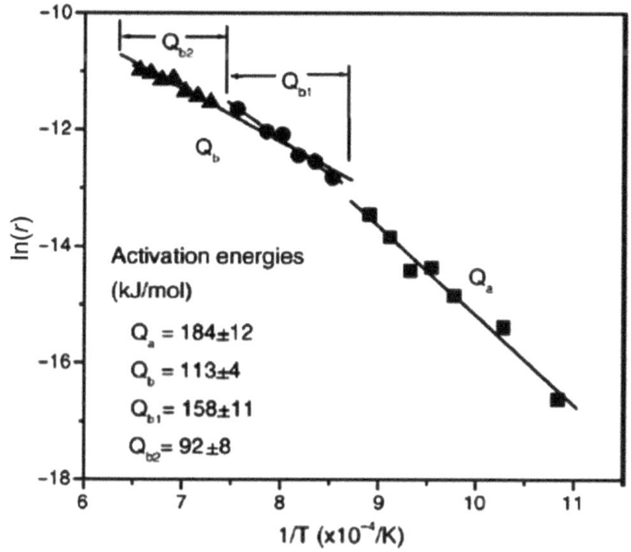

Figure 12.2 Arrhenius plots of shrinkage rate as a function of temperature for estimating the sintering activation energy of CP-Ti powder (<25 μm, about 0.25 wt% O) compacted at 300 MPa (70% TD) over the temperature range of 650–1250°C [20].

12.4 Sintering of Ti-10V-2Fe-3Al

Ti-10V-2Fe-3Al is a near beta titanium alloy with vanadium (V), a slow diffuser in titanium, as the principal alloying element. The decoupled sintering densification and solute homogenization issue during the sintering of titanium alloys containing slow diffusers is discussed below via the sintering of Ti-10V-2Fe-3Al.

12.4.1 Densification mechanism of Ti-10V-2Fe-3Al

HDH Ti powder and two MAPs, 66.7V-13.3Fe-20Al (equivalent to 10V-2Fe-3Al) and 85V-15Al (all in weight percentages), were used to fabricate Ti-10V-2Fe-3Al. Isothermal sintering was performed at 1300°C in vacuum. The equilibrium solidus temperatures of Ti-10V-2Fe-3Al, 10V-2Fe-3Al, and 85V-15Al alloys are about 1608°C, 1710°C, and 1860°C, respectively [11]. No liquid formation was detected in the Ti-10V-2Fe-3Al compacts during heating to, and isothermal holding at, 1300°C (Figure 12.3). Sintering occurred via solid-state diffusion.

Figure 12.4 shows the sintering response of Ti-10V-2Fe-3Al with the use of each MAP. As with the sintering of CP-Ti powder, shrinkage started at about 690°C and accelerated from about 880°C. The major difference noted is that the use of the 10V-2Fe-3Al MAP resulted in faster shrinkage than did the use of the 85V-15Al MAP above 1200°C. Figure 12.5 plots the experimental data on $\ln(r)$ and $1/T$ over the temperature

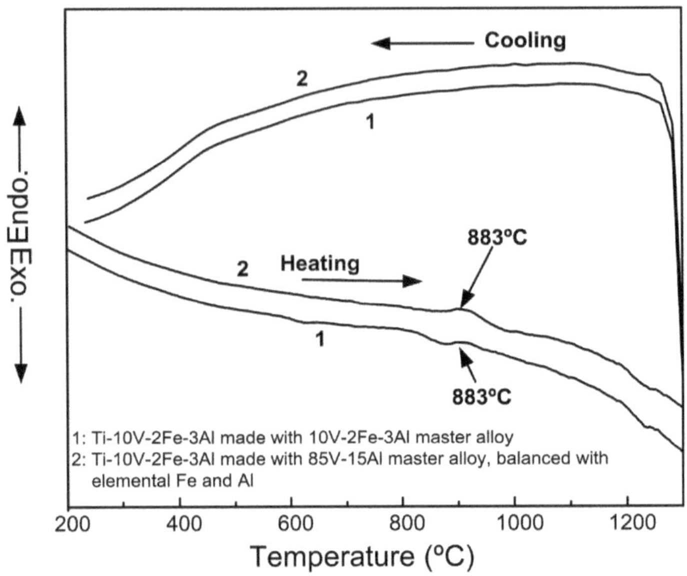

Figure 12.3 DSC curves of loose powder mixes in the composition of Ti-10V-2Fe-3Al during heating to 1300°C at 5°C/min and 15 min of subsequent holding at 1300°C, followed by cooling at 10°C/min to room temperature in high-purity argon [11]. The powder mixes are made of HDH Ti powder (75–145 μm) and 10V-2Fe-3Al or 85V-15Al master alloy powder (48–75 μm).

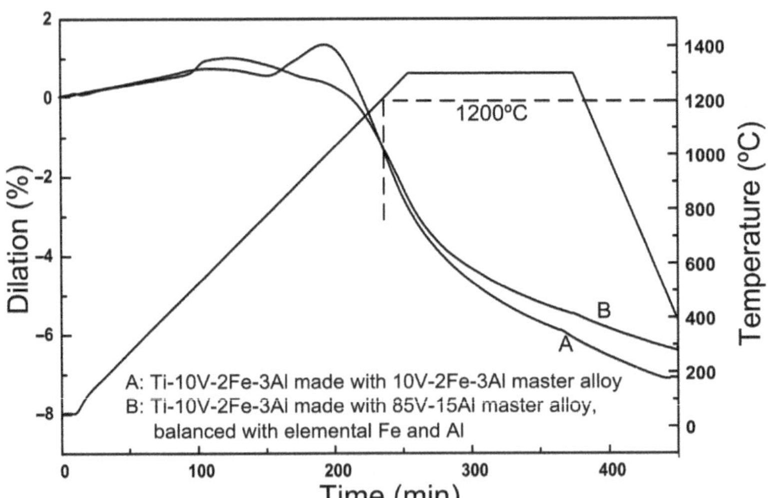

Figure 12.4 Dilatometer curves of Ti-10V-2Fe-3Al samples compacted from powder mixes of 75–145 μm HDH Ti powder and 48–75 μm master alloy powder at 400 MPa [11].

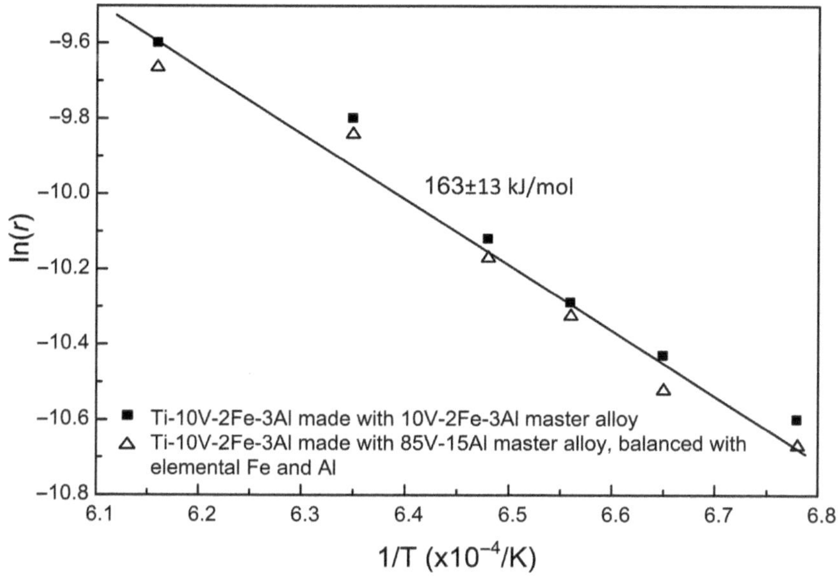

Figure 12.5 Logarithmic shrinkage rate ln(r) versus $1/T$ for the sintering of Ti-10V-2Fe-3Al compacted from powder mixes of 75–145 μm HDH Ti powder and 48–75 μm master alloy powder at 400 MPa (heating rate: 5°C/min) [11].

range of 1200–1350°C. The resulting linear relationship is essentially independent of the master alloy type. The Q value was determined to be 163 ± 13 kJ/mol, falling in the range of 130.6–251.2 kJ/mol for self-diffusion of titanium over the temperature range of 898–1540°C [25–28]. Data fitting using the Master Sintering Curve approach suggested a Q value of 160 kJ/mol for the sintering of CP-Ti powder over 1100–1300°C [21]. This suggests that the sintering densification of Ti-10V-2Fe-3Al over the temperature range of 1200–1350°C is dictated by self-diffusion of titanium, rather than the diffusion of vanadium.

The observation is not surprising considering that about 90% of the particle–particle contacts in Ti-10V-2Fe-3Al green compacts are titanium–titanium contacts. It also explains why the Q values are similar for the sintering of Ti-10V-2Fe-3Al and CP-Ti.

The sintering densification is affected by both the HDH Ti powder size and the MAP size. Figure 12.6 shows that achieving greater than 98% TD of Ti-10V-2Fe-3Al requires the use of fine titanium powder (≤38 μm), irrespective of master alloy selection. Fine MAP is preferred but the influence of the master alloy type was noticeable only when relatively coarse HDH Ti powder (≥75 μm) was used (Figure 12.6). Finally, the use of warm die (200°C) compaction increased the green density of Ti-10V-2Fe-3Al pressed from the same powder mixes by 4.9–6.4% TD over the compaction pressure range of 400–800 MPa and the sintered density by 1.5–2.7% TD after sintering at 1300°C in vacuum for 2 h [30].

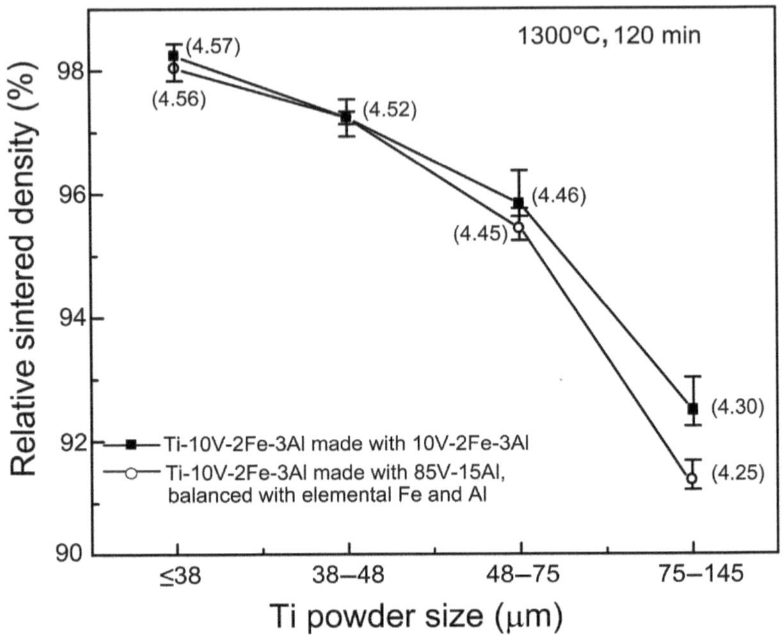

Figure 12.6 Effect of HDH Ti powder size and master alloy powder type (powder size: 48–75 μm) on the sintered density of Ti-10V-2Fe-3Al compacted from powder mixes at 400 MPa, sintered at 1300°C for 120 min (heating rate: 4°C/min) [11]. The theoretical density of Ti-10V-2Fe-3Al is 4.65 g/cm^3.

12.4.2 *Effect of alloying elements on titanium self-diffusion and sintering densification*

The introduction of a faster diffuser in general tends to enhance the self-diffusion of the base atoms and vice versa [31]. As a slow diffuser in β-Ti [22,32], vanadium (V) slows down the self-diffusion of titanium. For instance, at 1300°C, the self-diffusion coefficient of titanium decreases from $\sim 2 \times 10^{-8}$ cm^2/sec in pure titanium to $\sim 1.49 \times 10^{-8}$ cm^2/sec in Ti-10V (wt%), and the influence increases with increasing V content [32]. On the other hand, the diffusion of titanium from titanium particles into the neighboring V-enriched master alloy particles will enhance the diffusion of V. For example, at 1300°C, the self-diffusion coefficient of V increases from 2×10^{-11} cm^2/sec in pure vanadium to 3.57×10^{-9} cm^2/sec in Ti-50V (wt%) and further to 1.92×10^{-8} cm^2/sec in Ti-10V [32]. Aluminium (Al) is reported to be a slow diffuser in β-Ti [22,32], but experimental results indicate that Al becomes essentially uniform even before reaching the isothermal sintering temperature [11]. Iron (Fe) is a fast diffuser in β-Ti and enhances the self-diffusion of titanium [22,33]. However, increasing Fe content leads to noticeable grain coarsening during sintering [34]. Consequently, Fe has been introduced at a small quantity in most PM Ti alloys. Ti-1Al-8V-5Fe (wt%) is an exception and achieves a remarkable yield strength (>1600 MPa) when fabricated by PM routes [35]. Oxygen is a fast diffuser in β-Ti too and it enhances sintered density [36]. In general, alloying elements that enhance the self-diffusion of titanium tend to enhance sintering densification, and the opposite holds too.

Assuming the self-diffusion of titanium occurs via the first-nearest-neighbor monovacancy mechanism, the activation energy for the self-diffusion of titanium can be described as the sum of monovacancy formation energy (E_f) and migration energy (E_m). Density functional theory (DFT) modeling [33] has revealed that the introduction of Fe to β-Ti at the level of 1.89 at% exerts little influence on the monovacancy formation energy but it substantially decreases the monovacancy migration energy for the self-diffusion of titanium. Conversely, the introduction of V at the level of 1.89 at% to β-Ti only has a small influence on the E_f but it substantially increases the E_m. These DFT results offer a fundamental understanding of how Fe and V may affect the self-diffusion of titanium in the beta region, although the real situation can be more complicated.

Based on the discussion above, the densification of Ti-10V-2Fe-3Al compacted from HDH Ti and vanadium-enriched MAP mixes includes the following diffusion events [11]:

- the diffusion of aluminium from the master alloy particles to titanium particles, which slows down titanium self-diffusion;
- the diffusion of iron from the master alloy particles to titanium particles, which enhances titanium self-diffusion;
- the diffusion of vanadium from the master alloy particles to titanium particles, which decreases titanium self-diffusion; and
- the diffusion of titanium from the titanium particles to master alloy particles, which increases vanadium self-diffusion.

Compared to the Q value for Ti self-diffusion (130.6–251.2 kJ/mol), the Q value obtained (163 ± 13 kJ/mol) for the sintering of Ti-10V-2Fe-3Al suggests that the combined effects of the various diffusion events listed above resulted in little net influence on the self-diffusion of titanium in Ti-10V-2Fe-3Al. This helps explain why the densification of Ti-10V-2Fe-3Al at 1300°C is essentially similar to that of CP-Ti. In addition to its influence on titanium self-diffusion, the presence of each alloying element affects the solidus temperature of the resulting titanium alloy. For example, both Fe and Al decrease the solidus temperature while Mo and W offer the opposite. V first decreases but then increases the solidus temperature, depending on its content. A decrease in solidus temperature in general favors titanium self-diffusion. The influence is reflected in the apparent activation energy determined for sintering.

12.4.3 Solute homogenization during sintering

Apart from achieving a near-pore-free density, solute homogenization has proved to be another key issue that affects the microstructure and mechanical properties of an as-sintered titanium alloy. Figure 12.7 shows the as-sintered microstructures of Ti-10V-2Fe-3Al compacted from HDH Ti powder and 10V-2Fe-3Al or 85V-15Al MAP mixes after sintering at 1300°C for 2 h. The sintered densities (98.3% TD vs. 98.1% TD)

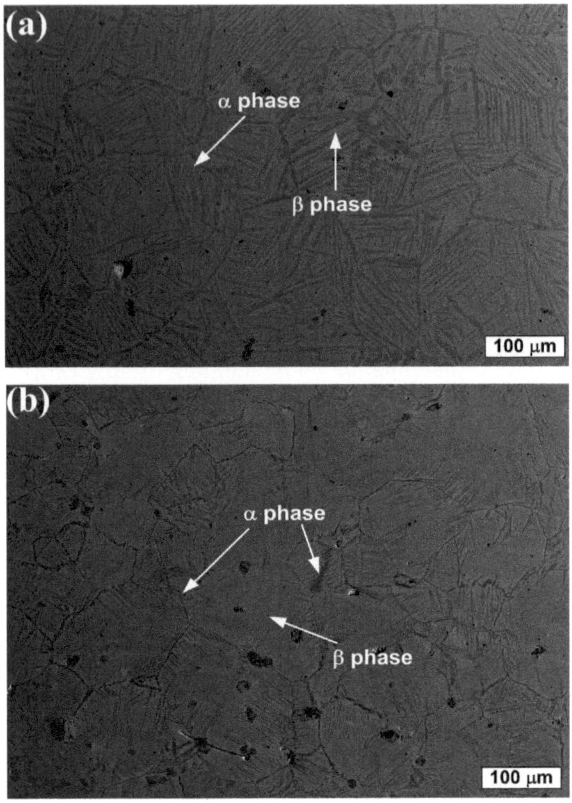

Figure 12.7 Scanning electron microscope–backscattered electron images of as-sintered microstructures of Ti-10V-2Fe-3Al compacted from powder mixes of HDH Ti powder (≤38 μm) and master alloy powder (≤38 μm) at 400 MPa: (a) 10V-2Fe-3Al master alloy and (b) 85V-15Al master alloy, sintered at 1300°C for 120 min in vacuum [11].

are similar. The use of the 10V-2Fe-3Al MAP produced a reasonably uniform microstructure. However, the use of the 85V-15Al MAP resulted in a distinctly nonuniform microstructure, with some β-Ti grains showing massive precipitation of acicular α-Ti (which formed during cooling below the beta transus temperature) while other β-Ti grains being essentially free of α-Ti precipitates (all the microstructures shown are at room temperature). The nonuniform distribution of V, shown in Figure 12.8a and b, was identified to be the principal reason because both Fe and Al were found to be essentially uniform even before reaching the sintering temperature of 1300°C (Figure 12.8c). Ivasishin et al. [10] investigated the diffusional phenomenon during heating of Ti-10V-2Fe-3Al compacted from TiH_2 powder and 10V-2Fe-3Al MAP mixes. They found that Al diffused faster than Fe into the titanium matrix, and this raised the α→β transformation temperature, thereby slowing down the diffusion of Fe from the master alloy particle. However, Fe will quickly diffuse into the titanium matrix from the master alloy particles when the α→β transformation completes [10]. The uniform distribution of

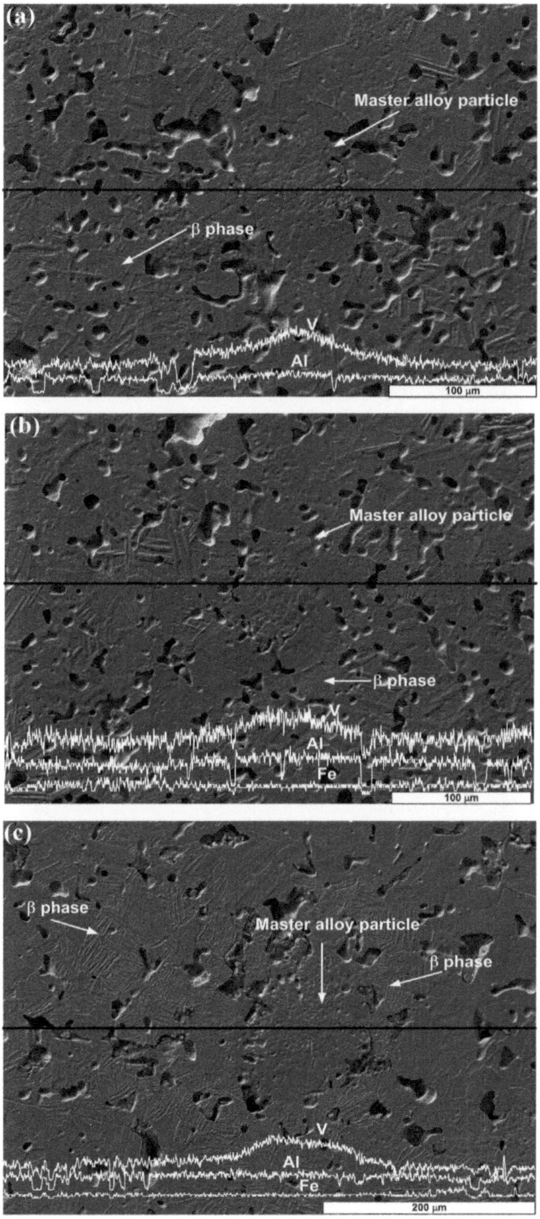

Figure 12.8 Scanning electron microscope microstructures and energy dispersive spectroscopy (EDS) line analyses across master alloy particles of 85V-15Al (a) and 10V-2Fe-3Al (b) in a Ti-10V-2Fe-3Al sample heated to 1300°C at 5°C/min without an isothermal hold [11]. Micrograph (c) shows the EDS line analyses across a 10V-2Fe-3Al master alloy powder particle in a Ti-10V-2Fe-3Al sample heated to 1200°C at 5°C/min. Samples of Ti-10V-2Fe-3Al were compacted at 400 MPa from powder mixes of HDH Ti powder (75–145 μm) and 10V-2Fe-3Al master alloy powder (48–75 μm) or 85V-15Al master alloy powder (48–75 μm), balanced with elemental Fe and Al powders. All samples were cooled at 40°C/min to room temperature.

Table 12.1 Tensile properties of as-sintered Ti-10V-2Fe-3Al fabricated with ≤ 38 μm titanium powder and ≤ 38 μm master alloy powder in the form of 10V-2Fe-3Al or 85V-15Al [11]

Alloy	Ultimate tensile strength (MPa)	Yield strength (MPa)	Elongation (%)	Sintered density (%)	Oxygen content (wt%)	Nitrogen content (wt%)
Ti-10V-2Fe-3Al (with 10V-2Fe-3Al master alloy), sintered at 1300°C for 120 min	952 ± 17	869 ± 13	8.2 ± 0.8	98.3 ± 0.3	0.5	0.02
Ti-10V-2Fe-3Al (with 85V-15Al master alloy, balanced with elemental Fe and Al), sintered at 1300°C for 120 min	931 ± 12	861 ± 12	3.3 ± 1.2	98.1 ± 0.4	0.53	0.02
Ti-10V-2Fe-3Al (with 85V-15Al master alloy, balanced with elemental Fe and Al), sintered at 1300°C for 180 min	958 ± 11	872 ± 9	8.0 ± 0.7	98.7 ± 0.1	–	–

Fe and Al shown in Figure 12.7c was reached during heating to 1200°C. This means that what happens during heating may not be that important as the densification occurs predominantly during isothermal sintering in the β phase region.

Table 12.1 lists the tensile properties of the as-sintered Ti-10V-2Fe-3Al with respect to the use of each master alloy. Similar sintered densities, ultimate tensile strengths, and yield strengths were obtained but the tensile ductility showed a distinct difference (8.2 ± 0.8% vs. 3.3 ± 1.2%). Chemical analyses revealed similar interstitial contents (0.5% O and 0.02% N vs. 0.53% O and 0.02% N; Table 12.1). The much lower tensile ductility arising from the use of the 85V-15Al MAP is thus attributed to the nonuniform microstructure caused by solute (vanadium) inhomogeneity. To address this issue, tensile samples of Ti-10V-2Fe-3Al compacted from fine HDH Ti powder (≤38 μm) and fine 85V-15Al MAP (≤38 μm) mixes were sintered at 1300°C for a further 60-min hold (i.e., from 120 to 180 min) to allow solute homogenization. The increased isothermal sintering time proved to be adequate to produce a microstructure similar to that shown in Figure 12.7a (see Figure 12.9). Accordingly, the tensile properties including tensile elongation became similar to those obtained with

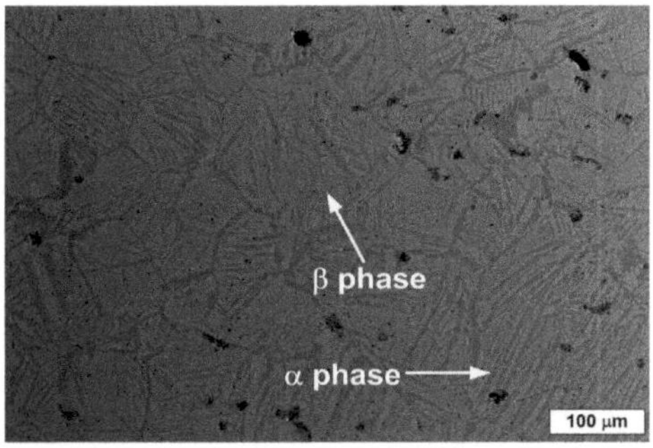

Figure 12.9 Scanning electron microscope–backscattered electron image of the as-sintered microstructure of Ti-10V-2Fe-3Al compacted from powder mixes of HDH Ti powder (≤38 μm) and 85V-15Al master alloy powder (≤38 μm) at 400 MPa, sintered at 1300°C for 180 min in a vacuum of 10^{-3}–10^{-2} Pa.

the use of the 10V-2Fe-3Al MAP (see Table 12.1). Fine 85V-15Al MAP (≤38 μm) was used in this case. The use of coarse 85V-15Al MAP may need a longer solute homogenization time.

Achieving high sintering densification is thus only the first step in the fabrication of titanium alloys containing slow diffusers such as Ti-10V-2Fe-3Al because of the decoupled sintering densification and solute homogenization. In order to achieve the desired microstructure and mechanical properties, it is important to also realize solute homogenization through appropriate master alloy selection and/or design of the sintering pathways.

12.5 Sintering of Ti-6Al-4V

The sintering densification mechanism of Ti-6Al-4V is expected to be similar to that of Ti-10V-2Fe-3Al, that is, controlled by the self-diffusion of titanium. However, no data seem to exist on the apparent activation energy (Q) for the sintering of Ti-6Al-4V compacted from HDH Ti powder and V-Al MAP mixes. The only data available are those reported recently by Crosby [37] for the sintering of PA Ti-6Al-4V powder (45–200 μm), which was ball-milled into angular Ti-6Al-4V powder (24 ± 2 μm in size, containing 4 wt% stearic acid) and uniaxially pressed into pellets of 12 mm diameter at 300 MPa (55.6% TD). Different heating rates (4–16°C/min) and isothermal sintering temperatures (900–1300°C) were employed. Using the Master Sintering Curve approach, Crosby [37] estimated a Q value of about 130 kJ/mol, agreeing with the reported Q values of 92.5–158 kJ/mol for self-diffusion of titanium over the β-Ti range from 900°C to 1250°C [22,26–28]. This supports the hypothesis that the sintering densification of Ti-6Al-4V is controlled by the self-diffusion of titanium. It also implies that decoupled sintering densification and solute homogenization will occur because of the involvement of the 4% V. Ivasishin et al. [9] compared the use of fine (≤20 μm) and coarse (120 μm) 60Al-40V MAPs for the fabrication of Ti-6Al-4V.

The use of fine alloying particles is critical to achieve solute homogenization. In addition, similar to the fabrication of Ti-10V-2Fe-3Al, Al diffused fast into the surrounding titanium matrix from the 60Al-40V master alloy particle during heating, leading to the formation of stable α phases, which can exist well above the α→β transformation temperature (882°C) for pure Ti [9]. Such stable α phases act as barriers to the diffusion of V from the master alloy particles during heating until the α→β transformation completes at a higher temperature [9]. However, it should be pointed out that both sintering densification and solute homogenization occur predominantly during the subsequent isothermal sintering and the influence of the diffusional phenomenon during heating is limited.

12.6 Enhanced densification with sintering aids

A small addition of boron (≤0.5 wt%) has proved to be effective in enhancing the sintering densification of CP-Ti, Ti-6Al-4V, Ti-10V-2Fe-3Al (see Figure 12.10), and Ti-Ni alloys as well as being able to significantly refine both the β-Ti grains

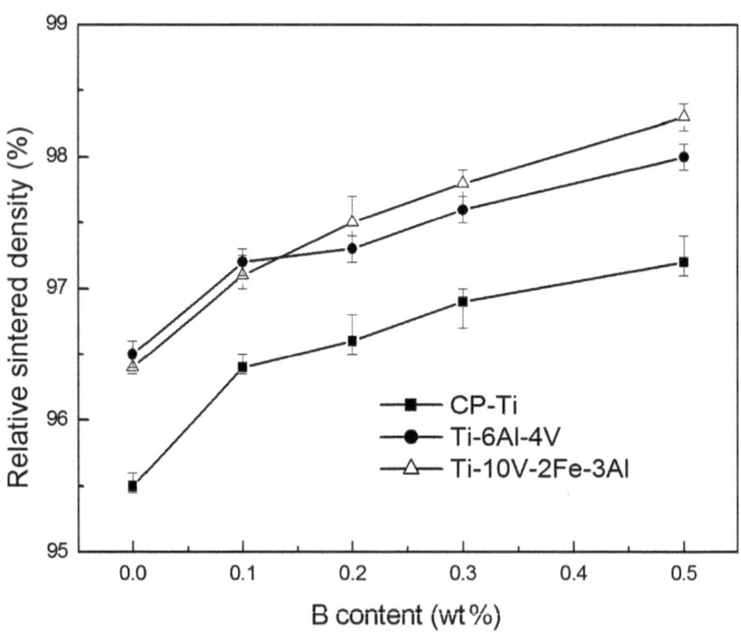

Figure 12.10 Sintered densities of CP-Ti, Ti-6Al-4V, and Ti-10V-2Fe-3Al compacted from powder mixes of HDH Ti powder (particle size <63 μm), 58V-42Al (particle size <45 μm), and 66.7V-13.3Fe-20Al master alloy powder (particle size <45 μm) at 600 MPa, with the addition of boron (<1 μm) as a sintering aid. Sintered at 1350°C for 120 min in a vacuum of 10^{-3}–10^{-2} Pa.

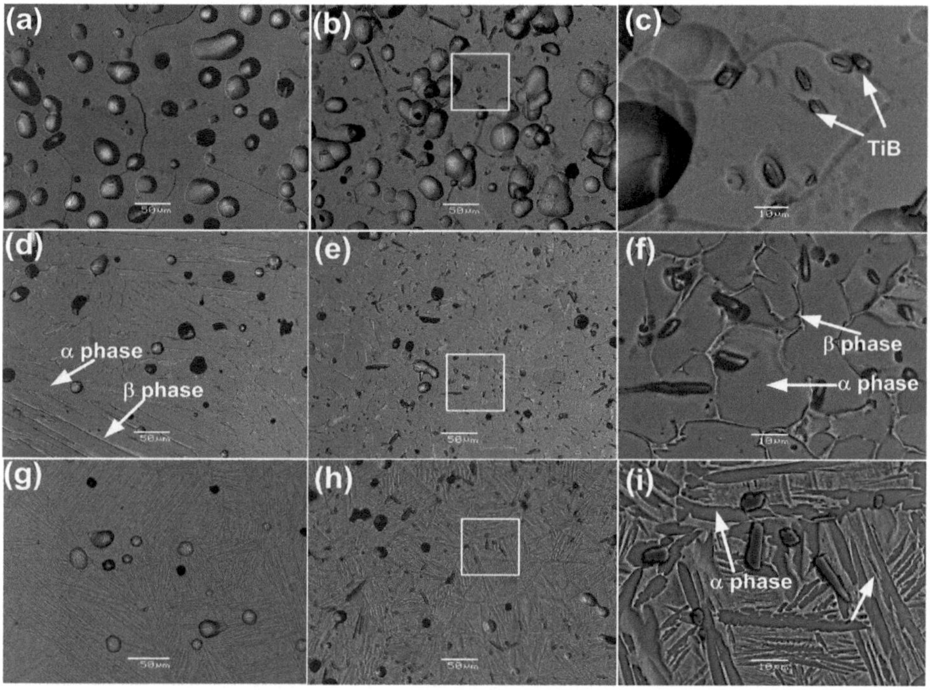

Figure 12.11 As-sintered microstructures of CP-Ti, Ti-6Al-4V, and Ti-10V-2Fe-3Al compacted from powder mixes of HDH Ti powder (particle size <63 μm), 58V-42Al (particle size <45 μm), and 66.7V-13.3Fe-20Al master alloy powder (particle size <45 μm) at 600 MPa, with and without an addition of 0.3%B (<1 μm): (a) CP-Ti; (b) CP-Ti-0.3B; (c) enlarged view of (b); (d) Ti-6Al-4V; (e) Ti-6Al-4V-0.3B; (f) enlarged view of (e); (g) Ti-10V-2Fe-3Al; (h) Ti-10V-2Fe-3Al-0.3B; (i) enlarged view of (h). Sintered at 1350°C for 120 min in a vacuum of 10^{-3}–10^{-2} Pa [38].

and the α-Ti laths and modify the α-lath morphology to near-equiaxed grains (see Figure 12.11) [38–39]. Silicon is another effective sintering aid for titanium alloys [34,40]. In addition, a small addition (≤1.5 at%) of cobalt, nickel, or copper enables the sintering of PA Ti-48Al-2Cr-2Nb powder to near-pore-free densities [1,41–42]. The mechanism by which each of the sintering aids works is due to the formation of a transient or persistent sintering liquid or the decreased solidus temperature and increased self-diffusion of titanium [2,38–42]. Exceptions seem to exist. Ivasishin and coworkers [43] sintered Ti-2.5Fe and Ti-5Mn alloys using TiH_2 powder with elemental Fe and Mn powders or 44Ti-56Fe and 31Ti-69Mn MAPs. The formation of transient eutectic liquid was found to be detrimental to the sintering densification [43].

Finally, the use of TiH_2 powder to replace HDH Ti powder has proved to be beneficial to the sintering densification of CP-Ti and titanium alloys. This has been discussed in a separate chapter of this book.

12.7 Conclusion remarks

Understanding the sintering densification mechanism and solute homogenization process is essential to the fabrication of titanium alloys containing slow diffusers by the press-and-sinter approach. From a sintering perspective, titanium powder is easy to sinter in a low oxygen atmosphere as there is no need to disrupt the surface oxide film, which will dissolve into the underlying titanium matrix from about 615–674°C, because of the significant solubility of oxygen in both α-Ti and β-Ti. As with the sintering of CP-Ti powder, the sintering densification of titanium alloys such as Ti-10V-2Fe-3Al and Ti-6Al-4V is controlled by the self-diffusion of titanium according to the apparent activation energy obtained for the sintering of each alloy. Fast-diffuser alloying elements tend to enhance the sintering densification while slow-diffuser alloying elements tend to slow down the sintering densification. It is possible that the net effect may be negligible in some alloys, making the sintering densification process of these alloys essentially similar to that of CP-Ti.

In general, it is easy to achieve high sintering densification of both CP-Ti and titanium alloys using relatively fine titanium powder. Additionally, sintering densification can be effectively improved via the use of warm die compaction or sintering aids. However, unlike the sintering of CP-Ti and titanium alloys free of slow diffusers, achieving high sintering densification is only the first step in the sintering fabrication of titanium alloys containing slow diffusers because of the decoupled sintering densification and solute homogenization. To ensure good mechanical properties, it is necessary to achieve both high sintering densification and solute homogenization. The latter can be achieved through appropriate master alloy selection (both master alloy type and powder size) and/or design of appropriate sintering pathways.

Acknowledgments

The authors acknowledge the financial support from the Australian Research Council (ARC) through the ARC CoE for Design in Light Metals and the Ministry of Science and Technology China under the International Science & Technology Cooperation Program (2011DFA5290). Dr Ya Feng Yang further acknowledges the financial support from the Australian Research Council (ARC) for his DECRA fellowship.

References

[1] S. Abkowitz, J.M. Siergiej, R.D. Regan, Titanium P/M preforms, parts and composites, Mod. Dev. Powder Metall. 4 (1971) 501–511.

[2] Y. Xia, G.B. Schaffer, M. Qian, The effect of a small addition of nickel on the sintering, sintered microstructure, and mechanical properties of Ti-45Al-5Nb-0.2C-0.2B alloy, J. Alloys. Comp. 578 (2013) 195–201.

[3] D.W. Rostron, Method of sintering titanium and like metals, U.S. Patent 2,546,320 (Priority date: 4 November 1948; Publication date: 27 March 1951) (1951).

[4] M. Yan, M. Qian, C. Kong, M.S. Dargusch, Impacts of trace carbon on the microstructure of as-sintered biomedical Ti-15Mo alloy and reassessment of the maximum carbon limit, Acta Biomaterialia 10 (2014) 1014–1023.

[5] D. Zhao, K. Chang, T. Ebel, M. Qian, R. Willumeit, M. Yan, et al., Titanium carbide precipitation in Ti-22Nb alloy fabricated by metal injection moulding, Powder Metall. 57 (2014) 2–4.

[6] D. Zhao, K. Chang, T. Ebel, M. Yan, M. Qian, R. Willumeit, et al., Microstructure and mechanical behavior of metal injection moulded Ti-Nb binary alloys as biomedical material, J. Mech. Behav. Biomed. Mater. 28 (2013) 171–182.

[7] M. Qian, Metal injection moulding (MIM) of titanium and titanium hydride reviewed at PM titanium 2013, Powder Inj. Mould. Int. 8 (2) (2014) 65–72.

[8] T. Saito, T. Furuta, Sintered powdered titanium alloy and method of producing the same, U.S. Patent, 5,409,518 (Priority date: 8 November 1991) (1995).

[9] O.M. Ivasishin, D. Eylon, V.I. Bondarchuk, D.G. Savvakin, Diffusion during powder metallurgy synthesis of titanium alloys, Defect Diffus. Forum 277 (2008) 177–185.

[10] O.M. Ivasishin, D.G. Savvakin, The impact of diffusion on synthesis of high-strength titanium alloys from elemental powder blends, Key Eng. Mat. 436 (2010) 113–121.

[11] Y.F. Yang, S.D. Luo, G.B. Schaffer, M. Qian, Sintering of Ti-10V-2Fe-3Al and mechanical properties, Mater. Sci. Eng. A. 528 (2011) 6719–6726.

[12] F.H. Froes, D. Eylon, Powder metallurgy of titanium alloys, Int. Mat. Rev. 35 (1) (1990) 162–184.

[13] I.M. Robertson, G.B. Schaffer, Review of densification of titanium based powder systems in press and sinter processing, Powder Metall. 53(2) (2010) 146–162.

[14] M. Qian, Cold compaction and sintering of titanium and its alloys for near-net-shape or preform fabrication, Int. J. Powder Metall. 46 (2010) 29–44.

[15] J.L. Murray, H.A. Wriedt, The O-Ti (Oxygen-Titanium) System, Phase Diagrams of Binary Titanium Alloys, ASM International, Metals Park, OH, 1987.

[16] T. Watanabe, Y. Horikoshi, The sintering phenomenon of titanium powders—a discussion, Inter. J. Powder Metall. 12 (1976) 209–214.

[17] W. Mo, G.Z. Deng, F.C. Luo, Titanium Metallurgy, second ed., Metallurgical Industry Press, Beijing, 2007.

[18] M. Marty, H. Octor, A. Walder, Process for forming a titanium base alloy with small grain size by powder metallurgy, U.S. Patent, 4,601,874 (Priority date: 6 July 1984) (1986).

[19] L.C. Pathak, S.K. Mishra, P.G. Mukunda, M.M. Godkhindi, D. Bhattacharya, K.L. Chopra, Sintering studies on submicrometre-sized Y-Ba-Cu-oxide powder, J. Mater. Sci. 29 (1994) 5455–5461.

[20] B.B. Panigrahi, M.M. Godkhindi, K. Das, P.G. Mukunda, P. Ramakrishnan, Sintering kinetics of micrometric titanium powder, Mater. Sci. Eng. A 396 (2005) 255–262.

[21] I.M. Robertson, G.B. Schaffer, Some effects of particle size on the sintering of titanium and a master sintering curve model. Metal. Mater. Trans. A 40A (2009) 1968–1979.

[22] Y. Mishin, C. Herzig, Diffusion in the Ti-Al system, Acta Mater. 48 (2000) 589–623.

[23] C. Herzig, R. Willecke, K. Vieregge, Self-diffusion and fast cobalt impurity diffusion in the bulk and in grain boundaries of hexagonal titanium, Phil. Mag. A 63 (1991) 949–958.

[24] M. Koppers, C. Herzig, M. Friesel, Y. Mishin, Intrinsic self-diffusion and substitutional Al diffusion in α-Ti, Acta Mater. 45 (1997) 4181–4191.

[25] C. Herzig, T. Wilger, T. Przeorski, F. Hisker, S. Divinski, Titanium tracer diffusion in grain boundaries of α-Ti, α2-Ti3Al, and γ-TiAl and in α2/γ interphase boundaries, Intermetallics 9 (2001) 431–442.

[26] M.C. Naik, R.P. Agarwala, Anomalous diffusion in beta zirconium, beta titanium and vanadium, J. Phys. Chem. Solids. 30 (1969) 2330–2334.
[27] A.E. Pontau, D. Lazarus, Diffusion of Titanium and Niobium in bcc Ti-Nb Alloys, Phys. Rev. A 19 (1973) 4027–4037.
[28] G. Neumann, V. Tolle, C. Tuijn, On the impurity diffusion in beta-Ti, Physica B 296 (2001) 334–341.
[29] M. Yan, S.D. Luo, G.B. Schaffer, M. Qian, Impurity (Fe, Cl, and P)-induced grain boundary and secondary phases in commercially pure titanium (CP-Ti), Metall. Mat. Trans. A 44(8) (2013) 3961–3969.
[30] S.D. Luo, Y.F. Yang, G.B. Schaffer, M. Qian, Warm die compaction and sintering of titanium and titanium alloy powders, J. Mater. Process. Technol. 14 (2014) 660–666.
[31] J.F. Murdock, C.J. Mchargue, Self-diffusion in body-centered cubic titanium-vanadium alloys, Acta Metall. 16 (1968) 493–500.
[32] G. Lütjering, J.C. Williams, Titanium, Second ed., Springer, Berlin, 2007.
[33] P. Yao, Density functional theory study of the self-diffusion in pure beta titanium and titanium alloys, Master's thesis, The University of Queensland, Queensland, 2012.
[34] Y.F. Yang, S.D. Luo, G.B. Schaffer, M. Qian, The sintering, sintered microstructure and mechanical properties of Ti-Fe-Si alloys, Metall. Mat. Trans. A 43 (12) (2012) 4896–4906.
[35] V.V. Joshi, C. Lavender, V. Moxon, V. Duz, E. Nyberg, K.S. Weil, Development of Ti-6Al-4V and Ti-1Al-8V-5Fe alloys using low-cost TiH_2 powder feedstock, J. Mat. Eng. Perform. 22 (4) (2013) 995–1003.
[36] O.M. Ivasishin, D.G. Savvakin, M.M. Gumenyak, O.B. Bondarchuk, Role of surface contamination in titanium PM, Key Eng. Mat. 520 (2012) 121–132.
[37] K.D. Crosby, Titanium-6Aluminum-4Vanadium for functionally graded orthopedic implant applications, Doctoral thesis, University of Connecticut, 2013.
[38] Y.F. Yang, S.D. Luo, G.B. Schaffer, M. Qian, Modification of the α-Ti laths to near equiaxed α-Ti grains in as-sintered titanium and titanium alloys by a small addition of boron, J. Alloys Comp. 579 (2013) 553–557.
[39] S.D. Luo, Y.F. Yang, G.B. Schaffer, M. Qian, The effect of a small addition of boron on the sintering densification, microstructure and mechanical properties of powder metallurgy Ti-7Ni alloy, J. Alloys Comp. 55 (2013) 339–346.
[40] Y.F. Yang, S.D. Luo, C.J. Bettles, G.B. Schaffer, M. Qian, The effect of Si additions on the sintering and sintered microstructure and mechanical properties of Ti-3Ni alloy, Mater. Sci. Eng. A 528 (2011) 7381–7387.
[41] Y. Xia, G.B. Schaffer, M. Qian, Cobalt-doped Ti-48Al-2Cr-2Nb alloy fabricated by cold compaction and pressureless sintering, Mater. Sci. Eng. A 574 (2013) 176–185.
[42] Y. Xia, S.D. Luo, X. Wu, G.B. Schaffer, M. Qian, The sintering densification, microstructure and mechanical properties of gamma Ti-48Al-2Cr-2Nb alloy with a small addition of copper, Mater. Sci. Eng. A 559 (2013) 293–300.
[43] O.M. Ivasishin, D.G. Savvakin, X.O. Bondareva, in: H. Danninger and R. Ratzi (Eds), PM synthesis of low porous titanium alloyed with eutectic-forming elements. Proceedings of World Congress on Powder Metallurgy (PM-2004), Published by EPMA, 2004, vol. 4, 2004, pp. 719–724.

Spark plasma sintering and hot pressing of titanium and titanium alloys

13

Ya F. Yang, Ma Qian
RMIT University, School of Aerospace, Mechanical and Manufacturing Engineering, Centre for Additive Manufacturing, Melbourne, Victoria, Australia

13.1 Introduction

Pressureless sintering offers a cost-effective approach to the fabrication of titanium and titanium alloys in near-net-shape product forms [1]. However, it requires the use of a high sintering temperature (≥1200°C) and a lengthy isothermal hold (≥120 min) for both densification and solute homogenization [2,3]. Even so, it may still be challenging to achieve a pore-free homogenous microstructure, particularly for titanium alloys that contain low diffusors [2].

Spark plasma sintering (SPS) is a pressure-assisted pulsed-current process in which the powder samples are loaded in an electrically conducting die and sintered under a uniaxial pressure [4–8]. The die itself acts as a heating resource when a pulsed direct current (DC) passes through it and the powder samples if conductive. Hence metal powder samples can be heated from both the outside and inside and this leads to fast heating, enhanced mass transfer, and rapid consolidation. SPS can now be used to fabricate large preforms up to 300 mm in diameter from which finished parts can be machined. This makes SPS a practical option for the fabrication of speciality and/or reactive alloys. To date, SPS of titanium and titanium alloys is still limited to laboratory studies [9–16].

Similar to SPS, metal powder can be sintered in a die under a uniaxial pressure hot pressing (HP) but in the absence of electric current [17–19]. Electrical resistance heating is often used in HP. Other heating methods include induction and electric convection/radiant/conduction [17–19] but they all differ from the heating mechanism used in SPS. HP has been used to consolidate titanium and titanium alloys from powder as a useful titanium powder metallurgy (PM) approach [20–25].

Since HP is an established PM process, the rest of this chapter will start with the fabrication of PM CP-Ti and Ti-6Al-4V by HP and then focus on the fabrication of PM CP-Ti and Ti-6Al-4V by SPS.

13.2 HP of CP-Ti and Ti-6Al-4V

Particle deformation is responsible for consolidation of metal powders by HP. For titanium, this can be inferred from the nearly linear decrease in the yield strength of titanium with temperature shown in Figure 13.1. The same mechanism applies to

Titanium Powder Metallurgy. http://dx.doi.org/10.1016/B978-0-12-800054-0.00013-7

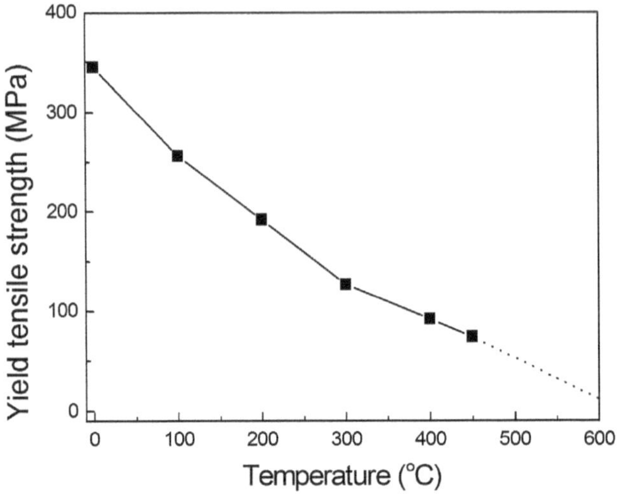

Figure 13.1 Yield strength of unalloyed titanium (ASTM Grade II) with temperature [25,26,31].

warm die compaction of titanium and titanium alloys [26]. Prealloyed (PA) Ti-6Al-4V powder has higher strength than unalloyed Ti powder at low temperatures but it will lose most of its strength at temperatures above 1000°C (Figure 13.2) [20]. For instance, the strength of Ti-6Al-4V decreases from about 140 MPa at 750°C to just about 12 MPa at 1050°C. Accordingly, it makes no difference essentially to consolidate either hydrogenated–dehydrogenated (HDH) Ti powder or PA Ti-6Al-4V powder by HP at temperatures above 1000°C.

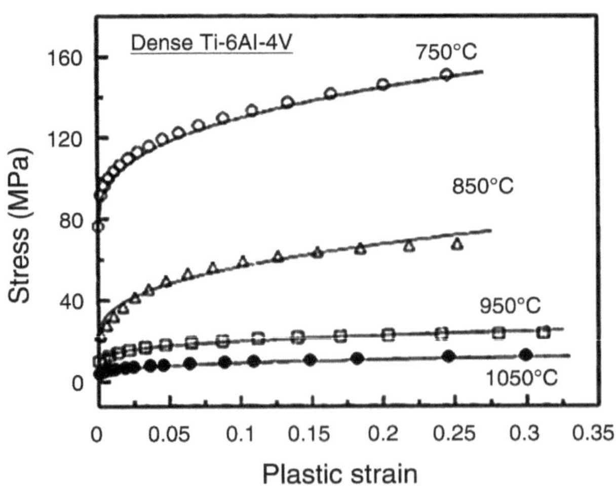

Figure 13.2 Compression strength of Ti-6Al-4V alloy at elevated temperatures [20].

Table 13.1 **Conditions for obtaining fully dense CP titanium by hot pressing [22]**

Powder	Powder size (μm)	Oxygen content (wt.%)	Consolidation temperature (°C)	Isothermal hold (min)	Pressure (MPa)
CP titanium	<26	0.45	1100	15	50
CP titanium	<75	0.272	1100	15	50
Titanium hydride	<26	0.67	1300	15	50

CP-Ti has been fabricated by HP using HDH Ti powder or TiH_2 powder [20–24,26]. Table 13.1 summarizes the HP conditions required for producing pore-free CP-Ti from HDH Ti powder [22]. The sintered density showed no dependence on powder size as densification was realized by particle deformation rather than diffusion. HP of TiH_2 powder seems to require a higher sintering temperature (1300°C) for full densification. However, it should be noted that dehydrogenation can be complete before 1000°C on slow heating. Hence full densification should be achievable at 1100°C for TiH_2 powder too. In addition, Ti-6Al-4V can be fabricated by HP using either elemental and master alloy (EMA) powder mixtures or PA Ti-6Al-4V powder under similar conditions (1100°C, 15 min, 50 MPa) used for HDH Ti (Table 13.2).

The microstructures of Ti-6Al-4V fabricated by HP of PA powder or EMA powder mixtures at 1100°C for 15 min are shown in Figure 13.3 [22,23]. The use of PA powder resulted in a microstructure (Figure 13.3a) similar to that fabricated by pressureless sintering, consisting of α–β colonies while the use of EMA powder mixtures produced an inhomogeneous microstructure with residual master alloy particles (Figure 13.3a). However, increasing HP temperature to 1300°C and isothermal hold to 30 min turned the inhomogeneous microstructure into a homogeneous microstructure (Figure 13.3c) because of the largely improved diffusion kinetics and the longer

Table 13.2 **Hot pressing of Ti-6Al-4V PA powders and EMA powder mixtures at 1100°C**

Powder	Powder size (μm)	Oxygen (wt%)	Consolidation temperature (°C)	Relative pore-free density (%)
PA Ti-6Al-4V	<45	0.39	1100	99.8
PA Ti-6Al-4V	<75	0.418	1100	99.9
EMA powder mixtures	<106 for Ti powder	0.428	1100	99.7

The isothermal hold was 15 min and the pressure used was 50 MPa [22,23]

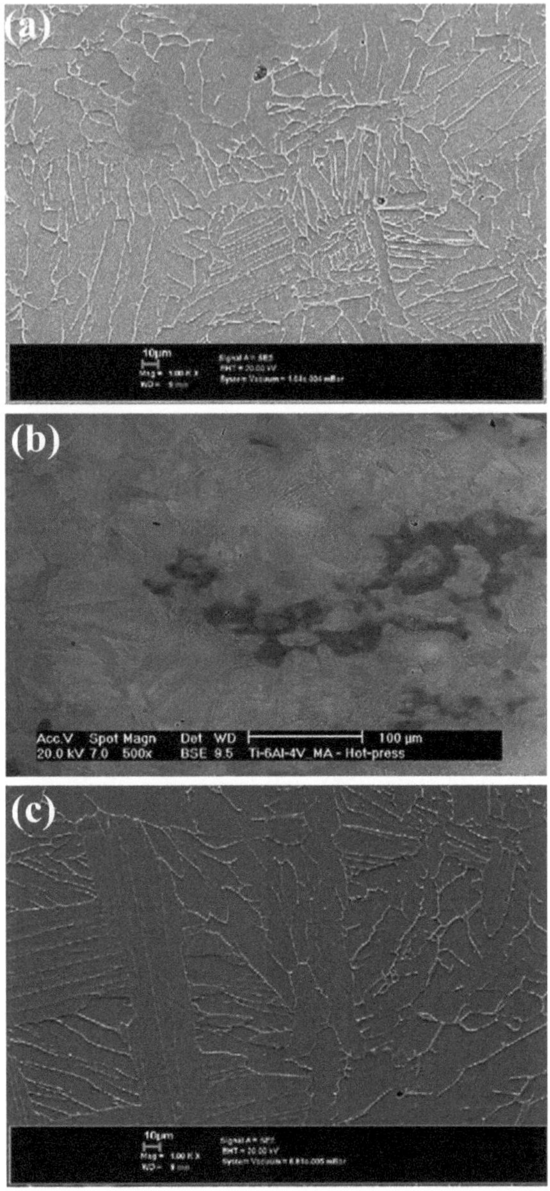

Figure 13.3 Microstructures of Ti-6Al-4V made from PA powder (75 μm) by hot pressing at 1100°C for 15 min (a), and EMA powder mixtures by hot pressing at 1100°C for 15 min (b) and 1300°C for 30 min (c) [22,23]. EMA powder mixtures: HDH titanium powder (<75 μm) and 40V-60Al master alloy powder (<106 μm). The applied pressure during HP: 30 MPa; heating rate: 10°C/min and vacuum pressure: 10^{-1} mbar.

isothermal hold [22,23]. Owing to the rapid consolidation process, the oxygen and nitrogen contents remained essentially unchanged after HP [22,23].

13.3 SPS of CP-Ti

13.3.1 Effect of sintering conditions

Both sintering temperature and applied pressure affect the sintered density of CP-Ti until full densification is achieved. Their influences are interrelated. Figure 13.4 shows the consolidation of gas-atomized Grade 1 and HDH Grade 3 CP-Ti powders (<45 μm) by SPS as a function of sintering temperature in low vacuum (2 Pa) under the pressure of 60 MPa [10]. The isothermal hold at each temperature was 5 min. Increasing sintering temperature was effective before reaching 950°C but not afterwards because of the rapid achievement of full densification (Figure 13.4). Figure 13.5 shows the effect of applied pressure over the range of 10–100 MPa on the sintered density of CP-Ti powder (<45 μm) by SPS [9], where samples were heated to 950°C at 50°C/min without an isothermal hold. Densification started to pick up from about 480°C, irrespective of the applied pressure, but full densification at 950°C was achieved only with the assistance of 100 MPa pressure. The combination of 60 MPa and a 5-min isothermal hold at 950°C proved to be capable of producing a pore-free microstructure of CP-Ti using <45 μm powder. This demonstrates the remarkable effectiveness of SPS in consolidating CP-Ti powder. Figure 13.6a shows a pore-free CP-Ti microstructure fabricated by SPS. A fine grain size was obtained because of rapid consolidation. However,

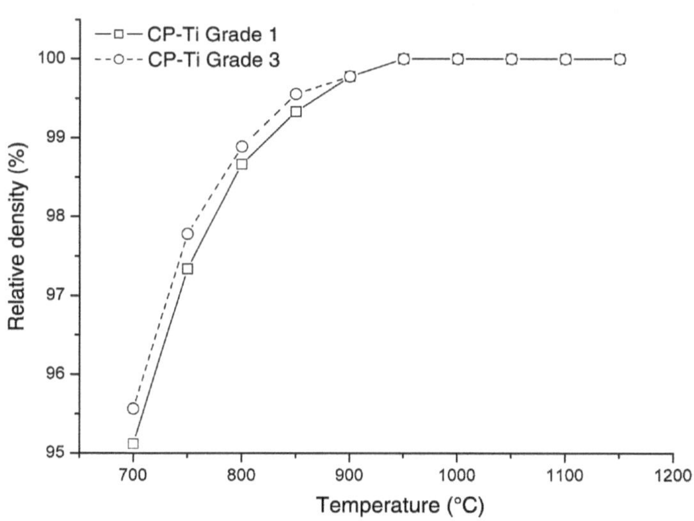

Figure 13.4 Consolidation of CP-Ti powders (gas-atomized Grade 1 and HDH Grade 3; <45 μm) by SPS vs. sintering temperature [10]. Applied pressure: 60 MPa, vacuum condition: 2 Pa and isothermal hold: 5 min at each temperature.

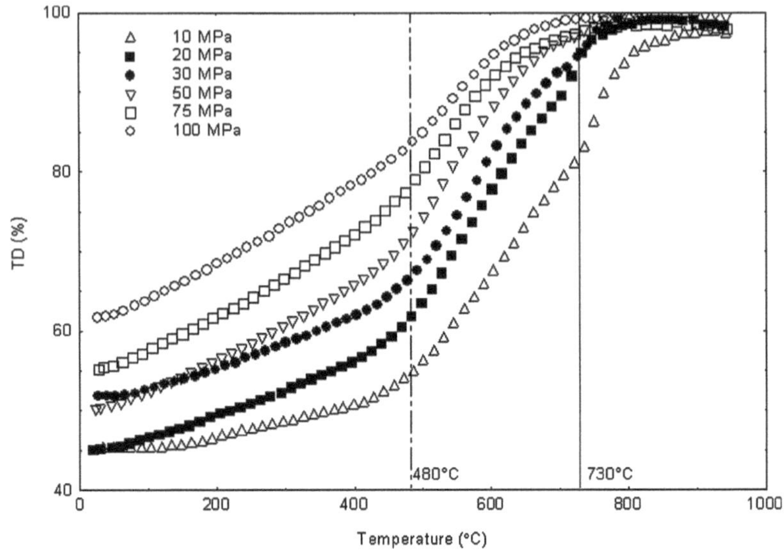

Figure 13.5 Consolidation of HDH CP-Ti powder (<45 μm) by SPS versus applied pressure [9]. Samples were heated to 950°C at 50°C/min without an isothermal hold.

significant grain growth will occur when sintered at temperatures above 900°C because of fast self-diffusion in the β region, accompanied by a change from equiaxed grains to irregular grains (Figure 13.6b) [10].

Heating rate was found to affect the densification process but not necessarily the final sintered density (Figure 13.7) [9]. This differs from the fabrication of bulk aluminium metallic glasses by SPS, where the use of a high heating rate was found to be essential for the removal of the surface oxide film on powder particles [27]. The effect of pulse sequence on sintering density, shown in Figure 13.8, is similar to that of the heating rate.

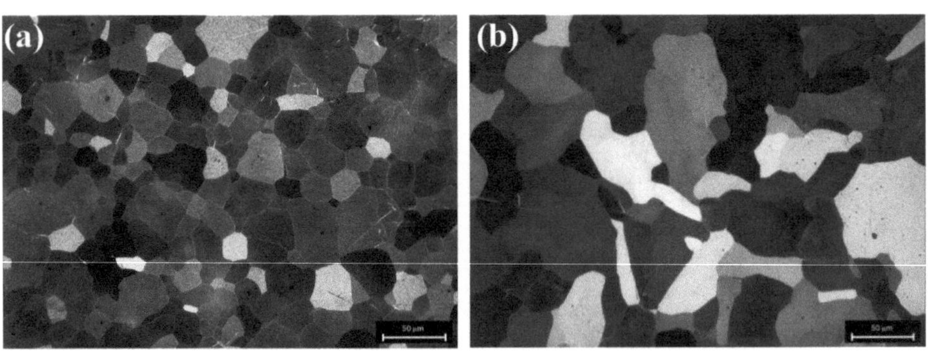

Figure 13.6 As-sintered microstructures of CP-Ti (gas-atomized ASTM Grade 1, <45 μm) by SPS: (a) 900°C for 5 min and (b) 950°C for 5 min. Applied pressure: 60 MPa and vacuum pressure: 2 Pa [10].

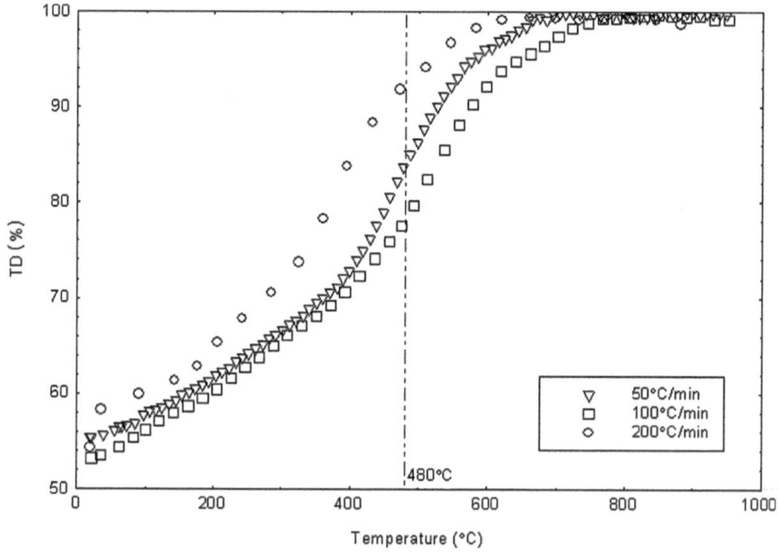

Figure 13.7 Consolidation of HDH CP-Ti powder (<45 μm) by SPS as a function of heating rate [9]. Samples were heated to 950°C at three heating rates under the pressure of 50 MPa without an isothermal hold.

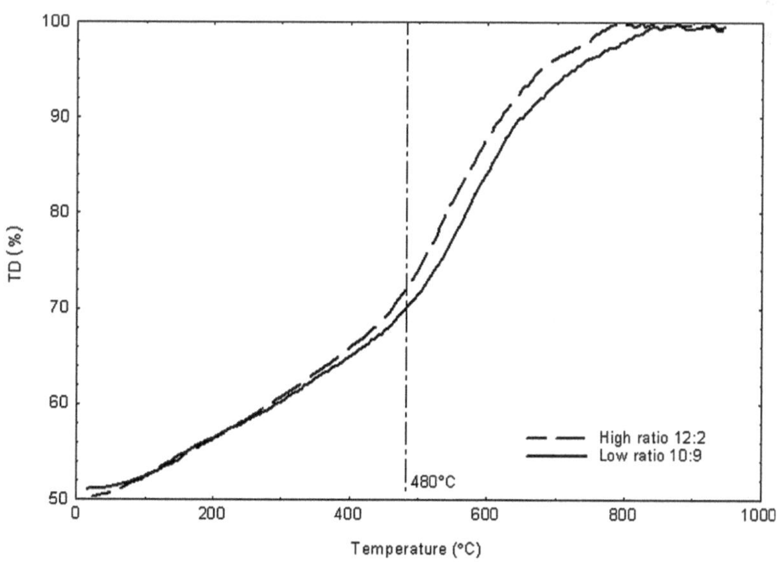

Figure 13.8 Consolidation of HDH CP-Ti powder by SPS versus pulse sequence [9]. Samples were heated to 950°C at 50°C/min under the pressure of 50 MPa without an isothermal hold [9].

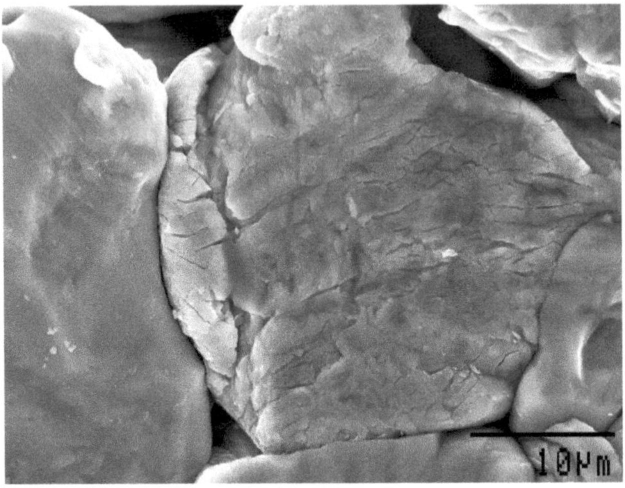

Figure 13.9 Particle deformation seen in the microstructure obtained from interrupting a CP-Ti powder sample heated to 500°C under the pressure of 50 MPa during SPS [9]. Titanium powder size: <45 µm and heating rate: 50°C/min.

13.2.2 Densification mechanism of CP-Ti

Two main hypotheses have been proposed to explain the fast densification of CP-Ti powder by SPS [9,25,28–30]. One assumes that the spark discharges generated between powder particles can clean and activate the surfaces of powder particles, thereby promoting mass transport for sintering [28,29]. The other suggests that particle deformation is responsible for densification as the yield strength of the powder particles decreases substantially with increasing temperature [9,25,30]. The latter mechanism is supported by the microstructure shown in Figure 13.9, obtained from interrupting the SPS of CP-Ti samples heated to 500°C under the pressure of 50 MPa [9]. This is essentially similar to HP and can be understood from the nearly linear decrease in yield strength of unalloyed titanium (ASTM Grad II) with temperature (Figure 13.1) [25,26,31]. Deformation of titanium particles is easy to occur at temperatures above 400°C because of their low yield strength and the applied pressure. Accordingly, the role of spark discharge in consolidation, even if it occurs to electrically conducting metal powders including titanium, may be negligible. This will be further discussed subsequently.

13.2.3 Impurities and mechanical properties of CP titanium

Owing to rapid consolidation, the impurity contents of carbon, nitrogen, and oxygen in the SPS-consolidated samples remained similar to their starting levels [10]. This is also supported by the high tensile elongation obtained in the as-sintered state, Table 13.3 [10,11]. In general, the tensile properties obtained under optimal SPS conditions compare favorably with those of the wrought CP-Ti materials. In particular,

Table 13.3 **Tensile properties of CP titanium fabricated by SPS [10,11]**

Materials and SPS conditions	UTS (MPa)	YS (MPa)	Elongation (%)
CP Ti (Grade 1) powder, 900°C, 5 min under 60 MPa	445	340	38.5
CP Ti (Grade 3) powder 900°C, 5 min under 60 MPa	720	595	18
Cryomilled nanocrystalline CP Ti (Grade 2) powder 850°C, 3 min under 80 MPa	840	770	27
Wrought titanium grade 4	655–690	480–635	20–25

nanostructured CP-Ti fabricated by SPS using nanocrystalline titanium powder attained tensile properties much superior to the wrought properties of CP-Ti: ultimate tensile strength (UTS) 840 MPa, yield strength (YS) 770 MPa, and elongation 27%, Table 13.3 [11]. This demonstrates the advantages of SPS.

13.4 SPS of Ti-6Al-4V from EMA powder mixtures and PA powder

13.4.1 Sintering densification

SPS of Ti-6Al-4V EMA powder mixtures was found to be more effective than SPS of PA Ti-6Al-4V powder in terms of densification (Figure 13.10). This may be attributed to the higher strength of the PA powder than the elemental Ti powder in the EMA powder mixtures, particularly at temperatures below 1000°C [25]. On the other hand, unlike pressureless sintering, the sintered density of Ti-6Al-4V was found to be essentially independent of the titanium powder size and master alloy powder size (Figure 13.11) [25]. This implies that the mechanism of densification by SPS is not diffusion controlled.

The dilation curve is informative for understanding the densification process. Figure 13.12 shows the dilation curve of Ti-6Al-4V EMA powder mixtures, heated to 1000°C at 20°C/min for an isothermal hold of 60 min [25]. The initial irregular shrinkage (<485°C) was caused by rapid compaction of the loose powder mixtures. Significant shrinkage occurred at ~485°C and lasted until 900°C, beyond which shrinkage was limited, consistent with the sintered density results.

The shrinkage rate derived from the dilation curve, shown in Figure 13.13 [25], displayed two peaks: one at ~570°C and the other at ~850°C. Also shown in Figure 13.13 is the shrinkage rate curve for PA Ti-6Al-4V powder, which showed only one peak at ~845°C starting from ~580°C. Fast densification of PA Ti-6Al-4V powder started much later than that of the EMA powder mixtures. The yield strength of unalloyed

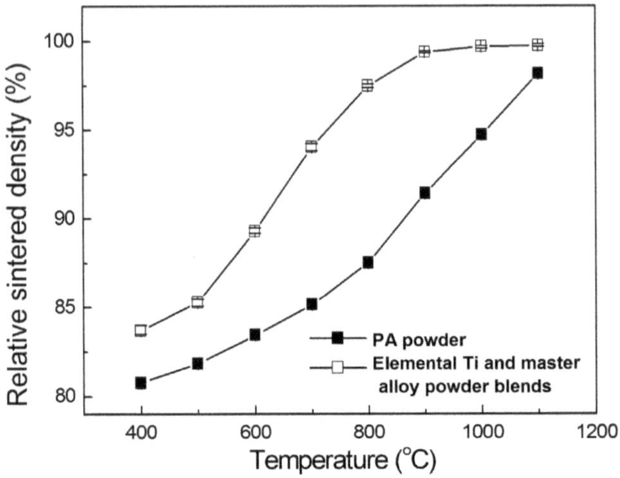

Figure 13.10 Effect of the SPS temperature on the sintered density of Ti-6Al-4V made from PA powder and EMA powder mixtures. The isothermal hold time at each temperature was 5 min. PA powder and HDH Ti powder size: 48–75 μm and 40V-60Al master alloy powder size: ≤38 μm [25]. Heating rate: 20°C/min, and vacuum pressure: <4 Pa.

titanium at 570°C is estimated to be around 40 MPa by extrapolation from Figure 13.1. No significant diffusion is expected at 570°C. Hence the rapid shrinkage recorded at 570°C for Ti-6Al-4V EMA powder mixtures can be attributed to titanium particle deformation under the applied pressure of 30 MPa. Apart from particle deformation,

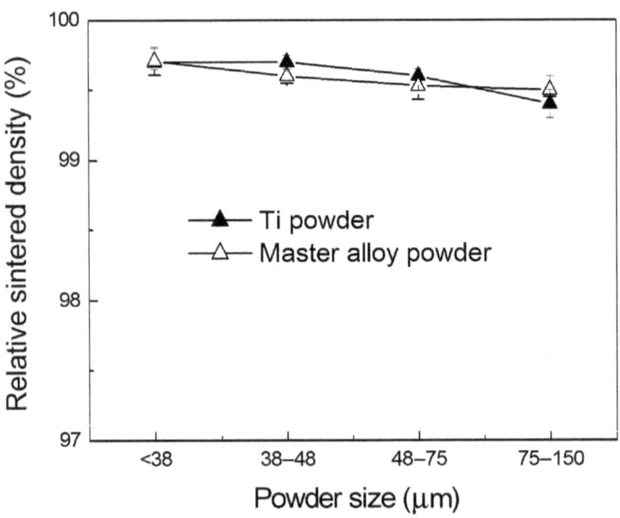

Figure 13.11 Effect of HDH Ti powder size (40V-60Al master alloy powder size: ≤38 μm) and master alloy powder size (titanium powder size: 48–75 μm) on the sintered density of Ti-6Al-4V. The sintering temperature was 1000°C and the isothermal hold time was 5 min. Applied pressure: 30 MPa, heating rate: 20°C/min, and vacuum level: <4 Pa [25].

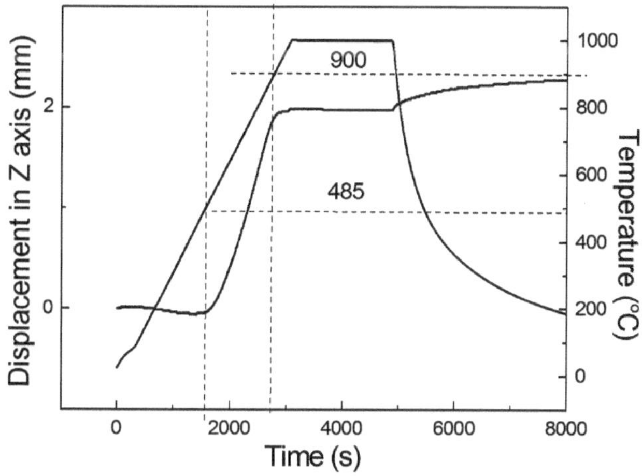

Figure 13.12 Dilatometer curves of Ti-6Al-4V fabricated using 48–75 μm HDH Ti powder and ≤38 μm 40V-60Al master alloy powder mixtures during SPS (heated to 1000°C at 20°C/min and held for 60 min). Applied pressure: 30 MPa, heating rate: 20°C/min, and vacuum level: <4 Pa [25].

the second peak recorded at 850°C can be partially attributed to the enhanced self-diffusion of titanium in the β region because there is always a temperature difference existing from the surface to the center of the powder compact. However, particle deformation played the decisive role in densification. This is also supported by the two

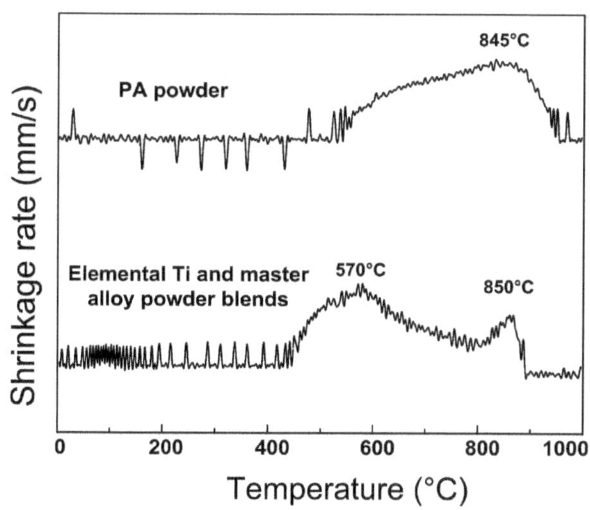

Figure 13.13 Shrinkage rate of Ti-6Al-4V fabricated from PA powder (48–75 μm) and HDH Ti (48–75 μm) and 40V-60Al master alloy powder (≤38 μm) mixtures derived from the dilation curves. Applied pressure: 30 MPa, heating rate: 20°C/min, and vacuum level: <4 Pa [25].

observations made from Figures 13.10 and 13.11: (i) the relative density achieved after SPS at 800°C for 5 min had already reached about 97.6% of the pore-free density before significant diffusion occurred; and (ii) the powder size showed little influence on densification. The late start of significant shrinkage for PA Ti-6Al-4V powder can be attributed to its high remaining strength before 850°C (Figure 13.2) [20], in relation to the applied pressure (30 MPa).

13.4.2 Microstructure homogenization and mechanical properties

Despite the near full densification achieved at 900°C for 5 min, the as-sintered microstructure made from the EMA powder mixtures still showed distinguishable 60Al-40V master alloy particles (Figure 13.14a). Both isolated α and β phases as

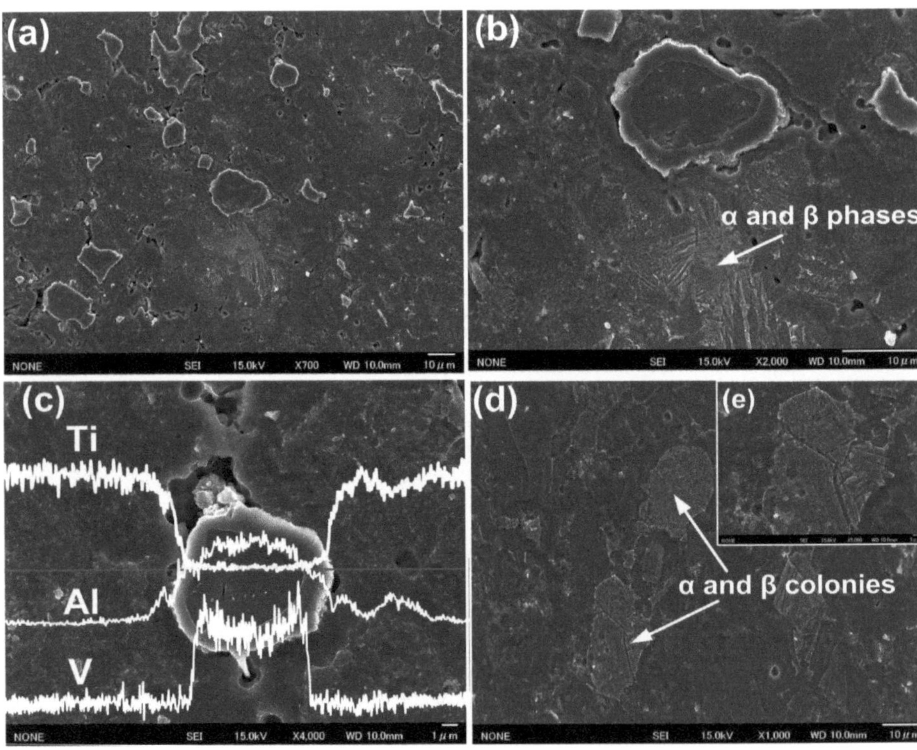

Figure 13.14 Microstructures of Ti-6Al-4V fabricated from HDH Ti (48–75 μm) and 40V-60Al master alloy powder (≤38 μm) mixtures after SPS at 900°C for 5 min: (a) overview, (b) an enlarged view of (a) showing the existence of α–β colonies, and (c) line analysis across a master alloy particle observed in (a). For comparison, the microstructure obtained after SPS at 1100°C for 5 min is shown in (d) (an enlarged view in (e)) [25]. Applied pressure: 30 MPa, heating rate: 20°C/min, and vacuum level: <4 Pa.

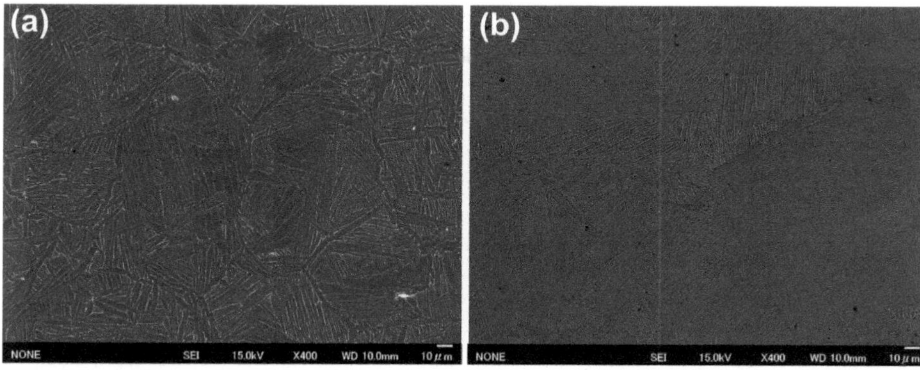

Figure 13.15 Microstructures of Ti-6Al-4V after SPS at 1100°C for 15 min: (a) from HDH Ti (48–75 μm) powder and 40V-60Al master alloy powder (≤38 μm) mixtures, and (b) from PA powder (48–75 μm) [25]. Applied pressure: 30 MPa, heating rate: 20°C/min, and vacuum level: <4 Pa.

well as α–β colonies were observed in the vicinity of these master alloy particles (Figure 13.14b–c). In addition, it was noted that each remaining master alloy particle was surrounded by a boundary layer. Energy dispersive spectroscopic line analysis across one such particle (Figure 13.14c) indicated that diffusion of aluminum and vanadium from the 60Al-40V master alloy particle into the titanium matrix was essentially limited to this layer.

Increasing SPS temperature to 1100°C with an isothermal hold of 5 min reduced the number of the remaining master alloy particles but the resulting microstructure remained noticeably inhomogeneous (Figure 13.14d–e). The α–β colonies were observed mainly in areas where the original master alloy particles resided. Increasing isothermal hold time from 5 to 15 min at 1100°C eventually produced a homogeneous as-sintered microstructure (Figure 13.15), typical of well-sintered Ti-6Al-4V consisting of uniform α–β colonies. This is indicative of solute homogenization. Compared to pressureless sintering at ≥1200°C for more than 120 min, SPS is thus very effective in the fabrication of Ti-6Al-4V from EMA powder mixtures in terms of both densification and solute homogenization. Since near full densification occurred in the first 5 min at 1100°C prior to solute homogenization, there were few pores available to induce spark discharges during the subsequent SPS from 5 to 15 min. The accelerated homogenization of vanadium is thus attributed to the improved diffusivity caused by the intense Joule heating effect when the direct current passes through the sample, together with the effect of the applied pressure. Rapid solute homogenization is another major advantage of SPS over HP. This is important for the fabrication of titanium alloys containing slow diffusers. PA powder does not have this homogenization problem.

Table 13.4 cites the tensile properties of as-sintered Ti-6Al-4V fabricated from PA powder and EMA powder mixtures by SPS at 1100°C for 15 min together with the ASTM Grade 5 specification [25,32]. Also listed are the oxygen contents before and

Table 13.4 Tensile properties of Ti-6Al-4V fabricated by SPS at 1100°C for 15 min and the oxygen contents before and after SPS [25]

Material	UTS (MPa)	YS (MPa)	Elongation (%)	Oxygen (wt%)
HDH titanium powder	–	–	–	0.2
Ti-6Al-4V sintered from EMA powder mixtures	1037	929	21.8	0.2
Ti-6Al-4V PA powder	–	–	–	0.19
Ti-6Al-4V sintered from PA powder	1014	818	17.8	0.19
ASTM Grade 5 specification	950	880	14	≤0.2

after SPS. The SPS process resulted in a negligible pickup of oxygen. The as-sintered Ti-6Al-4V from EMA power mixtures achieved a UTS of 1037 MPa, YS of 929 MPa, and tensile elongation of 21.8%, much superior to the ASTM Grade 5 specifications. In addition, the YS and tensile elongation are clearly higher than the respective values obtained from PA Ti-6Al-4V powder. The difference was attributed to the fine microstructure obtained from the use of EMA powder mixtures. The tensile fractographs shown in Figure 13.16 confirmed the difference: the samples sintered from EAM powder mixtures after SPS at 1100°C for 15 min showed finer and more dimples than did the samples sintered from PA Ti-6Al-4V powder.

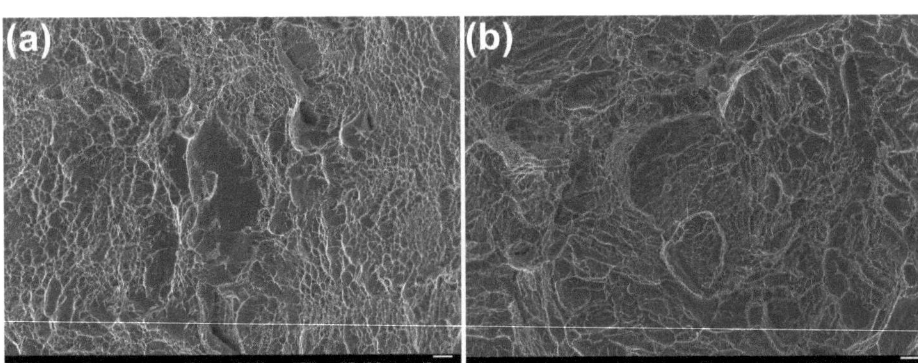

Figure 13.16 Tensile fractographs of Ti-6Al-4V after SPS at 1100°C for 15 min: (a) from HDH Ti (48–75 μm) powder and master alloy powder (≤38 μm) mixtures, and (b) from PA powder (48–75 μm) [25]. Applied pressure: 30 MPa, heating rate: 20°C/min, and vacuum level: <4 Pa.

Table 13.5 Minimum consolidation conditions for obtaining >97% relative pore-free density for Ti-6Al-4V by SPS and HP [22,23]

Material	HP	SPS
Ti-6Al-4V sintered from EMA powder mixtures	Temperature: 900°C; pressure: 30 MPa; isothermal hold: 30 min	Temperature: 900°C; pressure: 30 MPa; isothermal hold: 5 min
Ti-6Al-4V sintered from PA powder	Temperature: 1050°C; pressure: 60 MPa; isothermal hold: 0	Temperature: 1100°C; pressure: 30 MPa; isothermal hold: 15 min

13.5 Comparison of SPS and HP

Table 13.5 summarizes the minimum consolidation conditions of obtaining >97% of the pore-free density for Ti-6Al-4V by SPS and HP [22,23,25]. There is essentially no difference in terms of densification. The pulsed electrical current showed little influence on the densification of Ti-6Al-4V, despite several hypotheses that have suggested the other way around [27–29]. For instance, it has been proposed that the Joule heat generated when the pulse current passes through the powder particles can accelerate the densification through enhancing localized plastic flow around the connecting necks [29,30].

The major difference noticed between SPS and HP is their ability in realizing solute homogenization. SPS is much more effective than HP in producing a homogenous Ti-6Al-4V microstructure using EMA powder mixtures, as pointed out earlier. A homogeneous microstructure is indicative of a generally homogenous solute distribution, which depends on the diffusion rate [25]. It appears that the intensive Joule heating effect may have raised the local temperature and therefore enhanced the diffusion rate. This may be the reason why SPS is much more effective in achieving microstructure homogeneity than HP.

13.6 Conclusion remarks

SPS and HP are both effective in producing pore-free product forms of titanium and titanium alloys from powder. Powder particle deformation is responsible for densification by both SPS and HP, and the pulsed electrical current exerts little influence on densification. As a result, the density is essentially independent of powder size. No difference has been observed in the fabrication of CP-Ti by SPS and HP. In addition, there is little difference in consolidating either HDH Ti powder or PA Ti-6Al-4V powder by SPS and HP at temperatures above 1000°C because PA powder will lose most of its strength at such temperatures. The major difference noticed between SPS and HP is their ability in realizing solute homogenization of titanium alloy. The SPS process leads to much faster solute homogenization in the fabrication of Ti-6Al-4V from EMA powder mixtures than the HP process. The Joule heat generated during SPS is proposed to be the major reason. Achieving solute homogenization is critical to the fabrication of titanium alloys containing slow diffusers. In this regard, SPS is clearly advantageous over HP.

Acknowledgments

The authors acknowledge the financial support from the Australian Research Council (ARC) under the Linkage Projects programme. Dr. Ya Feng Yang further acknowledges the financial support from the ARC for his DECRA fellowship.

References

[1] M. Qian, Cold compaction and sintering of titanium and its alloys for near-net-shape or preform fabrication, Int. J. Powder Metall. 46 (2010) 29–44.
[2] Y.F. Yang, S.D. Luo, G.B. Schaffer, M. Qian, Sintering of Ti-10V-2Fe-3Al and mechanical properties, Mat. Sci. Eng. A 528 (2011) 6719–6726.
[3] O.M. Ivasishin, D.G. Savvakin, The impact of diffusion on synthesis of high-strength titanium alloys from elemental powder blends, Key Eng. Mat. 436 (2010) 113–121.
[4] V. Mamedov, Spark plasma sintering as advanced PM sintering method, Powder Metall. 45 (4) (2002) 322–328.
[5] Z.A. Munir, U. Anselmi-Tamburini, M. Ohyanagi, The effect of electric field and pressure on the synthesis and consolidation of materials: a review of the spark plasma sintering method, J. Mat. Science 41 (3) (2006) 763–777.
[6] G. Xie, O. Ohashi, K. Chiba, N. Yamaguchi, M. Song, K. Furuya, et al., Frequency effect on pulse electric current sintering process of pure aluminum powder, Mat. Sci. Eng. A 359 (1) (2003) 384–390.
[7] G. Xie, O. Ohashi, K. Wada, T. Ogawa, M. Song, K. Furuya, Interface microstructure of aluminum die-casting alloy joints bonded by pulse electric-current bonding process, Mat. Sci. Eng. A 428 (1) (2006) 12–17.
[8] S.W. Wang, L.D. Chen, Y.S. Kang, M. Niino, T. Hirai, Effect of plasma activated sintering (PAS) parameters on densification of copper powder, Mat. Res. Bull. 35 (4) (2000) 619–628.
[9] M. Eriksson, Z. Shen, M. Nygren, Fast densification and deformation of titanium powder, Powder Metall. 48 (3) (2005) 231–236.
[10] M. Zadra, F. Casari, L. Girardini, A. Molinari, Microstructure and mechanical properties of CP-titanium produced by spark plasma sintering, Powder Metall. 51 (1) (2008) 59–65.
[11] O. Ertorer, T.D. Topping, Y. Li, W. Moss, E.J. Lavernia, Nanostructured Ti consolidated via spark plasma sintering, Metall. Mat. Trans. A 42 (4) (2011) 964–973.
[12] T. Yoshimura, T. Thotsaphon, H. Imai, K. KONDOH, Microstructural and mechanical properties of Ti composite reinforced with TiO_2 additive particles, Trans. JWRI 38 (2) (2009) 37–41.
[13] H. Izui, G. Kikuchi, Sintering performance and mechanical properties of titanium compacts prepared by spark plasma sintering, Mat. Sci. Forum 706 (2012) 217–221.
[14] C.G. Goetzel, V.S. de. Marchi, Electrically activated pressure sintering (spark sintering) of titanium powders, Powder Metall. Int. 3 (1971) 80–87.

[15] R. Orru, R. Licheri, A.M. Locci, A. Cincotti, G. Cao, Consolidation/synthesis of materials by electric current activated/assisted sintering, Mat. Sci. Eng. R. 63 (**4**) (2009) 127–287.
[16] C.G. Goetzel, D. Marchi, Electrically activated pressure-sintering/spark sintering of titanium-aluminum-vanadium alloy powders, Mod. Dev. Powder Metall. 4 (1971) 127–150.
[17] W. Schatt, K.P. Wieters, Powder Metallurgy: Processing and Materials, EPMA-European Powder Metallurgy Association, Shrewsbury, 1997.
[18] A. Bose, W.B. Eisen, Hot Consolidation of Powders and Particulates, Metal Powder Industries Federation, Princeton, 2003.
[19] R.K. Malik, Vacuum hot pressing of titanium alloy powders, Prog. Powder Metall. 31 (1975) 277–288.
[20] K.T. Kim, H.C. Yang, Densification behavior of titanium alloy powder during hot pressing, Mat. Sci. Eng. A 313 (1) (2001) 46–52.
[21] K. Akechi, Z. Hara, Electrical resistance of resistance-sintered titanium and its alloys, J. Jpn. Soc. Powder Metall. 26 (5) (1979) 180–186.
[22] L. Bolzoni, E.M. Ruiz-Navas, E. Neubauer, E. Gordo, Inductive hot-pressing of titanium and titanium alloy powders, Mat. Chem. Phys. 131 (3) (2012) 672–679.
[23] L. Bolzoni, I.M. Meléndez, E.M. Ruiz-Navas, E. Gordo, Microstructural evolution and mechanical properties of the Ti–6Al–4V alloy produced by vacuum hot-pressing, Mat. Sci. Eng. A 546 (2012) 189–197.
[24] W.H. Kao, D. Eylon, C.F. Yolton, F.H. Froes, Effect of temporary alloying by hydrogen (HYDRO-VAC) on the vacuum hot pressing and microstructure of titanium alloy powder compacts, Prog. Powder Metall. 37 (1982) 289–301.
[25] Y.F. Yang, H. Imai, K. Kondoh, M. Qian, Comparison of spark plasma sintering of elemental and master alloy powder mixtures and prealloyed Ti-6Al-4V powder, Int. J. Powder Metall. 50 (1) (2014) 41–47.
[26] S.D. Luo, Y.F. Yang, G.B. Schaffer, M. Qian, Warm die compaction and sintering of titanium and titanium alloy powders, J. Mat. Proc. Tech. 14 (2014) 660–666.
[27] X.P. Li, M. Yan, H. Imai, K. Kondoh, G.B. Schaffer, M. Qian, The critical role of heating rate in enabling the removal of surface oxide films during spark plasma sintering of Al-based bulk metallic glass powder, J. Non-Cryst. Solids 375 (2013) 95–98.
[28] M. Omori, Sintering, consolidation, reaction and crystal growth by the spark plasma system (SPS), Mat. Sci. Eng. A 287 (2) (2000) 183–188.
[29] J.R. Groza, A. Zavaliangos, Sintering activation by external electrical field, Mat. Sci. Eng. A 287 (2) (2000) 171–177.
[30] M. Tokita, Trends in advanced SPS spark plasma sintering systems and technology, J. Soc. Powder Tech. Jpn. 30 (1993) 790–804.
[31] TIMETAL Datasheets, TIMETAL® 50A, http://www.timet.com (accessed 07.04.14).
[32] TIMETAL Datasheets, TIMETAL® 6–4, http://www.timet.com (accessed 07.04.14).

Microwave sintering of titanium and titanium alloys

14

Shudong D. Luo, Ma Qian**, M. Ashraf Imam†*
*The University of Queensland, School of Mechanical and Mining Engineering, Brisbane, Australia
**RMIT University, School of Aerospace, Mechanical and Manufacturing Engineering, Centre for Additive Manufacturing, Melbourne, Victoria, Australia
†Materials Science and Technology Division, Naval Research Laboratory, Washington DC, USA

14.1 Introduction

Microwaves (MWs) refer to electromagnetic waves over the frequency range from 300 MHz to 300 GHz. The most commonly used MW frequencies for materials processing are 2.45 GHz and 915 MHz. When exposed to MWs, some materials can interact with the waves, converting the electromagnetic energy into heat within the materials. This unique heating mechanism imparts a number of advantages to MW processing and the processed materials, including rapid and volumetric heating, reduced energy consumption and processing time, enhanced sintering, refined microstructure, and improved physical and mechanical properties [1–4].

Microwave sintering began in the early 1970s with ceramics [5]. Nowadays microwave sintering of ceramics has been realized in industrial production. However, microwave sintering of metals is still in its infancy, hindered by the perception that metals are strong MW reflectors. This is true for bulk metals but not necessarily the case for metal powders as fine metal powder can be close to the skin depth of MWs (typically less than 10 μm for a 2.45 GHz frequency at room temperature). In addition, metal powder particles are always enveloped with a surface oxide film, which usually decreases the MW reflectivity [6]. In 1988, Walkiewicz et al. [7] reported that metal powders including Al, Co, Cu, Fe, Mo, and Zn were quickly heated up to elevated temperatures (e.g., 768°C for Fe, in 7 min) when exposed to 1-kW, 2.45-GHz MWs. A decade later, for the first time bulk metals/alloys (Fe-Cu/Ni-C and Co) were fabricated from metal powders via MW heating and sintering [8]. The sintered density and mechanical properties were superior to the respective properties achieved by conventional sintering. This study has subsequently encouraged some interest in microwave sintering of metals [9–11].

Microwave sintering of titanium (Ti) is a recent academic development. Researchers have looked at the heating response of hydride-dehydride (HDH) Ti powder to MW radiation [6,12–14], MW sintering densification [14–22], and mechanical properties [14–16] of MW-sintered commercially pure Ti (CP-Ti) and titanium alloys since 2004. MW sintering of titanium hydride (TiH_2) powder is a new effort [23]. This

Titanium Powder Metallurgy. http://dx.doi.org/10.1016/B978-0-12-800054-0.00014-9

chapter reviews the current status of MW fabrication of titanium and titanium alloys based on the use of HDH Ti powder and TiH_2 powder.

14.2 Heating of metal powders by microwaves

Bulk metals cannot be heated up effectively by MWs because most of the incident power is reflected, for example, copper slab reflects as high as 99.98% power of 2.45-GHz MW [24]. The rest, accounting for 0.02% of the incident power, is converted into heat owing to the Joule loss by eddy current, induced by the alternating magnetic (*H*) field of the MWs. Even so, the conversion is confined to the skin layer, typically less than 10 μm. The heat generated increases with increasing volume of the skin layer [25]. This mechanism suggests that the *H*-field of the MWs tends to be more effective than the electrical (*E*) field in heating metal powders [26,27]. In addition to eddy current heating, magnetic loss plays an important role in heating ferromagnetic metals because the reorientation of magnetic domains in response to the *H*-field can generate heat [25]. For instance, Walkiewicz et al. [7] and Buchelnikov et al. [6] showed that metal powders of Fe, Co, or Ni exhibited faster MW heating than nonferromagnetic metal powders. Powder size affects the heating efficiency. For metal powder smaller than the MW skin depth, volumetric MW heating is expected [28]. In contrast, for coarse metal powder, MW heating will be restricted to the skin depth of each powder particle and the heat is then conducted to the rest of the particle [29–30].

14.3 Heating of Ti powder by microwaves

14.3.1 Heating by pure microwaves

Sato et al. [12] studied the heating of Ti powder (50 μm, 99.5% pure) in a TE103 waveguide cavity running with a 2.45-GHz MW source. The temperature reached about 100°C in the *H*-field and 200°C in the *E*-field after 2 min radiation at 300 W power. Buchelnikov et al. [6] showed that the temperature of Ti powder (<50 μm) reached only 40°C after radiation by 1000-W MWs for 1 min in a multimode cavity and changed little after another 1 min radiation. In contrast, Fe powder of the same particle size was heated to 200°C in 1 min under the same radiation conditions.

The weak response of Ti powder to MW radiation is generally attributed to the paramagnetism property of Ti [6]. However, other factors may play a role too such as spark (or plasma) discharging in the form of hot spot or electrical arc, which is observable both inside and outside the insulation package [14,19,31]. Spark discharges occasionally and uncontrollably leading to erratic heating behaviors (see Figure 14.1). Oxidation of Ti powder can be another influential factor, particularly when oxygen-deficient titanium oxides form on the surface, which are good MW absorbers such as TiO [12,32]. Hashiguchi and Sueyoshi [33] studied the effect of oxidation on the response of Ti powder (<45 μm) to MW heating in Ar or N_2. They observed slow heating to ~400°C but an abrupt rise to ~1700°C afterwards. Oxidation appears to be the major reason, although a small amount of TiN also formed.

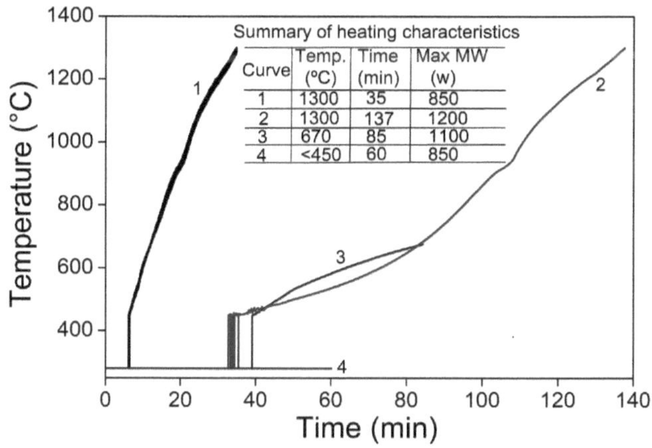

Figure 14.1 Heating behavior of powder compacts by pure MW radiation. The heating was conducted in a 2.45-GHz multimode MW chamber in a vacuum of 10^{-3} Pa [14].

14.3.2 Hybrid heating assisted by MW susceptors

In order to ensure effective and consistent heating of Ti powder compacts by MWs, the use of MW susceptors has been practiced [14–17,19–21,31]. MW susceptors are MW-absorbing ceramic materials such as SiC and $MoSi_2$. It has been suggested that "once a material is heated to its critical temperature, MW absorption becomes sufficient to cause self-heating" [1]. MW heating with the assistance of susceptors is thus also referred to as hybrid MW heating.

Figure 14.2 shows representative heating profiles of HDH Ti powder compacts radiated inside a cylindrical SiC susceptor [14]. Steady and controllable heating rates varying from 17°C/min to 50°C/min were realized by adjusting the incident MW

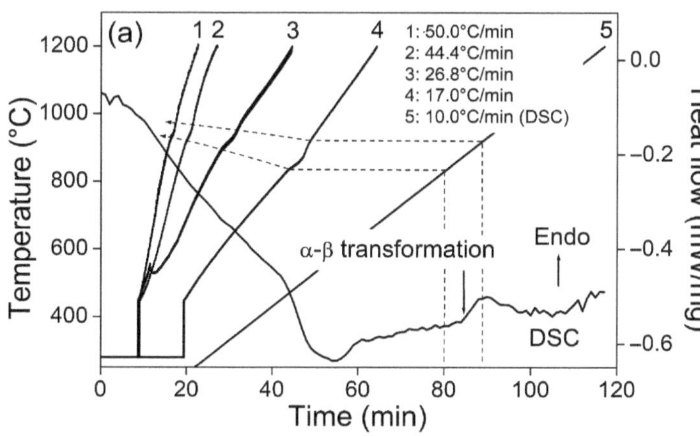

Figure 14.2 Heating profiles of 100 mesh HDH Ti powder compacts in the presence of a SiC susceptor. The temperature fluctuation between 850°C and 950°C is due to endothermic α→β transformation, confirmed by differential scanning calorimetry (DSC) analysis [14].

power. Moreover, the heating behavior was reproducible and insensitive to sample mass over the range studied (from 24 g to 56 g).

14.4 Sintering densification

14.4.1 Uniaxially pressed Ti powder compacts

The sintering densification of uniaxially pressed Ti powder compacts was affected by sintering temperature [16], isothermal holding time [14], compaction pressure and powder particle size [19]. Higher sintering temperature, finer Ti powder and/or higher compaction pressure produce higher sintered densities. Figure 14.3a shows the sintered density

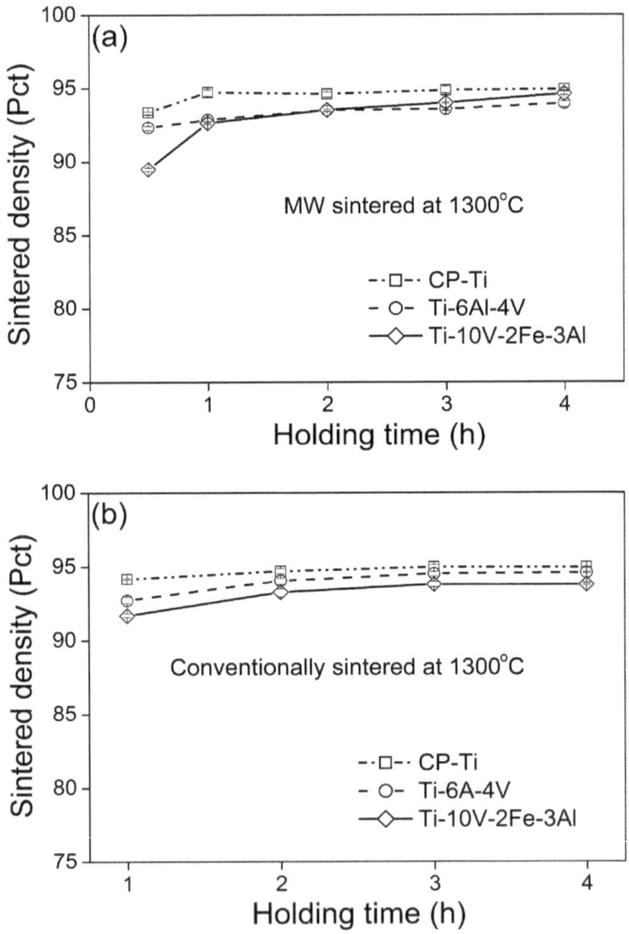

Figure 14.3 Sintered densities of CP-Ti, Ti-6Al-4V, and Ti-10V-2Fe-3Al by (a) MW sintering with the assistance of SiC susceptors, heating rate: 30–32°C/min, vacuum: $\leq 4.0 \times 10^{-3}$ Pa; and by (b) conventional vacuum sintering, heating rate: 4°C/min, vacuum: 10^{-2} Pa. Ti powder, Ti-6Al-4V, and Ti-10V-2Fe-3Al powder mixes were uniaxially compacted at 600 MPa [14].

of CP-Ti, Ti-6Al-4V, and Ti-10V-2Fe-3Al as a function of isothermal sintering time at 1300°C [14]. In each case, the sintered density increased notably in the initial 60 min and slightly thereafter toward reaching about 95% of the theoretical density (TD) after 240 min. The variation of sintered density with isothermal holding time is similar to vacuum sintering (less than 1% TD, see Figure 14.3b). No distinct difference was observed in sintering densification between MW sintering and conventional vacuum sintering.

14.4.2 Cold isostatic pressed Ti powder compacts

Different sintering behaviors were reported with MW radiation of Ti powder compacted by cold isostatic pressing (CIP). Sintered densities close to theoretical values were achieved for both Ti-6Al-4V and CP-Ti in one study reported [20], which used Armstrong Ti powder and prealloyed Ti-6Al-4V powder. Samples were first uniaxially pressed at 69 MPa and then CIPed at 690 MPa, resulting in a final green density of ~87% TD. MW sintering was conducted in an S-Band 2.45-GHz MW chamber in a vacuum of 1 Pa. Isothermal sintering at 1300°C for 10 min was able to produce sintered densities of greater than 98% TD for CP-Ti and ~100% TD for Ti-6Al-4V (see Figure 14.4a).

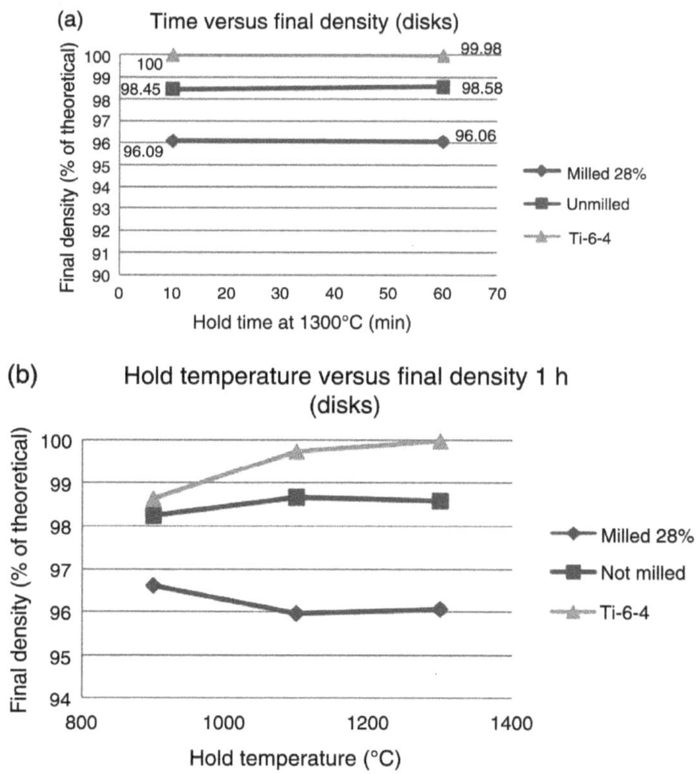

Figure 14.4 Sintered density of cold isostatically pressed Ti and prealloyed Ti-6Al-4V powder by MW radiation, affected by (a) isothermal holding time at 1300°C, and (b) sintering temperature when held for 60 min. Milled 28% refers to that as-received Ti powder is milled and the tap density of the milled powder is 28% TD [20].

Increasing isothermal holding time to 60 min resulted in little improvement. Near full densification of CIPed Ti was also achieved at 900°C (Figure 14.4b). In addition, milling the as-received Ti powder appears to be detrimental to densification (Figure 14.4).

14.5 Mechanical properties

14.5.1 Effect of contaminations

Titanium has a high affinity for interstitials, so contamination is a concern for the sintering of PM Ti. In MW sintering, Ti powder compacts are placed inside an insulation package that consists of porous alumina fiberboard and lossy susceptors [15,19,31]. The residual gas and volatile substances in the insulation package including those triggered by the intense MW radiation during sintering could contaminate the Ti powder compacts being sintered.

Table 14.1 summarizes the contamination observed during MW sintering of Ti powder compacts. Naked sintering without the protection of Ti sponge increased the oxygen content by 50–125% and the carbon content by 350–600%. The amount of interstitials absorbed varies with sample locations; samples close to the insulation-susceptor package picked up substantially more interstitials, justifying contamination from the insulation package. Interstitial contamination deteriorates tensile elongation ε_f (Figure 14.5a).

Titanium sponge is an effective impurity getter. Sintering under the protection of titanium sponge reduced the pickup of oxygen to 100–200 ppm compared to 1000–2500 ppm in naked sintering (Table 14.1). Accordingly, the ductility improved

Table 14.1 Comparison of interstitial concentrations in MW-sintered Ti with and without the protection of Ti sponge [14]

Package design	Sample location	O (ppm)	C (ppm)	N (ppm)	Si (ppm)	Mass gain (wt %)
No protection	1 and 4 in Figure 14.5a	4500	560	210	350	0.54
	2 and 3 in Figure 14.5a	3000	360	140	280	0.23
Protection	1 and 4 in Figure 14.5b	2200	140	96	80	−0.05
	2 and 3 in Figure 14.5b	2100	160	93	100	−0.06
HDH Ti powder	2000	80	77	73		

All samples were sintered from <150 μm Ti powder at 1300°C for 120 min at a heating rate of 30–32°C/min.

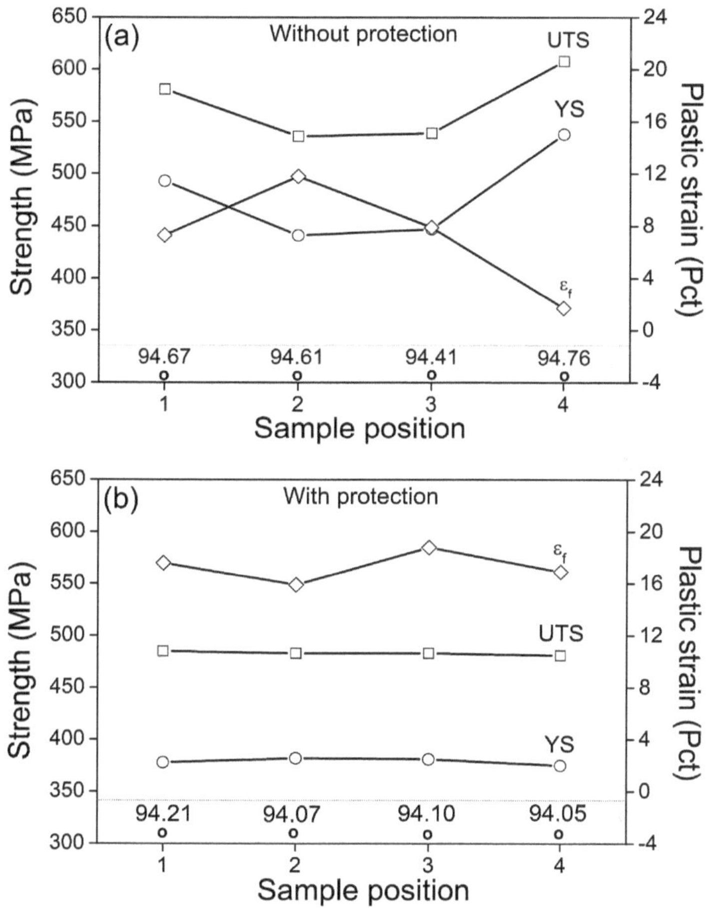

Figure 14.5 The effect of interstitial contamination on tensile properties of MW-sintered CP-Ti: (a) naked sintering, and (b) protected sintering. Ti rectangular bars were pressed at 600 MPa and sintered at 1300°C for 120 min. Four bars were positioned side by side in the insulation-susceptor package. Samples 1 and 4 were close to the package; samples 2 and 3 were in the central region of the package [14].

noticeably (Figure 14.5b). More importantly, the tensile properties including ultimate tensile strength (UTS) and yield strength (YS) are consistent.

14.5.2 Tensile properties by protected sintering

Microwave sintering under the protection of Ti sponge produces encouraging mechanical properties. Figure 14.6a shows the UTS versus ε_f for CP-Ti, sintered at 1300°C for 120–180 min in a vacuum of $\leq 4.0 \times 10^{-3}$ Pa [14]. The tensile properties obtained from 100 mesh HDH Ti powder varied between 480 and 512 MPa in UTS, and 15.0%

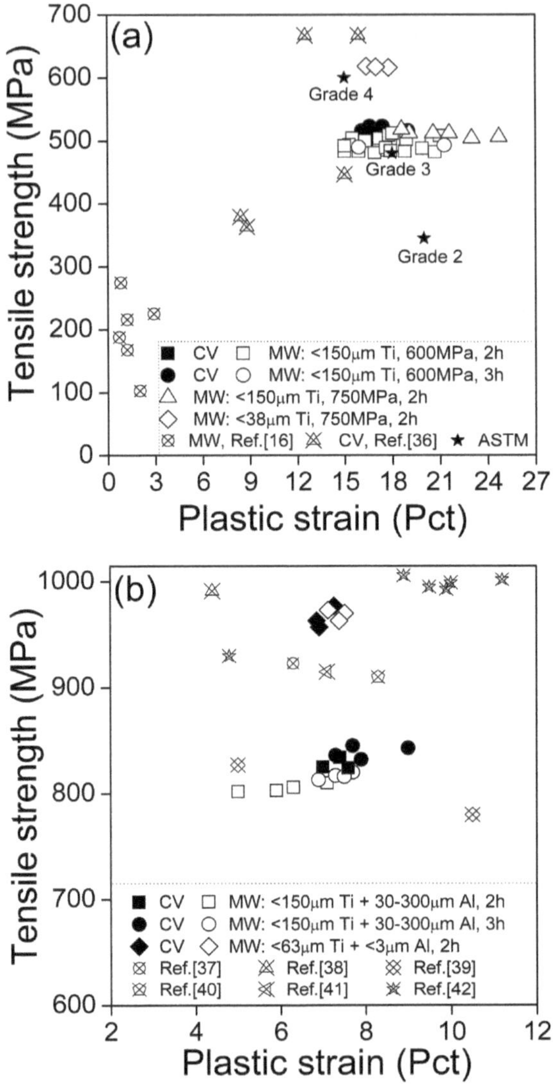

Figure 14.6 Comparison of tensile properties between MW-sintered and conventionally (CV) vacuum-sintered samples. (a) CP-Ti, (b) Ti-6Al-4V, compacted from elemental Ti, Al, and 42Al-58V master alloy (<150 μm) powders at 600 MPa, and (c) Ti-10V-2Fe-3Al, compacted from elemental Ti and 66.7V-13.3Fe-20Al master alloy powders at 600 MPa. Isothermal sintering was performed at 1300°C for 120–180 min with the protection of Ti sponge for both sintering processes. Some literature data are included for CP-Ti, Ti-6Al-4V, and Ti-10V-2Fe-3Al fabricated by the conventional press-and-sinter PM approach [14].

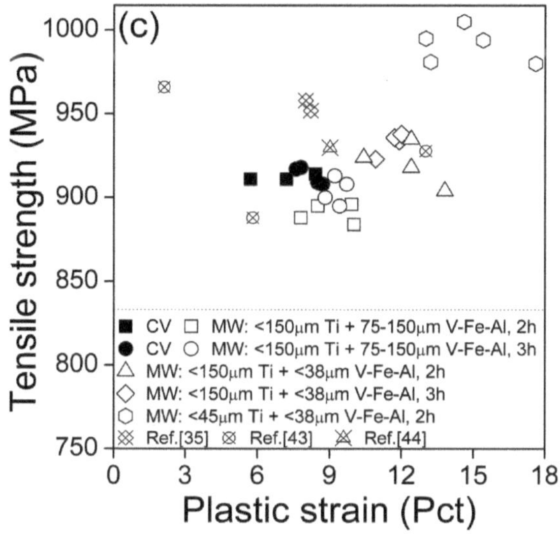

Figure 14.6 *(Cont.)*

and 21.3% in ε_f. The properties fall in the range of ASTM Grades 2–4 (superimposed in the plot). For Ti-6Al-4V (Figure 14.6b), the tensile properties obtained from MW and conventional vacuum sintering are essentially similar (UTS of 957–977 MPa, ε_f of ~7%) and typical of press-and-sinter Ti-6Al-4V [34].

MW sintering of Ti-10V-2Fe-3Al produces more encouraging properties. Figure 14.6c shows the comparison together with literature data. MW-sintered Ti-10V-2Fe-3Al, compacted from the powder blends of <45 μm Ti and <38 μm V-Fe-Al at 600 MPa, achieved UTS = 980–1005 MPa and ε_f = 13.0–17.6%. The tensile elongation is noticeably better than those obtained from conventional vacuum sintering. This implies that a homogeneous microstructure, which is dictated by the diffusion of vanadium [35], has been achieved in MW sintering.

14.6 Microwave heating and sintering of titanium hydride powder

Titanium hydride (TiH_2) is an intermediate product of HDH Ti powder. The use of TiH_2 powder offers a few notable advantages over HDH Ti powder in conventional sintering, including enhanced sintering and improved mechanical properties [45–47]. TiH_2 powder interacts effectively with MWs, leading to a potent heating response to pure MW radiation without the assistance of MW susceptors. For example, TiH_2 powder was heated to ~327°C in 3.5 min when radiated at 500 W in

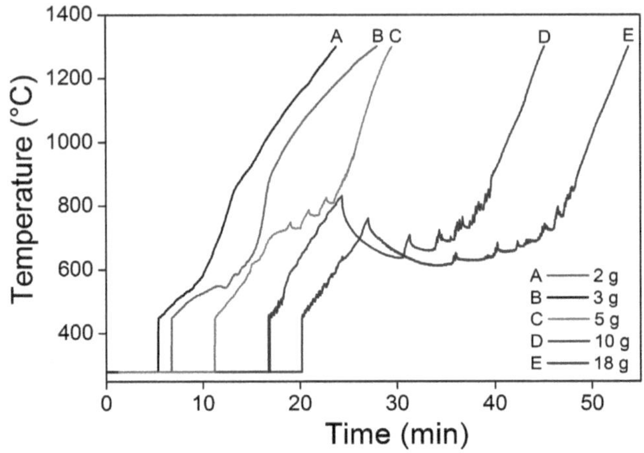

Figure 14.7 Representative heating profiles of TiH_2 powder compacts of different masses by pure MW radiation. Curves were shifted for clarity [23].

a multimode cavity [48], or ~370°C in 2 min when radiated at 400 W in a single-mode cavity [49]. A recent study [23] has shown that TiH_2 powder compacts can be heated to 1300°C by pure MW radiation. However, the heating response was found to depend on sample mass; samples of 2–3 g exhibited steady heating from 450°C to 1300°C while samples ≥5 g exhibited cyclic temperature fluctuations before 850°C (Figure 14.7). More recent experimental work suggests that the heating response of TiH_2 to pure MW radiation is far more complicated than expected; variables such as the cavity geometry of the insulation package and powder characteristics can significantly influence the heating behavior too, leading to ineffective heating of TiH_2 powder compacts.

The cyclic temperature fluctuation observed before 850°C occurred along with the dehydrogenation of TiH_2 [23]. Figure 14.8a shows the variations in temperature, vacuum and reflected MW power ratio ($Refl._{MW}$) recorded from a 3 g TiH_2 powder pellet heated in low vacuum (10^{-1} Pa). The temperature, vacuum, and $Refl._{MW}$ all fluctuated in a cyclic manner at the same time. Meanwhile, electric arc or plasma repeatedly discharged and distinguished, coinciding with the crest and trough of the $Refl._{MW}$ oscillation, respectively (see Figure 14.8b and c). A mechanism involving MW-induced hydrogen plasma has been proposed to understand the cyclic and synchronous variations of these parameters [23].

Isothermal sintering of TiH_2 powder compacts at 1300°C for 30 min resulted in 97.0% TD compared to 93.4% TD sintered from HDH Ti powder under the same sintering conditions (see Table 14.2) [23]. Extending the isothermal hold by 30 min slightly increased the sintered density to 97.5% TD. CP-Ti sintered from TiH_2 powder also exhibited better tensile properties than those sintered from HDH Ti (Table 14.2). This can be attributed to the higher sintered density and finer residual pores, confirmed

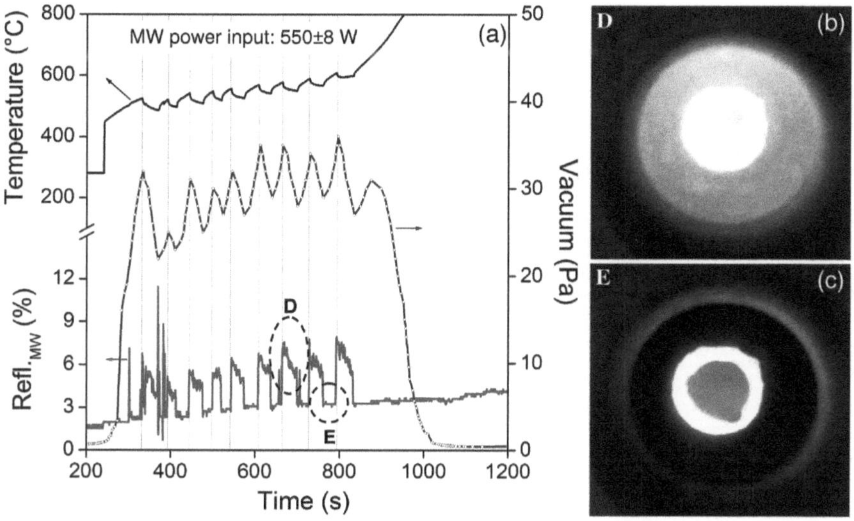

Figure 14.8 Temperature, vacuum, and MW power reflection ratio (Refl.$_{MW}$) vary synchronously as TiH_2 dehydrogenates. The 3-g sample was heated by pure MW radiation under vacuum of 10^{-1} Pa. (b) and (c) are images of the sample when plasmas are discharging (D) and extinguishing (E) corresponding to the positions of D and E indicated on the curve in (a) [23].

by metallographic analysis (see Figure 14.9). The short isothermal sintering time (30 min) at 1300°C and the resulting good combination of sintered density (97.0% TD) and tensile properties (643 MPa UTS, 512 MPa YS, and 16.3% elongation) demonstrate the advantage of MW sintering of TiH_2 powder over conventional vacuum sintering.

Table 14.2 Sintered density, tensile properties, and impurity levels of CP-Ti sintered from TiH_2 and HDH Ti powder

			Tensile properties			Impurity	
Powder	Holding time at 1300°C (min)	Sintered density (%TD)	UTS (MPa)	YS (MPa)	Elongation (%)	O (ppm)	H (ppm)
HDH Ti	30	93.4±0.1	466±4	374±14	13.1±1.2	2200	10
	60	94.7±0.3	479±4	384±5	15.6±1.9	2200	10
TiH_2	30	97.0±0.0	643±7	512±12	16.3±0.9	3000	10
	60	97.5±0.1	625±10	522±6	19.0±2.1	2900	10

TiH_2 was sintered by MWs with no susceptor while the sintering of HDH Ti was assisted with a SiC susceptor [23].

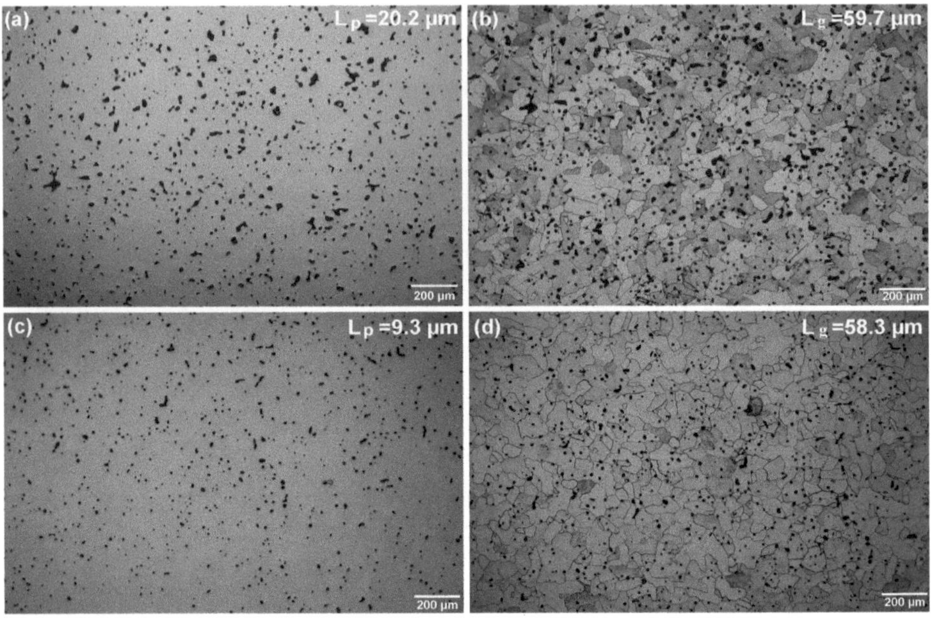

Figure 14.9 Microstructures of as-sintered CP-Ti: (a) and (b) from HDH Ti powder, and (c) and (d) from TiH_2 powder. Both powders were 100 mesh, compacted at 750 MPa, and sintered at 1300°C for 60 min under vacuum of 10^{-3} Pa. Average pore length (L_p) and grain size (L_g) were indicated [23].

14.7 Summary

Titanium powder interacts weakly with MWs, thus exhibiting an erratic heating response to MW radiation with no susceptor. The use of MW susceptors is essential to ensure effective and consistent MW heating and isothermal sintering of titanium powder compacts. Titanium and titanium alloys isothermally sintered by MW radiation under the protection of titanium sponge exhibited tensile properties comparable to or slightly better than their counterparts sintered conventionally. Titanium hydride powder appeared to be more responsive to MW radiation than HDH Ti powder but noticeable temperature fluctuations may accompany the hydrogenation process stimulated by MW-induced hydrogen plasma.

Acknowledgments

The first two authors acknowledge the financial support from the Australian Research Council (ARC) through the ARC Linkage Projects (LP) Programme and the Baosteel-Australia Joint R&D Centre.

References

[1] D.V. Clark, W.H. Sutton, Microwave processing of materials, Annu. Rev. Mater. Sci. 26 (1996) 299.

[2] Yu.V. Bykov, K.I. Rybakov, V.E. Semenov, High-temperature microwave processing of materials, J. Phys. D: Appl. Phys. 34 (2001) R55.

[3] D. Agrawal, J. Cheng, Y. Fang, R. Roy, Microwave processing of ceramics, composites and metallic materials, in: D.E. Clark, D.C. Folz, C.E. Folgar, M.M. Mahmoud (Eds.), Microwave Solutions for Ceramic Engineers, The American Ceramics Society, Inc, Westerville, OH, 2005, pp. 205.

[4] M. Oghbaei, O. Mirzaee, Microwave versus conventional sintering: a review of fundamentals, advantages and applications, J. Alloy Compd. 494 (2010) 175.

[5] J.C. Willams, Microwave Processing of Materials, National Academy Press, Washington D.C, 1994, pp. 81.

[6] V.D. Buchelnikov, D.V. Louzguine-Luzgin, G. Xie, S. Li, N. Yoshikawa, M. Sato, A.P. Anzulevich, I.V. Bychkov, A. Inoue, Heating of metallic powders by microwaves: experiment and theory, J. Appl. Phys. 104 (2008) 113505.

[7] J.W. Walkiewicz, G. Kazonich, S.L. McGill, Microwave heating characteristics of selected minerals and compounds, Miner. Metall. Proc. 5 (1988) 39.

[8] R. Roy, D. Agrawal, J.P. Cheng, S. Gedevanishvili, Full sintering of powdered-metal bodies in a microwave field, Nature 399 (1999) 668.

[9] M. Gupta, W.W. Leong, Microwaves and Metals, Wiley (Asia), Singapore, 2007.

[10] A. Mondal, Microwave Sintering of Metals, LAP Lambert Academic Publishing, Saarbrücken, 2011.

[11] D. Agrawal, Microwave sintering of metal powders, in: I. Chang, Y. Zhao (Eds.), Advances in Powder Metallurgy, Woodhead Publishing, Ltd, 2013, pp. 361.

[12] M. Sato, H. Fukusima, F. Ozeki, T. Hayasi, Y. Satito, S. Takayama, Experimental Investigation of Mechanism of Microwave Heating in Powder Metals, 2004 Joint 29th International Conference on Infrared and Millimeter Waves and 12th International Conference on Terahertz Electronics, Karlsruhe, 2004, pp. 831.

[13] S.D. Luo, Y.F. Yang, G.B. Schaffer, M. Qian, Characteristics of microwave sintering of titanium powder compacts, Proceedings of the 12th World Conference on Titanium, Beijing, 2011, pp. 1826.

[14] S.D. Luo, C.L. Guan, Y.F. Yang, G.B. Schaffer, M. Qian, Microwave heating, isothermal sintering, and mechanical properties of powder metallurgy titanium and titanium alloys, Metall. Mater. Trans. A 44 (2013) 1842.

[15] M.G. Kutty, S. Bhaduri, S.B. Bhaduri, Gradient surface porosity in titanium dental implants: relation between processing parameters and microstructure, J. Mater. Sci. Mater. Med. 15 (2004) 145.

[16] T. Hayashi, Microwave Sintering of Metal Matrix Alloys, Reports of Research Institute of Industrial Products Technology, Research Institute Industrial Products Technology, Gifu, 2005.

[17] T. Marcelo, J. Mascarenhas, F.A.C. Oliveira, Microwave sintering-a novel approach to powder technology, Mater. Sci. Forum 636-637 (2010) 946.

[18] S.D. Luo, C.J. Bettles, M. Yan, G.B. Schaffer, M. Qian, Microwave sintering of titanium, Key Eng. Mater. 436 (2010) 141.

[19] S.D. Luo, M. Yan, G.B. Schaffer, M. Qian, Sintering of titanium in vacuum by microwave radiation, Metall. Mater. Trans. A 42 (2011) 2466.

[20] A.W. Fliflet, S.L. Miller, M.A. Imam, Evaluation of microwave-sintered titanium and titanium alloy powder compacts, Ceramic Trans. 234 (2012) 83.
[21] P. Yu, G. Stephani, S.D. Luo, H. Goehler, M. Qian, Microwave-assisted fabrication of titanium hollow spheres with tailored shell structures for various potential applications, Mater. Lett. 86 (2012) 84.
[22] M.A. Imam, A. Fliflet, Sintering of metal and alloy powders by microwave/millimetre-wave heating, U.S. Patent 8,431,071 B2 (April 30, 2013).
[23] S.D. Luo, Q. Li, J. Tian, C. Wang, M. Yan, G.B. Schaffer, M. Qian, Novel fabrication of titanium by pure microwave radiation of titanium hydride powder, Scripta Mater. 69 (2013) 69.
[24] I. Maxim, M. Tanaka, Numerical analysis of the microwave heating of compacted copper powders in single-mode cavity, Jpn. J. Appl. Phys. 50 (2011) 097302.
[25] M. Tanaka, H. Kono, K. Maruyama, Selective heating mechanism of magnetic metal oxides by a microwave magnetic field, Phys. Rev. B 79 (2009) 104420.
[26] J. Cheng, R. Roy, D. Agrawal, Experimental proof of major role of magnetic field losses in microwave heating of metal and metallic composites, J. Mater. Sci. Lett. 20 (2001) 1561.
[27] J. Cheng, R. Roy, D. Agrawal, Radically different effects on materials by separated microwave electric and magnetic fields, Mater. Res. Innovat. 5 (2002) 170.
[28] J. Ma, J.F. Diehl, E.J. Johnson, K.R. Martin, N.M. Miskovsky, C.T. Smith, G.J. Weisel, B.L. Weiss, D.T. Zimmerman, Systematic study of microwave absorption, heating, and microstructure evolution of porous copper powder metal compacts, J. Appl. Phys. 101 (2007) 074906.
[29] P. Mishra, G. Sethi, A. Upadhyaya, Modeling of microwave heating of particulate metals, Metall. Mater. Trans. B 37 (2006) 839.
[30] A. Mondal, D. Agrawal, A. Upadhyaya, Microwave heating of pure copper powder with varying particle size and porosity, J. Microwave Power E.E. 43 (2009) 5.
[31] R.W. Bruce, A.W. Fliflet, H.E. Huey, C. Stephenson, M.A. Imam, Microwave sintering and melting of titanium powder for low-cost processing, Key Eng. Mater. 436 (2010) 131.
[32] J.P. Cheng, D.K. Agrawal, S. Komarneni, M. Mathis, R. Roy, Microwave processing of WC-Co composites and ferroic titanates, Mater. Res. Innovat. 1 (1997) 44.
[33] T. Hashiguchi, H. Sueyoshi, Effect of atmosphere on microwave heating of titanium powder, Powder Metall. 54 (2011) 537.
[34] H.T. Wang, Z.Z. Fang, P. Sun, A critical review of mechanical properties of powder metallurgy titanium, Int. J. Powder Metall. 46 (2010) 45.
[35] Y.F. Yang, S.D. Luo, G.B. Schaffer, M. Qian, Sintering of Ti-10V-2Fe-3Al and mechanical properties, Mater. Sci. Eng. A 528 (2011) 6719.
[36] D.F. Heaney, R.M. German, Advances in the Sintering of Titanium Powders, in: H. Danninger, R. Ratzi (Eds.), Proceedings of the PM 2004 Powder Metallurgy World Congress, European Powder Metallurgy Association, Shrewsbury, 2004, pp. 222.
[37] T. Saito, A cost-effective P/M titanium matrix composite for automobile use, Adv. Perform. Mater. 2 (1995) 121.
[38] Y. Yamamoto, J.O. Kiggans, M.B. Clark, S.D. Nunn, A.S. Sabau, W.H. Peter, Consolidation process in near net shape manufacturing of Armstrong CP-Ti/Ti–6Al–4V powders, Key Eng. Mater. 436 (2010) 103.
[39] S. Abkowitz, J.M. Siergiej, R.D. Regan, Titanium P/M, preforms, parts and composites, in: H.H. Hausner (Ed.), Modern Developments in Powder Metallurgy, Metal Powder Industries Federation, Princeton, 1971, pp. 501.
[40] A.D. Hanson, J.C. Runkle, R. Widmer, J.C. Hebeisen, Titanium near net shapes from elemental powder blends, Int. J. Powder Metall. 26 (1990) 157.

[41] F.H. Froes, S.J. Mashl, V.S. Moxson, J.C. Hebeisen, V.A. Duz, The technologies of titanium powder metallurgy, JOM 56 (2004) 46.
[42] O.M. Ivasishin, D.G. Savvakin, I.S. Bielov, V.S. Moxson, V.A. Duz, R. Davies, C. Lavender, Microstructure and Properties of Titanium Alloys Synthesized from Hydrogenated Titanium Powders, Proceedings of Conference on Science and Technology of Powder Materials: Synthesis, Consolidation and Properties, Pittsburg, MS&T, 2005, pp. 151.
[43] N.R. Moody, W.M. Garrison Jr., J.E. Smugeresky, J.E. Costa, The role of inclusion and pore content on the fracture toughness of powder-processed blended elemental titanium alloys, Metall. Trans. A 24 (1993) 161.
[44] H. Guo, Z. Zhao, C. Duan, Z. Yao, The powder sintering and isothermal forging of Ti-10V-2Fe-3Al, JOM 60 (2008) 47.
[45] O.M. Ivasishin, Cost-effective manufacturing of titanium parts with powder metallurgy approach, Mater. Forum 29 (2005) 1.
[46] V.A. Duz, O.M. Ivasishin, V.S. Moxson, D.G. Savvakin, V.V. Telin, Cost-effective titanium alloy powder compositions and method for manufacturing flat or shaped articles from these powders, U.S. Patent 7993577 B2 (August 9, 2011).
[47] Z.Z. Fang, P. Sun, H.T. Wang, Hydrogen sintering of titanium to produce high density fine grain titanium alloys, Adv. Eng. Mater. 14 (2012) 383.
[48] Y. Nakamori, S. Orimo, T. Tsutaoka, Dehydriding reaction of metal hydrides and alkali borohydrides enhanced by microwave irradiation, Appl. Phys. Lett. 88 (2006) 112104.
[49] M. Matsuo, Y. Nakamori, K. Yamada, S. Orimo, Effects of microwave irradiation on the dehydriding reaction of the composites of lithium borohydride and microwave absorber, Appl. Phys. Lett. 90 (2007) 232907.

Scavenging of oxygen and chlorine from powder metallurgy (PM) titanium and titanium alloys

15

Ming Yan, H.P. Tang**, Ma Qian**
*RMIT University, School of Aerospace, Mechanical and Manufacturing Engineering, Centre for Additive Manufacture, Melbourne, Victoria, Australia
**State Key Laboratory of Porous Metal Materials, Northwest Institute for Nonferrous Metal Research, Xi'an, China

15.1 Introduction

Oxygen and chlorine are two persistent issues for the fabrication of low-cost, strong, and ductile powder metallurgy (PM) Ti materials, especially when the inexpensive hydride-dehydride (HDH) Ti powder made directly from Kroll sponge is used. In most cases, oxygen is present in titanium as an interstitial impurity. However, even down to a trace level (e.g., 1200 ppm) [1], oxygen can still radically change the phase selection, microstructure, and mechanical properties of titanium and titanium alloys. In the context of Ti PM, recent studies have established that there exists a critical level of oxygen for as-sintered Ti-6Al-4V (wt%), which is around 0.33 wt% (Figure 15.1) [2–4], beyond which the tensile ductility drops dramatically making the as-sintered Ti-6Al-4V unsuited to structural applications. Ref. [5] gives an up-to-date review of the effect of oxygen on the room-temperature ductility of both PM and non-PM titanium materials.

Chlorine, which often exists in the form of residual chlorides, is an impurity in titanium powders made from Kroll or Hunter sponge or other forms of titanium materials (including scrap) manufactured from titanium sponge. Unlike oxygen, which can increase the strength of titanium despite being detrimental to ductility, chlorine does not seem to be beneficial to titanium in any way. Instead, it can impose a variety of negative impacts from degrading the quality of titanium welds to reducing sintered density. Table 15.1 summarizes the chlorine level of titanium powders made from different processes [6–11].

Much effort has been made to develop low-cost or more cost-affordable titanium powders over the last two decades (see Chapters 2–9). However, real commercial success is still to be achieved. Consequently, it would appear that the HDH Ti powder (see Chapter 7) made from Kroll sponge will continue to be the primary source of affordable titanium powder for PM Ti in the foreseeable future. In fact, the use of HDH Ti powder produced from Kroll sponge is the key to the commercial success

Titanium Powder Metallurgy. http://dx.doi.org/10.1016/B978-0-12-800054-0.00015-0

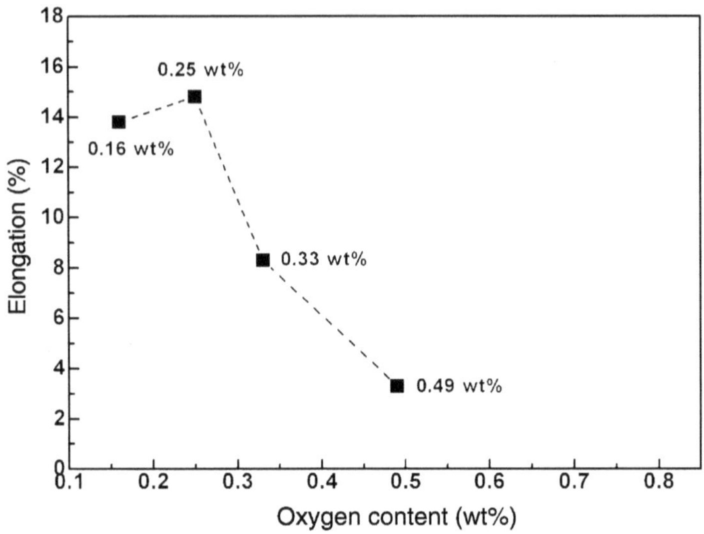

Figure 15.1 Effect of oxygen content on tensile ductility of as-sintered Ti-6Al-4V [4].

Table 15.1 Impurity levels of chlorine in titanium powders made by various processes [6–11]

Type of Ti	Reaction involved	Major elemental impurities	Impurity level, Cl
Ti (Kroll)	$TiO_2(s)+2Cl_2(g)+C(s) \rightarrow TiCl_4(g)+CO_2(g)$ $TiCl_4(g)+Mg(s) \rightarrow Ti(s)+MgCl_2(s)$	O, Cl, C, Mg	1200–1500 ppm
Ti (Hunter)	$TiO_2(s)+2Cl_2(g)+C(s) \rightarrow TiCl_4+CO_2(g)$ $TiCl_4+4Na(s) \rightarrow Ti(s)+4NaCl(s)$	O, Cl, C, Na	1200–1500 ppm
Ti (HDH)	$Ti(s)+H_2(g) \rightarrow TiH_2(s)$ $TiH_2(s) \rightarrow Ti(s)+H_2(g)$	H (according to reaction)	Can ~ 10 ppm, generally <800 ppm
TiH_2	$Ti(s)+H_2(g) \rightarrow TiH_2(s)$	H (according to reaction)	From 60 ppm to 500 ppm
Ti (Cambridge)	$TiO_2(s)+C(s) \rightarrow Ti(s)+CO(g)+CO_2(g)$	O, C	–
Ti (Armstrong)	$TiO_2(s)+2Cl_2+C(s) \rightarrow TiCl_4+CO_2(g)$ $TiCl_4+4Na(s) \rightarrow Ti(s)+4NaCl(s)$	O, Cl, C, Na	Can <50 ppm

Oxygen concentration depends on many factors and is therefore not listed in the table.

of Dynamet's elemental blend sinter PM process (see Chapter 17). This process has produced a wide range of affordable PM near-net-shape preformed components [12]. Another source of affordable powder for Ti PM can be titanium hydride powder (see Chapter 8).

HDH Ti powder can be made to contain low oxygen (e.g., <0.15 wt%) but with high production costs. Instead, there are a variety of inexpensive, decent HDH powder products available on the market that contain typically more than 0.25 wt% O, particularly for powders finer than 100 mesh (e.g., 250 mesh or 325 mesh). On the other hand, the powder handling and sintering process can readily add an extra 0.1 wt% O even with the use of TiH_2 powder [13–15], which is known to be advantageous over HDH Ti powder in terms of the control of oxygen (see Chapters 8 and 10). This will make the oxygen content of the as-sintered Ti-6Al-4V products exceed the critical level of 0.33 wt%. Being able to mitigate the detrimental effect of oxygen on ductility thus holds the key to the fabrication of low-cost, strong, and ductile PM Ti materials from inexpensive HDH Ti powders. The chlorine or chloride content in such inexpensive HDH Ti powders is often around 0.02 wt%, which can adversely affect the densification process during sintering. Hence, it is desirable to be able to also effectively mitigate the detrimental effect of chlorine on sintering densification. This chapter reviews the developments to date in scavenging of oxygen and chlorine from PM Ti and Ti alloys.

15.2 The effect of oxygen on ductility of Ti materials

Oxygen can decrease the ductility of titanium and titanium alloys via a number of mechanisms [5]:

1. Oxygen can promote the precipitation of α from β leading to acicular or grain boundary α precipitates, which are generally detrimental to ductility.
2. Oxygen may increase the tendency of martensitic transformation in Ti. The martensitic phases, typically referred to as α' in low β-stabilizer alloys, and α'' in high β-stabilizer alloys, are usually associated with reduced ductility, although they may increase strength.
3. Oxygen can contribute to the stabilization of α phase over ω phase and this favors suppressing or eliminating the potential negative impact of the ω phase on ductility.
4. Oxygen, which increases the Al equivalent (Equation 15.1) [1], can promote the formation of the α_2 phase (Ti_3Al) in Al-containing Ti alloys, which is highly detrimental to the ductility of Ti alloys.

$$[\mathrm{Al}]_{\mathrm{EQ}} = 1[\mathrm{Al}] + 1/3[\mathrm{Sn}] + 1/6[\mathrm{Zr}] + 10[\mathrm{O}+\mathrm{C}+2\mathrm{N}]\ (\mathrm{wt}\%) \quad (15.1)$$

5. Oxygen can induce oxygen-enriched clusters or ordering, and such microstructure inhomogeneities can obstruct plastic deformation and therefore reduce ductility.
6. Oxygen can suppress the activation of deformation twinning or change the deformation mode from twinning to slip in Ti. Either development will tend to lower ductility.
7. Oxygen can obstruct plastic deformation by slip through facilitating the precipitation of α from β.

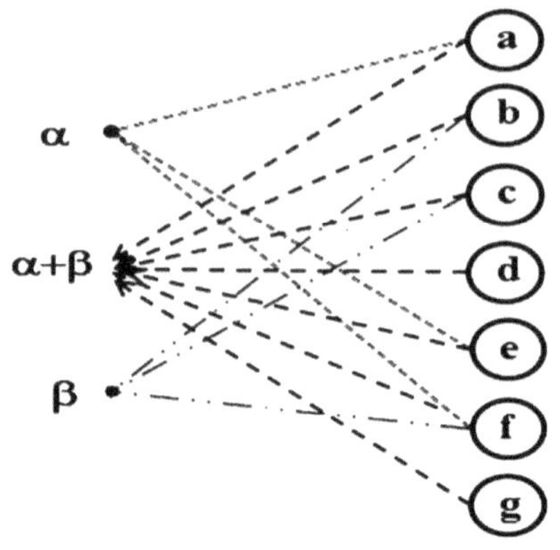

Figure 15.2 (a–g) represent different mechanisms by which oxygen may affect the ductility of titanium (see text). The diagram shows the relevance of each of the proposed mechanisms to α-, (α + β)-, and β-Ti alloys [5].

The relevance of each of these mechanisms to α-, (α + β)-, and β-Ti alloys is shown in Figure 15.2 [5]. These mechanisms suggest that it is important to control the oxygen content in titanium.

15.3 Scavenging of oxygen from PM Ti and Ti alloys

The discussion below will focus on scavenging of oxygen from PM Ti and Ti alloys during high-temperature isothermal sintering. There are a variety of approaches which may be used to lower the oxygen content in titanium powders. Alkaline earth metals, especially Ca [16,17], can be used to deoxidize Ti powders because of their higher affinity for oxygen than titanium (see Figures 15.3 and 15.4). In fact, it was proposed that titanium may be sintered while submerged in molten calcium or other alkaline earth metals or in the atmosphere of such a metal vapor to reduce the oxygen content and/or avoid oxidation during sintering [18]. However, it should be noted that the use of alkaline earth metals at high temperatures (typical isothermal sintering temperatures for titanium are 1200–1350°C) is technically challenging. In this regard, the approaches discussed in the following can be easily applied at high temperatures.

15.3.1 By rare earth (RE) element–based scavengers

Owing to the strong chemical affinity of titanium for oxygen, only a few options are available to allow scavenging of oxygen from titanium solid solutions. The use of RE element–based scavengers appears to be the most practical option. In the 1970s and 1980s, RE elements were introduced into a number of titanium alloys through rapid solidification processing (RSP) [19]. The purpose was to form fine RE oxide dispersoids from the supersaturated titanium matrix (consolidated from powder, flake, ribbon, etc.)

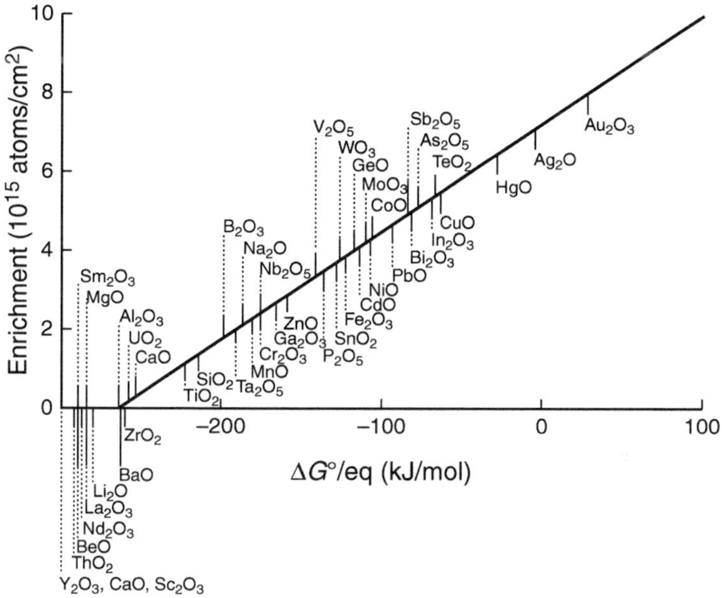

Figure 15.3 The Gibbs free energy (ΔG) of formation of various oxide materials, showing their thermodynamic stability as well as their affinity for oxygen [27]. Their elemental enrichment in the earth is also shown (y axis).

during subsequent heat treatment for dispersion strengthening and enhanced elevated temperature performance [19]. Scavenging of oxygen was not the purpose because of the low oxygen content in these vacuum-melted alloys. Driven by the need to develop low-cost PM Ti alloys, the last two decades have seen an increasing interest in the use of RE-based materials to mitigate the detrimental effect of oxygen on ductility. The following observations are notable.

- A variety of RE-based materials have proved to be effective in scavenging oxygen from titanium solid solutions. These include RE metals (e.g., Er and Ce) [20,21], oxygen-deficient RE oxides (i.e., Y_2O_3) [22], RE hydrides (e.g., YH_2) [23], RE borides (e.g., LaB_6) [24], RE silicides (e.g., $CeSi_2$) [25], and RE-containing master alloys (e.g., $NdAl_2$ and NdAl) [26].
- Research has shown that the oxygen scavenging capability of RE elements follows the following sequence: Y > Er > Dy > Tb > Gd (see Figure 15.3 [27] and Figure 15.4 [28]). The high potency of yttrium in scavenging oxygen from titanium alloys is demonstrated in Figure 15.5 [29], where the uniformly distributed Y_2O_3 dispersoids result from an addition of 0.1 wt% Y to a titanium alloy that contained only 0.07 wt% O.
- The oxygen-scavenging effect of RE oxides is insignificant unless the RE oxides introduced are largely oxygen-deficient [22]. The addition of pure RE metals has proved to be problematic too as they are essentially RE oxides. Accordingly, adding RE elements in the form of master alloys, hydrides, and/or borides are preferred.
- The cost of scavenging oxygen should be taken into account. According to Table 15.2, Y, La, Ce, and Sm are relatively low-cost RE metals. In addition, RE master alloys can be more cost effective [30,31].

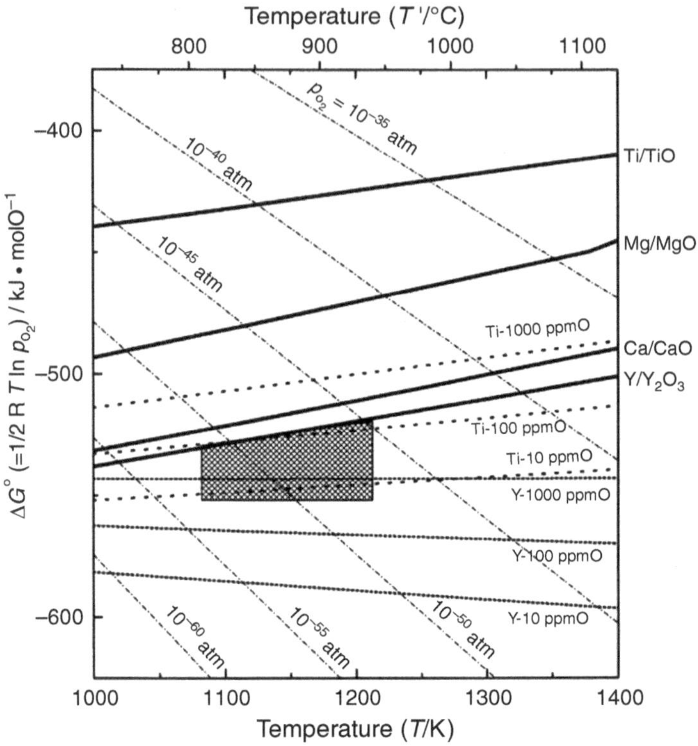

Figure 15.4 The Gibbs free energy (ΔG) of formation between metals (Ti, Mg, Ca, and Y) and oxygen as a function of temperature [28].

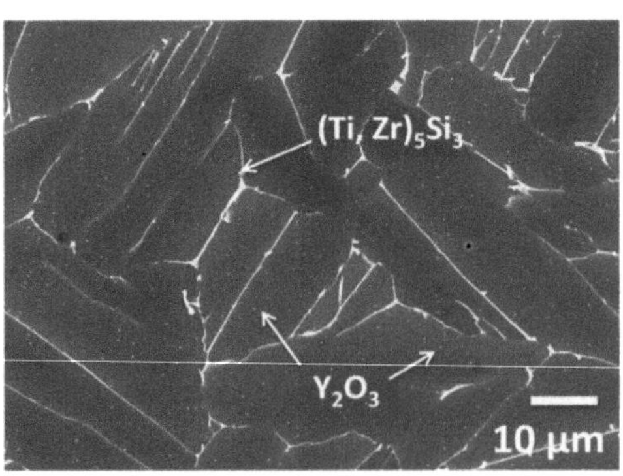

Figure 15.5 Precipitation of fine Y_2O_3 dispersoids (due to the addition of 0.1 wt% Y) in the α-Ti matrix of Ti-6Al-2.7Sn-4Zr-0.4Mo-0.45Si-0.1Y (wt%) which contained 0.07 wt% O and 0.1 wt% Y. The alloy was additively manufactured by selective electron beam melting [29].

Table 15.2 Price of rare earth (RE) metals (upper position in the table) and RE oxides (lower position in the table) per kg [30]

Sc														
$15,500														
$7,000														
1541°C														
2836°C														
Y														
$74														
$20														
1526°C														
2730°C														
La	**Ce**	**Pr**	**Nd**	**Pm**	**Sm**	**Eu**	**Gd**	**Tb**	**Dy**	**Ho**	**Er**	**Tm**	**Yb**	**Lu**
$13	$12	$175	$98	–	$30	–	$95	$1,900	$750	–	$225	–	–	–
$5.8	$5.5	$130	$72	–	–	$980	$41	$800	$525	–	$72	–	–	–
920°C	795°C	935°C	1024°C	1042°C	1072°C	826°C	1312°C	1356°C	1407°C	1461°C	1529°C	1545°C	824°C	1652°C
3464°C	3443°C	3130°C	3074°C	3000°C	1900°C	1529°C	3000°C	3123°C	2562°C	2600°C	2868°C	1730°C	1196°C	3402°C

The melting (upper position) and boiling temperatures (lower position) of the RE metals are also given in the table [31].

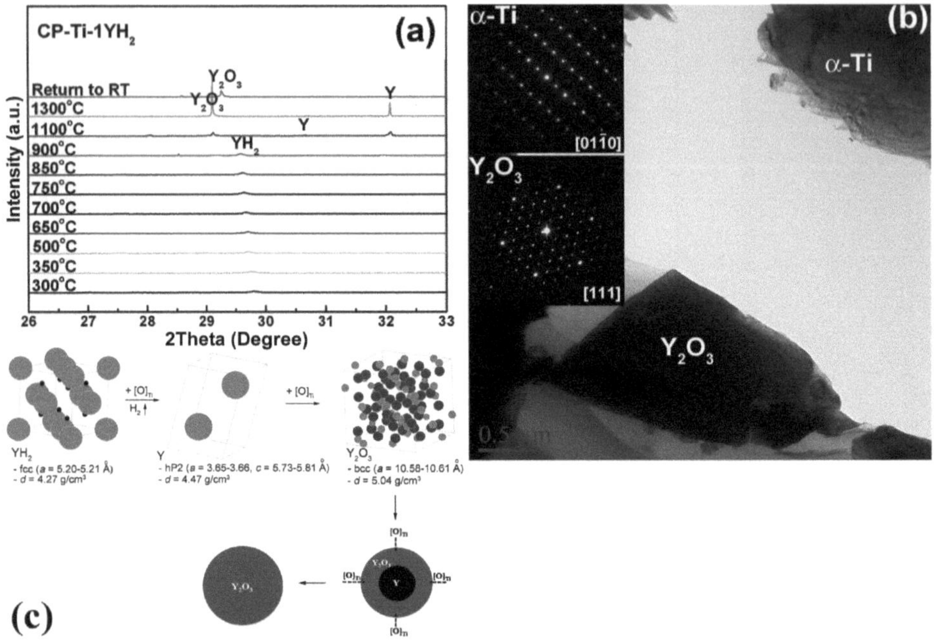

Figure 15.6 (a) *In situ* synchrotron x-ray radiation to show the decomposition of YH_2 in CP-Ti from room temperature to sintering temperature (1300°C). The corresponding as-sintered microstructure is presented in (b), showing as-formed Y_2O_3 by transmission electron microscopy bright-field image and selected area electron diffraction patterns for Y_2O_3 and α-Ti matrix. Schematic graph of the decomposition process as well as oxygen gettering procedure is shown in (c) [33].

- Boron [32] is an effective grain refiner for Ti by pinning grain boundary motion. The use of RE borides as oxygen scavengers has proved to be effective and can realize grain refinement at the same time.
- RE hydrides (e.g., YH_2) seem to have an extra advantage compared with other forms of RE-based materials because of the release of hydrogen (Figure 15.6) [33]. The hydrogen released has the potential to react with surface oxides and therefore produces surface cleaning, in a way similar to the use of TiH_2 as the starting powder material. Additionally, the chemical activity of the surrounding Ti particles may be enhanced by H_2 too and this may contribute to better densification.
- As a result of the reduced oxygen content due to the addition of RE oxygen scavenger, the ductility of the as-sintered Ti and Ti alloys can be substantially increased, e.g., by 60–90%, when merely 0.3–0.6 wt% of RE are used [23,25].
- However, the mechanical properties, particularly ductility as well as sintered density, do not necessarily show linear increase with increasing addition of the RE-based scavengers. For example, an optimum addition level is often seen around 0.5 wt% for the $CeSi_2$-doped CP-Ti, Ti-6Al-4V, and Ti-10V-2Fe-3Al [25] (Figure 15.7).

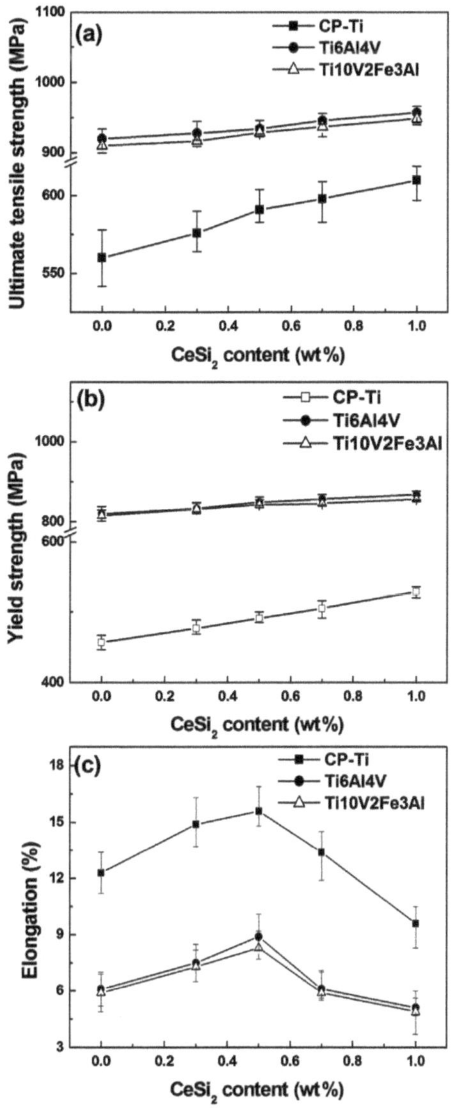

Figure 15.7 Ultimate tensile strength (a), yield strength (b), and elongation (c) of as-sintered CP-Ti, Ti-6Al-4V, and Ti-10V-2Fe-3Al versus the addition of $CeSi_2$ [25].

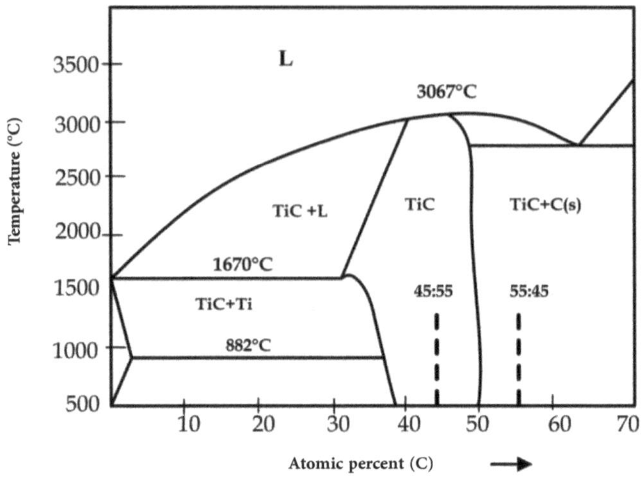

Figure 15.8 Ti-C binary phase diagram showing that the TiC is a stable phase while Ti_2C is not in the system [38].

15.3.2 By oxygen-stabilized compounds

15.3.2.1 Ti_2C-type compounds

Some compounds were found to be capable of scavenging oxygen from Ti solid solutions via forming metastable or substoichiometric, oxygen-stabilized phases. Ti_2C is one such example [34–37]. The Ti-C phase diagram suggests that TiC is the equilibrium phase while Ti_2C is not (Figure 15.8) [38]. It has been found that the Ti_2C phases observed in C-doped Ti materials are normally associated with oxygen (Table 15.3) [34–37]. The following observations and discussions are based on this finding.

- The formation of Ti_2C is closely related to oxygen in most cases, suggesting that local oxygen can be gettered via forming oxygen-stabilized Ti_2C. As a result of the reduced local oxygen concentration, the formation of other oxygen-sensitive phases such as grain boundary α, ω, and/or α_2 can be suppressed or avoided; see Figure 15.9 for example [37].
- The solubility of oxygen in Ti_2C is found to be generally below 10 at%, suggesting that the oxygen-scavenging capability of Ti_2C is less than that of RE-based materials such as YH_2, which leads to the formation of Y_2O_3 containing up to 60 at% O.
- The scavenging of oxygen via the formation of Ti_2C appears to be alloy dependent. For example, the Ti_2C phase can form in the Ti-15Ta alloy but it was found to contain less oxygen than the surrounding Ti matrix [37]. It was proposed that this is due to the enlarged lattice parameters of the Ti matrix by the alloying element Ta, which in turn increases the tendency for oxygen to occupy interstitial vacancies in the Ti matrix rather than in the Ti_2C phase.
- Controversy still exists. For instance, in a study of PM Ti-15Mo [35], the Ti_2C identified contains more oxygen than the Ti matrix, in contrast to the observation of Ref. [37]. It should be noted that light elements such as carbon and oxygen are difficult to be quantified and this might be the reason for the discrepancy.

Table 15.3 Oxygen concentration in Ti alloys containing carbon [34–37]

Ti alloy	Overall carbon content (wt%)	Ti-C phase	Carbon concentration in Ti-C phase	Oxygen concentration in Ti-C phase	References
Ti-15-3(Ti-15V-3Al-3Sn-3Cr)	0.2–0.7	Ti_2C or Ti(CO)	~0.5 in C/Ti ratio	Higher than matrix	34
Ti-15Mo (PM)	0.032	Ti_2C	~ 0.54 in C/Ti	Higher than matrix	35
Ti-13Cr	0.15	Ti_2C	–	–	36
Ti-15Cr	0.2	Ti_2C	~0.56 in C/Ti ratio	Higher than matrix	37
Ti-15Ni	0.2	Ti_2C	0.98	Higher than matrix	37
Ti-15Mo	0.2	Ti_2C	0.5	Lower than matrix	37

The Ti-C phase in these alloys was identified to be the Ti_2C phase, which normally contains higher oxygen than does the surrounding Ti matrix.

15.3.2.2 *$Ti_4Fe_2O_x$-type compounds*

The second type of oxygen-stabilized compounds is a group of intermetallics, abbreviated as Ti_2MO_x (M = Fe, C_O, Ni; the corresponding value of x is up to 0.5) or $Ti_4M_2O_y$ (M = Fe, Co, Ni) [39–42]. These are the so-called electronic concentration (*e*/*a*) phases. Taking $Ti_4Fe_2O_x$ as an example, this intermetallic phase can form in the presence of TiFe even though the oxygen concentration is on a trace level (e.g., 0.1 at% ≈0.03 wt%). The key to the formation of $Ti_4Fe_2O_x$ seems to be the availability of TiFe in the microstructure, which, however, does not normally form in PM Ti materials as Fe is a potent β-stabilizer, which tends to stay within β-Ti phases [43]. On the other hand, a recent observation of the $Ti_4Fe_2O_x$ phase (Figure 15.10) [44] in the PM Ti materials suggests that this intermetallic phase can somehow exist in PM Ti to enable scavenging of oxygen from PM Ti. The underlying mechanism is not clear as yet and is under investigation. However, the overall oxygen-scavenging efficiency of this compound is expected to be higher than that of the oxygen-stabilized Ti_2C.

Finally, it should be pointed out that the formation of coarse RE oxides, oxygen-stabled Ti_2C or $Ti_4Fe_2O_x$-type compounds can be detrimental to the ductility of the as-sintered PM Ti materials. The size and distribution should be controlled where possible.

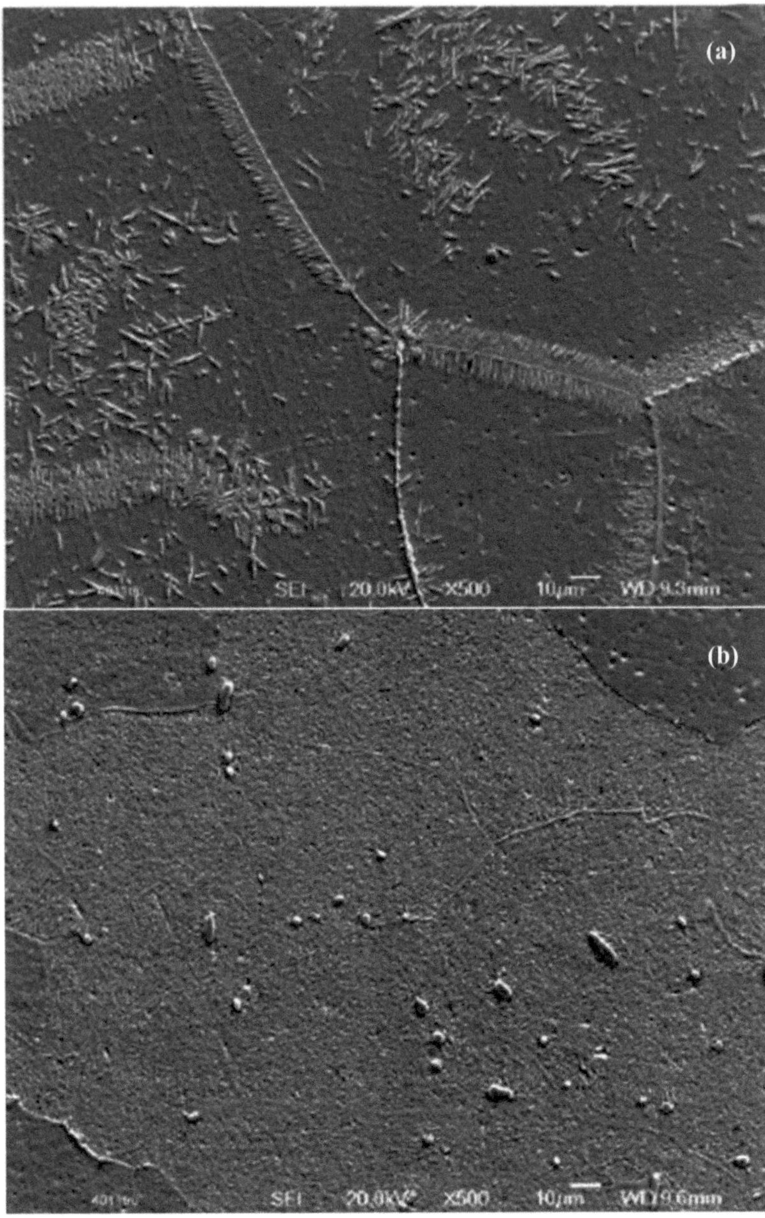

Figure 15.9 Scanning electron micrographs to reveal that due to the doping of 0.2 wt% C into Ti-13 wt% Cr, the grain boundary α phases in (a) were eliminated in the as-sintered microstructure (b) [37].

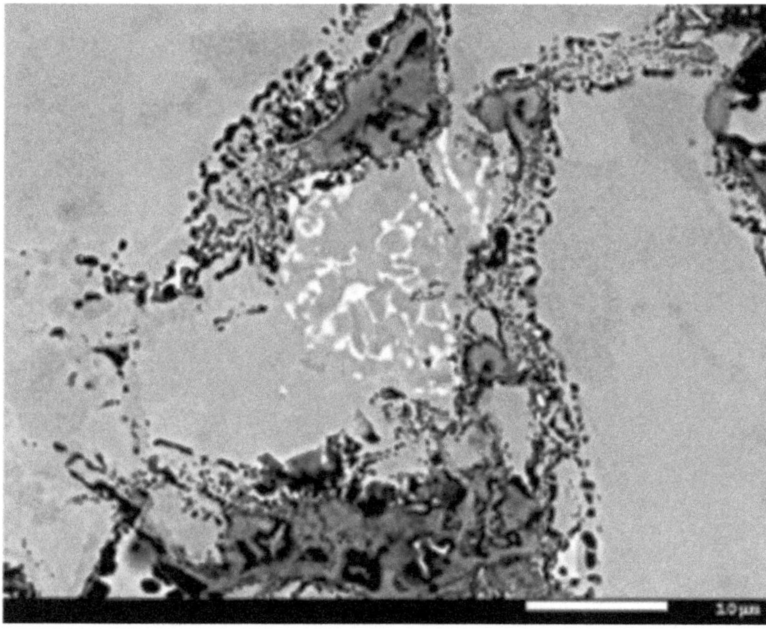

Figure 15.10 Scanning electron microscope–backscattered electron image to show bright-contrasted $Ti_4Fe_2O_x$ phase in a PM Ti-6Al-4V alloy [44].

15.4 Impact of chlorine on PM Ti materials

The chlorine impurity in titanium powder, often in the form of residual chlorides, originates from the Kroll or Hunter process. The two major impacts of Cl or Cl-containing phases include the following [7,45–48]:

- The volatilization of Cl-containing impurity phases can result in the formation of both macroporosity and microporosity in the as-sintered microstructure (Figure 15.11) [45]. Hence, they can be a killer to the sintering fabrication of strong and ductile PM Ti materials.
- The Cl-containing impurity phases can adversely affect the welding process of Ti alloys including PM Ti alloys. Owing to volatilization, the Cl-containing impurity phases can cause splashing-like effects during welding. In addition, the remaining Cl-containing impurity phases (e.g., NaCl and $MgCl_2$) may become the source of weakness in terms of mechanical properties and/or corrosion resistance.

15.5 Scavenging of chlorine from PM Ti and Ti alloys

15.5.1 Overall strategy

Four possible strategies may work together Cl from PM Ti:

- Vacuum soaking [49]. The Cl-containing impurity phases (e.g., $MgCl_2$ and NaCl) have low melting points (Table 15.4) [50]. For example, $MgCl_2$ can start to volatize around

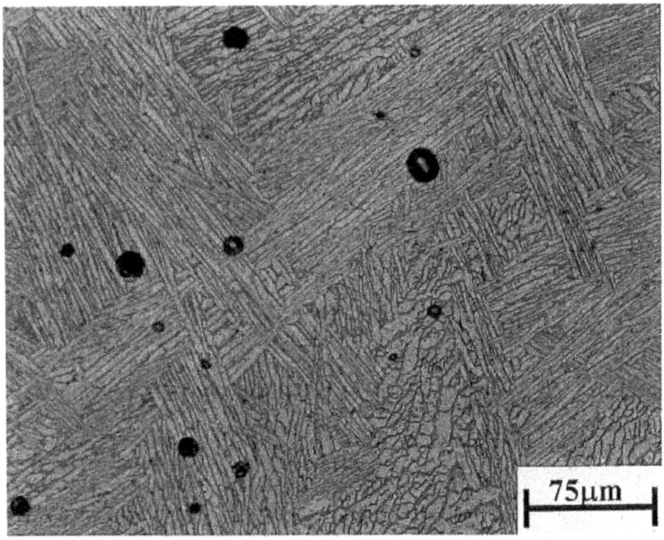

Figure 15.11 Microporosity (~1–50 μm) in an as-sintered Ti-TiB composite due to the presence of ~200 ppm Cl [45].

Table 15.4 Selected chlorides and their standard heat of formation, melting temperature, boiling temperate, and solubility in water and ethanol [49,50]

Material	Heat of formation (ΔH_f) (kJ/mol)	Melting temperature (°C)	Boiling temperature (°C)	Solubility (water)	Solubility (ethanol)
$TiCl_4$	−815	−24	136	Soluble	Soluble
$MgCl_2$ (Kroll)	−642	300–714	1412	54.3 g/100 mL (20°C)	7.4 g/100 mL (20°C)
$CaCl_2$	−795	772	1935	74.5 g/100 mL (20°C)	–
NaCl (Hunter)	−411	801	1413	35.9 g/100 mL (20°C)	1.49/100 mL (20°C)

900°C [51]. Vacuum soaking at higher temperatures can remove such chlorides, although the process may be slow depending on sample size. In comparison, the volatilization rate of $MgCl_2$ is much faster than that of NaCl [49].

- Powder leaching. Because of the high solubility of $MgCl_2$ and NaCl in water or ethanol (see Table 15.4), (water or ethanol) leaching is another approach to lowering the concentration of Cl or chlorides in the powder material. However, there can still be a considerable amount of Cl left in the powder (e.g., up to 800 ppm Cl in HDH powder, Table 15.1) after leaching or vacuum soaking. Further actions to lower the Cl concentration are necessary for most applications.

- Use of TiH_2 powder [11]. This is based on the assumption that the atomic hydrogen released from TiH_2 can react with Cl-containing compounds such as $MgCl_2$ to form gaseous HCl:

$$MgCL_2\,(s) + H_2\,(\text{decomposed from } TiH_2)(g) \rightarrow Mg\,(s) + HCL\,(g) \tag{15.2}$$

- Doping with Cl scavengers [7,23]. Suitable scavengers include RE metals and RE-containing compounds. This will be the focus of the discussion below.

15.5.2 By RE metals

It has been found [20] that metallic Y can scavenge Cl leading to the formation of a $Y_{75}Cl_{18}O_5Ti_2$ (at%) compound. The source of the metallic Y is from the decomposition of the metastable YH_2. The following observations are notable from Ref. [23].

- No Na or Mg was detected in the Y-Cl particles analyzed. However, we may still assume that the source of Cl has most likely come from the decomposition or volatilization of chlorides (NaCl or MgCl based) during sintering.
- Metallic Y has a high heat of formation (ΔH_f) with Cl. Table 15.5 lists the heat of formation between RE metals and Cl [52]. Again, Y is the best Cl-scavenger in terms of ΔH_f except Sc, which is much more expensive than Y (Table 15.2) and therefore less attractive.

Table 15.5 A summary of rare earth (RE) chlorides in terms of their melting temperature, heat of formation, crystal symmetry (H for hexagonal and M for monoclinic) and lattice parameters [52]

Compound	Melting point (°C)	Heat of formation (kJ/mol)	Symmetry	*a* (Å)	*b* (Å)	*c* (Å)
$ScCl_3$	960	−925.1	H	6.38		17.79
YCl_3	721	−1000	M			
$LaCl_3$	852	−421	H	7.48		4.36
$CeCl_3$	802	−252.8	H	7.45		4.32
$PrCl_3$	786	−252.1	H	7.42		4.28
$NdCl_3$	760	−245.6				
$SmCl_3$	678	−243	H	7.38		4.17
$EuCl_3$	774, 623	−233	H	7.37		4.13
$GdCl_3$	609, 602	−240	H	7.36		4.11
$TbCl_3$	588	−241				
$DyCl_3$	654	−236	M	6.91	11.97	6.40
$HoCl_3$	720	−233	M	6.85	11.85	6.39
$ErCl_3$	776	−211.4	M	6.80	11.79	6.39
$TmCl_3$	821	−229	M	6.75	11.73	6.39
$YbCl_3$	854	−228.7, −223	M	6.73	11.65	6.38
$LuCl_3$	892	−227.9	M	6.72	11.60	6.39

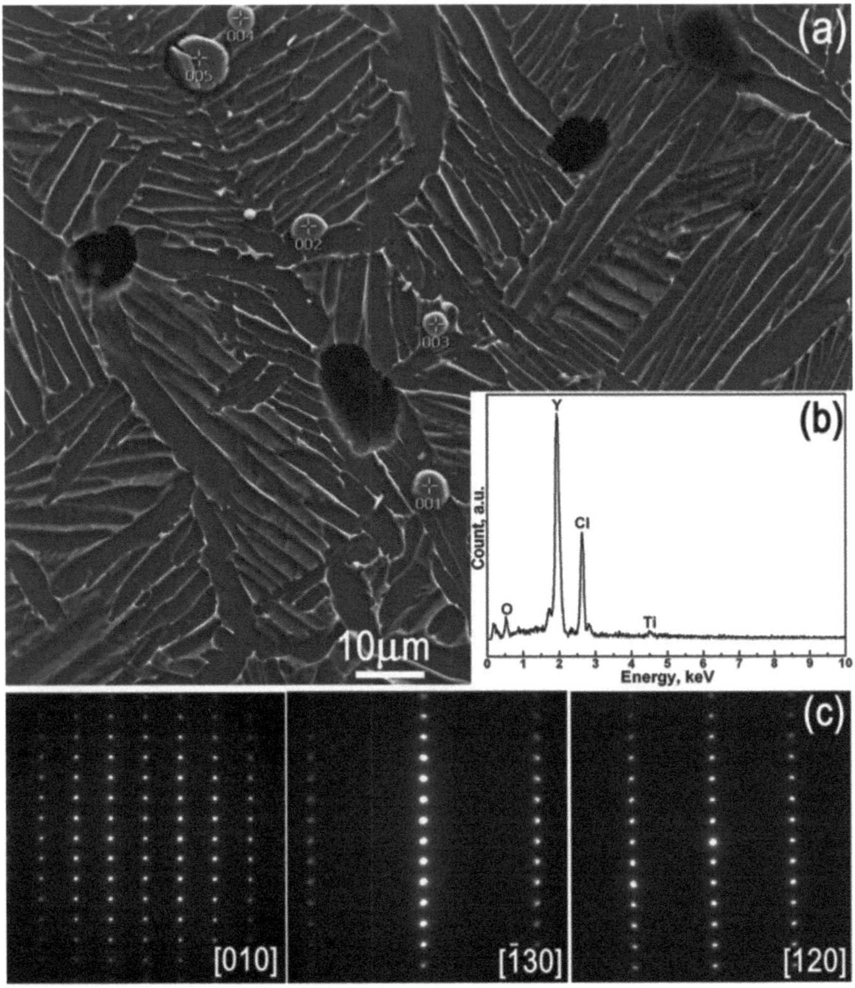

Figure 15.12 The Y-Cl phase in the as-sintered Ti-2.25Fe-1.5Mo (α+β) alloy due to the addition of 0.6 wt% Y [23]. SEM image (a), phase constitution determined by transmission electron microscopy (TEM)–energy-dispersive spectroscopy (b), and TEM–selected area electron diffraction results of the phase (c).

- The Y-Cl compound, which is close to Y_4Cl in nominal composition (at%), is not in a stoichiometric state (Figure 15.12) [23]. Research has shown that the stable phase YCl_3 in the Y-Cl binary system can be reduced by the presence of additional Y (see Equation 15.3) [51]. This could be the reason for the nonstoichiometric Y-Cl compound observed in the as-sintered microstructure, rather than the stable YCl_3.

$$YCl_3 + Y \rightarrow Y_xCl_y \tag{15.3}$$

15.5.3 By RE-based compounds

- A number of RE-based compounds including Y_2O_3 [53], YH_2 [23], and $CeSi_2$ [22] have proved to be effective Cl-scavengers for PM Ti. Among these, Y_2O_3 is used as a stable compound itself, while both YH_2 and $CeSi_2$ will decompose during sintering.
- The proposed reactions involving Y_2O_3 are shown as follows according to the thermodynamic data [53–55]:

$$Y_2O_3 + MgCl_2 \rightarrow YOCl + MgO(\Delta G \sim -49.3\,\text{kJ/mol at } 25\,°\text{C}) \tag{15.4}$$

$$Y_2O_3 + NaCl_2 \rightarrow YOCl + Na_2O(\Delta G \sim +320\,\text{kJ/mol at } 25\,°\text{C}) \tag{15.5}$$

$$Y_2O_3 + Cl \rightarrow YOCl + \tfrac{1}{2}O_2(\Delta G \sim -67\,\text{kJ/mol at } 25\,°\text{C}) \tag{15.6}$$

- It was proposed that the mechanism for the Y_2O_3 to scavenge Cl is based on the reaction of Equation 15.4. Yet an unanswered question is that there is no evidence of the formation of MgO or MgO in the reaction products (e.g., Ti-Mg-O) along with the YOCl phase (Figure 15.13) [53]. Instead, the reaction of Equation 15.6 is more likely to be responsible. However, one should also note that the Gibbs free energy for this reaction may be positive at high temperature, suggesting that thermodynamically it is less favorable to occur at high temperatures (i.e., at sintering temperatures) than at low temperatures [55].

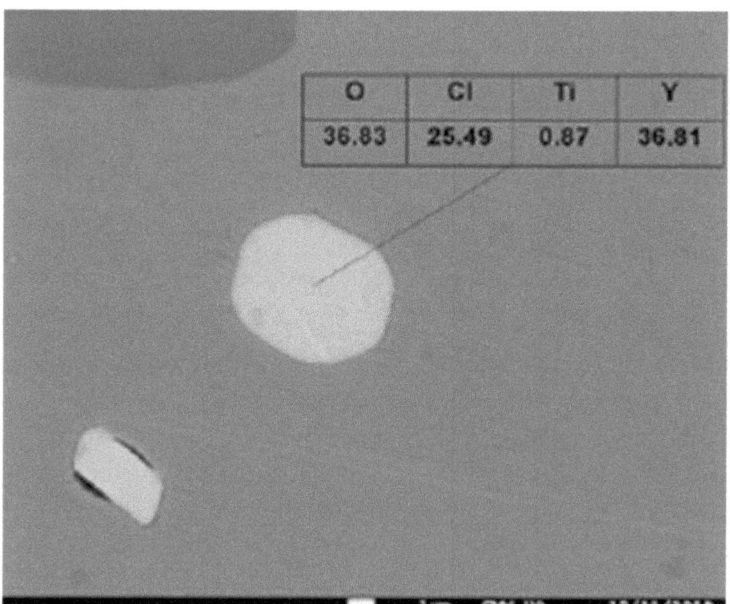

Figure 15.13 Cl-containing Y_2O_3 particles in a PM Ti-4Si alloy [53]. The phase constitution shows no presence of Na or Mg but enriched in Y, Cl, and O.

15.6 Scavenging of oxygen in additively manufactured Ti alloys and reaction kinetics

As shown earlier (Figure 15.5), scavenging of oxygen can also occur in additively manufactured Ti alloys (e.g., Ti-6Al-2.75Sn-4Zr-0.4Mo-0.45Si-0.1Y), despite the low oxygen content (0.07 wt%), low yttrium addition (0.1 wt%), and fast cooling rate (up to $\sim 10^4$ K/s) [56]. There is direct microscopic evidence of the formation of Y-O-Ti oxide particles (Figure 15.14), implying that the small addition of Y has effectively reacted with O to reduce the oxygen content in solid solution. Additionally, the resulting nanometric Y-O-Ti oxides are expected to offer dispersion strengthening too.

Similar to the additive manufacturing process, studies performed in the 1980s and 1990s on rapidly solidified Ti alloys have also demonstrated the formation of RE oxide dispersoids in the rapidly solidified microstructure [57]. The fast reaction rate between RE and O may be attributed to the fact that oxygen is a fast diffuser in Ti (e.g., $\sim 10^{-9}$ m^2/s at 1300°C) [58,59].

The only data available on the reaction kinetics about the scavenging of Cl is reported in a study of the chlorination of yttrium oxide over the temperature range of 575–975°C [55]. A global reaction rate, *R*, is determined and shown in Equation 15.7.

$$R = 10^5\,\text{kpa}^{-1} \cdot \exp(-187\,\text{kJ} \cdot \text{mol}^{-1} / (R/T)) \cdot p\text{Cl}_2 \cdot 1.51 \cdot (1-\alpha) \cdot [-\text{In}(1\text{-}\alpha)]^{0.34} \quad (15.7)$$

where R is gas constant, T is temperature, pCl_2 is the partial pressure of Cl_2, and α is conversion degree.

Three findings are noticeable from this study, which can also be interpreted according to Figure 15.15 [55]:

- The reaction occurs at faster rates at higher temperatures;
- Mass transfer rate of Cl is determinative to the overall reaction at high temperatures such as 850–975°C;
- It is a progressive process to achieve the maximum weight gain, that is, the completion of the reaction ($Y_2O_3 + Cl \rightarrow YOCl$). At 950°C, the time required is about 120 seconds and the resulting weight gain is about 25%.

Based on this study, the reaction rate seems to be fast enough for the use of Y_2O_3 as a Cl-scavenger in the conventional Ti PM processes. However, the reaction may not be able to complete in the case of additive manufacturing.

15.7 Concluding remarks

Both oxygen and chlorine can be effectively scavenged from inexpensive HDH Ti powders made directly from Kroll sponge. As a result, their detrimental effect can

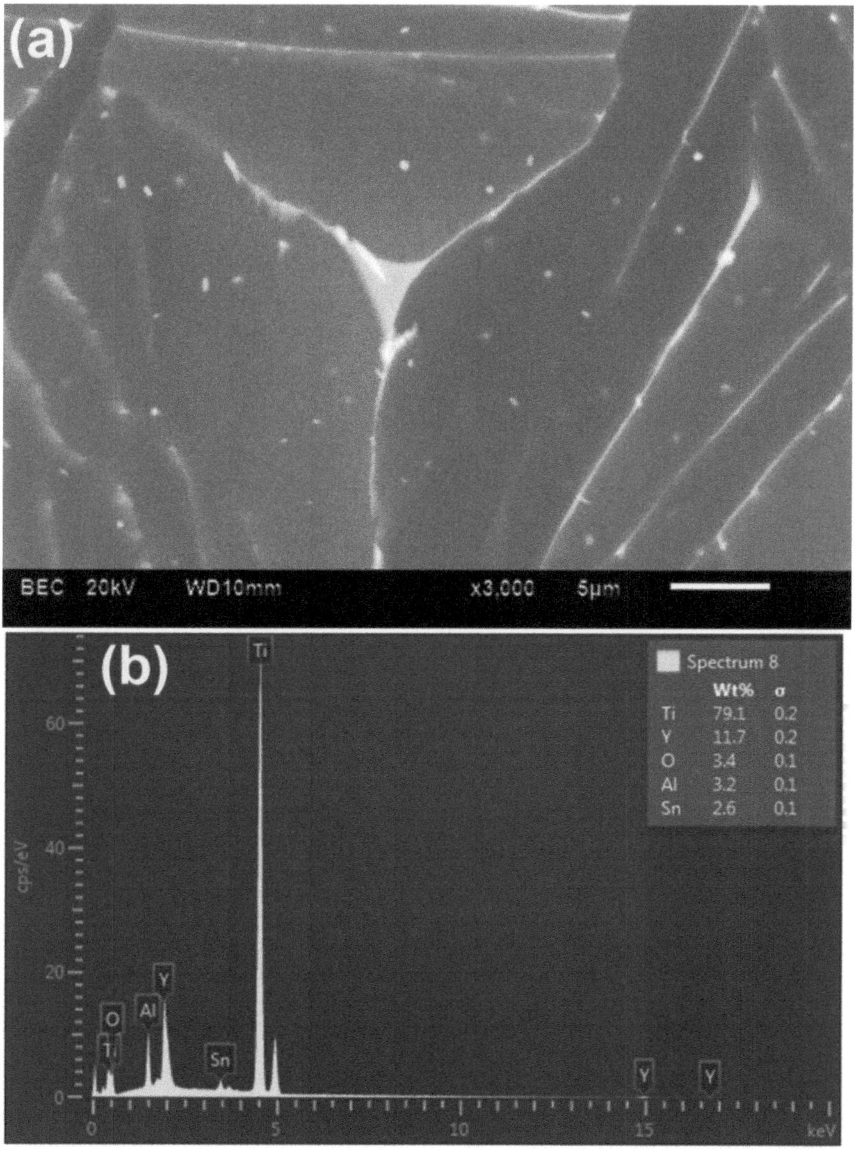

Figure 15.14 (a) Scanning electron microscopic image of additively manufactured Ti-6Al-4V showing nanoparticles containing Y and O (b) [56]. The results suggest that Y is capable to scavenge oxygen even under high cooling rate conditions (cooling rate up to 10^4 k/s).

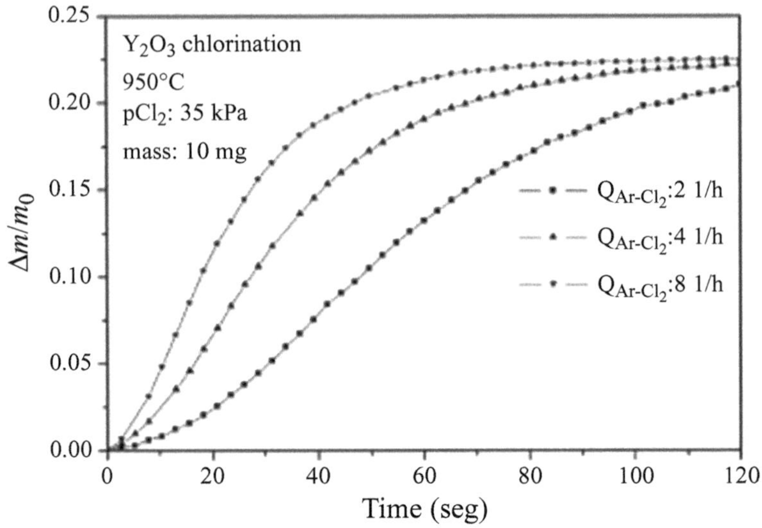

(a)

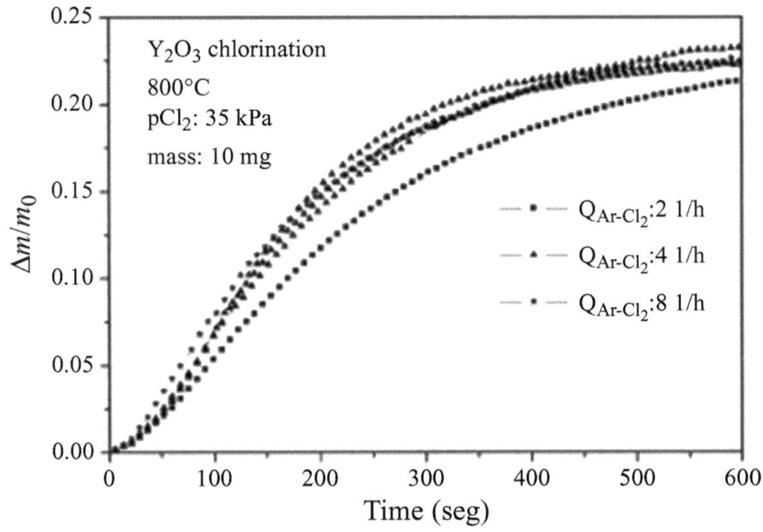

(b)

Figure 15.15 Reaction kinetics (weight gain $\Delta m/m_0$ vs. time) of Y_2O_3 with Cl_2 as a function of flow rate of Ar-Cl_2 mixture, partial pressure of Cl_2, and temperature [55].

be mitigated or controlled. This provides a practical avenue for the development of low-cost, strong, and ductile PM Ti materials for near-net-shape component production that aims for nonfatigue critical applications.

Acknowledgments

Support from the Australian Research Council (ARC) through the ARC CoE for Design in Light Metals is acknowledged. H.P. Tang acknowledges support from the Ministry of Science and Technology China and the National High Technology Research Program (No. 2013AA031103).

References

[1] J.Y. Lim, C.J. McMahon, D.P. Pope, J.C. Williams, The effect of oxygen on the structure and mechanical behavior of aged Ti-8 wt% Al, Metall. Trans. A 7 (1) (1976) 139–144.

[2] H. Miura, Y. Itoh, T. Ueamtsu, K. Sato, The influence of density and oxygen content on the mechanical properties of injection molded Ti-6Al-4V alloys, Adv. Powder Metall. Particulate Mater. (2010) 46–53.

[3] T. Ebel, O. Milagres Ferri, W. Limberg, M. Oehring, F. Pyczak, F.P. Schimansky, Metal injection moulding of titanium and titanium-aluminides, Key Eng. Mater. 520 (2012) 153–160.

[4] M. Yan, M.S. Dargusch, T. Ebel, M. Qian, A transmission electron microscopy and three-dimensional atom probe study of the oxygen-induced fine microstructural features in as-sintered Ti–6Al–4V and their impacts on ductility, Acta Mater. 68 (2014) 196–206.

[5] M. Yan, W. Xu, M.S. Dargusch, H.P. Tang, M. Brandt, M. Qian, Powder Metall. 57 (4) (2014) 251-257.

[6] F.H. Froes, D. Eylon, Powder metallurgy of titanium alloys, Int. Mater. Rev. 35 (1) (1990) 162–184.

[7] R.J. Low, Sintering of titanium for cost-effective net-shape manufacturing: scavenging of chlorine and sulphur with rare earth oxide additions and the sintering of titanium for surface densification and containerless HIP, PhD thesis, The University of Queensland, 2010.

[8] M.A. Imam, F.S. Froes, Low cost titanium and developing applications, J. Metall. 62 (5) (2010) 17–20.

[9] M. Qian, Cold compaction and sintering of titanium and its alloys for near-net-shape or preform fabrication, Int. J. Powder Metall. 46 (5) (2010) 29–44.

[10] I.M. Robertson, G.B. Schaffer, Review of densification of titanium based powder systems in press and sinter processing, Powder Metall. 53 (2) (2010) 146–162.

[11] O.M. Ivasishin, D.G. Savvakin, V.A. Duz, M.V. Matviychuk, V.S. Moxson, Extra low impurity content powder metallurgy titanium and titanium alloys, Titanium 2012, Hilton Atlanta, 2012.

[12] S. Abkowitz, S. Abkowitz, H. Fisher, Titanium alloy components manufacture from blended elemental powder and the qualification process, in: M. Qian, F.H. Froes (Eds.), Titanium Powder Metallurgy, Elsevier, 2014, (Chapter 17).

[13] O.M. Ivasishin, D.G. Savvakin, F.H.S. Froes, K.A. Bondareva, Synthesis of alloy Ti–6Al–4V with low residual porosity by a powder metallurgy method, Powder Metall. Metal Ceram. 41 (7–8) (2002) 382–390.

[14] S.L.G. Petroni, M.S.M. Paula, V.A.R. Henriques, Interstitial elements in Ti-13Nb-13Zr alloy produced by powder metallurgy suing hydride powder, Powder Metall. 56 (2013) (2013) 202–207.
[15] S.D. Luo, Y.F. Yang, G.B. Schaffer, M. Qian, Novel fabrication of titanium by pure microwave radiation of titanium hydride powder, Scripta Mater. 69 (1) (2013) 69–72.
[16] C.G. McCracken, J.W. Robison, C.A. Motchenbacher Jr., Manufacture of HDH low oxygen titanium-6aluminium-4vanadium (Ti-6-4) powder incorporating a novel powder deoxidation step, Adv. Powder Metall. Particulate Mater. (2009); Proceedings of the 2009 International Conference on Powder Metallurgy and Particulate Materials, June 20–July 1, Las Vegas, Nevada, The Metal Powder Industries Federation (MPIF), 2009.
[17] C.G. McCracken, C. Motchenbacher, D.P. Barbis, Review of titanium. Powder-production methods, Int. J. Powder Metall. 46 (5) (2010) 19–26.
[18] D.W. Rostron, Method of sintering titanium and like metals. U.S. Patent 2,546,320 (Priority date: 4 November 1948; Publication date: 27 March 1951) (1951).
[19] D.G. Konitzer, B.C. Muddle, H.L. Fraser, A comparison of the microstructures of As-cast and laser surface melted Ti-8Al-4Y, Metall. Trans. A 14 (10) (1983) 1979–1988.
[20] M.F.X. Gigliotti, A.P. Woodfield, The roles of rare earth dispersoids and process, Metall. Trans. A 24 (8) (1993) 1761–1771.
[21] S.M.L. Sastry, P.J. Meschter, J.E. O'neal, Structure and properties of rapidly solidified dispersion-strengthened titanium alloys: part I. Characterization of dispersoid distribution, structure, and chemistry, Metall. Trans. A 15 (7) (1984) 1451–1463.
[22] V. De Castro, T. Leguey, M.A. Monge, A. Muñoz, R. Pareja, D.R. Amador, M. Victoria, Mechanical dispersion of Y_2O_3 nanoparticles in steel EUROFER 97: process and optimisation, J. Nucl. Mater. 322 (2) (2003) 228–234.
[23] M. Yan, Y. Liu, Y.B. Liu, C. Kong, G.B. Schaffer, M. Qian, Simultaneous gettering of oxygen and chlorine and homogenization of the β phase by rare earth hydride additions to a powder metallurgy Ti–2.25 Mo–1. 5 Fe alloy, Scripta Mater. 67 (5) (2012) 491–494.
[24] I. Bilobrov, V. Trachevsky, Approach to modify the properties of titanium alloys for use in nuclear industry, J. Nucl. Mater. 415 (2) (2011) 222–225.
[25] Y.F. Yang, S.D. Luo, G.B. Schaffer, M. Qian, Impurity scavenging, microstructural refinement and mechanical properties of powder metallurgy titanium and titanium alloys by a small addition of cerium silicide, Mater. Sci. Eng. A 573 (2013) 166–174.
[26] Y. Liu, L. Chen, W. Wei, H. Tang, B. Liu, B. Huang, Improvement of ductility of powder metallurgy titanium alloys by addition of rare earth element, J. Mater. Sci. Technol. 22 (4) (2006) 465–469.
[27] G.E. Thompson, P. Skeldon, X. Zhou, K. Shimizu, H. Habazaki, C.J.E. Smith, Improving the performance of aerospace alloys, Aircraft Eng. Aerospace Technol. 75 (4) (2003) 372–379.
[28] T.H. Okabe, K. Hirota, E. Kasai, F. Saito, Y. Waseda, K.T. Jacob, Thermodynamic properties of oxygen in RE–O (RE= Gd, Tb, Dy, Er) solid solutions, J. Alloys Compd. 279 (2) (1998) 184–191.
[29] H.P. Tang, M. Qian, Unpublished result.
[30] http://www.mineralprices.com (recorded on the April 1, 2014).
[31] http://en.wikipedia.org.
[32] S. Tamirisakandala, R.B. Bhat, J.S. Tiley, D.B. Miracle, Grain refinement of cast titanium alloys via trace boron addition, Scripta Mater. 53 (12) (2005) 1421–1426.
[33] M. Yan, Y. Liu, G.B. Schaffer, M. Qian, *In situ* synchrotron radiation to understand the pathways for the scavenging of oxygen in commercially pure Ti and Ti–6Al–4V by yttrium hydride, Scripta Mater. 68 (1) (2013) 63–66.

[34] Z.Q. Chen, M.H. Loretto, Effect of carbon additions on microstructure and mechanical properties of Ti-15-3, Mater. Sci. Technol. 20 (3) (2004) 343–349.
[35] M. Yan, M. Qian, C. Kong, M.S. Dargusch, Impacts of trace carbon on the microstructure of as-sintered biomedical Ti–15Mo alloy and reassessment of the maximum carbon limit, Acta Biomater. 10 (2) (2014) 1014–1023.
[36] M. Chu, I.P. Jones, X. Wu, Effect of carbon on microstructure and mechanical properties of a eutectoid β titanium alloy, J. Mater. Eng. Perform. 14 (6) (2005) 735–740.
[37] Z.Q. Chen, Y.G. Li, M.H. Loretto, Role of alloying elements in microstructures of beta titanium alloys with carbon additions, Mater. Sci. Technol. 19 (10) (2003) 1391–1398.
[38] H. Okamoto (Ed.), Desk Handbook: Phase Diagrams for Binary Alloys, ASM International, 2000.
[39] M.H. Mintz, Z. Hadari, M.P. Dariel, Hydrogenation of oxygen-stabilized Ti_2MO_x (M= Fe, Co, Ni; $0 < x < 0.5$) compounds, J. Less Common Met. 74 (2) (1980) 287–294.
[40] L. Schlapbach, T. Riesterer, The activation of FeTi for hydrogen absorption, Appl. Phys. A 32 (4) (1983) 169–182.
[41] V.G. Chuprina, I.M. Shalya, I.I. Karpikov, Reaction of the hydrogen-absorbing intermetallic compound TiFe with oxygen III. Phase composition of the scale formed on TiFe, Powder Metall. Metal Ceram. 34 (11–12) (1996) 663–668.
[42] O. Heinen, D. Holland-Moritz, D.M. Herlach, Phase selection during solidification of undercooled Ti–Fe, Ti–Fe–O and Ti–Fe–Si–O melts – influence of oxygen and silicon, Mater. Sci. Eng. A 449 (2007) 662–665.
[43] M. Yan, S.D. Luo, G.B. Schaffer, M. Qian, Impurity (Fe, Cl, and P)-induced grain boundary and secondary phases in commercially pure titanium (CP-Ti), Metall. Mater. Trans. A 44 (8) (2013) 3961–3969.
[44] M. Yan, M.S. Dargusch, C. Kong, J. Kimpton, S. Kohara, M. Brandt, M. Qian, Unpublished data.
[45] C.Z. Yu, M.I. Jones, Investigation of chloride impurities in hydrogenated-dehydrogenated Kroll processed titanium powders, Powder Metall. 56 (4) (2013) 304–309.
[46] Z. Fan, H.J. Niu, B. Cantor, A.P. Miodownik, T. Saito, Effect of Cl on microstructure and mechanical properties of *in situ* Ti/TiB MMCs produced by a blended elemental powder metallurgy method, J. Microsc. 185 (2) (1997) 157–167.
[47] M. Kumagai, K. Shibue, M.S. Kim, M. Yonemitsu, Influence of chlorine on the oxidation behavior of TiAl–Mn intermetallic compound, Intermetallics 4 (7) (1996) 557–566.
[48] K. Majima, T. Hirata, K. Shouji, Effects of purity of titanium powder and porosity on static tensile properties of sintered titanium specimens, J. Jpn. Inst. Met. 51 (12) (1987) 1194–1200.
[49] S.D. Hill, R.V. Mrazek, Vacuum evaporation of salt from titanium sponge, Metall. Trans. 5 (1) (1974) 53–58.
[50] http://en.wikipedia.org.
[51] J.D. McKinley, Mass-spectrometric investigation of the high-temperature reaction between nickel and chlorine, J. Chem. Phys. 40 (1) (1964) 120–125.
[52] Standard Thermodynamic Properties of Chemical Substances, CRC Press LLC, http://www.update.uu.se/~jolkkonen/pdf/CRC_TD.pdf, 2000.
[53] R.J. Low, M. Qian, G.B. Schaffer, Sintering of titanium with yttrium oxide additions for the scavenging of chlorine impurities, Metall. Mater. Trans. A 43 (13) (2012) 5271–5278.
[54] L. Rycerz, M. Gaune-Escard, Lanthanide (III) halides: thermodynamic properties and their correlation with crystal structure, J. Alloys Compd. 450 (1) (2008) 167–174.

[55] J.P. Gaviría, A.E. Bohé, The kinetics of the chlorination of yttrium oxide, Metall. Mater. Trans. B 40 (1) (2009) 45–53.
[56] H.P. Tang, S.L. Lu, W.P. Jia, G.Y. Yang, M. Qian, Selective electron beam melting of titanium and titanium aluminide alloys, Int. J. Powder Metall. 50 (2014) 57–64.
[57] S. Naka, M. Marty, H. Octor, Oxide-dispersed titanium alloys Ti-Y prepared with the rotating electrode process, J. Mater. Sci. 22 (3) (1987) 887–895.
[58] C. Leyens, M. Peters, Titanium and Titanium Alloys, Wiley-VCH, Weinheim, 2003, p. 187.
[59] G. Lütjering, J.C. Williams, Titanium, vol. 2, Springer, Berlin, 2003.

Titanium metal matrix composites by powder metallurgy (PM) routes

16

Katsuyoshi Kondoh
Osaka University, Joining and Welding Research Institute (JWRI), Ibaragi, Osaka, Japan

16.1 Introduction

Many researchers have worked on metal matrix composites (MMCs) by powder metallurgy (PM) routes to improve the mechanical, thermal, and electrical properties of the monolithic alloys. Titanium metal matrix composites (TMCs) have pronounced potential for high strength, excellent heat resistance, and corrosion protection. The conventional ingot metallurgy process is not suitable for the fabrication of TMCs reinforced with *ex situ* additive particles because of the high chemical reactivity of Ti. Therefore, PM processes are commonly employed to fabricate TMC components used for automotive, motorcycle, and aerospace industries. The reinforcement dispersion is essential. Hence, the powder mixing, dispersion, and coating techniques for particles/fibers are important to ensure a good performance of TMCs. In particular, nanoscale fine particles can be easily segregated, and this will result in the formation of material defects at the primary powder boundary (PPB) of TMCs. Not only the selection of dispersed reinforcements, their size, morphology, and content but also the interfacial bonding between reinforcements and the matrix are important to obtain desired properties. The metallurgical bonding incident to the reaction behavior between the reinforcements and titanium matrix dominates the mechanical performance of TMCs. In addition, from a viewpoint of the interface control and bonding strength improvement, the process conditions in the fabrication of TMCs such as powder consolidation pressure, sintering temperature, and atmosphere must be optimized. Secondary treatments like hot rolling, forging, and extrusion as well as heat treatment are also important for full density consolidation and strength improvement. Furthermore, porous metal materials can be easily produced by PM processes such as the space holder method, use of pore-forming agents, and powder-based rapid-prototyping (3D printing) for high-energy shock absorption and artificial bone implant applications. In general, these porous materials show inadequate mechanical strength; however, the introduction of dispersed reinforcements can effectively improve the strength of porous structured components. Some implications of materials design and processing of TMCs are introduced and discussed below based on previous and recent studies on PM.

Titanium Powder Metallurgy. http://dx.doi.org/10.1016/B978-0-12-800054-0.00016-2

Figure 16.1 Samples produced by compressive induction heating at 850°C for 10 min, showing sintered powders, but porosity at the SiC particle clusters formed during mixing [19].

16.2 Materials design and processing of TMCs

Two types of PM routes have been successfully established for the fabrication of TMCs: *ex situ* and *in situ* processing for reinforcements dispersion in the matrix. The additive particles selected as effective reinforcements of TMCs include: Cr_3C_2 [1], TiC [2], TiN [3], TiO_2 [4], Si_3N_4 [5], SiC [6], TiB_2, TiB [7,8], Al_3O_2 [9], Zr_2O_3, R_2O_3 (R stands for rare earth element) [10,11], Ti_3Al or TiAl [12], and Ti_5Si_3 [13] intermetallic compounds. Boron (B) particles [14] and carbon nanoparticles such as nanodiamond (ND) [15], carbon nanotube (CNT) [16], and vapor-growth carbon fiber (VGCF) [17] are also effective single elements to form reinforcing compounds via the partial or complete reaction between these additive elements and the titanium matrix.

16.2.1 *Ex situ processing*

Thermodynamically stable ceramics such as SiC, TiC, TiB, and ZrC have high micro-indentation hardness and specific Young's modulus, but no or little reactivity with titanium matrix. Then no compound is synthesized during sintering and consolidation of the elementally blended raw Ti powders with these ceramics particles. This means that there is no change in both the particle size and morphology of the additive ceramics before and after sintering. TiB is thermally stable and has high hardness and Young's modulus [18]. In addition, the maximum solubility of B in titanium matrix is small (<0.001 at%); that is, TiB additive particles are stable in the matrix of TMCs after high temperature sintering and hot consolidation processing. β-titanium alloy (Ti-7Mo-4Fe-2Al-2V) reinforced with 20 vol% TiB particles is a typical composite having excellent static and dynamic strength, Young's modulus, and wear resistance. In particular, the high-temperature mechanical strength is remarkably improved by TiB. The modified TMCs of this material have been used to make automotive components (exhaust valves) by Toyota Motor Corp.

Table 16.1 Selected properties of CermeTi-C-10 and Ti-6Al-4V [21]

Property	CermeTi-C-10 (10 wt% TiC/Ti-6Al-4V)	Ti-6Al-4V*
Yield strength (MPa)	1014	862
Tensile strength (MPa)	1083	965
Elastic modulus (GPa)	135	116
Fracture toughness (MPa $\sqrt{m}$)	40	80

* HIP processed.

Coarse SiC particulates have also been used as additive reinforcements of PM Ti-6Al-4V (Ti-64) composites, where the elementally blended Ti-64 alloy powder and SiC particles are consolidated by ECAP, a severe plastic deformation process. However, macro- and micro-cracks occurred at SiC particles as shown in Figure 16.1 [19]. Pores also formed around the SiC particulates because of their poor interfacial bonding strength with titanium matrix. As a result, higher-temperature sintering or secondary operation such as hot extrusion and hot pressing are necessary to obtain such dense PM TMCs [19].

The characteristics of raw additive particulates used as reinforcements and their distribution uniformity in the matrix of TMCs have a significant influence on the mechanical properties of TMCs. For example, TMCs with 10 wt% TiC (sintered at 1230°C) show superior mechanical properties compared to the monolithic PM Ti-64 alloy as shown in Table 16.1. In addition, they show excellent tensile strength (TS) at elevated temperatures up to about 700°C [20,21]. However, it has been reported that pure titanium and TiC are not stable at high temperatures; for example, TiC_{1-x} layers form around TiC particles during sintering of the elementally blended Ti-TiC powders at 1375°C [22]. Consequently, the additive TiC particles are completely consumed to form TiC_{1-x} dispersoids at the end of this reaction.

On the other hand, from a tribological performance point of view, these additives are effective to control the seizure and sticking phenomena due to their poor reactivity with the counter metal materials under dry sliding conditions, and to improve the wear resistance and stabilize the friction coefficient [23]. For example, as shown in Figure 16.2, the wear coefficient of PM Ti composites reinforced with TiC particulates gradually decreases and their tribological property improves with increasing TiC content. However, they are inferior to other TMCs (Ti + TiB_2 and Ti + Si_3N_4). This is because *in situ*–formed TiB and Ti_5Si_3 compounds, which originated from raw TiB_2 and Si_3N_4 particles, respectively, have good interfacial bonding with the matrix and are therefore more effective for wear protection in sliding wear test.

16.2.2 In situ reactive processing

Because of the high reactivity of titanium matrix with the additive elements such as B, carbon (C), TiB_2, B_4C, Cr_3C_2, and Si_3N_4, stable particulates or needle-like reinforcements can form *in situ* and are dispersed in the matrix through solid-state reaction.

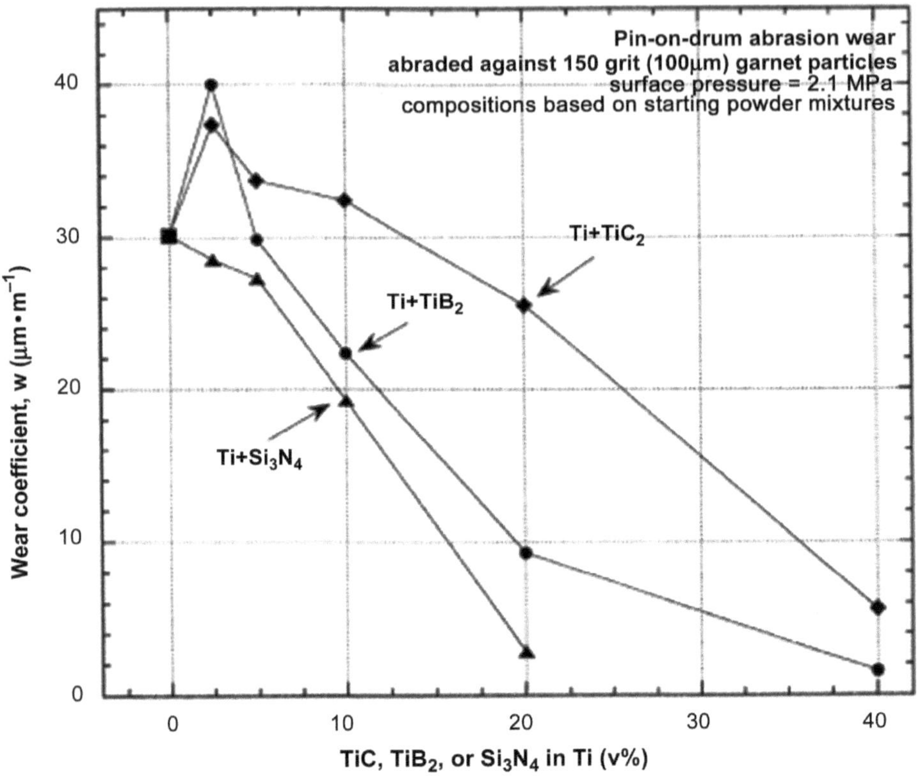

Figure 16.2 Wear coefficients versus Ti-MMC composite composition [23].

For example, when B_4C particles are used as the additives of TMCs, raw titanium powder can easily react with B_4C particles during sintering, and then two phases of TiB_w and TiC_p (with subscripts w and p indicating whisker and particle, respectively) are synthesized and dispersed as the reinforcements in the matrix [24–28]. They are highly effective to enhance the mechanical properties of TMCs, including elevated-temperature strength. The effect of additive B_4C particle size on the microstructural and mechanical properties of TMCs has been discussed in detail. Table 16.2 indicates that the use of finer B_4C particulates results in the improvement of both of UTS and elongation because of the absence of segregated TiB_w clusters and homogeneous dispersion of fine TiB_w in the matrix [26]. Furthermore, for β-Ti alloy composites with TiB_w/TiC_p, β-grains and secondary α-laths are refined after solution and aging heat treatment, which improves strength [29].

In case of boron powders and TiB_2 additives, whisker-like TiB_w reinforcements also form in TMCs via solid-state reaction during sintering and hot consolidation process [14,30–34]. TiB_w has high Young's modulus (467 GPa) and can result in the remarkable increment of Young's modulus of pure titanium (110 GPa) by its uniform dispersion in the matrix [33]. In general, the morphology of TiB whiskers is hard to

Table 16.2 **Mechanical properties of composites at room temperature [26]**

Sample	B_4C particle size (μm)	Mechanical properties			
		UTS (MPa)	Elongation (%)	Modulus (GPa)	Vickers hardness
1	3.5	817	0.55	140	452
2	0.5	950	0.64	142	581

control because the growth of TiB_w occurs anisotropically within a short time during high temperature sintering. In addition, TiB_w with a low aspect ratio can form during higher temperature sintering because of the high reaction speed of boron particles with titanium powder. The morphology of TiB_w reinforcements is more effective for strength improvement when they are arranged in one direction. As mentioned above, the whiskers grow anisotropically and are randomly dispersed in the matrix [31]. Secondary operations such as hot extrusion and die compaction are available for microstructure control, including whiskers rearrangement by severe plastic deformation [14,32]. In fact, the extruded TMCs with TiB_w show whisker reinforcements that are completely realigned along the extrusion direction, and this improves both tensile strength and ductility. *In situ* formation of TiB_w reinforcements from TiB_2 additives can be used to develop new PM TMCs (matrix; Ti–6.8Mo–4.2Fe–1.4Al–1.4V) by combination with plastic forming process (hot rolling and hot forging), which show high TS of 1500–1800 MPa and high Young's modulus of 159 GPa [34].

Carbon particles have a high reactivity with titanium powders during sintering to allow *in situ* synthesis of TiC_p as useful reinforcements of TMCs. The use of finer carbon particulates will lead to the formation of finer TiC_p reinforcements and therefore improved mechanical properties when they are uniformly blended with titanium powder, free of segregation [35–38]. Vapor growth carbon nanotubes (VGCFs) have been also used to blend with pure titanium powder, and TMCs reinforced with TiC_p are fabricated by *in situ* reactive sintering process. High-resolution transmission electron microscopic analysis indicates that *in situ*–formed TiC_p has good coherence with the titanium matrix as shown in Figure 16.3 [35], and is effective for load transfer from TiC_p reinforcements to the titanium matrix. In addition, TiC particle size is constant (0.5–1.5 μm) and does not depend on the VGCF content. This means that VGCFs are uniformly blended with pure titanium powder and no segregation exists in the starting mixture materials. For example, spark plasma sintering (SPS) and hot extrusion are applied to the elementally blended pure titanium powder and VGCFs to fabricate dense TMC rods with TiC_p. Figure 16.4 suggests that both the UTS and YS of TMCs increase noticeably with increasing VGCF content while the elongation gradually decreases [35]. In particular, in case of the 1.0 mass% VGCF addition, its YS of 1179 MPa is about 2.5 times that of PM pure titanium without reinforcement. However, TiC_p-reinforced pure titanium composite has 15% elongation, which is enough for many structural applications.

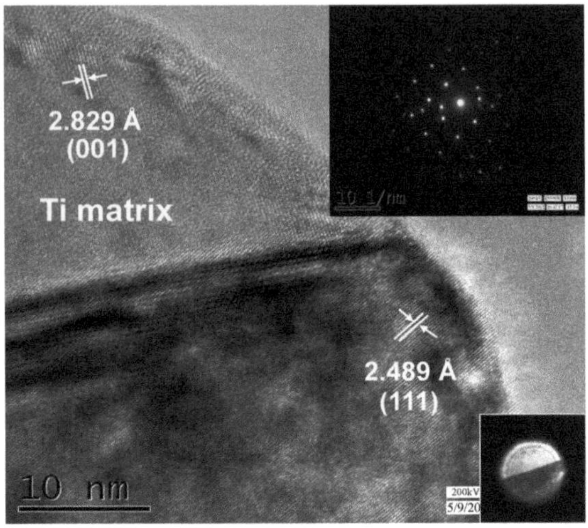

Figure 16.3 HR-TEM image of Ti/TiC interface in extruded Ti-1.0VGCFs PBM composite [35].

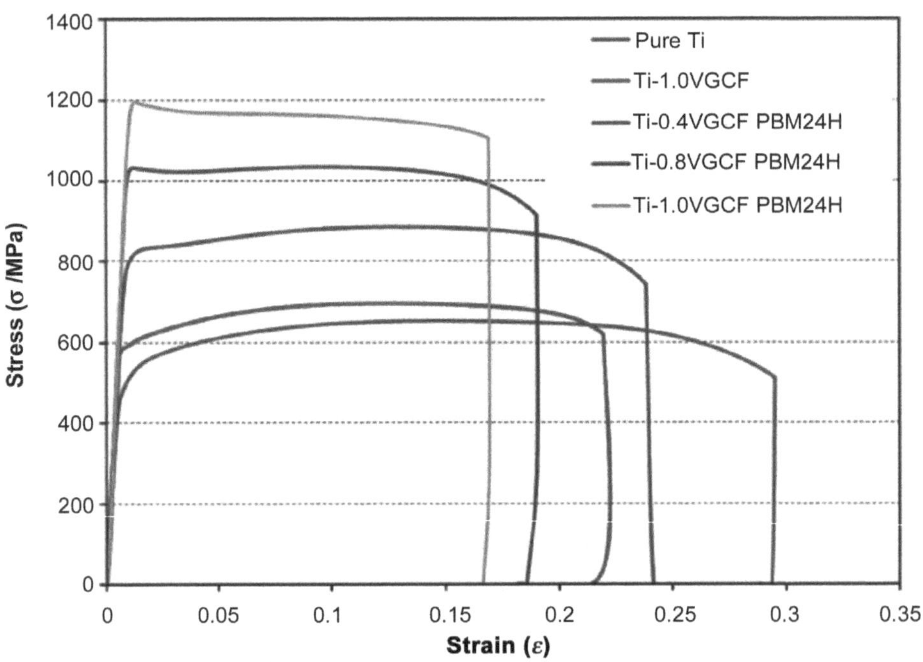

Figure 16.4 Stress–strain curves of the as-extruded Ti-VGCFs composites [35].

Carbon can also be introduced through coating titanium metal powder with resol nanospheres (10–30 nm) [39]. TiC nanoplatelet-reinforced TMCs fabricated by pressureless sintering of the coated titanium powder have outstanding compressive properties, such as 2.54 GPa US, 1.52 GPa YS, and strain to fracture of 44.4% [39]. These mechanical properties are superior to all other advanced Ti matrix materials, although they are just Ti-C binary alloys with no other alloying elements.

Multiwalled carbon nanotubes are smaller than VGCFs, and suitable for *in situ* reactive sintering to synthesize finer TiC_p in the matrix of TMCs. These thermally stable TiC dispersoids exist at α-Ti grain boundaries, and play an important role to obstruct grain growth (pinning effect). For example, as shown in Figure 16.5, the mean α-Ti grain size decreases from 10.6 μm (pure titanium) to 7.6 μm (+3.0 wt% CNTs) with increasing CNTs content [36]. In addition, TMCs with a larger amount of TiC_p reinforcements show higher microindentation hardness and YS at 573 K compared to PM pure titanium without reinforcement. This means that the pinning effect by TiC_p is also effective to improve the elevated temperature mechanical properties [36].

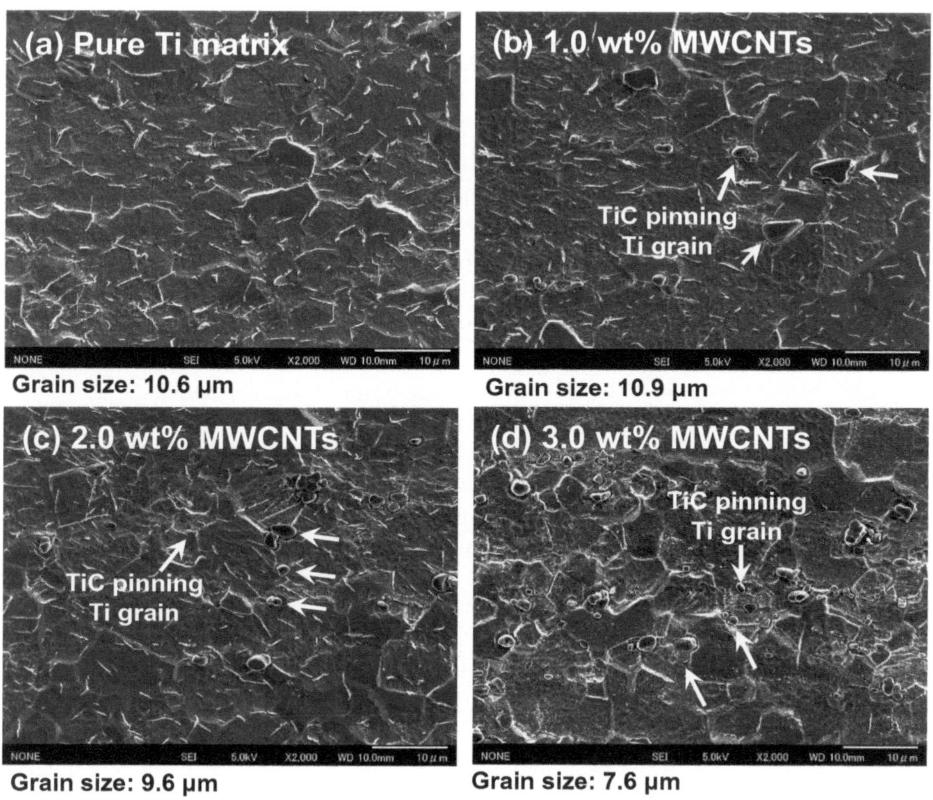

Figure 16.5 SEM images of (a) recrystallized grain of extruded pure Ti material and unchanged Ti grain size due to TiC pinning effect of the samples coated by (b) 1.0 wt%, (c) 2.0 wt%, and (d) 3.0 wt% CNTs/zwitterionic solution after annealing at 673 K for 360 ks [36].

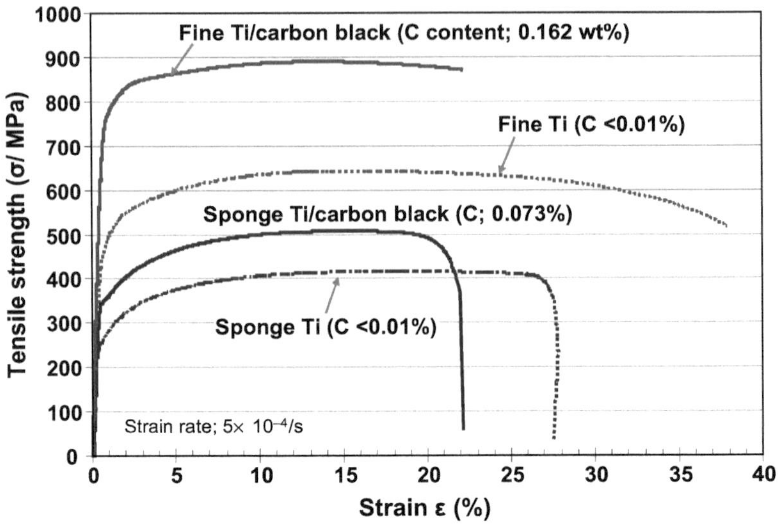

Figure 16.6 Stress–strain curves for PM extruded pure titanium composites reinforced with TiC particles in using different pure Ti raw powder [37].

X-ray diffraction (XRD) analysis indicates that the lattice constant of TMCs in the c-axis becomes larger than that of pure titanium because the solubility of carbon in α-Ti matrix at room temperature is 0.05 wt% and carbon solid solution occurs [37]. Commercial carbon black fine particles of 290 nm diameter, which are cheaper than VGCFs and MWCNTs, can be employed as raw additive elements to *in situ* synthesize TiC_p reinforcements in the matrix. Both sponge pure Ti powder and gas-atomized pure Ti fine powder can be used as the matrix materials. Figure 16.6 shows that extruded PM TMCs containing TiC_p particles of 1.8–3 μm diameter have higher UTS and YS than PM pure titanium with no reinforcement [38].

As mentioned earlier, B_4C, boron, TiB_2, and carbon additive elements can be selected as raw materials to control the volume fraction ration of TiB_w and TiC_p reinforcements dispersed in the matrix of TMCs, in order to optimize their microstructural, mechanical, thermal, and tribological properties.

16.2.3 Rapid solidification

Atomization is a unique method to produce metal powder with fine microstructures effective for mechanical property improvement [40–43]. As commercially economic processes, water atomization and gas atomization are often used for mass production. From a viewpoint of rapid solidification, the former is desired and much suitable in the production of powder with refined grains. However, metal powder surfaces can be easily oxidized by contact with water flow. Therefore, inert-gas atomization is generally employed for reactive metals such as aluminum, magnesium, and titanium [42–43]. When a molten metal containing some reinforcing particles or fibers, which must be thermally stable, is heated in the crucible and atomized through a nozzle, metal matrix composite powders with the above reinforcements are produced. For example, Ti-6Al-4V + 20 vol% TiC composite ingot is remelted using the induction heating process, and atomized by argon gas flow [43]. Gas-atomized Ti alloy composite powder is spherical, and TiC particles are uniformly dispersed in the matrix of the composite powder. However, it is necessary to note some demerits of the atomization process in the production of Ti composite powders. When high-density reinforcements are used, they can readily aggregate, leading to nonuniform distribution of reinforcements in the atomized powder. Therefore, the selection of suitable reinforcing materials and optimization of the atomization process such as heating and stirring conditions of the molten metal are important to the production of high-quality Ti composite atomized powders. Carbon and/or boron particulates are used as raw additives into molten Ti-6Al-4V for gas atomization, and Ti-64 alloy powders dispersed with TiC and/or TiB particles are produced by *in situ* reaction of these additive elements with titanium matrix. They are then consolidated by hot isostatic pressing (HIP) or direct extrusion [42]. In the extruded Ti alloys, TiC and TiB reinforcements are aligned in the extrusion direction. Table 16.3 shows the significant increase in TS and YS of TMCs compared to PM monolithic Ti-64 alloy. However, the ductility of TMCs decreases noticeably because of the brittle ceramic particulates dispersed in the matrix. Shape forming of TMC bars is achievable by hot forging or rolling process at the same temperature as conventional Ti alloys. In general, complicated shape components and sheet, rod, wire, and pipe products can be produced using conventional metal working facilities.

Table 16.3 Room-temperature tensile properties of PM Ti-6Al-4V with C and B additions [42]

Composition	Consolidation method (°C)	Tensile strength (MPa)	0.2% Yield strength (MPa)	Elongation (%)	Reduction in area (%)	Elastic modulus (GPa)
Ti-6Al-4V-0.5C	Extrude 1095	1306	1183	6	7	127
Ti-6Al-4V-0.9C	Extrude 1095	1237	1124	5	10	123
Ti-6Al-4V-0.9B	Extrude 1095	1312	1160	5	9	127
Ti-6Al-4V-1.4B	HIP 1000	1148	1027	1	4	123
	Extrude 1095	1297	1135	7	17	140
Ti-6Al-4V-1.7B	HIP 1000	1209	1080	1	4	128
	Extrude 1095	1359	1202	3	8	136
Ti-6Al-4V-2.2B	Extrude 1095	1470	1315	3	5	144
Ti-6Al-4V-1.3B-0.5C	HIP 1000	1206	1134	1	2	126
	Extrude 1095	1424	1257	7	12	138
Ti-6Al-4V	HIP 950	965	827	17	40	110

16.3 Carbon fiber–reinforced TMCs

Carbon fiber is one of the effective reinforcements of MMCs. Carbon fiber–reinforced titanium matrix composites could be manufactured through the adaptation of the "powder cloth" method while using specific conditions to take into account the chemical reactivity between titanium and carbon [44–47]. In one example, pure Ti powders with a particle size of less than 38 μm were mixed with carbon fibers having a 10 μm diameter, and then consolidated by hot press. The volume fractions of the fibers (V_f) were 7–11%. Figure 16.7 shows the microstructures of continuous carbon fiber–reinforced titanium matrix composite (CFRTMC) containing 8 vol% carbon fibers. There is no bundle of carbon fibers in the matrix, indicating that the reinforcements were effectively dispersed. Table 16.4 lists the mechanical performances of CRFTMCs prepared by the hot pressing method versus the hot pressing temperature and V_f of carbon fibers. The properties are affected by both the hot pressing temperature and the V_f of carbon fibers and are well below theoretical predictions. For example, with

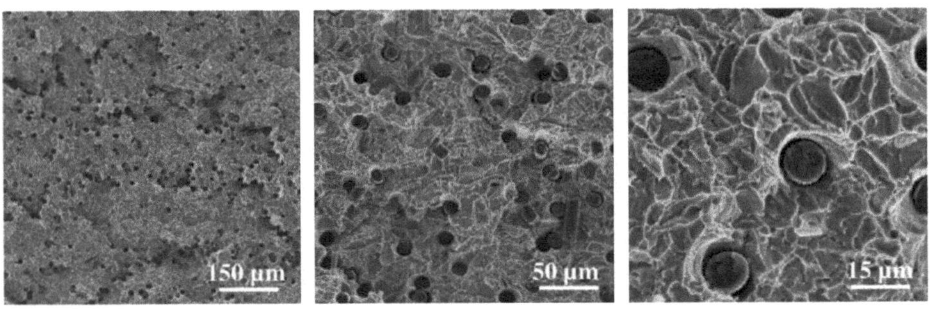

Figure 16.7 Fractography of CRFTMC containing 8 vol% carbon fibers (failure strain 13%) [46].

Table 16.4 **Mechanical performances of hot-pressed CRFTMCs with different V_f of carbon fibers [46]**

Specimen	Hot press temperature (°C)	Young modulus (GPa)	0.2% Yield strength (MPa)	Ultimate tensile strength (MPa)	Rupture strain (%)	V_f (%)
Matrix	700	105	321	581	34	0
CRFTMCs	600	114.5	350	580	1.5	8.5
	600	111	300	563	3.5	8
	650	110	313	586	3.5	7.5
	650	111	345	583	13	7.5
	700	118	380	580	2.5	11
	700	114	384	599	22	7

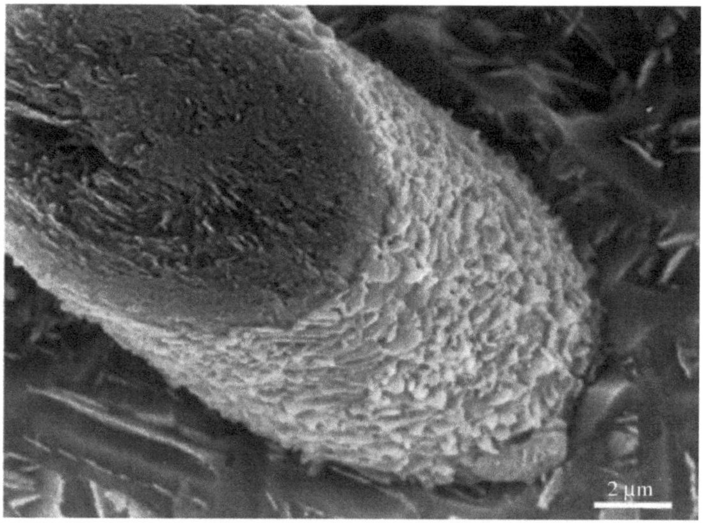

Figure 16.8 SEM micrograph of TiC crystallites formation at carbon fiber-matrix interface after 1 h of compression at 700°C [46].

a fiber volume fraction of 11%, the rigidity increased only by 12% whereas the rule of mixture predicts a rigidity increase of 45%. As regards their ductility, the presence of brittle carbon fibers within the ductile Ti matrix (30% of elongation) was clearly a source of embrittlement, although the matrix remains ductile, as illustrated in Figure 16.7. In this group of TMCs, the interface reaction between carbon fibers and the matrix depends on the processing temperature [48,49]. The fiber–matrix interaction is limited to the pyrolytic carbon precoating layers on carbon fiber surfaces as shown in Figure 16.8. The small pyrolytic carbon layer is transformed into TiC crystallites forming a thin discontinuous TiC layer of about 100 nm in average thickness whose corresponding notch effect on fibers should not be dramatic and noneffective. The creep deformation behavior and fatigue failure of CRFTMCs were found to be superior to conventional titanium alloys [47].

16.4 Atomic-scale reinforced TMCs with solute light elements

Rare metals such as vanadium (V), molybdenum (Mo), niobium (Nb), and zirconium (Zr) are limited mineral resources. They are very effective for improvement of mechanical properties of titanium alloys but expensive. Therefore, rare-metal-free titanium materials with high strength should be developed to reduce the materials' cost for wider industrial applications. In this regard, low-cost ubiquitous light elements such as oxygen, nitrogen, carbon, and hydrogen are available to be used for strengthening titanium materials. In particular, oxygen solution strengthening of titanium is well

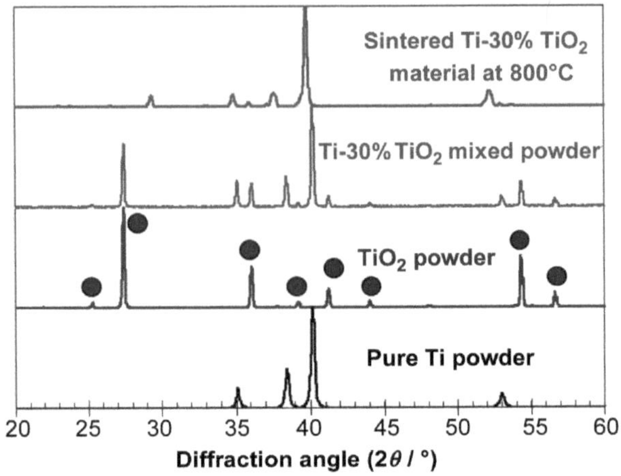

Figure 16.9 XRD profiles of raw Ti and TiO_2 powders, elementally blended Ti-30 mass% TiO_2 mixture powder and sintered Ti-30% TiO_2 powder material [51].

known because of the large solubility of oxygen atoms in titanium (~33 at%) [50]. In other words, these light elements may be used as effective "atomic-scale reinforcements" of PM titanium materials. For example, when the elementally blended pure Ti powder and TiO_2 particles (Ti-30 mass% TiO_2) are sintered at 800°C by SPS, XRD profiles indicate that the sintered Ti specimen shows no TiO_2 peak and the Ti diffraction peaks clearly shift to lower diffraction angles as shown in Figure 16.9 [51]. In addition, Bragg's law reveals that the lattice parameter in the c-axis increases proportionally with the TiO_2 content. It is concluded that TiO_2 raw particles of the elemental mixture decomposed during SPS and the oxygen atoms originated from TiO_2 become interstitial solute into the vacancies of titanium. Elementally blended Ti-TiO_2 powder mixtures containing 0, 0.6, 1.0, and 1.5 mass% TiO_2 are consolidated by SPS (800°C) and hot extruded at 1000°C to fabricate PM pure titanium rods containing various amounts of oxygen solute atoms.

Figure 16.10a shows that the ultimate tensile strength (UTS) and YS increase with increasing TiO_2 additive content while a small decrease of elongation occurs. As shown in Figure 16.10b, the experimental data of YS after removing the effect of grain refinement on the YS increment correspond well to the results calculated using the Labusch limit [52] regarding solid solution strengthening of metals described by equation (16.1).

$$\Delta\sigma_{0.2\%YS} = \frac{\tau_0}{S_F} = \frac{1}{S_F}\left(\frac{F_m^4 C^2 w}{4Gb^9}\right)^{\frac{1}{3}} \tag{16.1}$$

where τ_0 is the shear stress, S_F is the Schmid factor, F_m is the normalized maximum interaction force, c is the concentration of solute, w (~$5b$, b is the Burgers vector) is the distance between atoms and dislocation, and G is the shear modulus of Ti.

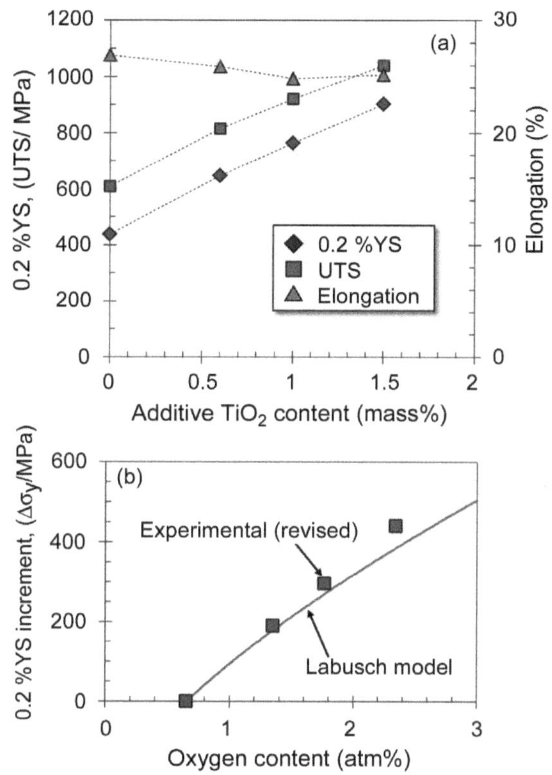

Figure 16.10 Dependence of tensile properties of PM extruded pure Ti materials with solute oxygen on TiO_2 content (a) and comparison of 0.2% YS increment by experimental and calculation by Labusch model [51].

The dynamic strength of PM Ti-O materials is evaluated using a rotating bending fatigue testing machine. Figure 16.11 reveals S-N curves for PM extruded pure Ti with 0.32 wt% oxygen solution (UTS; 702 MPa) and JIS grade 2 pure Ti with 0.10 wt% O fabricated by cast and hot extrusion (UTS; 432 MPa). The former shows a 480 MPa fatigue limit (σ_f) at 10^7 cycles, much higher than the conventional JIS grade 2 pure Ti material ($\sigma_f = 290$ MPa).

In case of nitrogen element, nitrogen solid solution occurs by heat treatment of pure titanium powder at suitable temperatures under nitrogen gas flow in furnace. DTA profiles of pure titanium powder heated under nitrogen gas flow suggest that the suitable heat treatment temperature for nitrogen solid solution is about 530–700°C. The nitrogen content of titanium powder can be quantitatively controlled by changing the heat treatment time. For example, pure titanium powder with 28 μm mean diameter is heated at 600°C for 1 and 2 h under 5 L/min gas flow in muffle furnace. XRD profiles of these powders indicate only Ti diffraction peaks; no TiN peaks are detected while Ti peaks shift to lower diffraction angles compared to

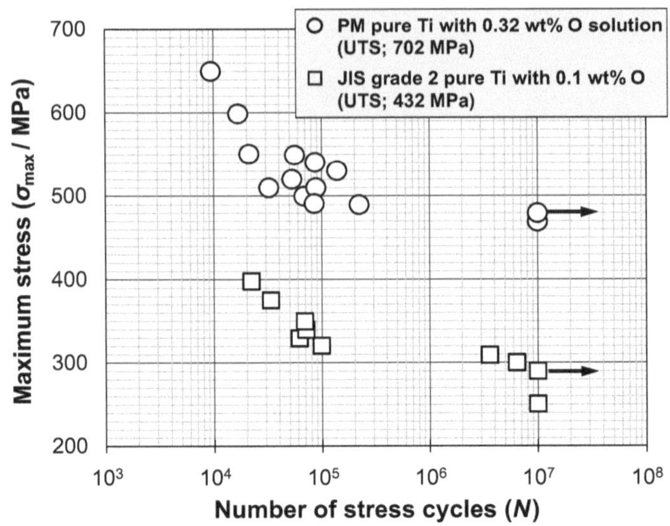

Figure 16.11 Dynamic properties of PM pure Ti with solute oxygen and JIS grade 2 pure Ti materials evaluated by rotating bending fatigue testing machine [51].

as-received raw pure titanium powder. The lattice constant in the c-axis also increases with increasing nitrogen content by heat treatment of titanium powder. Therefore, nitrogen solid solution in the Ti crystal occurs through the above processes, and results in the strength and hardness improvement of PM pure Ti material as shown in Table 16.5 [53].

In general, many deformation twins occur locally during tensile testing of pure titanium materials, and they finally result in fracture. Figure 16.12 shows typical evidence concerning microstructures of a pure Ti specimen (with 0.004 wt% hydrogen) after tensile testing, where (a) is near the fractured surface with local deformation and (b) is far from the fractured point and uniform deformation area. The fractured area (a) has a high density of large twins going through α-Ti grains;

Table 16.5 Mechanical properties of PM extruded pure Ti with solute nitrogen compared to monolithic pure Ti material [53]

Specimen	0.2% Yield strength (σ_y/MPa)	Ultimate tensile strength (σ/MPa)	Elongation, ε (%)	Hardness (Hv)
Pure Ti	479 ± 8.1	653 ± 6.6	28 ± 1.7	264 ± 26.3
Heated for 1 h	903 ± 17.4	1008 ± 6.1	24 ± 1.5	479 ± 34.2
Heated for 2 h	1045 ± 13.6	1146 ± 7.1	11 ± 2.3	539 ± 45.5

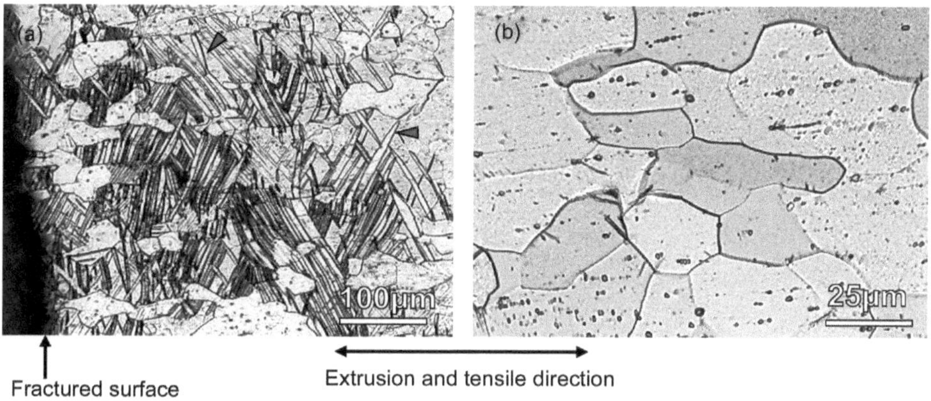

Figure 16.12 Optical microstructures of cross-section near fractured surface (a) and far from fractured surface (b) of pure Ti tensile specimen (with 0.004 wt% hydrogen) after tensile test [57].

however, no deformation twins but some elongated Ti grains along the tensile direction are observed at the uniformly deformed area in (b). Direct oxygen solid solution into Ti powder using oxygen gas can be made effective to improve the mechanical properties of PM Ti materials as well as in the use of TiO_2 additive particles [54].

Titanium easily reacts with hydrogen to form needle-like compounds of titanium hydride (TiH_2), resulting in poor ductility of Ti materials because of their brittleness [55,56]. However, with a small presence of TiH_2 compounds, the tensile elongation of pure titanium showed a significant increase at ambient temperature compared to the commercial pure Ti material containing less than 0.001 wt% hydrogen [57]. This suggests that TiH_2 dispersoids may be useful reinforcements of TMCs when the H content is small. PM pure Ti materials with different hydrogen contents are prepared by sintering TiH_2 raw powder, followed by hot extrusion. The hydrogen content can be controlled by the post–heat treatment time at 900°C in vacuum. Figure 16.13a shows the S-S curves of PM Ti materials with various hydrogen contents, and Figure 16.13b gives the relationship between elongation and hydrogen content. "0 ks" means no post–heat treatment of PM extruded Ti material, and its hydrogen content is 0.33 wt%. With increasing heat treatment time in vacuum, the hydrogen content quickly decreases because of the thermal decomposition of TiH_2 compounds. Figure 16.13b shows that the tensile elongation increases noticeably with increasing H content of PM Ti materials in the form of brittle TiH_2 dispersoids in the matrix.

Figure 16.14 shows the microstructures of cross-sectional area near the fractured surface after tensile testing. The number of deformation twins gradually decreases with increasing post–heat treatment time. In Figure 16.14b of pure Ti with 0.15% H, small twins were observed around α-Ti laths cut through Ti grains. In addition, the

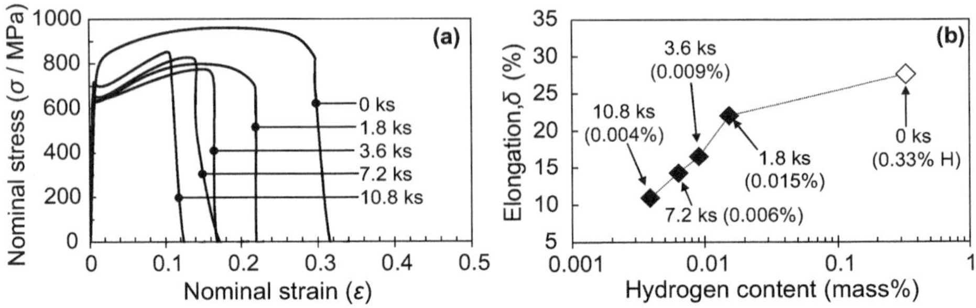

Figure 16.13 S-S curves of PM Ti materials with various hydrogen contents in (a) and relationship between elongation and hydrogen content of specimens (b). "0 ks" means no post–heat treatment of extruded PM Ti material and its hydrogen content is 0.33 wt% [57].

length of twins becomes shorter, indicating that the propagation of twins is obstructed with increasing H content in Ti materials.

To understand the effect of TiH_2 dispersoids on deformation twins propagation, *in situ* tensile testing was carried out in a scanning electron microscope (SEM) system to observe microstructural changes during tensile deformation of PM Ti material with TiH_2 compounds (H content = 0.33 wt%). Tensile testing was also temporarily stopped to observe the specimen surfaces by SEM at points A–D on the S-S curve, and

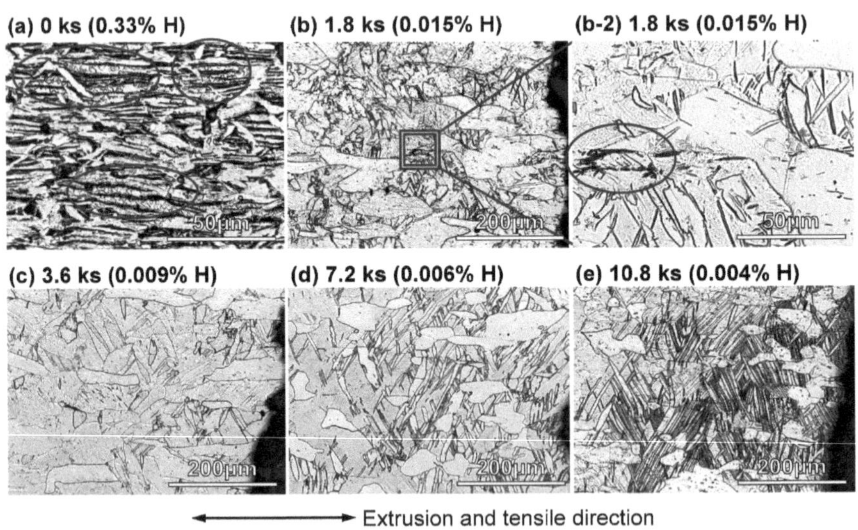

Figure 16.14 Optical microstructures of cross-sectional area near fractured surface after tensile testing of each PM pure Ti material. The number of deformation twins gradually decreases with increase in post–heat treatment time [57].

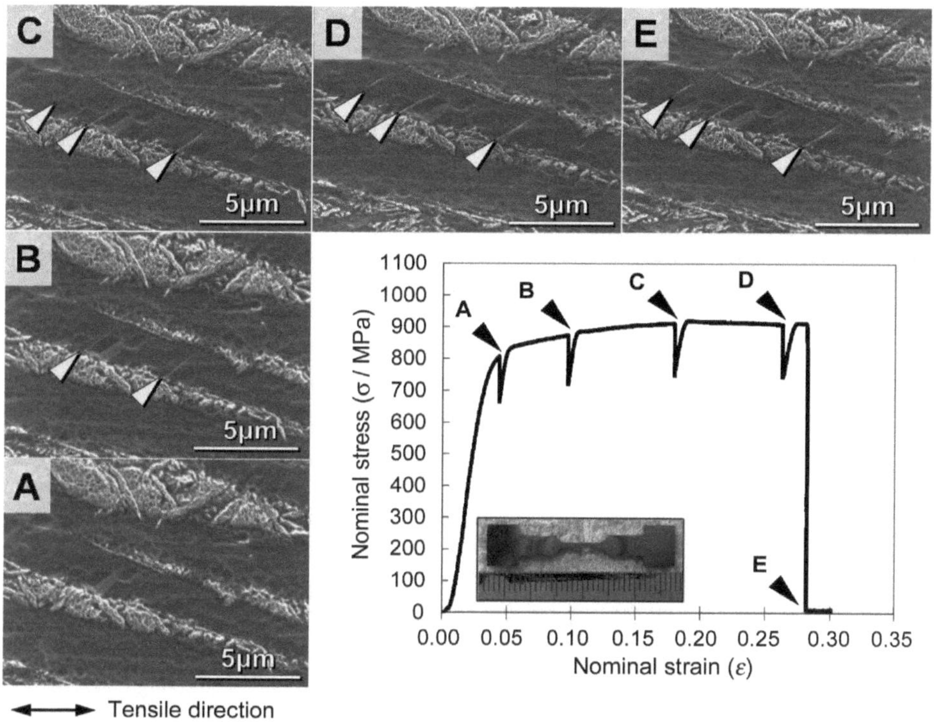

Figure 16.15 Microstructural changes during tensile deformation of PM Ti material with TiH_2 compounds (H content: 0.33 wt%) by *in situ* SEM observation system. Tensile test was temporarily stopped to observe the specimen surface by SEM at points A–D on the stress–strain curve, and restarted after that [57].

restarted after that (Figure 16.15). Each SEM micrograph shows white bands that correspond to aggregated TiH_2 compounds. As shown at point B, small deformation twins (indicated by arrows) initiate from local areas, and they gradually grow and propagate with increasing tensile load. After the deformation twins reached TiH_2 dispersoids; however, they failed to pass the TiH_2 bands as shown at points D and E. Detailed SEM observations and EBSD analysis of twins formation at fractured surface areas after tensile testing were performed. As shown in Figure 16.16a, PM Ti with no TiH_2 compound (H content = 0.004 wt%) clearly shows some twins propagation that cuts through α-Ti grains. On the other hand, PM Ti materials with layer-structured TiH_2 dispersoids, corresponding to the black areas in Figure 16.16b, show short deformation twins interrupted by TiH_2 layers. This is similar to the observation made with *in situ* tensile testing shown in Figure 16.15. In conclusion, a small number of TiH_2 dispersoids in PM Ti materials can effectively obstruct the propagation of deformation twins and their penetration through Ti grains. As a result, this can result in a significant improvement of tensile elongation due to uniform deformation in tension as shown in Figure 16.13b.

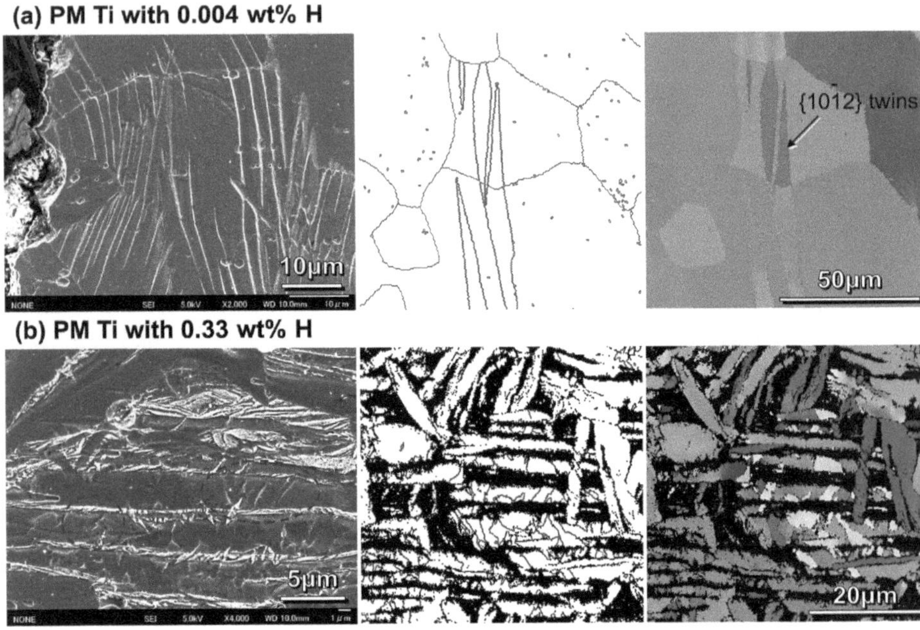

Figure 16.16 SEM-EBSD analysis of twins formation at fractured surface area of (a) PM Ti with no TiH_2 compound (H content: 0.004 wt%) and (b) PM Ti material with layer-structured TiH_2 dispersoids after tensile testing [57].

References

[1] Y. Liu, L.F. Chen, H.P. Tang, C.T. Liu, B. Liu, B.Y. Huang, Design of powder metallurgy titanium alloys and composites, Mat. Sci. Eng. A 418 (2006) 25–35.

[2] Y.J. Kim, H. Chung, S.J.L. Kang, Processing and mechanical properties of Ti–6Al–4V/TiC *in situ* composite fabricated by gas–solid reaction, Mat. Sci. Eng. A 333 (2002) 343–350.

[3] Z.D. Cui, S.L. Zhu, H.C. Man, X.J. Yang, Microstructure and wear performance of gradient Ti/TiN metal matrix composite coating synthesized using a gas nitriding technology, Surf. Coat.Tech. 190 (2005) 309–313.

[4] P.B. Joshi, G.R. Marathe, N.S.S. Murti, V.K. Kaushik, P. Ramakrishnan, Reactive synthesis of titanium matrix composite powders, Mater. Lett. 56 (2005) 322–328.

[5] D.E. Alman, J.A. Hawk, The abrasive wear of sintered titanium matrix–ceramic particle reinforced composites, Wear 225-229 (1999) 629–639.

[6] M.S. Selamat, L.M. Watson, T.N. Baker, XRD and XPS studies on surface MMC layer of SiC reinforced Ti–6Al–4V alloy, J. Mat. Proc. Tech. 142 (2003) 725–737.

[7] T. Saito, T. Furuta, T. Yamaguchi, Recent Advances in Titanium Metal Matrix Composites, in: F.H. Froes, J. Storer (Eds.), TMS, Warrendale, PA, 1995, pp. 33–44.

[8] T.M.T. Godfrey, P.S. Goodwin, C.M. Ward-Close, Production of titanium particulate metal matrix composite by mechanical milling, Mat. Sci. Tech. 16 (2000) 753–758.

[9] J. Lapin, L. Ondrus, O. Bajana, Effect of Al_2O_3 particles on mechanical properties of directionally solidified intermetallic Ti–46Al–2W–0.5Si alloy, Mat. Sci. Eng. A 360 (2003) 85–95.

[10] V. Castro, T. Leguey, A. Muñoz, M.A. Monge, R. Pareja, Microstructure and tensile properties of Y_2O_3-dispersed titanium produced by arc melting, Mat. Sci. Eng. A 422 (2006) 189–197.
[11] K. Geng, W. Lu, Z. Yang, D. Zhang, *In situ* preparation of titanium matrix composites reinforced by TiB and Nd_2O_3, Mater. Lett. 57 (2003) 4054–4057.
[12] E. Zhanga, S. Zenga, B. Wang, Preparation and microstructure of *in situ* particle reinforced titanium matrix alloy, J. Mat. Proc. Tech. 125-126 (2002) 103–109.
[13] M. Sumida, K. Kondoh, *In-situ* synthesis of Ti matrix composite reinforced with dispersed Ti_5Si_3 particles via spark plasma sintering, Mater. T. 46 (2005) 2135–2141.
[14] H. Lu, D. Zhang, B. Gabbitas, F. Yang, S. Matthews, Synthesis of a TiB_w/Ti6Al4V composite by powder compact extrusion using a blended powder mixture, J. Alloys Compd. 606 (2014) 262–268.
[15] I.M. Melendez, N. Neubauer, P. Angerer, H. Danninger, J.M. Torralba, Influence of nano-reinforcements on the mechanical properties and microstructure of titanium matrix composites, Compos. Sci. Technol. 71 (2011) 1154–1162.
[16] K. Kondoh, T. Threrujirapapong, H. Imai, J. Umeda, B. Fugetsu, Characteristics of powder metallurgy pure titanium matrix composite reinforced with multi-wall carbon nanotubes, Compos. Sci. Technol. 69 (2009) 1077–1081.
[17] S. Li, B. Sun, K. Kondoh, T. Mimoto, H. Imai, Influence of carbon reinforcements on the mechanical properties of Ti composites via powder metallurgy and hot extrusion, Mater. Sci. Forum 750 (2013) 40–43.
[18] T. Saito, The automotive application of discontinuously reinforced TiB-Ti composites, JOM 56 (2004) 33–36.
[19] C. Poletti, M. Balog, T. Schubert, V. Liedtke, C. Edtmaier, Production of titanium matrix composites reinforced with SiC particles, Compos. Sci. Technol. 68 (2008) 2171–2177.
[20] S. Abkowitz, P.F. Weihrauch, F.H. Abkowitz, L.H. Heussi, The commercial application of low-cost titanium composites, JOM 47 (1995) 40–41.
[21] S. Abkowitz, S.M. Abkowitz, H. Fisher, P.J. Schwartz, CermtTi® discontinuously reinforced Ti-matrix composites: manufacturing, properties, and applications, JOM 56 (2004) 37–41.
[22] J.B. Fruhauf, J. Roger, O. Dezellus, S. Gourdet, N. Karnatak, N. Peillon, S. Saunier, F. Montheillet, C. Desrayaud, Microstructural and mechanical comparison of Ti + 15%TiC_p composites prepared by free sintering HIP and extrusion, Mat. Sci. Eng. A 554 (2012) 22–32.
[23] D.E. Alman, J.A. Hawk, The abrasive wear of sintered titanium matrix–ceramic particle reinforced composites, Wear 225-229 (1996) 629–639.
[24] D.R. Ni, L. Geng, J. Zhang, Z.Z. Zheng, Fabrication and tensile properties of *in situ* TiBw and TiCp hybrid-reinforced titanium matrix composites based on Tin $_4$Chy, Mat. Sci. Eng. A 478 (2008) 291–296.
[25] X.N. Zhang, W.J. Lu, D. Zhang, R. Wu, Y.J. Bian, P.W. Fang, *In situ* technique for synthesizing (TiB + TiC)/Ti composites, Script. Mater. 41 (1999) 39–46.
[26] D.R. Ni, L. Geng, J. Zhang, Z.Z. Zheng, Effect of B_4C on microstructure of *in situ* titanium matrix composites prepared by reactive processing of Ti-B_4C system, Script. Mater. 55 (2006) 429–432.
[27] J. Lu, J. Qin, W. Lu, Y. Liu, J. Gu, D. Zhang, *In situ* preparation of (TiB + TiC + Nd_2O_3)/Ti composites by powder metallurgy, J. Alloys Compd. 469 (2009) 116–122.
[28] I.Y. Kim, B.J. Choi, Y.J. Kim, Y.Z. Lee, Friction and wear behavior of titanium matrix (TiB + TiC) composites, Wear 271 (2011) 1962–1965.

[29] Z.X. Du, S.L. Xiao, P.X. Wang, L.J. Xu, Y.Y. Chen, H.K.S. Rahoma, Effects of trace TiB and TiC on microstructure and tensile properties of β titanium alloy, Mat. Sci. Eng. A 596 (2014) 71–79.
[30] Z. Xinghong, X. Qiang, H. Jiecai, V.L. Kvanin, Self-propagating high temperature combustion synthesis of TiB/Ti composites, Mat. Sci. Eng. A 348 (2003) 41–46.
[31] S.I. Lieberman, A.M. Gokhale, S. Tamirisakandala, R.B. Bhat, Three-dimensional microstructural characterization of discontinuously reinforced Ti64–TiB composites produced via blended elemental powder metallurgy, Mater. Charact. 60 (2009) 957–963.
[32] S. Gorsse, D.B. Miracle, Mechanical properties of Ti-6Al-4V/TiB composites with randomly oriented and aligned TiB reinforce, Acta Materialia 51 (2003) 2427–2442.
[33] S. Gorsse, Y.L. Petitcorps, S. Matar, F. Rebillat, Investigation of the Young's modulus of TiB needles *in situ* produced in titanium matrix composite, Mat. Sci. Eng. A 340 (2003) 80–87.
[34] T. Saito, H. Takamiya, T. Furuta, Thermomechanical properties of P/M β titanium metal matrix composite, Mat. Sci. Eng. A 243 (1998) 273–278.
[35] S. Li, B. Sun, H. Imai, K. Kondoh, Powder metallurgy matrix composite *in situ* reactive processing of Ti-VGCFs system, Carbon 61 (2013) 216–228.
[36] K. Kondoh, T. Threrujirapapong, J. Umeda, B. Fugetsu, High-temperature properties of extruded titanium composites fabricated from carbon nanotubes coated titanium powder by spark plasma sintering and hot extrusion, Compos. Sci. Technol. 72 (2012) 1291–1297.
[37] S. Li, B. Sun, H. Imai, T. Mimoto, K. Kondoh, Powder metallurgy titanium metal matrix composites reinforced with carbon nanotubes and graphite, Composites Part A 48 (2013) 57–66.
[38] T. Threrujirapapong, K. Kondoh, H. Imai, J. Umeda, B. Fugetsu, Mechanical properties of a Titanium Matrix Composite Reinforced with Low Cost Carbon Black via Powder Metallurgy Processing, Mater. T. 50 (2009) 2757–2762.
[39] S.D. Luo, Q. Li, J. Tian, C. Wang, M. Yan, G.B. Schaffer, M. Qian, Self-assembled, aligned TiC nanoplatelet-reinforced titanium composites with outstanding compressive properties, Script. Mater. 69 (2013) 29–32.
[40] E. Klar, J.W. Fesko, Gas and water atomization, Metals Handbook, in: E. Klar (Ed.), vol. 7, Powder Metallurgy, American Society for Metals, Metals Park, OH, 1984, pp. 25–39.
[41] S. Lagutkin, L. Achelis, S. Sheikhaliev, V. Uhlenwinkel, V. Srivastava, Atomization process for metal powder, Mat. Sci. Eng. A 383 (2004) 1–6.
[42] C.F. Yolton, The pre-alloyed powder metallurgy of titanium with boron and carbon additions, JOM 56 (2004) 56–59.
[43] D. Hu, M.H. Loretto, Microstructural characterisation of a gas atomised Ti6A14V/TiC composite, Scripta. Metall. Mater. 31 (1994) 543–548.
[44] S.T. Mileiko, A.M. Rudnev, M.V. Gelachov, Low cost PM route for titanium matrix carbon fibre composites, Powder Metall. 39 (1996) 97–99.
[45] L.V. Kovalenko, V.I. Antipov, A. Afanas'ev, L.V. Vinogradov, Interphase interaction in production of the titanium-carbon fibre composite, J. Adv. Mater. 3 (1996) 215–217.
[46] C. Even, C. Arvieu, J.M. Quenisset, Powder route processing of carbon fibres reinforced titanium matrix composites, Compos. Sci. Technol. 68 (2008) 1273–1281.
[47] S.C. Tjong, Y.W. Mai, Processing-structure-property aspects of particulate- and whisker-reinforced titanium matrix composites, Compos. Sci. Technol. 68 (2008) 583–601.
[48] C. Arvieu, J.P. Manaud, P. Chadeyron, J.M. Quenisset, The design of an ephemeral interfacial zone for titanium matrix composites, Composites Part A 29A (1998) 1193–1201.
[49] C. Arvieu, J.P. Manaud, J.M. Quenisset, Interaction between titanium and carbon at moderate temperatures, J. Alloys Comp. 368 (2004) 116–122.

[50] H. Okamoto, O-Ti (Oxygen-Titanium), J. Phase Equilib. Diff. 32 (2011) 473–474.
[51] K. Kondoh, S. Li, B. Sun, T, Mimoto, H. Imai, J. Umeda, Pure Titanium with High Strength and Excellent Ductility by Solid Solute Oxygen Strengthening via Powder Metallurgy Route, Proceedings of MS&T 2013, Montreal, Canada, 2013, pp. 3004–3008.
[52] L. Labusch, A statistical theory of solid solution hardening, Phys. Stat. Solid. (b) 41 (1970) 659–669.
[53] S. Li, K. Kondoh, B. Sun, H. Imai, J Umeda, Investigation of Powder Metallurgy Titanium Matrix Composites by Planetary Ball-milling of Ti Powder Dispersed with Vapour Grown Carbon Nanofibers, Proceedings of MS&T 2013, Montreal, Canada, 2013, pp. 3009–3016.
[54] B. Sun, S. Li, H. Imai, T. Mimoto, J. Umeda, K. Kondoh, Fabrication of high-strength Ti materials by in-process solid solution strengthening of oxygen via P/M methods, Mat. Sci. Eng. A 563 (2013) 95–100.
[55] E. Nyberg, M. Miller, K. Simmons, K.S. Weil, Microstructure and mechanical properties of titanium components fabricated by a new powder injection molding technique, Mat. Sci. Eng. C 25 (2005) 336–342.
[56] K.H. Baik, Microstructural evolution and tensile properties of Ti-Al-V alloys manufactured by plasma spraying and subsequent vacuum hot pressing, Mater. T. 47 (2006) 1198–1203.
[57] T. Mimoto, K. Kondoh, J. Umeda, Phase Transformation and Orientation in Direct Consolidation of TiH_2 Powder and Their Effects on Tensile Behavior of P/M Extruded Ti Material, Proceedings of TMS2013, San Antonio, USA, 2013.

Titanium alloy components manufacture from blended elemental powder and the qualification process

17

Stanley Abkowitz, Susan Abkowitz, Harvey Fisher
Dynamet Technology, Inc. (now RTI Advanced Powder Materials a unit of RTI International)

17.1 Introduction

The elemental blend sinter (EBS) powder metallurgical (PM) method involves the blending of elemental powders along with master alloy or other desired additions, cold pressing the blended powders into shape, and sintering the shape to higher density and uniform chemistry. The sintered shape may subsequently be hot isostatically pressed to full density.

This method provides the flexibility to select the titanium powder from the wide range of powders now available and for the blending of various powders in order to develop target properties. Powders can be selected for compatibility or affordability and can be blended to achieve the desired level of interstitials and to develop specific combination of mechanical properties. The elemental method is typically lower cost and elemental powders are more easily processed than prealloyed powder. Dynamet Technology, Inc.,[1] has been able to achieve full density components with properties comparable to ingot metallurgy (IM) produced material.

Table 17.1 shows the characteristics of the different types of titanium powders that are either available or under development today. This table is based in part on a recent review of powder production methods coauthored by McCracken [1]. The new powders that are under development, such as the Armstrong, Frey, and MER process powders, may provide an opportunity for reducing the costs of PM product when they become available, that is, if they can be processed to the necessary density levels with properties equivalent or superior to baseline PM and wrought titanium. Finally, the cost of producing *components* from those powders must be competitive. The development of new titanium production methods shown in the table is aimed at lowering the cost of PM titanium powder. However, these powders will not be available for some time and their relative cost as well as their shape-making characteristics are yet to be established.

The EBS method using hydride–dehydride (HDH) titanium powder produced from Kroll sponge is a key to the commercial success of Dynamet Technology PM process. This process is producing a wide range of affordable PM near-net-shape preformed components. Dynamet has developed critical specifications for its titanium and master

[1] Dynamet Technology Inc. is now RTI Advanced Powder Materials, a unit of RTI International.

Titanium Powder Metallurgy. http://dx.doi.org/10.1016/B978-0-12-800054-0.00017-4

Table 17.1 **Characteristics of different types of titanium powders**

Type/process	Elemental or prealloyed	Advantages	Status/disadvantages
Hunter process (pure sodium)	Elemental	• Low cost • Excellent cold compactibility	• Limited availability • High chloride
HDH Kroll process (pure magnesium)	Elemental	• Lower cost • Good compactibility • Readily available • Low chloride	–
HDH powder produced from ingot, sheet or scrap	Prealloyed	Readily available	• High cost • Fair compactibility
Atomized	Prealloyed	• High purity • Available	• High cost • Not cold compactable
REP/PREP	Prealloyed	High purity	• High cost • Not cold compactable
Armstrong	Both	• Compactable • Potential for low cost	• Processibility/quality • Production scale-up
Fray	Both?	TBD	Developmental
MER	Both?	TBD	Developmental

Note: HDH, hydride–dehydride; REP/PREP, rotating electrode powder/plasma rotating electrode powder; MER, MER Corp., Tucson, AZ.

alloy powders that control for morphology, particle size, particle distribution, and chemistry have been developed.

17.2 The CHIP PM process

The PM CHIP[2] process, shown in Figure 17.1, is a green manufacturing technology [2] that has proven to be an acceptable process for producing aerospace, military, industrial, and medical components. This advanced PM process uses titanium powder, typically Kroll process HDH powder, blended with master alloy powder such as aluminum–vanadium master alloy powder. The blended powder is compacted to shape by cold isostatic pressing (CIP) in reusable elastomeric tooling. With proper selection of powders, well-designed CIP tooling, and appropriate pressing conditions, a shaped powder compact can be produced and readily extracted from the PM tooling with sufficient "green strength" for handling. It must also have sufficient uniformity and intimate contact of the powder particles for densification and homogenous alloying in the subsequent sintering process.

This CIP-Sinter or CHIP (CIP-Sinter-HIP) process is used to produce near-net-shape parts for finish machining to high-tolerance configurations. These processes

[2] CHIP is an acronym for *c*old isostatic pressing/*h*ot *i*sostatic *p*ressing.

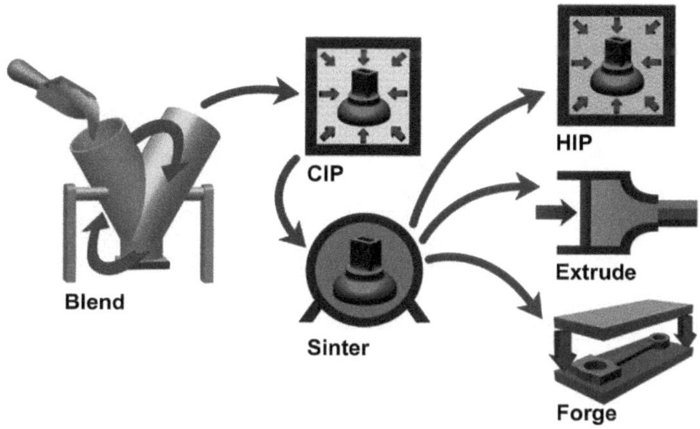

Figure 17.1 The PM CHIP process.

can also be used to make forging preforms or mill product shapes for subsequent processing such as billet for casting, extrusion, or hot rolling. In the case of as-sintered material, full density is achieved during subsequent processing.

A wide range of shapes has been produced with size only limited by the capacity of the equipment. The size of the CIP is usually the limiting factor since vacuum furnaces and HIP units are available in larger sizes than are high-pressure CIP units. Size capability also depends on the powder fill characteristics, product configuration, and by tooling parameters. Successful products can range from a few grams to hundreds of kilograms.

The sintering process was historically established to reach a minimum density level at which the material had no interconnected porosity. At this density threshold, the material could be hot isostatically pressed (HIP) without the processing expense of HIP encapsulation, making the HIP process economically viable. Through recent developments, the capability to reach greater than 98% sintered density has been achieved. This can result in as-sintered tensile properties that are suitable to compete with wrought properties and are superior to castings. This reduces the need for the HIP operation and further strengthens the economic advantage of this PM CIP-Sinter manufacturing technology.

Figure 17.2 compares the CHIP process to the VAR double melt process typically used to produce IM titanium. The major cost benefit of the powder process is that it uses relatively low-cost raw materials, avoids costly melt processes, and results in relatively little material lost during processing. The cold-pressed consolidated preforms are sintered in vacuum to high or nearly full density. Alloying of the titanium with the desired other elements is accomplished by solid state diffusion during the sintering process. By selecting the proper powders and sintering parameters, a homogeneous alloyed material with sufficiently high density, free of interconnected porosity, is achieved.

The CHIP process results in a fine-grain, anisotropic, lamellar structure as shown in Figure 17.3. The color-attributed micrograph (left) shows the field percentage of the porosity (0.7%), alpha phase (71.7%), and beta phase (27.6). The optical micrograph shows the typical lamellar structure of Ti-6Al-4V. In general, the microstructure and

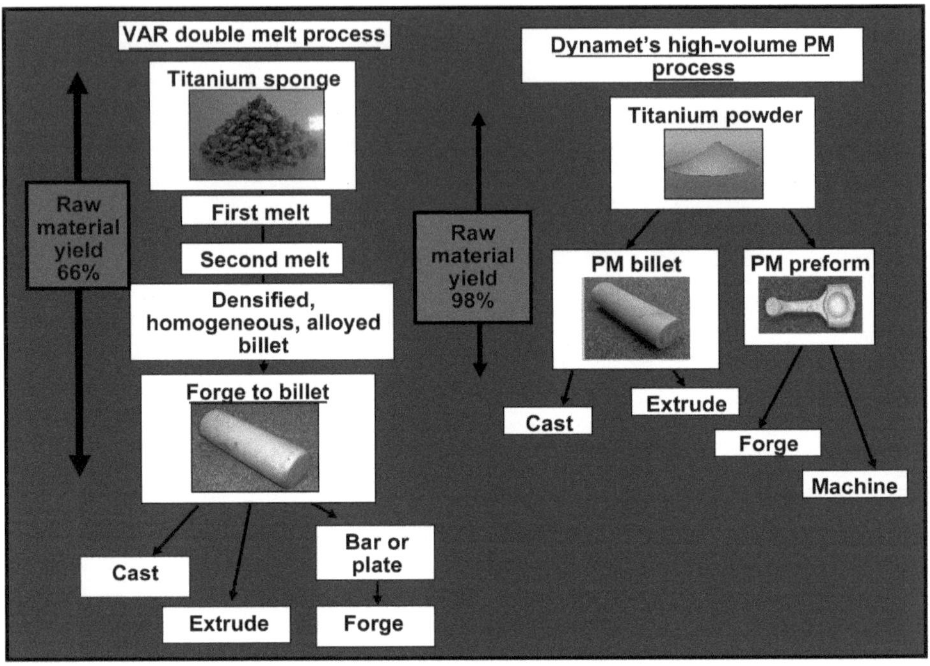

Figure 17.2 The PM CHIP process versus ingot metallurgy.

phase composition are typical of Ti-6Al-4V. Figure 17.4 shows that there is virtually no porosity in the CHIP structure.

If desired, hot isostatic pressing (HIP) can be employed to further densify the sintered part. HIP closes the residual fine porosity in the sintered shape by a combination of argon gas pressure and temperature. The result is a fully dense titanium alloy component with properties that are comparable to wrought material.

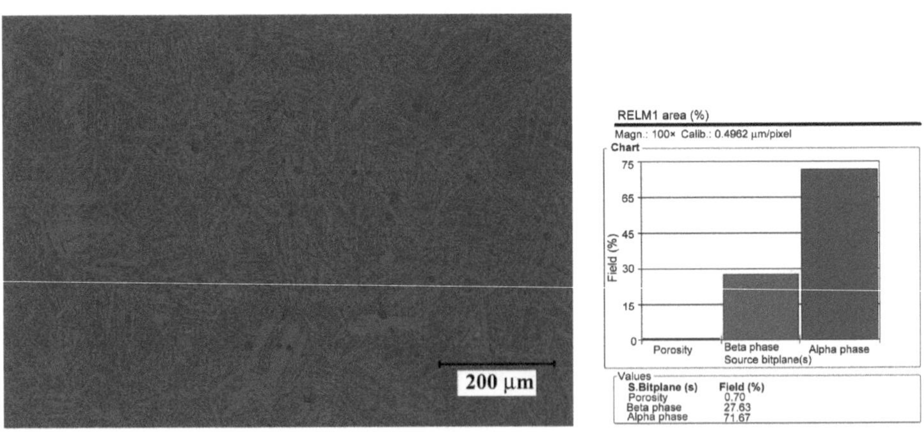

Figure 17.3 Color attributed as-sintered PM Ti-6Al-4V microstructure with image analysis insert.

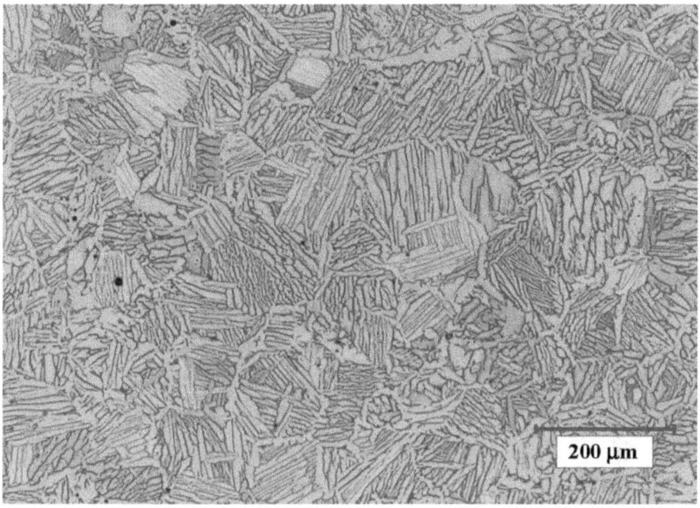

Figure 17.4 Microstructure of PM Ti-6Al-4V alloy produced by CHIP.

Table 17.2 compares the minimum tensile properties of wrought Ti-6Al-4V as specified in AMS 4928 and the typical tensile properties of wrought Ti-6Al-4V compared to the Ti-6Al-4V produced by the PM CHIP process. As shown in Table 17.2, PM CHIP meets the minimum tensile properties of AMS 4928 and can equal the typical properties of wrought Ti-6Al-4V, thus permitting a near-net shape, free of alpha case, to be finish machined to final configuration without further working. The properties of the 98% dense CIP-sinter are superior to cast and comparable to wrought. The fatigue properties of CIP-sinter and CIP-sinter-HIP are sufficient for static parts. Additional development is ongoing to achieve fatigue properties equivalent to wrought.

17.3 Titanium metal matrix composites

Dynamet Technology has applied the PM CHIP process to the production of its CermeTi family of titanium metal matrix composites (titanium MMCs). These MMCs consist of a titanium alloy matrix reinforced with a fine uniform dispersion of ceramic particulates. The ceramic particulate dispersions significantly improve the elastic

Table 17.2 Ti-6Al-4V alloy: ASTM E-8 tensile properties

	Theoretical density (%)	Ultimate tensile strength (MPa)	Yield strength (MPa)	Elongation (%)
AMS 4928 (min)	–	896	827	10
Typical wrought	–	965	896	14
Typical PM CIP-sinter	98	951	841	15
Typical PM CHIP	100	965	854	16

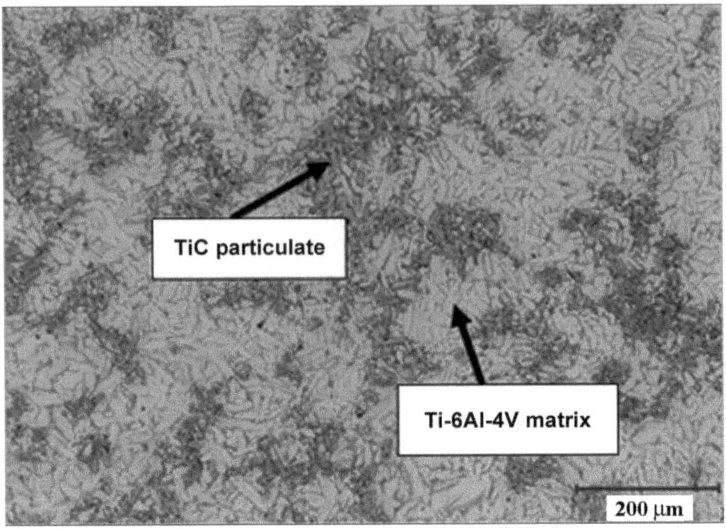

Figure 17.5 The microstructure of CermeTi® Ti-based MMC with TiC reinforcement.

modulus and wear resistance of titanium. The process also enables titanium MMCs reinforced with titanium carbide (TiC) or titanium diboride (TiB_2) particulates.[3] The particulates are added to a blend of titanium and master alloy and then processed by an innovative PM CHIP process to produce fully dense, high-quality, discontinuously reinforced CermeTi titanium-based MMCs. The microstructure of the resulting MMC material is shown in Figure 17.5. The structure consists of a Ti-6Al-4V alloy matrix reinforced with particles of TiC. The density of TiC is very close to that of the titanium alloy matrix; thus, the density of this titanium MMC is virtually the same as the density of Ti-6Al-4V. Therefore, the stiffness to weight ratio of the titanium MMCs is significantly better than the matrix alloy.

The particulate addition results in a metal matrix composite with excellent wear resistance, increased strength, and elastic modulus with useful ductility as shown in Table 17.3. Dynamet's extensive experience in the production of blended elemental titanium alloy and MMC billet and the extrusion of these billets has enabled high-quality dense commercial products ranging from a wear-resistant medical orthopedic device to a shot sleeve liner for die-casting aluminum alloys.

17.3.1 Titanium–tantalum alloys for implantable medical devices

Titanium–tantalum (Ti–Ta) alloys offer a range of desirable properties that make them of interest for implantable medical devices. However, Ti–Ta alloys are not produced commercially because they are very difficult to prepare as homogeneous alloys by conventional IM because of the disparity in melting point and density between

[3] CermeTi with TiC is referred to as CermeTi-C and CermeTi with TiB is referred to as CermeTi-B.

Table 17.3 Typical properties of CermeTi versus Ti-6Al-4V

	Ultimate tensile strength (MPa)	Yield strength (MPa)	Elongation (%)	Elastic modulus (GPa)	Hardness (Rc)
Ti-6Al-4V PM	965	896	14	110	36
CermeTi®-C MMC (Ti-64 + TiC)	1034	965	3	130	42

titanium and tantalum.[4] Dynamet Technology has developed a radically different PM method of producing these alloys. The properties and wire drawing characteristics of these PM Ti-Ta alloys have been investigated in the range of 30–95% Ta.[5] Several compositions have been found to be of interest for medical devices. Specific titanium-based alloys possess a unique combination of properties that makes them promising for a variety of implant applications, including medical wire, orthopedic implants, and stents.

17.3.2 Titanium alloys containing tungsten

The PM process has led to the development of the Abkolloy® family of PM titanium–tungsten alloys that cannot be produced economically by IM processing because of the high disparity in melting point and density between titanium and tungsten. The high-strength Abkolloy PM titanium–tungsten alloys possess significantly higher strength than Ti-6Al-4V with good ductility and as much as 25% higher hardness. The results show significant improvement in wear behavior, permitting the expanded use of titanium in demanding applications. By the PM process, these materials can be tailored to the specific requirements of individual applications. Ultimate strength can range from about 1034–1379 MPa (150–200 ksi) and hardness values from about 40–48 HRc. These properties, in conjunction with maintaining titanium's innate characteristics of lightweight, corrosion resistance, biocompatibility, imaging-friendly, and nonmagnetic properties, results in a suite of materials that provide solutions to a variety of technical challenges. The Abkolloy alloys are being evaluated for a variety of applications in aerospace, industrial, defense, and medical device markets. These materials are at various stages of testing from initial to mature by design engineers and end-users for products in these markets and are subjects of a US patent and trademark.

Table 17.4 gives examples of how powder metal processing and subsequent hot working and heat treatment influence the properties of Abkolloy materials. For example, Alloy 1 exhibits the unusual combination of the strength of Ti-6Al-4V with the high ductility of CP-Ti. Alloy 2 has higher strength than annealed Ti-6Al-4V with similar ductility. Alloy 3 reaches yield strength levels on the order of 1350 MPa (200 ksi) with useful (5%) ductility. The tungsten-containing alloys also have improved wear resistance

[4] Ti has a density of 4.5 g/cm^3 and melts at 1670°C while Ta has a density of 16.6 g/cm^3 and melts at 3020°C.
[5] This work was conducted in part under a National Science Foundation SBIR grant, award number: IIP-0724433.

Table 17.4 **The properties of W-containing Abkolloy materials**

Material	Ultimate tensile strength (MPa)	Yield strength (MPa)	Elongation (%)	Reduction in area (%)
CP titanium	517	414	24	46
Ti-6Al-4V	951	862	14	28
Alloy 1	910	855	27	58
Alloy 2	951	862	24	46
Alloy 3	1462	1351	5	24

and high temperature strength compared with Ti-6Al-4V. The properties of these alloys continue to be investigated. The processing-microstructure-property relationships of a range of alloys continue to be evaluated.

Applications for the tungsten-containing materials include medical implants, munitions, ballistic armor, and automotive powertrain components. Examples of products under development for various Abkolloy materials are shown in Figures 17.6 and 17.7.

17.4 Commercial products

17.4.1 Shot sleeve liners for aluminum die castings

One of the most successful applications of CermeTi is as shot sleeve liners for aluminum die casting.[6] In aluminum die casting, molten aluminum is forced under high pressure into a die cavity through a shot sleeve. A "shot" of molten aluminum is ladled

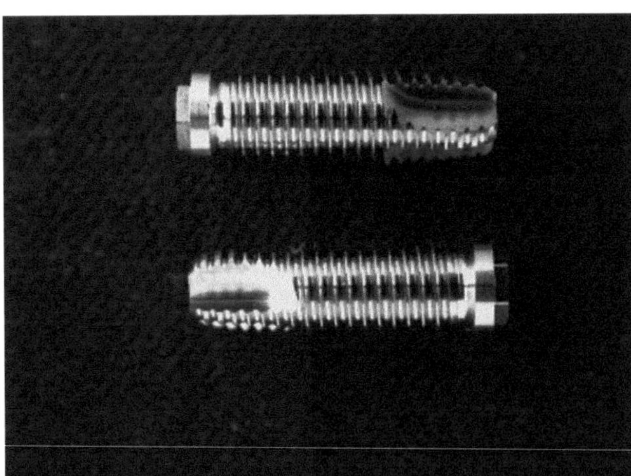

Figure 17.6 Prototype dental implants machined from PM advance composition bar.

[6] This application earned the 2002 Grand Prize in Advanced Particulate Materials at the International P/M Design Competition sponsored by the Metal Powder Industries Foundation.

Figure 17.7 PM preform for advanced munitions produced to near-net shape.

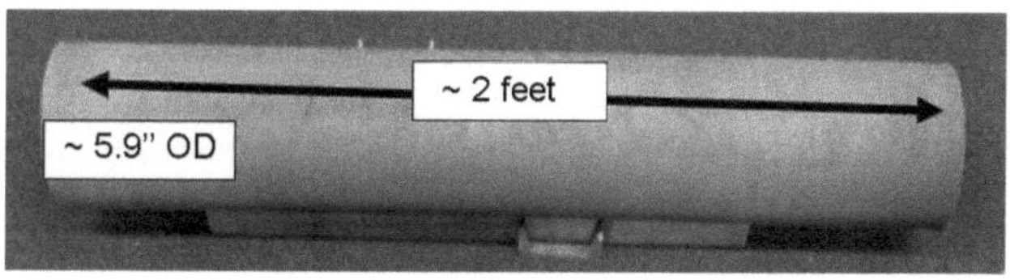

Figure 17.8 CermeTi PM shot sleeve liner preform with a 76 mm inside diameter, 7.6 mm wall thickness, by 610 mm length, weighing 9 kg.

into the shot sleeve typically made from H-13 tool steel. A plunger then forces the shot of molten aluminum through the shot sleeve into the die cavity. Shot sleeves are subject to severe conditions and must be replaced frequently. Shot sleeve replacement results in lost production time and wasted energy. Inserting a CermeTi liner into the H-13 tool steel shot sleeve dramatically improves the life of the shot sleeve, reducing down time and saving energy. The CermeTi does not erode under the high-pressure impingement of molten aluminum, withstands the shock of the sudden increase in temperature, and is virtually impervious to the constant wear of the moving plunger. In addition, CermeTi has insulating properties that reduce heat loss and improve product quality [3]. Figure 17.8 shows the shot sleeve liner preform and Figure 17.9 is a schematic of the shot sleeve assembly with the liner in place.

17.4.2 Artificial articulating replacement disc

The wear resistance of CermeTi is far superior to that of commercial titanium alloys. This material combines high wear resistance with the excellent imaging characteristics of titanium. This permits titanium to be used for metal-on-metal wear surfaces. This material is being used by a major medical device manufacturer for an artificial cervical disc eliminating the less desirable use of steel. The use of steel is less desirable

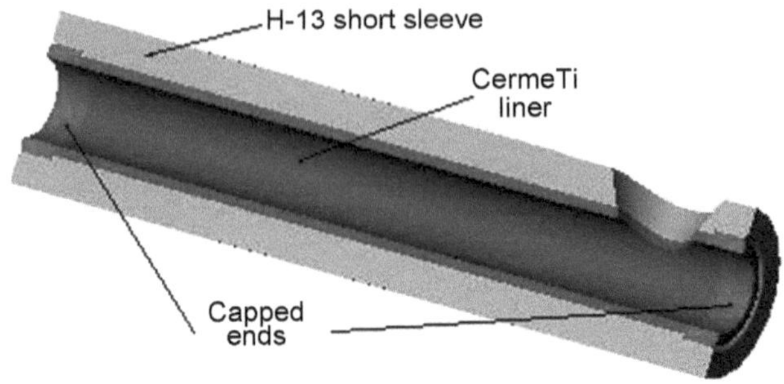

Figure 17.9 Schematic of the liner placed in the shot sleeve.

because of its poor imaging characteristics, making X-ray difficult to read and causing distortion in magnetic resonance imaging (MRI) or computed tomographic (CT) scans. The device is in commercial use outside of the United States and has recently received FDA approval for use in the United States.

17.4.3 Automotive components

Dynamet Technology has demonstrated the technical and economic viability of its titanium alloy and MMCs for lightweight titanium powertrain components.[7] The use of advanced titanium-based metal matrix composite materials can significantly improve automotive fuel efficiency and reduce greenhouse-gas emissions. Reduction in vehicle weight results from the substitution of lighter-weight titanium for steel and also from the significantly reduced vibration management system needed for reciprocating weight. The major barriers to the adoption of titanium for automobile engine components has been its poor wear resistance and high cost. Through PM processing, titanium components can be produced at lower cost than by conventional methods. Dynamet's novel alloys and metal matrix composites retain the benefits of titanium (light weight, high specific strength, and corrosion resistance) while overcoming other major barriers to the use of titanium.

The PM near-net-shape capability and the good forgeability of the innovative titanium alloy and MMCs have been demonstrated in the manufacture of prototype forged titanium MMC connecting rods for high-performance automotive engines. Conventional titanium alloys are light weight and have less rotating mass than steel but they do not have the desired wear resistance, stiffness, and creep resistance for optimal performance.

17.4.4 Powder metal preforms and improved titanium alloys for aircraft hydraulic tubing

Dynamet's novel alloys combined with its advanced powder metal technology offer the potential for producing affordable, higher-strength, damage-resistant, damage-tolerant, and cold-formable titanium alloys. These alloys hold promise for producing improved titanium thin-wall, high-pressure hydraulic tubing that could operate at the higher pressures required for future aircraft. Figure 17.10 shows the CHIP preforms, the machined CHIP preforms, and the hydraulic tube sections. The material has been

[7] NSF SBIR Phase I Grant; Award Number IIP-1045207.

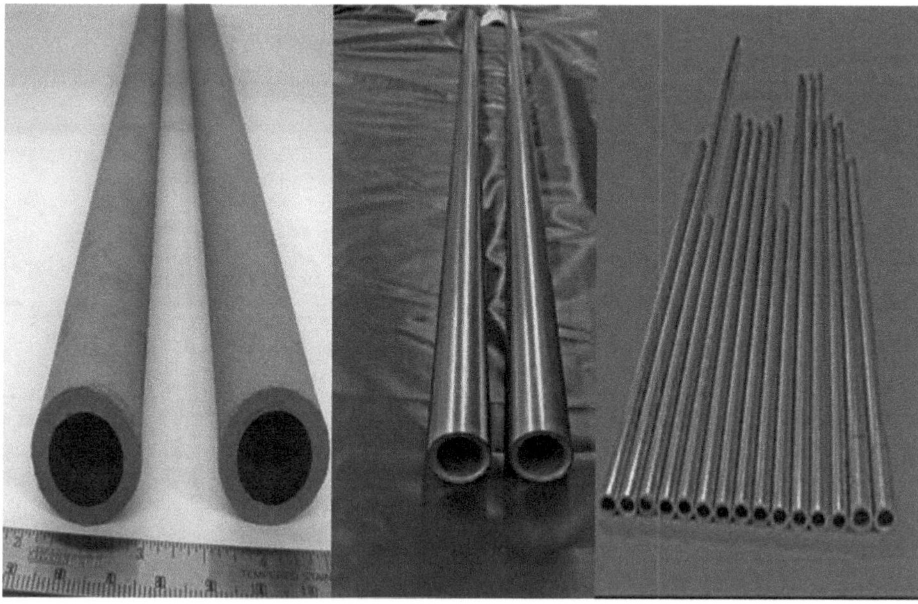

Figure 17.10 PM titanium alloy tube preforms (left) tube hollows (center), titanium alloy hydraulic tube sections produced by cold rolling from the PM tube hollows (right).

processed successfully through successive cold rolling operations. Property evaluation is being conducted on the resulting tube sections that have reached various sizes and wall thicknesses typical for aircraft hydraulic tubing.

17.5 The Boeing qualification process

PM Ti's near-net-shape advantage and the increased demand for titanium product because of titanium's compatibility with graphite composites have prompted Boeing to consider alternatives to the traditional supply of titanium products. This has resulted in heightened interest in qualifying PM Ti-6Al-4V alloy for commercial aircraft applications. Boeing has worked closely with Dynamet Technology for the last several years evaluating PM Ti-6Al-4V to enable its utilization on Boeing civil aircraft. The Boeing Research and Technology effort included extensive property testing and analysis to establish the design allowables for the novel PM titanium materials necessary to demonstrate acceptability of the material for aerospace use [4]. Dynamet Technology collaborated with Boeing in generating a Boeing Material Specification for the PM Ti-6Al-4V material. As a result of this effort, Boeing Material Specification [5] "BMS7-393B, Titanium 6Al-4V, Pressed and Sintered Powder Compacts" was released on January 30, 2012, with Dynamet Technology, Inc., as the sole qualified manufacturer on the qualified products list (QPL) for this specification. This breakthrough will enable RTI Advanced Powder Materials PM titanium products to be used as a substitute for a variety of wrought titanium airframe components on commercial aircraft.

Table 17.5 Comparison of the minimum tensile properties for IM Ti-6Al-4V and Dynamet PM Ti-6Al-4V

	Ultimate tensile strength (MPa)	Yield strength (MPa)	Elongation (%)	Reduction in area (%)	Oxygen (%)	Density (g/cm^3)	Theoretical density (%)
IM Ti-6Al-4V*	896	827	10	20	0.2	4.3	100
PM Ti-6Al-4V, CHIP	896 (130)	827	10	20	0.3	4.3	≥99

* Minimum properties for Ti-6Al-4V Alloy, Grade 5 per AMS 4906, ASTM B348, and MIL-T9046.

The inability of the PM titanium producers to generate the substantial data required for qualification has been a major barrier to the acceptance of PM Ti. As evidenced by the qualification of Dynamet's PM Ti-6Al-4V to Boeing's specification after thorough test and evaluation, Dynamet can produce PM Ti-6Al-4V product with tensile properties comparable to the properties of conventionally processed Ti-6Al-4V. In IM Ti, oxygen is considered one of the most problematic impurities. Oxygen has a significant interstitial solubility in titanium. While interstitial oxygen offers a substantial strengthening effect, it also degrades ductility. For this reason, the oxygen content of IM Ti-6Al-4V is limited to 0.2% maximum. Oxygen above that level is considered too deleterious [6]. Dynamet has found that oxygen levels of up to 0.3% actually enhance the properties of its PM Ti-6Al-4V resulting in tensile properties that are comparable to conventional Ti-6Al-4V. Table 17.5 shows that Dynamet's PM Ti-6Al-4V meets the minimum tensile properties of IM Ti-6Al-4V. Note that these properties are obtained in the CHIP near-net-shape condition without the need for hot working operations such as rolling, extrusion, or forging as required for conventional Ti-6Al-4V mill products.

17.6 Industry specification for PM titanium alloys

The heightened commercial interest in PM Titanium and the improvement in achievable properties have led to the need to develop an ASTM Specification that would provide a universal standard for PM titanium to facilitate the use of PM titanium product by industry. Dynamet served on the ASTM committee that recently issued ASTM B988-13, Standard Specification for Powder Metallurgy (PM) Titanium and Titanium Alloy Structural Components. The new standard includes PM equivalents for unalloyed titanium (Grades 1, 2, 3, and 4), Ti-6Al-4V (Grade 5), Ti-3Al-2.5V (Grade 9), Ti-6Al-4V LI (low interstitial), and Ti-6Al-6V-2Sn. This standard will facilitate broad industrial, commercial, military use of PM titanium, and thereby increase the commercial market for PM Ti-6Al-4V near net shape and basic product forms for a wide range of applications.

The influence of oxygen on PM titanium may be related to the fact that the negative effect of oxygen on ductility is offset by the inherently larger grain size of PM Ti. Figure 17.11 shows the typical microstructure of hot worked IM Ti-6Al-4V, the

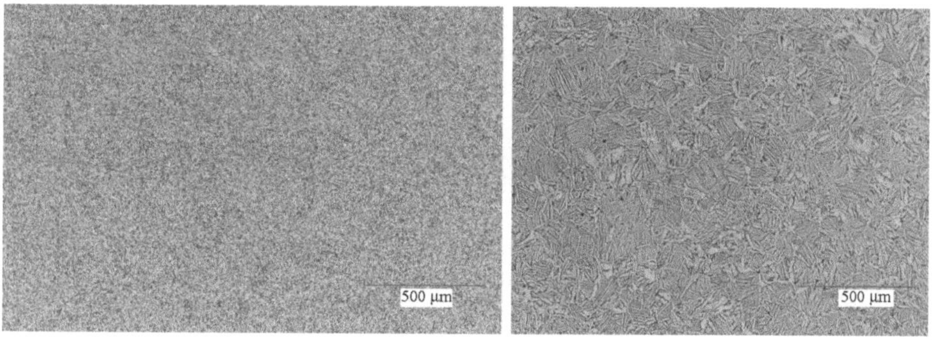

Figure 17.11 Comparison of the microstructures of IM Ti-6Al-4V and PM Ti-6Al-4V CHIP.

microstructure of PM Ti-6Al-4V as-sintered, and the microstructure of PM Ti-6Al-4V sintered and HIPed. Also shown is the grain size of each material. The grain size of the PM Ti-6Al-4V materials is approximately twice that of the mill-produced Ti-6Al-4V. Despite having much larger grains than IM Ti-6Al-4V, the PM materials possess strength and ductility equivalent to wrought titanium.

17.7 The shape-making capability

As shown in Figure 17.12, the PM process has the capability of producing a wide range of shapes and sizes. Shapes have been produced with sizes ranging from a few grams to hundreds of kilograms.

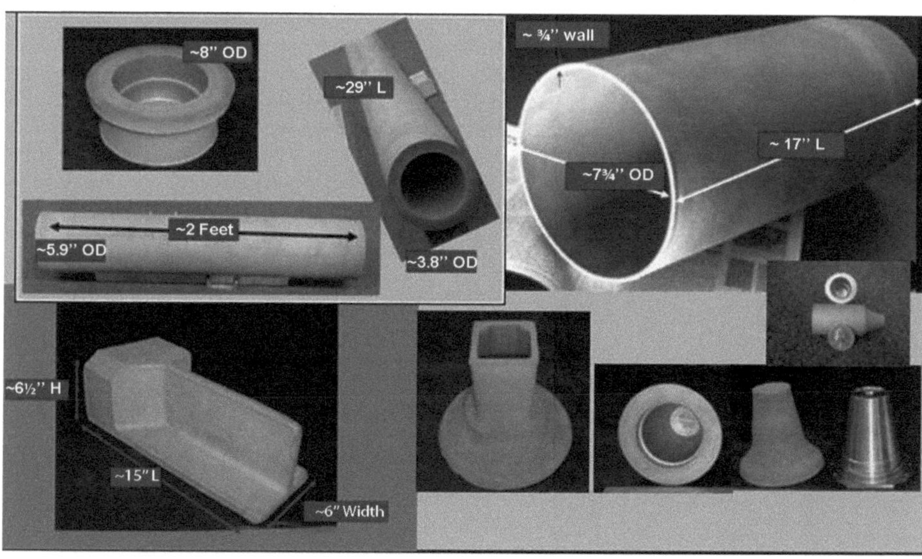

Figure 17.12 Examples of near-net shapes produced by Dynamet Technology.

17.8 Conclusions

There are expanding commercial opportunities for affordable PM titanium based on the demonstrated capability to produce high-quality PM Ti-6Al-4V near-net-shape product with reproducible tensile properties equivalent to those of conventional Ti-6Al-4V. The adoption of PM titanium by industry is being facilitated by the issuance of company-specific specifications and industry standards. Specialty PM titanium alloys and metal matrix composites developed specifically to take advantage of this PM processing route are providing new titanium material choices that offer enhanced properties and performance capability. The flexibility of PM manufacturing technology to produce a variety of near-net-shape components in sizes from a few grams to hundreds of kilograms has spurred the interest of industry. These recent developments are now leading to a number of breakthrough applications that will result in PM titanium becoming an important and expanding segment within the titanium industry. The recent acquisition of Dynamet Technology, Inc., by RTI International will enable an accelerated application of blended elemental powder technology to commercial aircraft, medical, industrial, and military titanium alloy components [7].

References

[1] C.G. McCracken, C. Mothenbacher, D.P Barbis, Review of titanium powder production methods, Int. J. Powder Metall. 46(5) (2010).

[2] R. James R. Dale, Sustainability manufacturing within the PM industry, Int. J. Powder Metall. 47(1) (2011).

[3] S.M. Abkowitz, S.M. Abkowitz, H. Fisher, Improvement of aluminum casting quality and productivity with use of titanium composite lined shot sleeves, in: Die Casting Engineer, North American Die Casting Association, vol. 56, 2012, pp. 44–48.

[4] S. Abkowitz, S.M. Abkowitz, D.H. Main, The New Era in Titanium Manufacturing – Powder Metal Components, Dynamet Technology, Inc., R.R. Boyer, T. Morton, The Boeing Company, 22nd Advanced Aerospace Materials and Processes Conference and Exposition (AeroMat 2011) Long Beach, CA, May 2011.

[5] BMS7-393B, Titanium 6Al-4V, Pressed and Sintered Powder Compacts, January 30, 2012.

[6] H.R. Ogden, R.I. Jaffee, The Effects of Carbon, Oxygen, and Nitrogen on the Mechanical Properties of Titanium and Titanium Alloys, Titanium Metallurgical Laboratory, Battelle Memorial Institute, Columbus 1, Ohio, TML Report No. 20, October 19, 1955.

[7] RTI International Metals increases advanced powder materials capabilities with acquisition of Boston-based Dynamet Technology, http://investor.rtiintl.com/press-release/rti-international-metals-increases-advanced-powder-materials-capabilities-acquisition.

Fabrication of near-net-shape cost-effective titanium components by use of prealloyed powders and hot isostatic pressing

18

V. Samarov, D. Seliverstov*, Francis H. (Sam) Froes***
*President, LNT PM Inc. (Laboratory of New Technologies)
**Consultant to the Titanium Industry, Tacoma, WA, USA

18.1 Introduction

The prealloyed (PA) powder metallurgy (PM) approach involves use of prealloyed powder, generally spherical in shape, that has been produced by melting, either by a technique such as the plasma rotating electrode processing (PREP) or gas atomization (GA) (see Chapter 2 on titanium powder production), followed by hot consolidation (generally by hot isostatic pressing) to full (100%) density, with greatly reduced "buy-to-fly" ratios compared to conventional forgings. The technique is used to fabricate complex near-net shapes (NNSs) with initial shape making using ceramic molds and in the past 20 years utilizing carefully shaped metal cans and mild steel inserts.

18.2 The ceramic mold process

Initial complex shape-making work was carried out using ceramic molds (Figure 18.1) [1–3]. Typical parts produced by this technique are shown in Figure 18.2. All mechanical properties met or exceeded cast and wrought (ingot metallurgy) levels. However, partly because of concerns (a) by engine and airframe manufacturers that ceramics or other particles could get into the titanium parts from the mold material during fabrication with a degradation in mechanical properties and (b) problems in precise shape making, this process has been discontinued. Degradation in fatigue resistance due to foreign particles was confirmed in a research program (Figure 18.3) [4].

18.3 The metal can process

The PM HIP technology for titanium using a metal can was developed in the 1970s by several companies, such as VILS (the All Russian Institute for Light Alloys), Crucible Research in the USA, or Tecphy in France, which had produced titanium parts for jet and rocket engines. The technology was further enhanced in 1992–2000 on the basis of

Titanium Powder Metallurgy. http://dx.doi.org/10.1016/B978-0-12-800054-0.00018-6

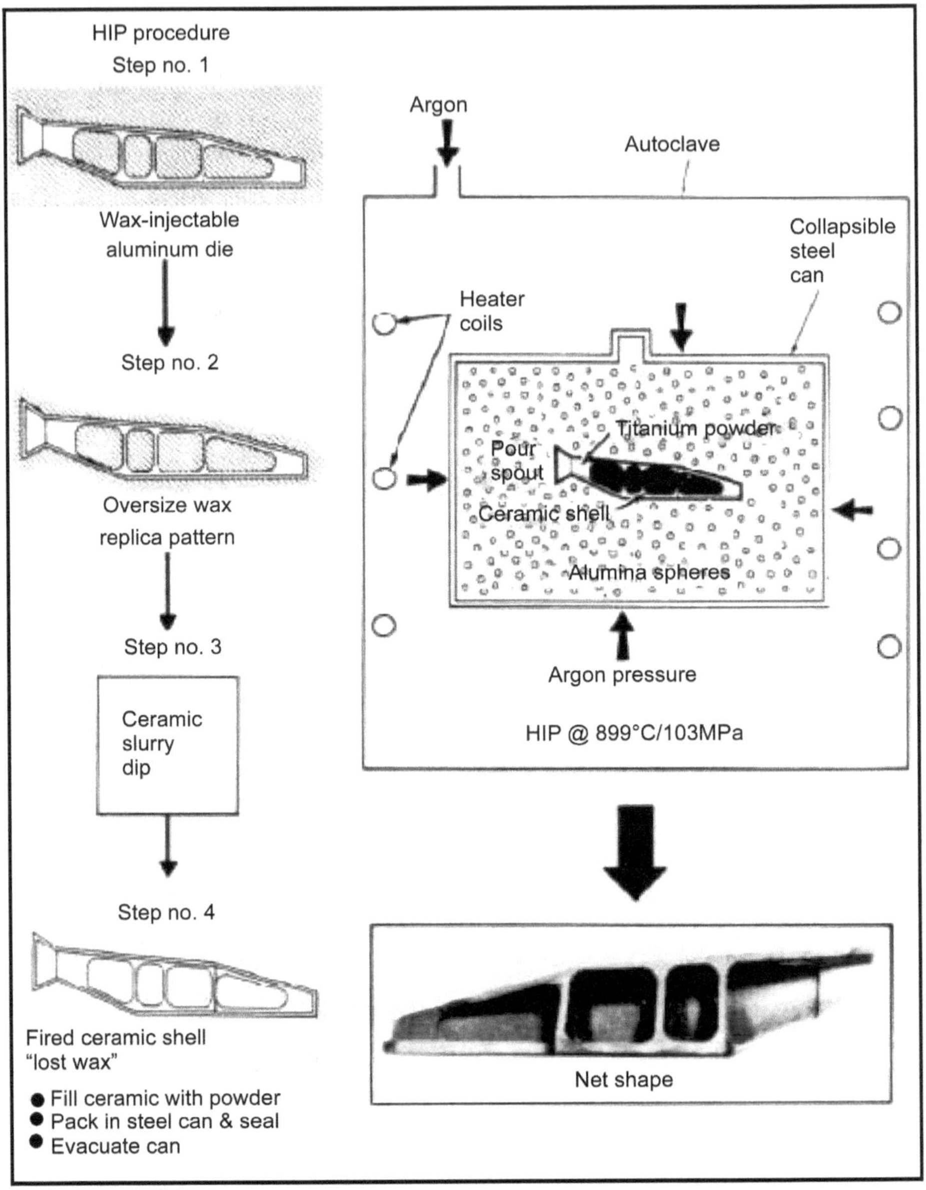

Figure 18.1 The sequence of operations to produce a part using the ceramic mold process [1–3].

computer modeling of HIP process for complex Net Shape and NNS technology by LNT (Laboratory of New Technologies). This enhanced technology has been used for the past 10 years by Synertech PM (founded in 2000), which has produced titanium parts for a variety of rocket engines, gas compressors, and test parts for jet engines and airframes,

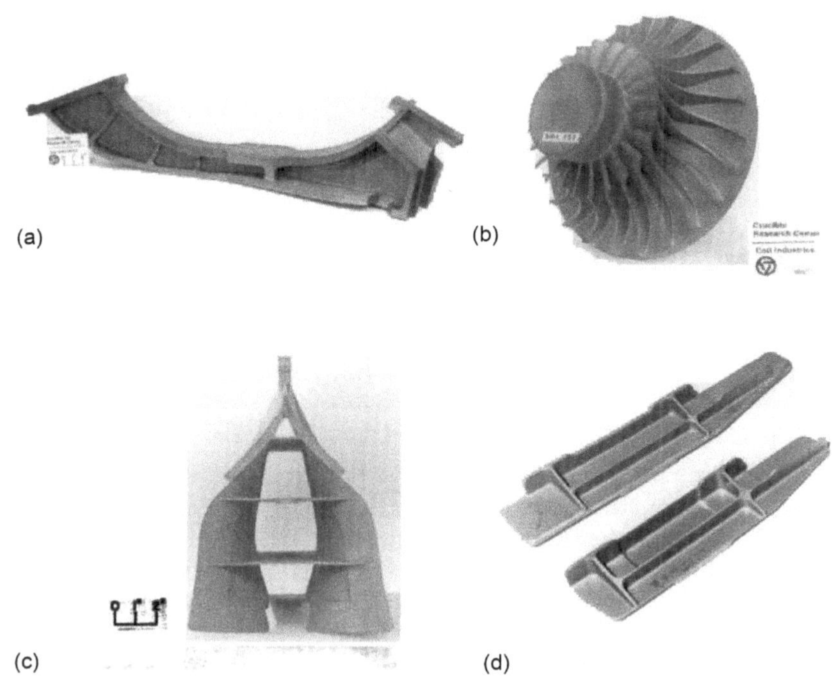

Figure 18.2 Components produced from prealloyed titanium powder, using HIPing and the ceramic mold process: (a) a nacelle frame for F14A, Ti-6Al-6V-2Sn; (b) radial impeller for F107 cruise missile engine, Ti-6Al-4V; (c) a complex airframe component for the stealth bomber, Ti-6Al-4V; and (d) engine mount support, Ti-6Al-4V. (Courtesy Crucible Materials Corporation.)

and has opened up the possibilities of significant cost reduction and introduction of Ti for various cost-efficient applications in oil and gas exploration and power generation.

Parts are produced using the shaped metal cans and removable mild steel inserts (which are removed by chemical dissolution technique).

Despite the 30–35% volume shrinkage (typical for HIP of PA powders), advanced process modeling and HIP capsule design allow to achieve nonmachined "net surfaces" and minimal machining stock on the "near-net surfaces." Also, these NNS titanium parts can be made up to the size of existing HIP furnaces, that is, up to 2 m, which is considerably larger than the capabilities of other PM technologies [1–3]. The parts produced exhibit nondirectional mechanical properties allowing an increase in design allowables (see later) and improved machinability beyond conventional cast and wrought levels.

The advanced manufacturing technology practiced by Synertech PM is based on LNT's mathematical models of consolidation and shrinkage of powder in the complex shape HIP tooling, advanced canning, and out-gassing techniques enabling high degree of control for the shape, properties, and surface quality of material.

These new advances in HIP tooling design and manufacturing techniques have enabled to reach the shape complexity of castings and the properties of the wrought

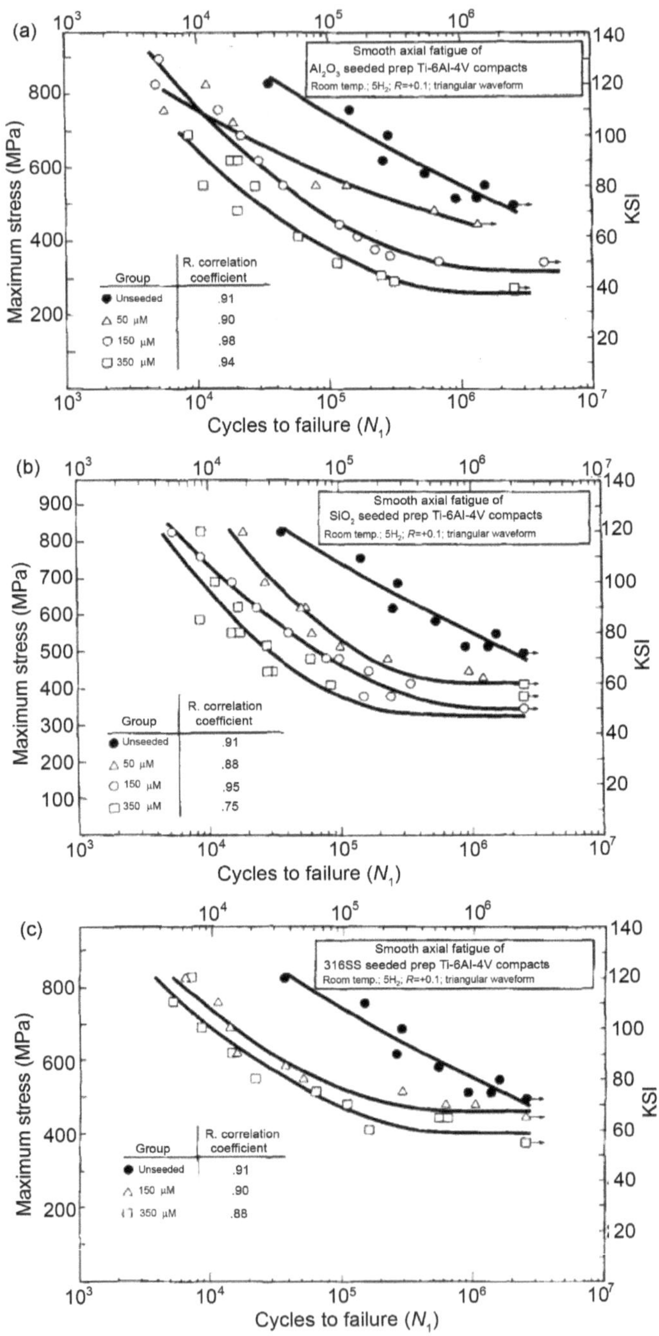

Figure 18.3 Data points and computer-generated S-N curves for all fatigue specimens initiated at seeded (a) Al_2O_3, (b) SiO_2, and (c) 316 SS contaminants, in comparison to contaminant-free unseeded baseline [4].

material and to accomplish the transfer of HIP from elaborate aerospace process to an industrial technology for a wide spectrum of Ti alloys.

The factors that allow for the cost-efficient use of HIP of PA Ti powders in the critical applications include finer powders (small inclusions), no porosity and prior particle boundary (PPB) concerns, a uniform and homogeneous microstructure, high mechanical properties with the minimal scatter, no oxygen pickup, excellent ultrasonic testing (UT) inspectability for complex shapes and better machinability.

18.3.1 General principles and advantages of HIP for complex shape parts compared to other consolidation processes for the PA Ti powders

Physically, five basic processes take place before and during HIP of Ti powder alloys that must be understood and implemented for the successful fabrication of complex shape parts: random packing of powders under vibration in a complex shape can, vacuum desorption during outgassing, densification along the given HIP trajectory, diffusion bonding, and formation of as-HIP microstructure, shaping of the part, and formation of external and internal surfaces.

Figure 18.4 presents the technological approach to manufacturing large-size complex-shape parts (Rocket Engine Housing) from an advanced Ti alloy, using the metal can approach, starting from computer-aided design (CAD) of the part and appropriate HIP tooling, manufacturing, filling, and out-gassing of capsules with powder through HIP, and subsequent tooling removal by premachining and pickling.

New technological solutions for producing high-quality net-shape PM parts from titanium alloys via HIP are based on the following physical principles, tools, and mechanisms [5,6]:

- Plastic stiffness of the capsules controlling the nonuniform deformation patterns during HIP;

Figure 18.4 HIP of large-size Ti engine housing using the metal can approach.

- Mathematical models of powder densification during a HIP cycle accounting for plastic behavior, heat diffusion, and anisotropic rheology;
- Formation of the "net" surfaces by controlling the deformation and diffusion at the capsule interface.

18.3.2 Plastic stiffness of HIP capsules controls the nonuniform deformation patterns

Despite the isotropy (uniformity) of the powder bulk that fills a HIP capsule, uniform temperature field during HIP, and isostatic pressure on the surface of the HIP cans, their shaping (shrinkage) during HIP occurs essentially nonisotropically because of the plastic stiffness of the metal cans; for example, the radial and axial shrinkages for axisymmetric cans may differ even several times [7,8]. The deformation pattern during HIP becomes even more complicated if special inserts are placed inside a capsule to provide for the internal cavities or channels in the part. Figures 18.5 and 18.6 illustrate examples of the capsules and inserts used to HIP a relatively simple shape blank for a turbine disk (Figure 18.5) and a complex shape impeller with the net shape internal flow passage (Figure 18.6).

Mathematical description of deformations for complex shapes under complex loading requires modeling with multiaxial stresses, and this type of modeling is essentially limited to the numerical finite-element techniques [9–19].

Figure 18.5 Capsule element for HIP of a simple-shape blank of a turbine disk.

Figure 18.6 Insert for HIP of a selectively net shape PM Ti impeller.

An adequate HIP process modeling enabling the necessary dimensional accuracy after densification is based on the following tools: new engineering models for powder material behavior during HIPing, adequate databases for the rheological properties of the powder and HIP tooling materials, techniques for the solution of the reverse problem of HIP – determining the HIP capsule geometry by the results of modeling.

18.3.3 Influence of the HIP cycle on the deformation, diffusion bonding, and formation of the "net surfaces"

As a rule, powder particles of Ti base alloys indent the surface of the mild steel tooling. This causes certain problems of the surface finish for the net shape parts and additionally intensifies diffusion of iron and carbon into the powder materials (Figure 18.7). However, if the HIP trajectory allows for the strength of HIP tooling material to be high enough versus the Ti alloy, the deformation pattern can be radically changed and the powder particles can flatten against the surface of the tooling, improving the surface finish (Figure 18.8). In this case, diffusion is reduced to a minimum.

18.3.4 The niche of PM HIP and the main tasks of the technology

For titanium alloys where the cost of the PA powders is still 2–3 times higher than that of the cast ingot, the applications are still limited, and the advantages of HIP can be

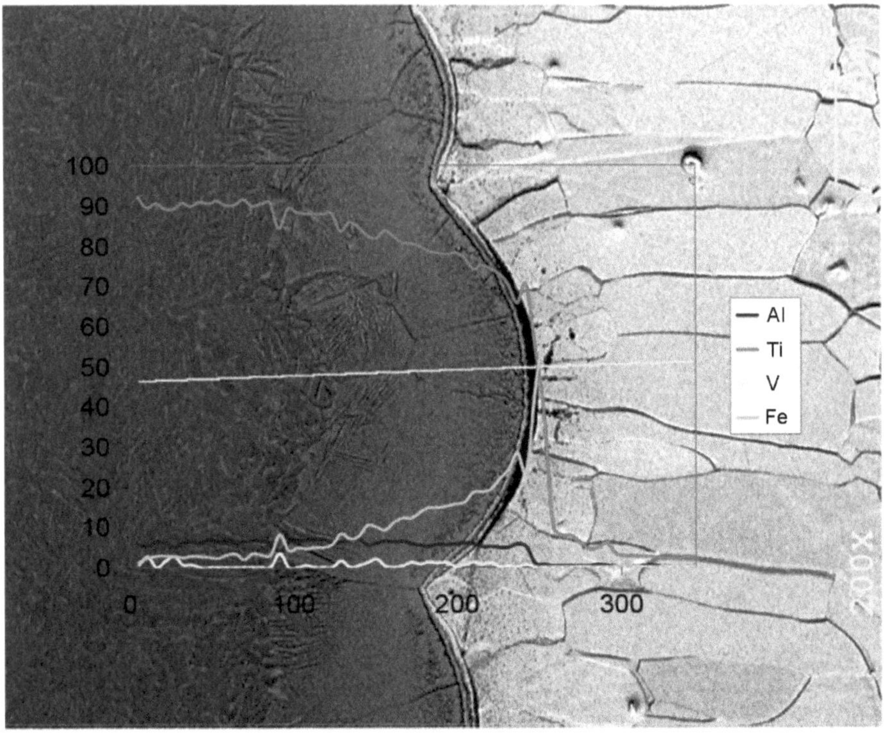

Figure 18.7 Interaction of Ti alloy with a low carbon steel during HIP.

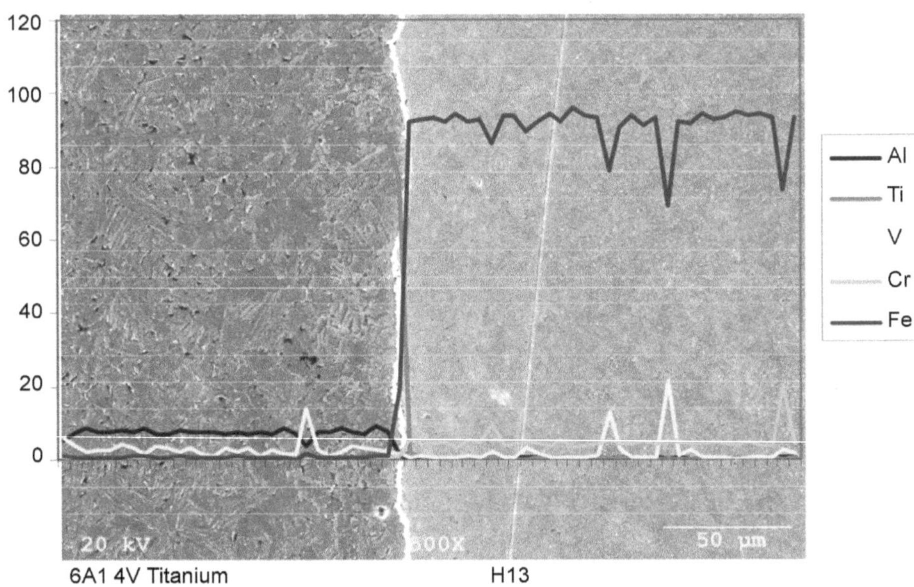

Figure 18.8 Interaction of Ti alloy with a carbon steel with higher strength during HIP.

revealed only by introducing its shaping potential through manufacturing complex, net- or near-net-shape parts, which eliminate the expensive and laborious operations of substantial machining, welding, and inspections.

Successful examples of the rather limited but technically very challenging markets such as aerospace and oil and gas include impellers, turbine and pump wheels, housings, manifolds and jackets, all made from existing advanced high-strength and environmentally compatible powder titanium alloys produced with the NNS or selectively net shape (SNS).

18.3.5 Preparation of powders for HIP (the issues and methods of filling and out-gassing)

There are two main operations with the powders prior to HIP that are an essential part of the NNS PM HIP technology: filling the HIP cans and out-gassing the powders in the capsules that to a major extend define the final quality of the HIPed products. Filling HIP capsules with powder is a very important technological step to bring the HIP can to its initial condition prior to HIP. Both under- and overfilling may be deteriorating for the dimensional future precision of the NNS parts, especially the large ones.

When the capsule dimensions and material and the filling parameters are "frozen," the reproducibility of filling density and of the final shape is very good for a given HIP cycle.

Out-gassing of Ti PA powders is a complicated process, combining the evacuation and desorption that are both extremely difficult in a powder bulk because of the thin and winding channels between the powder particles and "dead ends" and a large amount of adsorbed moisture on the powder surface once the powders are exposed to atmosphere. It has been experimentally shown [20–23] that the simple evacuation is not efficient and elevated temperatures must be carefully used for out-gassing, which is additionally complicated by the very low thermal conductivity of powders in vacuum.

18.3.6 Shaping of Ti powders during HIP (no rigid shaping tool) to the near-net and net shape using capsule as a plastically deformed tool

During HIP of powders in metal capsules, there is no rigid shaping tool, and the final geometry of the part comes out as a result of the 30–40% volumetric deformation (densification) of the powder in the capsule and of the plastic deformation of the non-compressible material of the HIP tooling (metal capsule and inserts).

As a result, despite the isostatic pressure, the deformation of the capsule with powder is not isotropic. Capsule partially shields the isostatic pressure, producing a nonuniform pressure on the surface of the powder. For Ti powders, the influence of the steel capsule on the deformation pattern is especially pronounced as the HIP temperatures are usually in the range of 800–1000°C when the capsule material may be still relatively strong.

The NNS or SNS hot isostatic pressing are based on computer modeling of HIP tooling using special numerical codes describing the joint plastic flow of the compressible (powder) and noncompressible (capsule and inserts) materials [15,16,18,19]. The capsule therefore acts as a plastically deformed shaping tool, giving the initial shape to the powder bulk and directing its further deformation till the material is fully densified.

Despite this large shrinkage (typical for HIP of PA Ti powders), this advanced process modeling allows to obtain "net surfaces" and minimal machining stock on the "near-net surfaces."

18.4 Problems and solutions

To realize through the shape control the potential advantages of PM HIP, the following technical problems have to be addressed and solved: development of the database of rheological properties for prealloyed Ti powders, advancement of the numerical models to fully account for the mass and heat transfer during HIP of large-size Ti components, modeling and study of the interaction of Ti powders with the HIP capsule material to enhance the surface finish and quality of the nonmachined surfaces, modeling and study of the mechanical properties as a function of the powder particle size, and powder processing parameters. Databases for every new Ti alloy are systematically built through special experiments in generating porous samples in the interrupted HIP cycles and testing them at the temperatures of this generation [8].

Design of the HIP capsules based on these models allows development and manufacture of complex NNS parts through one step and SNS parts – by using one intermediate iteration.

Figure 18.9 shows complex NNS housing from Ti-6Al-4V made in one development step.

The geometry of these parts is very close to the print, leaving only 2–3 mm for the final machining and enabling reduction of material 3–4 times compared to forgings. Considering the cost-efficient powder manufacturing processes producing Ti powders with a high packing density, >65%, and the very high cost of machining of Ti alloys, the NNS HIP process is highly competitive with forgings.

For the SNS parts such as impellers with the nonmachinable precise internal channels, usually an experimental iteration is needed to provide the necessary tolerances of 0.2–03 mm.

Figure 18.10 illustrates modeling, HIP tooling design, and the achieved dimensional accuracy within 0.2 mm for a rocket engine impeller fabricated from the Ti-5Al-2.5Sn alloy.

Nondestructuve (ultrasonic) inspection for complex shape parts is one of the key elements for the success of the NNS HIP technology [24,25].

Usually the requirements ultrasonic inspection for simplified ("sonic") external shape and surface quality builds barriers for enhancing the shape of the forgings and radical reduction of the "buy to fly" ratio. HIP allows to resolve this contradiction and to provide inspection for the very complex shape parts and even those with internal cavities and channels formed by steel inserts. This is done by performing such

Figure 18.9 Complex PM HIPed near-net-shape housings from Ti-6Al-4V.

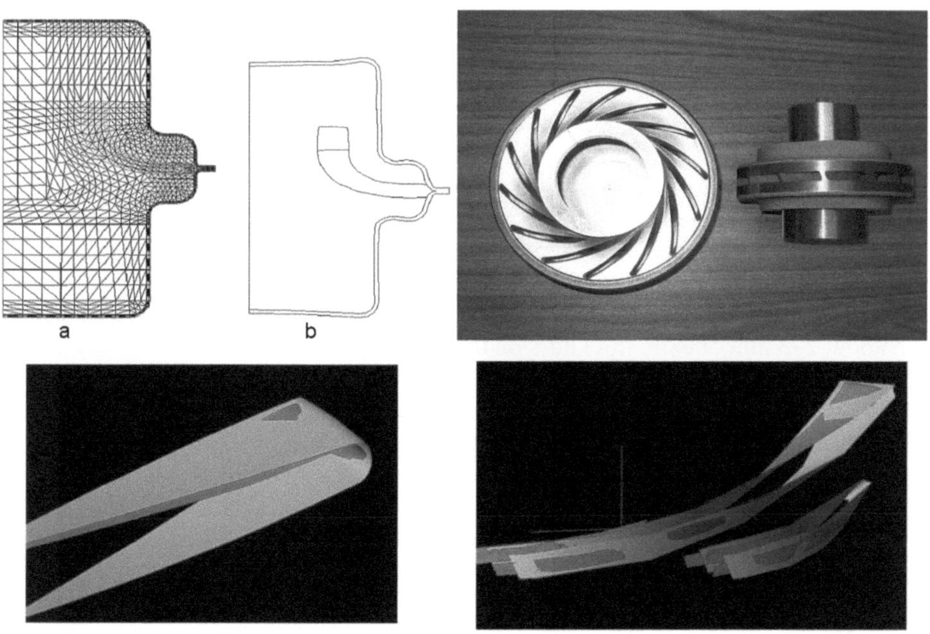

Figure 18.10 HIP modeling and dimensional precision for an upper-stage rocket engine impeller.

Figure 18.11 Rocket engine impeller premachined for ultrasonic testing inspection and acid leaching.

inspection through keeping some of the HIP tooling outside and inside the very complex shape parts, that is, machining them to the "sonic" shape and inspecting such solid "bimetallic" structures.

This approach to the UT inspection has been very successfully demonstrated while manufacturing complex shape shrouded impellers for rocket engines and gas compressors.

Figure 18.11 shows an impeller after premachining of the HIP capsule to a sonic shape and prior to UT inspection: the ring shiny layers of Ti alloy on the hub, shroud, and vanes and low-carbon steel surrounding it.

The advantage of this approach is that the net-shape geometries can be inspected, while the inaccessible "dead areas" typical for UT inspection of forgings and requiring about 2–3 mm of extra thickness on the surface, can stay within the steel tooling.

Figure 18.12 shows an airframe honeycomb structure made from Ti-6Al-4V to an SNS, that is, with substantial cost reduction compared to machining from a rolled plate.

In the case of critically loaded rotating parts, fine PA Ti powders (less than 100 μm) are used that is enabled by the vacuum induction melting and gas atomizing (VIGA), plasma atomizing, and electrode induction melting gas atomization (EIGA) powder production technologies.

The use of such fine powders removes all previously existing concerns regarding inclusions and their impact on the fatigue properties.

As for the oxygen pickup, the efficient out-gassing techniques allow the O_2 level of the produced powder to be maintained in the final HIPed product, see Table 18.1.

Figure 18.12 Ti-6Al-4V selectively net shape airframe honeycomb structure.

Table 18.1 **O_2 content in the initial Ti-6Al-4V powders and in the material HIPed in metal cans**

Powder production method	Powder size (mesh)	O_2 content in powder (wt%)	O_2 content in HIPed product (wt%)
VIGA	−140	0.14	0.13
VIGA	−325	0.16	0.16
Plasma atomizing	−140	0.09	0.08
Plasma atomizing	−325	0.10	0.10
Plasma atomizing	−140 + 325	0.065	0.07

18.4.1 HIP modeling and capsule design for HIP of complex NNS and SNS from Ti alloys

While the NNS parts are manufactured via HIP with some material stock (3–5 mm), the SNS parts must provide certain surfaces of the part with machining or casting tolerances that are usually in the range of 0.2–0.5 mm. Both tasks cannot be solved without an adequate process modeling as the linear displacements during HIP are nonuniform and large (usually 10–20%) of the corresponding dimension, and even a 1–2% error can substantially increase the necessary material stock for the NNS parts.

The new solutions in modeling and design for the NNS HIP of Ti alloys far surpass the traditional concept of a HIP capsule as just a shaped can transferring the isostatic pressure to the powder and providing its 100% density. The finite element codes specifically developed to model the densification of complex-shape powder parts in metal capsules account for the plastic deformation of a metal capsule and powder, viscous-plastic flow of powder material at high temperatures, diffusion bonding of powder particles, and pressure-activated diffusion sintering of the micropores at the final stages of HIP to form a 100% dense material typical of HIPing [6,8,13,14].

In the early decades of HIP development (approximately 1970–1989), a lot of efforts were directed toward increasing the precision of description of densification during HIP in order to optimize the main process parameters: temperature, time, and pressure [26–28]. The major modeling approach was based on the micromechanical models describing the behavior of powder at a particle scale imitating the real contacts between particles and utilizing the rheological properties of the powder particles obtained by special experiments [29,30]. This micromechanical modeling was acknowledged as a reliable and available modeling technique. However, these studies had not resolved the problem of shape control during HIP.

It was realized in the early 1980s [31] that the major factor defining deviations from isotropic deformation during HIP is the stiffness of the capsule. The thicker the capsule wall, and the stronger the capsule material relative to the powder material, the more control it has on the deformation pattern and the more nonuniformity is will introduce to the shrinkage. This is specifically the case for Ti alloys HIPed at much lower temperatures than steels or Ni base alloys. Mechanically, it means that the relative value of work to deform a capsule in a given direction increases versus the work needed to densify the powder [32]. Analytical solutions are possible only in the very general or special cases [33] and a detailed mathematical description of deformations for complex shapes under such complex loading requires modeling with multiaxial stresses, and this type of modeling is essentially limited to the numerical finite element techniques.

One more aspect is even more important in the theoretical study and in the practical modeling of HIP. While complex-shape parts are HIPed, capsule and insert may for a while substantially shield the outer isostatic stresses. Therefore, it is important to develop a numerical modeling technique that is able to account for the possibility of the rigid zones in powder and the effects caused by them.

Such accounting is especially important for the large bulky capsules where the temperature field inside the part is substantially nonuniform. As far as the coefficient of thermal conductivity grows substantially with temperature, and the value of the yield stress, on the contrary, decreases substantially with temperature, a kind of densification front can be formed [32], and this front will move inwards from the surface of the capsule. Modeling [33] shows that there can be colder and practically nondensified powder behind the front and hot and substantially densified material behind the front. The latter due to its higher strength can cause additional shielding of stresses.

The mathematical problem of design of the initial shape of the HIP tooling, capsule, and insert cannot be solved by direct recalculation from the final shape of the part because this problem is mathematically incorrect when the plastic flow law is considered. The mathematically correct and technically efficient approach to computer-aided CAD of the capsules consists of the solution of a set of direct problems of HIP modeling. At the first step, the initial shape of the HIP tooling is, for example, defined as the final shape of the part and HIP modeling is carried out. After this, using the deformation map obtained by mathematical modeling, the new shape is defined by "bloating" the geometry of the part. Then modeling is repeated for this new geometry and compared with the part definition. The new deformation map is then built accounting the residuals. The process is continued till it converges. Some practical criteria must be applied to optimize the external shape of the capsule from the technological point of view [34].

The precision of modeling is strongly influenced by the technological factors.

The value of the powder filling (tap) density, its uniformity within the capsule, and its stability for different powder heats is also extremely important: the changes (PREP, VIGA, EIGA, or Plasma) or even variations of atomization process for Ti P/M alloys can lead to radical changes of the filling density.

During HIP, at every increment of the pressure (that densifies the powder and makes it stronger) and of the temperature (that softens the capsule and the powder), the actual deformation vectors are defined by the minimum of the virtual work input, and therefore the dimensional stability for the NNS parts, especially, the large ones, requires almost identical HIP cycles to obtain the necessary dimensional accuracy of the NS and NNS parts.

For a thin wall capsule, the discrepancy of the final shape after HIP is directly a function of the same discrepancy at the capsule stage. Sheet forming and welding of the HIP cans are to be optimized in taking in account the cost issues to obtain a capsule as close as possible to the nominal definition.

The technological steps of HIP modeling and design for Ti PM alloys are illustrated by the process of development of a complex NNS Ti-6Al- 4V casing [35].

A Ti-6Al-4V engine casing component demonstrator has been modeled, designed, and manufactured into an NNS using PREP Ti-6Al-4V powder. The casing, as shown in Figure 18.13, consists of an axial-symmetric cylindrical body with the two bosses, two edge flanges, two body rings and two reinforced ribs. The whole design and manufacturing process involved initial shrinkage modeling, subsequent tooling design and fabrication, and the final HIP of the part with the dimensional analysis.

Figure 18.14 shows the results of modeling for the different cross-sections of the casing. The calculated deformation map was recorded and compared to the target NNS geometry of the casing. The magnitudes of the numerical discrepancies were then used to adjust the capsule geometry in the next iteration of calculation. For a general geometry as complex as this, three to four computing iterations are usually necessary to precisely define the necessary capsule dimensions. The results of modeling for the last iteration of the HIP capsule geometry are shown by the dotted lines. They provide the necessary minimal stock of 2–3 mm over the surface of the part

The set of capsule elements was then manufactured according to the design and assembled to encapsulate the powder during HIP. The powder was consolidated from the initial relative density of 67%, to the fully dense state. Figure 18.15 shows the HIPed casing demo after the removal of the capsule by pickling in Nitric acid.

The dimensions of the Ti-6Al-4V casing demo after HIP were measured using a coordinate measurement machine (CMM) and through high-energy X-ray tomography

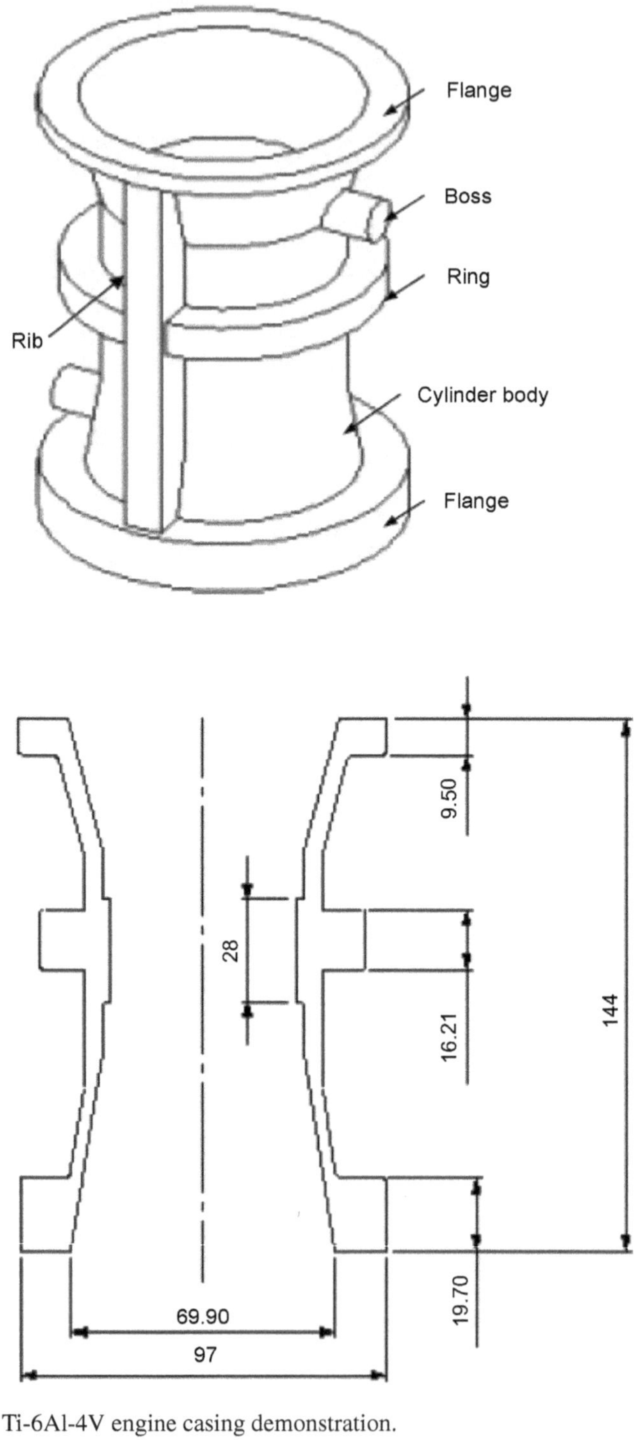

Figure 18.13 Ti-6Al-4V engine casing demonstration.

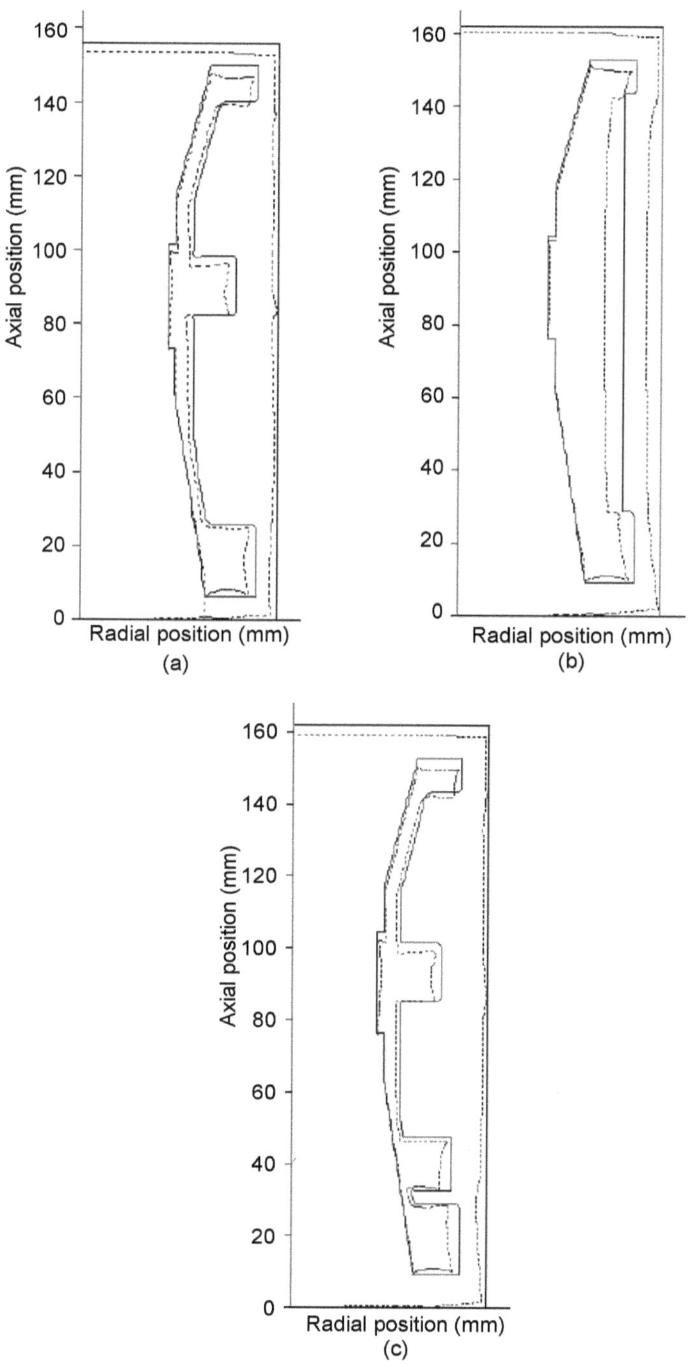

Figure 18.14 Results of HIP modeling of the NNS geometry for the different cross sections of the casing.

Figure 18.15 HIPed Ti-6Al-4V casing demonstration.

body scanner technology. The digitized surface data were imported into the CAD module and compared with the model predictions. The predictions and the measurements for the internal and external contours of the casing as well as the positions for each of the ring and bosses were in good agreement providing the dimensional accuracy within 1–2 mm and enabling manufacturing such components to the NNS.

All this modeling and results obtained for the demo had enabled to manufacture within the first shot a full size SNS Ti-6Al-4V compressor casing (Figure 18.21).

18.4.2 PA Ti powders for HIP (PREP, VIGA, EIGA, plasma), HIP parameters, examples, microstructures, and mechanical properties

There are at least four commercially available production techniques for the PA Ti powders. These include PREP (various versions), VIGA, and EIGA – another gas atomizing process of the PA Ti powder alloys utilizes vacuum induction melting of the slowly rotating rods of a Ti alloy and gas atomization of the melt and plasma atomization of wire. Further details on these powders can be found in Chapter 2, "Conventional Titanium Powder Production." Despite big differences in the production route, all these powders can be HIPed into shapes and parts with good mechanical properties and good shape control

HIP parameters for PA Ti powders are mainly determined by the beta transus temperature of each specific alloy and the necessary suite of mechanical properties in the final product – a compromise between the strength, ductility, and fatigue that is controlled by the microstructure. For the majority of the alloys requiring a fine grain structure, HIP is done below the beta transus at 900–930°C, while for some of the alloys such as Ti-5Al-5V-5Mo-3Cr near beta alloy, the HIP temperature can be as low as 800°C. It is important, however, that all these HIP temperatures provide the 100% density of the material and a well-controlled uniform microstructure.

Even when coarse (−500 μm, 35 mesh) powders are HIPed, the material already possesses high and uniform mechanical properties (Figure 18.16) quite comparable with the wrought material. Moreover, the minimum value of the 0.2% YS (that are used in the design allowables) for the PM material is above that for the conventionally fabricated material.

The same tendency is proved by the analysis of the Figure 18.17 that presents a comparison of ingot and powder metallurgy (HIPed) tensile properties for the 200-mm-diameter billet.

In this HIPed condition, the material already exhibits an excellent fracture toughness compared to the wrought (Figure 18.18) and fatigue properties of the same or better level than the wrought material (Figure 18.19).

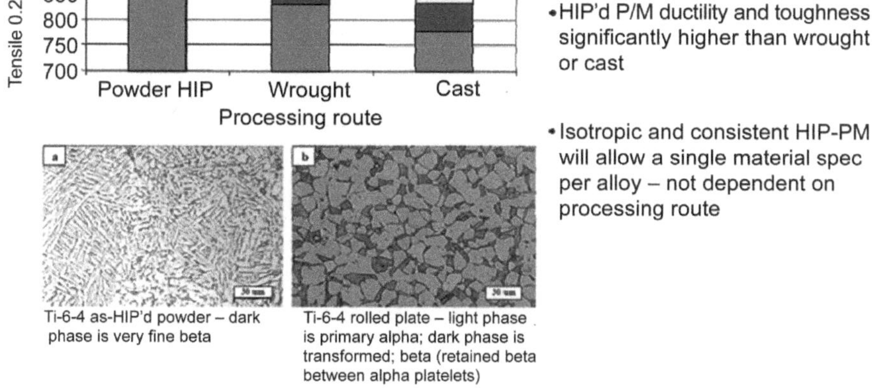

Figure 18.16 Comparative analysis of the PM HIPed Ti-6Al-4V powder with the cast and wrought material. (Courtesy Dr. Wayne Voice, Rolls-Royce, UK.)

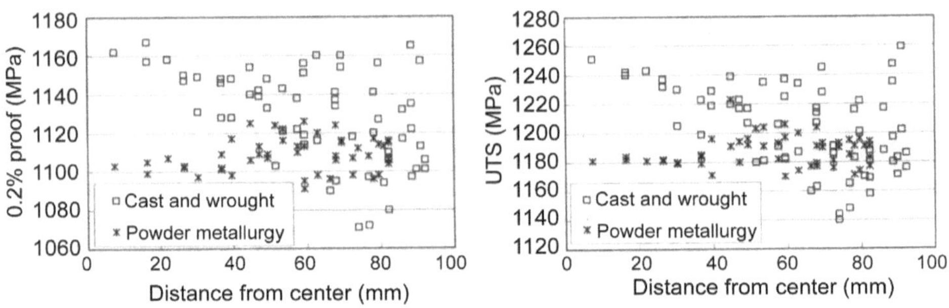

Figure 18.17 Comparison of the ingot and powder metallurgy (PM HIP) tensile properties for the 200-mm-diameter billet for a high-strength Ti alloy. (Courtesy Prof. Igor Polkin, VILS, Russia.)

HIP'd PM Ti-6Al-4V fracture toughness: 20°C

Specimen	K_{IC} (MNm$^{-3/2}$)
1	94.0
2	96.5
3	92.5

- Compact tension (CT) specimens
- BS 7488-1 (1991)
- Type 1 load vs. displacement
- Plane strain criteria confirmed
- Good consistency
- K_{IC} values relatively high

Forged Ti-6Al-4V K_{IC}= 55 MNm$^{-3/2}$

Figure 18.18 Fracture toughness of the PM HIPed Ti-6Al-4V. (Courtesy Dr. Wayne Voice, Rolls-Royce, UK.)

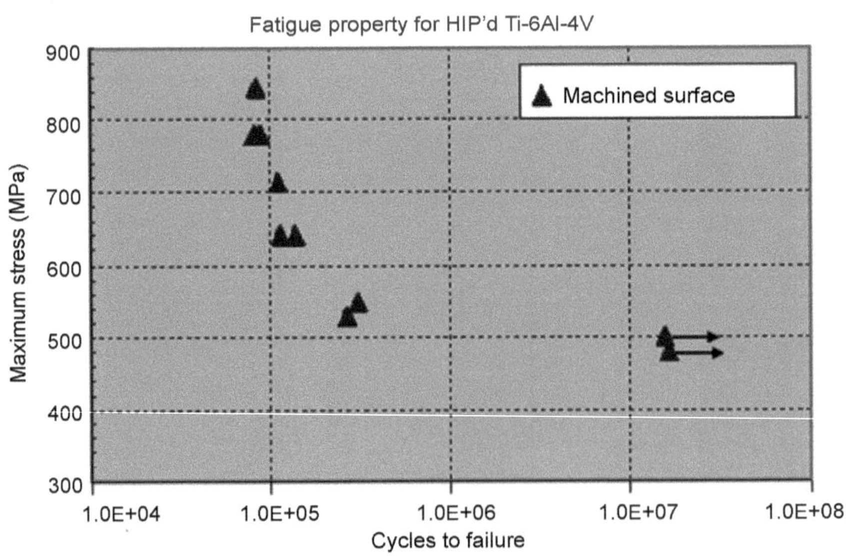

Figure 18.19 Fatigue properties for PM HIPed Ti-6Al-4V, equivalent to cast and wrought (ingot metallurgy) material. (Courtesy Prof. M. Loretto, University of Birmingham.)

Keeping in mind the issue of potential inclusions, many applications of PM HIP, such as the rotating turbomachinery, require finer PA Ti powders (less than 100 mesh, or -150 μm) for the critically loaded components The enhancements of the microstructure are becoming more uniform and homogenous and of the mechanical properties while the powder particle size is reduced, with the Ti-6Al-4V alloy serving as an example, are shown in Figure 18.20. In the case of PM HIPed PA Ti material, this increase of the strength is not accompanied with a drop in ductility [36].

Part No.:	6-4 Bar (150)
Material:	TI

Tensile						
Specification:	INFOR ONLY					
Test method:	ASTM E 8-04					
Sample no.	**Tensile strength**	**Yield strength at 0.2% offset**	**Elongation in 1"**	**Reduction of area**	**Diameter**	**Area**
	Min. (psi)	**Min. (psi)**	**Min. (%)**	**Min. (%)**	**(in)**	**(in)**
Requirements	**N/S**	**N/S**	**N/S**	**N/S**		
SAMPLE # 3-1	144,000	135,000	21	43	0.250	0.0491
SAMPLE # 3-2	144,000	135,000	21	47	0.250	0.0491

Charpy specimen Part no. 6-4 Bar (150) 100X

Figure 18.20 Microstructure and tensile properties of as HIPed Ti-6Al-4V powder of -100 mesh.

Figure 18.21 Selectively net shape Ti-6Al-4V compressor casing as HIPed (left) and after final limited machining (right). (Courtesy Rolls-Royce.)

SNS Ti-6Al-4V compressor casing made from this powder basing on the results of computer modeling, and testing of the prototype of Figure 18.15 is shown in Figure 18.21 as HIPed and after the final limited machining.

18.5 Analysis and conclusions

HIP of complex NNS and SNS parts is a synergetic technology that requires and involves exceptional engineering in design of the HIP tooling and of the entire process; high-quality powders, and elaborate powder handling technology, well-established and diversified canning technology, including out-gassing, continuous control of the powder bulk properties, surface and shape of powders, well-controlled and reliable HIPing, novel nondestructive inspection techniques, and attainment of mechanical properties (including S-N fatigue) at the cast and wrought (ingot metallurgy) levels.

With all these requirements answered, NNS and SNS HIP has become a cost-efficient alternative to forgings machined to the final complex shape.

References

[1] F.H. Froes, J. Smugeresky (Eds.), Proceedings of the Conference on "Powder Metallurgy of Titanium Alloys," TMS, Warrendale, PA, 1980.

[2] R.W. Witt, W.T. Highberger, "Hot Isostatic Pressing of Near Net Titanium Structural Parts," in Reference #1, 255.

[3] F.H. (SAM) Froes, Titanium powder metallurgy: developments and opportunities in a sector poised for growth, Powder Metall. Rev. 2 (4) (2013) 29.

[4] S.W. Schwenker, D. Eylon, F.H. Froes, Influence of foreign particles on fatigue behavior of Ti-6Al-4V prealloyed powder compacts, Metall. Trans. A 17A (1986) 271.
[5] V. Samarov, D. Seliverstov, G. Raisson, V. Goloveshkin "Physical Principles of Shape and Densification Control during HIP," Proceedings of the 2011 International Conference on Hot Isostatic Pressing, Kobe, Japan, 2011.
[6] A. Laptev, V. Samarov, S. Podlesny, Parameters of Hot Isostatic Pressing of Porous Materials, News of the USSR Academy of Sciences, Metals, 1989, N2.
[7] G. Garibov, V. Samarov, L. Buslavskii, Capsules deformation during the hot isostatic pressing of disks from powders, Powder Metall. Metal Ceram. 9 (9) (1980).
[8] D. Seliverstov, V. Samarov, V. Goloveshkin, P. Extrom, Capsule design for HIP of complex shape parts. Hot isostatic pressing'93, in: Proceedings of the International Conference on Hot Isostatic Pressing-HIP'93, Elsevier, 1993, pp. 555–561.
[9] V. Samarov, Industrial application of HIP for near net shape critical parts and components, Hot isostatic pressing'93, in: Proceedings of the International Conference on Hot Isostatic Pressing-HIP'93, Elsevier, 1994, pp. 171–185.
[10] HIP – theories and applications, in: Proceedings of the International Conference on Hot Isostatic Pressing, CENTEC, Sweden, 1988, p. 568.
[11] R.J. Green, A plasticity theory for porous solids, Int. J. Mech. Sci. 14 (4) (1972) 215–224.
[12] S. Shima, M. Oyane, Plasticity theory for porous materials, Int. J. Mech. Sci. 18 (1976) 285–291.
[13] M. Abouaf, J. Chenot, G. Raisson, Finite element simulation of the hot isostatic pressing of metal powders, J. Num. Methods Eng. 25 (1988) 191–212.
[14] A. Vlasov, D. Seliverstov, New phenomenological model of HIP simulation based on the development of plasticity theory, in: Proceedings of International Conference on Hot Isostatic Pressing HIP' 02, VILS, Russia, 2003.
[15] K. Zadeh, Finite element simulation of near net shape parts produced by hot isostatic pressing, Hot isostatic pressing, in: Proceedings of International Conference on Hot Isostatic Pressing, ASM, 1996, p. 297.
[16] D. Seliverstov, V. Samarov, HIP modeling of complex shape parts: experience trends and perspectives, 1994 Powder Metallurgy World Congress, Paris, 1994, Proceedings.
[17] G. Raisson, V. Samarov, Design rules for net shape or near net shape components produced by hot isostatic pressing of superalloy powders, 1994 Powder Metallurgy World Congress, Paris, 1994, Proceedings.
[18] C. Argento, D. Bouvard:, Modeling the effective thermal conductivity of random packing of spheres through densification, Int. J. Heat Mass Transfer 39 (7) (1996) 1343–1350.
[19] J. Palmkvist, Sandvik uses HIP modeling towards net shape HIP products, in: Proceedings of HIP 2011 Conference, Kobe, Japan, 2011, pp. 209–214.
[20] V.J. Koshelev, et al., Process investigation and industrial technology development of vacuum out-gassing of high-temperature nickel-based alloys granules, Light Alloys Technol. 5 (1989) 20–24.
[21] V. Samarov, R. Haykin, V. Nepomnyatschy, V. Koshelev, E. Khomyakov, Out-gassing of powders before HIP: problems and solutions, in: Proceedings of the International Conference HIP 2002, VILS, Moscow, May 20–22, 2002.
[22] A. Rot, Vacuum Sealing Techniques, Pergamon Press, 1979.
[23] N.V. Cherepnin, Sorption phenomena in vacuum technology, Russ. J. Phys. Chem. 58 (1984).
[24] F.H. Froes, M.A. Imam, D. Fray (Eds.), Cost Affordable Titanium, Symposium Organized by TMS, Warrendale, PA, 2004.
[25] M. Ashraf Imam, F.H. Froes, K.F. Dring, Spring Symposium on "Cost-Affordable Titanium III," Trans Tech Publications, Stafa-Zuerich, Switzerland, 2010.

[26] H.J. Frost, M.F. Ashby, Deformation Mechanism Maps, Pergamon Press, UK, 1982.
[27] M.F. Ashby, A first report of sintering diagrams, Acta Metall. 22 (1974) 275–284.
[28] E. Arzt, M.F. Ashby, K.E. Easterling, Practical application of hot isostatic pressing diagrams: four case studies, Metall. Trans. 14A (1983) 211–221.
[29] W. Kaysser, Present state of modeling of hot isostatic pressing, Hot isostatic pressing. Theory and applications, in: Proceedings of the International Conference HIP-1989, ASM International, 1991, pp. 1–14.
[30] D. Bovuard, M. Lafer, New developments in modeling of hot isostatic pressing, Ibid, pp. 15–23.
[31] G. Garibov, V. Samarov, L. Buslavskii, Capsules deformation during the hot isostatic pressing of disks from powders, Powder Metall. Metal Ceram. 9 (9) (1980) 25–30.
[32] B. Dryanov, V. Samarov, Isostatic densification of the powder material in a non-uniform temperature field, Poroshkovaya Metallurgia (Powder Metall.) 3 (1989) 25–29.
[33] V. Samarov, D. Seliverstov, E. Kratt, G. Raisson, HIP of complex shape parts through modeling, capsule design and demonstrators, in: Proceedings of the International Conference on Hot Isostatic Pressing-HIP'99, International Academic Publishers, Beijing, pp. 25–31, 1999.
[34] V. Goloveshkin, G. Raisson, A. Ponomarev, A. Bochkov, Accounting the non-stationary temperature field while modelling of HIP for large size components, in: Proceedings of International Conference on Hot Isostatic Pressing HIP, p. 231, 2011.
[35] YuanF W.X., J. Mei, V. Samarov, D. Seliverstov, X. Wu, Computer modelling and tooling design for near net shaped components using hot isostatic pressing, J. Mater. Process. Tech. 182 (1–3) (2007) 39–49.
[36] V. Samarov, E. Khomyakov, A Bisikalov, D. Seliverstov, HIP of complex shape parts from various Ti alloys. The AZO Journal of Materials OnLine, Presented at the 2011 International Conference on Hot Isostatic Pressing Kobe, Japan, April, 12–14, 2011.

Metal injection molding of titanium

19

Thomas Ebel, V. Friederici**, P. Imgrund**, T. Hartwig***
*Helmholtz-Zentrum Geesthacht, Zentrum für Material- und Küstenforschung GmbH, Geesthacht, Germany
**Fraunhofer-Institut für Fertigungstechnik und Angewandte Materialforschung IFAM, Bremen, Germany

19.1 The MIM process and market

MIM was developed in the 1970s and 1980s as a process for net-shape manufacturing of small and complex metallic parts. With this technology, the capabilities of processing a wide range of metals and alloys known from conventional powder metallurgy (PM) are combined with the freedom of shape forming and capabilities of series manufacturing known from polymer injection molding [1,2]. Figure 19.1 shows a schematic diagram of the MIM process: fine, preferably spherical metal powder (typical particle sizes range from d_{50} = 10 to 45 μm) is homogenized with an organic binder system in an appropriate mixing, kneading, and/or extrusion device. In this process, the binder is being molten and the powder-binder mixture (then called feedstock) is homogeneously blended by the operating shear forces. During the subsequent cool-down of the feedstock, the material is granulated, which can then be processed on an injection molding machine as known from thermoplastic polymer injection molding with the established modifications for MIM.

The binder, which usually makes up 40–50 vol% of the feedstock composition, has the function of a flowing agent for injection molding. It will be removed from the molded part (also called green part) in the subsequent debinding step. The methodology of debinding depends on the specific binder components. Binder components can be removed by solvent or thermal extraction or by combining both processes. In industrial production, catalytic debinding is also performed.

Following injection molding and debinding, the binder free part, so-called brown part, is sintered to a required density. For most materials, final density in the range of 95–99% of the theoretical density (TD) can be obtained in the as-sintered state [1]. Because of the densification during sintering, the part is subject to a linear, isotropic shrinkage of about 12–18% depending on the specific feedstock and its metal content. The sintered part may be used without any postprocessing. Hot isostatic pressing (HIP) may be performed on parts when full density is required.

The benefits of MIM include the following:

- Low-cost production, if a certain number of parts are exceeded
- High degree of freedom in terms of geometry
- Even hard and brittle materials can be processed
- High reproducibility

Titanium Powder Metallurgy. http://dx.doi.org/10.1016/B978-0-12-800054-0.00019-8

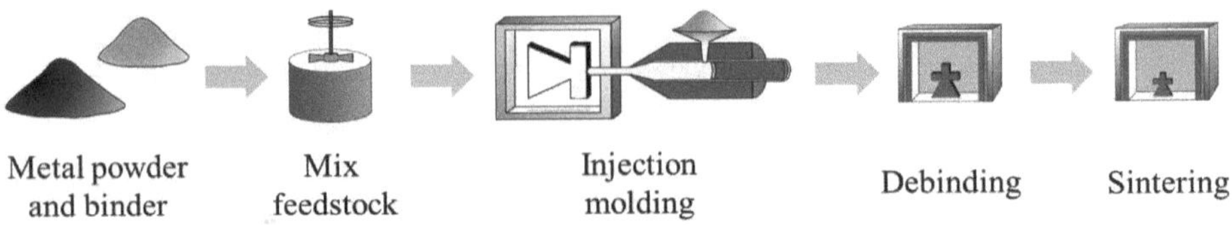

Figure 19.1 A schematic illustration of the MIM process.

Because of some limitations related to the injection molding process, the removal of the binder, and the risk of deformation during sintering, the size of MIM parts is preferentially rather small, typically in the range up to several centimeters. Wall thickness can be as small as 0.1 mm.

The MIM market was globally valued at $1.1 billion in annual sales in 2010, and the industry is currently growing at 14% per annum. Steels and stainless steels are by far the most common MIM materials, representing about 75% of the global market in terms of sales. Tungsten, nickel, and iron-nickel account for about 16%, while titanium and titanium alloys are still negligible as yet. The rest is electronic materials, copper, and tool steel grades. The most relevant market sectors for MIM are automotive, consumer goods, medical/dental, electronics, and industrial [3].

Applications of Ti MIM have covered all fields from consumer products, automobile and aerospace sectors, to medical devices and implants [4]. Figure 19.2 shows some examples of commercial products made from Ti-6Al-4V. Apart from economic

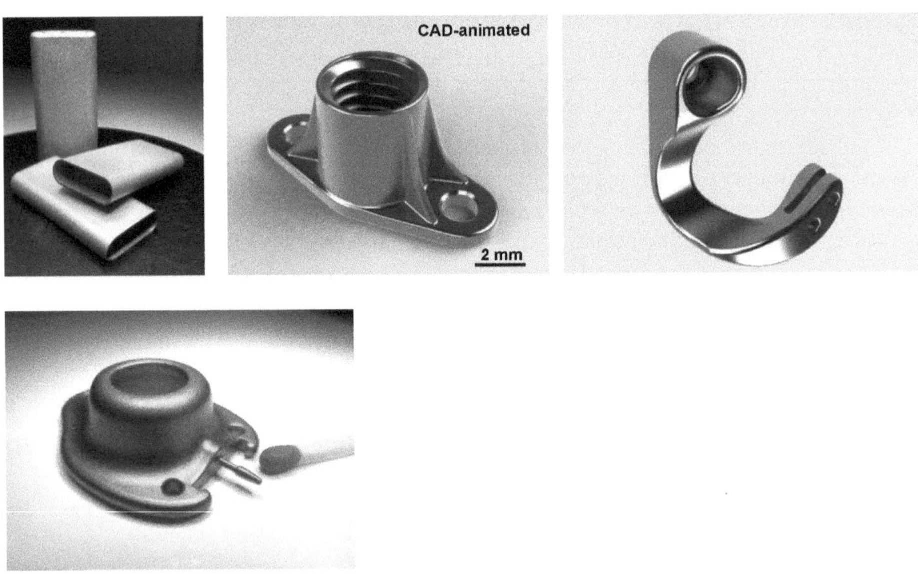

Figure 19.2 Examples for commercial products made from Ti-6Al-4V by MIM; (a) biocompatible thin-wall housings for implants, (b) airplane fastener, (c) lever; (d) housing of implantable port system. (a–c: Courtesy Element22 GmbH, Germany; d: Courtesy tricumed Medizintechnik GmbH, Germany.)

and design reasons, MIM is also applied for the realization of thin walls in the case of small housings. For instance, Ti-6Al-4V is hardly deep-drawable. Consequently, MIM is an attractive alternative.

19.2 Titanium MIM

According to Heaney and German [2,3], globally there are 19 companies that are processing or offering Ti MIM. A major technical challenge in Ti MIM processing is its high affinity for oxygen and carbon and their strong influence on mechanical properties. This issue has impeded market entrance for MIM manufacturers, as the specifications of Ti Grade 5 (Ti-6Al-4V) and especially those of Grade 23 (Ti-6Al-4V-ELI) can hardly be met. However, the MIM processing technology has improved recently, and ASTM material standards specifically for MIM of Ti-6Al-4V [5] and commercially pure titanium (CP-Ti) [6] have been established. For instance, ASTM F2885-11 for MIM Ti-6Al-4V, and ASTM F1472 for wrought Ti-6Al-4V both allow an oxygen level of 0.2 wt% while ASTM F136 for wrought Ti-6Al-4V-ELI limits oxygen to 0.13 wt% [7]. The ELI grade implies improved ductility and fracture toughness, with some reduction in strength. Table 19.1 exemplifies the maximum interstitial levels along with typical mechanical properties for wrought and MIM-processed Ti-6Al-4V.

Because of rather high costs of raw material and difficult machining of titanium alloys, near-net-shape techniques based on PM are especially attractive. Thus, interest in Ti MIM has been increasing steadily over the last decades, especially in the medical sector. The two standards mentioned earlier are intended to be applied for surgical purposes. Today, it is possible to purchase spherical Ti-6Al-4V powder smaller than 45 μm in diameter (suitable for MIM) at a price just about 30% more than that of semifinished wrought material. Considering that it is common to lose more than 80% of the starting material during machining, the advantage using powder is clear. If high processing costs and limitations in geometry of conventional techniques are taken into account, MIM can achieve an enormous economic benefit for suitable parts.

Nevertheless, mainly because of the sensitivity of titanium to interstitial elements and also owing to rather coarse powders when compared to MIM of stainless steel,

Table 19.1 Comparison of Ti-6Al-4V standards

	Max O (wt%)	Max N (wt%)	Max C (wt%)	UTS (MPa)	YS (MPa)	Min ε_f (%)	Min reduction of area (%)
Ti-6Al-4V-ELI ASTM F136	0.13	0.03	0.1	860	790	10	25
Ti-6Al-4V ASTM F1472	0.2	0.03	0.1	930	860	15	30
MIM Ti-6Al-4V ASTM F2885-11 (96% density)	0.2	0.05	0.08	780	680	10	15
MIM Ti-6Al-4V ASTM F2885-11 (98% density)	0.2	0.05	0.08	900	830	10	15

some specialties have to be considered when Ti MIM is applied. This concerns the composition of the binder, but also powder handling, injection molding machines, and the debinding and sintering equipment. In the following sections some of these issues will be discussed in more detail.

19.3 Powders and powder handling

Titanium and titanium alloy powders are available in all four grades of commercial purity, as well as according to the specification of Ti-6Al-4V grades 5 and 23. Other alloy powders may also be obtained from specialized manufacturers upon request. Generally, MIM requires fine powders for easier sintering and better surface quality. However, for the MIM of titanium a compromise is necessary as the finer the powder the more the contamination by oxygen it will have. While steel powders for MIM are usually −22 μm or −16 μm and even as fine as 6 μm, titanium powders are usually sieved to −45 μm.

The most common powder for Ti MIM is −45 μm and is produced by gas atomization or plasma atomization. In gas atomization, a rod of the given alloy is melted by induction melting and then atomized by an argon stream. In plasma atomization, a wire of the respective alloy is melted in a plasma and the resulting droplets cooled and transported to the collecting chamber by an argon stream. These powders are available from various companies in Europe, Canada, the United States, and some Asian countries. The advantage of these powders is their spherical shape, which allows a very high metal content in the feedstock. Also, the spherical shape presents the smallest surface area possible and thus picks up less contamination by carbon, oxygen, and nitrogen.

As the powder picks up some contamination during production and later during processing by MIM, the rod or wire used for atomization needs to contain clearly less oxygen and nitrogen than intended for the sintered MIM parts. An increase in oxygen of 0.08% is quite normal but can be reduced to 0.03–0.05% by careful handling in all production steps. Thus, in order to achieve a grade 5, a grade 23 powder has to be applied.

For the intended replication of small parts and improved surface finish, finer spherical powders with particle size −25 μm are available as well, or can be obtained by sieving from the −45 μm powder. Because of the increasing surface area of the finer powder, however, more complex measures need to be taken into account to prevent oxygen contamination during the process. Because of some industrial demand for finer powders, effort has been made to make Ti-6Al-4V powder available in particle sizes as fine as −15 μm [8].

If very fine titanium powder is to be processed, it should be handled in a protective argon atmosphere in all process steps from powder production to sintering. As this is an additional cost factor, it is not a common industrial practice.

Other possible powder sources for Ti MIM consist of titanium hydride (TiH_2) [9,10] or hydride–dehydride (HDH) powders [4]. For these, pure titanium stock is hydrided in a hydrogen atmosphere, which makes it brittle so that it can be milled to the projected particle size. Then it can be used directly as TiH_2 or dehydrided at an elevated temperature into HDH powder. TiH_2 powder can be dehydrided during sintering but will cause additional shrinkage depending on its hydrogen content. Milling of the brittle titanium hydride leads to angular particles, which necessitate larger binder amounts compared to spherical powders. Because of their larger surface area, they also easily pick up oxygen during production and during processing. Thus, sintered parts made from TiH_2 or HDH usually do not achieve oxygen levels necessary for CP-Ti grade 2 or even grade 3. In order to produce titanium alloys like titanium grade 5 from Ti powders, master alloy powders have to be added and long sintering times have to be applied for homogeneous distribution of the elements. All this has led to the preferred use of spherical powders for Ti MIM, especially for applications that have to comply with the standards.

A study concerning the influence of powder size and mold roughness was performed recently [8]. Ti-6Al-4V powder with different maximum particle sizes was utilized: powder A with a maximum diameter of 25 μm and powder B with 15 μm. Feedstock materials were prepared by mixing the powders with a binder system consisting of a wax–polymer combination. A test tool was developed providing four square-shaped cavities with different surface finishes and, thus, different roughness values of R_a. Furthermore, R_a was set to be 0.4, 0.8, 1.6, and 9.0 μm. Figure 19.3 displays a molded green part.

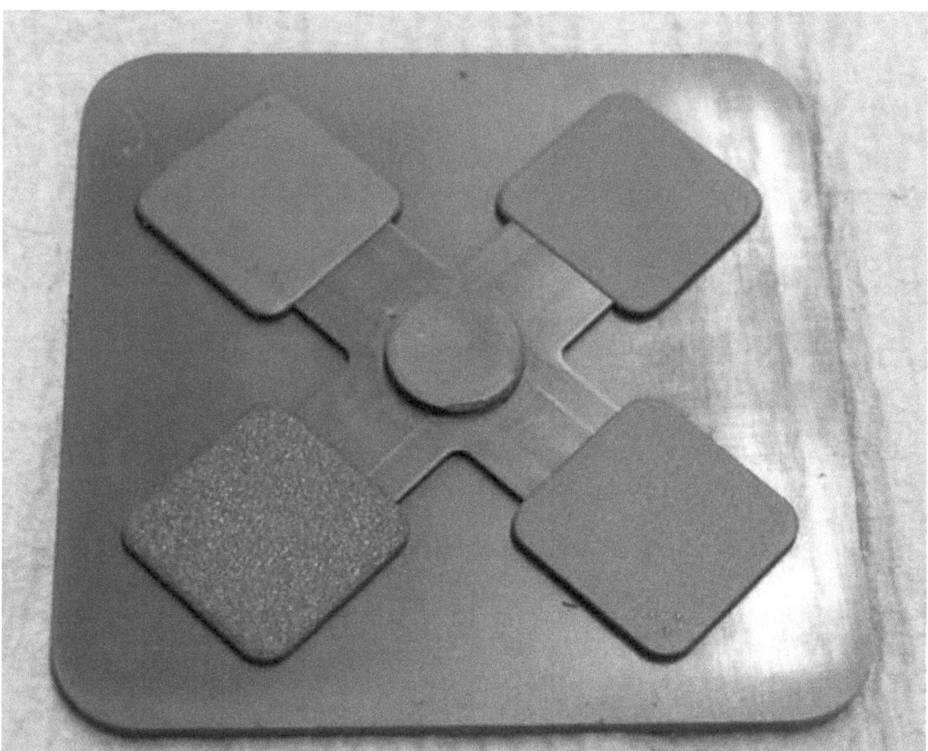

Figure 19.3 Moulded green parts with four areas of different surface roughness.

Table 19.2 **Results of surface roughness measurements (R_a in μm)**

			Green parts		Sintered parts	
Cavity number	Specification	Tool	Powder A	Powder B	Powder A	Powder B
1	0.4	0.90–1.00	1.81	0.91	1.20	1.18
2	0.8	1.40–1.46	1.93	1.19	1.16	1.17
3	1.6	2.37–2.49	2.64	1.86	1.81	1.89
4	9.0	7.41–11.74	9.94	9.42	8.63	8.56

Using a white-light profilometer, the surface roughness of cavities, green parts, and sintered parts was determined. The results are summarized in Table 19.2.

There is a clear relationship between the roughness of the tool and that of the green part. With regard to the effect of the powder, the finer powder (B in Table 19.2) follows rather accurately the roughness of the specific mold cavity, but the coarser powder (A in Table 19.2) tends to exceed the roughness of the cavity. However, after sintering, there is no difference in surface roughness between the two powders used.

19.4 Binder systems

A MIM binder is usually made up of at least three constituents. A polymer backbone provides the strength and is decomposed during thermal debinding. The largest volume fraction is made up by a plasticizer, which reduces the viscosity of the backbone and is removed in the first stage of debinding by solvent extraction, catalytic debinding, or evaporation. One or more surfactants in small amounts are added to enable good interactions between the powder surface and the binder and are removed either by the solvent or by the thermal treatment.

An ideal binder for the MIM of titanium allows perfect mixing and molding with good contact to the powder, and is then removed from the green part completely without interacting and reacting with the powder. No such ideal binders have been developed yet, as always small amounts of carbon from the binder are picked up and end up in the sintered part. There may also be some oxygen pickup from the binder if it contains a large fraction of oxygen groups in the binder constituents, but this usually plays a minor role as compared to the oxygen pickup from the sintering atmosphere. The amount of carbon and oxygen pickup depends on the binder and on the debinding method.

As the probability for the reaction between the powder and contaminants increases with temperature [4], the binder should be designed for removal at low temperature. This is partly achieved by extracting a major fraction of the binder, like paraffin or other low-molecular-weight waxes or polyethylene glycol (PEG) [11] by solvent debinding. Another possibility is to catalytically remove the major fraction at low temperature as is done for polyoxymethylene (POM) [12].

This leaves only the remaining polymer backbone to react with the powder, which cannot be completely prevented as the decomposition of typical polymers such as polyethylene and polypropylene only starts above 320°C. Here, the backbone should be chosen for completing the decomposition at as low a temperature as possible, as long as its properties do not compromise the flowability of the feedstock during molding.

There is a large number of possible binder systems that have been used successfully for Ti MIM [13–20] and there are also commercial feedstock materials available [10,11] with binders used for all kinds of materials. This shows that most of the binders being used for MIM at present can be used if the debinding and sintering cycle is carefully adjusted and care is taken to remove the binder decomposition products from the brown parts. There is a difference in carbon contamination caused by the different binders, but usually this can be kept inside the allowed values (Table 19.1). The powder should be handled completely under a protective argon atmosphere when fine or highly reactive alloy powder is used. The wax–polymer binders should be used exclusively in this case as the use of water as solvent or nitric acid as catalyst are prohibited.

19.5 Debinding and sintering

As far as solvent debinding is concerned, temperatures of immersion are generally low, varying between room temperature and 60°C. Depending on the soluble binder component, hexane, heptane, and water debinding are common. Hexane and heptane are used to dissolve, for example, paraffin waxes, while PEG-based binders can be extracted in warm water. Contamination is not a major concern in this processing step. However, explosion-safe equipment needs to be provided when choosing the wax debinding option. Drying of the parts is performed slowly to avoid cracks.

In the thermal debinding step following solvent extraction, use of an argon sweep gas process at reduced pressures of 200–600 mbar has proven most efficient for decomposing the polymers and at the same time safely removing the decomposition products. Peak debinding temperatures depend on the type of backbone polymer used. In practice, use of wax–polymer-based binder systems that can be debound at 450°C have led to the best results in this respect [20]. Usually, the thermal debinding cycle is integrated into the sintering cycle. However, the furnace has to be capable of trapping the binder residuals outside of the hot zone.

Sintering of titanium is mostly carried out either in high vacuum or in argon. In general, high vacuum sintering provides better results compared to argon in terms of the final density and oxygen contamination. This can be attributed to the fact that argon might be trapped in the pores of the sintering part, preventing full densification. Furthermore, impurities in the argon can lead to contamination of the titanium parts [20].

Especially if the furnace is not fully loaded, it is advisable to include getter materials as oxygen traps and to place them in the vicinity of the parts. Typical getter materials can be scrap titanium parts or titanium sponge. The latter is advantageous as it features a comparatively high surface area.

Concerning sintering parameters, typically temperatures are in the 1250°C range with dwell times of about 3 h. Higher sintering temperatures can generally lead to higher tensile strength (due to increased density). However, often a reduction of ductility is observed at temperatures around and above 1300°C, which may be due to either grain growth, or higher oxygen pickup, or both [20]. The ASTM F 2885-11 standard for sintered Ti-6A-4V by MIM outlines the mechanical properties sintered to 96% TD and 98% TD. A supplementary option is the use of HIP at moderate temperatures (850–1100°C) to optimize density, microstructure and properties. Provided that the density is higher than 95% TD (closed porosity), no encapsulation has to be performed. Thus, HIP is a rather simple and standard process for the mechanical improvement of MIM parts.

A major influence on sintering success of titanium alloys is known from the sintering substrates. Trays or plates made of alumina, as are common for sintering of MIM steel grades, are not suitable for titanium sintering. Instead, zirconia or yttria substrates need to be used. A cost-saving alternative may be the use of alumina substrates plasma spray coated with a yttria layer [20].

19.6 Properties of specific alloys processed by MIM

19.6.1 Ti-6Al-4V

Ti-6Al-4V is the most applied titanium alloy with well-balanced mechanical properties. Thus, it is also of special interest with regard to MIM processing. Most of the published studies on MIM of titanium deal with Ti-6Al-4V [16,19,21–45], Ti-6Al-7Nb [15,46–48], or modified Ti-6Al-4V [49–51] and since about 10 years ago, MIM of titanium has been applied commercially [52,53].

A typical microstructure of MIM-processed Ti-6Al-4V is shown in Figure 19.4. Sintering is performed at temperatures between 1200 and 1350°C in the single beta

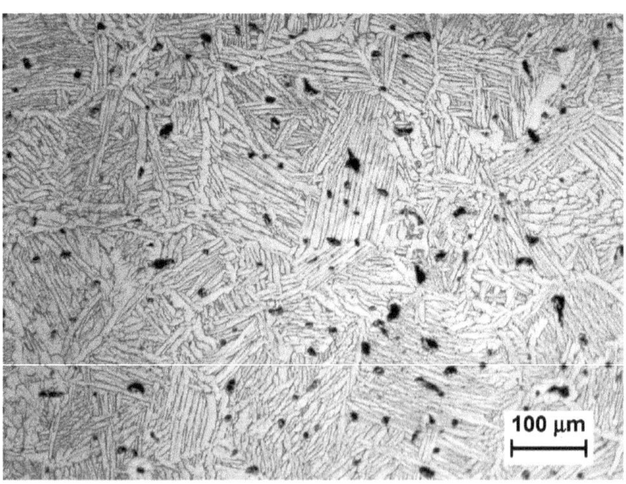

Figure 19.4 Typical microstructure of MIM-processed Ti-6Al-4V, revealing a coarse lamellar microstructure consisting of alpha and beta phase. Black features are pores.

Table 19.3 Comparison of Ti-6Al-4V ASTM standards for MIM-processed (F 2885-11) and wrought (B 348-02) material

Standard	Condition	Min YS (MPa)	Min UTS (MPa)	Min ε_f (%)	Max O (wt%)	Max C (wt%)	Max N (wt%)
F 2885-11	Sintered	680	780		0.20		0.05
	Densified	830	900	10		0.08	
B 348-02	Grade 5	828	895				
	Grade 23	759	828		0.13		0.03

phase region. During cooling, alpha phase forms by passing the beta transus at around 980°C (the precise value is among others dependent on the oxygen content), and the microstructure develops into alpha grains with beta lamellae. Often alpha phase can be observed at the colony boundaries. The amount of residual porosity is commonly in the range of 3–4%, but smaller values are possible, if finer powders or longer sintering times are utilized. Most of the pores are below 10 μm diameter.

As mentioned in Section 19.2, ASTM F 2885-11 describes the mechanical and chemical requirements for medical components from Ti-6Al-4V produced by MIM. In addition to Table 19.1, Table 19.3 compares the chemical and mechanical requirements with those of ASTM B 348-02 for wrought material intended for general application. In ASTM F 2885-11, besides the "sintered" state, a "densified" condition is distinguished, which usually is obtained by a subsequent HIP process. It is obvious that the MIM standard is guided by the chemical demands of Grade 5 and the mechanical properties of Grade 23.

As Table 19.3 reveals, basically, equivalent mechanical properties to wrought material can be achieved by applying MIM. However, one has to bear in mind that usually the actual tensile properties of thermomechanically treated wrought materials with a fine microstructure exceed the minimum standard values by far. Nevertheless, for most of the possible applications, the mechanical properties of MIM components are sufficient.

The actual properties of a Ti-6Al-4V MIM component depend on residual porosity, grain size, and interstitials content. In fact, results from different studies can hardly be compared because it is difficult to separate the specific influence of each factor on the mechanical properties. Porosity, colony size, and interstitial contents can be affected by the specification of the raw material (powders and binder system) as well as by processing parameters of debinding and sintering. In the following, some of these aspects are considered in more detail.

19.6.1.1 Influence of porosity

The amount of residual porosity has a strong influence on tensile properties as Table 19.3 reveals, when the sintered and the densified states are compared. The residual porosity should be at least smaller than 5% to avoid connected porosity. For best properties, an additional HIP process should be applied, but the exact

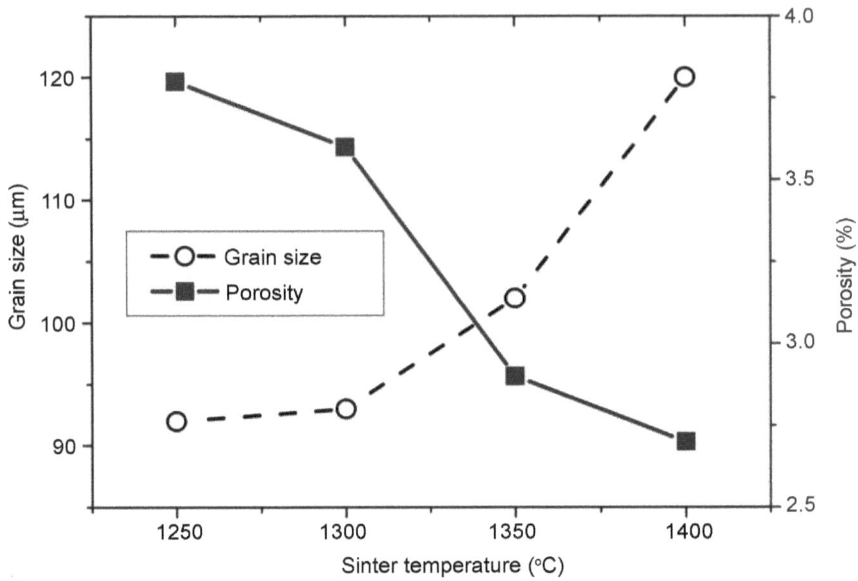

Figure 19.5 Influence of the sintering temperature on grain size and porosity of MIM-processed Ti-6Al-4V [26]. Sintering time was 2 h for all samples.

dependence of the tensile properties on porosity between 5% and 0% is not quite clear yet. Residual porosity depends on powder size and morphology as well as on sintering temperature and time. Sintering under vacuum leads to slightly higher density, because no gas is trapped in the closed pores, which hinders further shrinkage [54]. Figures 19.5–19.7 show the results of a study on processing parameters of gas-atomized PA Ti-6Al-4V powder processed by MIM [26]. Figure 19.5 displays the dependence of porosity and colony size on sintering temperature (all specimens were sintered for 2 h), while Figure 19.6 reveals their dependence on sintering time, when sintered at 1350°C. Based on these data, in Figure 19.7 the dependence of the tensile properties on residual porosity is presented with the 0% porosity being achieved by HIP.

From Figure 19.7, it appears obvious that there is a strong influence on tensile strength especially for small porosity. In studies by Zhang et al. [23] and Niinomi et al. [45], CP-Ti powder was mixed with prealloyed 60Al-40V powder and processed with long sintering times up to 8 h, so that very small porosity in the range 1–2% remained. The corresponding values are included in the figure for comparison, and a basically similar tendency is visible. However, the exact values are shifted compared to the results of Obasi et al. [26], and this shows the difficulty of comparing different studies. For example, it is very important to consider the specific oxygen concentration in the specimens. In the case of Obasi et al. [26], all samples show similar oxygen contents around 0.22 wt%. The results from the studies of Zhang et al. and Niinomi et al. tend to be higher in strength but lower in ductility. This may indicate an increased level of interstitials.

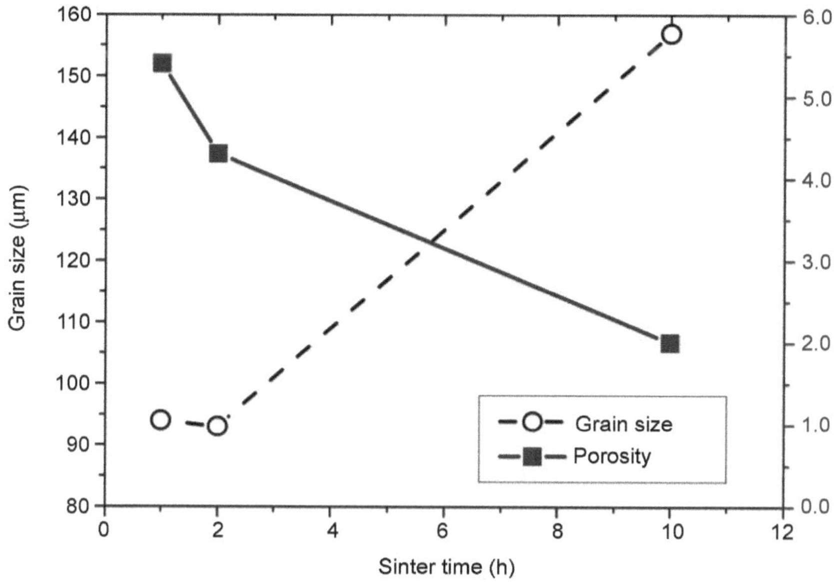

Figure 19.6 Influence of the sintering time on grain size and porosity of MIM-processed Ti-6Al-4V [26]. Sintering temperature was 1350°C for all samples.

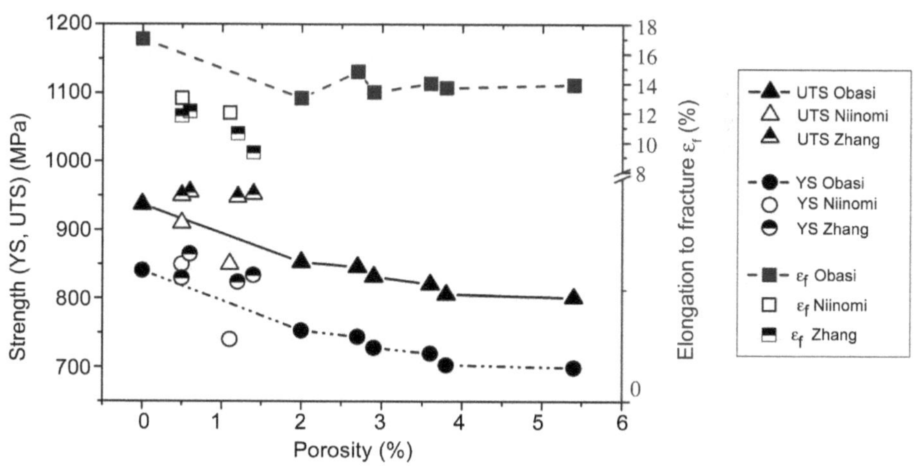

Figure 19.7 Influence of porosity of MIM-processed Ti-6Al-4V on strength and plastic elongation of MIM-processed prealloyed Ti-6Al-4V (Obasi [26]) and blended Ti and 60Al-40V powders (Zhang et al. [23], Niinomi [45]).

19.6.1.2 Influence of oxygen

Interstitials play a dominant role in terms of the influence on mechanical properties of MIM-processed Ti-6Al-4V, and Baril et al. pointed out the importance of low interstitial powders and binders [55]. In principle, the sum of oxygen, nitrogen, and carbon contents is responsible for strengthening and embrittlement processes based on solid solution hardening. For the discussion, it is helpful to introduce an oxygen equivalent according to Conrad [56] as:

$$O_{eq} = 2c_N + c_O + 0.5c_C \quad \text{(Eq. 19.1)}$$

c_N, c_O, and c_C mean the concentrations of nitrogen, oxygen, and carbon, respectively. In practice, oxygen is the most important element, because nitrogen is nearly not taken up if oxygen is present and the influence of carbon is firstly only the half of that of oxygen and secondly the amount is usually significantly lower. Thus, it is allowed to focus on the influence of oxygen on tensile properties as shown in Figure 19.8 [57]. All specimens showed porosity around 3.5%, a carbon content of 0.045 wt%, and a nitrogen concentration around 0.018 wt%.

The diagram shows clearly that the strength increases linearly with increasing oxygen content, while the elongation keeps more or less at a constant high value until it drops down suddenly at a certain limit. This behavior is confirmed by other studies [58,59]. It is important to note that this limit is more than 0.33 wt%, which is far away from the maximum value required according to the ASTM F 2885-11 standard. The fundamental reasons for the highly detrimental influence of oxygen on ductility beyond 0.33 wt% have been clarified recently via a detailed microstructure study [60]. In industrial MIM practice, it is difficult to realize oxygen content below 0.2 wt% to

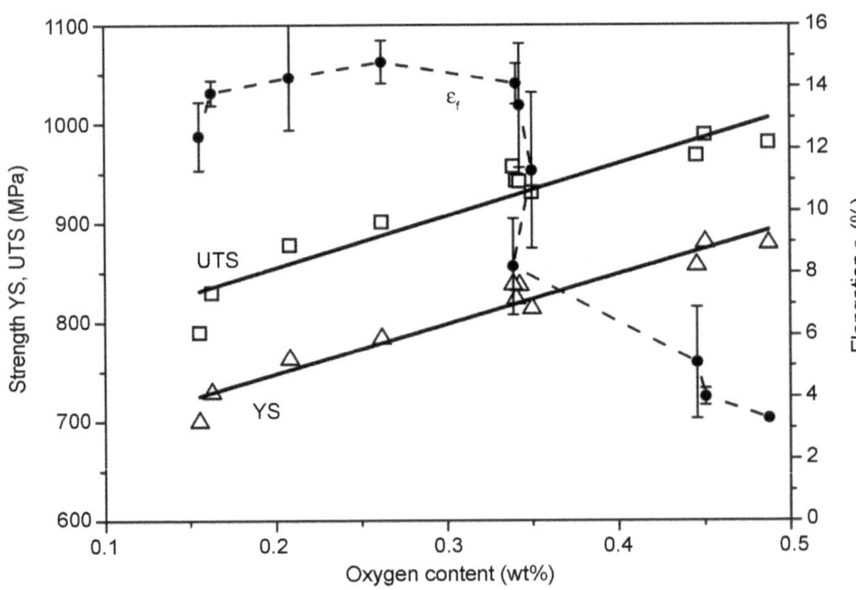

Figure 19.8 Dependence of tensile properties of MIM-processed Ti-6Al-4V on oxygen content [57].

satisfy the standard. On the other hand, Figure 19.8 shows that from the mechanical point of view, more oxygen is beneficial, at least for a quasi-static load.

16.6.1.3 Fatigue

For high fatigue resistance, commonly a material with high yield strength, fine and homogenous microstructure, and low impurity concentration is considered as advantageous, especially with regard to high cycle fatigue. In this case, each irregularity, for example, pores, oxide inclusions, or large and hard precipitates, can act as a site for crack initiation. Ductility can mitigate the effect of such flaws, but only to a certain limit. So, even if MIM allows geometrical freedom to avoid critical shapes, for example, sharp edges, the material itself basically shows all these features detrimental to sound fatigue properties. For instance, pores are an inherent part of the sintering process; reactions with oxygen or carbon can easily lead to the forming of oxides or carbides; during powder handling, foreign particles may be incorporated; and the microstructure tends to be rather coarse as a result of the high temperature sintering process where grain coarsening is expected. Because of these issues, the fatigue resistance of MIM titanium is expected to be somewhere between those of cast and wrought alloys, but maybe tending more toward cast materials.

To date, there are only few published studies on fatigue of MIM-processed titanium alloys, although many more tests may have been performed in institutes or companies. Three published studies on Ti-6Al-4V show similar results for the endurance limit in a range of 300–400 MPa for the as-sintered alloy. While Muterlle et al. [44] concentrated on the effect of shot peening on the fatigue resistance by introduction of residual stress, Ferri et al. [29,49] and Niinomi et al. [45] also investigated the dependence on porosity and microstructure. Their results were very similar, although Ferri et al. characterized the fatigue specimens by means of 4-point-bending tests while Niinomi et al. conducted testing under axial loading. In both studies, different powders and binders were used. In Table 19.4, the results for the 10^7 endurance limit in connection with porosity and tensile properties are summarized.

Considering the fact that different feedstocks, testing geometries, and test setups were applied, the results appear quite comparable. They reveal that as-sintered Ti-6Al-4V already shows sound properties exceeding the fatigue resistance of common cast material. However, applying a shot peening process is highly recommended, as Table 19.4 reveals. The surface of MIM components always shows rather large flaws due to the injection process [30]. The induced residual stress caused by a shot peening process [44] shift the crack initiation away from the surface into the more homogenous underlying material. Surface irregularities are the most important source for crack initiation and are also standardly taken into account in the case of conventionally manufactured components. An additional HIP process improves the fatigue behavior further as a result of the reduction of the number of pores, which can also serve as crack initiation sites.

Further optimization is only possible if the microstructure itself becomes more tolerable to crack initiation, and this is usually done by grain refinement. In the case of wrought material, thermomechanical treatments are applied, creating very fine

Table 19.4 Overview of mechanical properties of MIM-processed Ti-6Al-4V

Condition	Porosity (%)	YS (MPa)	ε_f (%)	Endurance limit (MPa)	Study
As sintered	1.1	740	12	380	Niinomi et al.
As sintered + solution treatment	1.1	920	5	280	Niinomi et al.
As sintered + HIP	0.5	850	13	420	Niinomi et al.
As sintered + hot rolling	0.4	900	3	200	Niinomi et al.
As sintered	n/a	750	10	180	Muterlle
As sintered + shot peening	n/a	810	9	330	Muterlle
As sintered	3.6	720	14	350	Ferri et al.
As sintered + shot peening	3.6	720	14	450	Ferri et al.
As sintered + HIP + shot peening	0.0	841	17	500	Ferri et al.

Data taken from three different studies: Niinomi et al. [45], Muterlle [44], and Ferri et al. [29,49].

microstructures by recrystallization processes. As MIM parts are in the final shape, mechanical treatments cannot be performed. Studies exist for grain refinement utilizing partly sintering under hydrogen atmosphere [61]. In this case, additional Ti-hydride phases and a shift of the beta-transus can lead to a finer microstructure. However, it was not successfully applied to MIM yet.

Another possibility to improve the fatigue resistance is to perform alloy modification. This was also done by Niinomi et al. [45] and Ferri et al. [49] and the results are shown in Table 19.5. While Niinomi et al. added 3 wt% Mo powder, Ferri et al. mixed the prealloyed Ti-6Al-4V powder with elemental boron powder.

Interestingly, the addition of Mo results in a significant increase of the yield strength, but not a corresponding growth in fatigue resistance in the same amount.

Table 19.5 Mechanical properties of MIM-processed or modified Ti-6Al-4V alloys

Alloy, condition	Porosity (%)	YS (MPa)	ε_f (%)	Endurance limit (MPa)	Study
Ti-6Al-4V-3Mo, as sintered	1.1	880	11	385	Niinomi et al. [45]
Ti-6Al-4V-0.5B, as sintered + shot peening	2.3	787	12	640	Ferri et al. [49]

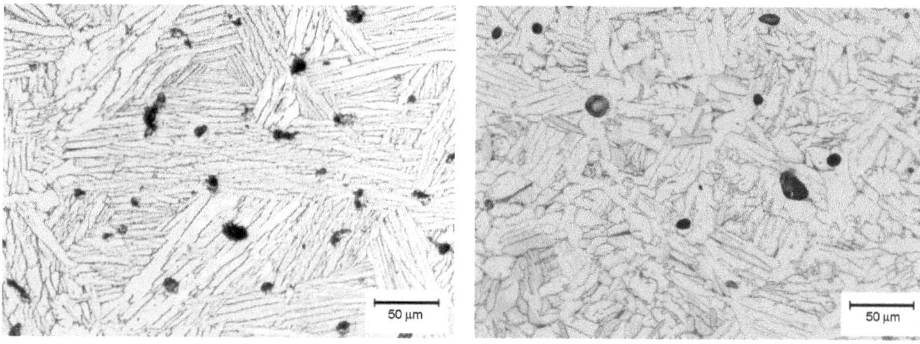

Figure 19.9 Comparison of typical microstructures of MIM-processed Ti-6Al-4V (left) and Ti-6Al-4V-0.5B (right).

In contrast, the addition of boron has a much stronger effect on the endurance limit than on tensile strength. Ferri et al. [49] showed that during heating the boron reacts with titanium, forming TiB. These borides hinder grain growth during sintering as it is known from casting. In addition, the particles serve as additional nucleation sites for alpha phase during cooling. Thus, more and smaller alpha colonies are formed compared to pure Ti-6Al-4V. Actually, the colony size is reduced from 120 to 18 μm. Figure 19.9 shows the microstructures of both alloys after sintering for comparison. It is clear that boron changes the morphology drastically. A recent detailed study of the role of boron in PM Ti and Ti alloys has reported similar observations and discussed the fundamental mechanisms behind it [62].

In summary, the fatigue resistance of MIM-processed Ti-6Al-4V amounts to sound values around 400 MPa. Shot peening is recommended, and HIP improves the properties further. In addition, rather simple alloy modifications can lead to an endurance limit even in the range of wrought material, although porosity is present. For example, the addition of 0.5 wt% boron improves the properties in such a manner that fatigue-loaded applications in aerospace automobile or even for medical implants can be taken into consideration [50].

19.6.2 MIM of beta-titanium alloys

One reason for using metastable beta-titanium alloys is their capability of optimizing mechanical properties like strength or fatigue resistance by heat treatments. Here, the basic mechanism is to control alpha phase forming after fast cooling from the pure beta-state. This makes the alloys suitable for highly loaded applications, for example, in the aerospace sector. Furthermore, beta-titanium alloys show a reduced Young's modulus compared to pure titanium or Ti-6Al-4V, and can be composed completely from biocompatible elements. Even aluminum can be avoided. Therefore, this alloy class is especially attractive for manufacturing permanent bone implants in order to reduce the risk of stress shielding and toxic reactions.

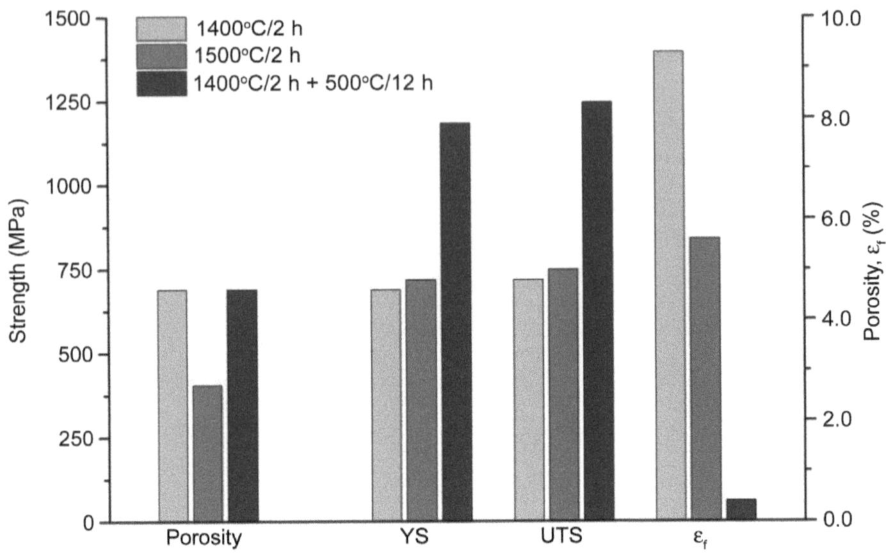

Figure 19.10 Typical tensile properties and porosity of Ti-15V-3Al-3Sn-3Cr processed by MIM under different sintering conditions [63].

There are only few published studies on MIM of beta-titanium alloys [63–66]. However, attempts were made on Ti-15V-3Al-3Sn-3Cr as an example for a technical alloy and Ti-24Nb-4Zr-8Sn and the binary system Ti-Nb as medical alloys. In the following, an overview of the results of these studies is given. In all cases, a binder system based on paraffin-wax, polyethylene-vinylacetate, and stearic acid was used, the same as those applied in Ref. [26] for Ti-6Al-4V feedstock.

Figure 19.10 displays the residual porosity and results from tensile tests applied to MIM-processed specimens made from gas atomized prealloyed Ti-15V-3Al-3Sn-3Cr powder with a particle size smaller than 45 μm [63]. The samples were sintered at two different temperatures under high vacuum for 2 h and then furnace cooled. Some samples were exposed to a subsequent heat treatment at 500°C for 12 h, enabling the partial transformation of beta to alpha phase. As visible in the diagram, an increase in the sintering temperature effects a decrease in residual porosity, resulting in higher strength as expected. However, the ductility decreases, which is probably caused by the higher interstitial content of these specimens. Applying the heat treatment leads to an enormous increase in strength, but also to strong embrittlement. However, the MIM processing leads to sound properties of the material and there is room for optimization by proper heat treatment.

Microscopic analyses showed carbide precipitations at the grain boundaries that are usually absent from MIM-processed Ti-6Al-4V. Such carbides were also observed in studies on beta-titanium alloys based on Nb [64] or Mo [67] as a beta stabilizing element. Table 19.6 shows the results from a study on MIM processing of prealloyed Ti-24Nb-8Sn-4Zr powder and of a study on the binary system Ti-Nb. In the latter

Table 19.6 Microstructural, chemical, and mechanical data of MIM-processed beta-titanium alloys based on the system Ti-Nb

	Porosity (%)	Oxygen (wt%)	Carbon (wt%)	YS (MPa)	UTS (MPa)	ε_f (%)	Young's modulus (MPa)
Ti-10Nb	3.6	0.20	0.056	552	638	10.5	85.2
Ti-16Nb	5.2	0.26	0.060	589	687	3.6	78.4
Ti-22Nb	5.8	0.23	0.059	649	754	1.4	70.9
Ti-22Nb + HIP	0.0	0.23	0.065	687	838	1.3	75.6
Ti-24Nb-4Zr-8Sn	3.9	0.32	0.067	636	656	9.2	53.5

case, elemental powders were blended to form alloys with different Nb concentrations ranging from 10% to 22 wt% [64]. All samples of this study were sintered at 1500°C for 4 h while the Ti-24Nb-8Sn-4Zr specimens were sintered at 1400°C for 4 h.

Obviously, for the binary system, porosity and strength increase with increasing Nb content while the Young's modulus decreases because of a larger amount of stabilised beta phase. However, the elongation to fracture is decreasing, too, and even an additional HIP process yields no improvement. The MIM-processed prealloyed Ti-24Nb-4Zr-8Sn material reveals sound mechanical properties and a very low Young's modulus. Comparing the final porosities of the different alloys, the prealloyed powder appears to be beneficial in terms of sintering compared to the blended elemental approach. Nevertheless, even in the case of the Ti-24Nb-4Zr-8Sn alloy, the ductility is not as high as might be expected [68].

Zhao et al. and Yan et al. [64,67] performed an in-depth investigation of the carbide formation. It is identified that titanium carbides formed, because the carbon solubility of the Ti matrix is reduced by the addition of Nb [64] or Mo [67]. This is the reason why the MIM-processed beta-titanium alloys show carbide precipitation while the Ti-6Al-4V does not, although the overall carbon content of all samples is nearly the same (around 0.06 wt%). The same study also showed that the titanium carbides caused the embrittlement.

In summary, MIM processing of beta-titanium alloys is feasible and the mechanical properties are principally sound; especially high strength is yielded, which can be further increased by heat treatments. However, for effective processing, relatively high sintering temperatures from 1400°C to 1500°C are needed, and the issue of carbide precipitation has to be taken into account. On the other hand, only few studies were performed yet, and there is significant room for optimization.

19.6.3 MIM of titanium aluminides

Titanium aluminides are intended for high-temperature applications, especially in gas turbines, jet engines, motor valves, and turbochargers as a significantly lighter replacement of Ni-based alloys. They consist mainly of two hard and brittle

Figure 19.11 SEM micrograph of the typical microstructure of MIM-processed TNB-V5 (Ti-45Al-5Nb-0.2B-0.2C).

intermetallic phases α_2 and γ. Although titanium aluminides have found commercial applications, as yet no standard technique for their processing has been established. Because of their processing difficulties by conventional techniques, PM techniques like MIM are especially interesting. However, MIM is also challenging, because very high sintering temperatures near the solidus are needed because of the low diffusivity. In addition, there is a strong dependence of the resulting microstructure on the sintering profile, so that processing is rather sophisticated. Nevertheless, over the last decade, sound results for the MIM processing of titanium aluminides are yielded [54,69–71]. Figure 19.11 shows the typical lamellar microstructure of Ti-45Al-5Nb-0.2B-0.2C (at%), also known as TNB-V5. The PA powder was sintered at 1500°C for 2 h under high vacuum. After sintering, the residual porosity is just around 0.2% and the colony size amounts to 80 μm.

Impurity control during MIM processing is rather difficult. However, the oxygen content could be limited to 0.12 wt% [54], and the corresponding tensile properties are shown in the Table 19.7. The tensile tests were performed at room temperature as well as at 700°C and are compared with typical values of cast materials. The results are nearly equivalent, although the cast samples were HIPed, machined to standard geometry, and polished, whereas the MIM-processed specimens were tested as-sintered.

Table 19.7 Results from tensile tests on MIM-processed and cast TNB-V5 alloy at room temperature (RT) and at 700°C [54]

Temperature	Sample	UTS (MPa)	ε_f (%)
RT	MIM	630	0.2
	Cast	745	0.1
700°C	MIM	650	1.0
	Cast	720	1.4

In Figure 19.12, the corresponding tensile test diagram of the MIM samples is shown, revealing the ductile behavior of the as-sintered material.

Compression creep tests of TNB-V5 specimens produced by MIM and arc melting, respectively, reveal that especially for low strain values, there is only a slight difference between sintered and melted materials (Figure 19.13). This shows the great potential of MIM for the production of fast-rotating components exposed to high temperatures, such as turbine blades or turbocharger wheels.

To sum up, MIM of titanium aluminides is possible; however, there are some issues, especially in terms of processing parameters, which have to be investigated in more detail before industrial introduction.

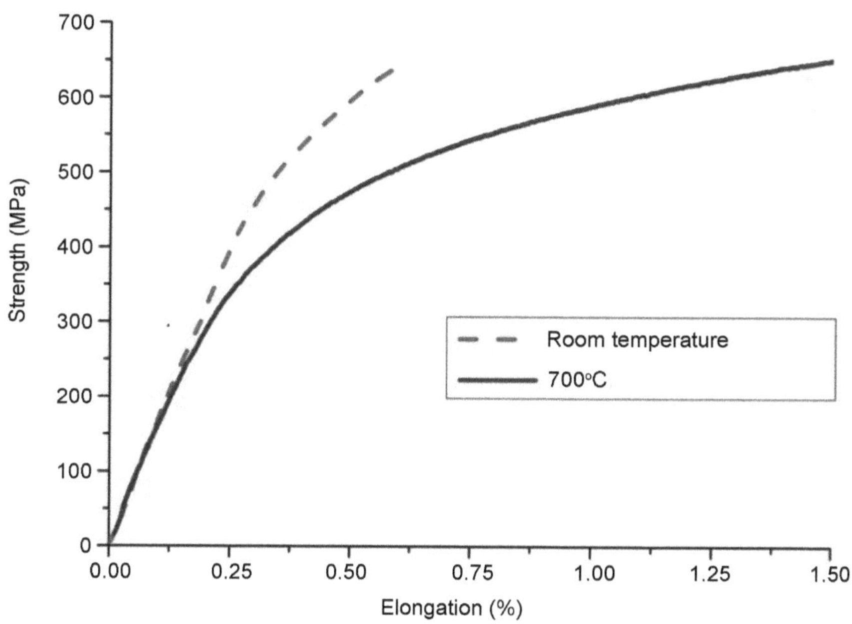

Figure 19.12 Tensile test diagram of TNB-V5, measured at room temperature and at 700°C.

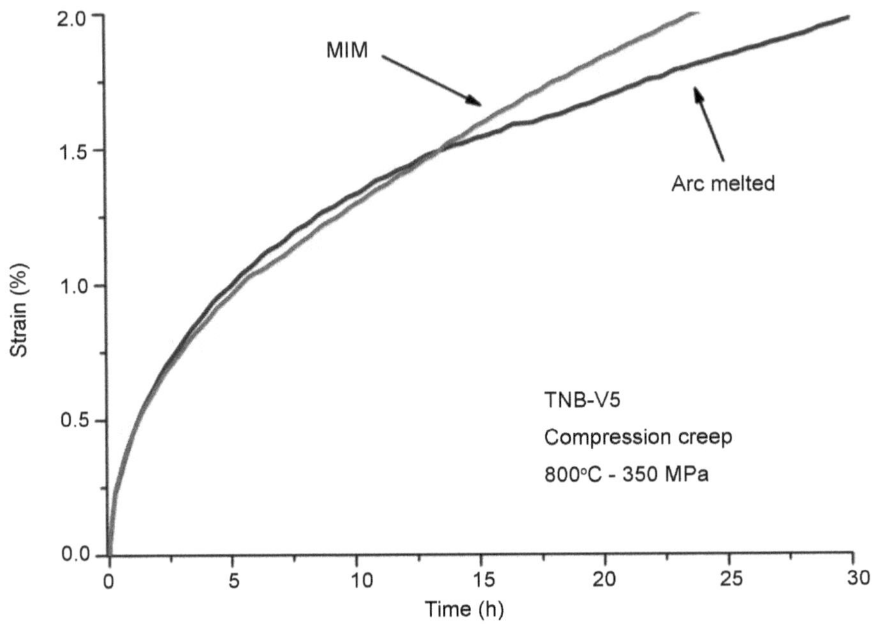

Figure 19.13 Comparison of TNB-V5 specimens made by MIM and arc melting in terms of compression creep properties.

19.7 Perspectives

MIM, even though still not a standard manufacturing technique for titanium and its alloys, is increasingly used commercially each year. Today, excellent mechanical properties are achievable and even fatigue properties are sound.

There are still concerns about applications for aerospace components, because of residual porosity and possible production flaws. However, the technique is used for medical components, even permanent implants, and the two existing ASTM standards target exactly this market. MIM components are already used for aero-engines, made from steel. MIM titanium components are already under testing and may be introduced in the near future for critical applications. The dramatically growing market for additive manufacturing could assist in such developments, because the use of powders could become more usual, and more experiences will be gathered in shorter time.

The growing usage of carbon fiber–reinforced polymer (CFRP) might be a new field for titanium MIM. Aluminum, used as the main metallic structural material for lightweight applications, tends to corrode if in contact with CFRP. Thus, connectors made from titanium may be a solution, and MIM might play an important role in this field.

There has been a lot of research performed, for example, on MIM-specific alloys with optimized properties, on porous components for improved medical implants and also related to the use of cheaper powders. It can be expected that a success of titanium MIM in the demanding applications of medical and aerospace industries and their standards will also accelerate the use in other applications. This would be a further impetus to supply good and maybe cheaper powder and allow more cost-effective applications. A significant boost of titanium MIM may occur within the next decade.

References

[1] R. German, A. Bose, Powder injection molding of metals and ceramics, Metal Powder Industries Federation, Princeton, NJ, 1997.

[2] D.F. Heaney (Ed.), Handbook of Metal Injection Moulding, Woodhead Publishing Limited, Sawston, Cambridge, UK, 2012.

[3] R. German, Markets applications, and financial aspects of global metal powder injection moulding (MIM) technologies, Metal Powder Report 67(1) (2012) 18–26.

[4] R. German, Progress in titanium metal powder injection molding, Materials 6 (2013) 3641–3662.

[5] ASTM F2885-11, Standard specification for metal injection molded titanium-6aluminium-4vanadium components for surgical implant applications.

[6] ASTM F2989-13, Standard specification for metal injection molded unalloyed titanium components for surgical implant applications.

[7] D.M. Brunette, et al., Titanium in Medicine, Springer Verlag Berlin, Heidelberg, 2001, p. 33.

[8] V. Frederici, M. Ellerhorst, P. Imgrund, S. Krämer, N. Ludwig, Metal Injection Moulding of Thin-Walled Titanium Parts for Medical Applications, in: Proceedings of EuroPM 2013, vol. 3, Gothenburg, Sweden, September 16–18, 2013, 265–27.

[9] E. Carreño-Morelli, W. Krstev, B. Romeira, M. Rodriguez-Arbaizar, J.-E. Bidaux, S. Zachmann, Powder Injection Moulding of Titanium from TiH_2 Powders, in: Proceedings of EuroPM 2009, vol. 2, Copenhagen, Denmark, October 12–14, 2009, pp. 9–14.

[10] E. Carreño-Morelli, J.-E. Bidaux, M. Rodriguez-Arbaizar, H. Girard, H. Hamdan, "Titanium Grade 4, by Powder Injection Moulding of Titanium Hydride," Proceedings of EuroPM2011, vol. 2, Barcelona, Spain, October 9–12, 2011, pp. 105–110.

[11] www.polymim.com.

[12] www.catamold.com.

[13] K.S. Weil, E. Nyberg, K. Simmons, A new binder for powder injection molding titanium and other reactive metals, J. Mat. Proc. Tech. 176 (1–3) (2006) 205–209.

[14] T. Kono, A. Horata, T. Kondo, Development of titanium & titanium alloy by metal injection molding process, Powder and Powder Metal 44 (11) (1997).

[15] E. Aust, W. Limberg, R. Gerling, B. Oger, T. Ebel, Advanced TiAl6Nb7 bone screw implant fabricated by metal injection moulding, Adv. Eng. Mater. 8 (2006) 265–370.

[16] Y. Itoh, T. Harikou, K. Sato, H. Miura, Improvement of Ductility for Injection Moulding Ti-6Al-4V Alloy, in: Proceedings of PM World Conference 2004, vol. 4, Vienna, Austria, October 17–21, 2004, pp. 445–450.

[17] H. Nakamura, T. Shimura, K. Nakabayashi, Process for production of Ti sintered compacts using the injection molding method, J. Jpn. Soc. Powder Metall. 46 (8) (1999).

[18] K. Simmons, K.S. Weil, E. Nyberg, Powder injection molding of titanium compounds, in: Industrial Heating, December 2005, p. 43.

[19] R. Ibrahim, M. Azmirruddin, M. Jabir, M.R. Ismail, M. Muhamad, R. Awang, S. Muhamad, Injection molding of titanium alloy implant for biomedical application using novel binder system based on palm oil derivatives, AJAC, 7 (6) (2010) 811–814.

[20] R. German, Titanium powder injection moulding, Powder Inj. Mould. Int. 3 (4) (2009) 21–37.

[21] B. Oger, T. Ebel, W. Limberg, The Manufacture of Highly-Ductile and Geometrically Complex MIM-Parts Based on TiAl6V4, in: Proceedings of EuroPM 2006, vol. 2, Gent, Belgium, October 23–25, 2006.
[22] S. Guo, X.U.J. Xiang, R. Ang, X. He, M. Li, S. Duo, W. Li, Effect of annealing processing on microstructure and properties of Ti-6A1-4V alloy by powder injection molding, Trans. Nonferrous Met. Soc. China 16 (2006) 701–704.
[23] R. Zhang, J. Kruszewski, J. Lo, A study of the effects of sintering parameters on the microstructure and properties of PIM Ti6Al4V alloy, PIM International 2 (2) (2008) 74–78.
[24] E. Nyberg, M. Miller, K. Simmons, K.S. Weil, Microstructure and mechanical properties of titanium components fabricated by a new powder injection molding technique, Mat. Sci. Eng. C 25 (2005) 336–342.
[25] G. Shibo, Q. Xuanhui, H. Xinbo, Z. Ting, D. Bohua, Powder injection molding of Ti-6Al-4V alloy, J. Mat. Proc. Tech. 173 (2006) 310–314.
[26] G.C. Obasi, O.M. Ferri, T. Ebel, R. Bormann, Influence of processing parameters on mechanical properties of Ti-6Al-4V alloy fabricated by MIM, Mat. Sci. Eng. A 527 (2010) 3929–3935.
[27] H. Miura, Y. Itoh, T. Ueamtsu, K. Sato, The Influence of Density and Oxygen Content on the Mechanical Properties of Injection Molded Ti-6Al-4V alloys, in: Proceedings of the 2010 International Conference on Advances in Powder Metallurgy and Particulate Materials, Hollywood, USA, vol. 4, June 27–30, 2010, pp. 46–53.
[28] A.T. Sidambe, W.L. Choong, H.G.C. Hamilton, I. Todd, Correlation of metal injection moulded Ti6Al4V yield strength with resonance frequency (PCRT) measurements, Mat. Sci. Eng. A 568 (2013) 220–227.
[29] O.M. Ferri, T. Ebel, R. Bormann, High cycle fatigue behaviour of Ti-6Al-4V fabricated by metal injection moulding technology, Mat. Sci. Eng. A 504 (2009) 107–113.
[30] O.M. Ferri, T. Ebel, R. Bormann, Influence of surface quality and porosity on fatigue behaviour of Ti-6Al-4V components processed by MIM, Mat. Sci. Eng. A 527 (2010) 1800–1805.
[31] M. Holm, T. Ebel, M. Dahms, Investigations on Ti-6Al-4V with gadolinium addition fabricated by metal injection moulding, Mater. Design 51 (2013) 943–948.
[32] W. Dietzel, T. Ebel, V. Heitmann, I. Scheider, Stress Corrosion Cracking Studies of MIM Ti-Al6-V4, in: Proceedings of Fracture Mechanics of Materials, 2nd Ukrainian-Greek Symposium, Lviv, Ukraine, October 03–07, 2011, pp. 57–60.
[33] S. Guo, X. Qu, X. He, Influence of sintering temperature on mechanical properties of Ti-6Al-4V compacts by metal injection molding, Mat. Sci. Forum 475–479 (2005) 2639–2642.
[34] K. Kusaka, T. Kono, A. Horata, T. Kondo, Tensile behavior of sintered Ti and Ti-6Al-4V alloy by MIM process, in: Advances in Powder Metallurgy and Particulate Materials-1996, MPIF: Princeton, NJ, vol. 19, 1996, pp. 127–131.
[35] H. Wohlfromm, M. Blomacher, D. Weinand, E. Langer, M. Schwarz, Novel Materials in Metal Injection Moulding, in: Powder Injection Moulding, Proceedings of the First European Symposium on Powder Injection Moulding, European Powder Metallurgy Association: Shrewsbury, UK, 1997, pp. 54–61.
[36] K. Maekawa, M. Takita, H. Nomura, Effect of MIM process conditions on microstructures and mechanical properties of Ti-6Al-4V compacts, J. Jpn. Soc. Powder Powder Metall. 46 (1999) 1053–1057.
[37] Y. Itoh, T. Uematsu, K. Sato, H. Miura, M. Niinomi, Fabrication of high strength alpha plus beta type titanium alloy compacts by metal injection molding, J. Jpn. Soc. Powder Powder Metall. 55 (2008) 720–724.

[38] T. Uematsu, Y. Itoh, K. Sato, H. Miura, Effects of substrate for sintering on the mechanical properties of injection molded Ti-6Al-4V alloy, J. Jpn. Soc. Powder Powder Metall. 53 (2006) 755–759.
[39] H. Miura, T. Takemasu, Y. Kuwano, Y. Itoh, K. Sato, Sintering behavior and mechanical properties of injection molded Ti-6Al-4V alloys, J. Jpn. Soc. Powder Powder Metall. 53 (2006) 815–820.
[40] S. Guo, X. Qu, X. He, T. Zhou, B. Duan, Powder injection molding of Ti-6Al-4V alloy, J. Mater. Proc. Technol. 173 (2006) 310–314.
[41] R.M. Gomes, L. Kowalski, L. Schaeffer, J. Duszczyk, Injection Molding of Blended Elemental Ti (6Al-4V) Powders, in: Proceedings of the 2000 Powder Metallurgy World Congress, Kyoto, Japan, November 12–16, 2000, pp. 324–327.
[42] H. Miura, Y. Itoh, T. Ueamtsu, K. Sato, The influence of density and oxygen content on the mechanical properties of injection molded Ti-6Al-4V alloys, in: Advances in Powder Metallurgy and Particulate Materials-2010; MPIF: Princeton, NJ, vol. 4, 2010, pp. 46–53.
[43] J.A. Sago, M.W. Broadley, J.K. Eckert, Metal injection molding of alloys for implantable medical devices, Int. J. Powder Metall. 482 (2012) 41–49.
[44] P.V. Muterlle, A. Molinari, M. Perina, P. Marconi, Influence of Shot Peening on Tensile and High Cycle Fatigue Properties of Ti6Al4V Alloy Produced by MIM, in: Proceedings of the PM2010 World Conference, Florence, Italy, vol. 4, Shrewsbury, EPMA, 2010, pp. 791–796.
[45] M. Niinomi, T. Akahori, M. Nakai, K. Ohnaka, Y. Itoh, K. Sato, T. Ozawa, Mechanical properties of α+β type titanium alloys fabricated by metal injection molding with targeting biomedical applications, in: M.N. Gungor, M.A. Imam, F.H. Froes (Eds.), Innovations in Titanium Technology, Wiley (2007 TMS Annual Meeting & Exhibition, Orlando, February 25–March 1, 2007) pp. 209–217.
[46] H. Miura, Y. Itoh, T. Uematsu, K. Asto, Advanced PIM Process of Ti-6Al-7Nb for Materials for Medical Applications, in: R.M. German, (Ed.), Proceedings of the Workshop on Medical Applications for Microminiature Powder Injection Molding [CD], Orlando, FL, March 2–5, 2009; MPIF: Princeton, NJ, 2009; pacts. J. Jpn. Soc. Powder Powder Metall. 46 (1999) 1053–1057.
[47] T. Osada, H. Miura, Y. Itoh, M. Fujita, N. Arimoto, Optimization of MIM process for Ti-6Al-7Nb alloy powder, J. Jpn. Soc. Powder Powder Metall. 55 (2008) 726–731.
[48] Y. Itoh, H. Miura, T. Uematsu, K. Sato, M. Niinomi, Improvement of the Properties of Ti-6Al-7Nb Alloy by Metal Injection Molding, in: Advances in Powder Metallurgy and Particulate Materials-2007; MPIF: Princeton, NJ, vol. 4, 2007, pp. 81–86.
[49] O.M. Ferri, T. Ebel, R. Bormann, The influence of a small boron addition on the microstructure and mechanical properties of Ti-6Al-4V fabricated by metal injection moulding, Adv. Eng. Mat. 13 (5) (2011) 436–447.
[50] T. Ebel, C. Blawert, R. Willumeit, B.J.C. Luthringer, O.M. Ferri, F. Feyerabend, Ti-6Al-4V-0.5B–A modified alloy for implants produced by metal injection molding, Adv. Eng. Mat. 13 (12) (2011) B 440 – B 453.
[51] Y. Itoh, T. Uematsu, K. Sato, H. Miura, M. Niinomi, Microstructural modification for injection molded Ti-6Al-4V alloys by addition of Mo powder, J. Jpn. Soc. Powder Powder Metall. 53 (2006) 750–754.
[52] M. Scharvogel, W. Winkelmueller, Metal injection molding of titanium for medical and aerospace applications, J. Met. Mater. Min. 63 (2011) 94–96.
[53] R.M. German, Market and Technology for Titanium Metal Powder Injection Moulding, in: Proceedings PM 2010 World Congress [CD], Florence, Italy, October 10–14, 2010; European Powder Metallurgy Association: Shrewsbury, UK, 2010.

[54] W. Limberg, T. Ebel, F. Pyczak, M. Oehring, F.P. Schimansky, Influence of the sintering atmosphere on the tensile properties of MIM processed Ti 45Al 5Nb 0.2B 0.2C, Mat. Sci. Eng. A 552 (2012) 323–329.
[55] E. Baril, L.P. Lefebvre, Y. Thomas, Interstitial elements in titanium powder metallurgy: sources and control, Powder Metall. 54 (3) (2011) 183–187.
[56] H. Conrad, The rate controlling mechanism during yielding and flow of titanium at temperatures below 0.4 TM, Acta Met. 14 (1966) 1631–1633.
[57] T. Ebel, O.M. Ferri, W. Limberg, M. Oehring, F. Pyczak, F.P. Schimansky, Metal injection moulding of titanium and titanium-aluminides, Key Eng. Mat. 520 (2012) 153–160.
[58] E. Baril, Titanium and titanium alloy powder injection moulding – matching application requirements, Powder Inject. Mould. Int. 4 (2010) 22–32.
[59] H. Miura, H. Kang, Y. Itoh, High performance titanium alloy compacts by advanced powder processing techniques, Key Eng. Mat. 520 (2012) 30–40.
[60] M. Yan, M.S. Dargusch, T. Ebel, M. Qian, A transmission electron microscopy and three-dimensional atom probe study of the oxygen-induced fine microstructural features in as-sintered Ti-6Al-4V and their impacts on ductility, Acta Mater. 68 (2014) 196–206.
[61] Z.Z. Fang, P. Sun, H. Wang, Hydrogen sintering of titanium to produce high density fine grain titanium alloys, Adv. Eng. Mat. 14 (6) (2012) 383–387.
[62] Y.F. Yang, M. Yan, S.D. Luo, G.B. Schaffer, M. Qian, Modification of the α-Ti laths to near equiaxed α-Ti grains in as-sintered titanium and titanium alloys by a small addition of boron, J. Alloys Compd. 579 (2013) 553–557.
[63] T. Ebel, O.M. Ferri, Processing of Ti-15V-3Al-3Sn-3Cr by Metal Injection Moulding, in: Proceedings of EuroPM 2011, vol. 2, Barcelona, Spain, October 9–12, 2011, pp. 265–270.
[64] D. Zhao, K. Chang, T. Ebel, M. Qian, R. Willumeit, M. Yan, F. Pyczak, Microstructure and mechanical behavior of metal injection molded Ti–Nb binary alloys as biomedical material, J. Mech. Beh. Biomed. Mat. 28 (2013) 171–182.
[65] J. Takekawa, N. Sakurai, Effect of processing conditions on density, strength and microstructure of Ti-12Mo alloy fabricated by PIM process, J. Jpn. Soc. Powder Powder Metall. 46 (1999) 877–881.
[66] J.-E. Bidaux, C. Closuit, M. Rodriguez-Arbaizar, D. Zufferey, E. Carreño-Morelli, Powder Inject, Mould. Int. 6 (2012) 72–75.
[67] M. Yan, M. Qian, C. Kong, M.S. Dargusch, Impacts of trace carbon on the microstructure of as-sintered biomedical Ti–15Mo alloy and reassessment of the maximum carbon limit, Acta Biomater. 10 (2) (2014) 1014–1023.
[68] L.C. Zhang, D. Klemm, J. Eckert, Y.L. Haod, T.B. Sercombe, Manufacture by selective laser melting and mechanical behavior of a biomedical Ti-24Nb-4Zr-8Sn alloy, Script. Mater. 65 (2011) 21–24.
[69] H. Zhang, X. He, X. Qu, L. Zhao, Microstructure and mechanical properties of high Nb containing TiAl alloy parts fabricated by metal injection molding, Mat. Sci. Eng. A 526 (2009) 31–37.
[70] Y.C. Kim, S. Lee, S. Ahn, N.J. Kim, Application of metal injection molding process to fabrication of bulk parts of TiAl intermetallics, J. Mater. Sci. 42 (2007) 2048–2053.
[71] R. Gerling, E. Aust, W. Limberg, M. Pfuff, F.P. Schimansky, Metal injection moulding of gamma titanium aluminide alloy powder, Mat. Sci. Eng. A 423 (2006) 262–268.

Powder-processing linkages to properties for complex titanium shapes by injection molding

20

Randall M. German
Professor, Mechanical Engineering, San Diego State University, San Diego, CA, USA

20.1 Introduction

Metal powder injection molding is a hybrid between powder metallurgy and plastic molding. Titanium component production by the metal powder injection molding process (termed Ti-MIM) has advanced considerably based on several reports detailing the powders, binders, debinding, sintering, and post-sintering details [1–150]. The intent here is to summarize those significant gains. The progress relies on commercially available clean, spherical (for good flow and packing characteristics) titanium alloy powders in the particle size required for molding, sintering densification, and minimized contamination. With major advances in powder fabrication and binder selection, Ti-MIM has converged on spherical alloy particles below 45 μm (325 mesh) to balance sintering densification (smaller particles are beneficial), impurity accumulation (large particles are desired), and component shape retention (small particles are required). With the advances in powder production customized to Ti-MIM, subsequent development evaluated how powders' characteristics affected feedstock rheology and component sintering. Progress in powder atomization provides an infrastructure enabling Ti-MIM. This allows qualification for medical or aerospace applications that require high-purity titanium components [114–143,150].

Depending on powder and process quality, the applications for Ti-MIM fall into three general categories:

1. Decorative items where mechanical and other properties are not demanding, such as in watch cases;
2. Mechanical components where mechanical and corrosion properties must exceed that of a stainless steel such as in medical surgical tools;
3. Life critical applications where titanium is needed for success such as aerospace components and biomedical implants.

Advances in powder synthesis were required for Ti-MIM applications in demanding fields. Accordingly, attention to the powder quality and process details leads to these three tiers. Decorative items are generally contingent on a marketing advantage as evident when Ti-MIM sunglass frames emerged using hydride–dehydride titanium powders. Other decorative successes occurred as cell phone and firearm components. Mechanical components were evident in high-value consumer products, such as

Titanium Powder Metallurgy. http://dx.doi.org/10.1016/B978-0-12-800054-0.00020-4

camera, automotive, firearm, and cellular telephone products. Subsequently, Ti-MIM moved into taxing applications for biomedical implants and aerospace components. Generally the components are small, since the powder and processing are expensive compared to investment casting; most economic successes are 10 g or less for slender components [107,139]. This maturation required clear links between the properties, powder, binder, and process to enable optimization as detailed below.

20.2 Powders for Ti-MIM

Quoted prices for Ti-MIM powders depend on the impurity level, alloying, particle size, and purchase quantity. Supply levels vary depending on other uses for titanium powders, resulting in swings from inadequate supply to excess inventory. High-quality spherical titanium alloy powders for MIM currently are selling for \$150–225 per kg. Lower quality powders are half that price, but carry higher interstitial levels and surface asperities that degrade packing. Low-cost powders, such as hydride–dehydride or sponge fines, are available for as little as \$35 per kg. A disadvantage of nonspherical powders is increased molding difficulty (packing density and flow characteristics).

Demanding applications require control over the impurities at every step in powder production and feedstock preparation. For example, Baril et al. [141] details the sources of oxygen contamination during MIM processing. Final oxygen levels reflect the starting powder purity. There is little opportunity to reduce oxygen from the starting powder level. Thus, the common dictate is to start with powder below the required final oxygen content. Oxygen increases with exposure to binders, debinding (process of binder extraction), and sintering. Generally with clean processing, the oxygen increase is about 0.04 wt%, some of which is traced to oxide contaminants in the binders. Most commonly, the increase in oxygen and carbon from the powder to the final sintered product is 0.15 wt% for oxygen and 0.05 wt% for carbon. Thus, starting with 0.2 wt% oxygen powder gives a realistic final oxygen content near 0.3 wt%. For a watch case, there is less concern over oxygen, and indeed the higher hardness improves wear resistance. On the other hand, medical biocompatible implant applications require low oxygen and carbon levels.

To minimize contamination, the decision is to use a larger particle size versus traditional systems, largely to reduce surface area and oxygen contamination. A few powder suppliers produce powders with 0.2 wt% interstitial levels at prices that depend on the quantity, particle size, tap density, and oxygen level – in the \$110–220 per kg range. This pricing is competitive with gas atomized cobalt-chromium powder at \$160 per kg. Note, on a volume basis, titanium is lower in cost (density of 8.4 g/cm^3 for cobalt-chromium vs. 4.5 g/cm^3 for Ti-6Al-4V). However, on a cost basis, titanium will not be able to displace biomedical stainless steels.

The powder offerings for Ti-MIM fall in the following categories; sponge fines, gas atomization, centrifugal atomization (rotating electrode), plasma atomization, hydride-milled-dehydride (HDH), mechanically spheroidized variants, and novel routes such as electrochemical, calcium reduction, and ultrasonic chemical reduction. In spite of much attention, there is difficulty ramping production to large quantities

Table 20.1 Characteristics of titanium powders used for metal injection molding

Powder type	Median particle size (μm)	Tap density of pycnometer (%)	Oxygen (wt%)	Carbon (wt%)
Sponge fines	38	48	0.35	0.05
Hydride–dehydride	38	38	0.25	0.04
Titanium hydride	35	40	0.20	0.02
Reactive	30	47	0.30	0.10
Gas atomized	32	60	0.15	0.03
Plasma atomized	60	62	0.15	0.04
Rotating electrode	130	72	0.15	0.02

without degrading quality. Table 20.1 provides a sampling of the current powders available for Ti-MIM showing key characteristics. The powders are specified by the following characteristics:

- Particle size distribution (quantified by the median particle size or D_{50});
- Particle shape (quantified by the tap density);
- Interstitial level (quantified by oxygen and carbon levels).

In addition, the powders must be free of internal voids and nonagglomerated. The common grades used in injection molding are spherical, below 45 μm, and alloyed (Ti-6Al-4V). Figure 20.1 shows an example of one newly offered powder made by plasma atomization of a hydride–dehydride raw powder.

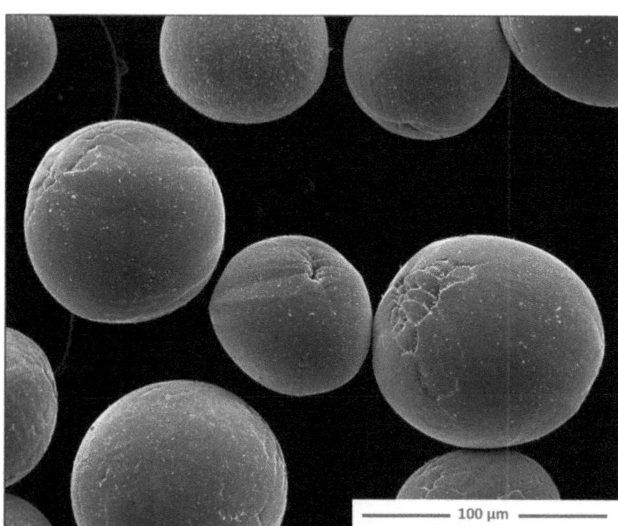

Figure 20.1 Titanium spherical powder formed using plasma atomization of a HDH powder.

One means to adjust rheology is by mixing different powder types. For example, HDH powders are mixed with gas atomized powders. The angular HDH powder adds interparticle friction to resist distortion during debinding. Solids loadings up to 72 vol% have been realized. Unfortunately, the HDH powders are higher in impurities, resulting in an increase in the final oxygen level as HDH is added to the feedstock. In experiments using gas atomized powder with 0.16 wt% oxygen and HDH powder with 0.23 wt% oxygen, after sintering the material derived from gas atomized powder alone produced 550 MPa tensile strength with 23% elongation, while the HDH powder by itself resulted in 710 MPa tensile strength and 8% elongation. The mixtures were intermediate in mechanical properties and interstitials.

To minimize impurities, powder selection favors a large particle size with less surface area. For example, Chen et al. [102] examined different particle sizes of HDH powder using a wax–polymer binder and two-step (solvent and thermal) debinding, followed by vacuum sintering at 1350°C for 90 min. The smaller particle size required a higher molding pressure, gave slower debinding, and resulted in a higher sintered density. The impurity level after sintering was high. The penalty with a larger powder is less dimensional precision, slower sintering densification, and a rough sintered surface.

A typical compromise in Ti-MIM is to use −45 μm, spherical or tumbled, powder with a high tap density starting below 0.20 wt% oxygen and below 0.05 wt% carbon. One option is to use hydride powders, where hydrogen liberated during sintering helps remove volatile impurities, especially those arising from the sintering atmosphere. The sintering of titanium in hydrogen has been used sporadically since the 1990s, with vacuum cooling to prevent hydriding the sintered body. One major disadvantage comes from the multiple temperature and gas steps required to effectively manipulate hydride, metal, and phase transformation. But properties for Ti-6-4 reported reach 940 MPa yield strength, 1035 MPa tensile strength, and 15% elongation, with about ±13 MPa data scatter on strength and ±4% on elongation.

Titanium powders do not react with oxygen during short duration air exposures at room temperature. Unfortunately, once titanium powder ignites, the reaction propagates with near-explosive speed reaching up to 100 MPa/s pressure rise. First oxidation occurs near 400°C, comparable to the peak temperature encountered during debinding. If a powder is sparked, then a fire or explosion is possible. Although not measured, it is anticipated that Ti-MIM powders are similar to aluminum and zirconium powders, which explode when dispersed at concentrations greater than 40 g/m^3. The greatest danger is during discharge from mixers, storage containers, or milling devices.

20.3 Key Ti-MIM success factors

Four parameters dominate the mechanical properties of sintered titanium – density, interstitial content, alloying, and microstructure. Corrosion and biomedical attributes are contingent on the same factors. Together, optimal properties arise when the powder is sintered to a high density, with little contamination, with alloying (such as Nb), and consolidated under conditions that avoid microstructure coarsening.

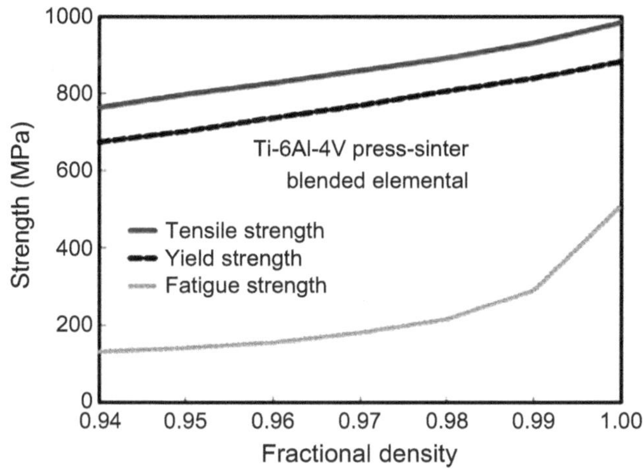

Figure 20.2 A plot of the mechanical properties for sintered Ti-6Al-4V from mixed powders versus fractional density showing the sensitivity of fatigue behavior as compared to tensile and yield strength.

The first point is final density. Residual pores degrade mechanical properties, so full density is desirable [114]. A classic demonstration for titanium is given in Figure 20.2 where tensile strength, yield strength, and fatigue strength are plotted for Ti-6Al-4V. Gains are evident with densification; in this plot, the yield strength changes 30% as the density increases from 94% to 100%, but the fatigue strength increases 400%. Ductility improves with density and densification and is also sensitive to the test geometry [125]; round bars give a higher ductility. Likewise, surface porosity degrades properties, so hot isostatic pressing (HIP) and shot peening are used to improve fatigue strength, usually by 100–200 MPa [127,148]. During sintering, residual gas generates pores in the sintered body, reaching pore diameters up to 80 μm [124]. Fracture toughness and fatigue strength are sensitive to these residual pores, more so than tensile strength. As an example, fatigue strength jumps 18% with the elimination of the last 2% porosity by HIP [126]. Containerless hot isostatic pressing is a common means to attain full density after sintering.

The second point with respect to Ti-MIM is the interstitial content. Interstitial oxygen, carbon, nitrogen, or hydrogen increase yield strength, tensile strength, and hardness, but decrease ductility and corrosion resistance. For sintered titanium, oxygen is the focus, leading to a sorting of alloy grades based on the oxygen content. Unalloyed commercially pure grade-1 titanium (CP Ti) must have below 1800 ppm oxygen, resulting in a tensile strength of 240 MPa with 24% fracture elongation. At higher oxygen levels, such as for grade-4 titanium, the oxygen ranges to 4000 ppm with tensile strength exceeding 550 MPa, but the ductility declines to 15% elongation. This latter level of properties is attainable using titanium hydride as the starting material.

In spite of much effort on applying advanced polymer science, the binders proving most successful are variants on the simple systems first patented in the 1970s – usually a wax phase (paraffin wax or polyethylene glycol), a backbone phase (polypropylene or ethylene vinyl acetate or a mixture of these), a surfactant/die release/plasticizer phase (stearic acid or similar small molecule) [111,114,118,147]. Solvent debinding where most of the binder is dissolved out at low temperatures is most successful in avoiding contamination [32,42,123]. The wax phase fills space and is easily extracted by solvents (heptane for paraffin wax or water for polyethylene glycol), and the backbone is intentionally evaporated during heating to the sintering temperature. Other systems are known [8,20,23,46–57]. Today several firms sell premixed feedstock using spherical prealloyed titanium powder in a compounded homogeneous mixture pelletized for direct molding.

The control of interstitials is a difficulty in sintered titanium. This is because the impurities are soluble at the sintering temperature and there are no effective reducing agents. Thus, contamination arriving with the powder is increased by furnace and substrate sources. Since oxygen is a major concern, it is common to denote an oxygen equivalent impurity level, where each impurity is assigned a weighting factor with respect to property changes when compared to oxygen. Strength increases linearly with the oxygen equivalent [126].

The third point with respect to Ti-MIM is alloying. The main compositions are copies of wrought alloys. Several alternative compositions exist with molybdenum, iron, chromium, niobium, zirconium, and boron alloying, but for commercial Ti-MIM the focus is on three alloys, CP Ti, Ti-6Al-4V, and Ti-6Al-7Nb. By far, the most common alloy is Ti-6-4. When taken to full density with less than 2000 ppm or 0.2 wt% oxygen, this alloy delivers tensile strength between 710 and 880 MPa with about 12% elongation (note the typical scatter in multiple samples is about 15 MPa on strength and 3% on elongation). Subsequent hot isostatic pressing delivers more than 1030 MPa tensile strength with 15% elongation [149]. Hot isostatic pressing is one means to lower the final temperature, limiting grain coarsening, while attaining full density. The higher sintering temperature coarsens the grain size, resulting in a lower strength and similar ductility. Alloying with boron is advocated to control microstructure during sintering [122,126,127], but this is outside the composition "equivalent" window and must be newly qualified for each application. Even so, Ti-6-4 products formed by injection molding reach yield strengths of 787 MPa with 0.5 wt% boron and 1400°C sintering. Most impressive is the fatigue endurance limit of 640 MPa. Although a few alloys have been developed just for Ti-MIM to compensate for the processing sensitivities, as yet they have not been put into production.

Clearly, the sintered tensile strength varies with the powder choice, density, impurity, and processing details. Without deviating from handbook of chemistries, generally the sintered material reaches a strength equivalent to wrought, near 950 MPa with 14% elongation. The Ti-MIM material reflects a coarse-grained microstructure concomitant with the time–temperature combinations required for sintering densification.

As noted already, the final factor is that the microstructure revolves around the grain size and phases present after sintering. The microstructure coarsens during sintering to reduce yield strength compared to wrought titanium. Accordingly, one option is to sinter to the closed-pore condition at about 95% density, then rely on lower-temperature HIP for final densification.

When Ti-MIM is fabricated with attention to these pivoting features including powder quality, binder, interstitial control, final densification, alloying, and microstructure coarsening, the mechanical properties approach the annealed wrought values. Unfortunately, without alloying and without optimal starting powder, the optimized properties are considerably degraded [138]. The balance of this chapter reviews the scientific steps needed to accomplish this success.

20.4 Optimized Ti-MIM processing

The processing science for titanium traces to the early availability of titanium in a powder form. The first reports in the 1950s relied on spark sintering by Lenel [144]. This was followed by hot isostatic pressing [145]. The HIP route required inert handling to deliver low interstitial levels for high-performance applications. Spherical, high-purity, and rather large rotating electrode powders played an important early role. Other efforts relied on lower cost powders, such as sponge fines and blended elemental powders, in die compaction. The progress was documented in 1980 [146]. The HIP Ti-6Al-4V product delivered a tensile strength of 975 MPa with 14% elongation, similar to wrought material. Post-consolidation heat treatments enabled strength-ductility manipulations ranging up to 1130 MPa tensile strength with 9% elongation. Without hot isostatic pressing, the best case corresponds to a tensile strength of 920 MPa with 11% elongation at 98% density for Ti-6Al-4V. However, the fatigue and fracture toughness were typically lower than the comparable wrought product. Without HIP full density was not attained, so the properties were lower. When subjected to shot peening, fatigue strength is about 410 MPa, but when subjected to HIP and shot peening, it is closer to 485 MPa [126].

Using the early base from press-sinter titanium sintering, Ti-MIM was demonstrated in 1988 [2]. Early reports showed 1000 MPa tensile strength, but just 2% elongation. In Japan, Ti-MIM reached production status for decorative uses about 1991. The most notable application was in running-shoe spikes: Leroy Burrell posted 100 m dash time of 9.88 s using ASICS shoes with low ductility but high strength Ti-MIM spikes. Dental orthodontic bracket efforts started soon after and the trials included surgical tools, automotive shifter knobs, toy components, model railroad train wheels, trigger guards, and eyeglass frames [139].

Subsequently, scientific study refined processing with a primary focus on oxygen to improve ductility and corrosion resistance. Several detailed studies emerged disclosing the processing cycles required for titanium and its alloys, as reviewed elsewhere [114–143,147–150]. Along the way, in spite of many industry skeptics, Ti-MIM reached production status for several mechanical components as shown in earlier reports [98,103,125,136,139]. One impressive Ti-MIM component is the tripod base shown in Figure 20.3.

Most recently, Ti-MIM has penetrated life critical applications in dental, aerospace, medical, and chemical devices [115,116,120–129,134,135,140–143]. Based on success at research institutions, the high-quality Ti-MIM process has been put into production with variants by several firms.

Figure 20.3 An example of a Ti-MIM component, in this case a tripod base.

In achieving this status, Ti-MIM was a favorite research topic in the research community, in part because the sensitive process provided a big challenge. Efforts pieced together new solutions within the Ti-MIM context. For example, El-Kadiri et al. [79] disclosed titanium alloys using lower-cost powder for automotive applications. It relies on −45 μm 99.8% pure sponge Ti powder alloyed using admixed Zr and Fe for liquid phase sintering densification. A composition with 7.5 wt% Fe and 5.0 wt% Zr, sintered at 1275°C for 1 h, gave 99% density with a high tensile strength but low ductility.

One study demonstrated that a cold isostatic pressing step between debinding and sintering was favorable [113], while another found that a polyacetal binder could be debound and sintered in a single step [72]. Miura et al. [31] produced Ti-6Al-7Nb using spherical gas atomized 325 mesh (−45 μm) Ti powder (0.008% C and 0.140% O) mixed with prealloyed additives. The powders and binder were formed into feedstock by mixing at 173°C for 2.5 h to give 65 vol% solids loading. After molding, the wax phase was removed via heptane immersion for 6 h, followed by vacuum sweep-gas treatment at 430°C [148]. Sintering densification occurs at 1350°C for 4 h, giving 97% density, with a tensile strength of 830 MPa and 11% elongation at 0.31% oxygen. As noted by Whittaker [149], the sintered mechanical properties are highly variable depending on the sintering approach, with yield strength from 140 to 940 MPa.

Cost reduction is the key issue ahead in Ti-MIM. Use of mixed atomized and HDH powders provide a means to control rheology and cost. Binder design needs to balance powder wetting, mixture viscosity, green strength, debinding, and contamination concerns. Mixing of the powder and binder at low temperatures prevents oxidation, and mixing under inert gas is generally most beneficial. Solvent debinding, including water and ethanol as solvents, opens the pore structure with minimized contamination. There is contamination from the backbone binders during the last portion of vacuum sweep-gas debinding. Accordingly, experimentation is required to isolate the optimal backbone polymer chemistry and concentration for minimized contamination.

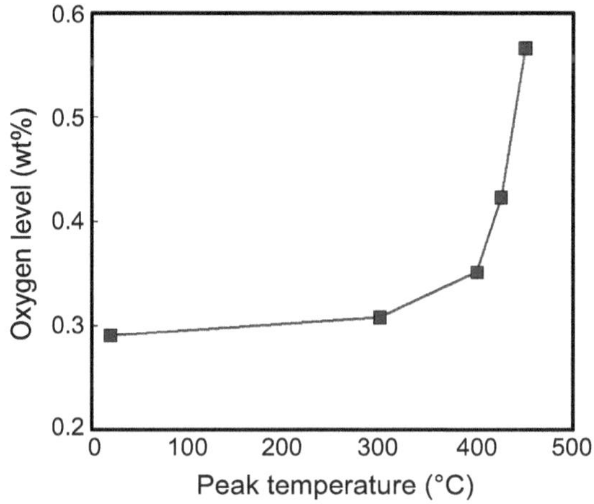

Figure 20.4 Oxygen content versus debinding temperature for titanium powder [136].

The rheology of highly loaded suspensions depends on the particle size distribution. Consequently, a broad size distribution is most desirable. However, oxygen concerns lead to removal of the smaller particles, giving a narrow and relatively coarse particle size distribution. Thus, the Ti-MIM feedstocks are different when compared to more traditional systems. The binder tends to be weaker and lower in molecular weight to ease mixing and debinding, but this leads to more molding defects and more required care in green component handling. The powder–binder interface adhesion needs to be improved to reduce separation in molding, a common difficulty with larger particle sizes and low-viscosity binders. Efforts with stronger polymers, such as polystyrene, result in increased contamination, so simple wax-like polymers remain the most successful in spite of the low strength and tendency toward separation.

Debinding is a sensitive aspect of Ti-MIM and requires two steps; solvent immersion followed by thermal pyrolysis under vacuum using the sweep-gas concept. The peak temperature, hold time, and other parameters are determined using analytical tools, including mass spectroscopy or similar *in situ* monitors. A good example of the increased oxygen uptake with debinding temperature is shown in Figure 20.4 for titanium powder held at various temperatures [136]. Some of the oxygen contamination traces to impurities such as titania (used to catalyze polymerization) carried in by the backbone polymer. Likewise, carbon pickup is common during debinding. In one variant, heating is accelerated during debinding to induce reactions between the titanium and carbon to give a composite of Ti and TiC. Graphite is another means to induce titanium carbide formation [64]. This produces a cermet of hard titanium carbide dispersed in a titanium matrix. The composite exhibits high tensile strength, hardness, and wear resistance, but low ductility.

The protocol is to isolate a sintering cycle that takes the titanium component to a closed pore condition at about 95% density. Curiously, too high a sintering temperature does damage, with loss of sintered strength due to either gas reactions or microstructure coarsening. This is evident in Figure 20.5 for Ti-12Mo sintered for 5 h at

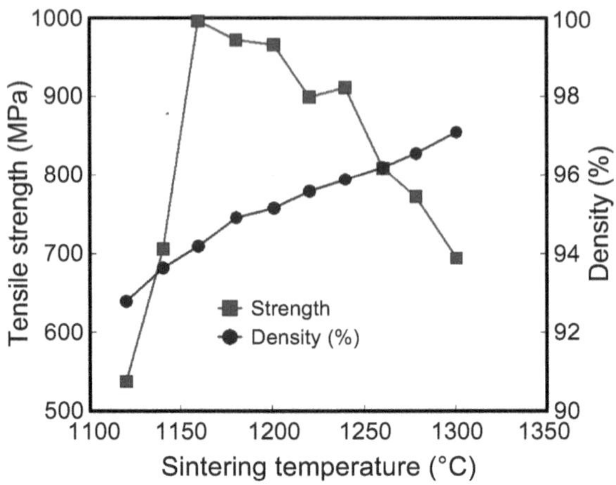

Figure 20.5 Sintered density and tensile strength for Ti-12Mo versus the sintering temperature [64].

various temperatures [64]. Subsequently, the component is densified using hot isostatic pressing. The difficult balance between densification in sintering, grain coarsening, gas generation, and contamination is without mathematical analysis, leading to a trial-and-error approach to development of sintering cycles. Higher sintering temperatures or longer sintering times potentially result in a higher final density, but lower strength because of enhanced grain and pore coarsening.

Research on demanding applications has made excellent progress [114–119,124,125, 128–143,147–150]. Much of this effort is focused on final properties, especially fatigue, as well as oxidation, corrosion, and various biocompatibility attributes. Mechanical properties at elevated temperatures are still largely missing.

With mechanical properties rivaling handbook values, the shift is to production scale-up. But the cost barrier is large still. New components are evaluated on a value-added basis, leading to focus on biomedical and aerospace applications. The early successes help convince designers to use MIM. As a metal, titanium components constitute the largest value in aircrafts (outside the turbine) and second largest value in biomedical components. Thus, the challenge is not in developing markets but in qualifying the Ti-MIM approach at a competitive cost.

For these successes, a baseline Ti-MIM process is defined in Table 20.2. This summarizes best practices for each of the processing steps. Changes in the powder or binder result in differences in mixing, impurities, sintering, and other steps. Hence, this is a demonstration of what works.

The need for the postsintering HIP treatment is evident by the sintered microstructure shown in Figure 20.6. In Figure 20.6, the black spots are pores that remain after sintering. Note the large grain size, indicative of considerable coarsening prior to full densification. In this case, the sintering temperature resulted in a mixture of alpha and beta phases, giving a desirable lamellar alpha structure. The grains are more than 100 μm in size, the pores are about 10–15 μm in diameter, and the lamellar plates

Table 20.2 Baseline Ti-MIM process

Steps	Key principle	Specific time, temperature, and such
Powder	Deagglomerated spheres	Gas or plasma atomized
	Typically -325 mesh	30–60 μm median particle size
	High tap density	60–62% of pycnometer density
	Low initial oxygen level	0.15 wt% O maximum
	Low initial carbon level	0.04 wt% C maximum
Binder	Majority low molecular polymer	65–75 wt% paraffin wax or polyethylene glycol
	Higher molecular weight backbone	15–25 wt% polypropylene or ethylene vinyl acetate
	Surfactant, lubricant, plasticizer	5 wt% stearic acid
Mixing	Mixing under protective conditions	Vacuum or argon cover gas
	Room temperature dry mix	At 65 vol% solids loading
	Heated, high shear mixing	Vacuum mix, 30 min at 120–185°C
	Target viscosity	At 500 s^{-1} of 150–250 Pa·s
Molding	Controlled nozzle temperature	120–180°C
	Slightly heated mold	30°C
	Injection temperature	160°C
	Injection pressure	30 MPa
	Green strength	10 MPa
Debinding	First stage solvent immersion	60°C; water for polyethylene glycol, heptane for paraffin wax
	Solvent penetration rate	2 mm/h
	Second stage thermal debinding	Slow heat, Ar sweep gas in vacuum
	Vacuum final step debinding	Heat slowly to 450°C, hold 1 h
	Presinter heating for strength	Hold near 900°C for 1 h, vacuum
Sintering	High temperature sintering	Vacuum, refractory metal furnace
	Support or substrate materials	Yttria or zirconia trays
	Peak temperature and time	1250°C for 120–180 min
	Sintered density	95% of theoretical, closed pores
Densification	Hot isostatic pressing	Ar without container
	Consolidation conditions	900°C, 100 MPa, 60 min
Properties	Final density and grain size	99.5–100%
	Grain size	40–100 μm
	Microstructure	Alpha and beta,10 μm platelets
	Final impurity level	0.20–0.22 wt% O, 0.04 wt% C
	Tensile strength	Tensile strength 900 MPa
	Tensile elongation	12%
	Fatigue endurance strength	Up to 500 MPa

are about 5–10 μm. Note the grain volume is equivalent to the volume of 30 starting particles, and the pore size is about 16 times the initial pore volume. Because of the porosity and microstructural coarsening during sintering, the mechanical properties are near cast material.

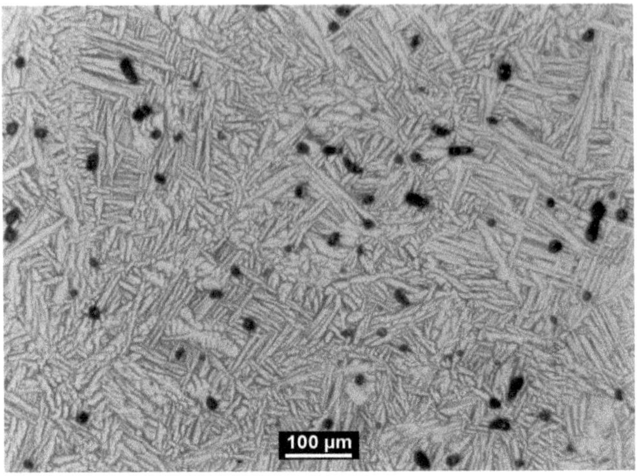

Figure 20.6 Sintered microstructure for Ti-6Al-4V produced by Ti-MIM prior to hot isostatic pressing.

After the HIP treatment, the grain size is similar but the pores are absent, and no artifact of the powder process remains.

20.5 Components design factors

Titanium metal powder injection molding is advancing, but is restricted to about 5% of the metal powder injection molding companies. Many components are formed using Ti-MIM, and a few feedstock companies are providing precompounded feedstock.

Even so, the market for Ti-MIM is small and dwarfed by the success of stainless steels, especially for cellular telephone and computer applications. Unfortunately, the use data for Ti-MIM are poorly organized and some of the trade associations ignore the activities by this segment. Since titanium is significantly higher priced versus stainless steel, the 10,000 kg of titanium powder used each year for metal injection molding produces substantial sales. This is because the typical Ti-MIM component sells for $8 each with an average mass of 10 g. In other words, Ti-MIM is probably about $10 million in component sales globally. Some of the early Ti-MIM shapes are shown in Figure 20.7, including heart valves, implants, and test geometries. As mentioned, other examples include dental implants, watch cases, trigger guards for firearms, model train wheels, decorative luggage buckles, toy components, surgical hand tools, automotive interior knobs, cellular telephone covers, and even artificial knees.

As noted above, a variety of products have been demonstrated using Ti-MIM. The more demanding applications require hot isostatic pressing to ensure full density, which adds to the expense. Unfortunately, the cost advantage is missing for some of the larger-scale applications, such as cellular telephone components. Thus, as powder costs decrease, there is ample opportunity for expanded applications.

Figure 20.7 Example Ti-MIM shapes produced by Element 22 GmbH, Kiel, Germany.

The median for MIM production tends toward 10 g and 25 mm maximum size while Ti-MIM is best suited to smaller components [107]. Costs drive Ti-MIM toward low mass components.

There are few reports on Ti-MIM tolerances, but in general the tolerance capabilities are not up to par with other MIM technologies. Whittaker [41] reports a case with gas atomized powder, wax-polymer binder, solvent debinding, and vacuum sintered at 1250°C for 2 h with a coefficient of variation (standard deviation divided by mean size) of 0.1%. Other reports with less detail on processing claim coefficients of size variation ranging from 0.08% to 0.65% and mass variations from 0.5% to over 1%. Thus, it appears Ti-MIM is closing in on the dimensional tolerance capabilities of other MIM approaches, but the sintered surface finish is rougher because of the larger particle size.

Recently, test data have been taken for life-critical Ti-MIM applications. These include standard mechanical properties, as well as a large set of challenge tests for cell toxicity, tissue irritation, systemic toxicity, hemocompatibility, and sensitization [4–9]. Such testing is complex and expensive, so it is no surprise that only a few MIM firms are taking up the challenge. Some of the often-mentioned targets are artificial bones, joints, heart valves, and dental implants, including scaffold structures for tissue ingrowth.

20.6 Summary

There is nothing routine about Ti-MIM. The field developed based on several methodical university research efforts coupled with advances in powder synthesis. The binders for Ti-MIM emphasize lower melting temperature ingredients that are easily extracted in solvents such as heptane, water, or ethanol. Curiously, the more successful binders are relatively inexpensive compared to the powder. Debinding by first-stage solvent immersion reduces contamination. All reports show contamination increasing from powder to product, so the strategy is to start with clean (low interstitial) powder and optimize each step in sequence to minimize any contamination increase. Higher-molecular-weight binder ingredients add more to the impurity burden, as do higher-temperature thermal cycles, process atmospheres, and oxide ceramic substrates. The Ti-MIM technology is well developed, and there are several vendors supporting the technology. Costs must be rationalized in the applications to justify the powder and processing costs. Decorative items, such as watch and toy components were early successes, but there was no compelling justification and these were transient applications. The new wave of applications in aerospace and medical devices are more justified since the properties are demanding and the cost reduction versus other manufacturing routes is impressive, making the life-critical areas permanent successes for Ti-MIM.

References

[1] F.H. Froes, Powder metallurgy of titanium alloys, in: I. Chang, Y. Zhao (Eds.), Advances in Powder Metallurgy, Woodhead, London, UK, 2013, pp. 202.

[2] Y. Kaneko, K. Ameyama, K. Saito, H. Iwasaki, M. Tokizane, Injection molding of titanium powder, J. Jpn. Soc. Powder Powder Metall. 35 (1988) 646–650 1988.

[3] K. Ameyama, Y. Kaneko, H. Iwasaki, M. Tokizane, Injection molding of titanium powders, Advances in Powder Metallurgy, vol. 3, MPIF, Princeton, NJ, 1989, pp. 121–126.

[4] Y. Kaneko, K. Ameyama, S. Sakaguchi, Application of injection molding to Ti-5wt.%Co and Ti-wt.%Al-4wt.%V mixed powders, J. Jpn. Soc. Powder Powder Metall. 37 (1990) 591–596.

[5] K. Kato, Y. Nozaki, A. Matsumoto, Properties of sintered TiAl by injection molding, J. Jpn. Soc. Powder Powder Metall. 39 (1992) 875–878.

[6] H. Kyogoku, T. Toda, K. Shinohara, Sintering behavior of titanium compacts with the addition of TiH powder by using injection moldings, J. Jpn. Soc. Powder Powder Metall. 40 (1993) 439–443.

[7] H. Kyogoku, S. Komatsu, K. Shinohara, H. Jinushi, T. Toda, Microstructures and mechanical properties of sintered Ti-4% Fe alloy compacts by injection moldings, J. Jpn. Soc. Powder Powder Metall. 41 (1994) 1075–1079.

[8] F. Petzoldt, H. Eifert, T. Hartwig, G. Veltl, Binder design and process control for high performance MIM materials, in: Advances in Powder Metallurgy and Particulate Materials, vol. 6, MPIF, Princeton, NJ, 1995, pp. 3–13.

[9] K. Kusaka, T. Kohno, T. Kondo, A. Horata, Tensile behavior of sintered titanium by MIM process, J. Jpn. Soc. Powder Powder Met. 42 (1995) 383–387.

[10] H. Kyogoku, S. Komatsu, I. Tsuchitori, T. Toda, Tensile properties of sintered Ti-4% Fe alloy compacts by injection moldings, J. Jpn Soc. Powder Powder Met. 42 (1995) 1052–1056.

[11] K. Kusaka, T. Kono, A. Horata, T. Kondo, Tensile behavior of sintered Ti and Ti-6Al-4V alloy by MIM process, in: Advances in Powder Metallurgy and Particulate Materials-1996, vol. 19, MPIF, Princeton, NJ, 1996, pp. 127–131.

[12] T. Deguchi, M. Ito, A. Obasta, Y. Koh, T. Yamagishi, Y. Oshida, Trial production of titanium orthodontic brackets fabricated by metal injection molding (MIM) with sinterings, J. Dental Res. 75 (1996) 1491–1496.

[13] R.M. German, R.G. Iacocca, Powder metallurgy processing, in: N.S. Stoloff, V.K. Sikka (Eds.), Physical Metallurgy and Processing of Intermetallic Compound, Chapman and Hall, New York, NY, 1996, pp. 605–654.
[14] K. Kato, A. Matsumoto, T. Ieki, Tensile properties at high temperature of sintered TiAl compacts by injection molding, J. Jpn. Soc. Powder Powder Metall. 44 (1997) 1029–1034.
[15] T. Kono, A. Horata, T. Kondo, Development of titanium and titanium alloy by metal injection molding process, J. Jpn. Soc. Powder Powder Metall. 44 (1997) 985–992.
[16] H. Wohlfromm, M. Blomacher, D. Weinand, E. Langer, M. Schwarz, Novel materials in metal injection moulding, Powder Injection Moulding, Proceedings of the First European Symposium on Powder Injection Moulding, European Powder Metallurgy Association, Shrewsbury, UK, 1997, pp. 54–61.
[17] T. Ieki, K. Katoh, A. Matsumoto, T. Masui, K. Andoh, Tensile properties of sintered Ti compacts by metal injection molding process, J. Jpn. Soc. Powder Powder Metall. 44 (1997) 448–452.
[18] H. Wohlfromm, M. Blomacher, D. Weinand, Metal injection molding of titanium and TiAl6V4, in: R.M. German, H. Wiesner, R.G. Cornwall (Eds.), Powder Injection Molding Technologies, Innovative Material Solutions, State College, PA, 1998, pp. 339–348.
[19] Y. Kim, N.J. Kim, T. Yoon, S. Ahn, Powder injection molding of Ti-48Al, in: Proceedings 1998 PM World Congress, Granada, Spain, European Powder Metallurgy Association, Shrewsbury, UK, CD, 1998.
[20] H. Eifert, T. Hartwig, State of the art in powder injection molding of titanium, in: Advances in Powder Metallurgy and Particulate Materials-1998, vol. 5, MPIF, Princeton, NJ, 1998, pp. 143–150.
[21] R.M. German, Powder injection molding applications to new materials, in: A. Khor, T.S. Srivatsan, J.J. Moore (Eds.), Processing and Fabrication of Advanced Materials IV, 2, Institute of Materials, London, 1998, pp. 1363–1375.
[22] H. Miura, T. Yokomizo, S. Sakaguchi, K. Tokumoto, K. Nakahara, Development of sintered titanium base hard alloy by metal injection molding, J. Jpn. Soc. Powder Powder Metall. 45 (1998) 892–895.
[23] S. Ito, N. Ogasawara, Y. Kusano, K. Ishiyama, Titanium feedstock for rapid catalytic debinding, in: A. Khor, T.S. Srivatsan, J.J. Moore (Eds.), Processing and Fabrication of Advanced Materials IV, 2, Institute of Materials, London, UK, 1998, pp. 1433–1441.
[24] K.H. Moyer, W.R. Jones, T. Loughridge, T. Rodzen, The application of metal injection molded metal parts to satisfy specific mechanical and physical properties, in: Advances in Powder Metallurgy and Particulate Materials-1998, vol. 5, MPIF, Princeton, NJ, 1998, pp. 85–91.
[25] I.R. Van Gestel, A new production method for high value parts, in: R.M. German, H. Wiesner, R.G. Cornwall (Eds.), Powder Injection Molding Technologies, Innovative Material Solutions, State College, PA, 1998, pp. 33–42.
[26] H. Nakamura, T. Shimura, K. Nakabayashi, Process for production of Ti sintered compacts using the injection molding method, J. Jpn. Soc. Powder Powder Metall. 46 (1999) 870–876.
[27] Y. Kim, N.J. Kim, T. Yoon, S. Ahn, Densification behavior of PIMedTiAl parts, J. Jpn. Soc. Powder Powder Metall. 46 (1999) 882–886.
[28] H. Wang, S.H.J. Lo, J.R. Barry, Development of high density (99%+) powder injection molded titanium alloys, P/M Sci. Tech. Briefs 1 (1999) 16–18.
[29] Y. Kato, Effect of sintering temperature on density and tensile properties of titanium compacts by metal injection molding, J. Jpn. Soc. Powder Powder Metall. 46 (1999) 865–869.

[30] K. Maekawa, M. Takita, H. Nomura, Effect of MIM process conditions on microstructures and mechanical properties of Ti-6Al-4V compacts, J. Jpn. Soc. Powder Powder Metall. 46 (1999) 1053–1057.

[31] H. Miura, Y. Itoh, T. Uematsu, K. Asto, Advanced PIM Process of Ti-6Al-7Nb for Materials for Medical Applications, in: R.M. German (Ed.), Proceedings of the Workshop on Medical Applications for Microminiature Powder Injection Molding, MPIF, Princeton, NJ, 2009.

[32] S.J. Park, Y. Wu, D.F. Heaney, X. Zou, G. Gai, R.M. German, Rheological and thermal debinding behaviors in titanium powder injection molding, Met. Mater. Trans. 40 (2009) 215–222.

[33] E. Baril, L.P. Lefebvre, Y. Thomas, F. Ilinca, Foam coated MIM gives new edge to titanium implants, Meter. Powder Rep. 63 (2008) 46–55.

[34] Y. Itoh, T. Uematsu, K. Sato, H. Miura, M. Niinomi, Fabrication of high strength alpha plus beta type titanium alloy compacts by metal injection molding, J. Jpn. Soc. Powder Powder Metall. 55 (2008) 720–724.

[35] L. Zhao, Z. Qu, X. He, S. Li, Sintering of TiAl injection molding, Chin. J. Rare Meter. 32 (2008) 180–184.

[36] T. Osada, H. Miura, Y. Itoh, M. Fujita, N. Arimoto, Optimization of MIM process for Ti-6Al-7Nb alloy powder, J. Jpn. Soc. Powder Powder Metall. 55 (2008) 726–731.

[37] Y.C. Kim, S. Lee, S. Ahn, N.J. Kim, Application of metal injection molding process to fabrication of bulk parts of TiAl intermetallic, J. Mater. Sci. 42 (2007) 2048–2053.

[38] Y. Itoh, H. Miura, T. Uematsu, K. Sato, M. Niinomi, Improvement of the properties of Ti-6Al-7Nb alloy by metal injection molding, in: Advances in Powder Metallurgy and Particulate Materials-2007, vol. 4, MPIF, Princeton, NJ, 2007, pp. 81–86.

[39] X. Zou, Y. Wu, G. Gai, D.F. Heaney, S.J. Park, R.M. German, The effects of spheroidizing treatment on the particle characteristics and feedstock rheologies of HDH Ti powders, Powder Met. Ind. 16 (2006) 1–6.

[40] S. Sunada, M. Kawagishi, K. Majima, Corrosion behavior of pure titanium by MIM process under SSRT conditions, J. Jpn. Soc. Powder Powder Metall. 54 (2007) 322–329.

[41] D. Whittaker, Developments in the powder injection moulding of titanium, Powder Inject. Mould. Int. 1 (2007) 27–32.

[42] Y. Li, X.M. Chou, L. Yu, Dehydrogenation debinding process of MIM titanium alloys by TiH_2 powder, Powder Met. 49 (2006) 236–239.

[43] Y. Itoh, T. Uematsu, K. Sato, H. Miura, M. Niinomi, Microstructural modification for injection molded Ti-6Al-4V alloys by addition of Mo powder, J. Jpn. Soc. Powder Powder Metall. 53 (2006) 750–754.

[44] Y. Wu, R. Wang, Y.S. Kwon, S.J. Park, R.M. German, Injection molding of HDH titanium powder, Int. J. Powder Met. 42 (2006) 59–66.

[45] T. Uematsu, Y. Itoh, K. Sato, H. Miura, Effects of substrate for sintering on the mechanical properties of injection molded Ti-6Al-4V alloy, J. Jpn. Soc. Powder Powder Metall. 53 (2006) 755–759.

[46] H. Miura, T. Takemasu, Y. Kuwano, Y. Itoh, K. Sato, Sintering behavior and mechanical properties of injection molded Ti-6Al-4V alloys, J. Jpn. Soc. Powder Powder Metall. 53 (2006) 815–820.

[47] K.S. Weil, E. Nyberg, K. Simmons, A new binder for powder injection molding titanium and other reactive metals, J. Mater. Proc. Technol. 176 (2006) 205–209.

[48] Y. Itoh, T. Uematsu, K. Sato, H. Miura, M. Niinomi, M. Ikeda, Sintering behavior and mechanical properties of injection molded Ti-4.3Fe-7. 1Cr alloys, J. Jpn. Soc. Powder Powder Metall. 53 (2006) 821–826.

[49] Y. Itoh, Y. Harikou, K. Satoh, H. Miura, Fabrication of near-alpha titanium alloy by metal injection molding, J. Jpn. Soc. Powder Powder Metall. 52 (2005) 43–48.
[50] K.S. Weil, E.A. Nyberg, K.L. Simmons, Use of a naphthalene based binder in injection molding net shape titanium components of controlled porosity, Mater. Trans. 46 (2005) 1525–1531.
[51] E. Nyberg, M. Miller, K. Simmons, K.S. Weil, Microstructure and mechanical properties of titanium components fabricated by a new powder injection molding technique, Mater. Sci. Eng. 25 (2005) 336–342.
[52] H. Miura, M. Uemura, Y. Kubota, High-temperature properties of injection molded Ti-Al intermetallic compounds, in: Advances in Powder Metallurgy and Particulate Materials-2004, vol. 4, MPIF, Princeton, NJ, 2004, pp. 102–105.
[53] A. Kitajima, T. Shimizu, A. Ito, Fabrication of bioactive titanium parts by MIM techniques using surface treatment, J. Jpn. Soc. Powder Powder Metall. 50 (2003) 739–744.
[54] E.S. Thian, N.H. Loh, K.A. Khor, S.B. Tor, Ti-6Al-4V/HA composite feedstock for injection molding, Mater. Lett. 56 (2002) 522–532.
[55] S. Guo, X. Qu, X. He, T. Zhou, B. Duan, Powder injection molding of Ti-6Al-4V alloy, J. Mater. Proc. Tech. 173 (2006) 310–314.
[56] H. Miura, T. Masuda, T. Ogasawara, Y. Kankawa, High performance injection molded Ti compacts, J. Jpn. Soc. Powder Powder Metall. 49 (2002) 825–828.
[57] S. Terauchi, T. Teraoka, T. Shinkuma, T. Sugimoto, Y. Ahida, Development of production technology by metallic powder injection molding for TiAl-type intermetallic compound with high efficiency, in: P. Rodhammer, H. Wildner (Eds.), Proceedings Fifteenth International Plansee Seminar, 1, Plansee Holding, Reutte, Austria, 2001, pp. 610–624.
[58] Y. Xu, H. Nomura, Corrosion behavior of biomedical titanium alloy Ti-5Al-2.5Fe processed by MIM, J. Jpn. Soc. Powder Powder Metall. 49 (2002) 382–389.
[59] Y. Xu, H. Nomura, M. Takita, H. Toda, Characteristics of metal injection processed Ti-5Al-2.5Fe alloy for implant material, J. Jpn. Soc. Powder Powder Metall. 48 (2001) 316–321.
[60] L.Y. Wang, X.L. Lin, Sintered titanium compacts by metal injection molding process, in: K. Kosuge, H. Nagai (Eds.), Proceedings of the 2000 Powder Metallurgy World Congress, 1, Japan Society of Powder and Powder Metallurgy, Kyoto, Japan, 2000, pp. 320–323.
[61] T. Shimizu, A. Kitajima, K. Kato, T. Sano, Supercritical debinding and its application to PIM of Ti-Al intermetallic compounds, in: K. Kosuge, H. Nagai (Eds.), Proceedings of the 2000 Powder Metallurgy World Congress, Japan Society of Powder and Powder Metallurgy, Kyoto, Japan, 2000, pp. 292–295.
[62] R.H. Froes, R.M. German, Cost reductions prime Ti PIM for growth, Meter. Powder Rep. 55 (2000) 12–21.
[63] L.P. Lefebvre, E. Baril, Effect of oxygen concentration and distribution on the compression properties on titanium foams, Adv. Eng. Mater. 10 (2008) 868–876.
[64] J. Takekawa, N. Sakurai, Effect of processing conditions on density, strength and microstructure of Ti-12Mo alloy fabricated by PIM process, J. Jpn. Soc. Powder Powder Metall. 46 (1999) 877–881.
[65] E. Ergul, H.O. Gulsoy, V. Gunay, Effect of sintering parameters on mechanical properties of injection moulded Ti-6Al-4V alloys, Powder Met. 52 (2009) 65–71.
[66] Y.X. Cai, Q. Chang, Y. Ding, Research of injection molding titanium parts, Powder Met. Technol. 23 (2005) 449–455.
[67] Y. Thomas, E. Baril, F. Ilinca, J.F. Hetu, Development of titanium dental implant by MIM: experiments and simulation, in: Advances in Powder Metallurgy and Particulate Materials-2009, vol. 4, MPIF, Princeton, NJ, 2009, pp. 81–93.
[68] J. Takekawa, Metal injection molding of Ti-Mo-Al mixed powders, J. Jpn. Soc. Powder Powder Metall. 41 (1994) 244–247.

[69] A. Matsumoto, K. Katoh, K. Andoh, Near net shaping of metal injection molded compacts by pulsed discharge sintering process under pseudo hot isostatic pressing, J. Jpn. Soc. Powder Powder Metall. 44 (1997) 1037–1042.
[70] F.H. Froes, Powder injection molding (PIM) of titanium alloys – ripe for expansion, Mater. Tech. 15 (2000) 295–299.
[71] T. McCabe, C. Vaidyanathan, Material advancement for PIM applications, in: Advances in Powder Metallurgy Particulate Materials-2007, vol. 4, MPIF, Princeton, NJ, 2007, pp. 87–92.
[72] M. Yoshimitsu, I. Kayo, S. Hiroshi, H. Nakamura, M. Iji, T. Masao, Y. Ippei, Single step MIM system with polyacetal binder without debinding process, in: K. Kosuge, H. Nagai (Eds.), Proceedings of the 2000 Powder Metallurgy World Congress, Japan Society of Powder and Powder Metallurgy, Kyoto, Japan, 2000, pp. 270–273.
[73] R.M. Gomes, L. Kowalski, L. Schaeffer, J. Duszczyk, Injection molding of blended elemental Ti (6Al-4V) powders, in: K. Kosuge, H. Nagai (Eds.), Proceedings of the 2000 Powder Metallurgy World Congress, Japan Society of Powder and Powder Metallurgy, Kyoto, Japan, 2000, pp. 324–327.
[74] B. Williams, Challenges for MIM titanium parts: Met, Powder Rep. 58 (2003) 30.
[75] H.S. Sun, The fabrication of Ti injection molded and sintered components, Titan Ind. Prog. 1 (2000) 19–20.
[76] J.A. Grohowski, B. Sherman, J.T. Strauss, Processing of titanium by metal injection molding, in: Advances in Powder Metallurgy and Particulate Materials-2003, vol. 8, MPIF, Princeton, NJ, 2003, pp. 273–281.
[77] L.Y. Wang, Metal injection molding of blended elemental titanium and zirconium powders, in: Advances in Powder Metallurgy and Particulate Materials-2002, vol. 10, MPIF, Princeton, NJ, 2002, pp. 301–314.
[78] T. Osada, S. Tanaka, K. Nishiyabu, H. Miura, Size effects of gas nitriding of injection molded titanium parts, J. Jpn. Soc. Powder Powder Metall. 53 (2006) 745–749.
[79] H. El-Kadiri, L. Wang, H.O. Gulsoy, P. Suri, S.J. Park, Y. Hammi, R.M. German, Development of a Ti-based alloy: design and experiment, J. Met. 61 (2009) 60–66.
[80] Y. Xu, H. Nomura, Homogenizing analysis for sintering of bio-titanium alloy (Ti-5Al-2.5Fe) in MIM process, J. Jpn. Soc. Powder Powder Metall. 48 (2001) 1089–1096.
[81] Y. Osafume, T. Uichida, Y. Nakamura, Y. Tanaka, Effects of debinding and sintering condition on distortion of sintered compact of injection molded TiAl intermetallic compound, J. Jpn. Soc. Powder Powder Metall. 41 (1994) 506–509.
[82] K. Kato, A. Matsumoto, Y. Nozaki, T. Ieki, Metal injection molding of prealloyed TiAl powders with various Ti/Al ratios, J. Jpn. Soc. Powder Powder Metall. 42 (1995) 1068–1072.
[83] R. Gerling, F.P. Schimansky, Prospects for metal injection moulding using a gamma titanium aluminide based alloy powder, Mater. Sci. Eng. 329 (2002) 45–49.
[84] S. Terauchi, T. Teraoka, T. Shinkuma, T. Sugimoto, Development of TiAl type intermetallic compounds by metal powder injection molding process, J. Jpn. Soc. Powder Powder Metall. 47 (2000) 1283–1287.
[85] S. Ishiyama, H.P. Buchkremer, D. Stover, The characterization of reinforced TiAl intermetallic with dispersed Cr particles consolidated by HIP, Mater. Trans. 43 (2002) 2331–2336.
[86] J.A. Sago, J.W. Newkirk, Metal injection molding of mechanically alloyed powders of advanced P/M alloys, in: Advances in Powder Metallurgy and Particulate Materials-1999, vol. 6, MPIF, Princeton, NJ, 1999, pp. 45–57.
[87] H. Kyogoku, S. Komatsu, Fabrication of TiNi shape memory alloy by powder injection molding, J. Jpn. Soc. Powder Powder Metall. 46 (1999) 1103–1107.

[88] E. Schuller, L. Krone, M. Bram, H.P. Buchkremer, D. Stover, Metal injection molding of shape memory alloys using prealloyed NiTi powders, J. Mater. Sci. 40 (2005) 4231–4238.
[89] L. Krone, J. Mentz, M. Bram, H.P. Buchkremer, D. Stover, M. Wagner, G. Eggeler, D. Christ, S. Reese, D. Bogdanski, The potential of powder metallurgy for the fabrication of biomaterials on the basis of nickel-titanium: A case study with a staple showing shape memory behaviour, Adv. Eng. Mater. 7 (2005) 613–619.
[90] H. Kyogoku, H. Kawasaki, S. Komatsu, Fabrication of Ti-Ni shape memory alloy by metal injection molding, in: Advances in Powder Metallurgy and Particulate Materials 2006, vol. 4, MPIF, Princeton, NJ, 2006, pp. 14–20.
[91] P. Imgrund, A. Rota, H. Schmidt, Micro MIM: Making the most of NiTi, Met. Powder Rep. 63 (2008) 21–24.
[92] M. Bram, A. Ahmad-Khanlou, A. Heckmann, B. Fuchs, H.P. Buchkremer, D. Stover, Powder metallurgical fabrication processes for NiTi shape memory alloy parts, Mater. Sci. Eng. 337 (2002) 254–263.
[93] P. Imgrund, F. Petzoldt, V. Friederici, Micro MIM for medical applications, in: Proceedings Europe PM 2008, European Powder Metallurgy Association, Shrewsbury, UK, 2008, CD.
[94] E.S. Thian, N.H. Loh, K.A. Khor, S.B. Tor, Microstructures and mechanical properties of powder injection molded Ti-6Al-4V/HA powder, Biomaterials 23 (2002) 2927–2938.
[95] Y. Cai, K. Luo, Q. Chen, $TiC_{0.7}N_{0.3}$ particulates reinforced titanium matrix composites and components prepared by powder injection molding, Mater. Sci. Eng. Powder Met. 10 (2005) 344–349.
[96] H. Ye, X.Y. Liu, H. Hong, Fabrication of metal matrix composites by metal injection molding – a review, J. Mater. Proc. Technol. 200 (2008) 12–24.
[97] S. Abkowitz, S.M. Abkowitz, H. Fisher, P.J. Schwartz, Cerme Ti discontinuously reinforced Ti-matrix composites: Manufacturing, properties, and applications, J. Met. 56 (2004) 37–41.
[98] R.M. German, Powder Injection Molding Design and Applications, Innovative Material Systems, State College, 2003.
[99] R.M. German, Global research and development in powder injection moulding, Powder Inject. Mould. Int. 1 (2007) 33–36.
[100] M. Kearns, P. Davies, Review of the MIM Industry: Recent Trends in Powder Size and Composition, in: Advances in Powder Metallurgy and Particulate Materials – 2007, vol. 4, MPIF, Princeton, NJ, 2007, pp. 1–7.
[101] M. Achikita, Development of MIM components for automobile and power tools, Ninth Case Studies on New Product Development, Japan Powder Metallurgy Association, Kyoto, Japan, 2000, pp. 25–34.
[102] Q. Chen, Y.T. Cai, Y.L. Yang, Effect of powder size and powder loading on the process of metal injection molding (MIM) for titanium, J. Guangdong Non-Ferrous Met. 9 (1999) 131–135.
[103] R.M. German, A. Bose, Injection Molding of Metals and Ceramics, MPIF, Princeton, 1996.
[104] S.B. Guo, X.H. Qu, B.H. Duan, X.B. He, M.L. Qin, Optimization of titanium alloy injection molding process parameters, Mater. Sci. Eng. 11 (2004) 32–36.
[105] J.W. Newkirk, J.A. Sago, G.M. Brasel, Metal injection molding of mechanically alloyed advanced materials, in: Advances in Powder Metallurgy and Particulate Materials-1998, vol. 5, MPIF, Princeton, NJ, 1998, pp. 93–105.

[106] R.F. Wang, Y.X. Wu, X. Zhou, C.A. Tang, Debinding and sintering processes for injection molded pure titanium, Powder Met. Technol. 24 (2006) 83–93.
[107] B. Smarslok, R.M. German, Identification of design parameters in metal powder injection molding, J. Adv. Mater. 37 (2005) 3–11.
[108] T. Oki, K. Matsugi, T. Hatayama, O. Yanagisawa, Microstructure and tensile properties of vacuum sintered/HIPed pure titanium, J. Jpn. Inst. Met. 59 (1995) 746–753.
[109] H. Conrad, Effect of interstitial solutes on the strength and ductility of titanium, Prog. Mater. Sci. 26 (1981) 123–403.
[110] Y. Ikeda, S. Takaki, Effects of pores and oxygen content on mechanical properties of a sintered Ti-4 mass % Cr alloy, J. Jpn. Soc. Powder Powder Metall. 42 (1995) 911–917.
[111] T. Gladden, Process for the manufacture by sintering of a titanium part and a decorative article made using a process of this type, U.S. Patent 5,441,695, 1995.
[112] T. Shimizu, K. Matsuzaki, Y. Ohara, Process of porous titanium using a space holder, J. Jpn. Soc. Powder Powder Metall. 53 (2006) 36–41.
[113] N. Sakurai, J. Takekawa, Effect of intermediate CIP treatment on density and strength of Ti-12Mo sintered alloy fabricated by MIM process, J. Jpn. Soc. Powder Powder Metall. 47 (2000) 653–657.
[114] R.M. German, Status of metal powder injection molding of titanium, Int. J. Powder Met. 46 (2010) 11–17.
[115] V. Friederici, A. Bruinink, P. Imgrund, S. Seefried, Getting the powder mix right for design of bone implants, Met. Powder Rept. 65 (2010) 14–16.
[116] H. Miura, Y. Itoh, T. Ueamtsu, K. Sato, The influence of density and oxygen content on the mechanical properties of injection molded Ti-6Al-4V alloys, in: Advances in Powder Metallurgy and Particulate Materials-2010, vol. 4, MPIF, Princeton, NJ, 2010, pp. 46–53.
[117] E. Baril, Titanium and titanium alloy powder injection moulding – matching application requirements, Powder Inject. Mould. Int. 4 (2010) 22–32.
[118] R.M. German, Conceptual optimization of titanium powder injection molding, in: Advances in Powder Metallurgy and Particulate Materials-2010, vol. 4, MPIF, Princeton, NJ, 2010, pp. 67–77.
[119] M. Scharvogel, W. Winkelmueller, Metal injection molding of titanium for medical and aerospace applications, J. Met. 63 (2011) 94–96.
[120] N. Muenya, A. Manonukl, Study of covering conditions for sintering of metal injection moulded commercially pure titanium, J. Met. Mater. Min. 20 (2010) 63–68.
[121] T. Osada, H. Miura, Nitriding resistance of microminiature powder injection molded titanium, Int. J. Powder Met. 46 (2010) 39–44.
[122] O.M. Ferri, T. Ebel, R. Bromann, Substantial improvement of fatigue behavior MIM using boron microalloying, in: Proceedings PM 2010 World Congress, European Powder Metallurgy Association, Shrewsbury, 2010 CD.
[123] Y. Thomas, E. Baril, Benefits of supercritical CO_2 debinding for titanium powder injection moulding, in: Proceedings PM 2010 World Congress, European Powder Metallurgy Association, Shrewsbury, 2010 CD.
[124] O.M. Ferri, T. Ebel, R. Bormann, Influence of surface quality and porosity on fatigue behaviour of Ti-6Al-4V components processed by MIM, Mater. Sci. Eng. 527 (2010) 1800–1805.
[125] E. Aust, W. Limberg, R. Gerling, B. Oger, T. Ebel, Advanced TiAl6Nb7 bone screw implant fabricated by metal injection moulding, Adv. Eng. Mater. 8 (2006) 265–370.
[126] T. Ebel, C. Blawert, R. Willumeit, B.J.C. Luthringer, O.M. Ferri, F. Feyerabend, Ti-6Al-4V-0.5B a modified alloy for implants produced by metal injection moulding, Adv. Eng. Mater. 13 (2011) B440–B453.

[127] O.M. Ferri, T. Ebel, R. Bormann, The influence of a small boron addition on the microstructure and mechanical properties of Ti-6Al-4V fabricated by metal injection moulding, Adv. Eng. Mater. 13 (2011) 436–447.
[128] C. Demangel, D. Auzene, M. Vayssade, J.L. Duval, P. Vigneron, M.D. Nagel, J.C. Puippe, Cytocompatibility of titanium metal injection molding with various anodic oxidation post-treatments, Mater. Sci. Eng. 32 (2013) 1919–1925.
[129] A.P.C. Barbosa, M. Bram, D. Stoever, H.P. Buchkremer, Realization of a titanium spinal implant with a gradient in porosity by 2-component metal injection moulding, Adv. Eng. Mater. 13 (2013) 510–521.
[130] R.M. German, Market and technology for titanium metal powder injection moulding, in Proceedings PM 2010 World Congress, Florence, Italy, October 10–14, 2010; European Powder Metallurgy Association, Shrewsbury, UK, 2010, on CD.
[131] A. Arockiasamy, R.M. German, D.F. Heaney, P.T. Wang, R.L. King, B. Alcock, Effect of additives on sintering response of titanium by powder injection moulding, Powder Met. 54 (2011) 420–426.
[132] G. Wen, P. Cao, D. Zhang, Development and design of binder systems for titanium metal injection molding: An overview, Met. Mater. Trans. 44 (2013) 1530–1547.
[133] R.M. German, Metal powder injection molding (MIM), key trends and markets, in: D. Heaney (Ed.), Handbook of Metal Injection Molding, Woodhead Publishing, London, UK, 2012, pp. 1–25.
[134] J.A. Sago, M.W. Broadley, J.K. Eckert, Metal injection molding of alloys for implantable medical devices, Int. J. Powder Met. 48 (2) (2012) 41–49.
[135] J.A. Sago, M.W. Broadley, J.K. Eckert, H. Chen, Manufacturing of implantable biomedical devices by metal injection molding, in: Advances in Powder Metallurgy and Particulate Materials, vol. 4, MPIF, Princeton, NJ, 2010, pp. 89–99.
[136] R.M. German, Infrastructure emergence for metal injection molded titanium medical devices, Int. J. Powder Met. 48 (2012) 33–38.
[137] H. Miura, M. Noda, H. Kang, Dynamic fracture characteristics of injection molded Ti alloy compacts, in: Advances in Powder Metallurgy and Particulate Materials-2011, vol. 4, MPIF, Princeton, NJ, 2011, pp. 58–64.
[138] A.T. Sidambe, L.A. Figueroa, H.G.C. Hamilton, I. Todd, Taguchi optimization of MIM titanium sintering, Int. J. Powder Met. 47 (2011) 21–28.
[139] R.M. German, Metal Injection Molding A Comprehensive MIM Design Guide, Metal Powder Industries Federation, Princeton, NJ, 2011.
[140] J.E. Bidaux, C. Closuit, M. Rodriguez-Arbaizar, E. Carreno-Morelli, Metal injection moulding of Ti-Nb alloys for implant application, Eur. Cells Mater. 22 (2011) 32.
[141] E. Baril, L.P. Lefebvre, Y. Thomas, Interstitial elements in titanium powder metallurgy: Sources and control, Powder Met. 54 (2011) 183–187.
[142] P. Ewart, S. Ahn, D. Zhang, Mixing titanium MIM feedstock – homogeneity, debinding, and handling strength, Powder Inject. Mould. Int. 5 (2011) 54–59.
[143] R.M. German, Powder injection moulding in the aerospace industry: opportunities and challenges, Powder Inject. Mould. Int. 5 (2011) 28–36.
[144] F.V. Lenel, Resistance sintering under pressure, Trans. TMS-AIME 203 (1955) 158–167.
[145] R. Widmer, Coarse powder techniques, in: J.J. Burke, V. Weiss (Eds.), Powder Metallurgy for High-Performance Applications, Syracuse University Press, Syracuse, NY, 1972, pp. 69–84.
[146] J.E. Smugeresky, D.B. Dawson, Effect of powder particle size and hot isostatic pressing temperature on the properties of Ti-6Al-6V-2Sn, in: F.H. Froes, J.E. Smugeresky (Eds.), Powder Metallurgy of Titanium Alloys, Metallurgical Society, Warrendale, PA, 1980, pp. 127–138.

[147] R.M. German, Titanium powder injection moulding: a review of the current status of materials, processing, properties, and applications, Powder Inject. Mould. Int. 3 (4) (2009) 21–37.

[148] H. Miura, Advanced powder processing techniques of Ti alloy powders for medical and aerospace applications, J. Kor. Powder Met. Inst. 20 (2013) 323–331.

[149] D. Whittaker, Powder processing, consolidation, and metallurgy of titanium, Powder Met. 55 (2012) 6–10.

[150] R.M. German, Progress in titanium metal powder injection molding, Materials 6 (2013) 3641–3662.

Titanium sheet fabrication from powder

21

G.M.D. Cantin, M.A. Gibson
CSIRO Process Science and Engineering, Clayton South MDC, Victoria, Australia

21.1 Introduction

Titanium is not easy to extract from its ore; it is also difficult to process and expensive to fabricate [1]. The conventional production of titanium mill products via the ingot route involves a large number of processing steps. As a result, the high cost of production compared to alternative materials has restricted it to critical applications in the aerospace, prosthetics, and chemical industries. Consequently, there is a general emphasis on lowering the cost of both primary production and fabrication of titanium through the application of innovative processes [2–5]. As highlighted in Figure 21.1 [6], innovative technologies in both the primary and secondary aspects of production will be needed to achieve the greatest cost reductions between powder feedstock and finished product. However, the most critical need is for a reduction of the price and improved availability of the feedstock powder [7]. New powder metallurgy (PM) techniques that have the potential to further reduce production costs with minimal or no loss in the mechanical properties of the resulting product [8–10] are also emerging. These may lead to a significant expansion in Ti powder derived component production into new applications in the automotive, architectural and sporting goods industries.

There are a number of processing routes that have been developed for the production of strip from powder by the direct rolling of various metals [11–16], and titanium specifically [17–23]. Roll compaction is one such technology and refers to the continuous consolidation of metal powder (elemental, blended elemental [BE], or prealloyed [PA] powder) by a standard rolling mill to produce a green strip. This strip undergoes further processing by sintering and rerolling (hot and/or cold) to produce a flat product with a tailored degree of porosity [24] or as fully dense sheet [5]. The general concept of the continuous production of strip from a metal powder originated early in the twentieth century but the process was not fully described until 1950 [25]. During the1950s and through the 1960s, there was a great deal of research and development activity in this field [26] but as yet the full commercial exploitation of the direct powder rolling process has not fulfilled its early potential. Success of the direct rolling process for sheet metal production, in general, depends largely on the availability of abundant low-cost high-quality powder, since as long as the metal in powder form demands a considerable premium over the price of the metal in ingot form, the economics of roll compaction for commodity metal products require careful evaluation. Nonetheless, the potential for the emerging technologies to eliminate several of these processing steps and capture a dramatic reduction in fabrication costs

Titanium Powder Metallurgy. http://dx.doi.org/10.1016/B978-0-12-800054-0.00021-6

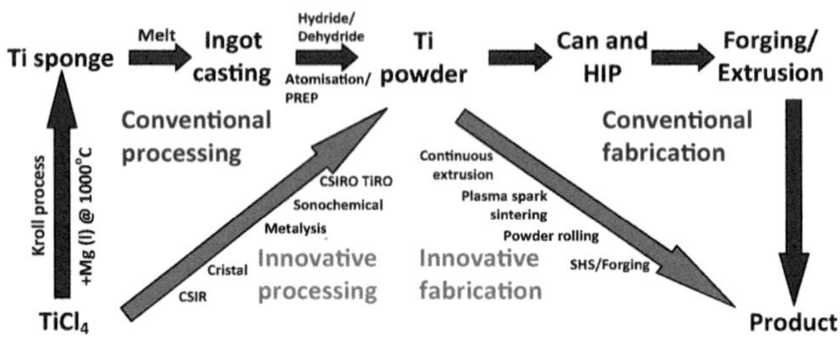

Figure 21.1 The current trend in R&D is to replace conventional processes in the production of titanium with new innovative processes in order to reduce production costs (after Ref. [6]).

is therefore significant. The main advantages of direct powder rolling technology over that of the conventional ingot/wrought route to strip production include the following:

- Reduction in capital equipment by minimizing the number of processing steps.
- Production of higher-purity strip, free from segregation and at a higher yield than can be obtained via conventional means.
- Possibility of producing fine-grained, high-strength strip with minimal preferred texture.
- Production of specialty materials, which are difficult to produce by conventional means, such as bimetallic strips, porous materials, composite bearing materials, functionally graded materials, clad materials, and those that are difficult to hot/cold work.

Methodologies that have been evaluated include bonded or free-flowing powder rolling, sintering, followed by cold rolling and annealing; direct hot rolling of the roll compacted sheet; and hot rolling of multiple layers of roll compacted sheet that are encapsulated in a steel can. Fabrication of fully consolidated sheet has been demonstrated using all three methods, and each processing route has the ability to produce a sheet that meets ASTM B265 specifications. However, not every method currently provides a sheet that can be formed adequately without tearing. The degree of sintering between powder particles, postprocessing density, and the particle-to-particle boundary layer where compositional variations may exist have a significant effect on the ability to form the sheet into useful components [27].

21.2 Direct powder rolling and consolidation

By the 1960s, DuPont [20] had developed a process for the direct rolling of Ti sponge fines into strip in commercial quantities. The mechanical properties were reported to be equivalent to the conventional wrought material; however, the fines contained small quantities of sodium chlorides (0.01–0.05%) trapped in the microstructure. These originated from the extraction process and reportedly made the material essentially unweldable. It has been demonstrated that this still remains an issue for the new low-cost Ti powders, [28] where the removal and minimization of both adsorbed water on the surface of titanium powder and the residues from the reduction process prior to consolidation are critical for achieving equivalent fusion welding performance similar to that obtained in conventional wrought titanium.

The production of strip via a slurry of powder containing an appropriate mixture of binder and plasticizer has been applied to the fabrication of intermetallic strip [29–31]. Hajaligol and coworkers, used both roll compaction of water atomized powder [29] and tape casting of gas atomized powder [30] techniques to produce green strips of FeAl that were then sintered and thermomechanically processed (cold rolled and annealed) into dense foil. It was determined that the water atomized powder, with binder, was much easier to roll compact than the gas atomized powder, also with a binder, because of the spherical nature of the latter and resultant lack of mechanical bonding. It was considered that the use of a binder that yields a low carbon residue after the debinding heat treatment was essential and they used aqueous binders such as methyl cellulose and polyvinyl alcohol to produce green strips. Rahaman et al. [31] applied the tape casting technique to Ti-33wt%Al-6wt%Nb-1.4wt%Ta PREP powder (<90 μm [<0.0035 in.]) and they provide a detailed description of the fabrication process, including the solvent–binder–plasticizer combination that they found to be most successful. The choice of the organic additives to produce the slurry is critical and must fulfill a number of criteria in that it must (1) facilitate the homogeneous packing of the powder particles into a uniformly thick strip and (2) must be able to be removed efficiently at a temperature that does not result in excessive oxidation of the powder nor lead to any contamination of the powder by either reaction products (carbides for example) or residue from the decomposition. The debound strip (~600 μm [~0.0236 in.] thick) was canned and then HIPed at 1373 K (2012°F) for 0.8 ks (15 min) at ~130 MPa (18.85 ksi) to full density (resulting in a 300–400 μm [0.0118–0.0157 in.] thick strip). Tape casting technology was applied successfully by Rak and Walter [32] to the fabrication of green Ti foil. They reported that a highly flexibile and strong foil was achieved for a tape manufactured from fine powder (25 μm [0.00098 in.]). The maximum sintering temperature was as low as 1273 K (1832°F). They also suggested that the crucial innovation was the use of a reducing agent, titanium hydride, in combination with vacuum, in a single sintering process. The organic components of the tape were first evaporated (i.e., toluene, ethylene) at temperatures below 423 K (302°F) and then the organic binder, plasticizer, and surfactant were pyrolyzed to an amorphous carbon phase up to 523 K (482°F). The residual carbon and products of decomposition of TiH_2 and TiO_2 reacted together forming gaseous products, which evaporated readily from the Ti surface and were removed from the system by a vacuum pump. Therefore, the surface of the sintered Ti was completely clean from all contaminants and allowed for a rapid sintering process between the neighbored Ti particles.

Direct powder rolling (DPR) has been applied by Park et al. [33] to make thin sheets from hydride–dehydride (HDH) powders of particle size -100 mesh. CP-Ti sheets of a thickness up to 1.5 mm (0.059 in.) and a width up to 300 mm (11.81 in.) were obtained without any rolling defects. The DPR-processed strips were sintered under vacuum at temperatures around 1273 K (1832°F) and 1523 K (2282°F), for intervals from 3.6 ks (1 h) to 14.4 ks (4 h), and postrolled with a thickness reduction of about 50% and annealed, at 1023 K (1382°F) for 7.2 ks (2 h), to produce about 0.5 mm (0.0197 in.) thick titanium sheets. Microstructural evaluation and mechanical testing indicated that the HDH titanium sheet material showed a near full density and

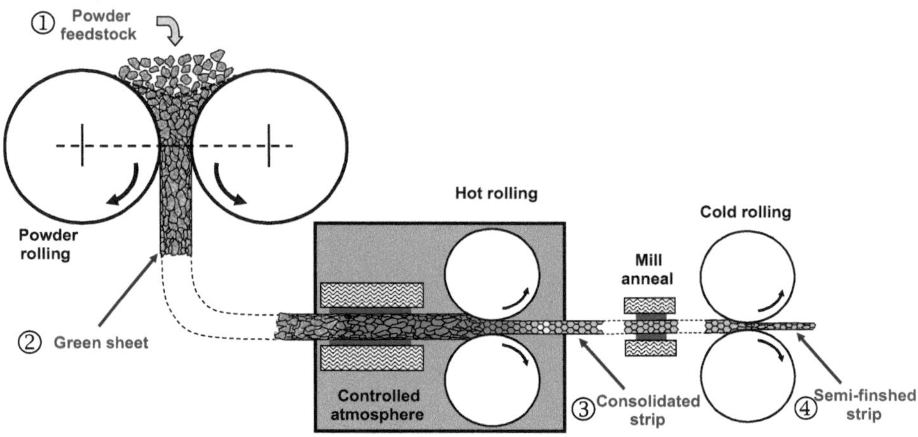

Figure 21.2 Schematic diagram of CSIRO process for continuous production of CP-Ti strip.

exhibit a good combination of tensile properties, depending on the annealing treatment after cold rolling. The optimal tensile properties were reportedly obtained by sintering at temperatures between 1323 K (1922°F) and 1423 K (2102°F).

Direct rolling into fully consolidated sheet appears as one of the most recent and economically adventurous consolidation techniques for metal powders [34]. CSIRO is working on the development of a continuous process for the manufacture of titanium and titanium alloy strip where powders are compacted by cold rolling to form a green strip; this strip is fed to a preheating station where it is rapidly heated in an argon atmosphere for up to a few minutes then transferred in a continuous manner to a hot rolling station (Figure 21.2).

The activities at CSIRO have also focused mainly on the production of thin gauge (1.5 mm [0.059 in.] or less) from CP-Ti powders, PA powders, and BE powders. Most of the development work has been conducted on binderless free-flowing (albeit classified as poor flowing powder) Grade 2 HDH powder with oxygen, nitrogen, and hydrogen contents of 0.14 wt%, 0.011 wt%, and 0.014 wt%, respectively (see Figure 21.3a). Cold metal powder is fed vertically into a set of horizontally positioned symmetrical rolls, which compact the powder into a green sheet that is between 75% and 90% of theoretical density (see Figure 21.3b). The green strips are rolled to thicknesses in the range of approximately 1.5–3 mm (0.059–0.118 in.), although greater thicknesses are possible. This thickness range takes into consideration the requirement for a degree of reduction of about 40–55% at the hot rolling densification (HRD) stage; the mechanisms and efficiency of densification of a porous preform by hot rolling and the implications for the required degree of thickness reduction are discussed in detail later. The achievement of uniformity of density and thickness across the width and along the length of the green strip is important for the success of subsequent rolling operations and the final properties of the wrought material. This is realized in part by controlling the powder rolling and material parameters including roll speed, roll gap, degassing, friction at the roll–powder interface, and particle size distribution.

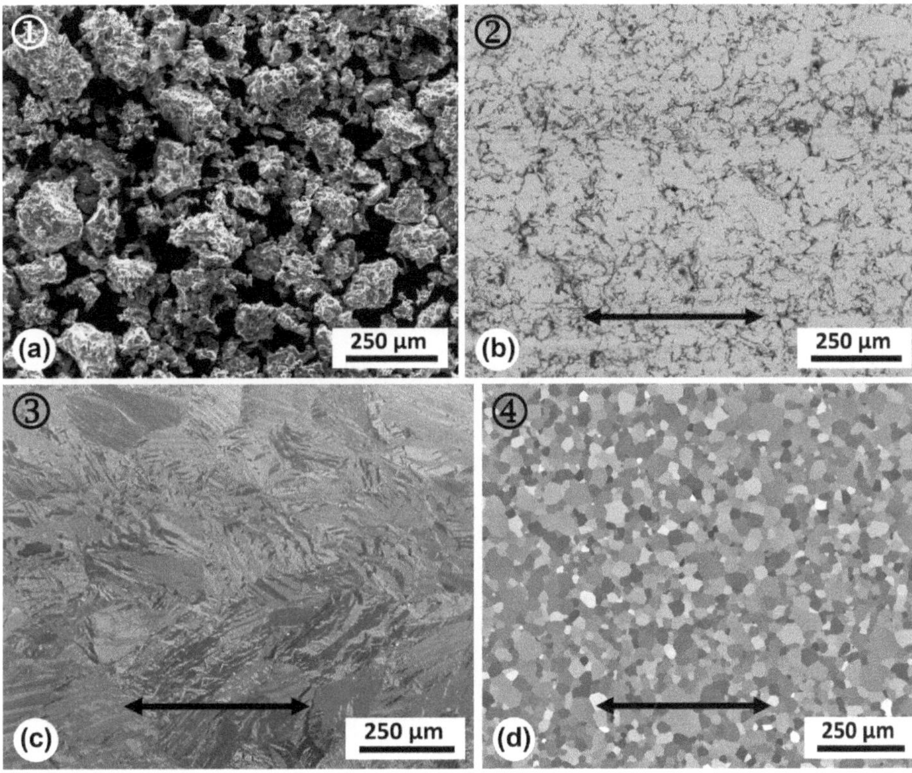

Figure 21.3 (a) Secondary electron SEM image of the powder morphology typical of hydride/dehydride (HDH) CP-Ti powder. (b) Backscattered electron SEM image the surface of a roll compacted 80% dense green sheet made from the same HDH powder shown in (a). (c) Optical image, under polarized light, of the as-hot rolled 100% dense consolidated strip. (d) Optical image, under polarized light, of the recrystallized CP-Ti strip after a standard mill anneal. The rolling direction is indicated by the double-headed arrows. The numbers in the top LHS of each micrograph corresponds to the various stages of the process indicated in Figure 21.2.

The influence of each of these parameters is discussed in detail in subsequent sections. Similarly, the rolling of other powders with various morphologies and properties, such as Armstrong powder [34] and TiRO™ [4], require the control of additional parameters such as powder feeding conditions.

For the purpose of illustration, one such green strip, with a thickness of about 1.8 mm (0.071 in.) and a green density of 88%, was passed through the atmosphere-controlled, inductively heated graphite furnace and then consolidated via a single HRD pass, which reduced the thickness by an amount of approximately 50%. The oxygen, nitrogen, and hydrogen contents of the strip after the HRD pass were 0.19 wt%, 0.02 wt%, and 0.010 wt%, respectively. A significant aspect of the combination of DPR and HRD is the short preheating time prior to hot rolling, which impacts directly on productivity. Another important advantage is that both CP-Ti grade and

Table 21.1 **Tensile properties of CP-Ti strip produced by DPR and HRD and commercial wrought CP-Ti Grade 2 sheet**

Material	Orientation with respect to rolling direction	UTS, MPa (ksi)	0.2% proof stress, MPa (ksi)	Elongation to failure (%)
HRD	Parallel	517 (75.0)	366 (53.1)	27
HRD	Perpendicular	520 (75.4)	372 (54.0)	20
CP-Ti Grade 2 (wrought sheet)	Parallel	524 (76.0)	385 (55.8)	37
CP-Ti Grade 2 (wrought sheet)	Perpendicular	523 (75.9)	450 (65.3)	38
ASTM B265	–	345 (50.0)	276 (40.0)	20

titanium alloy strip (e.g., Ti-6Al-4V) can be manufactured to thin gauges (<1 mm [<0.039 in.]) in one or two hot rolling passes, which significantly reduces material waste, especially the yield losses resulting from surface removal treatments. The strip exits from the hot consolidation rolls, fully dense, directly into a cooling chamber that is also purged with argon to minimize the pickup of atmospheric gases, and has a typical hot worked microstructure (see Figure 21.3c). The HRD strip was subsequently cold rolled (two passes) and annealed in vacuum at 923 K (1202°F) for 7.2 ks (2 h), after which all the strips were treated to remove 0.05 mm (0.002 in.) of material from each of the surfaces. After the mill anneal, the microstructure is fully recrystallized (see Figure 21.3d) and typical properties associated with wrought CP-Ti are obtained. The resulting tensile properties are shown in Table 21.1, together with the properties from a commercially available CP-Ti wrought sheet with comparable interstitial element contents (0.12 wt% O, 0.013 wt% N, and 10 ppm H).

The stress–strain behavior for the HRD strip is observed to be different for the two orientations tested; the strength parallel to the rolling direction tends to be lower than that in the perpendicular direction. This is evident in the proof stress values in particular, while the elongation to failure is higher in the direction of rolling. In comparison with the commercial wrought CP-Ti sheet, the UTS and proof stress values of the HRD strips are similar to those of the commercial wrought sheet. The anisotropy observed, in the yield stress in particular, also exists in the commercial wrought sheet and is a consequence of the rolling texture. However, the elongation to failure of the commercial wrought sheet is significantly higher than that for the HRD strip, but both satisfy the minimum standard. In terms of the microstructures of these materials, the average grain size was determined by electron backscatter diffraction; the commercial wrought sheet had an average grain size of 25 μm (0.00010 in.) and the average grain size of the HRD, cold-rolled, and annealed strip was 35 μm (0.00014 in.).

Each of the unit operations involved in the process (Figure 21.2) has a number of parameters and variables that affect the quality of the finished product and/or the productivity of the system as a whole. These are discussed in greater detail in the following sections.

21.2.1 Powder characteristics and properties

The processing characteristics of metal powders, such as flow rate, apparent density, compressibility, and sinterability are themselves all influenced by particle morphology, size distribution, and chemistry. It is therefore critical that these factors be carefully controlled to ensure consistency of the product. The preferred powder for roll compaction is characterized by generally fine, irregular-shaped particles, relatively low apparent density, and relatively poor flow properties [35,36]. Besides the particle morphology, there are a number of quantitative parameters that are used to characterize a powder, which include (1) particle size and its distribution; (2) particle shape and its variation with particle size; (3) surface area; (4) interparticle friction; (5) flow, packing, and segregation; (6) compressibility; (7) internal particle structure; (8) chemical gradients, surface films, and admixed materials; and (9) consistency of powder from batch to batch and particulate contamination [37–40]. All of the above characteristics have been demonstrated to be of importance to the roll compaction process to some degree or other. The green strength of a powder-compacted sheet results from the mechanical interlocking of the irregularities on the particle surfaces. During compaction, particle rearrangement, plastic deformation, and particularly surface deformation of the powder occur. Titanium powder is susceptible to interstitial element (particularly oxygen) contamination, which may lead to solid solution strengthening of the particles and/or the formation of nonmetallic compounds on the surface. The addition of oxygen or nitrogen has been found to decrease both the green density and the green strength of compacts [41]. It is, therefore, important to identify and control the sources of contamination by interstitials throughout the entire manufacturing process [42].

Moreover, these characteristics are, in turn, largely determined by the method of manufacture of the feedstock powder and its proper handling. There are a wide variety of techniques for powder production, and these involve a wide range of processing routes, which include mechanical comminution (see Figure 21.3a), electrolytic deposition, chemical precipitation, and melt disintegration. The technical aspects of all of these methods of production have been described in great detail in the literature (see, e.g., [37]). However, conventional preparation methods may not meet completely the demand for high purity and flexible controllability over particle morphology and as such new techniques for powder production continue to be developed [43].

A granular material is a complex system that exhibits nontrivial transitions between the static, the quasi-static, and the dynamical states. Understanding flowability is therefore important for multiple reasons [44]. First, automation in production requires powder to move efficiently without jamming and causing disruptions in the manufacturing process. Second, flowability also influences the mechanical properties of the sheet produced. If the particles do not have uniform flow characteristics, imperfections or nonuniformities in green density result and this leads to a poor-quality end product. Indeed, an assembly of grains can behave like a solid or a fluid according to the applied stress. Lumay et al. [45] have shown how three measurement techniques can be used to measure the physical properties of a powder and determine the influence of the different parameters (grain size, grain size distribution, grain shape) on the macroscopic properties of the assembly. Therefore, it is paramount to control the flow characteristics of the powder to obtain satisfactory strip properties. In general, a balance needs to be achieved such that the powder flows sufficiently freely to provide an uninterrupted supply of powder to the roll gap and to minimize any tendency for agglomerates or bridges to form, but not so excessive that the powder flows through the roll gap without being compacted [13,46]. The flow behavior of the powder depends on the size, shape and surface roughness of the particles (and indirectly on the speed of rolling). Spherical powders have good flow properties but lack the ability for particles to interlock during roll compaction, whereas an irregular shape assists in the interlocking of particles and imparts green strength [26]. The cohesion properties influence many characteristics of bulk powders including the apparent and tap densities, the compaction dynamics, friction properties, and flowability [45]. Such behavior typifies fine powders or powders with a fraction of fine particles. The angle of internal friction of a powder (obtained from shear tests) has been found to depend on both the particle size and shape such that asymmetric particles tend to increase friction and therefore decrease flow, and an increase in particle size decreases the angle of internal friction. The latter effect is more pronounced in the fine particle size range [47]. The effects of particle size on flowability have been studied and discussed by Liu et al. [48]. Although it is generally accepted that a larger particle size improves flowability, the size distribution is an equally important factor. By determining the flowability of sieved size fractions, the authors found that as the particle size distribution narrows the flowability improves for powders with the same median size. Poor flowability affects powder compaction by causing a number of issues such as stagnant zones in the powder body within the hopper, residual powder in the hopper, loss of flow and formation of powder bridges, and, due to their strong interaction with surfaces, a tendency to form residues on both tool and container surfaces [45].

The cohesion between powder particles and, therefore, the flowability can also be affected by the moisture content [26,49]. In general, increasing the moisture content of a powder decreases its ability to flow freely. For aluminium powders with a native oxide surface, Stevens et al. [49] have found from shear cell tests that increasing the humidity increases the bulk strength of the powder as a result of adsorbed water. This was attributed to the formation of liquid bridges between the particles such that particles adhere together due to capillary forces. The resulting effect in an increase in cohesion and particle strength [49]. Flowability of powders can be measured by relatively simple tests, such as the determination of the Hausner ratio, the Carr index, and the angle of repose, and by more complex tests such as shear tests. The latter tend to provide more comprehensive information on flow properties and with greater sensitivity and reproducibility [48].

The conditions and parameters that affect incipient flow and free flow of pharmaceutical powders have been modeled. The incipient flow criterion is based on the assumption that the gravity of the particles for a loosely packed powder exceeds the total adhesive forces. The models include the effect of the particle size and shape [48]. For example, the models can be used to show the influence of particle shape on flowability using a sphericity factor (ratio of the surface area of the equivalent volume sphere to the particle surface area) where powers with irregular particles have poorer flowability. Although, the experimental work of Liu et al. [48] was performed on pharmaceutical powders (with an elongated, irregular morphology), the trends that they observed are applicable to powders in general. Tap density and Hausner ratio (ratio of bulk tap to fill density) measurements provide simple but, in general, reasonable indications of powder flowability. The Hausner ratio tends to decrease with increasing particle size, meaning an increase in flowability with an increase in particle size. A bulk powder mixture with a smaller mean size and narrow size distribution may flow better than a coarser powder mixture with a wider size distribution and a larger mean particle size.

21.2.2 Cold roll compaction process

It is the main objective of a roll compaction mill to produce green strip in a continuous manner, with the required thickness and density to facilitate downstream processing into finished strip. Roll compaction operations require both the feeding (as powder is fed into the rolls prior to compaction) and the processing (compaction) conditions to be controlled carefully. Furthermore, the feeding conditions impact processing conditions and, thus, the interaction between feeding and processing factors also affects the physical qualities of the final compact. It has also been shown that the green sheet thickness needs to be approximately four times that required for the finished strip thickness, if similar mechanical properties to those of conventional wrought material are to be obtained in the end product. The roll geometry (i.e., roll diameter, barrel length, and roll gap), roll speed, the elastic modulus of the roll material, and the load-bearing capacity are all important process parameters [50]. The ability of the rolling mill to compact powder to a strip of a given thickness is influenced by the diameter of the rolls significantly. The roll diameter is one of the major process limitations and it has been reported in the literature that the maximum thickness of powder rolled strip varies from 0.33% to 1% of the roll diameter, depending on the type of powder [26].

During cold direct powder rolling, a free-flowing metal powder is transformed into a green strip of sufficient density and strength at room temperature that it will hold together during subsequent processing. In undergoing this transformation, the material passes through four stages of densification, as shown schematically in Figure 21.4a, for the case of "saturated" feeding conditions. In "saturated" feeding, an excess amount of powder is fed to the nip of the rolls, and the amount that passes through the mill is controlled by the frictional forces existing between the powder and roll surfaces, the

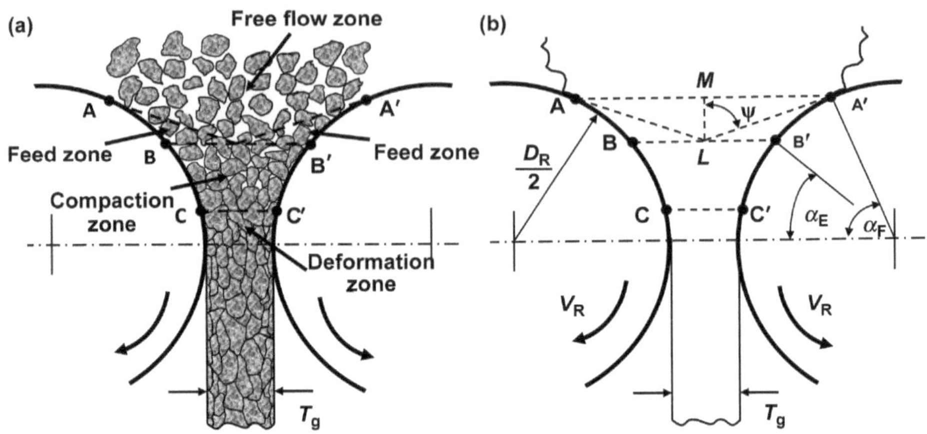

Figure 21.4 Schematic diagram of the "cold" direct rolling of a metal powder (after Refs [26,55]). NB: D_R is the roll diameter; V_R is the roll speed; Ψ is the slip line angle, which is a function of the angle of internal friction; α_E is the compaction angle; α_F is the feed angle, and T_g is the thickness of the green strip.

powder particles themselves, and the flow properties of the powder [11]. These stages are identified as follows:

1. The free-flow zone, in which the powder falls under the influence of gravity. According to Sturgeon et al. [13], variations of powder head within this zone have a negligible effect on strip properties. Dube [11], however, mentions that the head of powder in the hopper should be held constant in order to maintain a constant green density along the strip length.
2. The feed zone, in which the powder is dragged toward the roll bite under the action of friction between the particle/rotating roll interfaces. In this stage, densification occurs solely by particle rearrangement, and the medium remains noncoherent. The feed zone is defined by two angles α_F and α_E, where α_F is called the feed angle as it denotes the starting feed position on the roll surface. In this region, the relationship between the rolling pressure, σ, and rolling shear, τ, is defined by the friction coefficient at the roll–powder interface, μ, such that when $\tau = \mu.\sigma$, there is relative movement of the roll and the powder in contact with the roll [51].
3. The compaction zone, within which the rolling pressure becomes effective and the powder mass becomes coherent. The start of this zone is defined by the angle α_E, which is termed as the angle of compaction (alternatively referred to as the gripping angle, angle of rolling, nip angle, or angle of grip). Once the powder is gripped, the roll shear is less than the product of the coefficient of friction at the roll powder interface and the roll pressure [51].
4. The deformation zone, where deformation of the powder particles take place which leads to further increases in green density and strength. This zone often overlaps with the compaction zone [26,55].

The feeding of powder into the roll gap is the initial operation involved with roll compaction and the way in which this is achieved has a profound influence on the quality of the green strip. A number of configurations have been devised for the metering of feedstock powder into the roll bite of horizontal and vertical variants of a roll compaction mill and these have been described in the literature (see, e.g., [26,52]). There are two fundamental methods by which powder can be gravity fed (favored by a horizontal mill configuration) into the roll gap; these being "saturated" and "unsaturated" supply systems [16,52]. In a "saturated" feed system a surplus of powder is maintained at the roll bite and the amount of powder that passes through the mill is determined largely by the particle–particle and particle–roll surface frictional forces. A constant powder feed ensures that a constant density is obtained throughout the green strip. The surface of the rolls also needs to be maintained "constant" in order to keep the green density uniform. In an "unsaturated" feed system, feedstock metering is controlled solely by the dispensing apparatus, once an appropriate feed rate has been established. This usually involves a complicated control/feedback system. Both methods have been reported widely in the literature to produce strip from a variety of materials. Independent of method, precise control over the distribution of powder to the roll gap from edge to edge of the strip is required. Uniformity of green strip density eliminates defects that appear during subsequent processing of the finished sheet. Excessive powder feed to the edges of the strip will result in edge "frilling" (a sinusoidal profile of the sheet in the regions close to the edges) during finish cold rolling. This situation is compensated for by actively feeding more powder to the central portion of the strip. Overcompensation can be equally as detrimental and leads to "cupping" (the formation of depressions or mounds in the central section of the sheet) of the strip on finish cold rolling [7]. These effects become more acute with an increase in strip width, and control over the feed rate to various sections becomes progressively more critical. A variation on the above two powder supply methods is forced feeding through the application of an external pressure. One such means to achieve this is through the use of an Archimedean screw arrangement [11]. Chekmarev et al. [53] describe the use of an auger system to force-feed fine iron powder into the roll gap to enable the roll speed to be increased (up to 3.2 m/s [10.5 ft/s], which is an order of magnitude larger than generally achievable from gravity-fed machines) without a loss in quality of the green sheet produced and thereby increasing productivity significantly. This same system has apparently also been used to force-feed titanium powder, [54] although no processing details were provided. Both of these methods are influenced by the flow properties of the feedstock powder. The powder must flow smoothly through the feed hopper system to the compaction mill without the tendency to stick/slip or to bridge.

The process of free gravity rolling of metal powders is influenced by several parameters. The following parameters, which can be divided into two groups, influence the properties of the green strips [50].

1. Geometric parameters:
   - **(a)** Thickness of the green strip, T_g.
   - **(b)** Width of the powder feed zone, W_F.
   - **(c)** Width of the compaction zone, W_E, the starting point of densification/nip region.
   - **(d)** Diameter of the rolls, D_R.
   - **(e)** Compaction angle, α_E (related to the roll to powder friction/roll surface roughness).
   - **(f)** Width of the strip, L.
   - **(g)** Starting roll gap, g_s.

2. Physical, chemical, and mechanical parameters:
(a) Density of the rolled strip, ρ_S.
(b) Density of the powder, ρ_P (fill and tap densities can provide estimates).
(c) Powder particle size distribution and chemical composition.
(d) Powder particle shape and friability.
(e) Rolling rate, $V_R = \omega D_R/2$ (ω is roll angular velocity) cm/s.
(f) Viscosity of the gaseous medium in which rolling takes place.
(g) Roughness of the roll surfaces. Friction coefficient at the roll/powder interface, μ.
(h) Direction of rolling (vertical, horizontal, or angled).
(i) Rigidity of the operating stand of the roll mill.

The feed zone is contained within the angle, α_F, but outside the angle of compaction, α_E. By an analysis similar to that used to determine the gripping angle in conventional rolling theory, [13] it can be shown that a material will be drawn into the roll bite if the horizontal component of the friction force is larger than or at least equal to the opposing horizontal component of the normal force (for vertically positioned rolls). So the maximum angle of entry, α_F, cannot be greater than $\tan^{-1} \mu$, where μ is the static coefficient of friction between the material and the roll surface. The value of the feed angle, α_F, for various powders and roll surface finishes, has been estimated by Sturgeon et al. [13] by the use of a static test in which the powder is poured onto the top of a static roll and the angle of contact at which the powder slides off the roll surface is measured. Furthermore, Sturgeon et al. [13] have demonstrated that the strip thickness can be controlled by varying the feed angle, which was achieved by shielding the roll surface. The mass of the strip per unit area, which is controlled by the movement of the shielding plate, increases up to the angle that corresponds to the measured feed angle. It is also reported that typical values for the compaction angle are between 0.140 rad (8°) and 0.175 rad (10°) for most powders that have been successfully compacted into the strip [13]. Dube [26] reported that for most metals, the angle of compaction is about 0.122 rad (7°). The compaction angle influences both the density and the thickness of the strip. An increase in the compaction angle increases the green density at the exit of the roll gap and also increases the strip thickness [50].

During powder rolling, air is ejected and passes back through the powder mass, which can fluidize the powder feed, thereby restricting powder flow. Fluidization tends to increase as the rolling speed increases, typically above a "transition" speed [26]. The influence of the entrained gas that accompanies the particles is to restrict the amount of powder that is compacted and therefore results in both lower strip density and thickness. The volume of gas is inversely proportional to the apparent density of the feedstock powder. The majority of the entrained gas is expelled as the powder passes through the last three zones of the compaction process, and the direction in which the gas escapes is counter to the flow of the particles. As a consequence of this, the gas can create turbulence in the feed zone if it has a high velocity, high density, and high volume percent and if the particle mass has small average particle size and low solid density. It is also assumed that the escaping gas can affect both the coefficient of friction and the internal friction coefficient [55]. This can pose limits on the speed at which roll compaction can be conducted successfully, which influences productivity.

Some researchers have attempted to use a lower-viscosity gas, such as hydrogen, as the compacting atmosphere. The most effective method would be to carry out compaction under vacuum but this, as well as the use of a low-viscosity gas, has considerable practical implications [26]. The volume of powder between the rolls and feeder hopper needs to have good permeability so that the entrapped air can escape from the roll nip. Degassing becomes more difficult at higher rolling speeds. If the air is not properly released, the disturbance of the powder in the hopper can be sufficient to interfere with the smooth flow of powder to the roll bite and the strip produced will not be of uniform density and/or may contain discontinuities along the length. A change in the amount of fine powder particles in the powder feedstock can also affect the permeability of the powder, with the permeability decreasing the higher the percentage of fine particles in the powder feedstock. The routes for gas to escape during compaction can be between the hopper (feeder plate) and the rolls, and through the powder mass within the hopper.

The factors involved in roll compaction of metal powders have been analyzed in detail in a number of articles [56–58]. The entrainment of powder in the compaction zone depends on the friction coefficient at the roll/powder interface. The latter has a significant impact on the rolling process. Cunningham [51] has shown that, for pharmaceutical powders, the density of the strip at the exit of the rolls increases progressively while the corresponding maximum roll pressure increases slowly and then rapidly as the coefficient of friction at the roll/powder interface increases. Since the angle of compaction increases with roll friction, the influence of an increase in the friction coefficient also affects the angle of delivery of the powder. The powder should be delivered at a sufficiently high feed angle to take advantage of any increase in coefficient of friction at the roll/powder interface. Otherwise, if the powder sticks immediately to the rolls, the final density will depend only on the roll geometry and any increase in roll/powder friction will not increase the achievable densification [51]. The modeling work of Muliadi et al. [59] also, again for pharmaceutical powders, predicts an increase in the compaction angle as the friction coefficient increases, which is suggested to be consistent with the idea that rougher rolls can "grip" the powder better, resulting in powder being nipped closer to the feed zone. This behavior was explained in more detail in terms of the wall failure criterion for an ideal Mohr–Coulomb material, which further shows that there exists a limit to which an increase in the friction coefficient will lead to an increase in the compaction angle (NB: The Mohr–Coulomb criterion for yielding represents the linear envelope that is obtained from a plot of the shear strength, τ, of a material versus the applied normal stress, σ (compression is assumed to be positive), and is stated as $\tau = \sigma \tan\phi + c$ where c is the coefficient of cohesion (which is zero for a noncohesive material), and ϕ is the angle of internal friction [60,61]). The roll force (or roll separating force), which is the force between the centers of the rolls that acts to push the rolls apart, and strip density as a function of friction coefficient were also calculated by Cunningham; [51] at low friction coefficient in which the powder does not experience extensive densification, the roll force per unit length is low. For a given compaction ratio (or degree of densification), the thickness of the strip increases as the coefficient of friction increases [26]. In turn, the factors likely to affect the roll/powder friction include the nature of the two surfaces involved, surface roughness of the rolls, rolling speed, temperature, and powder morphology and hardness. High green strength brings several advantages such as producing longer strips and ease of handling and shipping in the green state.

Although the functioning principle of the roll compaction process is simple, its mechanism is not well understood. Efforts to quantitatively model the process have proven challenging because of complex material behavior in both the feeding and compaction zones. In efforts to better understand the roll compaction mechanism and to facilitate scale-up operations, numerous studies have attempted to describe explicit relationships between the various process parameters and the final physical properties of the resulting green sheet. The modeling of the powder roll compaction process dates to the 1960s, with much of the work conducted by Russian workers [62]. The relationships that were derived attempted to relate principally the thickness and density of the green sheet. Parameters affecting feed conditions (i.e., feed powder properties, feeding method, sealing systems, powder deaeration [63]) and processing conditions (i.e., roll surface, roll speed, roll pressure, roll gap) have been studied extensively. The key to a better understanding of the roll compaction process is not about determining individual relationships between the various individual process parameters but in determining how all of these input factors interact with one another. Towards this end, a dimensionless relationship between key process parameters and final compact properties has been elucidated by Rowe et al. [64]. Relating yield-to-viscous stresses during roll compaction delineates how the operating conditions impact green sheet properties, and thus provides guidance with regard to the design space for a roll compaction process. This dimensionless relationship also provides insight as to how roll compaction operating conditions should scale with equipment size in order to maintain desirable green sheet properties.

Analytical models were then devised to enable the prediction of the roll surface pressure, force, and torque from the physical characteristics of the powder; the geometry of the roll gap; and the dimensions of the rolls. Johanson proposed such a model in 1965 [65]. Later a model based on the "slab method" was applied to analyze the roll compaction of metal powder by Katashinskii [62] to predict the pressure distribution and roll-separating force [66]. More recently, finite element modeling approaches (one-, two-, and three-dimensional models) have been used to predict the density distribution of the resulting strip. Some of these models have been devised for pharmaceutical powders [66]. A review of the experimental validation for these models is summarized and discussed by Muliadi et al. [67]. This latter author also goes on to summarize the development of finite element methods (FEM) for modeling the roll compaction process using two- and three-dimensional FEM to predict the density distribution of roll compacted strips.

21.2.3 Powder consolidation and post processing

The green strip produced by cold direct powder rolling generally contains from 10% to 25% porosity, possessing minimal strength and essentially no ductility. In order to improve these mechanical properties, so that the strip can be handled more easily during subsequent processing, the strip is subjected to a sintering operation. Both batch-type and continuous sintering furnaces (either in-line or in parallel) have been considered for plant operation, but irrespective of the method employed it is generally accepted that economic constraints dictate the use of high temperatures (typically in excess of $0.75T_m$) and short cycle times or high throughput. There is also the added complication of needing to protect the green strip from excessive oxidation during the sintering process; this is particularly paramount for titanium-based materials. Increasing the sintering temperature is often favored over an increase in sintering time because of positive effects on final strip properties [11]. Use of increased temperature is a more practical solution for green strips that have been produced from single metal powders, or prealloyed feedstock.

In general, the strip exiting the sintering process still retains residual porosity and as a consequence requires further densification to attain optimum mechanical properties. This can be performed by hot rolling in single/multiple passes, or by repeated cold rolling/annealing cycles. During the rolling of porous strip, densification occurs under the action of a generalized stress field, which has both hydrostatic and shear components. The work of rolling collapses the pores and welds the particle surfaces together to increase the strip density. The hot rolling of porous strip has a number of processing advantages, which include shortening the sintering time and utilizing the heat from the sintering stage, and it has the capability to achieve near-full density in a single pass. As a consequence of these benefits, hot rolling is highly favored in a continuous process [46,68]. The rolling operation involves a variety of important parameters such as the initial density of porous strip, the rolling temperature, the degree of reduction, the rolling mill specifications (i.e., roll diameter, roll speed and roll temperature), and most significantly the rollability of the material being processed. Weaver et al. [69] have conducted a comprehensive investigation into the efficiency of densification of porous iron powder preforms by single-pass hot rolling. As the preform becomes denser with its passage through the roll gap, its compressive strength increases and the rolling pressure rises. Under these conditions, plastic deformation becomes the predominant densification mechanism, wherein some of the plastic flow goes into elongating the strip rather than in closing pores. The greater the proportion of this latter mechanism active during densification, the less efficient is the overall process. This has certain practical consequences in that the degree of thickness reduction to produce a fully dense strip is always greater than that which would be required theoretically to just close the pores present. In practice, the amount of reduction required could be as much as double the theoretical value [68]. This factor must therefore be taken into consideration when producing the initial green strip, if a specific gauge is targeted in the finished wrought product.

Research by Podrezov et al. [70] determined that in Ti powder consolidation, a physical bond forms at temperatures about 573 K (572°F) higher than the temperature at which mechanical bonding occurs. They showed that fracture analysis permitted the identification of factors that promote or hinder the formation of a physical bond. The observed fracture patterns testify that the fracture of a sintered titanium sample still occurs via an interparticle mechanism at temperatures much higher than that at which a mechanical contact forms. Further work by Podrezov et al. [71] has investigated the temperature dependency of the quality of contact bonds in samples produced by both dynamic hot pressing (DHP) and sintering. It was observed that mechanical, electrical, and physical contacts form in DHP well before they do in static sintering. For example, the bonds are 50% complete at the following temperatures: electrical contact at 648 K (707°F) and at 858 K (1085°F), mechanical contact at 723 K (842°F) and 1003 K (1346°F), and physical contact at 1133 K (1580°F) and 1120°C (2048°F) in DHP and in sintering, respectively. It was shown that during DHP, physical bonds of high quality are formed in the temperature range between 1073 K and 1223 K (1472 and 1742°F). Such thermomechanical treatment promotes the recrystallization of the material near interparticle boundaries and, thereby, prevents interparticle fracture. The precondition for such an improvement follows from the theory of hot pressing of compact materials, indicating that a high strain rate is one of the primary factors promoting dynamic recrystallization. Podrezov et al. [72] have, more recently, reported that DHP at an optimal temperature (1183 K [1670°F]) and strain rate (~200 s^{-1}) leads to a fine-grained structure with 2–5 μm (0.000079–0.000197 in.) grains due to dynamic recrystallization. These factors are also likely to be important in hot-rolling consolidation. A relationship between the fracture pattern and mechanical properties of a powder-derived titanium strip, by powder rolling, has been established by Gogaev et al. [73]. Samples with the highest true strain-to-failure (a parameter that characterizes the quality of the interparticle physical bonds) showed intracrystalline dimple fracture.

Irrespective of the route by which the strip was manufactured, it invariably requires additional cold rolling to clean up the surface finish and to bring the gauge into specification, and possibly a final heat treatment, depending on the temper required, and this is generally performed using conventional methods. The static mechanical properties determined for CP-Ti [74] or Ti-6Al-4V [21] are more than satisfactory, with only the ductility of the alloys derived from powder feedstock being generally lower than conventionally produced sheet.

Recently some alternative processing techniques have been demonstrated that could prove useful for enhanced mechanical property development. Kim et al. [75] have evaluated the effect of the roll speed ratio (RSR) on the microstructure, texture and mechanical properties of CP-Ti (Grade 2). The influence of differential speed rolling (DSR) was examined at 673 K (752°F) and also for a wide range of RSRs, between 1 and 5, to produce Ti sheets with an ultrafine-grained microstructure and exhibiting high strength. At a RSR of 5, a defect-free sheet, with shear bands and ultrafine grains distributed homogeneously over the entire thickness of the sheet, could be fabricated through a single rolling pass. Huang et al. [76] have demonstrated that the stretch formability of a CP-Ti sheet can be improved markedly by DSR. A basal texture can be obtained by the simultaneous action of shear deformation and high deformation temperature. Pure titanium sheets with ultrafine-grained (UFG) microstructures were produced by high-ratio DSR at different RSRs and different roll temperatures [77]. Significant strengthening was achieved after high-ratio DSR, and the sample processed at a higher speed ratio or a lower processing temperature yielded a higher strength. The UFG Ti exhibited high corrosion resistance in H_2SO_4 and HCl solutions. Randman et al. [78] have investigated the scaling up of the asymmetric rolling process to generate wider magnesium alloy sheet with greater formability for industries such as automotive and aerospace. It is envisaged that powder-derived titanium sheet would also benefit greatly from the application of such postprocessing.

21.3 Summary

In order to lower the cost of components made from titanium and titanium alloys, a number of innovative processes are being developed for both the production of powders and the consolidation of these powders into useful products. The combination of cold and hot direct rolling of Ti powder into fully consolidated strip and sheet represents one of the best means for rapid consolidation. The quality of the green sheets made by this process, as well as the uniformity of the green density and thickness across individual green strips, are governed largely by the properties of the powder, and the feeding and processing conditions during cold roll compaction. In turn, the quality of the green sheet has a profound influence on the hot rolling behavior and, as a consequence, the properties of the final product, be it fully dense strip or with a tailored amount of porosity. As a result, all the processing parameters and conditions must be carefully controlled. The nature and integrity of the interparticle bonding is paramount for the development of useful mechanical properties, ductility in particular.

References

[1] B.E. Hurless, F.H. Froes, Cutting the cost of titanium, Adv. Mater. Process. 160 (12) (2002) 37–40.
[2] Z.Z. Fang, P. Sun, Pathways to optimize performance/cost ratio of powder metallurgy titanium – a perspective, Key Eng. Mater. 520 (2012) 15–23.
[3] D.S. van Vuuren, S.J. Oosthuizen, J.J. Swanepoel, Development of a continuous process to produce Ti via metallothermic reduction of $TiCl_4$ in molten salt, Key Eng. Mater. 551 (2013) 16–24.
[4] C. Doblin, D. Freeman, M. Richards, The TiRO™ process for the continuous direct production of titanium powder, Key Eng. Mater. 551 (2013) 37–43.
[5] G.M.D. Cantin, P.L. Kean, N.A. Stone, R. Wilson, M.A. Gibson, M. Yousuff, D. Ritchie, R. Rajakumar, Innovative consolidation of titanium and titanium alloy powders by direct rolling, Powder Metall. 54 (2011) 188–192.
[6] F.H. Froes, B. Trindale, The mechanochemical processing of aerospace metals, J. Mater. Process. Technol. 153-154 (2004) 472–475.
[7] F.H. Froes, D. Eylon, Powder metallurgy of titanium alloys, Int. Mater. Rev. 35 (3) (1990) 162–182.
[8] V.S. Moxson, J.I. Qazi, S.N. Patankar, O.N. Senkov, F.H. Froes, Low-cost CP titanium and Ti-6Al-4V alloys, Key Eng. Mater. 230-232 (2002) 339–343.
[9] W. Peter, W. Chen, Y. Yamamoto, R. Dehoff, T. Muth, S. Nunn, J. Kiggans, M. Clark, A. Sabau, S. Gorti, C. Blue, J. Williams, Current status of Ti PM: progress, opportunities and challenges, Key Eng. Mater. 520 (2012) 1–7.

[10] M. Ashraf Imam, F.H. Froes, R.G. Reddy, Cost effective developments for fabrication of titanium components, Key Eng. Mater. 551 (2013) 3–10.
[11] R.K. Dube, Particle technology methods for making metal strip part 1, Powder Metall. Int. 13 (4) (1981) 188–190; Part 2. Powder Metall. Int. 14 (1) (1982) 45–48; Part 3. Powder Metall. Int. 14 (2) (1982) 108–111; Part 4. Powder Metall. Int. 14 (3) (1982) 163–165; Part 5. Powder Metall. Int. 15 (1) (1983) 36–40.
[12] G.M. Derkacheva, G. Ya Kalutskii, R.V. Minakova, Copper powder rolling technology (review), Sov. Powder Metall. Metal Ceram. 39 (3–4) (2000) 202–206.
[13] G.M. Sturgeon, G. Jackson, V. Barker, G.M.H. Sykes, The production of stainless-steel strip from powder, Powder Metall. 11 (22) (1968) 314–329.
[14] D.G. Hunt, R. Eborall, The rolling of copper strip from hydrogen-reduced and other powders, Powder Metall. (5) (1960) 1–23.
[15] D.K. Worn, The continuous production of strip by the direct rolling process, Powder Metall. (1/2) (1958) 85–93.
[16] A.F. Marshall, Some mechanical requirements of plant for the roll-compacting process, Powder Metall. (5) (1960) 24–31.
[17] G.F. Tikhonov, L.A. Pyryalov, V.K. Sorokin, A.N. Nikolaev, V.G. Khromov, L.S. Shmelev, Production of properties of porous rolled materials, Sov. Powder Metall. Metal Ceram. 12 (12) (1973) 1011–1014.
[18] L.S. Shmelev, Industrial production of porous strip from stainless steel and titanium powders, Sov. Powder Metall. Metal Ceram. 10 (1) (1971) 73–74.
[19] Y. Muramatsu, K. Tamura, Production of titanium strip by powder rolling, Trans. Natl. Res. Inst. Metals (Japan) 20 (3) (1978) 172–177.
[20] Titanium: past present and future, Report of the Panel on Assessment of Titanium Availability: current and future needs of the Committee on Technical Aspects of Critical and Strategic Materials, National Materials Advisory Board Commission on Engineering and Technical Systems National Research Council, Publication NMAB-392, National Academy Press Washington DC, USA, 1983, pp. 159–160.
[21] D.H. Ro, M.W. Toaz, V.S. Moxson, The direct powder rolling process for producing thin metal strip, JOM 35 (1) (1983) 34–39.
[22] K. Kusaka, T. Shimizu, Production of Ni-Ti alloy strip by powder rolling and its shape memory behaviour, Denki Seiko (Electr. Furn. Steel) 58 (4) (1987) 226–234.
[23] V.A. Duz, V.S. Moxson, The direct rolling process for producing titanium and titanium alloy foils, sheets and plates, in: L.L. Shaw, F.D.S. Marquis, E.A. Olevsky, I.E. Andderson, M.G. McKimpson, J.P. Singh, J.H. Adair (Eds.), Science and Technology of Powder Materials: Synthesis, Consolidation and Properties, Materials Science and Technology '05, 2005, pp. 45–52.
[24] G.F. Tikhonov, L.A. Pyryalov, V.K. Sorokin, A.N. Nikolaev, V.G. Khromov, L.S. Shmelev, Production of properties of porous rolled materials, Sov. Powder Metall. Metal Ceram. 12 (12) (1973) 1011–1014.
[25] C. Naeser, F. Zirm, Steel strip experimentally rolled from powder, Stahl und Eisen 70 (1950) 995–1004.
[26] R.K. Dube, Metal strip via roll compaction and related powder metallurgy routes, Int. Mater. Rev. 35 (5) (1990) 253–291.
[27] W.H. Peter, T. Muth, W. Chen, Y. Yamamoto, B. Jolly, N.A. Stone, G.M.D. Cantin, J. Barnes, M. Paliwal, R. Smith, J. Capone, A. Liby, J. Williams, C. Blue, Titanium sheet fabricated from powder for industrial applications, JOM 64 (5) (2012) 566–571.

[28] T.R. Muth, Y. Yamamoto, D.A. Frederick, C.I. Contescu, W. Chen, Y.C. Lim, W.H. Peter, Z. Feng, Causal factors of weld porosity in gas tungsten arc welding of powder-metallurgy-produced titanium alloys, JOM 65 (5) (2013) 643–651.
[29] M.R. Hajaligol, S.C. Deevi, V.K. Sikka, C.R. Scorey, A thermomechanical process to make iron aluminide (FeAl) sheet, Mater. Sci. Engi A258 (1998) 249–257.
[30] R.E. Mistler, V.K. Sikka, C.R. Scorey, J.E. McKernan, M.R. Hajaligol, Tape casting as a fabrication process for iron aluminide (FeAl) thin sheets, Mater. Sci. Eng. A258 (1998) 258–265.
[31] M.N. Rahaman, R.E. Dutton, S.L. Semiatin, Fabrication of dense thin sheets of γ-TiAl by hot isostatic pressing of tape cast monotapes, Mater. Sci. Eng. A360 (2003) 169–175.
[32] Z.S. Rak, J. Walter, Porous titanium foil by tape casting technique, J. Mater. Process. Technol. 175 (2006) 358–363.
[33] N.-K. Park, C.H. Lee, J.H. Kim, J.K. Hong, Characteristics of powder-rolled and sintered sheets made from HDH Ti powders, Key Eng. Mater. 520 (2012) 281–288.
[34] D. Mangabhai, K. Araci, M.K. Akhtar, N. Stone, D. Cantin, Processing of titanium powder into consolidated parts & sheet, Key Eng. Mater. 551 (2013) 57–66.
[35] D.K. Worn, The continuous production of strip from metal powder by the direct rolling process, Sheet Metal Ind. (1958) 615–619 August.
[36] T. Stevens Daugherty, Direct roll compacting sheet from particles, Powder Metall. 11 (22) (1968) 342–357.
[37] R.M. German, Powder Metallurgy Science, Metal Powder Industries Federation, Princeton, NJ, 1984.
[38] V.N. Eremenko, V.I. Nizhenko, The role of surface phenomena in powder metallurgy processes – a review, Sov. Powder Metall. Metal Ceram. 2 (4) (1963) 270–278.
[39] V.P. Katashinskii, G.A. Vinogradov, N.V. Rukhailo, G. Ya Kalutskii, Effects of particle size on the conditions of the strip shaping process in powder rolling – part I, Sov. Powder Metall. Metal Ceram. 14 (10) (1975) 791–794.
[40] V.P. Katashinskii, G.A. Vinogradov, N.V. Rukhailo, G. Ya Kalutskii, Effects of particle size on the conditions of the strip shaping process in powder rolling – part II, Sov. Powder Metall. Metal Ceram. 14 (11) (1975) 883–887.
[41] J.B. Lim, C.J. Bettles, B.C. Muddle, N.-K. Park, Effects of impurity elements on green strength of powder compacts, Mater. Sci. Forum 654–656 (2010) 811–814.
[42] E. Baril, L.P. Lefebvre, Y. Thomas, Interstitial elements in titanium powder metallurgy: sources and control, Powder Metall. 54 (3) (2011) 183–187.
[43] Q.G. Weng, R.D. Li, T.C. Yuan, Z.H. Zhou, Y.H. He, Investigation to impurity content and micromorphology of high purity titanium powder prepared by molten salt electrolysis, Mater. Res. Innovat. 17 (2013) 396–402.
[44] N.A. Pohlman, J.A. Roberts, M.J. Gonser, Characterization of titanium powder: microscopic views and macroscopic flow, Powder Technol. 228 (2012) 141–148.
[45] G. Lumay, F. Boschini, K. Traina, S. Bontempi, J.-C. Remy, R. Cloots, N. Vandewalle, Measuring the flowing properties of powders and grains, Powder Technol. 224 (2012) 19–27.
[46] G.M. Sturgeon, K.J. King, Development of powder route for production of stainless steel strip, Powder Metall. 25 (2) (1982) 57–61.
[47] F. Podczeck, Y. Miah, The influence of particle size and shape on the angle of internal friction and the flow factor of unlubricated and lubricated powders, Int. J. Pharmaceut. 144 (1996) 187–194.

[48] L.X. Liu, I. Marziano, A.C. Bentham, J.D. Litser, E.T. White, T. Howes, Effect of particle properties on the flowability of ibuprofen powders, Int. J. Pharmaceut. 362 (2008) 109–117.
[49] N. Stevens, S. Tedeschi, K. Powers, B. Moudgil, H. El-Shall, Controlling unconfined yield strength in humid environment through surface modification of powders, Powder Technol. 191 (2009) 170–175.
[50] G.A. Vinogradov, Application of the method of reference points for calculating the density of strips rolled from powders, Sov. Powder Metall. Metal Ceram. 1 (4) (1962) 275–282.
[51] J.C. Cunningham, Experimental studies and modeling of the roller compaction of pharmaceutical powders, PhD Thesis, Drexel University, 2005.
[52] W.V. Knopp, Roll compacting of metal powders, in: ASM Metals Handbook, 10th ed., Powder Metal Technologies and Applications, vol. 7, ASM, Materials Park, OH, 1998, pp. 389–395.
[53] A.P. Chekmarev, A.M. Musikhin, P.L. Klimenko, G.L. Lebedik, Use of sheet rolling mills for the rolling of metal powders, Sov. Powder Metall. Metal Ceram. 11 (2) (1972) 156–158.
[54] A.M. Musikhin, G.A. Vinogradov, High-speed rolling of metal powder, Sov. Powder Metall. Metal Ceram. 9 (6) (1970) 450–455.
[55] H.S. Nayar, Strip products via particle metallurgy, Powder Metall. Int. 4 (1) (1972) 30–36.
[56] A.M. Musikhin, Fundamental equations of metal powder dynamics and their application to the rolling process, Sov. Powder Metall. Metal Ceram. 12 (10) (1973) 785–789.
[57] S. Shima, M. Yamada, Compaction of metal powder by rolling, Powder Metall. 27 (3) (1984) 39–44.
[58] A.N. Nikolaev, The compacting and rolling of metal powders, Sov. Powder Metall. Metal Ceram. 2 (1) (1963) 30–34.
[59] A.R. Muliadi, J.D. Litster, C.R. Wassgren, Modeling the powder roll compaction process: comparison of 2-D finite element method and the rolling theory for granular solids (Johanson's model), Powder Technol. 221 (2012) 90–100.
[60] M. Massoudi, M.M. Mehrabadi, A continuum model for granular materials: considering dilatancy and the Mohr-Coulomb criterion, Acta Mechanica. 152 (2001) 121–138.
[61] S.M. Tahir, A.K. Ariffin, M.S. Anuar, Finite element modelling of crack propagation in metal powder compaction using Mohr-Coulomb and elliptical cap yield criteria, Powder Technol. 202 (2010) 162–170.
[62] V.P. Katashinskii, Analytical determination of specific pressure during the rolling of metal powders, Sov. Powder Metall. Metal Ceram. 5 (10) (1966) 765–772.
[63] V. Esnault, D. Heitzmann, M. Michrafy, D. Oulahna, A. Michrafy, Numerical simulation of roll compaction of aerated powders, Chem. Eng. Sci. 104 (2013) 717–726.
[64] J.M. Rowe, J.R. Crison, T.J. Carragher, N. Vatsaraj, R.J. McCann, F. Nikfar, Mechanistic insights into the scale-up of the roller compaction process: a practical and dimensionless approach, J. Pharmaceut. Sci. 102 (2013) 3586–3595.
[65] J.R. Johanson, A rolling theory for granular solids, ASME J. Appl. Mech. Series E 32 (4) (1965) 842–848.
[66] G. Bindhumadhavan, J.P.K. Seville, M.J. Adams, R.W. Greenwood, S. Fitzpatrick, Roll compaction of a pharmaceutical excipient: experimental validation of rolling theory for granular solids, Chem. Eng. Sci. 60 (14) (2005) 3891–3897.
[67] A.R. Muliadi, J.D. Litster, C.R. Wassgren, Validation of 3-D finite element analysis for predicting the density distribution of roll compacted pharmaceutical powder, Powder Technol. 237 (2013) 386–399.
[68] M.H.D.V. Blore, S. Silins, T.W. Romanchuk, V.N. Benz, Mackiw, Pure nickel strip by powder rolling, Metals Eng. Quart. (1966) 54–60 May.

[69] C.H. Weaver, R.G. Butters, J.A. Lund, Hot rolling behaviour of iron powder performs, Int. J. Powder Metall. 8 (1972) 3–15.
[70] Yu. N. Podrezov, V.A. Nazarenko, A.V. Vdovichenko, V.I. Danilenko, O.S. Koryak, Ya. I. Evich, Mechanical properties of powder titanium at different production stages. III. Contact formation in powder titanium based on examination of mechanical properties in sintering, Powder Metall. Metal Ceram. 48 (2009) 201–210.
[71] Yu. N. Podrezov, V.A. Nazarenko, A.V. Laptev, A.I. Tolochin, V.I. Danilenko, O.S. Koryak, Ya. I. Evich, Mechanical properties of powder titanium at different production stages. IV. Mechanical properties and contact formation in powder titanium produced by dynamic hot pressing, Powder Metall. Metal Ceram. 48 (2009) 295–301.
[72] Yu. N. Podrezov, V.A. Nazarenko, A.V. Laptev, A.I. Tolochin, Ya. I. Evich, N.I. Danilenko, O.M. Ivanova, Structural dispersion of powder titanium in the optimal conditions of dynamic hot pressing, Powder Metall. Metal Ceram. 51 (2012) 56–63.
[73] K.A. Gogaev, V.A. Nazarenko, V.A. Voropaev, Yu. N. Podrezov, D.G. Verbilo, O.S. Koryak, I. Yu. Okun, Mechanical properties of powder titanium at different production stages. V. Properties of a titanium strip produced by powder rolling, Powder Metall. Metal Ceram. 48 (2009) 652–658.
[74] Y. Muramatsu, K. Tamura, Production of titanium strip by powder rolling, Trans. Natl. Res. Inst. Metals (Japan) 20 (3) (1978) 172–177.
[75] W.J. Kim, S.J. Yoo, H.T. Jeong, D.M. Kim, B.H. Choe, J.B. Lee, Effect of the speed ratio on grain refinement and texture development in pure Ti during differential speed rolling, Scripta Mater. 64 (2011) 49–52.
[76] X. Huang, K. Suzuki, Y. Chino, Improvement of stretch formability of pure titanium sheet by differential speed rolling, Scripta Mater. 63 (2010) 473–476.
[77] H.S. Kim, S.J. Yoo, J.W. Ahn, D.H. Kim, W.J. Kim, Ultrafine grained titanium sheets with high strength and high corrosion resistance, Mater. Sci. Eng. A 528 (2011) 8479–8485.
[78] D. Randman, B. Davis, M.L. Alderman, G. Muralidharan, T.R. Muth, W.H. Peter, T.R. Watkin, Production of wide shear-rolled magnesium sheet for part forming, in: S.N. Mathaudhu, W.H. Siliekens, N.R. Neelameggham, N. Hort (Eds.), Magnesium Technology 2012, The Minerals, Metals & Materials Society, Warrendale, PA, 2012, pp. 23–28.

Cold-spray processing of titanium and titanium alloys

22

Phuong Vo, Dina Goldbaum**, Wilson Wong**, Eric Irissou*, Jean-Gabriel Legoux*, Richard R. Chromik**, Stephen Yue***

*National Research Council Canada, Boucherville, Québec, Canada

**Department of Mining and Materials Engineering, McGill University, Montréal, Québec, Canada

22.1 Introduction

Cold gas dynamic spray, or cold spray, is a process in which a layer-by-layer buildup of material is produced by successive high-speed impacts of solid-state, micron-sized particles onto a substrate or work piece. First developed in Russia in the mid-1980s followed by US and European patents in the mid-1990s [1,2], the process is gaining attraction within the traditional family of thermal spray technologies for coating deposition and is an emerging candidate in the family of metal additive manufacturing technologies. Cold spray exhibits various advantages over competing processes such as low process temperature relative to material melting point, high build rate, high deposition efficiency, etc. [3–5]. The low temperature relative to conventional thermal spray methods that produce molten- or semimolten-state particles minimizes adverse temperature effects such as oxidation, residual stress, phase transformations, or other undesired microstructural changes [6].

For titanium and its alloys, the capability to rapidly produce thick, unoxidized deposits (hereafter refers to coatings or free-formed parts) in open air with minimal adverse property effects makes cold spray particularly attractive. Cold-spray Ti coatings can provide protection of steel structures in harsh marine environments [7] or improvement in implant biocompatibility and integration [8]. Alternatively, additive manufacturing by cold spray is appealing for applications featuring relatively high material and manufacturing costs due to extensive machining (e.g., high buy-to-fly ratios of 33:1 for a bleed air leak detect bracket [9]). Various geometries produced using cold spray are shown in Figure 22.1.

In a conventional cold-spray process, particles with typical sizes of 5–50 μm are injected into a heated, pressurized gas stream and accelerated to velocities of 300–1200 m/s at a temperature sufficiently low to maintain a solid state during particle flight [1,5]. Particle deposition onto a substrate is obtained when the impact velocity is greater than a critical velocity [5,10]. An example of Ti powder in as-received and cold-sprayed states is shown in Figure 22.2, where to successfully spray a dense coating, velocities of 1173 m/s were used [11]. Within the Ti material class, commercially pure (CP-Ti) powder is by far the most investigated for cold spray. Alloy powders

Titanium Powder Metallurgy. http://dx.doi.org/10.1016/B978-0-12-800054-0.00022-8

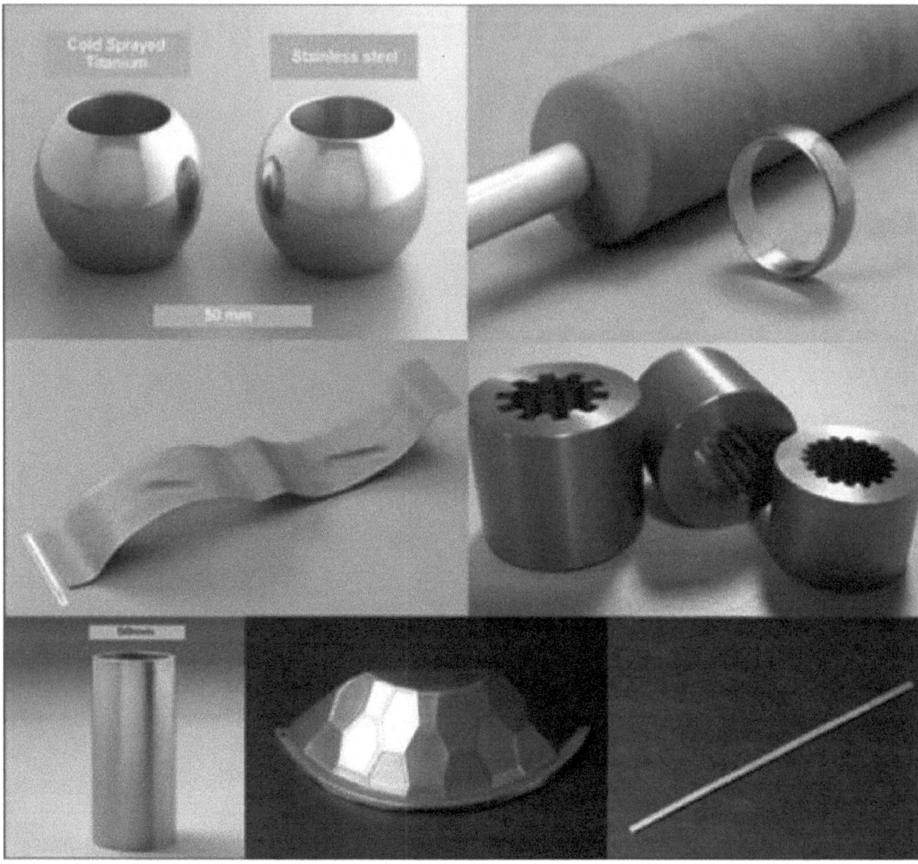

Figure 22.1 Components produced at CSIRO by cold-spraying commercially pure powder. (Source: From Ref. [9].)

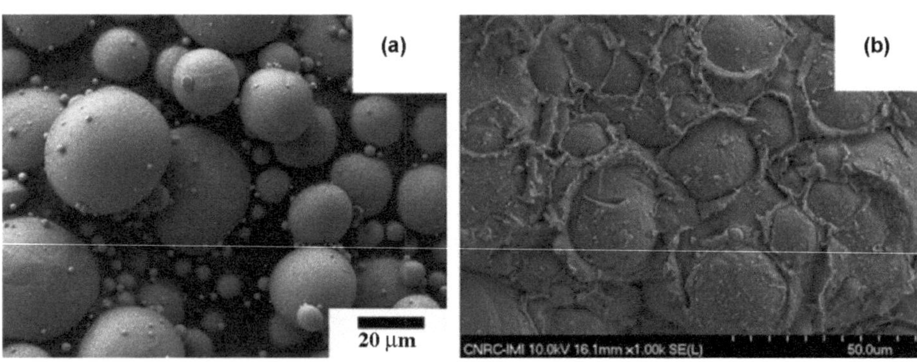

Figure 22.2 Example of CP-Ti (a) as-received powder and (b) top surface of cold-sprayed deposit. (Sources: (a) From Ref. [11] and (b) from Ref. [11].)

appears to be limited to the industry workhorse Ti-6Al-4V, although some powder mixtures have also been studied (e.g., to form TiAl [12]).

A number of reference books and review papers are available for an in-depth and comprehensive treatment of cold-spray processing [2–5,13,14]. Here, selected results from the available literature will be discussed to highlight key practical and fundamental aspects in cold-spray processing of titanium-base powders.

22.2 Process description

Figure 22.3 shows a schematic of a conventional cold-spray system: (1) a pressurized gas flows through a gas heater; (2) a powder is introduced into the gas flow; and (3) the gas–powder mix is accelerated through a de Laval type nozzle and directed onto a substrate. Additional in-process equipment such as He recycling [11,15], laser treatment [16,17], and machining [15] has also been developed to complement the base system. Detailed descriptions of cold-spray equipment are available in a number of books [3–5].

The basic process parameters are centered on gas conditions, nozzle position/movement, and powder feeding. With optimized parameters, dense coatings can be produced with a deposition efficiency greater than 95% (e.g., for CP-Ti [11,18]). Table 22.1 lists reported working ranges and general effects of increasing standard parameters for Ti cold spray. Broadly speaking, these parameters control the amount and properties of deposited material through their effects on particle velocity and temperature, substrate surface temperature, and amount of particles delivered. The selection of an appropriate nozzle is also a key consideration as the nozzle design and material will have a large impact on particle velocity and nozzle life, respectively. The modeling of gas flow has received significant attention, and Ti-specific studies are available [19].

Following spray, material properties can be improved through heat treatment, hot isostatic pressing (HIP), and/or surface modification. Ti and Ti alloy coatings have been heat treated using conventional equipment within vacuum [20,21], controlled

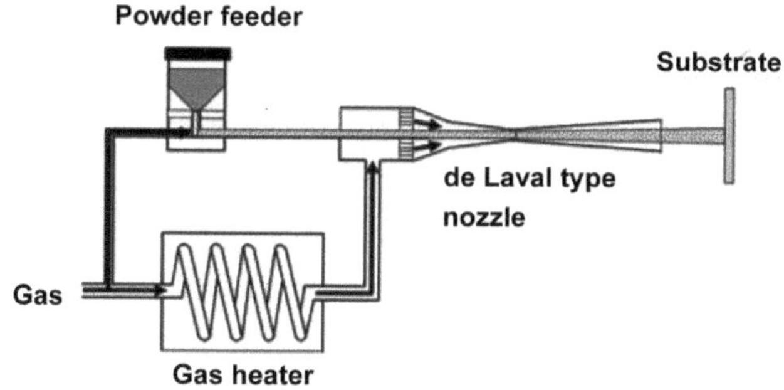

Figure 22.3 Schematic diagram of cold-spray system. (Source: From Ref. [14].)

Table 22.1 General effects of spray parameters on key deposition characteristics

Parameter	Working range	v_p*	T_p	T_s	Δd_{layer}
Gas	Air, N_2, He	+**			
Gas pressure	<5 MPa	+			
Gas temperature	<1000 °C	+	+	+	
Standoff distance	5–100 mm	+/−	−	−	
Spray angle	>70°	−†			−
Traverse speed	5–330 mm/s			−	−
Powder feed rate	<100 g/min			+	+

* v_p, particle impact velocity; T_p, particle temperature; T_s, substrate surface temperature; Δd_{layer}, layer thickness per pass.
** Air → N_2 → He.
† Increase in off-normal spray angle decreases normal component.

gas atmosphere [21–23], and air [24]. A comparison of Ti coatings heat treated within different environments revealed better coating density and purity using a vacuum instead of gas atmospheres [21]. In general, the effectiveness of heat treatment for improving coating properties is also dependent on the as-sprayed coating properties while a HIP treatment can be performed on higher-porosity coatings (e.g., ~20% for Ti-6Al-4V [22]). Surface modification by laser melting can also be performed to eliminate porosity near the coating surface without affecting the substrate [25].

22.3 Cold-spray principles

22.3.1 General particle-bonding mechanisms

Upon particle impact, cold-spray material is subject to high strain rate deformation, leading to localized viscoelastic material flow or jetting near the interface. The bonding mechanism can be partially explained by the mechanical interlocking of the interfacial asperities at the micro/nano level in addition to the physical intermixing of the material through roll-up and vortices [26]. Part of the kinetic energy of the impact is converted to heat, which can induce recrystallization, atomic diffusion, formation of the intermetallic phases, and localized melting [10,27–31]. The generally accepted view is that metallurgical bonding requires the formation of clean (e.g., oxide free) conformal contact surfaces at high pressure through the creation of adiabatic shear instabilities [10,28]. The phenomenon of adiabatic shear instability refers to the shear localization followed by a rise in the temperature observed at high strain rate deformation conditions [10,28,32]. In FEA modeling of single particle deformation [10,14,29], the predicted critical particle velocity for onset of localized adiabatic shear instabilities at the contact interface corresponded to an experimentally determined critical velocity for particle deposition. The critical velocity for successful particle bonding was then empirically related to the particle temperature and material properties of density, melting temperature, and strength [10].

22.3.2 Particle deformation for Ti and Ti alloys

The deposition mechanisms of Ti splats (i.e., deformed and adhered particles) vary depending on the particle impact velocity (often assumed equal to the measured in-flight particle velocity) and follow three generally accepted deposition mechanisms [5,27,29,33,34]. Figure 22.4 demonstrates all three deformation mechanisms observed for Ti and Ti-6Al-4V splats deposited on Ti and Ti-6Al-4V substrates, respectively [34]. At a deposition velocity below but approaching the critical velocity (see Figure 22.4a and d) development of the adiabatic shear instability is limited and results in weak bonding mainly through the mechanical interlocking of surface asperities [29,33]. As the deposition velocity reaches and surpasses the critical velocity (Figure 22.4b and e), the formation of the adiabatic shear instability leads to viscoelastic material flow in an outward direction from the substrate and, therefore, only a small fraction of the adiabatic shear contributes to splat adhesion [34]. The splat adhesion is counteracted by the elastic strain energy stored in the substrate. The elastic strain energy is typically highest at the center of the impact and contributes to particle

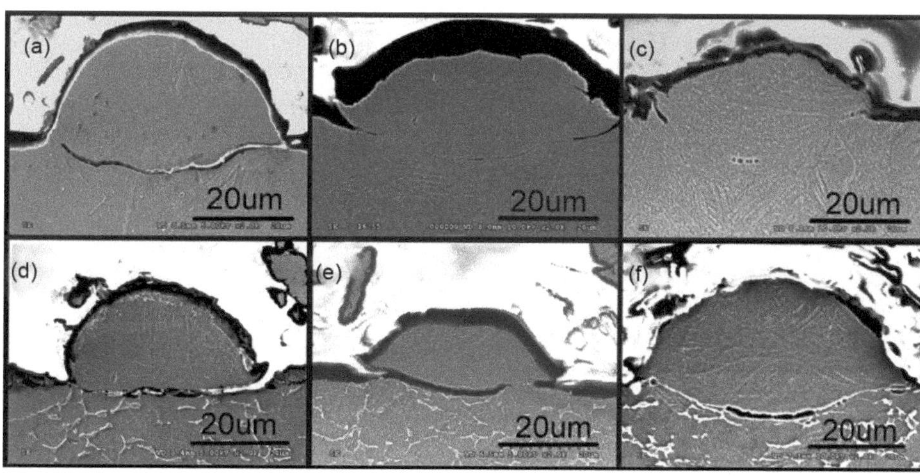

Figure 22.4 Etched SEM cross-sectional micrographs of: Ti splats deposited on Ti at (a) 724 m/s (N_2 at T_g = 500°C/P_g = 4 MPa); (b) 825 m/s (N_2 at 800°C/4 MPa); and (c) 1140 m/s (He at 350°C/4 MPa); and Ti-6Al-4V splats deposited on Ti-6Al-4V substrate at (d) 741 m/s (N_2 at 500°C/4 MPa); (e) 834 m/s (N_2 at 800°C/4 MPa); and (f) 1115 m/s (He at 350°C/4 MPa). The dark regions between the splat and the substrate indicate regions of poor bonding and void formation. (Source: From Ref. [34].)

de-bonding and void formation in that region [35,36]. A deformation in the substrate was shown to be necessary in order to retain greater surface contact between the substrate and adiabatic shear instability jet, as can be seen in Figure 22.4c and f [34]. The deformation in the substrate was obtained by increasing the particle deposition velocity (1140 m/s for Ti and 1115 m/s for Ti-6Al-4V) substantially above the critical velocity or by preheating the substrate (to 400°C). An increase in the bond strength of up to 100 MPa was measured by a modified ball bond shear testing technique [34,37]. Other studies have indicated that the selection of a softer substrate material can also contribute to substrate deformation, although with a resulting decrease in the adiabatic shear and jetting within the Ti splats [38,39].

TEM studies of cold-spray Ti splats revealed the formation of nanocrystalline grains, 50–500 nm in size [40–42], which was attributed to dynamic recrystallization within the adiabatic shear instability region [41]. The extent of recrystallization has been shown to highly depend on the temperature of impacting particles [40,43], with near full recrystallization reported for preheated Ti particles within the gas jet [40] or a preheated substrate [44]. The deformation within a Ti substrate was mainly through formation of twins, observed up to 30 μm into the substrate, with a narrow region, ~150 nm in size, near the impact demonstrating recrystallization [44]. Microstructural intermixing, between the splat and substrate nanograins, induced by high contact pressures and conformity of the interfaces, was reported to contribute to metallurgical bonding within the adiabatic shear instability region [44,45]. The morphology of the feedstock powder was shown to have an effect on the deformation mechanisms within the cold-sprayed Ti splats. The splats with angular morphology particles were observed to contain twins and the elongated grains, in addition to the recrystallized grains containing dislocations imparting retained strain energy [45,46]. Such microstructural variations were reflected in an inhomogeneous increase in hardness within the adiabatic shear instability region in the splats and the substrate [44,47] and to an overall increase in the hardness within the coatings [42,47,48]. The recrystallization may also contribute to the dimpled tensile fracture of cold-spray coatings [42,47].

22.4 Properties of deposited material

22.4.1 Microstructure

Figure 22.5 shows dense coatings of CP-Ti (≤0.5% porosity) successfully sprayed under optimized parameters with gas temperatures ranging from 350 to 1000°C [11,18,49,50]. Nearly dense coatings of Ti-6Al-4V (not shown) [23] have also been produced with identical spray parameters as the Ti coatings shown in (b). Although not shown, cold-spray coatings typically display some (minor) porosity in the surface layers because of the absence of tamping action induced by follow-up spray passes [11,51]. The optimized conditions reveal that relatively low gas temperature or pressure can be used with He because the lower the gas density the faster the gas velocity. At lower N_2 gas temperatures (≤800°C), coating porosity values of 1–20% were reported for pure Ti [13] and 5–12% for Ti-6Al-4V [17,23]. The coating–substrate

Figure 22.5 As-polished CP-Ti coatings produced with different optimized spray parameters: (a) on $AlMg_3$ substrate using N_2 gas at T_g = 1000°C / P_g = 4 MPa; and (b) on steel substrate using He gas at 350°C / 4 MPa. (Sources: (a) From Ref. [18] and (b) from Ref. [11].)

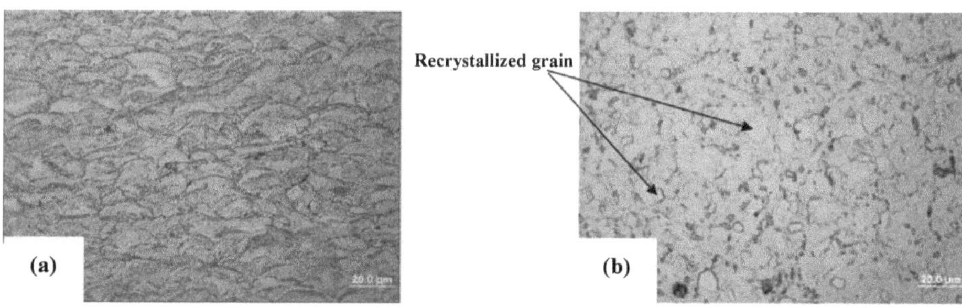

Figure 22.6 Etched CP-Ti coatings produced with optimized spray parameters on Al substrate (not shown) using He gas at T_g = 600°C and P_g = 1.5 MPa: (a) as-sprayed and (b) after heat treatment at 550°C for 2 h. (Source: From Ref. [24].)

interface in Figure 22.5a shows significant particle entrapment or penetration into the substrate compared to the relatively flat interface in Figure 22.5b, suggesting that mechanical anchorage will be less in the latter. This difference can be attributed to a combination of the harder substrate material and lower gas temperature in Figure 22.5b. Nevertheless, both interfaces display conformal contact with relatively few visible defects (e.g., grit blast media) between the deposited CP-Ti and the substrate.

For the etched, as-sprayed condition in Figure 22.6a, particle boundaries are clearly delineated and significant deformation can be observed [24]. Figure 22.7 shows that a fine recrystallized microstructure within the vicinity of the particle–particle interface of two bonded particles can be observed at higher magnification (10k ×) using electron channeling contrast imaging (ECCI) [44]. A heat treatment of CP-Ti at 550°C for 2 h in Figure 22.6b appeared to eliminate particle boundaries and develop fine recrystallized grains [24]. For Ti-6Al-4V, heat treatment produced only recovery at 600°C while recrystallization, phase transformation, and/or grain growth occurred at

Figure 22.7 ECCI micrograph showing the region near the particle-particle interface of two bonded CP-Ti particles cold-sprayed at T_g = 800°C/P_g = 4 MPa. (Source: From Ref. [44].)

temperatures of 840°C and higher [23]. The elimination of particle boundaries and formation of grain boundaries across these eliminated boundaries is an indication that improved metallurgical bonding has been obtained.

22.4.2 Tensile strength and ductility

Tensile properties have been investigated for cold-spray coatings of CP-Ti [13,18,22,24,44] and Ti-6Al-4V [22,23]. Table 22.2 shows the highest reported values of tensile strength and elongation for coatings in different conditions and ASTM B 265 [52] minimum values. As is generally the case with cold-sprayed metals (e.g., tensile results for Cu, Ni, and Al [23]), Ti and Ti alloy coatings in the as-sprayed condition exhibit limited ductility. The tensile strength varies widely depending on deposition conditions and the resulting coating microstructure, which also affects the coating response to annealing treatment [6]. Heat treatments can increase or decrease tensile strength, although the minimum tensile strength values of bulk material can be met for CP-Ti [24,44]. The ductility is improved after annealing and values for CP-Ti are near the minimum values of bulk material [24]. For cold-sprayed Ti-6Al-4V alloy, however, a HIP treatment has been necessary in order to obtain bulk equivalent strength and ductility [22].

22.4.3 Coating bond strength

The coating bond (adhesion) strength obtained using a variety of spray parameters, substrate materials, and substrate preparations has been investigated for Ti and Ti-6Al-4V. The limit of the ASTM C633 [53] (e.g., ~60 MPa [54] to 80 MPa [55]) or

Table 22.2 Highest reported tensile properties in as-sprayed, annealed, and HIP conditions for CP-Ti and Ti-6Al-4V cold-spray coatings produced using He or N_2 gas

		He-spray			N_2-spray			
Coating*	**Condition**	**UTS (MPa)**	**El. (%)**	**Ref.**	**UTS (MPa)**	**El. (%)**	**Ref.**	**HT/HIP parameters**
CP Ti	As-spray	800	0	[24]	450**	–	[18]	–
	HT	600	8	[24]	250	8	[44]	550°C/2 h (He) 800°C/1 h (N_2)
	HIP	920	3	[22]	–	–	–	900°C/2 h/103 MPa
Ti-6Al-4V	As-spray	450	4†	[23]	150	2†	[23]	–
	HT	760	6†	[23]	460	6†	[23]	600°C/2 h (He) 1000°C/4 h (N_2)
	HIP	900	13	[22]	–	–	–	900°C/2 h/103 MPa
Bulk		ASTM B 265						
CP Ti	Min	240–550	15–24	[52]				
Ti-6Al-4V	Min	895	10	[52]				

* All coatings with <1% porosity except N_2-sprayed Ti-6Al-4V.
** Tubular coating tensile test (TCT-test).
† Elongation at failure values reported in Ref. [23].

similar test type [33] has been reached for cold-spray coatings produced using optimized spray conditions (e.g., Ti on steel [33] or Ti [55] and Ti-6Al-4V on Ti-6Al-4V [17,55]). Bond strength results for Ti-6Al-4V coatings on Ti-6Al-4V measured by ASTM C633 and LASAT (Laser Shock Adhesion Test), although not directly comparable tests, revealed a similar trend of weak bond strength for substrate surface roughness within a specific range (2.56 μm < R_a < 3.21 μm) [17]. This was generally consistent with previous observations that Ti coatings on Ti-6Al-4V showed a reduction in bond strength with similar roughness values after grit blasting [8,51].

22.4.4 Machinability

Ti and Ti alloy cold-spray deposits can be effectively milled in the as-sprayed condition provided the quality of the coating is good and proper milling parameters are selected [56]. The as-sprayed coatings may not be suitable for machining if hardness and porosity are too high [22]. The compressive stresses produced during machining can result in densification in the machine-affected zone, which can be ~100 μm in depth with 2–3 times higher hardness than as-sprayed material [57]. Electric discharge machining has also been effectively used for Ti and Ti alloy cold-spray coatings [23,24].

22.4.5 Substrate corrosion protection and fatigue life

The corrosion performance of CP-Ti coatings has been evaluated in 3.5 wt% NaCl solution (Ti on 045M10 steel [20,25]), natural seawater (on 1Cr13 stainless steel [7]), and simulated body fluid (on Ti [58]). Results show that porosity degraded performance relative to bulk Ti [20]. A coating post treatment of laser surface melting to eliminate porosity on the top of the coating resulted in corrosion performance, as measured by open circuit potential, corrosion potential, and passive current density, nearly similar to bulk Ti [25].

The effect of cold-spray CP-Ti coatings on substrate fatigue life has been investigated for Ti-6Al-4V substrates [8,59]. Rotating bend tests showed a 15% reduction in the fatigue endurance limit relative to as-received material, which was attributed to an increase in substrate surface topography and induced tensile residual stresses within the substrate [8]. Cantilever-beam cyclic bend tests revealed a 9% reduction in fatigue life because of the initiation of a crack in the coating that was subsequently transferred to substrate [59]. The effect of grit blasting prior to coating on fatigue life is dependent on multiple factors (e.g., blasting parameters, substrate surface [59]) and can result in improving [59] or reducing fatigue performance [8].

22.5 Process–microstructure–property relationships

22.5.1 Spray parameter effects

Because of its fundamental role in bonding, the determination of particle velocity relationships (e.g., to gas conditions, nozzle geometry, coating properties) has been one of the central tenets of cold-spray optimization. Particle velocity data via modeling and/or measurement at various gas conditions and standoff distances are available for CP-Ti [11,16,19], with the velocity for Ti-6Al-4V similar to CP-Ti at comparable conditions [34]. The effect of spray parameters on particle/substrate temperature, which allows greater particle deformation by lowering material flow stress, has also received consideration [18,34,60]. Deposition thickness effects are generally of lesser concern within normal operating ranges and typically considered after other factors have been satisfied. The general effects of spray parameters were previously shown in Table 22.1.

The effect of gas conditions on gas velocity (and thus particle velocity through drag forces) can be understood through an analytical 1-D isentropic (adiabatic and frictionless) model; wherein gas velocity at the nozzle exit is increased with increasing gas temperature or pressure (i.e., $v_g = M(\gamma R T_g)^{0.5}$, where v_g is the gas velocity, M is the Mach number, γ is the ratio of specific heats, R is the specific gas constant, and

T_g is the gas temperature) [61]. The switch to He from N_2 gas results in an improvement because of its smaller molecular weight and higher specific heat ratio [61]. An optimum standoff distance (according to powder, nozzle geometry, gas type, etc.) exists because of the increase in particle velocity after the nozzle exit for some distance before starting to decrease at longer distances [19]. The shock wave developed in front of the substrate (termed the bow shock) due to the disruption of the supersonic flow by the substrate must also be considered, particularly at short standoff distances, because of possible deflection and/or slowing of particles [19]. In practice, this also means that velocity measurements do not accurately reflect impact speed on the surface of the substrate, particularly for small particles [49].

Cold spray at off-normal spray angles and/or at high feed rate, which are of practical concern, can also have significant effects on particle velocity because of a lower normal velocity component [62] and/or particle loading effect [63], respectively. Steep decreases in strength have been observed at spray angles of 70° compared to 90°, although relatively high deposition efficiency can be maintained [18]. The traverse speed controls the amount of deposition per spray pass and can influence the coating bond strength by changing substrate temperatures [63]. As the traverse slows, significantly more heat input and oxidation of steel substrates can be clearly observed, with potentially deleterious effects from oxidation and nitridation [11].

In order to provide more practical methods to select spray parameters for Ti and Ti alloys, there have been attempts to move beyond identifying one-to-one trends (e.g., strength vs. gas temperature [18] and porosity vs. velocity [50], standoff distance [49], or spray angle [18]) to account for interrelationships through factorial analysis [64] and optimization software [65]. Alternatively, Figure 22.8 shows empirical relationships of coating properties as functions of the ratio of particle velocity to critical velocity formulated to use with parameter selection maps [66]. Equations for this method are available for CP-Ti, with increasing particle velocity showing improvements in porosity, strength, flattening ratio, and deposition efficiency [18,66,67].

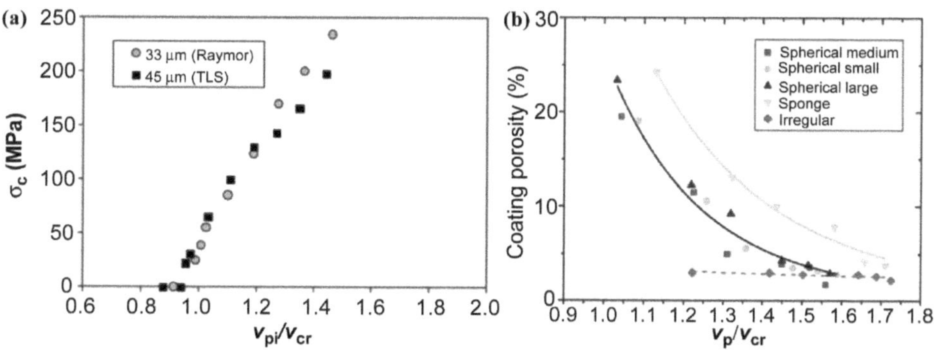

Figure 22.8 Relationships for Ti coating properties as a function of particle to critical velocity ratio: (a) cohesive strength for two particle size distributions and (b) porosity for three particle morphologies. (Sources: (a) From Ref. [66] and (b) from Ref. [67].)

22.5.2 Powder effects

The consideration of powder characteristics has received significantly less attention than spray parameters. Standard powder characteristics for cold spray include composition, particle morphology, particle size distributions (PSD), and packing and flowability. The latter are important practical measures (e.g., ensure powder feed stability), although a study of multiple measures for five commercial CP-Ti powders showed limited influence on coating properties [67]. Although any impurities may result in deleterious effects, the oxygen content is most often considered as high levels can translate into thick surface oxide layers and/or higher particle strength. This can result in increased potential for particle rebound and/or entrainment of oxides at the interface [68].

The effect of particle morphology is typically associated with the particle velocity, as irregular powders travel at higher velocity because of higher drag forces compared to spherical powders. Irregular particles at impact can also rotate or fold to conform to the surface, which may influence the contact surface and porosity development. Investigations with different CP-Ti [67] and Ti-6Al-4V [22] morphologies demonstrated the difficulty in isolating morphology effects through commercial powders as they typically feature a host of differences (e.g., size distribution, hardness, impurities). Nevertheless, empirical relationships based on the particle–critical velocity ratio revealed that deposition efficiency was relatively independent of morphology while porosity had distinct dependencies by morphology [67], shown in Figure 22.8b.

Although higher velocities can also be achieved with a decrease in PSD, small particles encounter greater difficulty in forming adiabatic shear instabilities (because of higher cooling rates, higher impurity content, etc.) and an optimum PSD balancing impact and critical velocities is necessary [29]. Smaller particles are also more likely to be decelerated or deflected by the bow shock, and loosely adhered particles can be removed by gas jet pressure during follow-up spray passes [11]. The porosity may show different PSD dependencies as a decrease in PSD can show relatively little effect [67], a decrease for smaller PSD [49], or an increase for larger PSD [51,64]. Porosity can be decreased by an increase in fines or in large particles, as the former may lead to better packing while the latter may lead to increased tamping, assuming the mechanical properties of particles are constant.

22.5.3 Microstructure effects

The microstructural effects for cold-spray deposits can generally be considered in terms of the porosity, interfacial bonding, and grain evolution. Microstructural characterization for cold-spray coatings has been predominately focused on porosity and establishing related trends. However, the amount of metallurgical relative to mechanical bonding obtained between particles has been shown to be significant in fully dense coatings (e.g., for Cu [6]) or coatings with comparable porosity levels (e.g., Ti-6Al-4V [23]). Fracture surface investigations for CP-Ti and Ti-6Al-4V [23,69] qualitatively showed low ductility was associated with brittle failure initiated at particle–particle interfaces compared to ductile inter- and transparticle failures. Although up to ~80%

of the fracture surface has been observed as ductile-type failure for as-sprayed copper coatings [6], this failure type is significantly more limited in the Ti and Ti alloy coatings.

In dense and well-bonded coatings, grain evolution can have significant effects on properties (e.g., grain size strengthening via the Hall–Petch relationship). The static recrystallization of CP-Ti during annealing has been investigated and recrystallization kinetics modeled using an Arrhenius-type equation [24]. For less dense, less well-bonded coatings produced using the stronger Ti-6Al-4V alloy, however, mechanical properties were more dependent on particle interfacial bonding than grain evolution associated with recrystallization or phase transformation [23].

22.6 Applications

22.6.1 Coatings

The benefits of cold spray as a general coating technology are well established (e.g., low temperature, high deposition efficiency, minimal masking due to a narrow particle jet, absence of hazardous chemicals such as used for electroplating) [3–5]. Two practical examples for which cold-spray Ti coatings are suitable are for corrosion protection and biomedical applications. For corrosion protection of steel structures in seawater, Ti coatings are applied using thermal spray processes such as shrouded or low-pressure plasma spray [7]. However, the need for atmosphere control to maintain phase and chemical purity in addition to post-sealing treatments to ensure airproof worthiness renders a relatively high production cost, which can be avoided by cold spray [7]. For biomedical applications, bulk implant materials such as cobalt chromium and Ti-6Al-4V are used for their mechanical properties despite less than ideal biocompatibility and integration with bone structures [8,70]. As a result, pure Ti [8,59] and composite hydroxyapatite–titanium (HAP-Ti) [70] coatings have been investigated to improve corrosion resistance, in vivo nontoxicity, and biocompatibility. In this case, a pure Ti coating acts as a bond coat to an HA top coat. These Ti bond coats are typically manufactured with high surface roughness and porosity, since such a porous bond coat structure is considered advantageous in promoting bone growth onto the implants. The use of cold spray in this application is advantageous as alternative high-temperature methods may result in oxide formation, deterioration of adhesion, and/or damage to hydroxyapatite [70].

22.6.2 Additive manufacturing

Cold spray has received attention as an additive manufacturing (AM) technology [15,56,71] primarily because of its high build rate and low process temperatures, with various secondary advantages also offered (e.g., open-atmosphere processing, multimaterial or functionally graded deposits). Figure 22.9 shows that within the AM group, cold spray features a very high build rate at the expense of shape fidelity [15]. The low temperatures allow temperature- and/or oxygen-sensitive materials to be

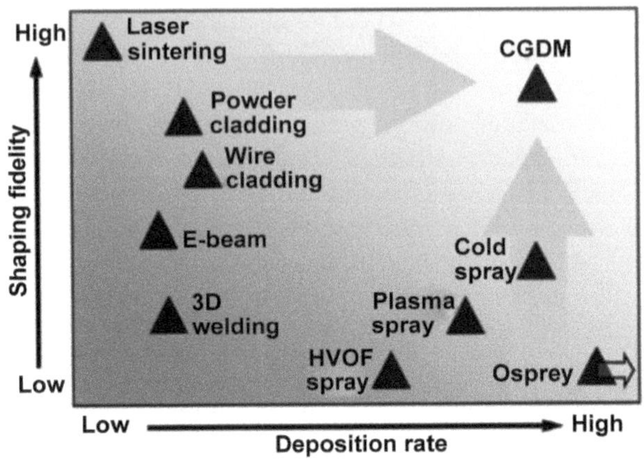

Figure 22.9 Schematic of shaping fidelity and deposition rate for additive manufacturing processes. (Source: From Ref. [15].)

sprayed without modifying underlying microstructure or composition. In addition to the pool of cold-spray materials and deposit properties discussed in previous sections, key considerations for cold-spray AM are spray resolution, build strategies, and process economics [15,56,71].

Because of limited dispersion of the gas jet following nozzle exit at typical standoff distances, the spatial resolution and profile obtained with cold spray can be controlled through nozzle design (e.g., micronozzles for spot diameters <1 mm or nonspherical nozzle exit geometries for custom deposit shapes) [56]. Alternatively, different deposit profiles can be achieved through in-process milling and/or build strategies (e.g., vertical wall by a triangular-tessellation technique) [15]. Multilayered build strategies can also be used to produce internal geometries by spraying sacrificial support structures of a second powder [15]. In addition, the solid-state nature and layer-by-layer buildup in cold spray allows a degree of flexibility in producing functionally graded [56] and multimaterial [15,56] forms without adverse reactions between layers of different material. This also presents the opportunity for cold spray to be used in hybrid technologies with other AM techniques (e.g., with electroplating or selective laser melting) [56].

The business case for fabricating Ti parts by cold spray was demonstrated using a value stream analysis that showed a cost advantage could be realized [71]. Developed as part of the US Air Force Forging Supplier Initiative for laser powder deposition of Ti, this analysis showed that the use of cold spray addressed multiple aspects in the value stream and, compared to a typical casting process, resulted in a reduction in raw material, elimination of mold and melt pour costs, reduction in rework and finishing, and increase in material utilization [71]. An analysis of the economics of using He instead of N_2 gas has also been performed for Ti [15]. The analysis, which used a 30:1 cost ratio for He:N_2 gas, revealed that coatings of high-value powders such as Ti can be cost-effectively produced with He gas provided that gas recycling, which necessitates a high capital cost, is used.

22.7 Status and future

To date, cold-spray processing for Ti and Ti alloy materials has largely focused on CP-Ti, with limited forays using the workhorse Ti-6Al-4V alloy. These investigations have revealed that dense deposits featuring acceptable service properties can be produced for select applications with present commercially available equipment, although not as easily as the more commonly used metals (e.g., aluminum). Processing currently requires operating at the temperature and pressure limits of current equipment when using N_2 gas while economic considerations with costly He gas may present a significant barrier to entry for many potential end users. In addition, deposits often cannot be employed in the as-sprayed condition as postprocessing is required to recover a measure of ductility and/or eliminate porosity. In order to obtain wider adoption, several critical areas need to be addressed.

The hardware development cycle has been rapid, with next-generation equipment constantly being introduced. Process boundaries continue to be extended through improvements to the base system and/or complementary in-process actions. Generally speaking, the main thrust has been increasing gas temperature to increase particle velocity, and this effort is sure to continue. With much of the focus on equipment, feedstock powder improvement to date has been limited. However, the inherent ceiling to process improvement through higher temperature before the process mimics warm spraying [40] means that powders will be studied in greater detail. The importance of powder properties for Ti [67] and in general [68] have been analyzed and future advances may involve the use of tailored powders [68]. The expansion of tested alloys beyond Ti-6Al-4V is also to be expected. Lastly, the cold-spray field is still relatively new, with few established protocols and/or standards. Although basic standards such as MIL-STD-3021 [72] can serve as an excellent starting point, continued refinement and additions will better promote industry acceptance. Efforts toward this goal will be aided by moving from the qualitative characterizations available to more quantitative relationships and the ability to incorporate these results into a unified form.

Despite the technical issues raised, however, the considerable advantages held by cold spray for processing Ti and Ti alloys promises that development efforts will not only continue but increase going forward.

References

[1] A.P. Alkhimov, A.N. Papyrin, V.F. Kosarev, N.I. Nesterovich, M.M. Shushpanov, U.S. Patent 5,302,414; WOWO9119016; EP0484533, 1994.

[2] E. Irissou, J.G. Legoux, A.N. Ryabinin, B. Jodoin, C. Moreau, Review on cold spray process and technology: part I – intellectual property, J. Therm. Spray Technol. 17 (4) (2008) 495–516.

[3] V.K. Champagne, The Cold Spray Deposition Process: Fundamentals and Applications, Woodhead Publishing Ltd, Cambridge, UK, 2007.

[4] R.G. Maev, V. Leshchynsky, Introduction to Low Pressure Gas Dynamic Spray: Physics & Technology, WILEY-VCH Verlag GmbH & Co. KGaA, Weinheim, Germany, 2008.

[5] A. Papyrin, V. Kosarev, K.V. Klinkov, V.M. Fomin, Cold Spray Technology, Elsevier Ltd, Oxford, UK, 2006.

[6] F. Gartner, T. Stoltenhoff, J. Voyer, H. Kreye, S. Riekehr, M. Kocak, Mechanical properties of cold-sprayed and thermally sprayed copper coatings, Surf. Coat. Technol. 200 (24) (2006) 6770–6782.

[7] H.-R. Wang, W.-Y. Li, L. Ma, J. Wang, Q. Wang, Corrosion behavior of cold sprayed titanium protective coating on 1Cr13 substrate in seawater, Surf. Coat. Technol. 201 (2007) 9–11, 5203–5206.

[8] T.S. Price, P.H. Shipway, D.G. McCartney, Effect of cold spray deposition of a titanium coating on fatigue behavior of a titanium alloy, J. Therm. Spray Technol. 15 (2006) 507–512.

[9] F.H. Froes, Titanium powder metallurgy: a review – part 2, Adv. Mater. Process. 170 (10) (2012) 26–29.

[10] H. Assadi, F. Gartner, T. Stoltenhoff, H. Kreye, Bonding mechanism in cold gas spraying, Acta Mater. 51 (15) (2003) 4379–4394.

[11] W. Wong, E. Irissou, A. Ryabinin, J.-G. Legoux, S. Yue, Influence of helium and nitrogen gases on the properties of cold gas dynamic sprayed pure titanium coatings, J. Therm. Spray Technol. 20 (1) (2011) 213–226.

[12] T. Novoselova, P. Fox, R. Morgan, W. O'Neill, Experimental study of titanium/aluminium deposits produced by cold gas dynamic spray, Surf. Coat. Technol. 200 (8) (2006) 2775–2783.

[13] T. Hussain, Cold spraying of titanium: a review of bonding mechanisms, microstructure and properties, Key Eng. Mater. 533 (2013) 53–90.

[14] T. Schmidt, H. Assadi, F. Gartner, H. Richter, T. Stoltenhoff, H. Kreye, T. Klassen, From particle acceleration to impact and bonding in cold spraying, J. Therm. Spray Technol. 18 (2009) 5–6, 794–808.

[15] J. Pattison, S. Celotto, R. Morgan, M. Bray, W. O'Neill, Cold gas dynamic manufacturing: a non-thermal approach to freeform fabrication, Int. J. Mach. Tool. Manuf. 47 (2007) 3–4, 627–634.

[16] M. Bray, A. Cockburn, W. O'Neill, The laser-assisted cold spray process and deposit characterisation, Surf. Coat. Technol. 203 (19) (2009) 2851–2857.

[17] M. Perton, S. Costil, W. Wong, D. Poirier, E. Irissou, J.G. Legoux, A. Blouin, S. Yue, Effect of pulsed laser ablation and continuous laser heating on the adhesion and cohesion of cold sprayed Ti-6Al-4V coatings, J. Therm. Spray Technol. 21 (6) (2012) 1322–1333.

[18] K. Binder, J. Gottschalk, M. Kollenda, F. Gärtner, T. Klassen, Influence of impact angle and gas temperature on mechanical properties of titanium cold spray deposits, J. Therm. Spray Technol. 20 (1) (2011) 234–242.

[19] J. Pattison, S. Celotto, A. Khan, W. O'Neill, Standoff distance and bow shock phenomena in the cold spray process, Surf. Coat. Technol. 202 (8) (2008) 1443–1454.

[20] T. Hussain, D. McCartney, P. Shipway, T. Marrocco, Corrosion behavior of cold sprayed titanium coatings and free standing deposits, J. Therm. Spray Technol. 20 (1) (2011) 260–274.

[21] J.S. Yu, H.J. Kim, I.H. Oh, K.A. Lee, Densification and purification of cold sprayed Ti coating layer by using annealing in different heat treatment environments, Adv. Mater. Res. (2013) 602–604, 1604–1608.

[22] R. Blose, in: E. Lugscheider (Ed.), Thermal Spray 2005", ASM International, Materials Park, OH, 2005, pp. 199–207.

[23] P. Vo, E. Irissou, J.G. Legoux, S. Yue, Mechanical and microstructural characterization of cold-sprayed Ti-6Al-4V after heat treatment, J. Therm. Spray Technol. 22 (6) (2013) 954–964.

[24] S.H. Zahiri, D. Fraser, M. Jahedi, Recrystallization of cold spray-fabricated CP titanium structures, J. Therm. Spray Technol. 18 (1) (2009) 16–22.
[25] T. Marrocco, T. Hussain, D. McCartney, P. Shipway, Corrosion performance of laser post-treated cold sprayed titanium coatings, J. Therm. Spray Technol. 20 (4) (2011) 909–917.
[26] M. Grujicic, J.R. Saylor, D.E. Beasley, W.S. DeRosset, D. Helfritch, Computational analysis of the interfacial bonding between feed-powder particles and the substrate in the cold-gas dynamic-spray process, Appl. Surf. Sci. 219 (2003) 3–4, 211–227.
[27] S. Guetta, M.H. Berger, F. Borit, V. Guipont, M. Jeandin, M. Boustie, Y. Ichikawa, K. Sakaguchi, K. Ogawa, Influence of particle velocity on adhesion of cold-sprayed splats, J. Therm. Spray Technol. 18 (3) (2009) 331–342.
[28] M. Grujicic, C.L. Zhao, W.S. De-Rosset, D. Helfritch, Adiabatic shear instability based mechanism for particles/substrate bonding in the cold-gas dynamic-spray process, Mater. Des. 25 (8) (2004) 681–688.
[29] T. Schmidt, F. Gartner, H. Assadi, H. Kreye, Development of a generalized parameter window for cold spray deposition, Acta Mater. 54 (3) (2006) 729–742.
[30] A.P. Alkhimov, A.I. Gudilov, V.F. Kosarev, N.I. Nesterovich, Specific features of microparticle deformation upon impact on a rigid barrier, J. Appl. Mech. Tech. Phys. 41 (1) (2000) 188–192.
[31] P.C. King, G. Bae, S.H. Zahiri, M. Jahedi, C. Lee, An experimental and finite element study of cold spray copper impact onto two aluminum substrates, J. Therm. Spray Technol. 19 (3) (2010) 620–634.
[32] C. Borchers, F. Gartner, T. Stoltenhoff, H. Kreye, Microstructural bonding features of cold sprayed face centered cubic metals, J. Appl. Phys. 96 (8) (2004) 4288–4292.
[33] G. Bae, S. Kumar, S. Yoon, K. Kang, H. Na, H.J. Kim, C. Lee, Bonding features and associated mechanisms in kinetic sprayed titanium coatings, Acta Mater. 57 (19) (2009) 5654–5666.
[34] D. Goldbaum, J. Shockley, R. Chromik, A. Rezaeian, S. Yue, J.-G. Legoux, E. Irissou, The effect of deposition conditions on adhesion strength of Ti and Ti6Al4V cold spray splats, J. Therm. Spray Technol. 21 (2) (2012) 288–303.
[35] K. Kim, M. Watanabe, K. Mitsuishi, K. Iakoubovskii, S. Kuroda, Impact bonding and rebounding between kinetically sprayed titanium particle and steel substrate revealed by high-resolution electron microscopy, J. Phys. D Appl. Phys. 42 (6) (2009) 65304–65308.
[36] P.C. King, C. Busch, T. Kittel-Sherri, M. Jahedi, S. Gulizia, Interface melding in cold spray titanium particle impact, Surf. Coat. Technol. 239 (2014) 191–199.
[37] R.R. Chromik, D. Goldbaum, J.M. Shockley, S. Yue, E. Irissou, J.-G. Legoux, N.X. Randall, Modified ball bond shear test for determination of adhesion strength of cold spray splats, Surf. Coat. Technol. 205 (5) (2010) 1409–1414.
[38] T. Hussain, D.G. McCartney, P.H. Shipway, Impact phenomena in cold-spraying of titanium onto various ferrous alloys, Surf. Coat. Technol. 205 (2011) 21–22, 5021–5027.
[39] S. Yin, X.-F. Wang, W.Y. Li, H.-E. Jie, Effect of substrate hardness on the deformation behavior of subsequently incident particles in cold spraying, Appl. Surf. Sci. 257 (17) (2011) 7560–7565.
[40] K. Kim, S. Kuroda, M. Watanabe, Microstructural development and deposition behavior of titanium powder particles in warm spraying process: from single splat to coating, J. Therm. Spray Technol. 19 (6) (2010) 1244–1254.
[41] K. Kim, M. Watanabe, J. Kawakita, S. Kuroda, Grain refinement in a single titanium powder particle impacted at high velocity, Scripta Mater. 59 (7) (2008) 768–771.
[42] G. Bae, K. Kang, J.-J. Kim, C. Lee, Nanostructure formation and its effects on the mechanical properties of kinetic sprayed titanium coating, Mater. Sci. Eng. A 527 (23) (2010) 6313–6319.

[43] G. Bae, K. Kang, H. Na, C. Lee, H.J. Kim, Thermal spray 2009: Expanding thermal spray performance to new markets and applications, in: B. Marple, M. Hyland, Y.-C. Lau, C.J. Li, R.S. Lima, G. Montavon (Eds.), Proceedings of the International Thermal Spray Conference, ASM International, Materials Park, OH 2009, pp. 290–295.
[44] D. Goldbaum, Micromechanical testing of cold sprayed Ti splats and coatings, PhD Thesis, McGill University, Montreal, QC, 2012.
[45] C.K.S. Moy, J. Cairney, G. Ranzi, M. Jahedi, S.P. Ringer, Investigating the microstructure and composition of cold gas-dynamic spray (CGDS) Ti powder deposited on Al 6063 substrate, Surf. Coat. Technol. 204 (23) (2010) 3739–3749.
[46] N. Cinca, J.M. Rebled, S. Estradé, F. Peiró, J. Fernández, J.M. Guilemany, Influence of the particle morphology on the cold gas spray deposition behaviour of titanium on aluminum light alloys, J. Alloys Compd. 554 (2013) 89–96.
[47] D. Goldbaum, R. Chromik, S. Yue, E. Irissou, J.-G. Legoux, Mechanical property mapping of cold sprayed Ti splats and coatings, J. Therm. Spray Technol. 20 (3) (2011) 486–496.
[48] J. Ajaja, D. Goldbaum, R.R. Chromik, Characterization of Ti cold spray coatings by indentation methods, Acta Astronaut. 69 (2011) 11–12, 923–928.
[49] S.H. Zahiri, C.I. Antonio, M. Jahedi, Elimination of porosity in directly fabricated titanium via cold gas dynamic spraying, J. Mater. Process. Technol. 209 (2) (2009) 922–929.
[50] D. Goldbaum, J. Ajaja, R.R. Chromik, W. Wong, S. Yue, E. Irissou, J.-G. Legoux, Mechanical behavior of Ti cold spray coatings determined by a multi-scale indentation method, Mater. Sci. Eng. A 530 (2011) 253–265.
[51] T. Marrocco, D.G. McCartney, P.H. Shipway, A.J. Sturgeon, Production of titanium deposits by cold-gas dynamic spray: numerical modeling and experimental characterization, J. Therm. Spray Technol. 15 (2) (2006) 263–272.
[52] ASTM B265 Standard Specification for Titanium and Titanium Alloy Strip, Sheet, and Plate, ASTM International, West Conshohocken, PA, 2006.
[53] ASTM C633 Standard Test Method for Adhesion or Cohesion Strength of Thermal Spray Coatings, ASTM International, West Conshohocken, PA, 2001.
[54] W.-Y. Li, C. Zhang, X. Guo, C.-J. Li, H. Liao, C. Coddet, Study on impact fusion at particle interfaces and its effect on coating microstructure in cold spraying, Appl. Surf. Sci. 254 (2) (2007) 517–526.
[55] S. Costil, E. Irissou, Y. Danlos Belfort, J. Legoux, W. Wong, S. Yue, Global solutions for future applications, in: B. Marple, A. Agarwal, M. Hyland, Y.C. Lau, C.J. Li, R.S. Lima, G. Montavon (Eds.), Thermal Spray 2010, DVS-Verlag, Dusseldorf, Germany, 2010, pp. 836–841.
[56] A. Sova, S. Grigoriev, A. Okunkova, I. Smurov, Potential of cold gas dynamic spray as additive manufacturing technology, Int. J. Adv. Manuf. Technol. 69 (2013) 9–12, 2269–2278.
[57] J. Karthikeyan, C.M. Kay, J. Lindeman, R.S. Lima, C. Berndt, in: C. Berndt (Ed.), Thermal Spray 2000, 2000, pp. 255–262.
[58] X. Zhou, P. Mohanty, Corrosion behaviour of cold sprayed titanium coatings in simulated body fluid, Corros. Eng. Sci. Technol. 47 (2) (2012) 145–154.
[59] J. Cizek, O. Kovarik, J. Siegl, K.A. Khor, I. Dlouhy, Influence of plasma and cold spray deposited Ti layers on high-cycle fatigue properties of Ti6Al4V substrates, Surf. Coat. Technol. 217 (2013) 23–33.
[60] A.N. Ryabinin, E. Irissou, A. McDonald, J.G. Legoux, Simulation of gas-substrate heat exchange during cold-gas dynamic spraying, Int. J. Therm. Sci. 56 (2012) 12–18.
[61] R.C. Dykhuizen, M.F. Smith, Gas dynamic principles of cold spray, J. Therm. Spray Technol. 7 (2) (1998) 205–212.

[62] S. Yin, X. Suo, J. Su, Z. Guo, H. Liao, X. Wang, Effects of substrate hardness and spray angle on the deposition behavior of cold-sprayed Ti particles, J. Therm. Spray Technol. 23 (2014) 1–2, 76–83.
[63] A. Rezaeian, W. Wong, S. Yue, E. Irissou, J.G. Legoux, MS&T'09, Association for Iron & Steel Technology, Warrendale, PA, 2009, pp. 2268–2278.
[64] N. Cinca, M. Barbosa, S. Dosta, J.M. Guilemany, Study of Ti deposition onto Al alloy by cold gas spraying, Surf. Coat. Technol. 205 (4) (2010) 1096–1102.
[65] P. Cavaliere, A. Silvello, Processing parameters affecting cold spay coatings performances, Int. J. Adv. Manuf. Technol. 71 (2014) 1–4, 263–277.
[66] H. Assadi, T. Schmidt, H. Richter, J.O. Kliemann, K. Binder, F. Gärtner, T. Klassen, H. Kreye, On parameter selection in cold spraying, J. Therm. Spray Technol. 20 (6) (2011) 1161–1176.
[67] W. Wong, P. Vo, E. Irissou, A.N. Ryabinin, J.G. Legoux, S. Yue, Effect of particle morphology and size distribution on cold-sprayed pure titanium coatings, J. Therm. Spray Technol. 22 (7) (2013) 1140–1153.
[68] M. Jeandin, G. Rolland, L.L. Descurninges, M.H. Berger, Which powders for cold spray? Surf. Eng. 30 (5) (2014) 291–298.
[69] W.Y. Li, C. Zhang, X. Guo, J. Xu, C.J. Li, H. Liao, C. Coddet, K.A. Khor, Ti and Ti-6Al-4V coatings by cold spraying and microstructure modification by heat treatment, Adv. Eng. Mater. 9 (5) (2007) 418–423.
[70] X. Zhou, P. Mohanty, Electrochemical behavior of cold sprayed hydroxyapatite/titanium composite in Hanks' solution, Electrochim. Acta 65 (2012) 134–140.
[71] J. Karthikeyan, Cold Spray Technology: International Status and USA Efforts. Report from ASB Industries Inc., Barbeton, OH, 2004, pp. 1–14.
[72] MIL-STD-3021, U.S. Department of Defense Manufacturing Process Standard: Materials Deposition, Cold Spray. www.arl.army.mil/www/pages/375/MIL-STD-3021.pdf, 2008 (accessed May 2014).

Thermal spray forming of titanium and its alloys

23

Jo Ann Gan[*,**], *Christopher C. Berndt*[*,†]
[*]Industrial Research Institute Swinburne, Faculty of Science, Engineering and Technology, Swinburne University of Technology, Hawthorn, Australia
[**]Research Services, La Trobe University, Melbourne, Australia
[†]Department of Materials Science and Engineering, Stony Brook University, NY, USA

23.1 Introduction to thermal spray

Thermal spray is a generic term referring to a group of coating processes in which a stream of finely divided metallic or nonmetallic particles is applied onto a substrate to form flattened lamellae that stack progressively to form a coating [1,2]. Upon impact with the substrate, the particles bond with the surface by means of mechanical interlocking. These processes can be used to deposit a wide range of materials, from ceramic to metallic, polymers and composites, because of their wide range of processing temperatures and velocities (Figure 23.1). The feedstock materials used for thermal spray processes can be in the form of powder, rod, wire, or in the form of a liquid solution or suspension.

Thermal spray coatings can be formed in one of two ways: (1) deposition of ductile metals in the solid state, such as in cold spray, and (2) deposition of molten or semimolten particles that have passed through heat sources, such as in flame spray, plasma spray, high-velocity oxygen fuel (HVOF), high-velocity air fuel (HVAF), and wire arc spray [3]. Thermal spray processes can be categorized according to their source of energy, that is, combustion, electric arc, and kinetic (Figure 23.2).

Thermal spray processes have high deposition rates, low processing and equipment costs, can be used on site, have simple disposal of waste streams, causes minimal degradation of substrates, and can be used to deposit a wide range of coating thicknesses for many applications. However, thermal spray is a line-of-sight process and the coatings have relatively low bond strength, low loading capacity, are porous, and have anisotropic properties. The ability to deposit porous coatings is sometimes beneficial in certain applications that require such a microstructure. For instance, porosity is known to (1) decrease thermal conductivity of thermal barrier coatings and (2) enhance bone growth in biomedical coatings. Other well-known applications of thermal spray coatings include wear and corrosion resistance coatings, abradable and abrasive coatings, dimensional restoration, and near-net-shape component manufacturing [1,2]. Thermal spray is considered as an additive manufacturing process because of its ability to form dense coherent deposits.

Titanium Powder Metallurgy. http://dx.doi.org/10.1016/B978-0-12-800054-0.00023-X

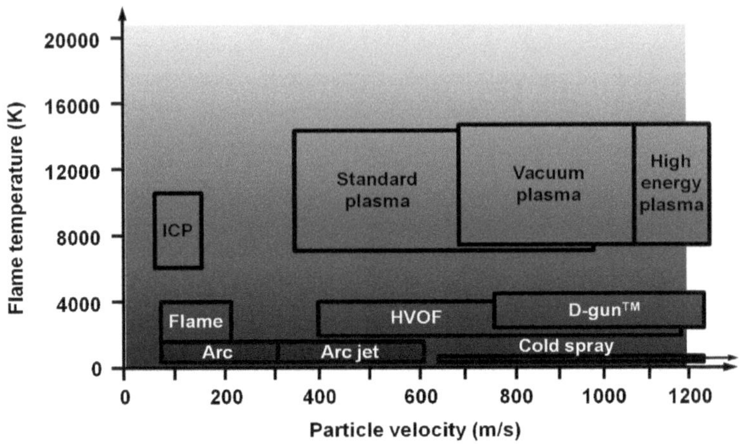

Figure 23.1 The temperature–velocity distribution envelope of thermal spray processes. HVOF, high-velocity oxygen-fuel; ICP, inductively coupled plasma; D-gun^{TM}, detonation gun.

23.2 Titanium and titanium alloy feedstock characteristics

The outstanding mechanical strength, corrosion properties, and biocompatibility of titanium have led to increased potential and interests in the manufacturing of titanium coatings. The most common titanium-based materials deposited by thermal spray processes include commercially pure (CP) titanium, titanium alloys (Ti-6Al-4V), and

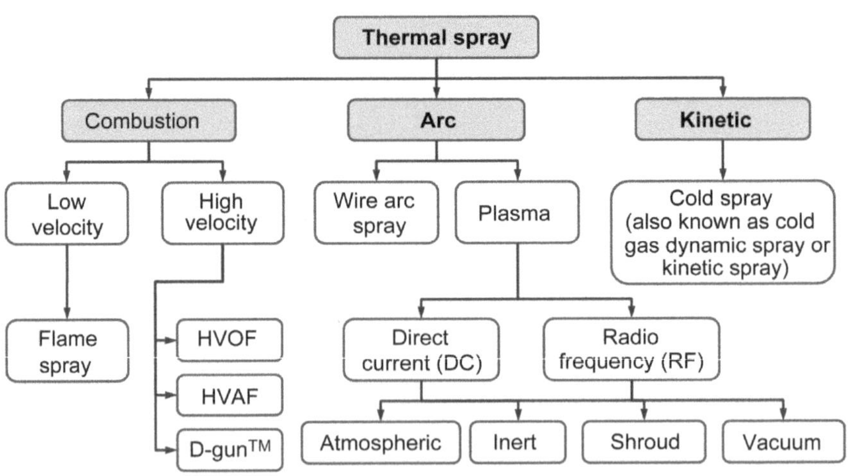

Figure 23.2 Classification of thermal spray processes. HVOF, high-velocity oxygen fuel; HVAF, high-velocity air fuel; D-gun^{TM}, detonation gun.

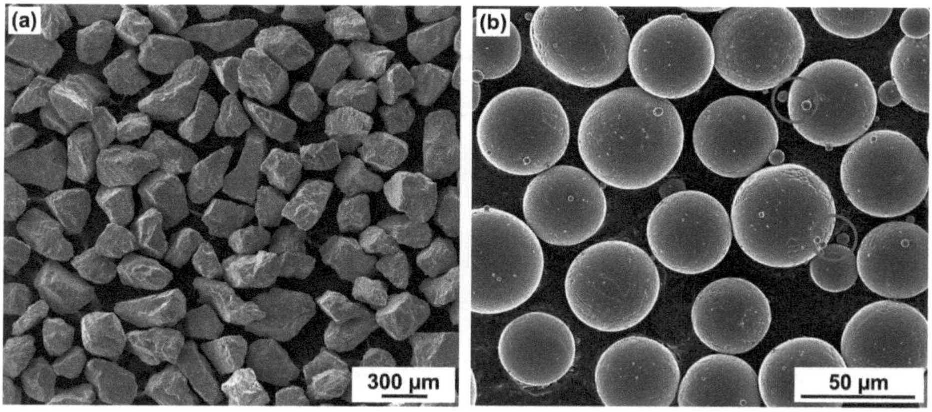

Figure 23.3 Titanium powder with (a) angular morphology manufactured by the hydride-dehydride (HDH) process, and (b) spherical morphology manufactured by the plasma atomization process. Note: several regions of Figure 23.3b are circled and depict features that incorporate satellites. ((a) Courtesy AMETEK Inc. (Reading Alloys), USA; (b) Courtesy Raymor Industries Inc. (AP&C), Canada.)

titanium aluminide (γ-TiAl, α-Ti_3Al, and $TiAl_3$). The main forms of feedstock used in the thermal spraying of titanium are powder and wire.

Ideally, the preferred particle feedstock morphology for thermal spray is spherical with smooth surfaces, mainly for their excellent flow characteristics (Figure 23.3). These spherical particles are manufactured through an atomization processes. However, the conventional atomization process is not appropriate for titanium since it tends to react with the crucible. Crucible-free or cold-crucible variations of the atomization process must be employed for titanium production: (1) electrode induction melting gas atomization (EIGA) [4], (2) plasma melting induction guiding gas atomization (PIGA) [5], (3) centrifugal atomization [6], and (4) plasma atomization [7]. Angular-shaped particles from titanium sponge or hydride–dehydride (HDH) process have also been employed as thermal spray feedstock [8,9]. The titanium powder manufactured via the HDH route has a variable size range between 5 and 300 μm [10]. The particle size distribution depends on parameters such as material feed rate as well as (1) gas pressure for gas atomization, (2) rotation speed for centrifugal atomization, and (3) plasma power and gas flow rate for plasma atomization.

The as-received powder is often sieved before being fed into thermal spray source to obtain a specific size range. A wide particle size range will result in inconsistent particle trajectories and irregular deposition. In addition, the optimum particle size distribution differs for each thermal spray process because of the difference in their temperature and velocity profiles, which will result in a variable degree of particle melting and flattening. Table 23.1 shows the particle size range of commonly used titanium powder for several thermal spray processes.

Table 23.1 Common particle size of titanium based feedstock used in several thermal spray processes

Process	Particle size (μm)	References
Atmospheric plasma spray	20–90	[11,12]
Vacuum plasma spray	10–75	[13,14]
High velocity oxy-fuel	1–45	[15,16]
Detonation spray	5–60	[17,18]
Cold spray	5–45	[19,20]

23.3 Deposition of titanium and titanium alloy coatings

This section will discuss the thermal spray processes that have been applied to deposit titanium-based coatings, from the technological aspects to the expected outcomes. It is important to identify the titanium phases required for the intended applications to select the appropriate spray process for the task.

23.3.1 Plasma spray

There are many renditions of plasma spray processes that have been used to deposit titanium-based coatings. In the atmospheric plasma spray (APS) process, for example, a plasma is generated by supplying a high-voltage discharge to ionize the plasma-forming gas. The energy sources used in thermal spray devices to generate thermal plasmas are usually DC electric arcs or high-frequency sources. A schematic of a plasma spray process is shown in Figure 23.4.

The plasma generator generally consists of a copper anode and a cathode that is usually made of thoriated tungsten because of its high melting point and good thermionic emission properties. Monoatomic gases, such as argon and helium, have lower gas enthalpy but the generated gas jet can reach high velocity. On the other hand, diatomic gases, such as hydrogen and nitrogen, generate greater enthalpy because

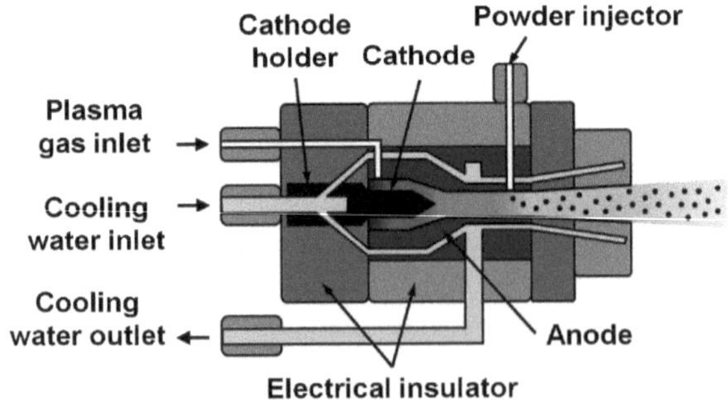

Figure 23.4 Schematic of a plasma spray process.

of dissociation of their molecular structure and have higher thermal conductivity. Combinations of monoatomic and diatomic gases are used to generate a plasma with high enthalpy and velocity.

A typical plasma spray system has a degree of ionization within the range of $0.5–4.0 \times 10^{-3}$ [21], which corresponds to a plasma efficiency of less than 10% to over 70% [22]. The degree of ionization depends on the temperature and pressure of the system [23] while the efficiency of a plasma jet is influenced by the plasma jet design, size of mixing chambers and nozzles, input power, secondary gas flow rate, and operating conditions [22].

23.3.1.1 *Atmospheric plasma spray*

The APS process operates under an ambient, atmospheric environment. This process has been used to deposit titanium aluminide or titanium composites rather than pure titanium because of the strong affinity of titanium for oxygen and nitrogen. Several studies [24–26] have used APS to deposit pure titanium feedstock. The coatings were formed with a porosity level up to 10.2%. Intra-splat cracks and elongated cracks along the coating–substrate interface were also found in these coatings. The cracks will severely impact the bond strength of the coatings. High contents of titanium oxide and titanium nitride are often found in the APS-deposited titanium coatings, which arise from reaction between the pure titanium feedstock and the oxygen and nitrogen in the working gases or surrounding atmosphere.

Titanium aluminide coatings deposited by APS also undergo phase and compositional changes during the process, resulting in titanium oxide and titanium nitride [11,27]. However, these coatings are denser and demonstrate better bond strength compared to coatings deposited from pure titanium feedstock, an outcome that arises due to the filling of cracks and pores by an aluminium phase [11]. Deposition of titanium tri-aluminide by the APS process has resulted in less than 1% oxide content in the coatings—all in the form of aluminium oxide [28]. Titanium tri-aluminide coatings were reported to be more resistant to thermal expansion–induced cracking, but remain prone to cracking after thermal cycling [28].

The APS process has been employed to deposit Ti-6Al-4V–hydroxyapatite composite coatings. The purpose of composite coatings was to impart the mechanical strength and toughness of titanium alloys while benefiting from the biocompatibility of hydroxyapatite. Oxidation of titanium was evidenced in some work [29] but not in others [30,31]. Formation of undesirable phases in APS coatings through oxidation and nitriding of titanium can be minimized by using (1) a gas shroud, (2) a controlled-atmosphere chamber, or (3) a modified plasma such as the Gator–Gard® process. In addition to these methods, the formation of titanium nitride can be minimized by substituting nitrogen with helium or hydrogen as the plasma forming gas.

Shrouding provides an inert gas to envelope the plasma jet, thus minimizing air entrainment and the amount of oxygen and reactive gases available to react with the sprayed particles. This process has been used effectively to reduce the oxide contents in titanium coatings by Kinos et al. [12]. A controlled-atmosphere chamber approach is more effective than shrouding, but at the expense of increased cost. The most commonly used controlled-atmosphere chamber for the plasma spray process is the vacuum plasma spray (VPS) process, which will be discussed in Section 3.1.3. Gator-Gard® [32,33] is a modified plasma spray process that operates on 100% helium gas. This process comprises a cooled extension tube that fits around the nozzle of a standard plasma torch. This process enables deposition of particles in a plasticized state instead of a molten state, and at a higher velocity compared to conventional APS. The lower particle temperature and higher velocity minimizes any potential in-flight oxidation reaction.

23.3.1.2 Reactive plasma spray

The reactive plasma spray (RPS) process is a variation of the plasma spray process that implements both *in-situ* synthesis and product deposition in a single process. The feedstock is injected into the plasma jet and heated in the same fashion as conventional APS. The particles then react with a reactive species, typically a hydrocarbon or nitrogen, to form carbides and nitrides. The introduction of reactive species into the system increases significantly the amount of carbon or nitrogen in the coatings [34]. The coatings formed are metal-matrix composites that consist of *in-situ* synthesized hard phases formed from the reaction between the molten metal particles and the reactive species. There are four essential components in a RPS system, Figure 23.5: (1) a plasma generator, (2) a method of injecting reactive precursors, (3) a reaction chamber/region, and (4) a means to transfer the reactants and products to the substrate [35].

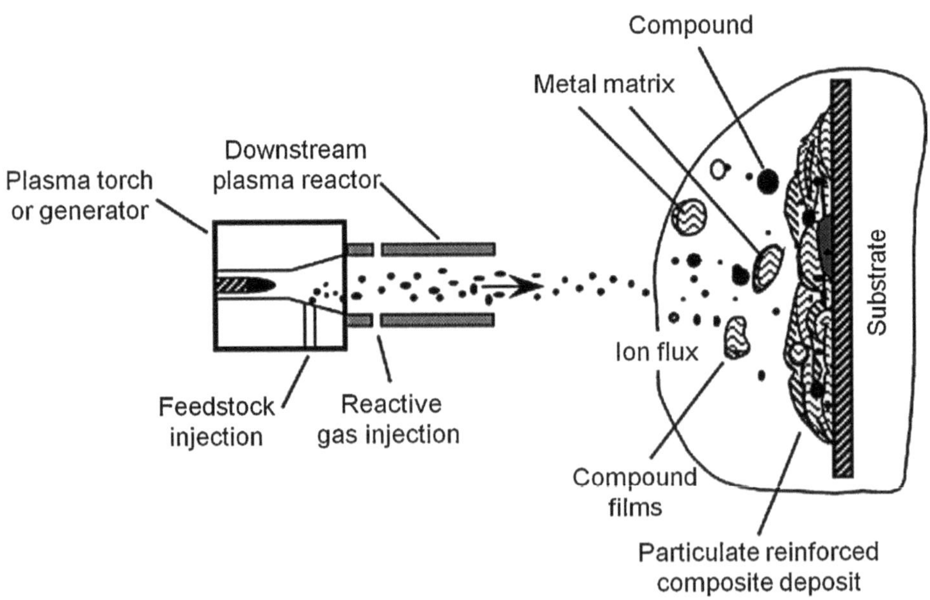

Figure 23.5 Schematic diagram of a reactive plasma spray process. (Source: From Ref. [35], reprinted with permission from Springer Science and Business Media.)

The earliest work on the RPS forming of Ti-based material was carried out by Tsantrizos et al. [36,37], where aluminium powder was fed into a plasma plume formed by titanium tetrachloride, $TiCl_4$, gas. The process formed titanium aluminide alloy, but at the expense of reduced cathode life due to corrosion by the high concentration of $TiCl_4$ [38]. The use of RPS for titanium-based materials has evolved to create titanium nitride and titanium carbide composites. The formation of titanium nitride and titanium carbide are exothermic reactions with standard heats of formation of -336.686 ± 1.130 and -183.468 ± 1.632 kJ/mol, respectively [39]. The formation of metal-matrix composites from a RPS process imparts enhanced mechanical properties and improved wear resistance.

The prerequisites for the formation of hard phases in the RPS process are (1) the dissociation of precursors into reactive species; e.g., CH_4 into CH_3^+ or N_2 into N^+, and (2) adequate processing temperature and residence time for the reaction to occur [35]. These prerequisites can be achieved by optimizing variables such as power, gas flow rate and spray distance. The extent of reaction further depends on the (1) spray system, (2) processing parameters, and (3) feedstock characteristics.

The use of a reactor increases significantly the concentration of reactive species, extends the reaction zone, increases the available reaction time and improves the precursor gas homogeneity [34], thus increasing the nitride or carbide contents in the coatings. The use of an alternative plasma generator such as the electromagnetic acceleration plasma generator developed by Tahara et al. [40,41], also affects the extent of reaction and coating formation since these devices form larger-area plasmas with higher velocity, temperature, and energy density than conventional plasma torches.

The flow rate of reactive gas determines the concentration of reactive species available in the reaction regions, thus affecting the rate of reaction and nitride/carbide content in the coatings. The chamber pressure increases the extent of reaction through two actions: (1) increase of reaction time due to the decrease of particle velocity and (2) increased particle temperature due to the increase of heat transfer from the plasma gas to the particles. The nitride content in the coatings increase by almost half by tripling the nitrogen flow rate and doubling the chamber pressure [34].

The spray distance increases the in-flight time, hence extending the available reaction time for hard phase formation [42]. However, when the spray distance exceeds an optimum value, the hard phase content in the coatings will decrease because this formation is suppressed by rapid cooling at long spray distances [43]. The power supply influences the hard phase content to a smaller extent by increasing the rate of heat transfer to the particles [44]. Thus, the particles are not completely molten at the low power and low heat transfer rate. Therefore, the reaction will only occur on the particle surface to form a layer of nitride or carbide that will act as a barrier that prevents further reaction during the residence time.

The most substantial feedstock characteristics that determine the extent of reaction is the particle size. Smaller particles have greater surface area per gross volume of material for reaction; thus, an increased reaction rate ensues. Smaller particles also require less energy to completely melt; hence the hard phases that were formed on the molten particle surface can be transported to the particles interior through Hill vortex motion [45].

23.3.1.3 Vacuum plasma spray

The VPS process is also known as the low-pressure plasma spray (LPPS^{TM}) process. Unlike the RPS process, the VPS process operates under a soft vacuum and, thereby, avoids formation of secondary phases in the coatings. The intent in using the VPS process is to produce a high-purity coating by retaining the original composition of feedstock, without undergoing reaction during in-flight to form secondary phases. This process operates at a higher cost compared to other plasma spray processes because of the need to maintain the low pressure condition, typically between 4 and 40 kPa (30–300 Torr or 0.04–0.40 atm), throughout the spray process.

The VPS system consists of a conventional plasma spray torch with a nozzle modified to withstand the high-pressure expansion from the plasma jet exiting into the low-pressure spray chamber. The plasma jet in a VPS process is wider and longer than for APS. This variation in spray patterns results in dense (up to ~98% of theoretical density) and more consistent coatings.

Spray forming by VPS has been used to manufacture Ti-6Al-4V alloy components [13,46,47]. Components from this alloy are conventionally fabricated through processes such as casting, forging, and powder metallurgy, which pose some disadvantages due to the reactivity of titanium alloys. The VPS forming process combines the processes of melting, rapid solidification, consolidation, and welding into a single step [46], thereby eliminating the need to repeatedly expose the reactive titanium alloys to high temperature. In addition, the VPS forming process offers several other advantages over conventional methods: (1) a finer-scale microstructure, (2) reduced segregation, and (3) ability to form complex shapes with layered or graded structures [47].

The manufacturing of a near-net-shape product using VPS forming begins with the preparation of a mold that has the shape of the desired product. The selection of the mold material, shape, and surface finish is a critical step since it defines the inner geometry of the product and controls the stress level within the spray-formed component [46]. The deposition process is then executed using optimized conditions until the required deposit thickness is achieved. The mold can be preheated to minimize stress buildup within the deposit. The spray-formed product is then cooled to room temperature in an inert environment, and detached naturally from the mold as a result of the difference in the coefficients of thermal expansion between the mold and deposit materials. The mold can be reused to form other components. The main challenges in fabricating near-net-shape components using VPS forming processes are the optimization and control of (1) phase composition and distribution, (2) microstructure such as grain size and porosity, and (3) residual stress [46].

Despite the fact that VPS operates under low-pressure conditions, it is not uncommon to have a small degree of oxide, nitride, or even hydride present in the coatings, as reported by several researchers [48,49] on VPS titanium coatings. The presence of secondary phases in these coatings is primarily because of (1) reaction with working gases such as nitrogen and hydrogen and (2) a residual amount of air because of the relatively soft vacuum environment. The content of secondary phases can be minimized by substituting hydrogen and nitrogen with other working gases and/or backfilling the chamber using an inert gas such as argon. There has also been a reported loss of aluminium content from spraying Ti-6Al-4V alloy and titanium aluminide [14,49] since aluminium has a lower vapor pressure compared to titanium. Titanium prealloyed with silicon, which forms a eutectic mixture with a lower melting point, has also been deposited by the VPS process [50].

23.3.2 Wire arc spray

The wire arc spray process is also known as the twin-wire arc process since it uses direct-current (DC) to form an electric arc in the gap between two consumable wire tips that are continuously fed into the arc zone (Figure 23.6).

Wire arc spray is a favorable technique to deposit reactive metals such as titanium because of the low specific surface area of the wire feedstock compared to powder and, therefore, a potentially lower amount of impurities. In-flight oxidation in the wire arc spray process is minimized by shortening the spray distance and increasing the atomizing gas flow rate. Low-pressure arc spray [8,51] or inert-gas shielded arc spray [52,53] has been developed as a more cost-effective alternative to VPS to produce high-purity titanium coatings. However, wire arc spray under low-pressure conditions poses complications since the low pressure condition causes expansion of the arc that melts and damages the contact tips [51,54]. Thus, this condition is prevented by adjusting the melting process to take place in an antechamber where the pressure is maintained at atmospheric conditions [51].

Despite its ability to deposit high-purity titanium coatings, low-pressure twin-wire arc spraying has several disadvantages: (1) it requires a high pressure behind the short circuit point, (2) different melting behavior between two wires can occur based on the applied potential and chamber pressure, and (3) the requirement for high argon consumption [55]. A new process that used only a single wire was based on arc generation

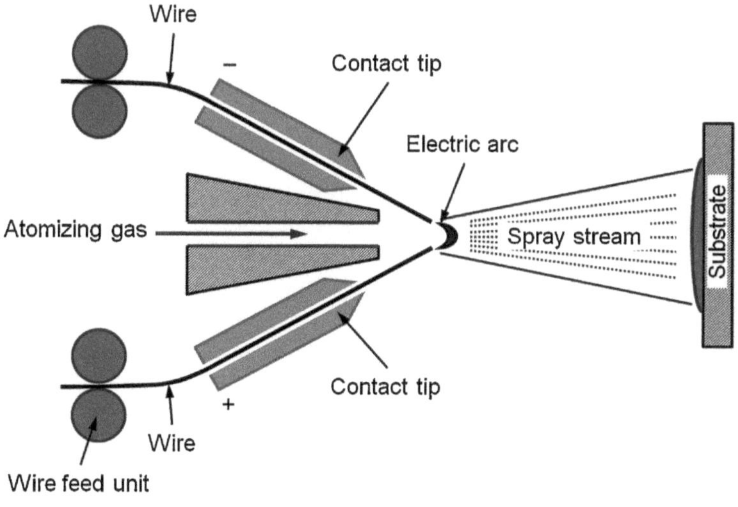

Figure 23.6 Schematic of wire arc spray process.

between a water-cooled, nonconsumable anodic nozzle and a cathodic poled consumable wire [54–56]. The arc was rotated around the wire tip by a disk behind the nozzle to prevent the nozzle wall from overheating at any one point. The arc generation for the single-wire arc process is based on high-voltage ignition instead of the short circuit ignition employed in a twin-wire arc process. The single-wire arc spray process is advantageous compared to the conventional wire arc spray because of (1) reduced gas consumption by 5 times, (2) better atomization, and (3) accurate control of energy input since it is independent of the wire feed rate [55,56].

The wire arc spray process has also been used to deposit titanium nitride and titanium aluminide coatings. Wear-resistant titanium nitride coatings were deposited by wire arc spray through reaction between nitrogen atomization gas and titanium or prenitrided titanium wire [57–59]. The coatings produced using at least one prenitrided titanium wire were harder and more wear resistant than those deposited without employing prenitrided titanium wire [57]. On the other hand, titanium aluminide coatings were produced by using titanium wire as the anode and aluminium wire as the cathode since the reverse polarity would result in arc instability because of uneven melting rates at electrode tips [60]. The key to obtaining a homogeneous coating is through uniform dispersion of titanium and aluminium droplets from their corresponding wires, which depends on the arc stability.

23.3.3 High-velocity oxygen fuel

The HVOF process (Figure 23.7a) works by confining combustion gases and particles in a high-pressure chamber to create a high-velocity jet. Fuel gas, such as propane, acetylene, and oxygen are fed into a combustion chamber and ignited. The jet travels through the nozzle and an extended confining barrel to generate a supersonic gas jet with high particle velocity. The feedstock powder can be fed into the system axially or radially, entrained into the high velocity jet, and accelerated through the barrel to deposit onto a substrate.

The lower average particle temperature of the HVOF process compared to APS, in addition to the high particle velocity, is expected to produce denser coatings with lower oxide content because of greater particle deformation and a lower degree of melting. Commercially pure titanium coatings deposited by HVOF at optimized parameters have approximately 4% porosity and an oxide content of 14% [62].

Modified HVOF systems have been developed to reduce the degree of oxidation of titanium coatings [15,63,64]. These modified HVOF systems are also referred to as *warm spray* in the literature. There are two variations of the warm spray process [61]: (1) HVOF with a gas shroud (Figure 23.7b) and (2) HVOF with an intermediate mixing chamber (Figure 23.7c). The gas shroud system reduces particle oxidation by introducing an inert gas into the system, thus restricting the amount of entrained air and oxygen available to react with the particles.

On the other hand, the second system uses a chamber to mix nitrogen gas with the combustion gas to lower the gas temperature. Titanium coatings sprayed using a modified HVOF system with an intermediate mixing chamber exhibit an oxide content as low as 0.25 wt% at a nitrogen flow rate of 1.5 m^3/min [61]. However, excessive

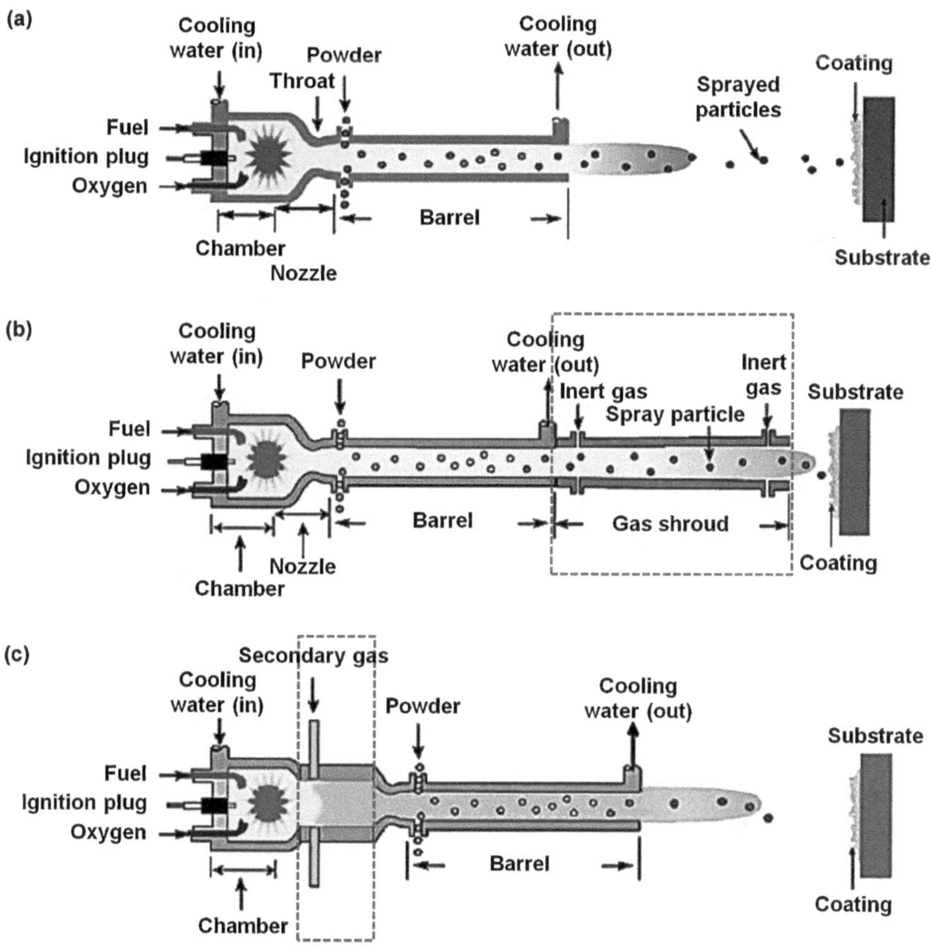

Figure 23.7 Schematic of (a) a standard HVOF system, (b) a modified HVOF system with gas shroud, and (c) a modified HVOF system with an intermediate mixing chamber. Note: the squares in Figure 23.7 b and c indicate the modified components. (Source: From Ref. [61], reprinted with permission from Elsevier.)

nitrogen may reduce the particle temperature to the point where the particles are not able to deform upon impact, which lead to increased coating porosity [63].

In order to counter the effects of increased porosity at high nitrogen flow rate in the modified HVOF systems, a secondary material that has a peening effect was added and mixed with the titanium feedstock [15]. A peening effect is often observed in HVOF coatings, even without the use of secondary peening material, because of the high-velocity impact of semimolten particles that cause plastic deformation of the previously deposited layers [65–67]. The peening effect is known to generate internal compressive residual stress that could improve the fatigue strength of the coatings [66,68].

Reactive HVOF spraying of titanium powder using nitrogen as the carrier gas has been used to deposit titanium nitride coatings [69,70]. The oxide content in HVOF sprayed titanium nitride coatings was higher than that in coatings deposited by the RPS process because of the presence of oxygen in the HVOF process [69].

23.3.4 Detonation spray

The components and principle of detonation spray, or detonation gun (D-gun™) [71], are similar to the HVOF process except that the D-gun™ involves explosion of fuel and oxygen to create a detonation-pressure wave rather than the continuous steady state in a HVOF process to heat and accelerate the feedstock particles. A long water-cooled barrel confines the combustion to generate a high kinetic energy jet. The high particle velocity contributed to good particle deformation and produced coatings with low porosity and excellent bond strength.

Detonation spray has found application for the manufacture of wear-resistant coatings because of its ability to deposit the most hard and dense thermal spray coatings. Deposition of titanium-based materials by detonation spray has mostly revolved around titanium aluminide, γ-TiAl [72–74] and Ti_3Al [17,75,76], although pure titanium feedstock [18] has been deposited by detonation spray. The coatings formed were reported to be composed of nanocrystalline grains, exhibit high density, and demonstrate high bond strength, hardness, and wear resistance [17,18,76].

Although the main phases of titanium and titanium aluminide were retained in the respective coatings, small amounts of titanium carbide, titanium nitride, and titanium oxide were detected. These secondary phases can be observed at the inter-splat boundaries or within the splats, depending on the mechanism of formation. Formation of such secondary phase inclusions are sometimes intentional to impart or enhance specific functions in the coatings. For instance, titanium nitride was included in TiAl coatings to enhance the tribological properties of these coatings [72].

Detonation sprayed titanium aluminide coatings deposited using hydrogenated feedstock (TiH_2–$TiAl_2$–$TiAl_3$) that were recombined from a destructive hydrogenation process [77] were distinguished to exhibit greater hardness and resistance to heat and oxidation compared to the coatings deposited from the starting titanium aluminide [17,75]. The particles were protected against oxidation by desorption of hydrogen from the titanium hydride during their in-flight transit [17,75].

23.3.5 Cold spray

The cold-spray process, also known as cold gas dynamic spray or kinetic spray, is part of the thermal spray family since the principles of particle heating and acceleration are valid, although cold spray operates at a lower temperature and higher velocity. The cold-spray process uses the supersonic velocity generated by heating and expansion of working gases to accelerate and deform particles instead of melting them prior to acceleration as in other thermal spray processes. This process has been a popular technique to deposit titanium-based materials because of its lower processing temperature;

hence, there is less tendency for secondary phase formation. More details on the cold spray forming of titanium-based materials are found in Chapter 22 of this book.

23.4 Microstructure of titanium coatings

The typical microstructural features observed in thermal-sprayed coatings include splats, oxide inclusions, unmelted particles, resolidified particles, and pores. Splats are thin lamellae that are formed from flattening of particles that are rapidly quenched upon impact with a surface, and are the basic building blocks of thermal spray coatings. Oxides are formed from (1) interaction of heated particles with the surrounding atmosphere (in-flight oxidation) or (2) heating of the coating surface during deposition (postimpact oxidation). The extent of these microstructural artifacts in coatings can be controlled through proper selection and optimization of the spray process and feedstock material, as well as postdeposition treatments.

Cross sections of titanium-based coatings deposited using several thermal spray methods are shown in Figure 23.8. Each method produced coatings with a different extent of unmelted particles, oxides, nitrides, and porosity because of the variation in processing temperature, particle velocity, working gases, and surrounding environment. Depending on the processing conditions, some coatings reveal a relatively "clean" microstructure while others consist of gray interlayer boundaries because of oxidation and other in-flight reactions.

Coatings deposited using techniques that operate in a reactive environment such as APS, RPS, and detonation spray produce coatings with clearly discernible gray boundaries of secondary phases (Figure 23.8a, b, and e). RPS has better control over the oxide, nitride, and carbide contents in the coatings since a controlled amount of reactive species are introduced into the RPS process [34] rather than relying on reactive species in working gases and entrained air within APS and detonation spray systems. Control of secondary phases in wire arc spray coatings can be achieved by varying the atomizing gases [59]. The more distinct intersplat boundaries in wire-arc sprayed coatings compared to other coatings arise because the particles have started solidifying prior to impact since they are immediately cooled by atomizing gas beyond the arc zone.

Titanium-based coatings sprayed by a regular, unmodified HVOF process [79] exhibited a microstructure similar to that of detonation-sprayed coatings (Figure 23.8e and f), under a comparable range of particle temperature and velocity conditions. The relatively low temperature and high velocity of these two processes rendered the particles impervious to formation of internal oxides by convective movement. The oxide layers observed results from either surface oxidation during in-flight or postimpact oxidation.

On the other hand, methods such as VPS and cold spray produced coatings that were relatively dense and oxide-free (Figure 23.8c and i). These methods restrict the formation of secondary phases in the coatings either by (1) control of the surrounding environment (i.e., VPS) or (2) deposition at low temperature (i.e., cold spray). These principles were adopted to create modified HVOF systems [61], for example,

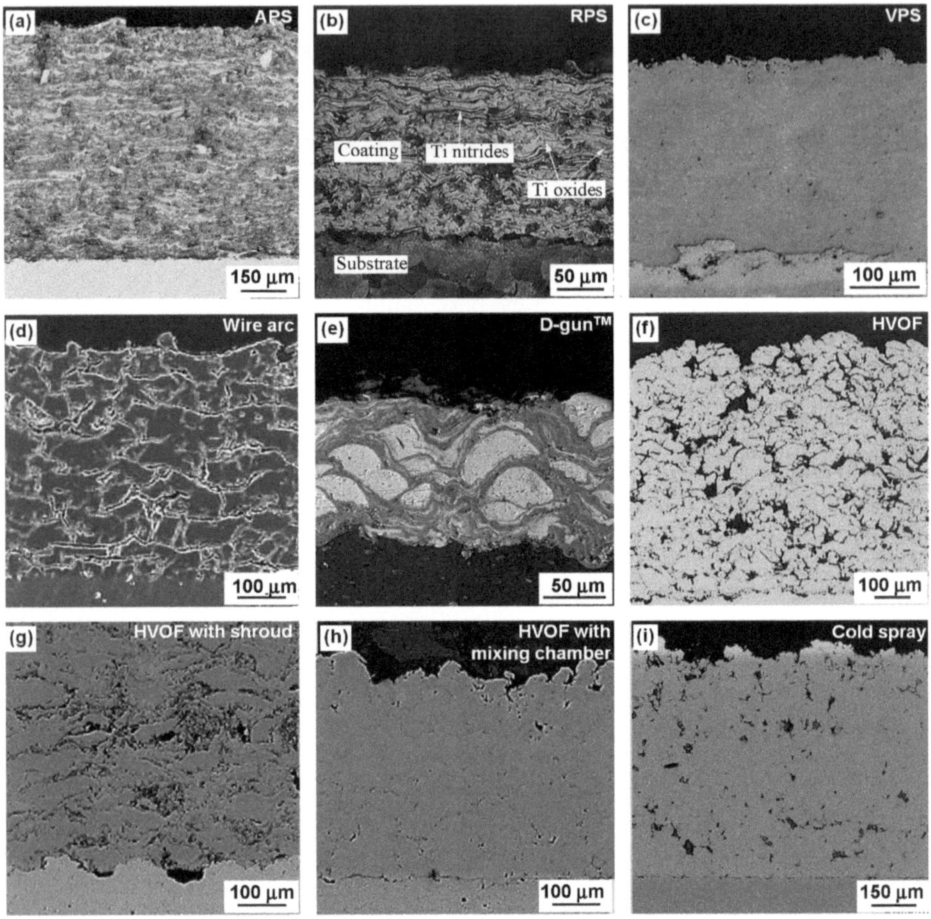

Figure 23.8 Cross sections of titanium-based coatings deposited by (a) atmospheric plasma spray, (b) reactive plasma spray, (c) vacuum plasma spray, (d) wire arc spray, (e) detonation spray, (f) high-velocity oxygen fuel, (g) modified HVOF with shroud, (h) modified HVOF with mixing chamber, and (i) cold spray. (Sources: (a) From Ref. [26], reprinted with permission from Elsevier; (b) from Ref. [42], reprinted with permission from Springer Science and Business Media; (c) from Ref. [78], reprinted with permission from Elsevier; (d) from Ref. [59], reprinted with permission from Elsevier; (e) from Ref. [18], reprinted with permission from Springer Science and Business Media; (f) from Ref. [62], reprinted with permission from ASM International; (g) from Ref. [61], reprinted with permission from Elsevier; (h) from Ref. [61], reprinted with permission from Elsevier; (i) from Ref. [26], reprinted with permission from Elsevier.)

(1) HVOF with a gas shroud and (2) HVOF with an intermediate mixing chamber. These modified HVOF systems created oxide-free coatings (Figure 23.8g and h).

Other factors such as power [80,81], type of working gases [82,83], gas flow rate [84,85], gas pressure and temperature [86,87], particle size [61,88], spray distance [80,81], and substrate temperature [87,89] also influence the coating microstructures. These factors directly influence the particle temperature, velocity, and rate of solidification, hence affecting splat formation and the microstructure of coatings.

23.5 Potential applications

Titanium and titanium alloys have been widely used in various industries because of its high strength-to-weight ratio, excellent corrosion resistance, and biocompatibility. The ability of thermal spray to produce either dense or porous coatings with reasonable control over the material composition, in addition to spray-forming near-net-shaped products, has enabled substantial progress in the development of titanium-based coatings since its first application in the 1980s [48,51].

The biocompatibility of titanium, which is its ability to perform appropriately in contact with a host material without experiencing any toxic, irritation, inflammatory, allergic, mutagenic, or carcinogenic effects, has been established. A porous plasma-sprayed titanium tapered cementless stem was proven clinically appropriate for all ages and femora types, with high long-term survivorship [90]. The rough surface and porous internal structure of thermal-sprayed titanium coatings promote osseointegration since they allow bone growth into the coating to form a mechanical interlock that improves osteoblast adhesion and shear strength between the bone and implant [91]. Similar successful survival rates were observed for thermal sprayed titanium dental implants [92], where the osseointegration of the implants was favored by the low thermal conductivity of titanium and the formation of a TiO_2 film that created neutrality in an oral environment [93].

Dense titanium-based coatings have exhibited excellent corrosion resistance [61]. Sealing of porous coatings with resins, such as epoxy or silica, is required to improve the corrosion resistance [94]. In a physiological fluid environment, a fitting structure with the right combination of dense and porous material is possible through the application of porous thermal sprayed titanium-based coatings on biocompatible and corrosion resistant metallic implants [95].

Another application of thermal sprayed titanium-based materials is the formation of near-net-shape products of complex geometry, which has been discussed in Section 3.1.3. One such example is the VPS forming of near-net-shape TiNi shape memory alloy that has comparable phase transformation behavior, superelastic behavior, shape memory effect, and mechanical properties as bulk materials [96]. Other applications of thermal-sprayed titanium-based coatings include: (1) *in situ*–formed composite coatings to improve the sliding wear resistance of industrial components [97], and (2) an intermediate layer to isolate and prevent toxic metal release by surgical implants, as well as (3) to reduce stresses that arise because of thermal expansion coefficient difference between metallic implants and ceramic hydroxyapatite coatings [24].

23.6 Summary

The flexibility of the thermal spray process provides a means to produce coatings or components with good control over the phase composition and microstructure. The characteristics of feedstock material, such as morphology and particle size distribution, have a significant influence on the coating microstructure. The APS produces coatings with significant secondary phases; thus, this process is more often used to deposit titanium aluminide or composites other than pure titanium. The RPS process offers better control over the oxide, nitride, and/or carbide content since the proportion of reactive species available to react with titanium is well regulated. VPS produces coatings that are free of secondary phases, but at the expense of cost since low-pressure conditions are required. The VPS process has been used to fabricate spray-formed, near-net-shape products.

Wire arc spray is able to deposit titanium-based coatings with higher purity than its powder counterpart since wire has a lower specific surface area per volume of material than powder. This process is unique compared to other thermal spray processes in the sense that the feedstock is melted before being propelled toward the substrate, instead of being accelerated then melted. Incorporation of a gas shroud to APS and wire arc spray is an economical alternative to the VPS process, where high-purity titanium coatings can be obtained. HVOF and detonation-sprayed titanium-based coatings exhibit a similar microstructure because of the similarities between the two processes. These coatings are dense with some secondary phases, albeit being lower than in APS and RPS coatings because of the high-velocity processing. A modified HVOF process known as warm spray permits the formation of secondary phases by introducing an inert gas into the system to reduce air entrainment or lower the combustion gas temperature.

Thermal spray processes are versatile in nature. Therefore, careful thought is needed to select the appropriate process that will produce the desirable microstructure and phases for the intended application. Considering the reactivity of titanium, it is also crucial to exercise control over the feedstock storage and handling, as well as parameter optimization to ensure that high purity is preserved in the manufactured product.

Acknowledgements

The authors acknowledge the financial support by a Swinburne University Postgraduate Research Award (SUPRA) for this work.

References

[1] J.R. Davis, Handbook of Thermal Spray Technology, ASM International, Materials Park, OH, USA, 2005.

[2] L. Pawlowski, The Science and Engineering of Thermal Spray Coatings, John Wiley & Sons, England, UK, 2008.

[3] P. Fauchais, Montavon, Thermal and cold spray: recent developments, Key Eng. Mater. 384 (2008) 1–59.
[4] M. Hohmann, N. Ludwig, Mechanism for making powders of metals, German Patent DE 4102101 A1, Leybold AG, issued Jul. 30, 1992.
[5] O.W. Stenzel, G. Sick, M. Hohmann, Procedure and device for the figuration of a casting jet, German Patent DE 4011392 A1, Leybold AG, issued Oct. 10, 1991.
[6] D.J. Klaphaak, L.G. Barnes, Method of centrifugal atomization, U.S. Patent 3,720,737, Atomization Systems Inc., issued Mar. 13, 1973.
[7] P.G. Tsantrizos, F. Allaire, M. Entezarian, Method of production of metal and ceramic powders by plasma atomization, U.S. Patent 5,707,419, Pegasus Refractory Materials, Inc. and Hydro-Quebec, issued Jan. 13, 1998.
[8] H.D. Steffens, E. Erturk, K.H. Busse, A Comparison of low-pressure arc and low-pressure plasma sprayed titanium coatings, J. Vac. Sci. Technol., A 3 (6) (1985) 2459–2463.
[9] J. Karthikeyan, C.M. Kay, J. Lindeman, P.S. Lima, C.C. Berndt, Cold Spray Processing of Titanium Powder, in: C.C. Berndt (Ed.), Thermal Spray: Surface Engineering via Applied Research, ASM International, Montréal, Québec, Canada, 2000, pp. 255–262.
[10] C. McCracken, Production of fine titanium powders via the hydride-dehydride (HDH) process, PIM Int. 2 (2) (2008).
[11] S. Adachi, K. Nakata, Improvement of adhesive strength of Ti-Al plasma sprayed coating, Surf. Coat. Technol. 201 (9–11) (2007) 5617–5620.
[12] T. Kinos, S.L. Chen, P. Siitonen, P. Kettunen, Densification of plasma-sprayed titanium and tantalum coatings, J. Therm. Spray Technol. 5 (4) (1996) 439–444.
[13] L. Leblanc, P. Tsantrizos, H.R. Salimijazi, T.W. Coyle, J. Mostaghimi, On vacuum plasma spray forming of Ti-6Al-4V, in: C. Moreau, B.R. Marple (Eds.), Thermal Spray 2003: Advancing the Science and Applying the Technology, ASM International, Materials Park, OH, Orlando, FL, 2003, pp. 603–609.
[14] K. Honda, A. Hirose, K.F. Kobayashi, Properties of titanium-aluminide layer formed by low pressure plasma spraying, Mater. Sci. Eng., A 222 (2) (1997) 212–220.
[15] J. Kawakita, S. Kuroda, S. Krebs, H. Katanoda, *In-situ* densification of Ti coatings by the warm spray (two-stage HVOF) process, Mater. Trans. 47 (7) (2006) 1631–1637.
[16] K.H. Kim, M. Watanabe, J. Kawakita, S. Kuroda, Effects of temperature of in-flight particles on bonding and microstructure in warm-sprayed titanium deposits, J. Therm. Spray Technol. 18 (3) (2009) 392–400.
[17] T.I. Bratanich, V.V. Skorokhod, L.I. Kopilova, A.V. Kotko, V.Y. Oliker, V.F. Gorban, Destructive hydrogenation and recombination of α_2-Ti_3Al alloy: production of solid nanocomposites and coatings with improved properties. II. Recombination of Ti_3Al and production of detonation-sprayed coatings, Powder Metall. Met. Ceram. 49 (9–10) (2011) 598–605.
[18] M. Kovaleva, Y. Tyurin, O. Kolisnichenko, M. Prozorova, M. Arseenko, Properties of detonation nanostructured titanium-based coatings, J. Therm. Spray Technol. 22 (4) (2013) 518–524.
[19] T. Marrocco, D.G. McCartney, P.H. Shipway, A.J. Sturgeon, Production of titanium deposits by cold-gas dynamic spray: numerical modeling and experimental characterization, J. Therm. Spray Technol. 15 (2) (2006) 263–272.
[20] H.R. Wang, B.R. Hou, J. Wang, Q. Wang, W.Y. Li, Effect of process conditions on microstructure and corrosion resistance of cold-sprayed Ti coatings, J. Therm. Spray Technol. 17 (5–6) (2008) 736–741.
[21] R. Ramasamy, V. Selvarajan, K. Ramachandran, Characterization of DC plasma spray torch using energy balance technique and thermo-fluid dynamical consideration, Plasma Devices Oper. 5 (3) (1997) 161–180.

[22] D.A. Gerdeman, N.L. Hecht, Arc Plasma Technology in Materials Science, in: V.D. Fréchette, H. Kirsch, L.B. Sand, F. Trojer (Eds.), Applied Mineralogy, Springer-Verlag, New York, 1972.

[23] B. Gross, B. Grycz, Miklóssy, in: R.C.G. Leckey (Ed.), Plasma Technology, Iliffe Books Ltd, London, 1968.

[24] H. Ji, P.M. Marquis, Characterization of plasma-sprayed titanium coatings on stainless steel, Surf. Coat. Technol. 45 (1–3) (1991) 121–127.

[25] Y. Yang, J.L. Ong, J. Tian, *In vivo* evaluation of modified titanium implant surfaces produced using a hybrid plasma spraying processing, Mater. Sci. Eng., C 20 (1–2) (2002) 117–124.

[26] J. Cizek, O. Kovarik, J. Siegl, K.A. Khor, I. Dlouhy, Influence of plasma and cold spray deposited Ti Layers on high-cycle fatigue properties of Ti6Al4V substrates, Surf. Coat. Technol. 217 (2013) 23–33.

[27] P. Cao, B. Gabbitas, D.L. Zhang, A. Salman, Fabrication of bulk titanium aluminides by thermal spray, Int. J. Mod. Phys. B 23 (6–7) (2009) 1777–1782.

[28] D. Dewald, M. Austin, E. Laitila, D. Mikkola, Cubic titanium trialuminide thermal spray coatings – a review, J. Therm. Spray Technol. 10 (1) (2001) 111–117.

[29] X. Zheng, M. Huang, C. Ding, Bond strength of plasma-sprayed hydroxyapatite/Ti composite coatings, Biomater. 21 (8) (2000) 841–849.

[30] K.A. Khor, C.S. Yip, P. Cheang, Ti-6Al-4V/hydroxyapatite composite coatings prepared by thermal spray techniques, J. Therm. Spray Technol. 6 (1) (1997) 109–115.

[31] C.S. Yip, K.A. Khor, N.L. Loh, P. Cheang, Thermal spraying of Ti-6Al-4V/hydroxyapatite composites coatings: powder processing and post-spray treatment, J. Mater. Process. Technol. 65 (1–3) (1997) 73–79.

[32] C.C. McComas, L.S. Sokol, E.M. Hanna, Thermal spray apparatus and method, U.S. Patent 4,235,943, United Technologies Corporation, issued Nov. 25, 1980.

[33] L.S. Sokol, C.C. McComas, E.M. Hanna, Plasma spray method and apparatus, U.S. Patent 4,256,779, United Technologies Corporation, issued Mar. 17, 1981.

[34] E. Lugscheider, L. Zhao, A. Fischer, Reactive plasma spraying of titanium, Adv. Eng. Mater. 2 (5) (2000) 281–284.

[35] R.W. Smith, R. Knight, Thermal spraying II: recent advances in thermal spray forming, JOM 48 (4) (1996) 16–19.

[36] P.G. Tsantrizos, L.T. Mavropoulos, B. Maher, J. Jurewicz, B. Henshaw, R. Lachance, K. Chen, Reactive spray forming process, U.S. Patent 5,217,747, Noranda Inc., issued Jun. 8, 1993.

[37] P.G. Tsantrizos, Reactive spray forming production of titanium aluminides in the tail flame of a D.C. plasma torch, International SAMPE Metals and Metals Processing Conference, F.H. Froes, W. Wallace, R.A. Cull and E. Struckholt (Eds.), Toronto, Canada, Society for the Advancement of Material and Process Engineering (SAMPE), Covina, CA, 1992, pp. 685–691.

[38] P. Tsantrizos, W.H. Gauvin, Characteristics of transferred-arc plasmas at high $TiCl_4$ concentrations, Plasma Chem. Plasma Process 10 (1) (1990) 99–113.

[39] G.L. Humphrey, The heats of combustion and formation of titanium nitride (TiN) and titanium carbide (TiC), J. Am. Chem. Soc. 73 (5) (1951) 2261–2263.

[40] H. Tahara, T. Shibata, T. Yasui, Y. Kagaya, T. Yoshikawa, Development of electromagnetic acceleration plasma generator for titanium nitride coatings, Vacuum 59 (1) (2000) 203–209.

[41] H. Tahara, Material spraying using electromagnetically accelerated plasma jet, Mater. Sci. Forum 449–452 (2004) 389–392.

[42] Y. Yao, Z. Wang, Z. Zhou, S. Jiang, J. Shao, Study on reactive atmospheric plasma-sprayed *in situ* titanium compound composite coating, J. Therm. Spray Technol. 22 (4) (2013) 509–517.

[43] A. Hirose, K. Honda, K.F. Kobayashi, Properties of *in-situ* nitride reinforced titanium-aluminide layers formed by reactive low pressure plasma spraying with nitrogen gas, Mater. Sci. Eng., A 222 (2) (1997) 221–229.

[44] T. Bacci, L. Bertamini, F. Ferrari, F.P. Galliano, E. Galvanetto, Reactive plasma spraying of titanium in nitrogen containing plasma gas, Mater. Sci. Eng., A 283 (1–2) (2000) 189–195.

[45] P. Fauchais, Understanding plasma spraying, J. Phys. D Appl. Phys. 37 (9) (2004) R86–R108.

[46] H.R. Salimijazi, T.W. Coyle, J. Mostaghimi, Vacuum plasma spraying: a new concept for manufacturing Ti-6Al-4V structures, JOM 58 (9) (2006) 50–56.

[47] H.R. Salimijazi, T.W. Coyle, J. Mostaghimi, L. Leblanc, P. Tsantrizos, Microstructural formation of vacuum plasma sprayed Ti-6Al-4V alloy, in: C. Moreau, B.R. Marple (Eds.), Thermal Spray 2003: Advancing the Science and Applying the Technology, ASM International, Materials Park, OH; Orlando, FL, 2003, pp. 611–616.

[48] E. Lugscheider, P. Lu, B. Häuser, D. Jäger, Optimized vacuum plasma-sprayed titanium coatings, Surf. Coat. Technol. 32 (1–4) (1987) 215–226.

[49] K.H. Baik, Microstructural evolution and tensile properties of Ti-Al-V alloys manufactured by plasma spraying and subsequent vacuum hot pressing, Mater. Trans. 47 (4) (2006) 1198–1203.

[50] C. Jaeggi, V. Frauchiger, F. Eitel, M. Stiefel, H. Schmotzer, S. Siegmann, The effect of surface alloying of Ti powder for vacuum plasma spraying of open porous titanium coatings, Acta Mater. 59 (2) (2011) 717–725.

[51] H.D. Steffens, E. Ertürk, Low pressure arc spraying of reactive materials, Thin Solid Films 121 (2) (1984) 143–150.

[52] A. Fujisawa, I. Noda, Y. Nishio, H. Okimatsu, The development of new titanium arc-sprayed artificial joints, Mater. Sci. Eng., C 2 (3) (1995) 151–157.

[53] Y. Nakashima, K. Hayashi, T. Inadome, K. Uenoyama, T. Hara, T. Kanemaru, Y. Sugioka, I. Noda, Hydroxyapatite-coating on titanium arc sprayed titanium implants, J. Biomed. Mater. Res. 35 (3) (1997) 287–298.

[54] H.d. Steffens, Z. Babiak, M. Wewel, Recent developments in arc spraying, IEEE Trans. Plasma Sci. 18 (6) (1990) 974–979.

[55] H.D. Steffens, K. Nassensteln, Recent developments in single-wire vacuum arc spraying, J. Therm. Spray Technol. 3 (4) (1994) 412–417.

[56] H.D. Steffens, M. Wewel, Recent developments in vacuum arc spraying, Mater. Manuf. Processes 7 (4) (1992) 573–591.

[57] Z. Zurecki, E.A. Hayduk, J.G. North, R.B. Swan, Method of producing titanium nitride coatings by electric arc thermal spray, U.S. Patent 5,066,513, Air Products and Chemicals, Inc., issued Nov. 19, 1991.

[58] Z. Zurecki, E.A. Hayduk, J.G. North, R.B. Swan, D.L. Mitchell, Method of forming titanium nitride coatings on carbon/graphite substrates by electric arc thermal spray process using titanium feed wire and nitrogen as the atomizing gas, U.S. Patent 5,254,359, Air Products and Chemicals, Inc., issued Oct. 19, 1993.

[59] N. Sakoda, M. Hida, Y. Takemoto, A. Sakakibara, T. Tajiri, Influence of atomization gas on coating properties under Ti arc spraying, Mater. Sci. Eng., A 342 (1–2) (2003) 264–269.

[60] T. Watanabe, T. Sato, A. Nezu, Electrode phenomena investigation of wire arc spraying for preparation of Ti-Al intermetallic compounds, Thin Solid Films 407 (1–2) (2002) 98–103.

[61] J. Kawakita, S. Kuroda, T. Fukushima, H. Katanoda, K. Matsuo, H. Fukanuma, Dense titanium coatings by modified HVOF spraying, Surf. Coat. Technol. 201 (3–4) (2006) 1250–1255.
[62] R. McCaw, R. Hays, Application of commercially pure titanium coatings using the high velocity oxy-fuel thermal spray process under atmospheric condition, in: C.C. Berndt (Ed.), Thermal Spray: International Advances in Coatings Technology, ASM International, Orlando, FL, 1992, pp. 881–886.
[63] T. Wu, S. Kuroda, J. Kawakita, H. Katanoda, R. Reed, Processing and properties of titanium coating produced by warm spraying, in: B.R. Marple, M.M. Hyland, Y.-C. Lau, R.S. Lima, J. Voyer (Eds.), Thermal Spray 2006: Building on 100 Years Success, ASM International, Materials Park, OH; Seattle, WA, 2006.
[64] J. Kawakita, H. Katanoda, M. Watanabe, K. Yokoyama, S. Kuroda, Warm spraying: an improved spray process to deposit novel coatings, Surf. Coat. Technol. 202 (18) (2008) 4369–4373.
[65] S. Kuroda, Y. Tashiro, H. Yumoto, S. Taira, H. Fukanuma, S. Tobe, Peening action and residual stresses in high-velocity oxygen fuel thermal spraying of 316L stainless steel, J. Therm. Spray Technol. 10 (2) (2001) 367–374.
[66] S. Sampath, X.Y. Jiang, J. Matejicek, L. Prchlik, A. Kulkarni, A. Vaidya, Role of thermal spray processing method on the microstructure, residual stress and properties of coatings: an integrated study of Ni-5 wt% Al bond coats, Mater. Sci. Eng., A 364 (1–2) (2004) 216–231.
[67] P. Bansal, P.H. Shipway, S.B. Leen, Effect of particle impact on residual stress development in HVOF sprayed coatings, J. Therm. Spray Technol. 15 (4) (2006) 570–575.
[68] A. Valarezo, G. Bolelli, W.B. Choi, S. Sampath, V. Cannillo, L. Lusvarghi, R. Rosa, Damage tolerant functionally graded WC-co/stainless steel HVOF coatings, Surf. Coat. Technol. 205 (7) (2010) 2197–2208.
[69] Z. Mao, J. Ma, J. Wang, B. Sun, Properties of TiN-matrix coating deposited by reactive HVOF spraying, J. Coat. Technol. Res. 6 (2) (2009) 243–250.
[70] Z. Mao, J. Ma, J. Wang, B. Sun, Dry abrasion property of TiN-matrix coating deposited by reactive high velocity oxygen fuel (HVOF) spraying, J. Coat. Technol. Res. 7 (2) (2010) 253–259.
[71] R.M. Poorman, H.B. Sargent, Method and apparatus utilizing detonation waves for spraying and other purposes, U.S. Patent 2,714,563, Union Carbide and Carbon Corporation, issued Aug. 2, 1955.
[72] V.E. Oliker, V.L. Sirovatka, E.F. Grechishkin, A.D. Kostenko, V.V. Lashneva, I.I. Maksyuta, Y.F. Anikin, A.F. Goncharenko, V.M. Derkach, Tribological properties of detonation coatings based on titanium aluminides and aluminum titanate, Powder Metall. Met. Ceram. 44 (11–12) (2005) 531–536.
[73] V.E. Oliker, V.L. Sirovatka, I.I. Timofeeva, E.F. Grechishkin, T.Y. Gridasova, Effect of properties of titanium aluminide powders and detonation spraying conditions on phase and structure formation in coatings, Powder Metall. Met. Ceram. 44 (9–10) (2005) 472–480.
[74] V.E. Oliker, V.L. Sirovatka, I.I. Timofeeva, T.Y. Gridasova, Y.F. Hrechyshkin, Formation of detonation coatings based on titanium aluminide alloys and aluminium titanate ceramic sprayed from mechanically alloyed powders Ti-Al, Surf. Coat. Technol. 200 (11) (2006) 3573–3581.
[75] T.I. Bratanich, V.V. Skorokhod, L.I. Kopylova, A.V. Kotko, V.Y. Oliker, Ti_3Al recombination and detonation coats on its basis with improved characteristics, Int. J. Hydrogen Energy 36 (1) (2011) 1330–1337.

[76] D.V. Dudina, M.A. Korchagin, S.B. Zlobin, V.Y. Ulianitsky, O.I. Lomovsky, N.V. Bulina, I.A. Bataev, V.A. Bataev, Compositional variations in the coatings formed by detonation spraying of Ti_3Al at different O_2/C_2H_2 ratios, Intermetallics 29 (2012) 140–146.
[77] T.I. Bratanich, V.V. Skorokhod, L.I. Kopylova, A.V. Kotko, M.O. Krapivka, Destructive hydrogenation and recombination of α_2-Ti_3Al alloy: production of solid nanocomposites and coatings with improved properties. I. Mechanism of destructive hydrogenation of Ti_3Al, Powder Metall. Met. Ceram. 49 (7–8) (2010) 484–494.
[78] L. Zhao, E. Lugscheider, Reactive plasma spraying of TiAl6V4 alloy, Wear 253 (11–12) (2002) 1214–1218.
[79] K. Kim, S. Kuroda, M. Watanabe, R. Huang, H. Fukanuma, H. Katanoda, Comparison of oxidation and microstructure of warm-sprayed and cold-sprayed titanium coatings, J. Therm. Spray Technol. 21 (3–4) (2012) 550–560.
[80] K.A. Khor, Y. Murakoshi, M. Takahashi, T. Sano, Plasma spraying of titanium aluminide coatings: process parameters and microstructure, J. Mater. Process. Technol. 48 (1–4) (1995) 413–419.
[81] W.M. Zhao, C. Liu, L.X. Dong, Y. Wang, Effects of arc spray process parameters on corrosion resistance of Ti coatings, J. Therm. Spray Technol. 18 (4) (2009) 702–707.
[82] T. Valente, F.P. Galliano, Corrosion resistance properties of reactive plasma-sprayed titanium composite coatings, Surf. Coat. Technol. 127 (1) (2000) 86–92.
[83] V.E. Oliker, V.L. Sirovatka, T.Y. Gridasova, I.I. Timofeeva, E.F. Grechishkin, M.S. Yakovleva, E.N. Eliseeva, Effect of gas media on the structural evolution and phase composition of detonation coatings sprayed from mechanically alloyed Ti-Al-B powders, Powder Metall. Met. Ceram. 48 (11–12) (2009) 620–626.
[84] D.J. Varacalle Jr., L.B. Lundberg, H. Herman, G. Bancke, Titanium carbide coatings fabricated by the vacuum plasma spraying process, Surf. Coat. Technol. 86–87 (PART 1) (1996) 70–74.
[85] X. Li, D.Q. Sun, X.Y. Zheng, Z.A. Ren, Effect of N_2/Ar gas flow ratios on the nitrided layers by direct current arc discharge, Mater. Lett. 62 (2) (2008) 226–229.
[86] C.J. Li, W.Y. Li, Deposition characteristics of titanium coating in cold spraying, Surf. Coat. Technol. 167 (2–3) (2003) 278–283.
[87] W. Wong, E. Irissou, A.N. Ryabinin, J.G. Legoux, S. Yue, Influence of helium and nitrogen gases on the properties of cold gas dynamic sprayed pure titanium coatings, J. Therm. Spray Technol. 20 (1–2) (2011) 213–226.
[88] D. Goldbaum, J.M. Shockley, R.R. Chromik, A. Rezaeian, S. Yue, J.G. Legoux, E. Irissou, The effect of deposition conditions on adhesion strength of Ti and Ti6Al4V cold spray splats, J. Therm. Spray Technol. 21 (2) (2012) 288–303.
[89] S. Sampath, X. Jiang, Splat formation and microstucture development during plasma spraying: deposition temperature effects, Mater. Sci. Eng. A 304–306 (1–2) (2001) 144–150.
[90] A.V. Lombardi Jr., K.R. Berend, T.H. Mallory, M.D. Skeels, J.B. Adams, Survivorship of 2000 tapered titanium porous plasma-sprayed femoral components, Clin. Orthop. Relat. Res. 467 (1) (2009) 146–154.
[91] L. Saldaña, J.L. González-Carrasco, M. Rodríguez, L. Munuera, N. Vilaboa, Osteoblast pesponse to plasma-spray porous Ti6Al4V coating on substrates of identical alloy, J. Biomed. Mater. Res. A 77 (3) (2006) 608–617.
[92] M. Vallecillo Capilla, M.N. Romero Olid, M.V. Olmedo Gaya, C. Reyes Botella, C. Zorrilla Romera, Cylindrical dental implants with hydroxyapatite- and titanium plasma spray-coated surfaces: 5-year results, J. Oral Implantol. 33 (2) (2007) 59–68.

[93] C. Leyens, M. Peters, Titanium and Titanium Alloys, Wiley-VCH GmbH & Co, KGaA, Weinheim, Germany, 2003.
[94] K. Ishikawa, T. Suzuki, Y. Kitamura, S. Tobe, Corrosion resistance of thermal sprayed titanium coatings in chloride solution, J. Therm. Spray Technol. 8 (2) (1999) 273–278.
[95] M. Simon, C. Lagneau, J. Moreno, M. Lissac, F. Dalard, B. Grosgogeat, Corrosion resistance and biocompatibility of a new porous surface for titanium implants, Eur. J. Oral Sci. 113 (6) (2005) 537–545.
[96] H. Nakayama, M. Taya, R.W. Smith, T. Nelson, M. Yu, E. Rosenzweig, Shape memory effect and superelastic behavior of TiNi shape memory alloy processed by vacuum plasma spray method, Mater. Sci. Eng. A 459 (1–2) (2007) 52–59.
[97] F. Borgioli, E. Galvanetto, F.P. Galliano, T. Bacci, Sliding wear resistance of reactive plasma sprayed Ti-TiN coatings, Wear 260 (7–8) (2006) 832–837.

The additive manufacturing (AM) of titanium alloys

24

B. Dutta, Francis H. (Sam) Froes***
*DM3D Technology, Auburn Hills, MI, USA
**Consultant to the Titanium Industry, Tacoma, WA, USA

24.1 Introduction

Titanium alloys are among the most important of the advanced materials that are key to improved performance in aerospace and terrestrial systems [1–5]. This is because of the excellent combinations of specific mechanical properties (properties normalized by density) and outstanding corrosion behavior [6–11] exhibited by titanium alloys. However, negating widespread use is the high cost of titanium alloys compared to competing materials (Table 24.1).

The high cost of titanium compared with the other metals shown in Table 24.1 has resulted in the yearly consumptions shown in Table 24.2.

In publications over the past few years [1–3], the cost of fabricating various titanium precursors and mill products has been discussed (very recently the price of TiO_2 has risen to \$2.00 per pound and $TiCl_4$ to \$0.55 per pound), and it has been pointed out that the cost of extraction is a small fraction of the total cost of a component fabricated by the cast and wrought (ingot metallurgy) approach (Figure 24.1). To reach a final component, the mill products shown in the figure must be machined, often with very high buy-to-fly ratios (which can reach as high as 40:1). The generally accepted cost of machining a component is that it doubles the cost of the component (with the buy-to-fly ratio being another multiplier in cost per pound) (Figure 24.2). This means that anything that can be done to produce a component that is closer to the final configuration will result in a cost reduction – hence the attraction of near-net-shape components.

The high cost of conventional titanium components has led to numerous investigations of various potentially lower cost processes [1–3], including powder metallurgy (PM) near-net-shape techniques [1–2,6–10,12,13]. In this chapter, one PM near-net-shape technique, additive manufacturing (AM), will be reviewed, with an emphasis on the "work horse" titanium alloy Ti-6Al-4V. This technique (described below) is receiving a lot of attention from the US Navy, who envision future use of the approach aboard a carrier where parts can be rapidly fabricated for immediate use on the battle group that the carrier is supporting [14]. The various approaches to AM are presented, followed by some examples of components produced by AM. The microstructures and mechanical properties of Ti-6Al-4V produced by AM are listed and shown to compare very well with cast and wrought product. Finally, the economic advantages to be gained using the AM technique compared to conventionally processed material are presented.

Titanium Powder Metallurgy. http://dx.doi.org/10.1016/B978-0-12-800054-0.00024-1

Table 24.1 Cost of titanium – a comparison*

Item	Material ($/pound)		
	Steel	Aluminum	Titanium
Ore	0.02	0.01	0.22 (rutile)
Metal	0.10	1.10	5.44
Ingot	0.15	1.15	9.07
Sheet	0.30–0.60	1.00–5.00	15.00–50.00

*Contract prices. The high cost of titanium compared to aluminum and steel is a result of (a) high extraction costs and (b) high processing costs. The latter relates to the relatively low processing temperatures used for titanium and the conditioning (surface regions contaminated at the processing temperatures, and surface cracks, both of which must be removed) required prior to further fabrication.

Table 24.2 Metal consumption

Structural metals	Consumption/year (10^3 metric tons)
Ti	50
Steel	700,000
Stainless steel	13,000
Al	25,000

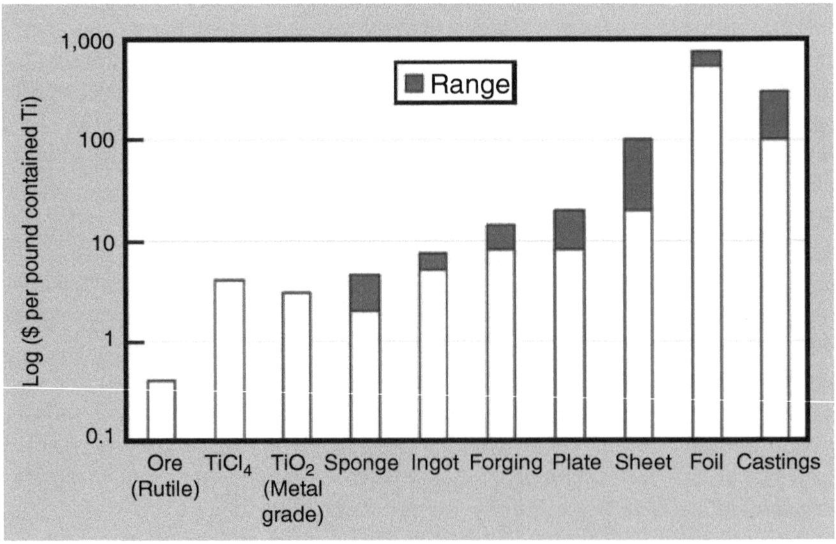

Figure 24.1 Cost of titanium at various stages of a component fabrication.

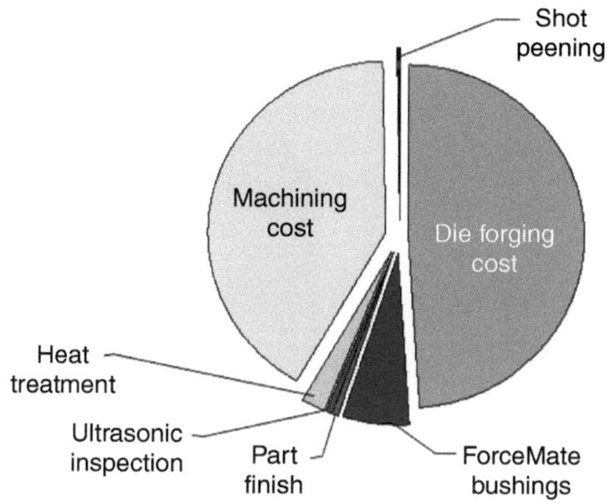

Figure 24.2 Boeing 787 side-of-body chord, manufacturing cost breakdown. (Courtesy Boeing.)

24.2 Technology overview

All the AM technologies are based on the principle of slicing a solid model in multiple layers to create a toolpath, uploading the data in the machine, and building the part up layer by layer following the sliced model data using a heat source (laser, electron beam, or electric arc) and feedstock (metal powder or wire). This section first addresses the creation of the computer toolpath and is followed by a discussion of the two broad classes of part-building technologies: powder bed fusion (PBF) and directed energy deposition (DED).

24.2.1 Software for AM

In principle, 3D printing is based on taking a 3D geometry, slicing it into multiple layers, and creating a toolpath that will trace the part layer by layer, one layer at a time. Three-dimensional printing of metals has its roots in stereo-lithographic process, invented by 3D Systems. Stereo lithography was built on a surface file format, called STL (Standard Tessellation Language) and widely used in rapid prototyping and computer-aided manufacturing. Many of the metal-based 3D printing technologies use STL files as input. Since STL represents the raw unstructured triangulated surface by the unit normal and vertices of the triangles using a 3D Cartesian coordinate system and does not contain any scale information, these files may not be suitable for complex operations and precision applications. Therefore, many AM technologies are using solid models as input. Remanufacturing and/or surface coating using deposition-based technologies (DED) poses additional challenges as it involves creating 3D layers as opposed to 2D layers for PBF technologies requiring five- or six-axis software for creation of toolpath. Figure 24.3 shows a typical deposition path simulated on a computer-aided design (CAD) model for a 5-axis deposition process.

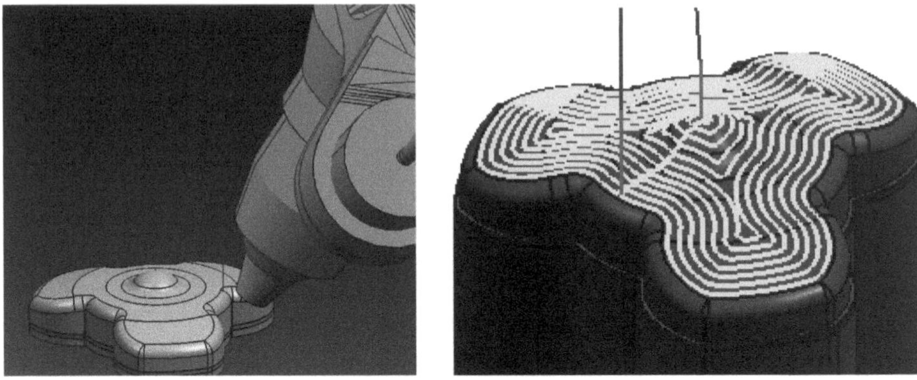

Figure 24.3 Left: CAD model of the part and process head. Right: Simulated toolpath for 5-axis deposition using DMDCAM software. (Courtesy DM3D Technology.)

24.2.2 Part-building technology

Following ASTM classification, AM technologies for metals can be broadly classified into two categories, DED and PBF (Table 24.3). There are several technologies under each category as branded by different manufacturers. While the PBF technologies enable building of complex features, hollow cooling passages and high precision parts, these are limited by build envelop, single material per build, and horizontal layer-building ability. In comparison, the DED technologies offer larger build envelop and higher deposition rate, while their ability to build hollow cooling passages and finer geometry is limited. DMD and LENS technology also offer the ability to deposit multiple materials in a single build and the ability to add metal on existing parts. Commercially available AM technologies are based on two types of heat sources, namely, laser and electron beam, for the purpose of melting the feedstock (powder or wire). Laser-based systems operate under inert atmosphere (for titanium processing) in contrast to the vacuum environment of the electron beam systems. While the vacuum systems are more expensive, they offer the advantage of low residual stress as compared to laser-based systems, and electron beam–processed parts can be used without any stress-relieving operation. The effect of the heat source on the microstructure and mechanical properties is discussed in more detail in sections 4.2 and 4.3.

24.2.2.1 Powder bed fusion

PBF technologies are based on the principles of laying down a layer of metal powder on the build platform and scanning the bed of powder with a heat source, such as laser or electron beam, that either partially or completely melts the powder in the path of the beam and the powder resolidifies and binds together as it cools off (ASTM specification F2924-12a and 13 for Ti-6Al-4V and Ti-6Al-4V ELI grade respectively). Layer-by-layer tool path tracing is governed by the CAD data of the

Table 24.3 Various AM technologies for processing of titanium and its alloys

AM category	Technology	Company	Description
Directed energy deposition (DED)	Direct metal deposition (DMD)	DM3D Technology LLC (Formerly POM Group)	Uses laser and metal powder for melting and depositing using a patented close loop process.
	Laser engineered net shaping (LENS)	Optomec, Inc.	Uses laser and metal powder for melting and depositing.
	Direct manufacturing (DM)	Sciaky, Inc.	Uses electron beam and metal wire for melting and depositing.
	Shaped metal deposition or wire and arc additive manufacturing (WAAM)	Not commercialized yet (patented by Rolls Royce Plc.)	Uses electric arc and metal wire for melting and depositing.
Powder bed fusion (PBF)	Selective laser sintering (SLS)	3D Systems Corp. (acquired Phenix Systems)	Uses laser and metal powder for sintering and bonding
	Direct metal laser sintering (DMLS)	EOS GmbH	Uses laser and metal powder for sintering, melting and bonding
	Laser melting (LM)	Renishaw Inc.	Uses laser and metal powder for melting and bonding
	Selective laser melting (SLM)	SLM Solutions GmbH	Uses laser and metal powder for melting and bonding
	LaserCUSING	Concept Laser GmbH	Uses laser and metal powder for melting and bonding
	Electron beam melting (EBM)	Arcam AB	Uses electron beam and metal powder for melting and bonding

part being built. Figure 24.4 shows a schematic diagram explaining the steps involved in this process:

- A substrate is fixed on the build platform.
- The build chamber is filled with inert gas (for laser processing) or evacuated (for electron beam processing) to reduce oxygen level in the chamber to the desired level.
- A thin layer of the metal powder (20–200 μm thick depending on the technology and equipment) is laid down on the substrate and leveled to a predetermined thickness using a leveling mechanism.

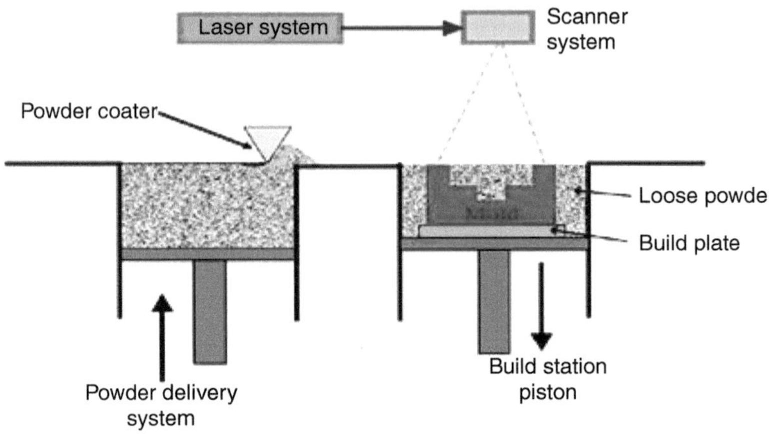

Figure 24.4 Schematic showing powder bed fusion technology. (Courtesy Jim Sears.)

- The laser or electron beam scans the powder bed surface following the toolpath precalculated from the CAD data of the component being built.
- The above process is repeated for the next and subsequent layer until the build is complete.

24.2.2.2 Directed energy deposition

DED technologies use material injection in to the meltpool instead of scanning on a powder bed (AMS specification 4999A for Ti-6Al-4V). Figure 24.5 shows a schematic of the DMD technology (laser-based metal deposition). The process steps for the DED are as follows:

- A substrate or existing part is placed on the work table.
- Similar to PBF, the machine chamber is closed and filled with inert gas (for laser processing) or evacuated (for electron beam processing) to reduce oxygen level in the chamber to the desired level (AMS 4999A specifies below 1200 ppm). The DMD process also offers local shielding and does not require inert gas chamber for less reactive metals than titanium, such as, steels, Ni alloys, Co alloys, etc.
- At the cycle start, the process nozzle with a concentric laser or electron beam is focused on the part surface to create a meltpool. Material delivery is in the form of powder through a coaxial nozzle (for laser) or through a metal wire with a side delivery (for electron beam). The nozzle moves at a constant speed and follows a predetermined toolpath created from the CAD data. As the nozzle (tooltip) moves away meltpool solidifies forming a layer of metal.
- Successive layers follow the same principle and build up the part layer by layer until completion.

24.2.3 Technology comparison

Table 24.4 below provides a comparison of capabilities, benefits, and limitations of various AM technologies that are used for producing titanium parts today.

Figure 24.6 below shows a comparison of PBF technologies with DED technologies in terms of deposition rate and surface roughness. Note that the layer thickness

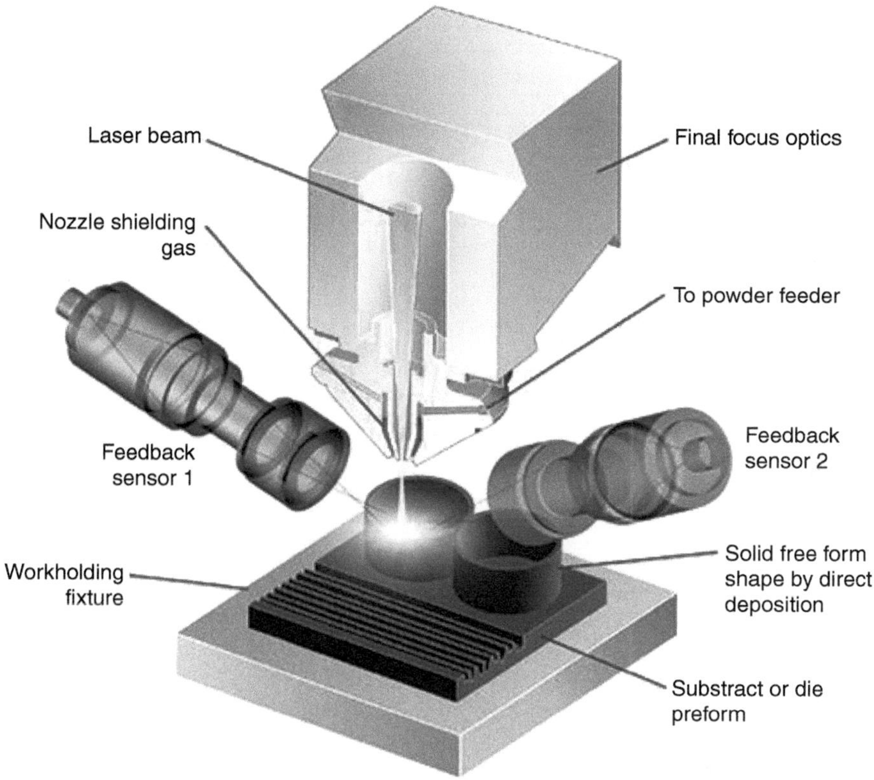

Figure 24.5 Schematic showing direct metal deposition (DMD) technology. (Courtesy DM3D Technology.)

has been used as a measure of roughness here as this determines the roughness of the vertical walls of the structure being built. Clearly, the PBF technologies offer better surface finish as these use smaller beam size (for both laser and electron beam) and smaller layer thickness as compared to DED technologies; however, as a consequence the deposition rate is also lower for these technologies. Therefore, PBF is more suitable for more accurate, complex, small-size objects, while DED is more suitable for building relatively larger parts at a high processing rate, but with coarser finish surface.

24.3 Titanium AM applications

Extensive exploration is currently underway for usage of AM titanium parts in aerospace and medical applications. Other applications for AM include applications in chemical, defense, and other industries. While PBF technologies are suitable for smaller, complex geometries, with hollow unsupported passages/structures, DED is better suited for larger parts with coarser features requiring higher deposition rates. Usage of finer powder grains combined with smaller laser/electron beam size leads to

Table 24.4 **Comparison of various technologies [15–18]**

Item	Laser based PBF (e.g., DMLS)	Electron beam based PBF (e.g., EBM)	Laser based directed energy deposition (e.g., DMD)
Build envelop	Limited	Limited	Large and flexible
Beam size	Small, 0.1–0.5 mm	Small, 0.2–1 mm	Large, can vary from 2–4 mm
Layer thickness	Small, 50–100 μm	Small, 100 μm	Large, 500–1000 μm
Build rate	Low, cc/h	Low, 55–80 cc/h	High, 16–320 cc/h
Surface finish	Very good, Ra 9/12 μm, Rz 35/40 μm	Good, Ra 25/35 μm	Coarse, Ra 20–50 μm, Rz 150–300 μm, Depends on beam size
Residual stress	High	Minimal	High
Heat treatment	Stress relieve required, HIPing preferred	Stress relieve not required, HIPing may or may not be performed	Stress relieve required, HIPing preferred
Chemistry	ELI grade possible, negligible loss of elements	ELI grade possible, loss of Al need to be compensated in powder chemistry	ELI grade possible, negligible loss of elements
Build capability	Complex geometry possible with very high resolution Capable of building hollow channels	Complex geometry possible with good resolution Capable of building hollow channels	Relatively simpler geometry with less resolution. Limited capability for hollow channels, etc.
Repair/ remanufacture	Possible only in limited applications (requires horizontal plane to begin remanufacturing)	Not possible	Possible (capable of adding metal on 3D surfaces under 5 + 1-axis configuration, making repair solutions attractive)
Feature/metal addition on existing parts	Not possible	Not possible	Possible. Depending on dimensions ID cladding is also possible
Multimaterial build or hard coating	Not possible	Not possible	Possible

a superior surface finish on the as-built parts from the PBF technologies as compared to DED technologies. However, majority of the AM parts need finish machining for most of the practical applications. Ability of the directed energy technologies to add metal on existing parts allow them to apply surface protective coatings, remanufacture and repair of damaged parts, and reconfigure or add features to existing parts, besides building new parts.

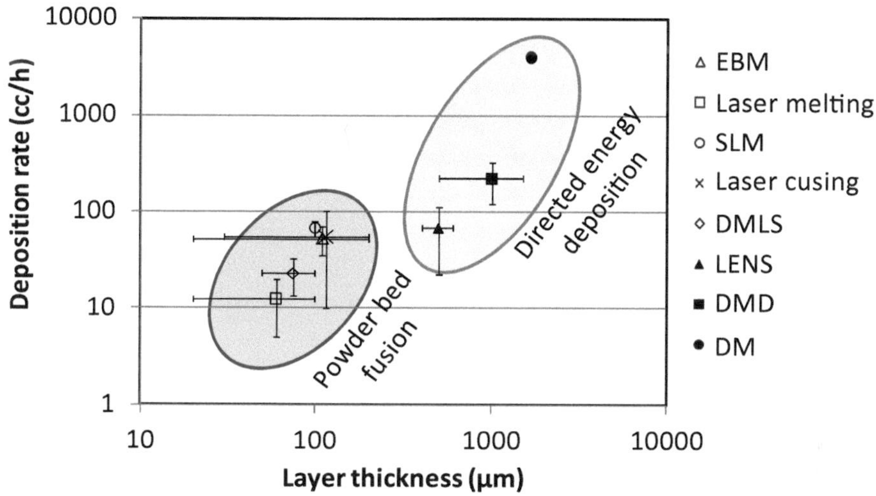

Figure 24.6 Comparison of PBF and DED technologies in terms of layer thickness and deposition rate [15,19–24].

24.3.1 Complex geometry

Small beam size and low layer thickness along with support of the powder bed allow PBF technologies, such as electron beam melting (EBM), direct metal laser sintering (DMLS), or selective laser sintering (SLS), to produce complex geometries with high precision and unsupported structures. Figure 24.7 shows one such example of

Figure 24.7 Hydraulic manifold built using EBM technology. The part was built at the MDF in Oakridge National Laboratory through an ONR-sponsored project. (Courtesy ORNL, TN.)

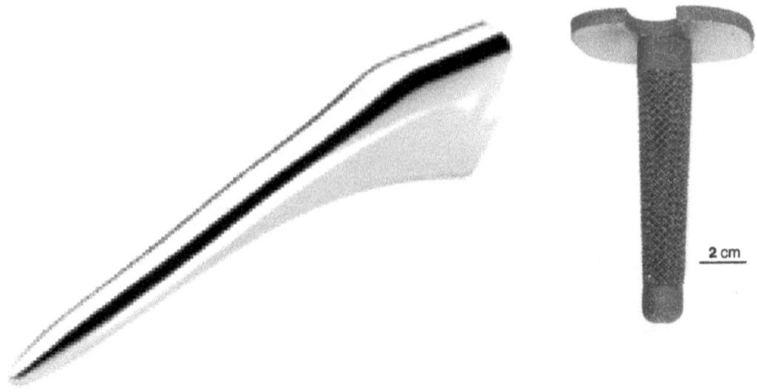

(a) (b)

Figure 24.8 Medical implant application using; (a) DMLS technology. (Courtesy Jim Sears.) (b) EBM technology [26].

a hydraulic manifold mount for an underwater manipulator built using EBM technology. Building the integrated mount and manifold with internal passageways in a single operation eliminates multiple part fabrication and effects significant cost savings. Good surface finish of the part eliminated finish machining needs on all surfaces, except seal surfaces and threading of screw holes. Generally the PBF technique gives a better surface finish than the DED approach; however, for demanding applications such as aerospace, finish machining is required [25].

Figure 24.8a shows an example of a biomedical implant built with Ti-6Al-4V alloy using DMLS technology, while Figure 24.8b shows a tibial (knee) Ti6Al4V stem built using EBM technology. These technologies also have the potential of building patient-specific custom implants to better suit the needs.

24.3.2 Feature addition

DED technologies, such as, DMD and/or LENS have the ability to add metal on 3D surfaces and, thus, allow addition of features on existing parts and/or blanks. This is not possible with the PBF approach. Adding features to a forged or cast preform as opposed to machining of such features can provide the most cost-effective manufacturing option, where a significant reduction of the preform size and weight can be effected through the elimination of the need for a machining allowance. Examples are various casings and housings in jet engines where flanges, bosses, etc. can be added on cast or forged cylindrical preforms. This is demonstrated for a feature addition on a titanium fan casing for an aircraft engine (Figure 24.9).

24.3.3 Remanufacturing

One of the best application areas suited for DED techniques is remanufacturing and repair of damaged, worn-out, corroded parts. Because of their ability to add metal on select locations on 3D surfaces, these technologies can be used to rebuild lost material

Figure 24.9 Fan case produced by adding features with AM (laser-aided directed energy deposition) to a forged perform. (Courtesy Jim Sears.)

on various components [27–29]. Closed-loop technologies, such as DMD, offer the particular benefit of a minimum heat-affected zone (HAZ) in the repaired part and helps to retain the integrity of the part. The closed-loop control allows DMD to repair parts with short HAZ and produce a high-quality repaired part. Figure 24.10 shows cross-sectional microstructures of the DMD area of a remanufactured turbine blade.

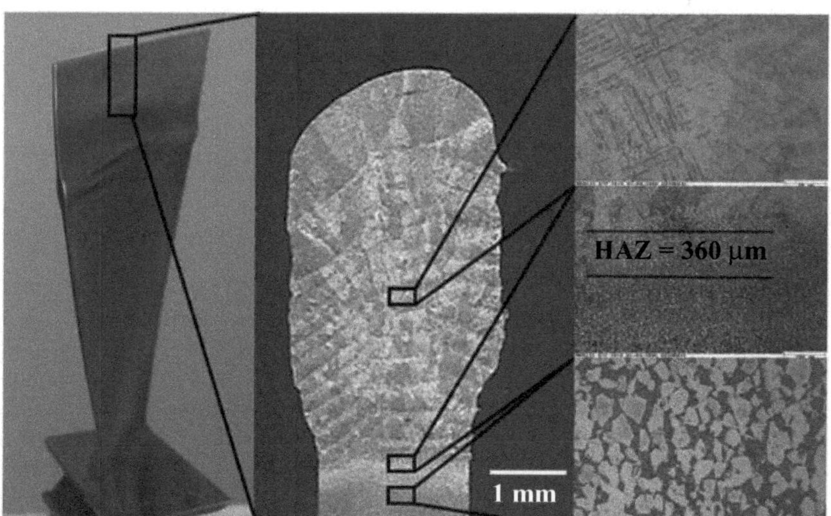

Figure 24.10 DMD repair of turbine components; left: repaired vane, middle: macro cross section, and right: microstructures (top to bottom shows the clad, interface, and base material). (Courtesy DM3D Technology.)

The excellent process control during DMD leads to a fully dense microstructure as observed in the vertical cross section. A layer thickness of about 0.1–0.2 mm has been applied in this case and a minimal HAZ is observed in the as-deposited blade. DMD vision system plays a significant role in this type of remanufacturing applications. A calibrated vision system integrated with the machine allows automatic identification of part location in the machine coordinate system and precision processing. Other titanium components that can be repaired include housings, bearings, casing flanges, seals, landing gears, etc.

24.4 Microstructure and mechanical properties

24.4.1 Specifications

The Aerospace Materials Specification SAE AMS4999A covers Titanium Alloy Direct Products Ti-6Al-4V Annealed. This calls for a postbuild annealing treatment of 550°C (1025°F) and if a HIPing treatment is used, it should be at not less than 100 MPa (14.5 ksi) within the range 899–954°C (1650–1750°F) for 2–4 hours followed by a slow cool to below 427°C (800°F). Minimum tensile properties shall be ultimate tensile strength (UTS) 855/889 MPa (124/129 ksi) (depending on direction), yield strength (YS) 758/800 MPa (110/116 ksi), and elongation of 6% [30].

24.4.2 Microstructures

Microstructure of the additively manufactured titanium is very different from that of a wrought product. With suitable post-processing, additively manufactured titanium can exhibit equivalent or better mechanical property than a conventionally manufactured titanium part.

Figure 24.11 shows typical microstructures of as-built Ti6Al4V alloy from various AM processes. Corresponding tensile properties (UTS and YS) are also plotted as a function of elongation to show the effect of microstructure on mechanical properties. Clearly, laser-based technologies offer higher strength and lower ductility because of the formation of α'-martensite as a result of fast cooling. Electron beam processed material exhibit α-β microstructure due to slower cooling in vacuum atmosphere and results in lower tensile strength and higher ductility. In comparison, microstructure morphology is coarser in the as-cast material and wrought material has equiaxed α-β microstructure. Arc-processed material (wire and arc AM [WAAM]) offers a microstructure similar to cast structure, though finer in length scale. The effect of these various microstructures is well demonstrated in their tensile behavior.

Figure 24.12 shows microstructure of as-built material using DMD process and after subsequent HIPing (hot isostatic pressing) and aging. The as-built microstructure shows the typical martensitic structure expected for Ti-6Al-4V cooled rapidly from the beta phase field, while the HIPed and aged material shows the expected grain boundary alpha and intergranular coarse alpha plates. This microstructural transition from as-deposited to HIPed-aged condition is also reflected through their tensile properties. While UTS and YS is a little lower after HIPing and aging, ductility improves significantly (see below) as a result of the microstructure changing from martensitic to transformed beta (precipitated alpha) structure.

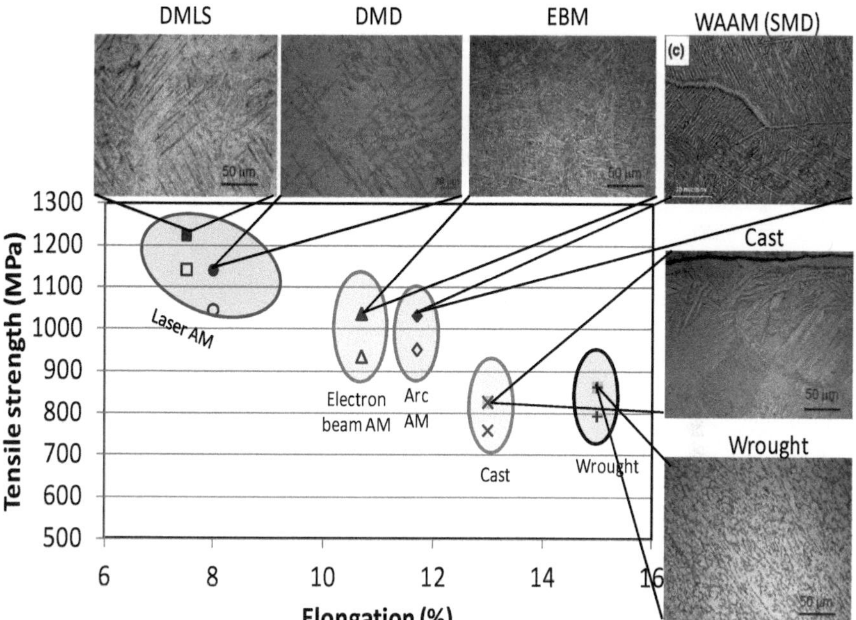

Figure 24.11 Room-temperature tensile behavior of AM Ti6Al4V alloy produced using various AM technologies and their comparison with cast and wrought material properties [18, 19, 31–33]. Typical microstructures are also included for comparison. Closed and open symbols represent UTS and YS respectively.

24.4.3 Mechanical properties

Tensile properties of Ti-6Al-4V fabricated by a number of AM techniques are shown in Figure 24.13. All of the processes show strength levels superior or comparable to conventional material (cast, forged, and wrought-annealed). As built materials in

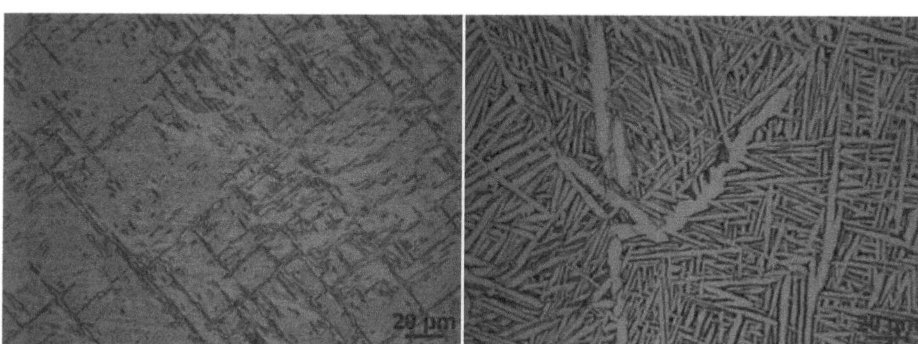

Figure 24.12 Microstructure of DMD built Ti6Al4V before and after HIPing. (Courtesy DM3D Technology.)

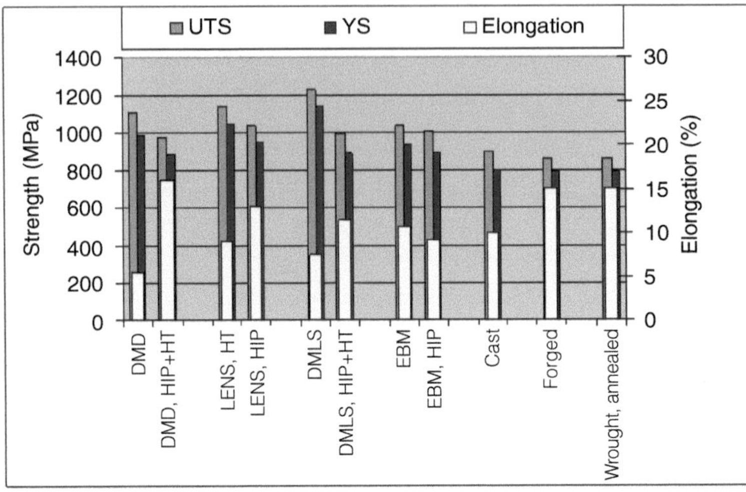

Figure 24.13 Tensile strength, yield strength, and elongation of Ti-6Al-4V alloy built using various AM processes. DMD, direct metal deposition [19]; LENS, laser engineered net shaping [21]; DMLS, direct metal laser sintering [18]; EB, electron beam melting [18]; HIP, hot isostatic pressing; HT, heat treatment.

laser-based processes, such as DMD, LENS, and DMLS, exhibit less ductility because of the formation of the martensite phase; however, the ductility can be improved through subsequent HIPing and/or heat treatment operation. As a result of reduced residual stress, EBM processed Ti-6Al-4V shows greater ductility when compared to laser processed Ti-6Al-4V. Fatigue properties have been tested using many different cycles. In general, as built Ti-6-4 offers fatigue resistance similar to cast and wrought material, even without a HIP treatment. (Figure 24.14).

DED technologies offer the benefit of repair and remanufacture of damaged parts as well as feature addition on existing parts and performs (such as castings, extrusions, forgings). Therefore, it is of considerable interest to examine properties of such repairs and feature additions. Figure 24.15 shows that "V" groove–repaired and "slot"-repaired Ti6Al4V samples exhibit comparable tensile strength with nonrepaired wrought Ti6Al4V sample. These samples were repaired using DMD technology and represent repairs done in the seal areas of jet engine components, such as casings and housings [19]. Ti6242 samples repaired using laser cladding has shown higher high cycle fatigue strength than wrought Ti6242 material [34].

24.5 Economics of AM

Figure 24.16 shows a typical cost breakdown of various steps involved in AM of titanium components based on the following assumptions: (a) small batch size (between 10 and 30 parts), (b) medium-size part, ~600–900 mm (2–3 ft) in size and relatively simple geometry. It is to be noted that any and all of these factors can significantly

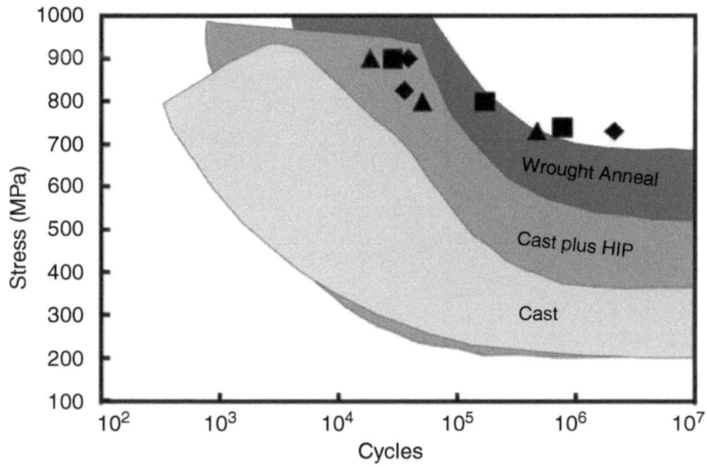

Figure 24.14 Comparison of room temperature fatigue properties of AM-fabricated Ti-6Al-4V and conventionally fabricated Ti-6Al-4V. ■, ◆, and ▲ represent properties in the three orthogonal directions *x*, *y*, and *z*, respectively. (Courtesy EADS/Jim Sears.)

influence the cost. The batch size plays a major role in costing for small batch sizes, while part size plays a more significant role in costing for larger batch sizes. Benefits of tool-less manufacturing make AM as an attractive manufacturing option for small batch sizes when compared with conventional manufacturing techniques, such as casting, forging, extrusion, etc. However, relatively lower throughput renders AM less attractive for high-volume manufacturing [35].

Success of expanding AM in the manufacturing industry depends on the selection of right applications. Value-added AM applications include long lead time complex components, weight savings, cost-effective remanufacturing, customized medical

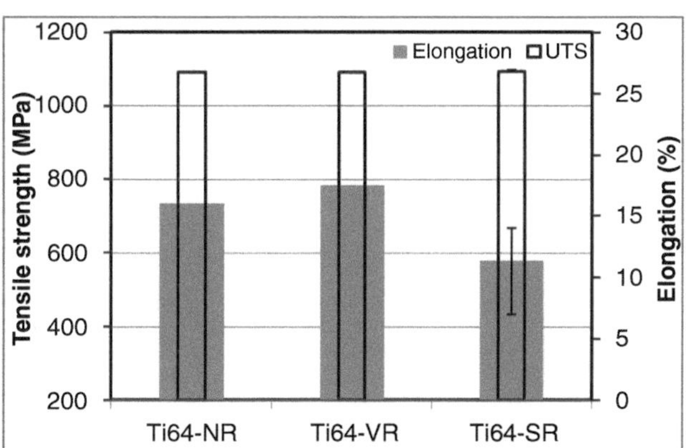

Figure 24.15 Mechanical behavior of DMD-repaired Ti6Al4V alloy. NR respresents nonrepaired material, while VR is a "V" groove–repaired material and "SR" is slot-repaired material.

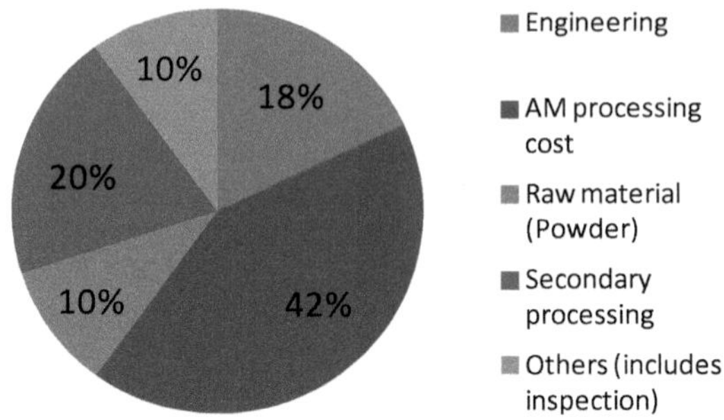

Figure 24.16 Typical cost breakdown of various steps involved in AM of titanium

implants, multimaterial components, etc. Below are few case studies that elaborate more on these aspects of AM.

One of the major benefits of the PBF processes is their ability to create hollow structures and, thereby, allow weight savings. Aerospace industry, where weight savings can have very significant impact, is actively looking into this ability of AM processes. A case study involving a seat buckle for commercial passenger jets is a perfect example of this [36]. A lightweight seat buckle with hollow structures was designed based on extensive FEA study to ensure enough strength against shock loading. The part was produced using DMLS Ti-6Al-4V alloy (Figure 24.17). Replacement of a conventional steel buckle with hollow AM titanium buckle causes 85 g weight saving per buckle (55% weight reduction). An Airbus A380 with 853 seats will result in a possible weight saving of 72.5 kg. According to the project sponsor, Technology Strategy Board, United Kingdom, this weight saving translates to 3.3 million liters of fuel saving over the life of the aircraft that is equivalent to £2 million ($3 million), while cost of making all the buckles using DMLS is only £165,000 ($256,000).

The direct manufacturing ability of AM technologies also helps to reduce manufacturing costs in the case of high buy-to-fly ratio parts. Researchers at the Oakridge

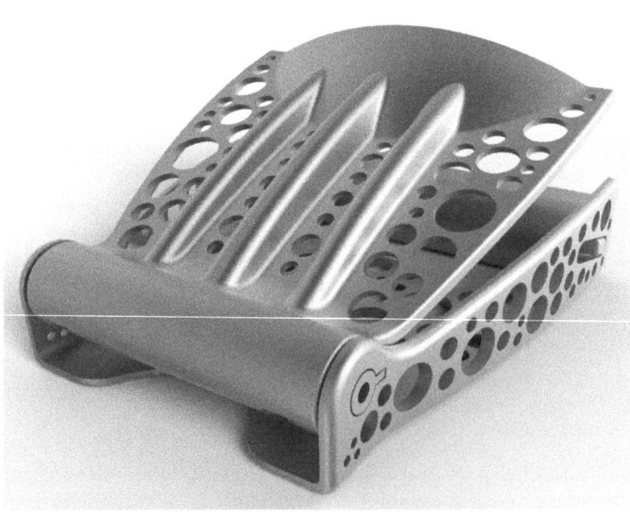

Figure 24.17 Seat buckle produced using DMLS technology.

Figure 24.18 BALD bracket for Joint Strike Fighter (JSF) built using EBM technology. (Courtesy ORNL, TN.)

National Laboratory built a Ti-6A-4V Bleed Air Leak Detect (BALD) bracket for the Joint Strike Fighter (JSF) engine using EBM technology (Figure 24.18)[25]. Traditional manufacturing from wrought Ti-6Al-4V plate costs almost $1000/lb because of a high (33:1) buy-to-fly ratio as opposed to just over 1:1 ratio for the AM built part. Estimated saving through AM is about 50%.

Direct deposition techniques such as DMD can not only be used to create parts, but these technologies can also be used for remanufacturing, repair, and/or feature building on existing parts. Damaged expensive aerospace titanium components, such as bearing housing, flanges, fan blades, casings, vanes, landing gears, etc., can be rebuilt using these technologies at 20–40% of the cost of the new parts [19]. Worn-out flanges in jet engine casings have been rebuilt using DMD at less than half of the cost of a new part. Extensive work is underway to investigate the feasibility of using such technologies to salvage components that are mismachined during conventional manufacturing. Successful realization of these efforts will have very significant impact on the titanium manufacturing industry. While most of the leading commercial activities in the AM industry is concentrated in USA and Europe, significant efforts are underway in many parts of the world, including China [37].

24.6 Research and development

The vast majority of Ti AM components have been fabricated using high-cost spherical gas atomized or plasma rotating electrode process (PREP) powder. As seen in Figure 24.16, the raw material (mostly powder) cost can be almost 10% of the overall component cost. As AM continues to evolve, focus is on driving the powder cost down

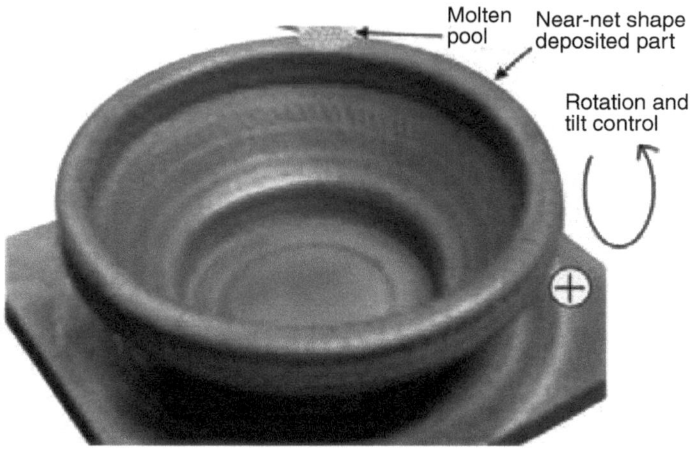

Figure 24.19 Schematic of the MER plasma transferred arc (PTA) additive manufacturing process [38].

to make it a competitive alternative solution. Recent work has demonstrated that titanium AM parts can be successfully produced using much lower cost angular powders [38, 39]. In one program [38], titanium sponge was blended with Al and V powder or Al/V master alloy to produce the Ti-6Al-4V composition. After processing this combination through a plasma transferred arc (Figure 24.19) the as-fabricated tensile properties were at cast and wrought levels: 980 MPa (142 ksi) UTS, 882 MPA (128ksi) YS, and 10% elongation, and S-N fatigue was at ingot metallurgy levels. Later work with ADMA Products TiH_2 powder was equally successful. In another work, angular Metalysis powder [39] was first converted to a spherical morphology and then fabricated by AM. Such an example is a turbocharger for an auto engine shown in Figure 24.20.

Figure 24.20 Automotive component produced from Metalysis powder after spheroidization of the powder [39].

Interestingly, primary work on low-cost titanium AM has targeted the automotive applications, low cost always being a primary concern with the auto industry.

Another major research focus in the AM is dedicated toward process monitoring, control, and in-line inspection. As AM involves layer-by-layer buildup, it allows unique opportunity for in-line inspection of each layer during the build. Various investigations are underway to develop in-line inspection tools for AM processes using ultrasonic, x-ray, and other tools. A combination of latest sensor technologies and predictive control algorithms have shown to be able to compensate for over- or underbuild of a layer caused by issues from toolpath overlap or powder catchment [27, 40]. On-line spectroscopic analysis of meltpool plasma has yielded promises of monitoring and possibly controlling chemistry of build material [27]. Advances in process monitoring and control are being well complemented by mathematical modeling of heat flow and microstructure evolution and predicting mechanical behavior of AM-built titanium [31, 41–42].

With the acceptance of AM in mainstream manufacturing, researchers are focusing on exploiting the added benefits of AM: (1) ability to build a single part with multiple materials or graded materials [43–44] and (2) applying wear-resistant coatings on titanium parts [44–46]. An example is graded coating of Rene 88 on Ti6Al4V [44] and Mo-WC coating on Ti6Al4V material [45] and surface alloying with N, C, or B to form hard coatings of TiN, TiC, or TiB [46]. Building multimaterial components require process and control development as well as developing software capability that will allow building a single object with multiple materials. Another area of research focus is software development for interpretation of the computed tomography/magnetic resonance imaging data, and subsequent translation into CAD data for building customized biomedical products, such as patient-specific implants for the orthopedic industry [47].

24.7 Summary

The past few years have shown significant advances in the AM technologies leading to the production of fully functional parts using titanium and its alloys. While PBF technologies offer the ability to build hollow near-net shapes with finer resolution, directed energy–based technologies offer the ability to add features on existing parts and remanufacture/repair damaged parts, besides building parts directly from CAD data. Most of the studies reveal that the mechanical properties of AM material are as good as or better than the conventionally fabricated titanium alloys. Selection of the right AM technology along with proper design optimization can lead to very significant savings through greatly reduced buy-to-fly ratios, overall weight reduction, scrap reduction, etc. Besides this, these technologies offer design freedom that conventional manufacturing does not. However, full exploitation of the benefits of AM depend largely on educating the manufacturing and design community and successful integration of these technologies in the manufacturing industry. The aerospace and medical industries have so far been the largest driver for the usage of titanium AM materials, while other industries, such as the automotive industry, are beginning to exploit the benefits of AM of titanium alloys. The recent push in low-cost titanium powders is expected to expand usage of AM in more cost-sensitive industries such as automotive.

Acknowledgments

The authors recognize inputs for this chapter and useful discussions with the following: Jim Sears, Ryan DeHoff, Richard Grylls, Jessica Nehro, Anders Hultman, Scott Thompson, Laura Kinkopf, Michael Cloran, David Whittaker, Karl D'Ambrosio, and Ma Quin.

References

[1] F.H. (Sam) Froes, Powder metallurgy of titanium alloys, in: Chang Isaac, Zhao Yuyuan (Eds.), Advances in Powder Metallurgy, Woodhead Publishing, Philadelphia, USA, 2013, p. 202.
[2] F.H. (Sam) Froes, Imam M. Ashraf, Fray Derek (Eds.), Cost Affordable Titanium, TMS, Warrendale, PA, 2004.
[3] M.N. Gungor, M.A. Imam, F.H.(Sam) Froes (Eds.), Innovations in Titanium Technology, TMS, Warrendale, PA, 2007.
[4] M.A. Imam, F.H.(Sam) Froes, K.F. Dring (Eds.), Cost-Affordable Titanium III, Trans Tech Publications Ltd, Switzerland, 2010.
[5] M.A. Imam, F.H.(Sam) Froes, R.G. Reddy (Eds.), Cost Affordable Titanium IV, Trans Tech Publications, Switzerland, 2013.
[6] F.H. (Sams) Froes, D. Eylon, H. Bomberger (Eds.), Titanium Technology: Present Status and Future Trends, TDA, Dayton, OH, 1985.
[7] F.H. (Sams) Froes, Y. Te-Lin, H.G. Weidenger Titanium, Zirconium and Hafnium Chapter 8, in: K.H. Matucha (Ed.), Materials Science and Technology – Structure and Properties of Nonferrous Alloys, vol. 2, VCH Weinheim, FRG, 1996, p. 401.
[8] F.H. (Sams) Froes, Titanium, chapters 3.3.5a – 3.3.5e Encyclopedia of Materials Science and Engineering, in: P. Bridenbaugh (Ed.), Elsevier, Oxford, UK, 2000.
[9] F.H. (Sam) Froes (Sam), Titanium Alloys, chapter 8 of the Handbook of Advanced Materials, in: Chief J.K. Weasel (Ed.), McGraw-Hill Inc., New York, NY, 2000.
[10] F.H. (Sam) Froes (Sam) Titanium Metal Alloys, Handbook of Chemical Industry Economics, Inorganic, in: Chief Jeff Ellis (Ed.), John Wiley and Sons Inc., New York, NY, 2000.
[11] R.R. Boyer, G. Welsch, E.W. Collings E W (Eds.), Materials Properties Handbook: Titanium Alloys, ASM Int., Materials Park, OH, 1994.
[12] F.H. (Sam) Froes, Eylon D, Powder metallurgy of titanium alloys, Int. Mater. Rev. 35 (1990) 162.
[13] F.H. (Sam) Froes, C. Suryanarayana, Powder Processing of Titanium Alloys, in: A. Bose, R.M. German, A. Lawley (Eds.), Reviews in Particulate Materials, vol. 1, MPIF, Princeton, NJ, 1993, p. 223.
[14] Defence News, Advanced Manufacturing Shows its Merits, June 10, 2013, p. 24.
[15] http://www.eos.info/additive_manufacturing/for_technology_interested.
[16] http://www.arcam.com/technology/additive-manufacturing/.
[17] http://www.dm3dtech.com/index.php/expertise-innovations/experticeandinnovations-dmddtechnology.

[18] http://www.morristech.com/Docs/Ti64ELI%20DataSheet.pdf.
[19] B. Dutta, Private communication, DM3D Technology, July 2013.
[20] S. Stecker, K.W. Lachenberg, H. Wang, R.C. Salo, Advanced electron beam free form fabrication methods & technology, AWS Conference (2006) 35–46.
[21] http://www.optomec.com/Additive-Manufacturing-Technology/Laser-Additive-Manufacturing.
[22] http://resources.renishaw.com/en/details/brochure-the-power-of-additive-manufacturing--57719.
[23] http://stage.slm-solutions.com/index.php?slm-500_en.
[24] http://www.industriallaser.com.au/pdf/X%20Line%201000R%20Brochure.pdf.
[25] R. Dehoff, C. Duty, W. Peter, Y. Yamamoto, W. Chen, C. Blue, C. Tallman, Case study: additive manufacturing of aerospace brackets, Adv. Mater. Process. 171 (3) (2013) 19–22.
[26] E.M. Lawrence, M.G. Sara, M. Edwin, M. Frank, B.W. Ryan B., Next generation orthopaedic implants by additive manufacturing using electron beam melting, Int. J. Biomater. 2012, 14 article ID 245727, doi:10.1155/2012/245727.
[27] B. Dutta, S. Palaniswamy, J. Choi, L.J. Song, J. Mazumder, Additive manufacturing by direct metal deposition, Adv. Mater. Proc. May (2011) 33–36.
[28] B. Dutta, S. Palaniswami, J. Choi, J. Mazumder, Rapid manufacturing and remanufacturing of DoD components using direct metal deposition, AMMTIAC Quarter. 6 (2) (2012) 5–9.
[29] B. Dutta, H. Natu, J. Mazumder, Near net shape repair and remanufacturing of high value components using DMD, TMS Proceedings of the Fabrication, Materials, Processing and Properties, vol. 1, 2009, pp. 131–138.
[30] Titanium Alloy Direct Deposited Products Ti-6Al-4V Annealed, SAE Aerospace Material Specification (AMS) 4999A, Sept 2009, http://www.sae.org/technical/standards/AMS4999A.
[31] F. Wang, S. Williams, P. Colegrove, A.A. Alphons, Microstructure and mechanical properties of wire and arc additive manufactured Ti-6Al-4V, Metall. Mater. Trans. A Sept. (2012) 1–10.
[32] K. Mari, G. Preston, O. Kelly, L. Guo, E. M. Lawrence, M. G. Sara, M. Edwin, T. Okabe, Evaluation of titanium alloys fabricated using rapid prototyping technologies – electron beam melting and laser beam melting, Materials 4 (2011) 1776–1792.
[33] ASM Handbook, vol. 2, Properties and Selection: Nonferrous Alloys and Special Purpose Materials, pp. 621, 637.
[34] R. K. Hermann, O. Sven, Nowotny, Steffen, Laser cladding of the titanium alloy TI6242 to restore damaged blades, Proceedings of the 23rd International Congress on Applications of Lasers and Electro-Optics, 2004 pp. 1–10.
[35] A. Eleonora, A. Salmi, Economics of additive manufacturing for end-usable metal parts, Int. J. Adv. Manuf. Technol. 62 (2012) 1147–1155.
[36] http://www.manufacturingthefuture.co.uk/_resources/case-studies/TSB-AirlineBuckle.pdf.
[37] M. Quin, Private communication, July 2013.
[38] J.C. Withers, V. Shapovalov, R. Storm, R.O. Loutfy, There is Low Cost Titanium Componentry Today, Reference #5, p. 11.
[39] P. Whittaker, posted on http://www.ipmd.net/news/002519.html, Metalysis' Titanium Powder used to 3D print Automotive Parts.
[40] L. Song, V. B. Singh, B. Dutta, J. Mazumder, Control of melt pool temperature and deposition height during direct metal deposition process, Int. J. Adv. Manuf. Technol. May (2011) 10.1007/s00170-011-3395-2.
[41] K. Makiewicz, S.S. Babu, M. Keller, A. Chaudhary, Microstructure evolution during laser additive manufacturing of Ti-6Al-4V alloy, Trends in Welding Research 2012 Proceedings of the 9th International Conference (ASM International), February 01, 2013, pp. 970–977.

[42] S.M. Kelly, S.L. Kampe, Microstructural evolution in laser-deposited multilayer Ti-6Al-4V builds: Part II. Thermal Modelling, Metall., Mater. Trans. A 35A (2004) 1869–1879.
[43] P.C. Collins, R. Banerjee, S. Banerjee, H.L. Fraser, Laser deposition of compositionally graded titanium/vanadium and titanium/molybdenum alloys, Mat. Sci. Eng. A 352 (2003) 118–128.
[44] X. Lin, T.M. Yue, H.O. Yang, W.D. Huang, Solidification behavior and the evolution of phase in laser rapid forming of graded Ti6Al4V-Rene88DT alloy, Metall. Trans. A 38A 2007, 127–137.
[45] W. Pang, H.C. Man, T.M. Yue, Laser surface coating of Mo–WC metal matrix composite on Ti6Al4V alloy, Mat. Sci. Eng. A 390 (2005), 144–153.
[46] R. Filip, Alloying of surface layer of the Ti-6Al-4V titanium alloy through the laser treatment, JAMME, 15(1–2) (2006) 174–180.
[47] L.B. Bourell, C.L. Leu, W.D. Rosen, Roadmap for additive manufacturing identifying the future of freeform processing, The University of Texas at Austin, Laboratory for Freeform Fabrication, Advanced Manufacturing Center, 2009.

Powder-based titanium alloys: properties and selection

25

Sami M. El-Soudani
Associate Technical Fellow, The Boeing Company, Huntington Beach, CA, USA

25.1 Mechanical properties of PM titanium alloys

Much of the database reported herein is based on Refs [1–19], specifically Refs [18,19].

25.1.1 Tensile properties and fracture toughness of blended elemental alloys

Tables 25.1a–25.1f provide a comprehensive listing of tensile properties and fracture toughness. With the exception of the powder used for extruded components examined in Table 25.1e, the BE compacts data for all other Tables 25.1a–25.1d and 25.1f was exclusively prepared by sampling titanium -100 mesh sponge fines, blended with master alloy powders. In these compacts, the chlorine content could be in the range of 500–1500 ppm. Such high chloride content could result in chlorine-associated voids and may preclude the attainment of 100% densification. By contrast, the extrusion billets used in generating the data for Table 25.1e were fabricated using ADMA's TiH_2 powder (Chapter 8), blended with master alloy powders then CIP-and-vacuum-sintered, followed by canless extrusion [19]. Chemical analysis reported chloride content <100 ppm, and in the as-sintered billets very low values (chloride <0.0010 wt%) [19]. These findings are critical, as Ref. [18] notes that the BE method must use chloride-free titanium powder [6]. One source for such powder is commercially pure titanium ingot material or machine turnings embrittled by hydrogenation that are subsequently crushed and dehydrogenated. Alternatively and with equal success, the ADMA process provides essentially chloride-free CIP-and-sintered compact billets for extrusion (Table 25.1e). For each processing pathway, the following conclusions may be drawn:

- *Pressing and sintering* (Table 25.1a): Both strength and ductility increased with higher densification values.
- *Pressing and sintering plus HIP* (Table 25.1b): HIP reduced strength and furthermore did not increase ductility over that of press-and-sinter only, but maximum benefit was attained at densification = 99.8%.
- *Pressing and sintering plus HIP, then hot rolling* (Table 25.1c): The best hot-rolling results of strength and ductility are those following mill annealing or beta annealing after rolling. Ductility somehow dropped with recrystallization annealing (RA).

Titanium Powder Metallurgy. http://dx.doi.org/10.1016/B978-0-12-800054-0.00025-3

Table 25.1a Tensile and fracture toughness properties of Ti-6Al-4V BE compacts pressed and sintered to varying degrees of densification

Condition*	0.2% yield strength		Ultimate tensile strength		Elongation (%)	Reduction in area (%)	K_{IC} or (K_Q)		Density (%)	Chlorine (ppm)	O_2 (ppm)
	MPa	ksi	MPa	ksi			MPa $\sqrt{m}$	ksi $\sqrt{in.}$			
Pressed and sintered (96% dense)	758	110	827	120	6	10	…	…	96	1200	…
Pressed and sintered (98% dense)	827	120	896	130	12	20	…	…	98	1200	…
Pressed and sintered (MR-9 process) (99.2% dense)	847	123	930	135	14	29	38	35	99.2	1200	…
Pressed and sintered (92% dense)	827	120	910	132	10	…	…	…	92	1500	2100

* HIP, hot isostatic pressing; CIP, cold isostatic pressing; ELCI, extra-low chlorine powder; BUS, broken-up structure; TCP, thermochemical processing; L, longitudinal; TL, transverse longitudinal; T, transverse; LT, longitudinal transverse (TL and LT per ASTM E 399), and STA, solution treatment and aging.

Table 25.1b Tensile and fracture toughness properties of Ti-6Al-4V BE compacts cold-pressed and sintered then HIPed (such a processing combination is sometimes referred to as CHIP)

Condition*	0.2% yield strength		Ultimate tensile strength		Elongation (%)	Reduction in area (%)	K_{IC} or (K_Q)		Density (%)	Chlorine (ppm)	O_2 (ppm)
	MPa	ksi	MPa	ksi			MPa $\sqrt{m}$	ksi $\sqrt{in.}$			
Pressed and sintered plus HIP	806	117	875	127	9	17	41	37	≥99	1500	2400
CIP and sintered plus HIP	690	100	793	115	9	15	85	77	>99	…	…
	793	115	896	130	10	20	83	76	>99	…	…
CIP and sintered plus HIP**	896	130	965	140	12	22	…	…	99.8	…	…

** Forged 1010°C + water quench.

Table 25.1c Tensile and fracture toughness properties of Ti-6Al-4V BE compacts cold-pressed and sintered then HIPed, followed by hot rolling of BE plate and heat treating as noted

Condition* Hot rolled plus noted below	0.2% yield strength		Ultimate tensile strength		Elongation (%)	Reduction in area (%)	K_{IC} or (K_Q)		Density (%)	Chlorine (ppm)	O_2 (ppm)
	MPa	ksi	MPa	ksi			MPa $\sqrt{m}$	ksi $\sqrt{in.}$			
Mill annealed (L or TL)	903	131	958	139	10	26	(72)†	65†	≥99	200	1600
Mill annealed (T or LT)	923	134	965	140	14	31	(71)†	(64)†	≥99	200	1600
Recrystallization annealed (L or TL)	888	129	916	133	4	8	(75)†	(68)†	≥99	200	1600
Recrystallization annealed (T or LT)	868	126	937	136	5	9	(67)†	(61)†	≥99	200	1600
β annealed (L or TL)	841	122	937	136	10	26	(89†	(81)†	≥99	200	1600
β annealed (T or LT)	875	127	958	139	7	20	(92)†	(84)†	≥99	200	1600
Minimum properties (MIL T-9047)	827	120	896	130	10	25	...	...	...	...	...

† Precracked Charpy, K_v.

Table 25.1d Tensile and fracture toughness properties of Ti-6Al-4V BE compacts pressed and sintered plus alpha-beta forged with added isothermal forging steps as noted

Condition*	0.2% yield strength		Ultimate tensile strength		Elongation (%)	Reduction in area (%)	K_{IC} or (K_Q)		Density (%)	Chlorine (ppm)	O_2 (ppm)
	MPa	ksi	MPa	ksi			MPa $\sqrt{m}$	ksi $\sqrt{in.}$			
Pressing and sintered plus α/β forged	841	122	923	134	8	9	...	...	≥99.4	1500	2400
Pressed and sintered plus α/β forged	951	138	1027	149	9	24	49	45	99	1200	...
Plus α/β 30% isothermally forged	841	122	930	135	30	...	...	...	99.7	1500	2100
Plus α/β 70% isothermally forged	896	130	999	145	30	...	...	...	99 8	1500	2100

Table 25.1e Tensile and fracture toughness properties of Ti-6Al-4V BE compacts extruded using 6-in.-diameter billets of ADMA TiH_2 hydrogenated powder CIP'ed and vacuum-sintered, then extruded into either L-shaped or five-flange profiles [19]

Condition*	0.2% yield strength		Ultimate tensile strength		Elongation (%)	K_{IC} or (K_Q) LT/TL		Density (%)	Chlorine content (ppm)	Oxygen content (ppm)
	MPa	ksi	MPa	ksi		MPa $\sqrt{m}$	ksi (in.)$^{1/2}$	Theoretical		
Beta-extruded	899	130	1000	145	16.3	58/43	53/39	≥99	<100	2500
Alpha-beta-extruded	944	137	1031	149	16	52/35	47/32	≥99	<100	3200
Beta-extruded (higher temperature)	960	139	1047	152	15.2	47/35	43/32	≥99	<100	3400
Beta-extruded	841	122	967	140	12.3	75/60	68/55	≥99	<100	2000
Alpha-beta-extruded	837	121	958	139	13.7	78/63	71/58	≥99	<100	1800
Beta-extruded (low oxygen)	900	130	990	144	13	76/66	69/60	≥99	<100	1700

Table 25.1f Tensile properties of Ti-6Al-4V BE compacts with microstructural modifications: (a) cold isostatically pressed and sintered then HIPed, (b) same as (a) but with extra-low chlorine ELCL), (c) broken-up structure or BUS, and (d) thermochemically processed, (TCP) consolidated, powder-based BE microstructure

Condition*	0.2% yield strength		Ultimate tensile strength		Elongation (%)	Reduction in area (%)	K_{IC} or (K_Q)		Density (%)	Chlorine (ppm)	O_2 (ppm)
	MPa	ksi	MPa	ksi			MPa $\sqrt{m}$	ksi $\sqrt{in.}$			
CIP and sintered plus HIP (low chlorine)	827	120	923	134	16	34	...	...	99.8	160	...
CIP and sintered plus HIP (ELCI)	882	128	985	143	11	36	...	...	100	<10	...
Plus BUS treated	951	138	1034	150	7	15	...	...	...	...	...
Plus TCP treated	1007	146	1062	154	14	20	...	...	...	...	...

- *Pressing and sintering plus alpha-beta forging* (Table 25.1d): Forging provided adequate strength level with moderate ductility. Isothermal forging substantially enhanced ductility with moderate strength, while alpha-beta forging maximized strength while reducing ductility.
- *CIP, sintering, followed by extrusion* (Table 25.1e): Extrusion provided superior strength and ductility values in all cases.
- *CIP and sintered plus HIP, broken-up structure, or thermo-hydrogen processed (THP) consolidated, powder-based BE microstructure* (Table 25.1f): Special processing provided adequate or superior strength levels and with good ductility except for the "BUS" condition.

However, more data are needed for these alloys before reliable parameters for property levels and optimum processes can be established. In the case of alloy Ti-10V-2Fe-3Al, a tensile strength of 1268 MPa with 9–10% elongation can be achieved with BE methods. The processing–property relationships for Ti-6Al-4V alloy are strongly dependent on the degree of densification, which in turn depends on the amount of total reduction, whether it is a hot rolling or forging operations.

Table 25.2 provides the more limited available information on the properties of several additional titanium BE alloys. Mechanical test data for BE alloys, such as Ti-6Al-2Sn-4Zr-6Mo, Ti-5Al-2Cr-1Fe, and Ti-4.5Al-5Mo-1.5Cr, cover a wider range of strength and ductility values. The most detailed work has been done on the Ti-10V-2Fe-3Al alloy, with some results reported at levels close to those for IM materials [9].

Over the past decades, studies of BE Ti-6Al-4V alloy using the -100 mesh sponge fines with possibly excessive chloride and oxygen contents rendered compacts with densities between 92% and 100% of the theoretical density (TD). The measured yield and tensile strength of these compacts were found to be proportional to the compact density. The fracture toughness over this densification range varied widely but was also increasing with compact density. Above 98% TD, the BE compacts exhibited K_{IC} values approaching the level of mill-annealed ingot-based IM materials. However, IM materials with coarse lenticular microstructures similar to those of BE compacts and those with extra-low interstitial (ELI) chemistry have exhibited normally much higher K_{IC} values (70–100 MPa $(m)^{1/2}$). The relatively lower K_{IC} level of the BE compacts was probably the result of their higher oxygen levels [7] and possibly also due to some residual porosity caused by excessive chloride content within the -100-mesh titanium sponge fines. The compact density will also be affected by the application of thermomechanical processing such as forging, rolling, or extrusion. The transition point in forging seems to occur at 30% forging reduction. A similar effect is expected for hot rolling reduction of BE billets. The total rolling reduction from the initial billet thickness to the final plate and/or sheet thickness must also exceed a certain minimum value for enhanced mechanical properties and this, as will be shown later in this section, is very important for enhancement of the fatigue endurance limit properties.

The reduced fracture toughness of BE compacts produced using titanium sponge fines in all prior investigations (Tables 25.1 and 25.2) was possibly due to high oxygen and chloride contents. To verify this hypothesis, we shall use in this section the BE extruded product optimization studies from Ref. [19], as the basis for such conclusions, and further verify whether current BE optimization can result in properties equivalent

Table 25.2 Tensile and fracture toughness properties of additional BE titanium alloy compacts processed under various conditions

Alloy and condition*	0.2% yield strength		Ultimate tensile strength		Elongation (%)	Reduction in area (%)	K_{IC} or (K_Q)		Density (%)	Chlorine (ppm)
	MPa	ksi	MPa	ksi			MPa $\sqrt{m}$	ksi $\sqrt{in.}$		
Ti-5Al-2Cr-lFe, pressed and sintered plus HIP	980	142	1041	151	20	39			≥99	310
Ti-4.5Al-5Mo-l.5Cr (Corona 5), pressed and sintered plus HIP	951	138	1000	145	17	39	(64)	(58)	≥99	310
Ti-6Al-2Sn-4Zr-6Mo, pressed and sintered, no STA or HIP	1068	155	1109	161	2	1	31	28	99	150
Ti-10V-2Fe-3A1										
Pressed and sintered, HIP (1650°C, or 3000°F), and STA (775–540°C, or 1425–1005°F)	1233	179	1268	184	9	...	30	27	99	1900
Pressed and sintered, HIP, and STA (750–550°C, or 1380–1020°F)	1102	160	1158	168	10	...	32	29	99	1900
Pressed and sintered, no STA or HIP	854	124	930	135	9	12	51	46	98	150
Ti 6Al -6V 2-Sn, CHIP	931	140	1035	150	15	35	78	71	100	...

to IM material. For this demonstration, the ADMA hydrogenated powder-based billets extruded was used to produce the full-scale extrusions.

The initial TiH_2 titanium hydride powder lots used for the extrusion process had initial oxygen contents in the range of 700–1000 ppm, chloride content in the range of 440–700 ppm, and hydrogen content of 33,500–38,800 ppm, which closely approached the 100% hydride stoichiometric composition of about 39,000 ppm. After blending with master alloy powders, CIP'ing, and vacuum-sintering of the 6-in.-diameter billets, followed by extrusion, the extruded part, had a billet center oxygen content increased

Table 25.3 Distribution profiles of oxygen, hydrogen, and chlorine gaseous elements [19] in one of the blended-elemental extrusion compact billets (prior to extrusion)

Location*	Billet center	Mid-radius	Outer surface
Oxygen (ppm)	2850	2790	2670
Hydrogen (ppm)	1001	564	484
Chlorine (ppm)	100	<100	100

* Chemical analysis location within a slice removed from one end of a 28-in.-long, 6-in.-diameter billet. All other elements met the Ti-6Al-4V AMS 4934 specification.

up to 2900 ppm. The contents of all three interstitial gaseous elements, of the billet at midradius and at the outer surface showed distribution profiles (Table 25.3) indicating where the extrusion processing optimization should be focused. It was apparent that (a) the powder milling and CIP processing must avoid introducing any oxygen and/or moisture into the BE powder mix and, furthermore, the efficiency of the sintering process had to be improved by reducing its duration considerably, as this was also identified as another contributing source of oxygen uptake. Table 25.3 also indicates that once oxygen contamination takes place, it cannot be removed by a vacuum anneal. In contrast, hydrogen and chloride concentrations were greatly reduced by a vacuum anneal.

Implementing these BE powder consolidation processing optimization measures, ADMA Products, Inc., were able to reduce the oxygen content within a new series of optimized-chemistry BE powder–based Ti-6Al-4V billets down to 1900 ppm oxygen, which met the AMS Specification (AMS 4935) requirements. In this latter series of BE extrusion billets, ADMA Products independently measured the chlorine and chloride contents in the as-sintered new (optimized) billets and reported very low values: chloride <0.0010 wt%, and chlorine = 12 ppm. The fracture toughness values were measured in both the unoptimized (Table 25.4) and optimized (Table 25.5) titanium BE billet series, for which measurements of the threshold for stress-corrosion resistance, K_{ISCC}, was determined (Table 25.6) per NACE TM0177-2005 Standard, and all these properties were then compared to similarly extruded double-arc-remelt billets (see Tables 25.4–25.6). The database in these tables shows clearly that when the blended elemental BE compacts were tested unoptimized as in Table 25.4, there was a large difference in fracture toughness values between the powder-based extrusions and the double-arc-remelt ingot-based extrusions. By contrast Table 25.5 shows that on optimization of the oxygen content to within the AMS Specification limits, and near total elimination of chloride content (down to 12 ppm chlorine), the fracture toughness gap between BE powder–based extrusions and IM extrusions was reduced to within the scatter of the fracture toughness measurements while Table 25.6 showed very closely matching K_{ISCC} values. The recently developed BE technology exhibits mechanical properties equivalent to IM material. In the following section it will be

Table 25.4 Comparison of ambient-environment fracture toughness values of ADMA-blended-elemental, powder-based, Ti-6Al-4V extrusions having oxygen content ≈3000 ppm with ingot-based same-alloy, with oxygen content ≈2000 ppm, and similarly processed extrusion R4 [19]

Extrusion identification	Longitudinal orientation (LT) fracture toughness K_Q (K_{IC}) (MPa. m$^{1/2}$)	Transverse orientation (TL) fracture toughness K_Q (K_{IC}) (MPa.m$^{1/2}$)	Averages in both orientations (LT/TL) fracture toughness K_Q (K_{IC}) (MPa. m$^{1/2}$)	Notes on extrusion processing
P-1 (B-E)	57 59.3	43.4 42.7	58.2/43.07	Beta-extruded
P-2 (B-E)	47.7 50.65	35.1 35.2	49.2/35.4	Extruded at beta transus
P-3 (B-E)	51.1 52.5	35.5 35.4	51.9/35.5	Alpha-beta-extruded
P-5 (B-E)	75.8 74.5	73.6 59	75.2/66.6	≈Beta-transus-extruded
R1 (B-E)	48.3 46.6	35.5 35.2	47.5/35.4	Highest extrusion temp.
R2 (B-E)	54.2 58.2 (valid K_{IC})	39.6 40.7	56.3/40.1	Same extrusion temp. as R4
Average of all blended-elemental powder-based extrusions			56.4/42.6	All temperatures, same extrusion ratio and strain rates
R4 extrusion (ingot based)	81.1 83.2	80.2 82.7	82.2/81.5	Same extrusion temp. as R2
General conclusion based on a total of 28 fracture toughness tests conducted (four per each extrusion) with powder-based extrusions exceeding AMS 4935 specification oxygen limit				≈ 45–60% knock-down in K_Q at higher oxygen content

shown that identical conclusions may also be drawn for fatigue crack initiation and growth in powder-based titanium compacts.

25.1.1.1 Tensile properties and fracture toughness of PA alloy compacts

It has been stated [18] that while BE compacts are produced and used in a wide range of densities, PA PM parts are acceptable only at 100% density [1–5,10]. This distinction may have been prompted by the fact that most of the prior PA PM applications were oriented toward aerospace applications, and other critical component manufacture. This does not relieve the BE powder process developers of optimizing their

Table 25.5 Comparison of ambient-environment fracture toughness values of ADMA-blended-elemental, powder-based, Ti-6Al-4V extrusions having oxygen content ≤2000 ppm with ingot-based same-alloy, with oxygen content ≈2000 ppm, and similarly processed extrusion R4 [19]

Extrusion identification	Longitudinal orientation (LT) fracture toughness K_Q (K_{IC}) (MPa. m^{½})	Transverse orientation (TL) fracture toughness K_Q (K_{IC}) (MPa. m^{½})	Averages in both orientations (LT/TL) fracture toughness K_Q (K_{IC}) (MPa.m^{½})	Notes on extrusion processing
P-4 (B-E)	75.6 74.5	59.6 61.3	75/60.4	Beta-extruded
P-6 (B-E)	75.2 80.3	61.8 64.8	77.8/63.3	Alpha-beta-extruded
R5 (B-E)	77.7 75.6	57.2 58.9	76.7/58.1	Beta-extruded
R6 (B-E)	76.5 75.5	76.4 56	76/66.2	Extruded at same temp. as ingot-based extrusion R4
R7 (B-E)	74.7 74.3	58.8 57.6	74.5/58.1	Highest extrusion temp. and 50% reduced strain rate
Average of all blended-elemental powder–based extrusions			76/61.2	All temperatures but same extrusion ratio and two strain rates
R4 extrusion (ingot based)	81.1 83.2	80.2 82.7	82.2/81.5	Same extrusion temp. as R6
General conclusion based on a total of 28 fracture toughness tests conducted (four per each extrusion) with powder-based extrusions not exceeding AMS 4935 specification oxygen limit.				B-E powder-based product closely matching ingot-based product form, albeit with some property directionality

densification processing so as to approach 100% densification. This is particularly important where fatigue resistance must meet minimum requirements, and specifically in order to match those of IM products. For PA applications, the earlier-developed titanium PA powders have been commercially available over the span of some three decades. The PA PM powders were supplied as fairly uniform size distribution of spherical particles. Because of their morphology, these spherical powder particles usually have high tap density (65%) and lend themselves to good powder flow and mold-fill characteristics.

Table 25.6 Correlation of stress-corrosion resistance, K_{ISCC}, of blended-elemental powder-based extrusions versus ingot-based extrusion, R4, prior to and after extruded powder–based product optimization [19]

Extruded product form	Extrusion identification	Optimization status	Average K_{ISCC}(MPa.m$^{1/2}$) T-L stress: corrosion resistance at 1000 h* (ksi in.$^{1/2}$)
B-E powder-based parts	P-1, P-3, R1, and R2	Before optimization	13.4 12.2
B-E powder-based parts	P4, P-6, R5, R6, and R7	After optimization	21.6 19.7
Ingot-based extrusion R4	R4	AMS-4935 specification-conformant	20 18.2

* Stress-corrosion tests conducted per NACE TM0177-2005 Standard.

In recent years, however, several emerging PA PM technologies have demonstrated the capability of producing prealloyed Ti-6Al-4V compositions yielding powder morphologies that are nonspherical. Data availability on the mechanical properties associated with these emerging technologies are still rather limited and will not be further reviewed in the remainder of this section.

More mechanical test data have been developed within the aerospace industry on PA compacts than on BE compacts, whereby once again the majority of PA work has been done on Ti-6Al-4V. In the remainder of this section we shall focus, however, not only on this alpha-beta alloy (Tables 25.7a–25.7d) but also on other near-beta and beta alloys (Tables 25.8a–25.8f).

Tables 25.7a–25.7d provide a comprehensive review of the tensile properties of Ti-6Al-4V PA compacts based largely on spherical powder processing pathways under various conditions. Table 25.8 provides limited information on the properties of additional alloys. When the alloy compacts are produced using HIP [14], VHP [12], or rapid omnidirectional compaction (ROC) [13] at higher pressures, but at lower temperatures, higher strength levels without losses in ductility are achieved. This result is due to the substantial microstructure refinement developed during high-pressure low-temperature powder processing. Similarly, postcompaction hot work, such as rolling [15] or forging [16], results in microstructural refinement that improves tensile strength and ductility.

25.1.2 Fatigue strength and crack propagation properties of Ti-6Al-4V BE and PA compacts

As with the tensile properties and fracture toughness analyses discussed in the previous section, investigations of the fatigue resistance of BE-based compacts were conducted using nonoptimized chlorine-containing BE materials consolidated largely from by-products of sponge fines. The fatigue life scatter band of chloride-containing

Table 25.7a Tensile and fracture toughness properties of Ti-6Al-4V PA compacts consolidated using a HIP cycle and also followed by heat treatments

Condition*	0.2% yield strength		Ultimate tensile strength		Elongation (%)	Reduction in area (%)	K_{IC} or (K_Q)		Titanium PA powder preparation**			
									Powder process	Compaction temperature		Other variables
	MPa	ksi	MPa	ksi			MPa $\sqrt{m}$	ksi $\sqrt{in.}$		°C	°F	
HIP	861	125	937	136	17	42	(85)	(77)	PREP	925	1695	...
HIP and annealed (700°C or 1290°F) (PREP)	882	128	944	137	15	40	(73)	(67)	PREP	955	1750	...
HIP (PSV) and β annealed	1020	148	1095	159	9	21	(67)	(61)	PSV	950	1740	975°C (1785°F) anneal
HIP and annealed (700°C, or 1290°F) (REP)	820	119	889	129	14	41	(76)	(69)	REP	955	1750	
HIP, annealed (700°C or 1290°F). and STA (955–480°C or 1750–855°F)	1034	150	1130	164	9	34	...	...	REP	955	1750	...
ELI; HIP (as-compacted)	855	124	931	135	15	41	(99)	(90)	REP	955	1750	1300 ppm O_2
ELI; HIP; and β annealed	896	130	951	138	10	24	93	85	REP	955	1750	1020°C (1870°F) anneal

** PREP: Plasma Rotating Electrode Process; REP: Rotating Electrode Process; PSV: Pulverization Sous Vide (Powder Under Vacuum).

Table 25.7b **Tensile properties of Ti-6Al-4V PA compacts consolidated using a HIP cycle followed by either of two special heat treatments, hydrogen broken up structure (BUS), or thermochemical (TCP) processing**

	0.2% yield strength		Ultimate tensile strength				K_{IC} or (K_Q)		Titanium PA powder preparation**			
									Powder process	Compaction temperature		Other variables
Condition*	MPa	ksi	MPa	ksi	Elongation (%)	Reduction in area (%)	$\text{MPa}\sqrt{\text{m}}$	$\text{ksi}\sqrt{\text{in.}}$		°C	°F	
HIP and BUS treated	965	140	1048	152	8	17	...	...	PREP	925	1695	...
HIP and TCP treated	931	135	1021	148	10	16	...	...	PREP	925	1695	...

** Definitions: Same as per Table 25.7a.

Table 25.7c **Tensile and fracture toughness properties of Ti-6Al-4V PA compacts consolidated using a HIP cycle followed by thermomechanical processing (rolling or forging) followed by heat treatments, as noted**

Condition*	0.2% yield strength		Ultimate tensile strength		Elongation (%)	Reduction in area (%)	K_{IC} or (K_Q)		Titanium PA powder preparation**			
									Powder process	Compaction temperature		Other variables
	MPa	ksi	MPa	ksi			$\frac{MPa}{\sqrt{m}}$	$\frac{ksi}{\sqrt{in.}}$		°C	°F	
HIP and rolled (955°C or 1750°F) (T)	958	139	992	144	12	35	…	…	REP	925	1695	75% rolling reduction
HIP, rolled (955°C or 1750°F), and β annealed												
L or LT	820	119	896	130	13	31	73	66	REP	925	1695	75% rolling reduction
T or TL	813	118	896	130	11	23	61	55	REP	925	1695	75% rolling reduction
HIP, rolled (950°C or 1740°F), and STA (960–700°C or 1760–1290°F)	924	134	1041	151	15	35	…	…	REP	950	1740	60% rolling reduction
HIP forged (950°C or 1740CF), and STA (960–700°C or 1760–1290°F)	1000	145	1062	154	14	35	…	…	REP	915	1680	56% forging reduction

** Definitions: Same as per Table 25.7a.

Table 25.7d Tensile properties of Ti-6Al-4V PA compacts consolidated using alternate compaction techniques, namely, rapid omnidirectional compaction (ROC) or vacuum hot pressing (VHP), or high-pressure low-temperature compaction (HPLT) followed by heat treatments, as noted

Condition*	0.2% yield strength		Ultimate tensile strength		Elongation (%)	Reduction in area (%)	K_{IC} or (K_Q)		Titanium PA powder preparation**			
									Powder process	Compaction temperature		Other variables
	MPa	ksi	MPa	ksi			$MPa\sqrt{m}$	$ksi\sqrt{in.}$		°C	°F	
ROC (900°C or 1650°F) (as-compacted)	882	128	904	131	14	50	...	...	PREP	900	1650	As-ROC
ROC (900°C or 1650°F) and RA (925°C or 1695°F)	827	120	882	128	16	46	...	...	PREP	900	1650	925°C(1695 °F) RA
ROC (650°C, or 1200°F) (as-compacted)	1131	164	1179	171	10	23	...	...	PREP	600	1110	As-ROC
ROC (600°C or 1100°F) and RA (815°C or 1500°F)	965	140	1020	148	15	43	...	...	PREP	600	1110	815°C(1500 °F) RA
VHP (830°C, or 1525°F) (as-compacted)	945	137	993	144	19	38	...	...	REP	830	1525	...
VHP (760°C or 1400°F) (as-compacted)	972	141	1014	147	16	38	...	...	REP	760	1400	...
HPLT and HIP (as-compacted)	1082	157	1130	164	8	19	...	...	PREP	650	1200	315 MPa (46 ksi)
HPLT, HIP, and RA (815°C or 1500°F)	937	136	1013	147	22	38	...	...	PREP	650	1200	315 MPa (46 ksi)

** Definitions: Same as per Table 25.7a; VHP: Vacuum Hot Pressed; HPLT: High Pressure / Low Temperature.

Table 25.8a Tensile properties of near-alpha titanium alloys PA compacts consolidated using a HIP cycle or rapid omnidirectional compaction (ROC) followed by heat treatments as noted

Alloy and condition*	0.2% yield strength		Ultimate tensile strength		Elongation (%)	Reduction in area (%)	K_{IC} or (K_Q)		Titanium PA powder preparation**			
									Powder process**	Compaction temperature		Other variables
	MPa	ksi	MPa	ksi			MPa $\sqrt{m}$	ksi $\sqrt{in.}$		°C	°F	
Ti-5.5Al-3.5Sn-3Zr-0.25Mo-1Nb-0.25Si (IMI 829)												
HIP and STA (1060–620°C) or 1940–1150°F	951	138	1089	158	18	22	...	...	PREP	1040	1905	...
ROC and STA (1060–620°C) or 1940–1150°F	909	132	1034	150	18	20	...	...	PREP	...	...	α+β ROC
Ti-6Al-5Zr-0.5Mo-0.25Si (IMI 685)												
HIP and STA (1050–550°C) or 1920–1020°F	970	141	1020	148	11	19	...	...	PREP	950	1740	...
Ti-6Al-2Sn-4Zr-2Mo												
HIP and STA (1050–550°C) or 1920–1020°F	924	134	1034	150	17	36	...	...	PREP	910	1670	...

** Definitions: Same as per Table 25.7a.

Table 25.8b **Tensile and fracture toughness properties of alpha-beta titanium alloy PA compacts consolidated using a HIP cycle followed by annealing heat treatment as noted**

Alloy and condition*	0.2% yield strength		Ultimate tensile strength		Elongation (%)	Reduction in area (%)	K_{IC} or (K_Q)		Titanium PA powder preparation**			
									Powder process**	Compaction temperature		Other variables
	MPa	ksi	MPa	ksi			$\frac{MPa}{\sqrt{m}}$	$\frac{ksi}{\sqrt{in.}}$		°C	°F	
Ti-6Al-6V-2Sn HIP and annealed (760°C or 1400°F	1008	146	1055	153	18	37	59	54	PREP	900	1650	...

** Definitions: Same as per Table 25.7a.

Table 25.8c Tensile and fracture toughness properties of near beta titanium alloys PA compacts consolidated using either a HIP cycle, or forging, or ROC'd, followed by solution and age heat treatments as noted

Alloy and condition*	0.2% yield strength		Ultimate tensile strength		Elon-gation (%)	Reduc-tion in area (%)	K_{IC} or (K_Q)		Titanium PA powder preparation**			
									Powder process**	Compaction temperature		Other variables
	MPa	ksi	MPa	ksi			MPa $\sqrt{m}$	ksi $\sqrt{in.}$	°C	°F		
Ti-6Al-2Sn-4Zr-6Mo HIP forged (920°C or 1690°F) and annealed (705°C or 1300°F)	1165	169	1296	188	11	37	…	…	REP	900	1650	920°C (1690°F), 70% forging reduction
Ti-5Al-2Sn 2Zn-4Cr-4Mo(Ti-17) HIP and STA (800–635°C or 1470–1175°F)	1123	163	1192	173	8	11	…	…	REP	915	1680	…
Ti-10V-2Fe-3Al HIP and STA (745–490°C or 1375–915°F)	1213	176	1310	190	9	13	…	…	PREP	775	1425	…
HIP, forged, and STA (750–495°C or 1380-925°F)	1286	186	1386	201	7	20	28	25	PREP	775	1425	750°C (1380°F), 70% forging reduction
HIP, forged, and STA (750–550°C or 1380-1020°F)	1065	155	1138	165	14	41	55	50	PREP	775	1425	750°C (1380°F), 70% forging reduction
ROC (as-compacted)	965	140	1007	146	16	54	…	…	PREP	650	1200	…
ROC and STA (760–510°C or 1400–950°F)	1296	188	1400	203	6	26	…	…	PREP	650	1200	…

** Definitions: Same as per Table 25.7a.

Table 25.8d Tensile and fracture toughness properties of beta titanium alloys PA compacts consolidated using a HIP cycle followed by aging or solution heat treat and age (STA) as noted

Alloy and condition*	0.2% yield strength		Ultimate tensile strength		Elongation (%)	Reduction in area (%)	K_{IC} or (K_Q)		Titanium PA powder preparation**			
									Powder process**	Compaction temperature		Other variables
	MPa	ksi	MPa	ksi			MPa $\sqrt{m}$	ksi $\sqrt{in.}$		°C	°F	
Ti-4.5A1-5Mo-1.5Cr (Corona 5) HIP and aged (705°C or 1300°F)	944	137	999	145	13	…	(75)	(68)	REP	845	1555	†
HIP and aged (760°C or 1400°F)	916	133	971	141	14	…	(79)	(72)	REP	845	1555	†
Ti-II.5Mo-6Zr-4.5Sn (Beta III) β HIP and STA (745–510°C or 1375–950°F)	1288	187	1378	200	8	I8	…	…	PREP	760	1400	…

** Definitions: Same as per Table 25.7a.

Table 25.8e **Tensile properties of meta-stable beta titanium alloy Ti185 PA compacts consolidated using either a beta-extrusion processing or a beta-HIP cycle followed by a solution and age heat treatment as noted**

	0.2% yield strength		Ultimate tensile strength		Elon-gation	Reduc-tion in area	K_{IC} or (K_Q)		Titanium PA powder preparation**			
									Powder process**	Compaction temperature		Other variables
Alloy and condition*	MPa	ksi	MPa	ksi	(%)	(%)	$\frac{MPa}{\sqrt{m}}$	$\frac{ksi}{\sqrt{in.}}$		°C	°F	
Ti-1.3Al-8V-5Fe												
β extruded and STA (705°C or 1300°F)	1392	202	1482	215	8	7	…	…	PREP	760	1400	…
β extruded and STA (770°C or 1420°F)	1461	212	1516	220	8	20	…	…	GA	760	1400	…
β HIP and STA (675°C or 1245°F)	1315	191	1414	205	5	10	…	…	GA	725	1335	…

** Definitions: Same as per Table 25.7a; GA: Gas Atomization.

Table 25.8f Tensile properties of intermetallic titanium-aluminum PA compacts consolidated using either a HIP cycle followed by a solution and age heat treatment as noted or rapid omnidirectional compaction (ROC) with no heat treatment

Alloy and condition*	0.2% yield strength		Ultimate tensile strength		Elongation (%)	Reduction in area (%)	K_{IC} or (K_Q)		Titanium PA powder preparation**			
									Powder process**	Compaction temperature		Other variables
	MPa	ksi	MPa	ksi			$\text{MPa}\sqrt{\text{m}}$	$\text{ksi}\sqrt{\text{in.}}$		°C	°F	
Ti-24A1-11Nb												
HIP (1065°C or 1950°F) and STA 1175°C or 2145°F)	510	74	606	88	2	2	...	...	PREP	1065	1950	...
HIP (925°C or 1700°F) and STA (1175°C or 2145°F)	696	101	765	111	2	2	...	...	PREP	925	1695	...
Ti-25A1-10Nb-3Mo-IV												
ROC (as-compacted)	710	103	854	124	5	6	...	...	PREP	1050	1920	...

** Definitions: Same as per Table 25.7a.

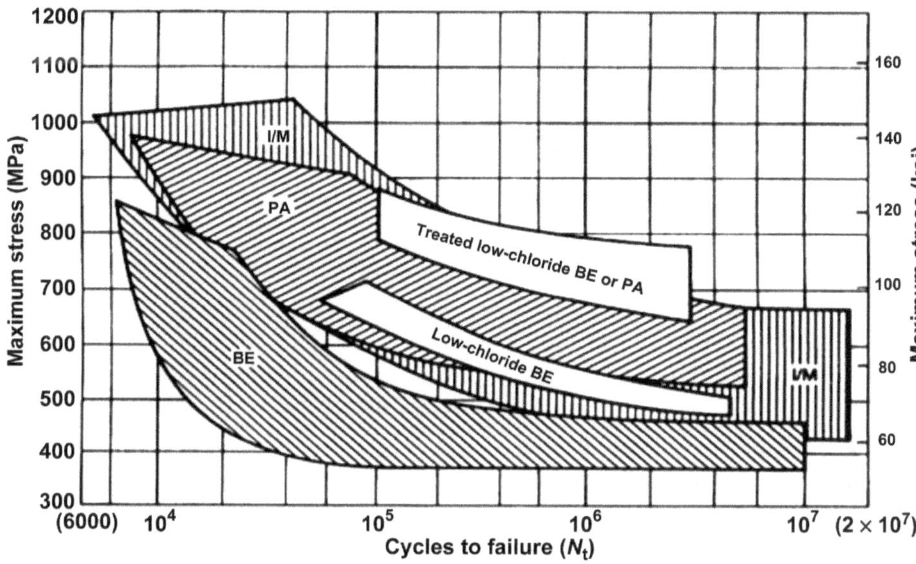

Figure 25.1 Showing comparison of the room-temperature fatigue life scatter bands of BE and PA Ti-6Al-4V compacts to that of a mill-annealed IM alloy. BE alloys were consolidated from chlorine-containing sponge fines blended with master alloy powders. The PA data were obtained by testing high-cleanliness REP and PREP compacts. (Source: From Ref. [18].)

Ti-6Al-4V BE compacts is compared in Figure 25.1 to a mill-annealed Ti-6Al-4V IM-based alloy. The effect of low chloride levels and postsintering treatments on fatigue strength is shown within the database scatter bands in Figure 25.1. In contrast, the smooth-bar fatigue life scatterband of Ti-6Al-4V PA compacts is compared to that of a mill-annealed IM material also in Figure 25.1 [3] and shows that the PA compacts are at equivalent levels to the best IM results. The PA data were obtained by testing high-cleanliness REP and PREP compacts that had undergone HIPing, with some of the compacts receiving a post-compaction heat treatment. It follows that powder contamination must be avoided in order to maintain a high fatigue strength in these materials. Figure 25.2 shows that upon full densification of BE compacts with extra-low chlorine content (which virtually eliminates residual void presence in these compacts), the fatigue endurance of these optimized BE compacts matched closely the endurance limit properties of the IM material scatter band, and with further treatments for microstructural refinement such as BUS and/THP, such BE compacts matched the best fatigue properties of the IM material scatter band (Figure 25.2).

The effect of reducing chlorine content, and increasing the BE compact density by the use of consolidation steps such as HIP processing is shown in Figure 25.3, which indicates that BE compacts must attain in fact 100% densification in order for fatigue properties to match exactly those of IM materials. HIP processing as an added

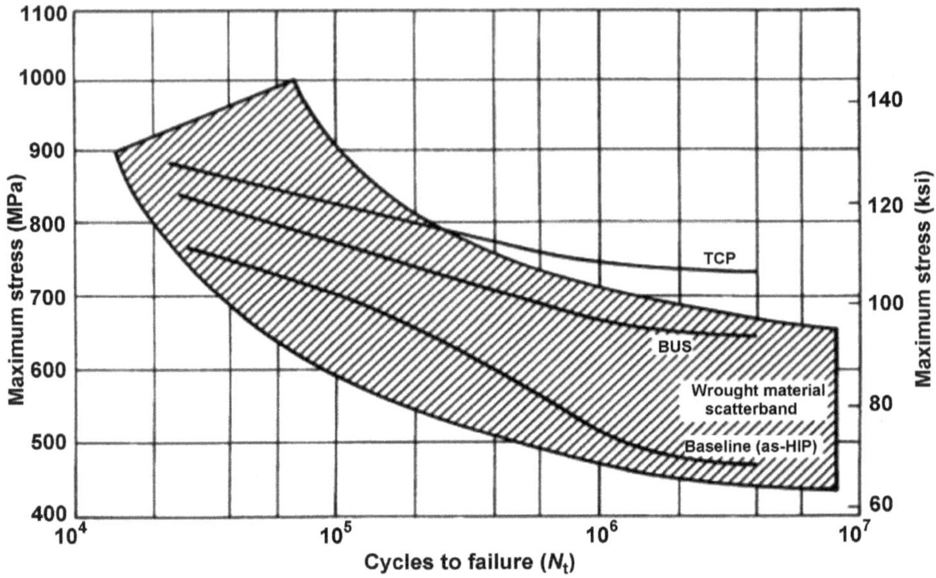

Figure 25.2 Showing comparison of the fatigue strengths of fully dense extra-low chloride Ti-6Al-4V BE compacts with the scatter band for the IM material. The BE compacts were tested in the as-HIP, BUS condition, and THP conditions. Smooth axial fatigue data were obtained at room temperature. Stress ratio *(R)*, 0.1; frequency *(f)*, 5 Hz with triangular waveform. (Source: From Ref. [8].)

step in the consolidation of BE compacts increases the cost of powder-based alloys, but a better alternative using thermomechanical processing by hot rolling, forging, and/or extrusion is essential for mechanical property optimization. Using the more recently developed emerging technology powders, it has been demonstrated that the fatigue properties match the fatigue endurance of IM and similarly extruded Ti-6Al-4V, Figure 25.4 [19].

The optimized blended-elemental BE compacts fatigue S/N data in Figures 25.1–25.3 along with the extruded BE compacts in Figure 25.4 and with the PA database for high-purity powder compacts using spherical powders in Figure 25.1 strongly confirm that powder contamination must be avoided in order to maintain high fatigue strength. In some PA materials where shape-making was attained using the ceramic mold process, there is a concern that other forms of contamination by particulates (such as metal or ceramic inclusions) may be present and the effect of such contaminants was studied by seeding the PA compacts with particles of different sizes. This study demonstrated the detrimental effects of such powder contaminants on the S/N curves and the compacted PA materials endurance limits.

Figure 25.5 shows the fatigue crack growth behavior of the blended elemental billets whose S/N behavior was examined in Figure 25.4. Figure 25.5 shows that the extruded ingot-based R4 billet is somewhat superior in crack growth, *da/dN*, relative

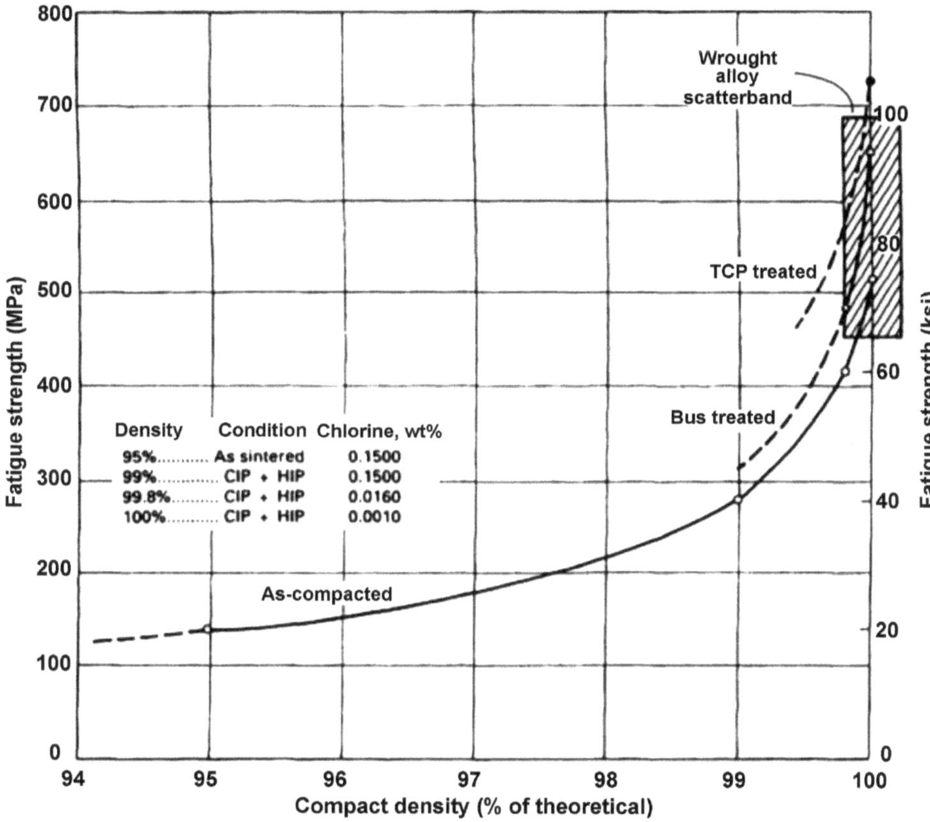

Figure 25.3 Effect of compact density on fatigue strength of CIP and sintered Ti-6Al-4V BE compacts. Note that the higher densities are only possible in the low-chloride material. BUS, broken-up structure; TCP, thermochemical processing. (Source: From Ref. [5].)

to the R6 BE billet. However, as the majority of part life is expended in fatigue crack initiation rather than crack growth, it may be concluded that both product forms are reasonably matched in terms of total fatigue resistance. The fatigue crack growth behavior of the PA materials is also shown for comparison with BE *da/dN* behavior in Figure 25.6.

25.1.2.1 Conclusions on fatigue behavior of BE, and PA compacts of the Ti-6Al-4V alloy

The database shows that with optimization of both BE and PA compacts, it is possible for powder-based materials to match the fatigue properties of IM materials, but close attention must be paid to reducing chloride content to values on the order of 10 ppm and the densification must approach very closely the theoretical value of 100% (i.e., essentially all residual void presence is to be eliminated).

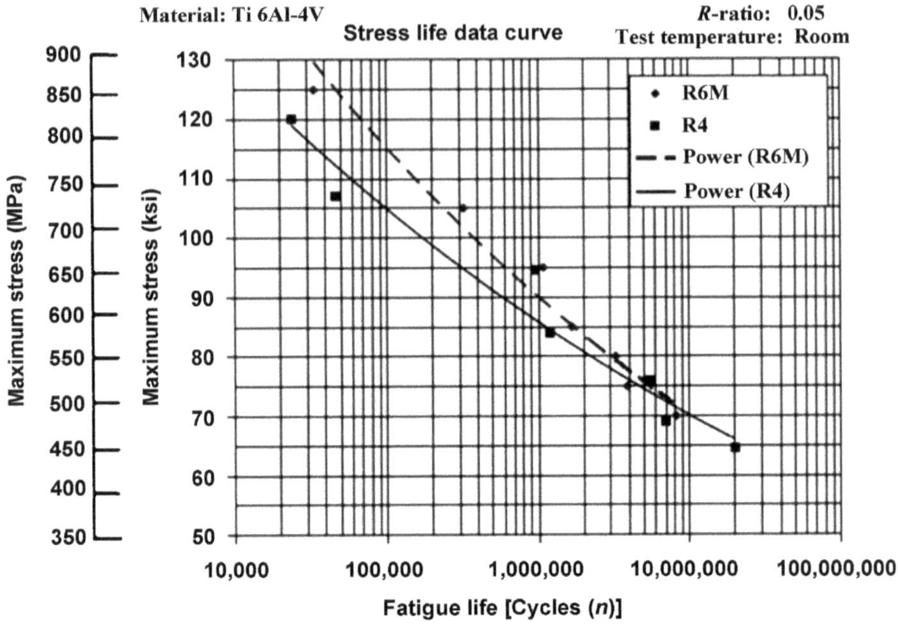

Figure 25.4 Showing the optimized blended-elemental powder–based extrusion S/N behavior to be superior to the back-to-back extruded IM material extruded at the same temperature, strain rate and extrusion ratio. (Source: From Ref. [19].)

25.2 Selection of powder processes and materials

1. *Titanium Powder Process Selection*
   There are five considerations for powder process selection:
   a. Cost of powder per unit weight: The powder process that will approach most closely the price of titanium sponge is most likely to be high on the list for process selection.
   b. Powder process amenability to scale up for powder production, and for meeting major customer and market demands (e.g., annual yield of powder reactors, and extraction process options for production volume increase with multiple-modular units)
   c. Process controllability and lot reproducibility of same product quality over many production lots as deliverables over time (lot after lot after lot).
   d. Ability to meet the final processed product specification requirements given specific alloy chemistry and required product form, etc. Especially critical in this regard is the supplier's ability to keep the production process under control, and to meet the customer procurement specification requirements for interstitial constituents.
   e. Minimum lead time. Note that the lead time for delivery of IM product from the time of order placement to the point of delivery of say a rolled plate or sheet product is on the order of 50–60 weeks. Powder producers must aim at reducing that lead time by far as their goal for final deliverables.

   The price of titanium sponge must be used as the benchmark for price comparisons of powder-based supply chain products. The two distinctively different titanium PM technologies, namely,

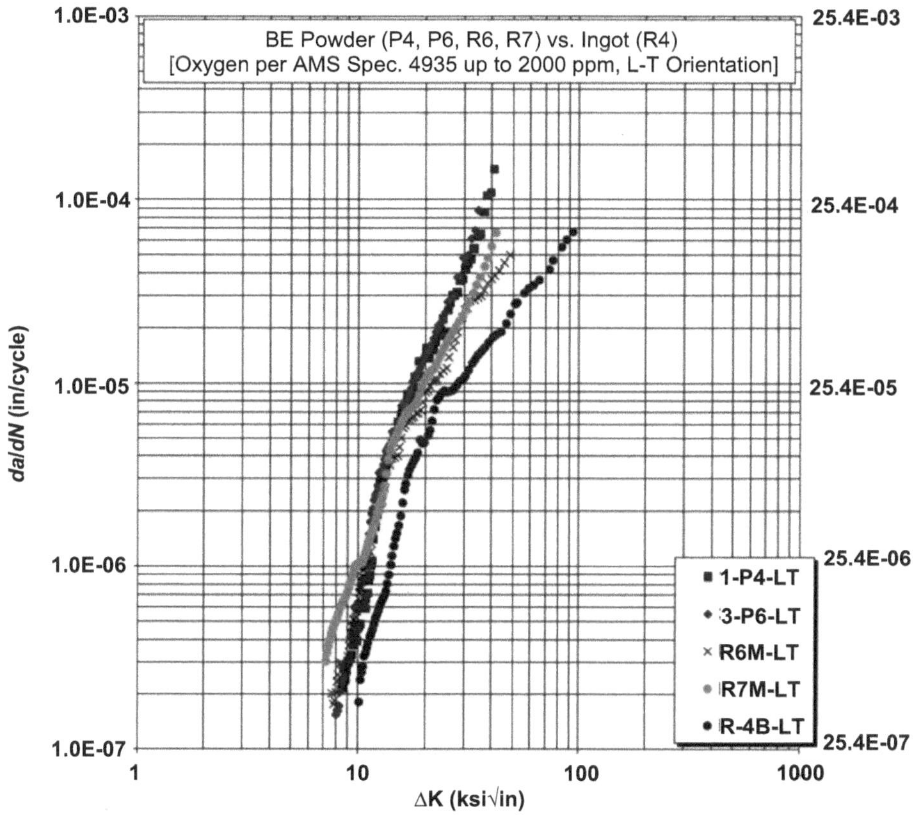

Figure 25.5 Showing the optimized Ti-6Al-4V BE powder-based extrusions (P4, P6, R6, and R7) *da/dN* behavior in comparison with back-to-back extruded IM-based Ti-6Al-4V material (R-4B-LT) extruded at the same temperature, strain rate, and extrusion ratio as all the other powder-based extrusions. Fatigue crack growth test parameters are: *R*-Ratio = 0.1, test frequency = 10 Hz, test environment: ambient temperature and laboratory air. (Source: From Ref. [19].)

the BE and the PA methods, not only produce compacts with different sets of properties but also with two different price ranges. At present, the most costly powders are those obtained by PREP, REP, and GA methods. These technologies deliver powders at a price upwards of $100/kg. This is because PA spherical powders for high-performance applications and 100% densification require an expensive melt stock, ultraclean handling, and expensive compaction tools. It has been claimed [11] that the introduction of GA powder is expected to lower powder costs and make spherical-particle powder-based PA products more cost competitive, but this has yet to happen. Over the past decade or so, the emerging titanium powder technologies briefly mentioned earlier in this section, and delivering randomly shaped particles are currently in tenacious competition for the future of low-cost titanium production. The deciding factor in selection of these new processes will be which one of them will deliver titanium powder at a cost as close as possible to the current cost of titanium sponge. But such an initial process selection will be also moderated by the other four factors (b) through (e) listed above.

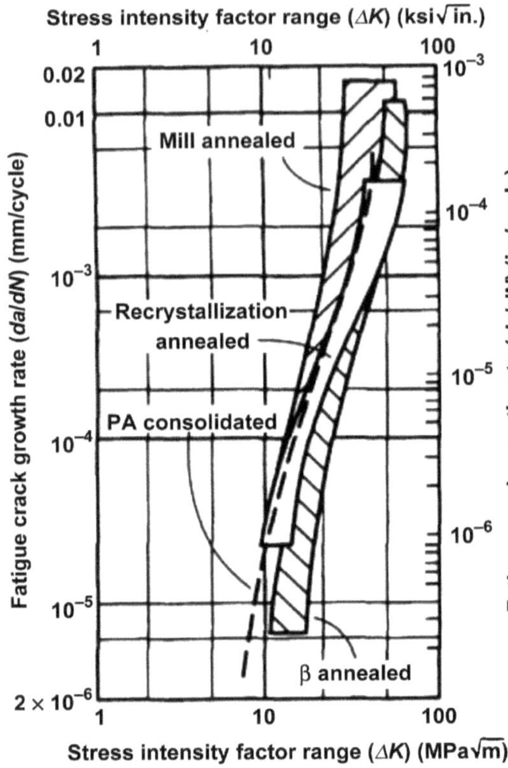

Figure 25.6 Comparison of the fatigue crack growth rate at room temperature in air of Ti-6Al-4V PA compacts with that of IM alloy material. Stress ratio *(R)*, 0.1; frequency *(f)*, 5–30 Hz (5 Hz for a PA compact). (Source: From Ref. [17].)

2. *Titanium Powder–Based Material Selection*

Alloy selection must be determined by the part design requirements and design allowables for the intended application, and this clearly falls outside the scope of this chapter. What may be derived from this chapter database, however, is the powder-consolidation processing pathway sequence that will deliver the best balance of mechanical properties, namely, tensile strength, ductility (elongation %), fracture toughness, and fatigue S/N endurance limit, as well as fatigue crack growth resistance (*da/dN*). Selection of the desired combination of processing steps must also keep in mind the final product cost. For example, HIPing may show improvements in certain properties and as such may be a desirable step following CIP and vacuum sintering, but if that step is to be followed by thermomechanical processing such as hot rolling, forging, or extrusion, then the optimum selection of the processing pathway must be optimized with the final product cost in view. For guidance in making specific optimum choices, the reader is referred at this point to our exhaustive analyses and concluding remarks on processing-microstructure–property relationships summarized earlier in this section for Tables 25.1, 25.2, 25.7, and 25.8 covering choices of processing step sequences for the baseline alloy Ti-6Al-4V as well as for a broad variety of near-beta, and beta titanium alloys.

References

[1] F.H. Froes, D. Eylon, G.E. Eichelman, H.M. Burte, Developments in titanium powder metallurgy, J. Met. 32 (2) (1980) 47–54.

[2] F.H. Froes, D. Eylon, in: F.H. Froes, D. Eylon (Eds.), Titanium Powder Metallurgy—A Review, Titanium Net-Shape Technologies, The Metallurgical Society of AIME, 1984, pp. 1–20.

[3] F.H. Froes, D. Eylon, Powder Metallurgy of Titanium Alloys—A Review, Titanium, Science and Technology, vol. 1, in: G. Lutjering, U. Zwicker, W. Bunk (Eds.), DGM, 1985, pp. 267–286; Powder Metall. Int., 17 (4) (1985) 163–167; continued in 17 (5) (1985) pp. 235–238; Titanium Technology: Present Status and Future Trends, in: F.H. Froes, D. Eylon, H.B. Bomberger (Eds.), Titanium Development Association, 1985, pp. 49–59.

[4] F.H. Froes, D. Eylon, Powder metallurgy of titanium alloys, Int. Mater. Rev., Private Communication.

[5] F.H. Froes, D. Eylon, Titanium powder metallurgy – a review, in: PM Aerospace Materials, vol. 1, MPR Publishing, 1984, pp. 39-1–39-19.

[6] D. Eylon, R.G. Vogt, F.H. Froes, Property Improvement of Low Chlorine Titanium Alloy Blended Elemental Powder Compacts by Microstructure Modification, Progress in Powder Metallurgy, vol. 42, compiled by E.A. Carlson, G. Gaines, Metal Powder Industries Federation, 1986, pp. 625–634.

[7] Y. Mahajan, D. Eylon, R. Bacon, F.H. Froes, Microstructure Property Correlation in Cold Pressed and Sintered Elemental Ti-6Al-4V Powder Compacts, Powder Metallurgy of Titanium Alloys, in: F.H. Froes, J.E. Smugeresky (Eds.), The Metallurgical Society of AIME, 1980, pp. 189–202.

[8] D. Eylon, R.G. Vogt, F.H. Froes, Property Improvement of Low Chlorine Titanium Alloy Blended Elemental Powder Compacts by Microstructure Modification, Progress in Powder Metallurgy, vol. 42, compiled by E.A. Carlson, G. Gaines, Metal Powder Industries Federation, 1986, pp. 625–634.

[9] R.R. Boyer, D. Eylon, C.F. Yolton, F.H. Froes, in: F.H. Froes, D. Eylon (Eds.), Powder Metallurgy of Ti-10V-2Fe-3Al, Titanium Net-Shape Technologies, The Metallurgical Society of AIME, 1984, pp. 63–78.

[10] F.H. Froes, H.B. Bomberger, D. Eylon, R.G. Rowe, in: R.J. Cunningham, M. Schwartz (Eds.), Potential of Titanium Powder Metallurgy, Competitive Advances in Metals and Processes, vol. 1, Society for the Advancement of Material and Process Engineering, 1987, pp. 240–254.

[11] C.F. Yolton, Gas Atomized Titanium and Titanium Aluminide Alloys, in Powder Metallurgy in Aerospace and Defense Technologies, Metal Powder Industries Federation, 1989.

[12] W.H. Kao, D. Eylon, C.F. Yolton, F.H. Froes, Effect of Temporary Alloying by Hydrogen (Hydrovac) on the Vacuum Hot Pressing and Microstructure of Titanium Alloy Powder Compacts, Progress in Powder Metallurgy, vol. 37, in: J.M. Capus, D.L. Dyke (Eds.), Metal Powder Industries Federation, 1982, pp. 289–301.

[13] D. Eylon, C.A. Kelto, A.F. Hayes, F.H. Froes, Low Temperature Compaction of Titanium Alloys by Rapid Omnidirectional Compaction (ROC), Progress in Powder Metallurgy, vol. 43, compiled by C.L. Freeby, H. Hjort, Metal Powder Industries Federation, 1987, pp. 33–47.

[14] D. Eylon, F.H. Froes, HIP compaction of titanium alloy powders at high pressure and low temperature (HPLT), Met. Powder Rep., 41 (4) (1986) 287–293; Titanium, Rapid Solidification Technology, F.H. Froes, D. Eylon (Eds.), The Metallurgical Society, 1986, pp. 273–289.

[15] R.F. Vaughan, P.A. Blenkinsop, in: F.H. Froes, J.E. Smugeresky (Eds.), A Metallurgical Assessment of Ti-6Al-4V Powder, Powder Metallurgy of Titanium Alloys, The Metallurgical Society of AIME, 1980, pp. 83–92.
[16] D. Eylon, F.H. Froes, D.G. Heggie, P.A. Blenkinsop, R.W. Gardiner, Influence of thermomechanical processing on low cycle fatigue of Ti-6Al-4V powder compacts, Metall. Trans. A 14 (1983) 2497–2505.
[17] S.W. Schwenker, A.W. Sommer, D. Eylon, F.H. Froes, Fatigue crack growth rate of Ti-6Al-4V prealloyed powder compacts, Metall. Trans. A 14 (7) (1983) 1524–1528.
[18] D. Eylon, F.H. (Sam) Froes, S. Abkowitz, Titanium powder metallurgy alloys and composites, in: Powder Metal Technologies and Applications, ASM vol. 7, 1998 ed., pp. 2191–2229.
[19] S.M. El-Soudani, K.O. Yu, E.M. Crist, F. Sun, M.B. Campbell, T.S. Esposito, J.J. Phillips, V. Moxson, V.A. Duz, Optimization of blended-elemental powder-based titanium alloy extrusions for aerospace applications, in: Metallurgical and Materials Transactions A, vol. 44A, ASM International, 2013, pp. 899–909.

A realistic approach for qualification of PM applications in the aerospace industry

26

R.R. Boyer*, J.C. Williams**, X. Wu†, L.P. Clark‡
*Retired Boeing Technical Fellow, Seattle, WA, USA
**Professor Emeritus, The Ohio State University, Columbus, OH, USA
†ARC Centre for Design in Light Metals, Monash University, Melbourne, Australia
‡Retired Boeing, Phoenix, AZ, USA

26.1 Introduction

Numerous studies have been and continue to be conducted throughout the aerospace community to reduce the cost of titanium components for aerospace applications. The rationale for using titanium is well documented, inclusive of specific strength, corrosion resistance, temperature capabilities, and composites compatibility. That notwithstanding, the primary restriction to greater use of titanium is cost. The aerospace industry can afford to pay more than most other industries, but cost is still a limitation. There are many instances where titanium is initially chosen, and later changed out as designers, faced with increasing cost of the structure, look for ways to modify the design to replace the Ti with lower-cost materials such as aluminum or steel. Thus, in order to obtain the superior structural efficiency of titanium, ways to reduce its cost have been and continue to be pursued.

One of the primary means of reducing the cost of titanium components is reduction of the buy-to-fly ratio (hereafter BTF), the amount of material procured relative to the weight of the final component. This is an excellent approach as it reduces the two highest cost factors in the price of Ti hardware, the raw material cost (through procurement of less material) and reduction of the amount of machining (machining costs are often as much as 50% of the component cost). There are many means of reducing the BTF, such as closed die forging, superplastic forming (SPF), superplastic forming plus diffusion bonding (SPF/DB), welding, and powder metallurgy (PM) methods. The PM methods include conventional press and sinter methods, cold and hot isostatic pressing, and more novel approaches such as additive manufacturing (AM) or 3-D printing.

The focus of this discussion will be on the most promising PM approaches, several of which have been developed and demonstrated to be capable of producing components with acceptable mechanical property performance at significantly lower cost than conventional wrought or cast processing. However, full production implementation has not yet been accomplished for a variety of reasons to be discussed. The processes most aggressively pursued have been conventional press and sinter utilizing low-cost blended elemental (BE) powder, utilizing cold isostatic pressing (CIP) and

Titanium Powder Metallurgy. http://dx.doi.org/10.1016/B978-0-12-800054-0.00026-5

hot isostatic pressing (HIP) using prealloyed powder. (Further details on the powder production processes can be found in Chapters 2 and 7.) Other techniques such as laser AM, electron beam consolidation, and metal injection molding (MIM) and a variety of roll consolidation methods for sheet manufacturing have also been pursued but will not be discussed in this article. Further information on these processes can be found in Chapters 8, 18, and 24.

Among the numerous reasons for utilization of titanium components are specific strength, corrosion resistance, elevated temperature capabilities, and electrochemical compatibility with carbon fiber polymer matrix composites [1]. The primary restriction with regard to more extensive use of titanium is cost. The aerospace industry can afford to pay more than most other industries because of the economic value created by lower weight and maintenance expense, but procurement cost (as opposed to life cycle cost) is still a limitation. There are instances where titanium is initially chosen, and designers immediately begin looking for ways to modify the design to replace the Ti with lower cost materials such as aluminum or steel. Ti PM is a possible avenue to reduce the cost of Ti hardware. Powder metallurgy is seen as a prime method of reducing the cost of manufactured shapes, but concerns about lower properties and inspection procedures continue to be a barrier to acceptance.

26.2 A brief history of Ti powder metallurgy in the United States

The first known titanium powder metallurgy developmental effort in the United States was conducted in the late 1940s by P.R. Mallory [2]. He and the Sharon Steel Corp. formed Mallory-Sharon Titanium Corporation under Navy sponsorship. That Corporation eventually evolved into RTI, a current major producer of titanium wrought products for the aerospace market and components for petroleum recovery. Prealloyed titanium powder was first produced in the United States in 1962 at Nuclear Metals (now Advanced Specialty Metals) using their novel rotating electrode process (REP). This led to the use of Ti PM particles on the surface of orthopedic knee and hip prostheses; the interstices between the powder particles provided bone ingrowth sites, increasing the bond strength between the bone and prostheses. The developer of that process, S. Abkowitz, left Nuclear Metals in 1972 to found Dynamet Corp., which eventually became Dynamet Technology, Inc. Dynamet is among the most active companies to pursue the blended elemental Ti PM approach to fabrication of titanium parts. Other companies, also pursuing applications of titanium press and sinter technology, were Imperial Clevite, Inc., and Gould Laboratories, neither of which exists today. Over the ensuing years, Dynamet has produced thousands of missile parts such as dome housings, warhead casings and wings [3], and continues to produce them today. ADMA Products is also engaged in this technology [4] and has been actively pursuing alternate powder making (hydrogenated powder) and consolidation techniques beyond conventional press and sinter such as those noted earlier.

Although work continued on applications of blended elemental titanium PM technology for small, expensive, complex, and difficult-to-make shapes such as those noted above, there was no significant thrust to take advantage of the technology for airframe or engine components. This was in part because the parts were significantly less than fully dense and, consequently, exhibited reduced mechanical property performance until the early 1970s. This is when the Air Force Materials Laboratory funded a program [5] with Pratt & Whitney (PWA) to apply the technology to advanced engines. The funding of this effort was the result of a major thrust by the Air Force Materials Laboratory (AFML) to refocus some of their effort on lowering the acquisition cost of primary airframe and engine components, which were escalating at an alarming rate. To achieve their goals of lowered acquisition cost of advanced aerospace systems, AFML convened a government–industry meeting at Lake Sagamore, New York [6], in September 1973. This meeting resulted in a major refocusing of their extensive R&D programs with a greater emphasis on acquisition cost reduction. Several major thrust areas were established. One of these was powder metallurgy with the intent to exploit the known potential of PM technology to reduce the acquisition cost of primary airframe and engine structures. Prime targets of the Laboratory's effort were structures made from expensive and difficult-to-process materials such as titanium and superalloys. In support of that objective, the Materials Laboratory, Manufacturing Technology division (Man-Tech), conducted an industry-wide Powder Metallurgy Conference [7] in June 1973 at Battelle's Columbus Laboratories. The conference was well attended by key university and manufacturing company technologists involved in advancing the state of the art in powder metallurgy. From that conference, a number of Air Force–sponsored programs emerged that advanced the application potential for titanium and superalloy PM technologies in both engines and airframes. The program with PWA noted above performed a program with powder supplier REM Metals together with press and sinter consolidation sources Gould Laboratories and Imperial Clevite to establish a process for the production of a near-net F100 engine compression rigid stator connecting link, Figure 26.1. The process produced parts with 91% theoretical density with static strength properties comparable to wrought material but with somewhat lower fatigue properties that were judged adequate for the intended application.

Economic analysis of the process showed that an acquisition cost reduction of 71% was possible in production lot quantities over the then current bill of material component that was machined from bar stock. It was recommended that the process be considered for application to the engine and that press and sinter technology be applied to other engine parts. This successful part making process also was selected as an IR100 award winner in 1981, but was never implemented into production.

A more challenging program [8] was undertaken by Crucible Materials Research Center (CMRC) to produce near-net-shape titanium alloy components for primary structural airframe and engine components by HIP of prealloyed powder. McDonnell Douglas participated in the effort to produce a keel splice fitting for the F-15 aircraft and General Electric to produce a compressor stub shaft for the F101 engine. This program utilized existing investment casting ceramic mold technology to create complex, but oversized, shapes from prealloyed Ti powder. The oversized parts were made

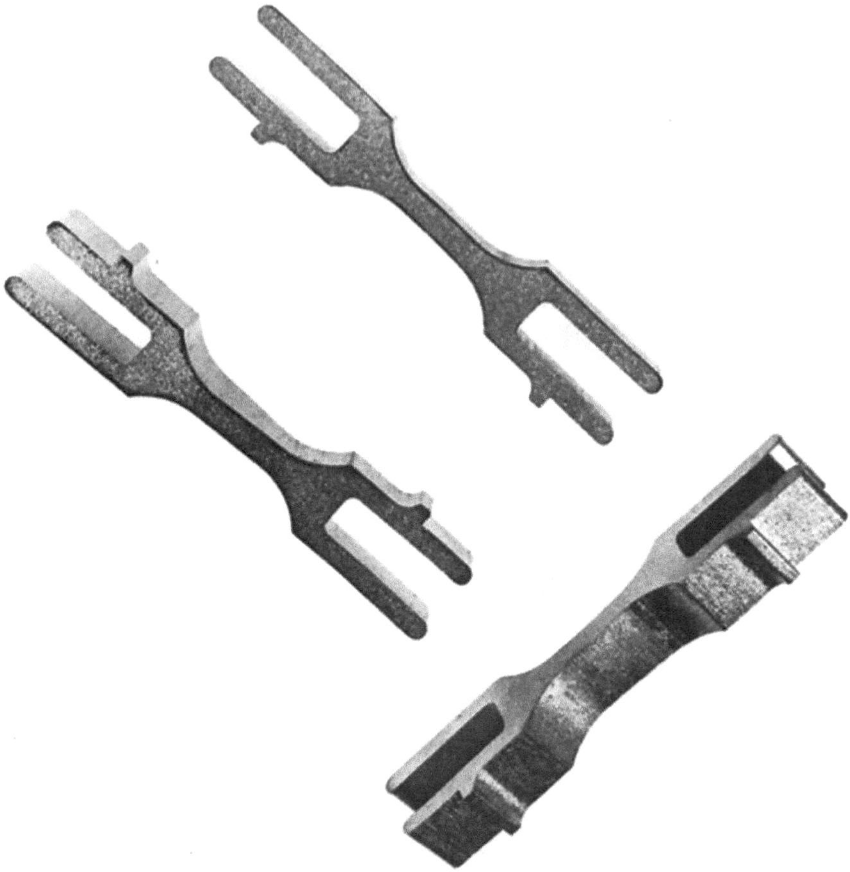

Figure 26.1 PWA F-100 engine compression rigid stator link produced by press and sinter of BE powder.

in shaped containers that were filled with powder, sealed, and then supported in secondary pressure transfer medium contained in a welded steel can. The cans would then be evacuated and exposed to high temperature (~915°C for the Ti-17 and 955°C for the Ti-6Al-4V) and isostatic pressure (~105 MPa) for 8 h during which the can would collapse around the container, resulting in consolidation of the powder to full density. Prealloyed powder made by the REP process developed by Nuclear Metals was selected as the powder source for the process because it exhibited the best flow and packing density characteristics among the limited powder types available at that time. Considerable characterization work was performed to determine these and other powder characteristics that were crucial to make the manufactured parts reproducible. Good flow was necessary to adequately fill complex-shape molds with high packing densities consistently; in general, densities of around 65% were necessary to ensure proper sizing of the molds to account for shrinkage during consolidation. The more consistent the packing density, the closer to net size the mold could be made. This

Figure 26.2 Keel Splice – F-15: from left: conventional 2.1-kg forging; 0.5-kg PM part; and 0.2-kg final shape.

was factored into the oversizing requirements and subsequent end product BTF, and ultimately cost. This work enabled CRC to produce complex, near-net-shape parts such as the F-15 keel splice fitting shown in the Figure 26.2. Extensive evaluation of microstructural and mechanical (static and dynamic) properties of the Ti-6Al-4V keel splice fitting showed comparable properties to the wrought product.

Additional parts were produced to further demonstrate reproducibility, five of which were then finish machined and structurally tested in side-by-side tests against finished wrought (forge and machined) parts. The test results are shown in Table 26.1 and demonstrated equivalency of the processes.

Similar results were obtained from consolidated products using General Electric's Ti-17 alloy (Ti-5Al-2Sn-2Zr-4Mo-4Cr), with the exception that the powder used was contaminated, resulting in lower than acceptable dynamic properties. This problem curtailed performing a full-up spin test on a finished part. Fracture analyses on early

Table 26.1 Spectrum fatigue data, F-15 horizontal tail spectrum, test limit load, 982 kg

Specimen no.	Fabrication method	Hours to failure	Average
2FF	Forged	11,255	
3FF	Forged	10,400	
4FF	Forged	6,960	
5FF	Forged	5,760	8,594
3HF	HIP	23,360	
5HF	HIP	17,840	
6HF*	HIP	14,240	18,480

* Oversized hole and fasteners (4.8 mm in lieu of 4.3 mm).

failed parts showed tungsten, iron, and nonmetallic inclusions at the crack initiation sites that contributed to the poor fatigue properties.

The powder contamination issue resulted in neither part being considered for production implementation until Nuclear Metals (or any other potential powder source) could reproducibly make contamination-free powder that met industry requirements. Since Nuclear Metals had the only viable process (REP) available at that time, AFWAL contracted with General Electric to work with Nuclear Metals [9] to improve their process, enabling them to produce an acceptable powder product to an acceptable specification. As part of the process improvement, Nuclear Metals altered their REP process by replacing the tungsten electrode and associated arc with a plasma energy heat source, thereby eliminating the source of tungsten inclusions. This process became known as plasma rotating electrode process (PREP). Nuclear Metals also built a dedicated facility with a clean room environment for processing powder. They utilized a number of sampling techniques to validate that the process produced clean powder that met specification requirements. The improved PREP powder met the user's requirements and was used in subsequent programs.

Naval Air Systems Command (NAVAIR) likewise saw the potential benefit of titanium PM technology in reducing the cost of titanium structure. Consequently, NAVAIR funded Grumman Aerospace [10] to establish manufacturing methods for producing a Ti-6Al-6V-2Sn near-net-shape fuselage brace fitting for the F-14 using Crucible's HIP process. The program demonstrated viability in producing the part to near-net shape with slightly reduced but acceptable property performance and potential acquisition cost savings of 30–40%. A follow-up contract [11] structurally tested the part to 24,000 equivalent flight hours (4 times design life) without failure (again, success but no implementation). This provided sufficient incentive for NAVAIR to collaborate on a more extensive effort with the AFML as discussed below.

Early on, the AFML and NAVAIR near-net-shape PM programs identified a number of follow-on actions that would enable full application of PM technology to the production of high-performance airframe and engine components. Paramount to any successful application of the technology to the near-net-shape production of aeronautical systems structural components is the availability of a clean, reasonable-cost powder with acceptable flow and packing density characteristics. With quality powder available and the prior success of the earlier near-net program, a follow-on program with Crucible Research Center (CRC) was cosponsored by AFML and NAVAIR [12] to resolve the manufacturing issues identified in the earlier programs.

Thus CRC, utilizing the same technology already established on the earlier program, expanded their reach to a wider aerospace industry base to resolve the technical issues from the earlier programs and assess its applicability to produce larger, more complex components. A phase I evaluation of four airframe manufacturers – Boeing (walking beam support fitting); General Dynamics (horizontal stabilizer pivot shaft); McDonnell Aircraft Co. (Drop-out link); Northrop Corp (engine mount support fitting; Figure 26.3) – and three engine manufacturers – Pratt & Whitney (third-stage fan disk), General Electric (3:8 compressor spool), and Williams International (HP centrifugal compressor rotor; Figure 26.4) – was conducted.

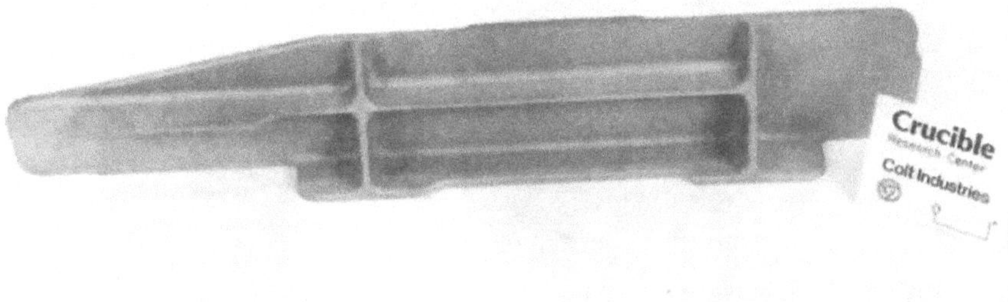

Figure 26.3 Engine mount support fitting (EMS) for an F-18A aircraft (Northrop Corporation).

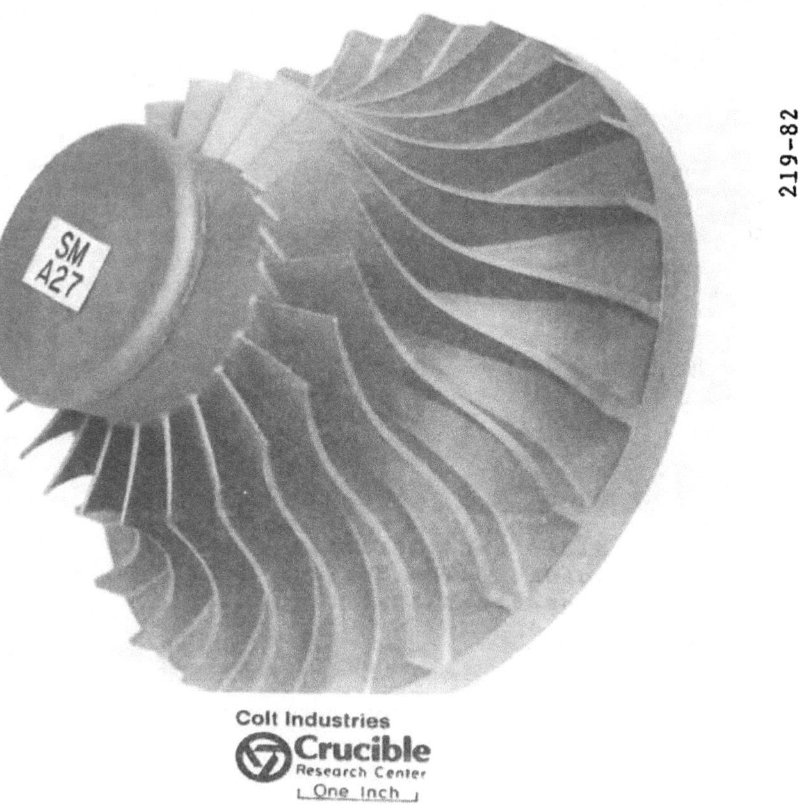

Figure 26.4 Williams International centrifugal compressor rotor, F-107 engine.

Selection of these parts demonstrated the wide industry interest and growing confidence in the PM processes as demonstrated in the earlier programs. This effort demonstrated the belief that use of PM had the capability to significantly impact the acquisition cost of major high-cost primary structures they were currently producing using conventional wrought product technology. The Parts selected by each of the users were made and extensively evaluated to demonstrate the viability of the ceramic mold HIP process to make acceptable parts in production at significantly reduced cost over more conventional wrought product processes. All participants in the Phase I effort were willing to remain involved with the program and to carry their particular part through the full qualification process and, if successful, commit the material/process to production.

The continuation (Phase II) was a production verification activity wherein one engine part and one airframe part was selected for pilot production. At the outset of Phase II, an extensive generic mechanical property database was established for Ti-6Al-4V product HIPed from Nuclear Metals' improved PREP powder that was used in support of both parts. Typical properties obtained were consistent with wrought product values as shown in Table 26.2.

The compressor rotor for the F107 engine (Figure 26.4) made by Williams Research was selected as the engine component. Eight rotors and four test blocks were made and provided to Williams to evaluate, which involved non-destructive inspection (NDI), mechanical property testing, and an overspeed spin test. All were successful and the process demonstrated the potential to reduce the BTF from 8:1 to 2:1.

The Airframe part selected was the engine mount support (EMS) shown in Figure 26.3 for the F-18A. A total of 72 parts and seven test blocks were produced, and extensively evaluated in 16 different static and dynamic mechanical tests by Northrop. As a result, 12 parts were fully qualified and installed on six F-18A aircraft. Northrop projected a manufacturing cost savings of 24% for full implementation of the PM parts in production. Although the six F-18A vehicles were identified at the time the parts were installed, no follow-up was performed to validate their performance. However, it is assumed that it has been successful since there is no known reporting to the contrary. Again, apparent success was achieved in the fabrication of powder components both from a material property and an economic point of view, but nothing implemented into production.

Table 26.2 Mechanical properties of Ti-6Al-4V made by HIP consolidation [12]

Property	Spec. min.	Min. value	Max. value	Avg. value	No. of tests
TYS (MPa)	828	869.4	917.7	890.1	64
UTS (MPa)	897	931.5	986.7	959.1	64
Elong. (% in 1.4")	10	12	17	14	64
Reduct. of area (%)	20	35	42	39	64
K_{IC} (MPa-m$^{0.5}$)	55	77	90.2	84.7	8
K_{Iscc} (MPa-m$^{0.5}$)		50.6	63.8	55	3

Finally, there has been continuing effort to further establish Ti PM as a viable production process, especially in the arena of alternate Ti powder production methods and processes. This has opened the door to an active pursuit within the industry to put into production the processes already developed, demonstrated, and flown on air vehicles. Additionally, alternate powder canning technologies have been pursued such as the use of steel cans originally demonstrated by Kelsey Hayes in the 1970s wherein an oversized shape is machined into a steel block that, when filled with Ti powder, evacuated and HIPed to produce near-net-size/shape parts with properties equivalent to wrought. This methodology is being actively pursued by Synertech [13] (see Chapter 18) and others. Applicability of the steel can methodology as with CMRC's ceramic mold process, are viable processes and should be considered for future use if they can compete cost-wise with conventional processing technology or other new innovations that are on the horizon.

For example, recent work by the group at the University of Birmingham and now at Monash University has led to an improved understanding the fundamental science governing HIP of Ti and TiAl prealloyed powders [14–17]. This work has found the powder HIPed Ti64 to exhibit mechanical properties, including tensile and fatigue strength, comparable to the forged Ti64 [14,16]. If high-quality Ti64 powder can be provided at a price lower than $150/kg, this process has the potential to manufacture load-bearing Ti structures with up to 30% cost reduction and up to 90% reduction in lead time from design to final component. This process also offers consistency and repeatability, so the uptake of this process in commercial aircraft or static parts in aero engines is very much dependent on the price of aerospace-graded powder. Some European aerospace companies have secured the source of such Ti powder and are therefore now actively investigating the potential of this process for commercial production.

This historical account of the trials of PM leading to apparent technical successes and disappointments is an excellent example of the effect nontechnical factors can have on actual implementation of a new technology. This is particularly true in the aerospace industry, where the consequences of inadequate risk analysis can be devastating. It also shows how undue conservatism can thwart thoughtful, low-risk introduction of new technology with resulting long-term benefits.

26.3 Assessment of the current status of Ti PM and its potential

Costs continue to rise today, and with the intense competition for commercial aircraft, there is a very strong effort, worldwide, to reduce costs. A lot of companies have been looking at PM as a potential means of reducing the component costs via reducing the BTF value. The BTF reduction must be substantial, at least for the PA approach, as the powder costs are significantly higher than that of a wrought product. For instance, a die forging will have a typical BTF in the range of 5–7:1 and could be significantly higher (some parts can have BTF value greater than 40:1) and forging costs could be in the range of \$100/kg or more. On the other hand, PA powder costs particularly in small quantities can range from \$100 to more than \$300/kg. The initial components to pursue would be parts machined from forged block or plate, where the BTF could be as high as 28:1, and even higher in certain circumstances. One would anticipate that the BE approach would be the lowest cost, though this may not always be the case, depending on the part and complexity. Most of these shapes are produced via an elastomeric mold, so these parts will usually have a somewhat higher BTF than the PA approach, and thus more machining would be required. Some smaller parts can be die-pressed using metal dies, and net shapes or very close to net shapes may be achieved using this approach because of the higher pressures and precision metal dies [4]. Components made via the BE approach will generally have static properties similar to that of wrought material, but the fatigue and damage tolerance properties will normally be less than that of wrought material.

With the PA approach, the powder is produced from wrought product, normally bar using the hydride–dehydride process, PREP or gas atomization processes, for example, and directly from the melt via gas atomization. Other approaches are being studied but those listed above have been the production processes used today. These PA powders are fairly expensive at this time and there are efforts to reduce those costs. Theoretically each particle has the proper chemistry at the outset. This will generally result in the full range of the PM product properties being comparable to wrought material. One of the primary caveats found during the 1970s in a large Air Force program to make large near-net-shape parts was the presence of a contaminant within a specimen, which had a dramatic impact on S-N fatigue performance (see Chapters 1 and 18). It is clear that the properties (static and dynamic) were comparable to those of the wrought parts when the powder is contamination free. Subsequently, component spectrum fatigue tests were run on two parts for four lifetimes; the part tested at the 100% operational level survived the required number of cycles. The second part was tested at 150% of the operational stress level and still survived four lifetimes.

As mentioned previously, the primary problems with the PA approach in those days was concern with contaminants, which had a very significant impact on durability properties. Contaminants were not much of an issue with the BE approaches as they were not 100% dense anyway, and so would not be candidates for fatigue-designed parts. As mentioned above, there is also a lot of work going on in reduction processes to lower the cost of titanium sponge or powder. This work will not be a part of this chapter, but substantial information is provided on some of these efforts elsewhere [18,19].

26.4 Qualification requirements

There are several requirements that must be met to supply product for aerospace applications. First and foremost, the supplier must be able to provide a near-net-shape (or complex) part with properties comparable to wrought with consistently high quality at a significantly reduced price, including machining (ready to install). The significantly reduced price is important for two reasons: (1) customers are going to consider a PM process high risk until there is a lot of data/experience indicating it is not, and (2) fatigue, durability, and damage tolerance properties allowables will have to be developed, which can easily cost over a million dollars. This cost can be substantially lower if a component is strictly static designed. In this case, the durability and damage tolerance testing, which are by far the most expensive part of the allowables, cost can be significantly reduced. Sufficient testing will be required to demonstrate that the fatigue properties are not so low that a static designed part will not become a potential fatigue problem. For example, porosity in a BE PM part could reduce the fatigue performance substantially; sufficient data must be generated to demonstrate that this will not become an issue for the parts in question.

In terms of quality control and processing, each company must be approved via audits to AS9100 and NADCAP in order to supply parts for the aerospace industry. The former is a rigorous review of the quality and management systems to help assure potential customers that you have the system in place to consistently produce a high-quality product with the potential for continuous improvements. The NADCAP audit will go through the processes requiring approval, such as heat treatment, to make sure that you follow all specification requirements. In addition, the customer will undoubtedly want to perform a technical audit of the entire process with the producer.

26.4.1 Properties evaluation

In order for Ti PM technology to be successfully applied to airframe and engine parts, they must fulfill form, fit, and function requirements of the parts they are to replace as well as be significantly lower in acquisition price. This was the primary objective of the foregoing programs, to produce near to net size (BTF ratios significantly less than the average 7:1 for airframe and engine parts) with equivalent mechanical property performance as discussed earlier.

Given that the shape-making capability has been established as discussed above, potential users must have confidence in achieving consistently uniform properties in each part. This can only be demonstrated through statistically based allowables evaluation of the mechanical properties of parts made by the process under consideration. The guidelines for attaining these allowables are contained in Metallic Materials Properties Development and Standardization (MMPDS) [20]. This document replaces the Military Handbook 5 (MIL-HDBK-5) and is the de facto standard for design data in the United States. The former document was maintained by the US Air Force through funding to Battelle Memorial Institute (BMI), Columbus, Ohio. BMI remains responsible for analysis of MMPDS data supplied by industry, but the funding of this effort is now through industry.

In instances where there is a unique alloy/heat treatment for a given application, a company may generate their own internal allowables, which still must be approved by the FAA (or another appropriate authority) for utilization on commercial aircraft. The de facto standard used by Europe and the United Kingdom is ESDU 00932. The approaches for these two documents are somewhat different; this is being worked on by the two organizations to reconcile the differences in an attempt to come up with a single standard. All of the testing must be done in accordance with "standard" aerospace test procedures, such as the ASTM specifications, or special company specifications.

The MMPDS also contains some durability and damage tolerance (D&DT) data, though at this stage it is not statistically based; it should be treated more as representative data. The goal for MMPDS is to eventually provide statistically based data of this nature, but the authors are not sure of the value of this endeavor. Different companies handle data of this nature in different fashions, so indications are that individual companies will require specific testing to satisfy their requirements. Data generated via standard tests using appropriate ASTM standards will still be useful for comparison purposes, but unique testing requirements could also be imposed. Generating these data is where the high costs come in.

In light of the above, in order for a new technology such as described herein to be approved for use in a form/fit/function replacement of an existing primary structural part, the allowables for all critical properties, usually both static and dynamic properties, must meet or exceed those of the part to be replaced. Static design allowables would involve testing for tensile, compression, shear, and bearing properties along with determination of precision tension, compression, and shear moduli; Poisson's ratio; density; coefficient of thermal expansion; specific heat; and thermal conductivity. Extensive testing of the physical properties would not probably be required as there is no reason to believe that these would not be the same as the wrought product. These properties must be determined over the temperature and thickness ranges of interest for the applications under consideration. Depending on the application, other properties, such as electrical conductivity, could be required. Additionally, where durability and damage tolerance performance is required, as discussed above, sufficient testing under service conditions would be performed to verify performance and in many cases full-scale structural testing would be done. This was done in several of the programs discussed earlier to provide an additional measure of assurance.

If there are deficiencies in a specific property and yet the benefit is sufficient to proceed forward, it could still be accomplished but would be significantly more expensive since drawing changes and stress analyses would be required before approvals would be granted. For new designs, if all the property performance requirements are met, the product, or product option, can be put on the drawing at the start. As mentioned above, application of the technology for static designs would be simpler, take less time, and require a smaller investment to gain approval. If the properties do not meet all the values in MMPDS or some other acceptable allowables source, the value of the technology diminishes greatly as the volume of potential parts will be significantly reduced. If it meets all the requirements of Ti-6Al-4V, for instance, then the material/process could potentially be used wherever Ti-6Al-4V is the bill of material. If not, it can only be used for components where the reduced value will not affect the function of the part. This could significantly reduce the production volume of the powder product.

26.4.2 Shape making capability

Clean, full-cut powder with reproducible flow characteristics and packing density is the major criteria for implementation of the PM processes requiring filling of some sort of metal or ceramic can (see Chapter 18). These characteristics are necessary to ensure that parts made are reproducibly close to net in shape, hence low BTF ratios and lower cost. The PREP powder process met these criteria and was thus chosen to demonstrate the inherent capabilities of the CMRC ceramic mold consolidation process. A disadvantage of the PREP process is the high cost of making the powder that can be more than $200/kg, significantly higher than mill products of forging stock. Alternate methods of making spherical powder have been established that has created a competitive environment and thus significantly reducing the cost of the powder. Regardless of the cost of the powder, the real cost savings potential comes from less material usage to make the finished product through reduced BTF ratios, which is accomplished with the near-net-shape-making processes. As noted earlier, the processes discussed have the capability of producing consolidated parts with BTF ratios as low as 2:1 and even lower, whereas most other conventional wrought metal processes produce parts on average with BTF ratios equal to 7:1 or greater. In some instances, they can be as high as 40:1 or more.

The other cost savings benefit comes in subsequent finishing operations on the parts, particularly machining to dimensionally precise requirements. Machining titanium is expensive and the closer to net a part can be made the less finish machining is required. Most parts will require some machining, and the supplier must work closely with the customer to provide a suitable configuration. One wants to provide as many surfaces as possible that would not require machining, and this would depend on load requirements. If the part is designed by fatigue, will the surface finish of the as-manufactured part be satisfactory? Some effort has been directed at producing parts that can be finish sized through chemical milling or other non-machining actions as well as design considerations where nonmating surfaces can either be left as produced or surface finished by chemical milling, grit blasting, shot peening, or some other inexpensive method, thereby reducing final processing costs.

Maximum reduction of the BTF is not as important for the BE approach except for the die-pressed components; the surface finish and shape reproducibility will not be as consistent. Most of the shapes will be produced using an elastomeric type mold to provide "green" compacts that must be sintered at temperatures exceeding something on the order of 1200°C. Sintered densities 98% of the theoretical density are achievable. For static designed parts this is usually acceptable. The BTF will probably be on the order of 1.5–2 and will require finish machining to achieve the desired dimensions. Because of the fact that the powder costs are lower than that of prealloyed powder, this may very well be acceptable.

26.4.3 Process controls

Suppliers will need to establish a range of process parameters, possibly for each part, at least early in the production phase. These parameters would include things such as approved powder suppliers, powder size range, powder flow rate, and packing density for HIPed parts.

For parts produced via AM (3-D printing), additional factors such as traverse speed, power settings, the path of the energy source, and any parameter that could affect the part quality will need to be carefully controlled (see Chapter 24). The supplier will need to establish an acceptable range of each of these parameters that provides consistent acceptable quality for each part that will account for normal tolerances, variations that could be required because of geometry, etc. Once established for a given part, these parameters will be entered into some type of process control document. Once approved by the customer, the process will be considered frozen.

26.4.4 Quality assurance

Another consideration prior to acceptance of new processes is inspection of the parts to ensure that contamination or porosity is within satisfactory limits (see Chapter 18). These types of defects will probably be quite small and may be difficult to inspect. It is anticipated that a rigorous inspection plan and testing will probably be required in early production. As production and flight experience is gained and no problems are encountered, it is likely that the amount of testing/inspection can be reduced to a sampling plan or development of a test procedure that the user can be assured will provide him the confidence in the product.

It would be recommended that extensive use of etched macro-slices be utilized, particularly in the early stages of development. They would provide a good idea of areas where porosity could be a problem, indicating areas where process parameters might need adjustment. New inspection processes are being developed but they may not yet be affordable. This is an area that needs further development.

26.5 Other development areas

Extensive working of the BE product to significantly enhance fatigue performance has been demonstrated by El-Soudani et al. [21]. They produced wrought and pressed and sintered BE Ti-6Al-4V billets and extruded them. The extrusion process significantly reduced the grain size and eliminated virtually all of the residual porosity in the extruded product. Evaluation of the extruded product showed that the mechanical, fatigue, and crack growth rate properties were equivalent to that of the wrought product, though the toughness values were reduced. Joshi et al. [22] exhibited a similar finding regarding static properties with extensive rolling of Ti-1Al-8V-5Fe above the beta transus. They exhibited yield strengths on the order of 1655 MPa with elongations of 12–14%. This was an alloy developed by Mallory-Sharon Metals in the 1950s, but with the high Fe content, the alloy was virtually impossible to melt with a production-sized ingot. Although this is not the subject of this paper, it brings out another important attribute of PM: the possibility of developing new alloys that cannot be produced via ingot metallurgy because of melting problems such as solute segregation.

26.6 Additive manufacturing (AM)

Recently, there also has been a lot written, mainly in the popular, semitechnical press [23], about various methods for making titanium shapes at lower cost (see Chapter 24). These articles largely ignore the effects of lot size on cost and even imply that these methods will produce across-the-board cost reductions. This is not the case for large lot sizes with relatively simple shapes that require minimal final machining. For these applications, forging is still the most competitive method. The low deposition rates characteristic of the laser, electron beam and other printing methods cause them to be more costly as the part size increases and for large production runs.

The general term that encompasses these methods is "additive manufacturing," which means the desired component is created incrementally by adding materials in a variety of means as opposed to starting with a larger piece and removing material to get the desired shape. In principle, this is "subtractive manufacturing" but this term is never used.

Among the range of AM methods, some of the most promising methods is the use of powder, which is selectively fused by a computer-controlled intense heat source, such as a laser or an electron beam.

- Components manufactured by laser and EB melting methods will have microstructures similar to components made from forgings or plate of the commonly used alloy Ti-6Al-4V (Ti-64) in the beta annealed condition, albeit on a finer scale. The microstructure and properties of material deposited by AM powder–based methods is not well characterized at the present time, but since it also starts with powder, it is logical to suggest the microstructures will be similar.
- *Feedstock:* The powder-based AM processes almost all use spherical PA as the raw material source, with one known exception [24]. Powder quality and cost are of central importance for these processes also.
- *Shape Making:* The shape-making capability of the powder-based AM methods is nothing short of spectacular, as can be seen from Figure 26.5.

Figure 26.5 Example of complex shape made by laser powder bed AM.

- *Capital Cost:* The contribution of the cost of capital to each component made via AM is significant, but is largely offset by the cost saving gained because no tooling is required. In principle, these methods also have the capability of much faster turnaround time (TAT) because the lead time for a forging die can be as long as 6 months. The laser and EB melting machines cost at least $500K. As in most cases, the actual cost depends on the size and configuration of the machine. At present, essentially all of the laser and EB machines have been installed in R&D facilities and would be overconfigured for a focused production activity.
- *Cost per Component:* However, the real issue that dominates the contribution of the cost of capital to each piece produced is the time required to make a component by any powder-based AM method (see Chapter 24). The real benefit of powder-based AM is the near-net-shape capability and the intricate shapes that can be created. When combined with the use of true design synthesis, part count can be reduced and manufacturability is no longer a design constraint. Components made by AM can be lighter also because the material can be placed only where it is needed to meet design intent. A non-aerospace example is non-load-bearing orthopedic devices such as cranial closures and facial reconstruction prosthetics, where custom dimensions are required and can be obtained directly from a CT scan.
- *Turnaround Time:* The time to deliver a new part for a new production program will be much reduced for 3-D printing compared to a forging. This can be particularly important if there are questions about the applied stresses and size of the part for long delivery items such as die forgings. It may take a year or longer to make the die for a forging. If there are design questions, the part will undoubtedly be made oversize to be certain a usable part can be machined from it. The part can be made via 3-D printing in a much shorter time. In addition, if the design needs to be modified, there would be no need to remachine the die or fabricate a new one; all one needs to do using 3-D printing is make alterations to the program and remake the part to the proper configuration.

The major challenge for broad acceptance of components made by AM is developing methods for rapidly qualifying them. This will almost certainly be accomplished through the use of high-fidelity physics-based computer models. There is a large effort underway to create such models. It is a challenging task to reach the level of confidence required. Concurrently, it will be essential to improve the control and stability of the AM machines. This need also has been recognized and is being addressed. Success will add a major dimension to the market for PA spherical powder. This is the main reason this short discussion is included here.

26.7 Summary

AM and near-net shapes produced via HIP, using both prealloyed and blended elemental powders, offer the potential to reduce the cost of titanium components through the manufacture of near-net shapes. Titanium has many attributes that make it a highly desirable material, but its utilization has always been limited by the high cost. Consequently, it is used only when its unique properties can justify the cost; if feasible, the manufacturers would rather use aluminum or steel because of their significantly lower cost. Providing near-net shapes offers the potential for reducing the cost of titanium components through two factors – reducing the manufactured weight of the part, which means less input of this expensive material, and reducing the machining costs, which is an important factor. Machining of titanium components can be half of the part cost. It is felt that a substantial cost savings must be offered to be successful as designers consider that powder metallurgy (or components produced by additive manufacturing using wire) will be high risk based on historical information – residual porosity or contamination. For these reasons, it would seem that the best approach would be to start with static designed parts. Data indicate that parts with densities on the order of 98% of the theoretical density will have static properties comparable to those of wrought product, but the dynamic properties will be reduced.

Once successful experience with static designed parts has been achieved, and additional data are generated indicating that the fatigue properties consistent with those of wrought material can be achieved and reliable inspection procedures are available, then components designed by durability may be pursued.

There are data showing that additional working of a powder product, that is, extrusion of a powder billet, that the durability of a powder part, even with the blended elemental approach, can be brought up to that of wrought material. This does, however, add to the cost, which reduces the potential cost savings. Obviously, there are instances where this may be justified.

Powder cost and quality is an issue. There are efforts worldwide to reduce titanium raw powder costs. Any gains made in this area would obviously improve the cost effectiveness of an additive manufacturing or HIP approach to further reduce the cost.

The authors perceive a market for products of this nature but for those working this area, the size of the market will be dependent on demonstration of a consistent high-quality product at a reasonable cost.

References

[1] R.R. Boyer, New titanium applications on the Boeing 777 aircraft, JOM 44 (1992) 23–25.

[2] R.A. Wood, Lesson 1 – History and Extractive Metallurgy, in: R.A. Wood (Ed.), ASM MEI Course 0142. Titanium and Its Alloys, ASM, Materials Park, 1994.

[3] S.M. Abkowitz, S. Abkowitz, H. Fisher, D.H. Main, Affordable PM titanium – microstructures, properties and products, in: I.V. Anderson, T.W. Pelletiers (Eds.), Advances in Powder Metallurgy and Particulate Materials – 2011, MPIF, Princeton, 2011 70-117-70-127.

[4] A. Abakumov, V. Duz, O. Ivasishin, V. Moxson, D. High. Savvakin, Performance titanium powder metallurgy components produced from hydrogenated powder by low cost blended elemental approach, in: L. Zhou, H. Chang, Y. Lu, D. Xu (Eds.), Ti-2011, Proceedings of the 12th World Conference on Titanium, vol. 2, Science Press Beijing, Beijing, 2011, pp. 1639–1643.

[5] AFML-TR-79-4028, in: J.A. Miller, G. Brodi (Eds.), Consolidation of Blended Elemental Ti 6Al-4V Powder to Near Net Shapes, United Technologies Corp, Pratt & Whitney Aircraft Group, W. Palm Beach, FA, 1979.

[6] Technical Memorandum AFML-TM-LT-73-1, Summary of Air Force/Industry Cost Reduction Conference, Wright Patterson Air Force Base, Ohio, 1973.

[7] Technical Memorandum AFML-TM-LT-74-4, Summary Report on the Powder Metallurgy Seminar, Wright Patterson Air Force Base, Ohio, 1974.

[8] V.C. Petersen, V.K. Chandhok, C.A. Kelto, Hot isostatic pressing of large titanium shapes, in: F.H. Froes, J.E. Smugeresky (Eds.), Powder Metallurgy of Titanium Alloys, A.I.M.E., Warrendale, pp. 243–254.
[9] AFWAL-TR-82-4113, in: R.E. Peebles (Ed.), Advanced Manufacturing Techniques for High Quality Low Cost Titanium Powder Production, General Electric Company, Evendale, OH, November 1982.
[10] R. Witt, J. Magnuson, Navy Contract No. N00019-74-C-0301, C/N 5018-002-4, April 1975, Manufacturing of Titanium Airframe Components by Hot Isostatic Pressing, Grumman Aerospace Corporation, Bethpage, NY, April 1975.
[11] R.H. Witt, Navy Contract No. N00019-74-C-0301, C/N 5018-002-4, April 1975, Manufacturing of Titanium F-14A Components Manufactured by Hot Isostatic Pressing (HIP), Grumman Aerospace Corp., Bethpage, NY, June 1977.
[12] AFWAL-TR-85-4120, in: P.M. Shapes, V.C. Peterson, V.K. Chandhok, J.H. Moll (Eds.), Manufacturing Process for the Hot Isostatic Pressing of Large Titanium, Crucible Research Center, Pittsburgh, PA, 1985.
[13] C. Bampton, T.V. Daam, G. Creeger, M. Jacinto, L. Arenas, V. Samarov, Progress in net-shape HIP'd powder metal for rocket engines, in: Proceedings of the 2008 International Conference on Hot Isostatic Pressing, Garden Grove, CA, 2008, pp. 191–197.
[14] X. Wu, Near-Net Shape Forming of Aeroengine Components. Encyclopedia of Aerospace Engineering, vol. 4, part 19, John Wiley and Sons, Hoboken, 2010, p. 197.
[15] K. Zhang, J. Mei, N. Wain, X. Wu, Effect of hot-isostatic-pressing parameters on the microstructure and properties of powder Ti-6Al-4V hot-isostatically-pressed samples, Met. Matls. Trans. 41A (2010) 1033–1045.
[16] C. Yang, D. Hu, X. Wu, A. Huang, M. Dixon, Microstructure and tensile properties of hot isostatically pressed Ti4522XD powders, Mater. Sci. Eng. A 534 (2012) 268–276.
[17] X. Wu, Falcon Tech, Unpublished Public Information 6, 2013.
[18] F.H. Froes, J.E. Smugeresky, Powder Metallurgy of Titanium Alloys, AIME, Warrendale, 1980.
[19] F.H. Froes, PM in Aerospace and Defense Technologies, MPIF, Princeton, 1990.
[20] MMPDS-07, Metallic Materials Properties Development and Standardization. Federal Aviation Administration, Washington DC, 2012.
[21] S.M. El-Soudani, K.-O. Yu, E.M. Crist, F. Sun, M.B. Campbell, T.S. Esposito, J.J. Phillips, V. Moxson, V.A. Duz, Optimization of blended-elemental powder-based titanium alloy powder based extrusions for aerospace applications, Met. Trans. A. 44A (2013) 899–910.
[22] V.V. Joshi, C. Lavender, V. Moxon, V. Duz, E. Nyberg, K.S. Weil, Development of Ti-6Al-4V and Ti-1Al-8V-5Fe alloys using low cost TiH_2 powder feedstock, J. Mater. Eng. Perf. 22 (4) (2013) 995–1003.
[23] The Printed World, The Economist, February 10, 2011.
[24] J.C. Withers, et al. in: M.A. Imam, F.H. (Sam) Froes, R.G. Ramana (Eds.), There is low cost titanium components today in cost affordable titanium IV, Trans Tech Pubs, Switzerland, 2013, pp. 11–15.

Powder metallurgy titanium aluminide alloys

27

Bin Liu, Yong Liu
State Key Lab of Powder Metallurgy, Central South University, Changsha, P.R. China

27.1 Introduction

Titanium aluminide (TiAl)-based alloys have attracted much attention for high-temperature aerospace and automobile applications over the last two decades because of their attractive properties, such as low density, high strength, high stiffness, and good corrosion, creep, and oxidation resistance [1–4]. Because of their high specific strength, TiAl-based alloys have the potential to increase the thrust-to-weight ratio of aircraft engines. They are particularly suited to the applications of low-pressure turbine blades and high-pressure compressor blades, which are traditionally made up of Ni-based superalloy (nearly twice as heavy as TiAl-based alloys). Recently, General Electric has announced that γ-TiAl low-pressure turbine blades made by Precision Castparts Corp. have been used in GEnx engines, which equipped Boeing 787 and Boeing 747-8 aircraft [5]. This is the first large-scale use of TiAl-based alloys on a commercial jet engine. In the automobile area, commercial use of TiAl-based alloys was realized in high-performance turbochargers and exhaust valves for Formula One and other sports cars [6,7]. In addition, TiAl-based alloys have found applications in military aviation and nuclear industry [8].

However, because of the long range order of intermetallics, TiAl-based alloys lack room temperature ductility and fracture toughness, which make their processing and machining difficult, thereby restraining their wider applications [9]. PM offers the potential for minimizing many of the problems associated with large ingot production and hot working, as well as reducing the overall cost of the final TiAl components. These problems include center-line porosity, chemical inhomogeneity, and regions of varying densities and microstructure [10–12]. With regard to γ-TiAl-based alloys, two PM approaches have been reported, namely, PA and elemental powder metallurgy. The PA powder approach relies on relatively expensive powder, but the properties of the processed material are generally far superior to those made using the inexpensive elemental powder technique. Additionally, the microstructures are chemically homogeneous. This chapter will focus on the PA PM approach of TiAl-based alloys, including powder preparation, compaction, hot deformation, and mechanical properties.

27.2 Preparation of PA TiAl powder

PA TiAl powder can be fabricated by various atomization techniques, including plasma rotating-electrode process (PREP) [13–15], gas atomization (GA) [16,17], and rotating disc atomization [18]. Each powder production technique has advantages and

Titanium Powder Metallurgy. http://dx.doi.org/10.1016/B978-0-12-800054-0.00027-7

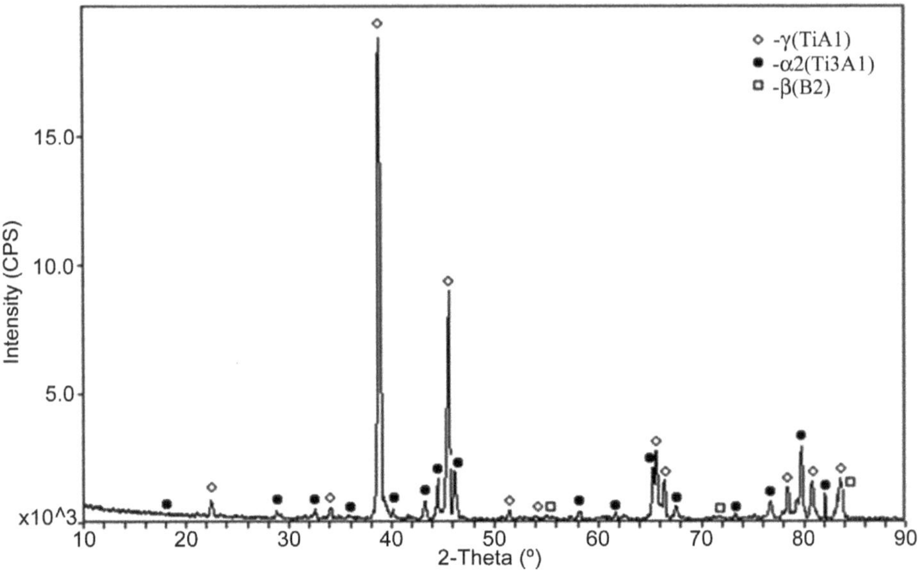

Figure 27.1 X-ray diffraction patterns of prealloyed Ti-47Al-2Nb-2Cr powder.

disadvantages. Entrapped gas pores can form within some gas-atomized TiAl powder particles during GA and rotating disc atomization, which will significantly affect the mechanical properties, such as fatigue and superplastic properties. In contrast, TiAl powder made by the PREP contains no internal porosity. Additionally, compared with GA, PREP is capable of producing highly spherical powder with few satellites. Such powder particles are desirable because they show the best flow behavior and highest packing densities. Therefore, PREP was selected for preparing the PA TiAl powder discussed in this chapter. Specific information of PREP can be found in Chapter 2 of this book.

The nominal composition of the TiAl powder is Ti-47Al-2Nb-2Cr (at%). The average particle size is about 90 μm. Figure 27.1 shows the X-ray diffraction patterns of the PREP-TiAl powder. The powder contains α_2 phase, γ phase, and a small amount of B2 phase, which results from the high cooling rate during atomization.

Figure 27.2 shows the interstitial (O, H, and N) content in the PREP-TiAl powder with respect to different particle sizes. The O content increases with decreasing particle size while the N and H contents remain unchanged. When the particles size is coarser than 150 μm, the O content is 600–700 wt ppm, while when the particle size is smaller than 46 μm, the O content increases to 1000 wt ppm. The average contents of O, H, and N are 800, 35, and 60 wt ppm, respectively. The powder is suited to producing high-purity PM TiAl components with proper protections during the subsequent processing. Figure 27.3 shows the distribution of interstitial elements in a PA powder particle. The interstitial content at the powder surface is much higher than that inside the powder because of reactions between the powder surface and the environment atmosphere. Oxygen has significant influences on the mechanical properties

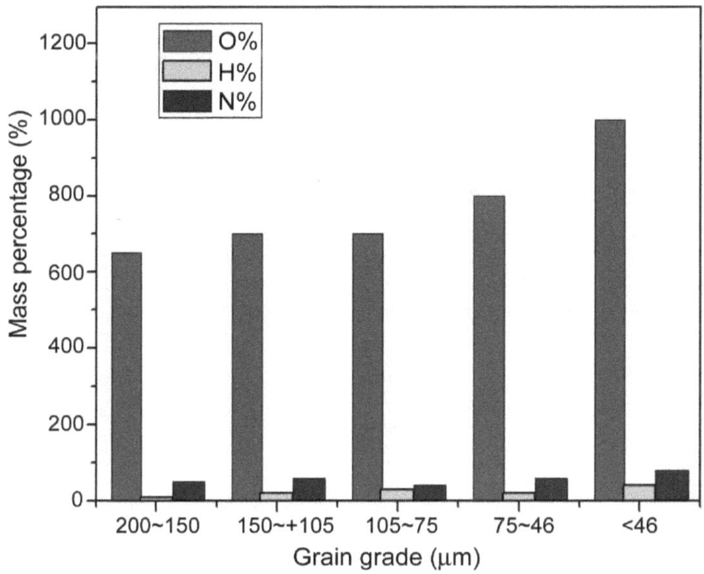

Figure 27.2 Contents of interstitial elements (O, H, and N) in the Ti-47Al-2Nb-2Cr powder with different particle sizes.

of PM TiAl-based alloys. Tönnes et al. [19] have shown that oxygen levels between 700 and 1600 wt ppm have little effect on the tensile strength of as–hot isostatically pressed (HIPed) Ti-48Al-2Cr-2Nb alloys made from PA powder. However, the room temperature tensile ductility of a duplex microstructure decreased from 2.1% to 0.5%, and that of a fully lamellar microstructure decreased from 0.5% to 0.2% when increasing the oxygen level from 1050 to 1600 wt ppm. It was thus proposed that the oxygen level should be limited to less than 1050 wt ppm in order to ensure adequate ductility.

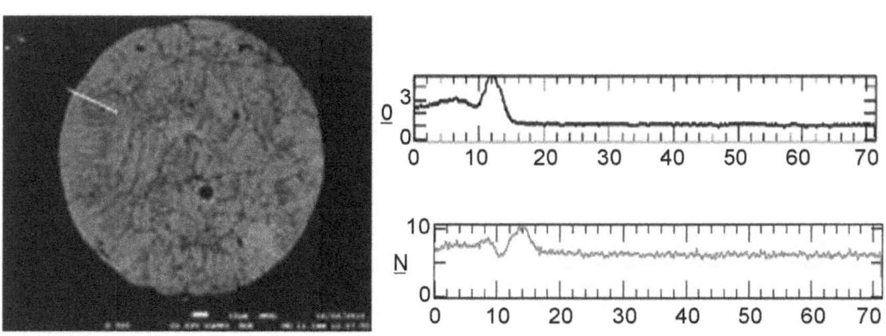

Figure 27.3 Interstitial elements distribution (O and N) in the prealloyed Ti-47Al-2Nb-2Cr powder.

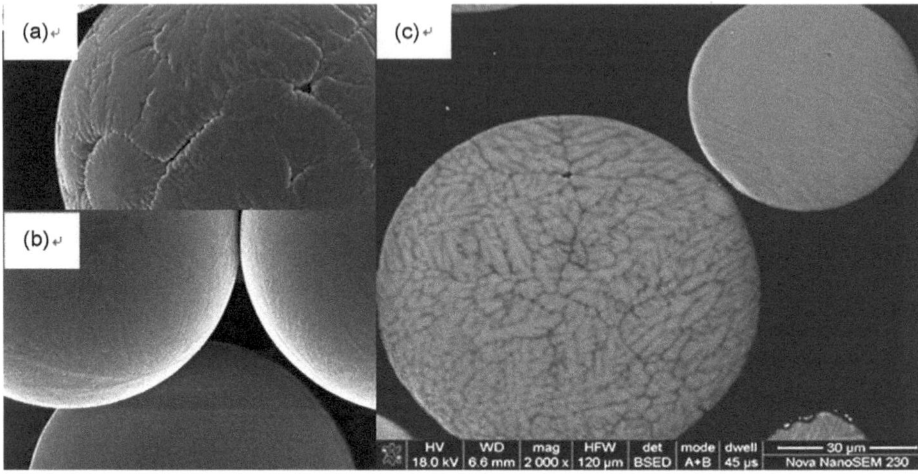

Figure 27.4 Images of Ti-47Al-2Nb-2Cr powder with particle size of (a) 80 μm and (b) 35 μm made by PREP under helium gas; (c) sections of the powder.

Figure 27.4 shows the SEM images of PA Ti-47Al-2Nb-2Cr powder. Two different types of particle surface morphologies are observed, namely, dendritic and featureless. Coarse particles with diameters more than 105 μm exhibit dendritic structures, suggesting that the disordered hexagonal α phase solidified as the primary phase. Small particles with diameters less than 70 μm show a featureless surface. Depending on the composition, about 20–35% of the powder particles exhibit a featureless surface. Examination of the microstructure of these particles indicates that no segregation or second phases are present, indicating that the undercooling might have been large enough to suppress the diffusion during solidification. It was speculated that these particles in fact consisted of disordered α, and not martensitic α′, as no characteristic β dendritic structure was observed. Hence, fine PREP-TiAl powder particles show nonequilibrium microstructures [20–22]. The relationship between the cooling rates and particle size can be described by [23]:

$$\upsilon_c = 6h\left[\frac{T - T_A}{\rho_1 D C_1}\right] \tag{27.1}$$

where v_c is the cooling rate, h is the heat transfer coefficient, T is the droplet temperature (K), T_A is the atmosphere temperature (K), ρ_1 is the density (g/cm^3), D is the particle size (μm) and C_1 is the specific heat (J/g·K). The cooling rate for the PREP powder prepared was calculated to be 2.83×10^4–6.6×10^5 K/s. Figure 27.5 gives the relationship between the particle size and cooling rate. It has been reported that the cooling rates of PREP IMI-829 titanium alloy particles ranging from 420 to 37 μm in diameter lie in 10^4–10^6 K/s [24]. Cooling rates as high as 10^6 K/s have also been reported for PREP atomized (under helium) TiAl-based alloy powder [25]. Detailed

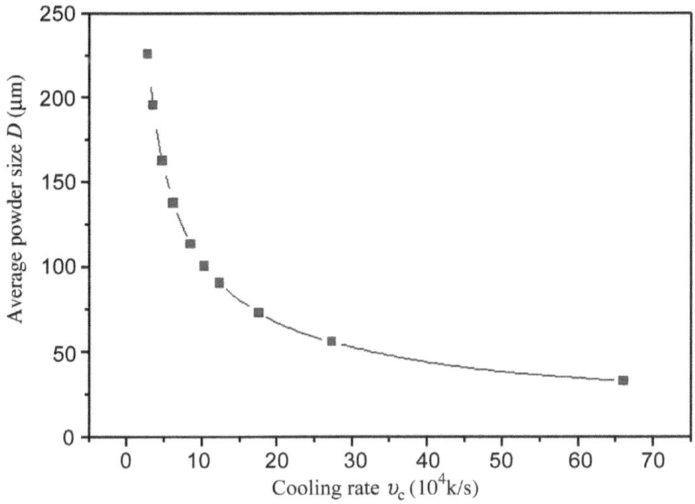

Figure 27.5 Dependence of the cooling rate on the particle size of the powder.

discussion concerning the effect of rapid cooling conditions on the microstructure of TiAl powder can be found elsewhere [26,27].

27.3 Consolidation of TiAl powder

HIP of powder compact is a popular method of producing TiAl billets [20,28–30]. Figure 27.6 shows Ti-47Al-2Nb-2Cr billet samples HIPed at 1220°C and 140 MPa for 4 h. The oxygen content of the as-HIPed compact is about 850 wt ppm, slightly higher than that of the PA powder. Figure 27.7 shows the SEM and EBSD microstructures of the as-HIPed Ti-47Al-2Nb-2Cr alloy. It is a fine $\alpha_2 + \gamma$ microstructure with the average grain size of around 15 μm and α_2 volume fraction of about 12%. The EBSD IPF map (Figure 27.7c) indicates that there is no apparent texture in the as-HIPed Ti-47Al-2Nb-2Cr alloy. The microstructure is fine-grained and shows an improved homogeneity in terms of grain size, phase distribution, and distribution of alloying elements compared to that of a traditional ingot. In this alloy, microstructural defects, such as porosity, coarse lamellar particles, and ceramic inclusions, are limited. Nevertheless, the microstructural defects sometimes still exist in as-HIPed TiAl-based alloys [31–33], which may have a negative impact on mechanical properties. During thermomechanical processing, for example, hot-forging and rolling, these microstructural defects can be eliminated, and no apparent degrading effect on the mechanical properties (e.g., tensile ductility at room temperature) has been found [31].

The pressure during HIPing may affect both the microstructure and phase constitution. In as-HIPed TiAl-based alloys, the volume fraction of the α_2 phase is clearly

Figure 27.6 Ti-47Al-2Nb-2Cr alloy compacts HIPed at 1220°C and 140 MPa for 4 h. The dimensions of the (a) cylinder and (b) rectangle are *d*30 mm × 70 mm and 600 mm × 400 mm ×30 mm, respectively.

lower than that of pressure-free sintered alloys. In an attempt to explain a lower than expected volume fraction of the α_2 phase in HIPed TiAl materials, Zhang et al. [34] speculated that under high pressure, α transus (T_α) may be increased and the kinetics of the γ to α transformation is much slower than that under atmospheric pressure. Huang et al. [35] have shown that the volume fraction of γ in as-HIPed TiAl-based alloys is higher than expected, which results from its lower atomic volume compared to the α_2 phase. The powder fraction used can also affect the microstructure. It is observed that HIPed TiAl compacts made from fine powder contained more α_2 than those made from coarser powder [34].

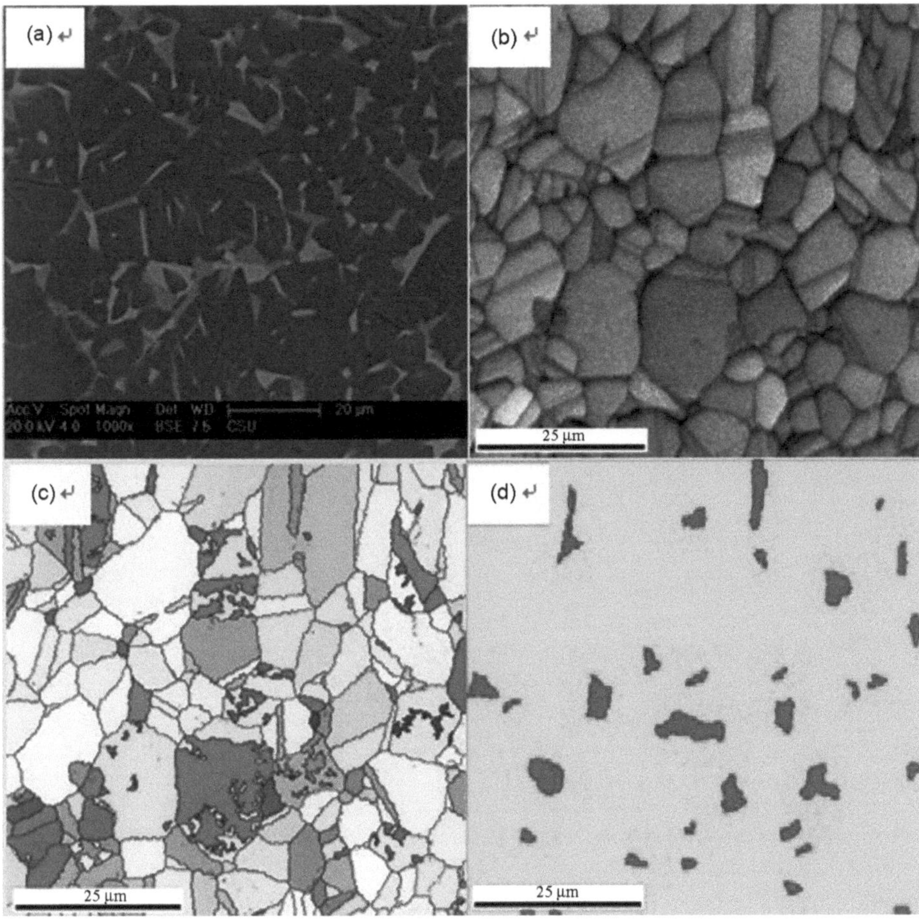

Figure 27.7 Microstructures of the Ti-47Al-2Cr-2Nb alloy HIPed at 1220°C and 140 MPa: (a) SEM taken in the backscattering electron mode; (b) image quality map; (c) orientation imaging (IPF) map; (d) phase map (red-α_2 phase, green-γ phase).

27.4 Hot deformation of PM TiAl-based alloys

The true strain–true stress (σ–ε) curves of PM Ti-47Al-2Nb-2Cr alloy deformed at 1000–1150°C to a 50% reduction with different strain rates are shown in Figure 27.8. All the curves exhibit a peak flow stress at a relatively low strain (<0.2%), followed by flow softening to a steady-state stress region. Such behaviors are believed to be typical for dynamic recrystallization described by Sakai et al. [36]. In addition, the flow stress decreases noticeably with the increase of temperature and the decrease of strain rate. These results indicate that the alloy is sensitive to the deformation temperature and strain rate, especially the latter. Remarkably, nearly all the deformed

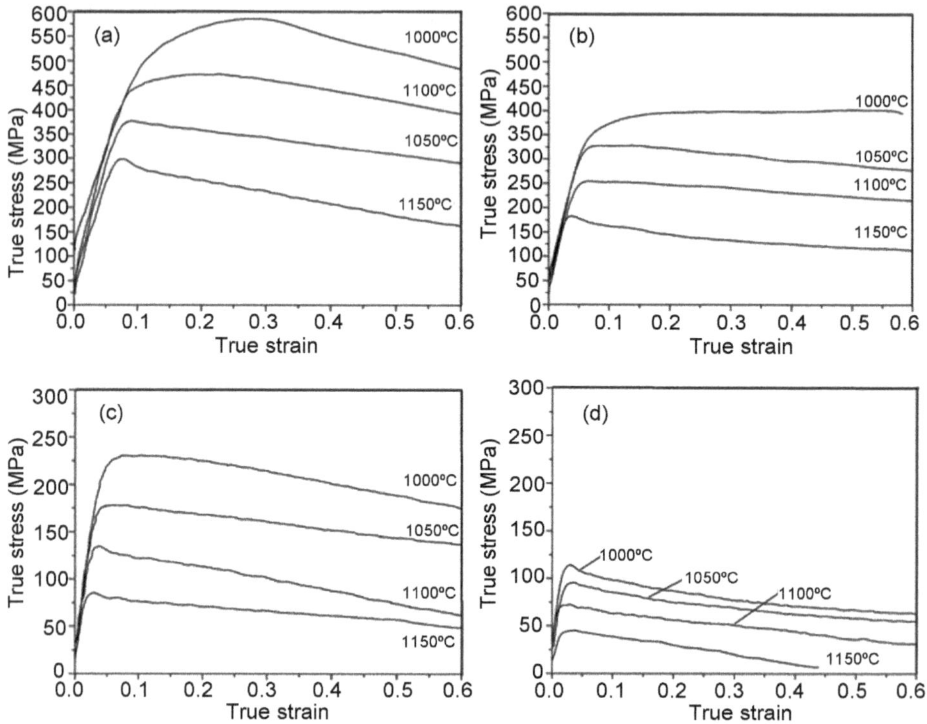

Figure 27.8 Flow curves of P/M Ti-47Al-2Nb-2Cr alloy deformed at different temperatures and strain rates. (a) 0.001 s^{-1}; (b) 0.01 s^{-1}; (c) 0.1 s^{-1}; (d) 1 s^{-1}. These curves were determined using cylindrical specimens of 10 mm in diameter and 15 mm in height.

samples appear to be sound and did not show any penetrating cracks on the surface, indicating that the PM Ti-47Al-2Nb-2Cr alloy prepared with PREP powder has a good deformability.

The flow stress data were analyzed using the processing map technique to evaluate the constitutive behavior during hot deformation. The power dissipation efficiency η in the processing map is given by [37]:

$$\eta = \frac{2m}{(m+1)} \tag{27.2}$$

where m is the strain rate–sensitive coefficient, which represents the pattern in which the input power is dissipated by the plastic deformation through microstructural evolution rather than by heat. The three-dimensional variations in η with temperature and strain rate constitute a power dissipation map that identifies the characteristic domains of various deformation and damage processes. Simultaneously, the microstructural instability regimes during the plastic flow are evaluated using Zilgler's criteria [37]:

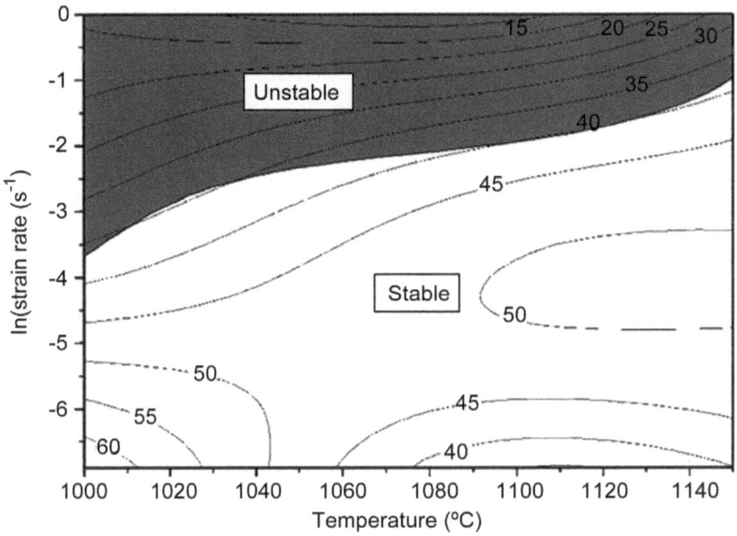

Figure 27.9 Processing map of P/M Ti-47Al-2Nb-2Cr alloy at a temperature range of 1000–1150°C and strain rate of 0.001–1 s^{-1}.

$$\xi = \frac{\partial \log[m/(m+1)]}{\partial \log \dot{\varepsilon}} + m \leq 0 \tag{27.3}$$

A detailed introduction to processing map can be obtained from Ref. [37]. The processing map of the PM Ti-47Al-2Nb-2Cr alloy obtained at a strain of 0.7 (at steady-state flow conditions) is shown in Figure 27.9. An instability map constructed with Equation (27.3) is also shown in Figure 27.9. From this equation, an instability regime occurs at strain rates higher than approximately 0.01 s^{-1}, and this is labeled as an unstable region in Figure 27.9. Thus, the high strain rate regime is not favorable for hot working.

The map exhibits a low efficiency (less than 25%) domain over the temperature range of 1000–1100°C and the strain rate range of 0.5–1 s^{-1}. In the samples deformed at 1000°C and 1 s^{-1}, cracks can be observed (Figure 27.10a). This is because stress concentration cannot be relieved or accommodated by the dislocation motion, but by nucleation of cracks, and this area corresponds to unstable area. Two domains with power efficiency peaks of 63% (at 1000°C and 0.001 s^{-1}) and 50% (at 1150°C and 0.01 s^{-1}) appear in the processing map, which show fine microstructures (Figure 27.10b and c). These domains indicate a feature of dynamic recrystallization and thus corresponds to the optimum hot working conditions. In addition, the efficiency of dissipation at 1150°C and 0.001 s^{-1} is relatively low and decreases sharply with the strain. It is suggested that the decrease in efficiency of dissipation in this region is caused by abnormal grain growth, which is due to long-time exposure at high temperatures. This observation is consistent with the deformation microstructure shown in Figure 27.10d.

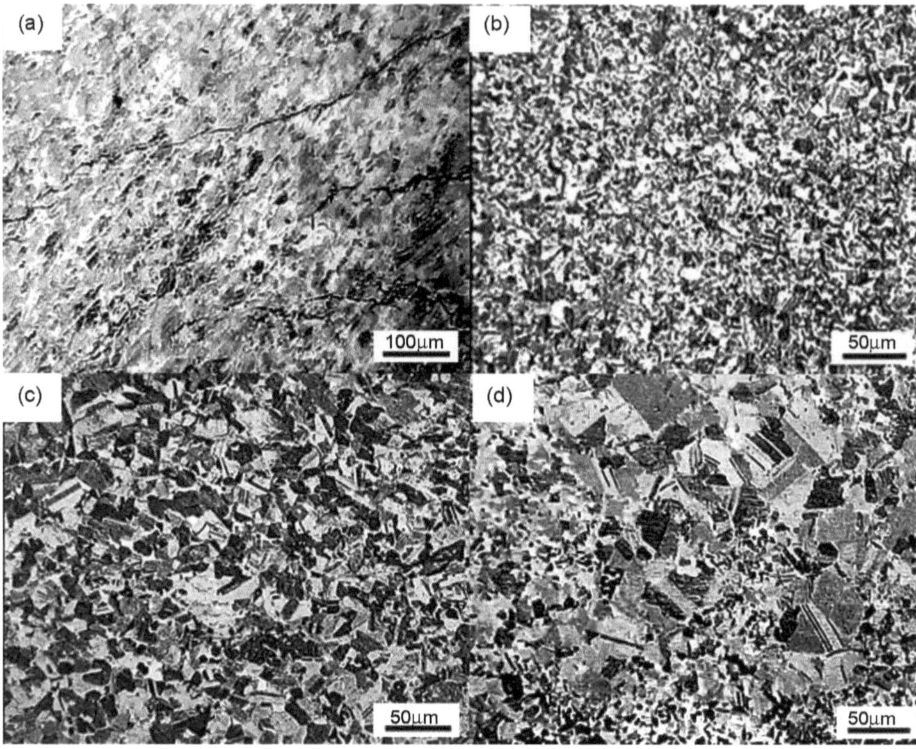

Figure 27.10 Microstructures of the P/M Ti-47Al-2Nb-2Cr alloy deformed at conditions of (light micrographs): (a) 1000°C, 1 s^{-1}; (b) 1000°C, 0.001 s^{-1}; (c) 1150°C, 0.01 s^{-1}; and (d) 1150°C, 0.001 s^{-1}.

Figure 27.11 shows the TEM microstructure of PM Ti-47Al-2Nb-2Cr alloy deformed at 1000°C and 0.001 s^{-1}. Dense dislocation structures, dislocation walls, and subgrains are observed. These observations suggest that the recrystallization of the present TiAl material is not a typical discontinuous dynamic recrystallization process, in which the nucleation and growth of new recrystallized grains occur during hot deformation. Instead, it is a continuous recrystallization process, which often occurs in high stacking fault energy (SFE) materials such as Al alloys and Ti alloys. In continuous dynamic recrystallization, as the strain increases, dislocations produced by strain hardening accumulate progressively in subgrain boundaries, leading to an increase in their misorientation angle and then the formation of high angle grain boundaries when a critical value of the misorientation angle is reached. Finally, the microstructure is an intermediate structure between a subgrain and a grain structure, like the microstructure shown in Figure 27.11. In continuous dynamic recrystallization, the true stress–true strain curves exhibit a single smooth maximum, followed by a slow but significant softening stage (Figure 27.8).

The processing map defines the appropriate deformation parameters, which are located in the low strain rate region. However, in industrial production, higher strain

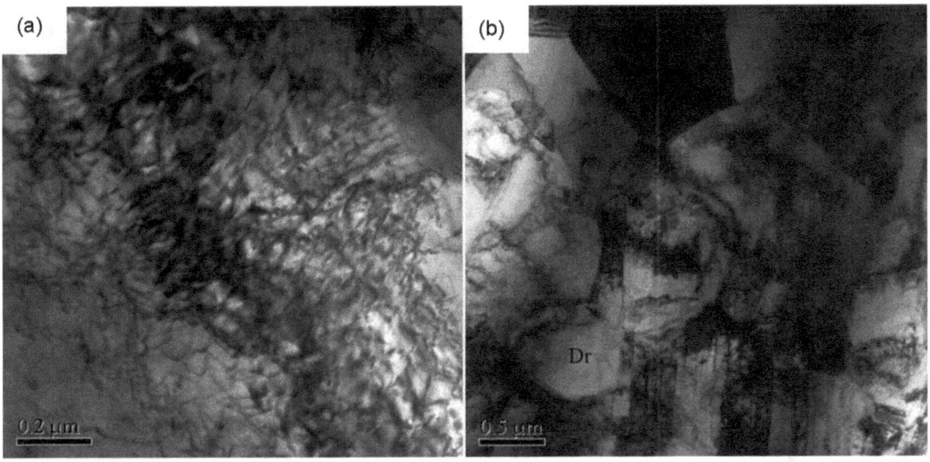

Figure 27.11 TEM microstructures of the PM Ti-47Al-2Nb-2Cr alloys deformed at 1000°C, 0.001 s^{-1}: (a) dislocations; (b) dynamic recrystallized grains.

rates are usually preferred for cost reductions. Therefore, a higher strain rate of 0.1 s^{-1} was chosen for industrial deformation, where the deformation condition is located in the stable-flow region in the processing map. Canned billets were processed using the multipass quasi-isothermal deformation technology. Figure 27.12 shows a PM Ti-47Al-2Nb-2Cr deformed pancake with dimensions of 500 mm × 30 mm and a sheet with dimensions of 440 mm × 300 mm × 2 mm, after a total deformation ratio of over 80% (strain = 1.61) for both products. Sound deformation was achieved free of surface cracks. However, areas of microporosity were observed that limit the strength of the material. Similar results have been reported in Ref. [36]. Hence, it is essential that the process of powder synthesis should be kept very clean to inhibit the presence of impurities.

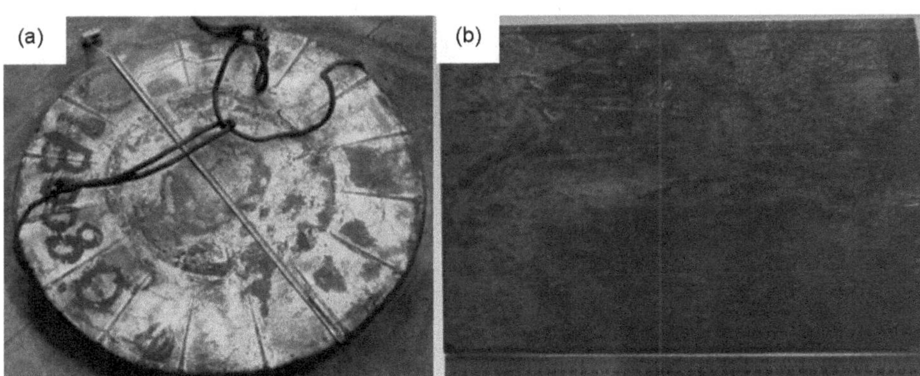

Figure 27.12 Deformed specimens of PM Ti-47Al-2Nb-2Cr alloy: (a) pancake with dimension of *d*500 mm × 30 mm, and (b) sheet with dimension of 440 mm × 300 mm × 2 mm.

The deformation behaviors of the PM TiAl-based alloys are different from those of the IM TiAl-based alloys. Semiatin et al. [38] compared the hot compression behaviors of cast ingot and as-HIPed powder compact of Ti-48Al-2Cr-2Nb and found that the peak flow stress was much lower in the PM material because of its fine grain size than in the IM material. This suggests that the PM route may facilitate a reduced load operation during isothermal forging. Similarly, Fuchs et al. [39] studied IM and PM Ti-48Al-2Nb-2Cr alloys and reached similar conclusions. Liu et al. and Beddoes et al. [40–43] studied the high-temperature compression behaviors of PM TiAl-based alloys and found that the PM-based alloys can be deformed over a wide range of temperatures and strain rates. With regard to deformation at lower temperatures, successful forging at 850°C to an 80% reduction has been reported for a PM TiAl-based alloy [12].

PM route also offers an attractive alternative for sheet rolling. HIPed TiAl-based alloys can be rolled to sheets without prior homogenizing heat treatments and forging. Bartolotta et al. [8] compared the TiAl-based alloy sheets produced by PM and IM routes. The yield from the IM process was found to be much lower than that from the PM process, and the PM route was more cost effective than the IM route. Furthermore, the PM route enables production of large TiAl-based alloy sheets because HIPing of material is possible in large dimensions. So, PM TiAl-based alloys seem to be suitable for hot working, such as pack rolling and closed die forging.

27.5 Properties of PM TiAl-based alloys

PM route offers attractive properties because of the fine-grained microstructure and improved homogeneity in terms of grain size, phase distribution, and distribution of alloying elements. For HIPed PM TiAl-based alloys, their mechanical properties are usually comparable to those of the cast ingots. Tönnes et al. [19] obtained a yield strength of around 370 MPa, tensile strength of 460 MPa, and plastic elongation of around 2% for a HIPed PM Ti-48Al-2Cr-2Nb alloy with a microstructure containing 30–40% lamellar colonies with a size of 100 μm. These properties compare favorably with those given by Austin et al. [44] for a cast Ti-48Al-2Cr-2Nb alloy.

For hot-worked PM TiAl-based alloys, their mechanical properties are usually close to or even higher than those of the IM TiAl-based alloys. Semiatin et al. [45] reported that HIPed + hot-worked PM TiAl products can exhibit mechanical properties superior to those of IM-based counterparts. Thomas et al. [46] and Malaplate et al. [47] found that the PM approach led to more homogeneous microstructures and higher strength and ductility than did the IM approach. In addition, the property variability is smaller for PM TiAl-based alloys because of improved microstructural homogeneity, which can be controlled through adjusting particle size fractions with a narrower size range. Liu et al. [48] reported yield strength values of around 971 MPa and tensile elongation of 1.4% for an extruded PM Ti-47Al-2Cr-2Nb alloy, the highest yield strength value for Ti-47Al-2Cr-2Nb alloys [49–52]. The outstanding properties were attributed to the extremely fine colony size and the fine interlamellar spacing (0.1 μm) combined with the unique ultrafine morphology of the α_2 lamellae. Table 27.1 summarizes

Table 27.1 Room temperature properties of PM TiAl-based alloys

Alloy	Processing	Microstructure	$\sigma_{0.2}$ (MPa)	σ_{max} (MPa)	Elongation (%)	References
Ti-47Al-4(Cr, Mn, Nb, Si, B)	HIP	NG	–	550	–	[12]
	MIM + HIP	DP	410	430	0.6	[53]
Ti-45Al-2Cr-Nb-1B-0.5Ta	Milling + HIP	Nano-grained	200–300	200–300	–	[54]
Ti-48Al-2W	HIP	DP	376	436	1.4	[55]
		FL	346	384	0.5	[55]
Ti-48Al-2Mn-2Nb-0.8C	HIP	NG	419	425	0.2	[56]
	HIP	FL	368	422	0.8	[56]
Ti-48Al-2Cr-2Nb	HIP	DP	370	460	2	[19]
	HIP	NG	460	550	1.8	[19]
Ti-46Al-2Cr-2Nb	HIP	NG	715	715	0.2	[28]
	HIP	FL	445	590	1.7	[28]
Ti-47Al-2Cr-2Nb	HIP + HP	FL	340	380	0.5	[49]
	HIP + HP	NFL	510	597	2.9	[49]
	HIP	NG	445	465	2	[50]
	Forged, 1175°C, 75%, HT 1350°C	FL	380	433	1.6	[51]
	Extruded, 1300°C, 16:1, HT 1350°C	NFL	486	512	1.2	[52]
	Extruded 1350°C, 16:1	NFL	582	793	3.8	[52]
	Extruded in α field, 16:1	FL	971	1005	1.4	[48]
Ti-47Al-4(Cr, Mn, Nb, Si, B)	HIP + deformed	Fine-grained	–	800	–	[12]
Ti-46Al-5Nb	HIP + rolled	DP	800	850	–	[57]
Ti-46Al-5Nb-0.5C	HIP + rolled	DP	1070	1100	–	[57]

Note: HT, heat treatment; DP, duplex; NG, near gamma; NFL, near full lamellar; FL, full lamellar.

literature data on the strength and ductility properties of PM TiAl-based alloys along with processing methods and microstructures.

Distinct superplastic behaviors of TiAl-based alloys processed via the PM routes have been reported in several studies [29,58–60]. Wegmann et al. [29] reported that, because of the good deformability, PM TiAl-based alloys can be isothermally forged at a relatively low temperature of 850°C, which result in an extremely fine near-γ microstructure with an average grain size of 0.9 μm. The forged material exhibits excellent superplastic properties at temperatures as low as 900–1100°C, and quite high deformation rates in the range of 0.0001–0.001 s^{-1}. At 1000°C with a nominal strain rate of 0.0015 s^{-1}, a tensile elongation of 920% was achieved with a low level of cavitation, which could be attributed to improved accommodation processes because of the very fine grain size. The excellent superplastic properties at low temperatures are important because superplastic forming facilities that are in use for Ti alloys can also be used for superplastic forming of TiAl-based alloys.

27.6 Summary

In recent years, considerable progress has been achieved in the development of PM TiAl alloys, which now exhibit significantly increased room and elevated temperature mechanical properties. The room mechanical properties of hot-worked PM TiAl alloys can be comparable to those of ingot-based TiAl alloys when processing is optimized. The distinct advantage of the PM route over the IM route is the much higher chemical and macrostructural homogeneity, which leads to better property reproducibility and thus improved component consistency and reliability. Through near-net processing, the PM approach also offers the advantage of increasing the material yield of the final products. Despite recent acceptance by the aerospace and automobile industries, wider industrial applications of TiAl-based alloys are still elusive largely because of their low ductility and workability. Development of advanced PM manufacturing techniques, which can negotiate the low workability of TiAl-based alloys, may act as an important avenue to widen the applications of these alloys.

Acknowledgments

The authors would like to thank financial supports of National Key Fundamental Research and Development Project of China (2011CB605505), National Natural Science Foundation of China (51301203), and fundamental research funds of Central South University.

References

[1] J.J. Bertin, R.M. Cummings, Fifty years of hypersonics: where we've been, where we're going, Progr. Aerospace Sci. 39 (2003) 511.

[2] G. Das, H. Kestler, H. Clemens, P.A. Bartolotta, Sheet gamma TiAl: status and opportunities, J. Mater. (2004) 42.

[3] W. Voice, The future of gamma-titanium aluminides by Rolls Royce, Aircraft Eng. Aerospace Technol. 71 (1999) 337.
[4] F.H. Froes, C. Suryanarayana, D. Eliezer, Review synthesis, properties and applications of titanium aluminides, J. Mater. Sci. 27 (1992) 5113.
[5] GE Aviation. GE – Aviation: GEnx, GE [Online]. http://www.geae.com/engines/commercial/genx/, 2007.
[6] D. Dimiduk, P.L. Martin, R. Dutton, Accelerated insertion of materials: the challenges of gamma alloys are really not unique, in: Y.W. Kim, H. Clemens, A.H. Rosenberger (Eds.), Gamma Titanium Aluminides 2003, TMS, San Diego, 2003.
[7] A.W. Sommer, G.C. Keijzers, Gamma TiAl and the engine exhaust valve, in: Y.W. Kim, H. Clemens, A.H. Rosenberger (Eds.), Gamma Titanium Aluminides 2003, TMS, San Diego, 2003.
[8] P. Bartolotta, D.L. Krause, Titanium aluminide applications in the high speed civil transport, in: Proceedings of International Symposium on Gamma Titanium Aluminides, 1999, p. 1.
[9] G. Lutjering, J. Williams, Titanium: Engineering Materials and Processes, Springer, New York, 2007.
[10] J.H. Moll, B.J. McTierman, PM TiAl alloys: the sky's the limit, MPR 1 (2000) 18–22.
[11] Y. Liu, B.Y. Huang, K.C. Zhou, Recent progress in PM γ-TiAl base alloy, J. Aeronaut. Mater. 24 (2001) 50.
[12] R. Gerling, H. Clemens, F.P. Schimansky, Powder metallurgical processing of intermetallic gamma titanium aluminides, Adv. Eng. Mater. 6 (2004) 23.
[13] M. Tokizane, T. Fukami, T. Inaba, Structure and mechanical properties of the hot pressed compact of Ti-rich TiAl powder produced by the plasma rotating electrode process, ISIJ Int. 31 (1991) 1088.
[14] B. Champagne, R. Angers, Fabrication of powder by the rotating electrode process, Int. J. Powder Metall. Powder Technol. 16 (1980) 359.
[15] X.Z. Ma, J. Shen, X.M. Qi, J. Jia, Cooling rates of PREP Ti48Al alloy, J. Mater. Sci. Technol. 17 (1) (2001) 91.
[16] P.I. Gouma, N. Saunders, M.H. Loretto, Microstructural evolution and microsegregation of gas atomized powders of a TiAl based alloy, Mater. Sci. Technol. 12 (1996) 823.
[17] V.M.A. Leo, G.R. Ramana, Processes for production of high-purity metal powders, JOM (2003) 314.
[18] D.J. Larson, C.T. Liu, M.K. Miller, The alloying effects of tantalum on the microstructure of an $\alpha_2 + \gamma$ titanium aluminide, Mater. Sci. Eng. A 270 (1999) 1.
[19] C. Tönnes, J. Rösler, R. Baumann, M. Thumann, Influence of microstructure on the tensile and creep properties of titanium aluminides processed by powder metallurgy, in: R. Darolia, J.J. Lewandowski, C.T. Liu, P.L. Martin, D.B. Miracle, M.V. Nathal (Eds.), Structural Intermetallics, TMS, Warrendale, PA, 1993, pp. 241.
[20] G.E. Fuchs, S.Z. Hayden, Microstructural evolution of as-solidified and heat-treated γ-TiAl based powders, Mater. Sci. Eng. A 152 (1992) 277.
[21] X.Z. Cai, X. Zhang, D. Eylon, Features of the solidified microstructure of PREP TiAl base alloy powders, Rare Metal Mater. Eng. 23 (1994) 41.
[22] J. Shen, X.Z. Ma, G. Wang, J. Jia, Calculation of cooling rates of rapidly-solidified TiAl alloy, Rare Metal Mater. Eng. 30 (2001) 273.
[23] H. Jones, Rapid Solidification of Metals and Alloys, Monograph No. 8, The Institution of Metallurgists, London, (1982).
[24] N.R. Osborne, D. Eylon, F.H. Froes, , in: T.G. Gasbarre, W.F. Jandeska (Eds.), in: Advances in Powder Metallurgy, vol. 3, MPIF, Princeton, New Jersey, 1989, pp. 213.

[25] X.Z. Cai, D. Eylon, Features of the solidified microstructure of PREP TiAl base alloy powder plasma rotating electrode processing, in: P.A. Blenkinsop, W.J. Evans, H.M. Flower (Eds.), Titanium '95: Science and Technology, IOM, London, UK, 1996, pp. 455.

[26] E.L. Hall, S.C. Huang, Microstructures of rapidly-solidified binary TiAl alloys, Acta Metall. Mater. 38 (1990) 539.

[27] C. McCullough, J.J. Valencia, C.G. Levi, R. Mehrabian, Microstructural analysis of rapidly solidified Ti-Al-X powders, Mater. Sci. Eng. A 124 (1990) 83.

[28] U. Habel, B.J. McTiernan, HIP temperature and properties of a gas-atomized gamma-TiAl alloy, Intermetallics 12 (2004) 63.

[29] G. Wegmann, R. Gerling, F.-P. Schimansky, H. Clemens, A. Bartels, High temperature mechanical properties of hot isostatically pressed and forged gamma titanium aluminide alloy powder, Intermetallics 10 (2002) 511.

[30] R.J. Schaefer, G.M. Janowski, Phase transformation effects during HIP of TiAl, Acta Metall. Mater. 40 (7) (1992) 1645.

[31] H. Clemens, H. Kestler, N. Eberhardt, W Knabl, Processing of γ-TiAl based alloys on an industrial scale, in: Y.W. Kim, D.M. Dimiduk, M.H. Loretto (Eds.), Gamma Titanium Aluminides, TMS, Warrendale, PA, 1999, pp. 209.

[32] H. Clemens, Intermetallic γ-TiAl based alloy sheet materials – processing and mechanical properties, Z. Metallkd. 86 (12) (1995) 814.

[33] H. Clemens, A. Lorich, N. Eberhardt, W. Glatz, W. Knabl, H. Kestler, Technology, properties and applications of intermetallic γ-TiAl based alloys, Z. Metallkd. 90 (8) (1999) 569.

[34] G. Zhang, P.A. Blenkinsop, M.L.H. Wise, in: P.A. Blenkinsop, W.J. Evans, H.M. Flower (Eds.), Titanium '95: Science and Technology, IOM, London, UK, 1996, pp. 542.

[35] A. Huang, D. Hu, M.H. Loretto, J. Mei, X. Wu, The influence of pressure on solid-state transformations in Ti–46Al–8Nb, Scripta Mater. 56 (2007) 253.

[36] T. Sakai, J.J. Jonas, Overview no. 35-dynamic recrystallization: mechanical and microstructural considerations, Acta Metall. 32 (1984) 189.

[37] Y.V.R.K. Prasad, S. Sasidhara, Hot Working Guide: A Compendium of Processing Maps, ASM International, Metal Park, OH, USA, 1997.

[38] S.L. Semiatin, G.R. Cornis, D. Eylon, Hot-compression behavior and microstructure evolution of pre-alloyed powder compacts of a near-γ titanium aluminide alloy, Mater. Sci. Eng. A 185 (1994) 45.

[39] G.E. Fuchs, The effect of processing on the hot workability of Ti-48Al-2Nb-2Cr alloys, Metall. Mater. Trans. A 28 (1997) 2543.

[40] B. Liu, Y. Liu, W. Zhang, J.S. Huang, Hot deformation behavior of TiAl alloys prepared by blended elemental powders, Intermetallics 19 (2011) 154.

[41] J. Beddoes, L. Zhao, J.P. Immarigeon, W. Wallace, The isothermal compression response of a near γ-TiAl + W intermetallic, Mater. Sci. Eng. A 183 (1994) 211.

[42] J. Beddoes, L. Zhao, P. Au, W. Wallace, The brittle-ductile transition in HIP consolidated near γ-TiAl + W and TiAl + Cr powder alloys, Mater. Sci. Eng. A (1995) A192–193 324.

[43] J. Beddoes, L. Zhao, W. Wallace, High temperature compression behaviour of near γ-titanium aluminides containing additions of chromium or tungsten, Mater. Sci. Eng. A 184 (1994) L11.

[44] C.M. Austin, T.J. Kelly, Development and implementation status of cast gamma titanium aluminide, in: R. Darolia, J.J. Lewandowski, C.T. Liu, P.L. Martin, D.B. Miracle, M.V. Nathal (Eds.), Structural Intermetallics, TMS, Warrendale, PA, 1993, pp. 143.

[45] S.L. Semiatin, J.C. Chesnutt, C. Austin, V. Seetharaman, Processing of intermetallic alloys, in: M.V. Nathal, R. Darolia, C.T. Liu, P.L. Martin, D.B. Miracle, R. Wagner, M. Yamaguchi (Eds.), Structural Intermetallics, TMS, Warrendale, PA, 1997, pp. 273.

[46] M. Thomas, J.L. Raviart, F. Popoff, Cast and PM processing development in gamma aluminides, Intermetallics 13 (2005) 944.

[47] J. Malaplate, M. Thomas, P. Belaygue, M. Grange, A. Couret, Primary creep at 750°C in two cast and PM $Ti_{48}Al_{48}Cr_2Nb_2$ alloys, Acta Mater. 54 (2006) 601.

[48] C.T. Liu, P.J. Maziasz, D.R. Clemens, J.H. Schneibel, V.K. Sikka, T.G. Nieh, J. Wright, L.R. Walker, Room and elevated temperature mechanical properties of PM TiAl alloy Ti-47Al-2Cr-2Nb, in: Y.W. Kim, R. Wagner, M. Yamaguchi (Eds.), Gamma Titanium Aluminides, TMS, Warrendale, PA, 1995, pp. 679.

[49] J.H. Moll, E. Whitney, C.F. Yolton, U. Habel, Laser forming of gamma titanium aluminide, in: Y.W. Kim, D.M. Dimiduk, M.H. Loretto (Eds.), Gamma Titanium Aluminides, Warrendale, PA, TMS, 1999, pp. 255–273.

[50] M. Thomas, O. Berteaux, F. Popoff, M.P. Bacos, A. Morel, B. Passilly, V. Ji, Effects of exposure at 700°C on RT tensile properties in a PM γ-TiAl alloy, Intermetallics 14 (2006) 1143.

[51] G.E. Fuchs, in: Y.W. Kim, R. Wagner, M. Yamaguchi (Eds.), Gamma Titanium Aluminides, Warrendale, PA, TMS, 1995, pp. 563.

[52] G.E. Fuchs, Supertransus processing of TiAl-based alloys, Metall. Mater. Trans. 29A (1998) 27.

[53] R. Gerling, F.P. Schimansky, Prospects for metal injection moulding using a gamma titanium aluminide based alloy powder, Mater. Sci. Eng. A 329 (2002) 45.

[54] H.B. Yu, D.L. Zhang, Y.Y. Chen, P. Cao, B. Gabbitas, Synthesis of an ultrafine grained TiAl based alloy by subzero temperature milling and HIP, its microstructure and mechanical properties, J. Alloys Compd. 474 (1–2) (2009) 105.

[55] L. Zhao, J. Beddoes, D. Morphy, W. Wallace, Microstructure and mechanical properties of a PM TiAl-W alloy processed by hot isostatic pressing, Mater. Sci. Eng. A 192/193 (1995) 957.

[56] P.I. Gouma, S.J. Davey, M.H. Loretto, Microstructure and mechanical properties of a TiAl-based powder alloy containing carbon, Mater. Sci. Eng. A 241 (1–2) (1998) 151.

[57] R. Gerling, F.P. Schimansky, A. Stark, A. Bartels, H. Kestler, L. Cha, C. Scheu, H. Clemens, Microstructure and mechanical properties of Ti-45Al-5Nb (0–0.5°C) sheets, Intermetallics 16 (5) (2008) 689.

[58] C. Koeppe, A. Bartels, H. Clemens, P. Schretter, W. Glatz, Optimizing the properties of TiAl sheet material for application in heat protection shields or propulsion systems, Mater. Sci. Eng. A201/201 (1995) 182.

[59] H. Clemens, H. Kestler, Processing and applications of intermetallic γ-TiAl-based alloys, Adv. Eng. Mater. 2 (9) (2000) 551.

[60] H. Clemens, I. Rumberg, P. Schretter, S. Schwantes, Characterization of Ti-48Al-2Cr sheet material, Intermetallics 2 (1994) 179.

Porous titanium structures and applications

28

H.P. Tang*, J. Wang*, Ma Qian**
*State Key Laboratory of Porous Metal Materials, Northwest Institute for Nonferrous Metal Research, Xi'an, China
**RMIT University, School of Aerospace, Mechanical and Manufacturing Engineering, Centre for Additive Manufacturing, Melbourne, Victoria, Australia

28.1 Introduction

Porous titanium, including titanium foam and cellular titanium, refers to titanium materials that contain a large number of pores. Porous titanium inherits an array of the outstanding properties of dense titanium materials such as high specific strength, stiffness, excellent corrosion resistance, and biocompatibility. In addition, porous titanium can be made to have high specific surface, high permeability, and super low density. These attributes make porous titanium a material of choice for many important applications, either structural or functional.

Titanium sponge produced by the Hunter process (invented in 1910) represents the earliest form of porous titanium materials made by humankind. However, it was not until 1948 that titanium was established as a metal of industrial importance, marked by the commercial production of titanium sponge using the Kroll process [1]. In either process, the production of titanium sponge involves light sintering of the titanium crystals reduced from titanium tetrachloride, particularly during the lengthy vacuum distillation stage.

A detailed literature survey has been made of the publications and patents available from 1910 on porous titanium, titanium foam, and cellular titanium using Scopus, Web of Science, Springer Link, Google Scholar, and Google Patent Search. Porous titanium received little attention before the industrial importance of titanium was established in 1948. An additional reason was the availability issue of titanium prior to 1948. Interestingly, in an early patent [2], the inventors suggested the use of a selective corrosion approach, referred to as dealloying today, to make a porous titanium skeleton from a Ti-Al alloy for the production of a porous titanium oxide structure by subsequent oxidation. Information on porous titanium remained scarce even over the period from 1948 to 1960. A patent filed on May 5, 1955 [3], discussed the importance of interposing a porous titanium barrier in the fused salt bath between the anode and cathode in the electrolytic production of titanium. The patent covered the fabrication of a porous "titanium" barrier by *in situ* deposition of titanium on a perforated metallic structure (e.g., nickel or stainless steel) placed in the fused salt bath between the anode and cathode.

Titanium Powder Metallurgy. http://dx.doi.org/10.1016/B978-0-12-800054-0.00028-9

It appears that research on porous titanium based on powder metallurgy (PM) routes started around 1960 in the former USSR, evidenced by several systematic studies published at that time [4–7]. The Kroll process was established in the former USSR in 1954 following the United States (1948), the United Kingdom (1951), and Japan (1952) [1]. China began small batches of production of titanium sponge in 1958, and the Northwest Institute for Nonferrous Metal Research (NIN) China started their R&D of porous titanium about a decade later. To date, porous titanium has found good commercial applications in China.

Figure 28.1 summarizes the results of the literature survey by focusing on the post-1970 period. The R&D activities on porous titanium showed a healthy pickup from the early 1990s. In particular, the last decade has shown a strong uptrend. For instance,

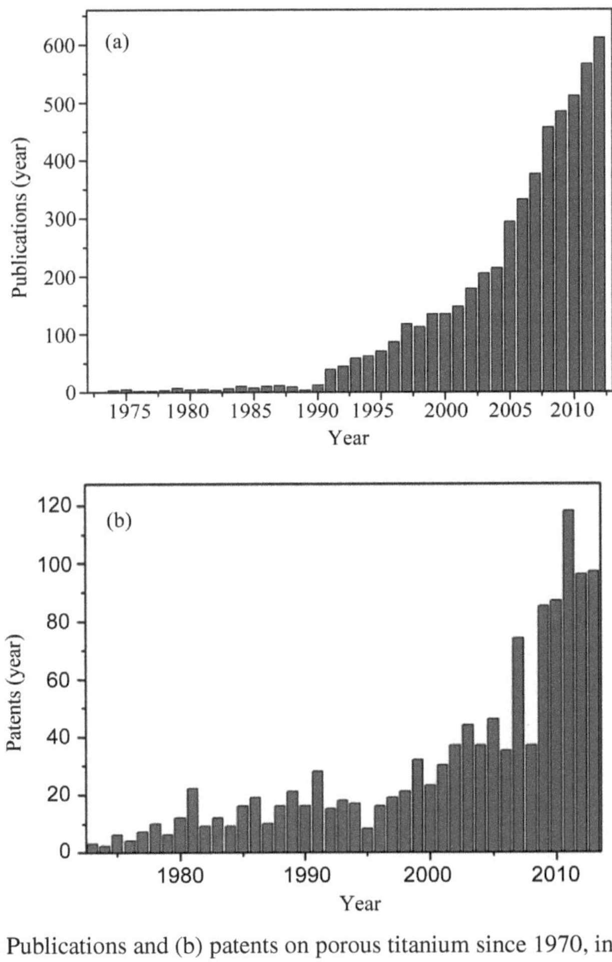

Figure 28.1 (a) Publications and (b) patents on porous titanium since 1970, including titanium foam and cellular titanium. (Sources: Web of Science for (a) and http://patft.uspto.gov, http://worldwide.espacenet.com and http://www.pss-system.gov.cn for (b).)

more than 600 research papers and 100 patent applications were documented in 2012 alone. Porous titanium has evolved into an important class of metallic materials, with new applications being introduced at a pace much than before. The rest of this chapter reviews the characteristics and properties of porous titanium materials and their commercial applications.

28.2 Porous titanium structures

Pore structure is a general term used to describe the porosity, pore size, pore size distribution, and pore morphology of a porous medium. As with other porous materials, porous titanium can have various pore structures. For instance, the porosity can be made up to 98% using hollow titanium spheres [8]; the pore size distribution can be produced to follow a normal, bimodal, or gradient distribution; and there can be open or closed pores, spherical or nonspherical pores, and smooth or rough surface pores, distributed either randomly or regularly. Discussed below are five common pore structures for porous titanium.

28.2.1 Uniform pore structures

A uniform pore structure means that the pore size distribution is limited to a narrow range following a normal distribution and the pores are randomly distributed. They can be produced via conventional PM approaches including loose powder sintering, press and sinter [9], powder rolling [10] or powder extrusion followed by sintering, using angular or spherical titanium powder, titanium fibers, or hollow titanium spheres. The porosity and pore size can be controlled by changing preform formation and sintering conditions, as well as the feedstock characteristics. In general, porous titanium sintered from powder achieves less than 60% porosity with the average pore size being around 1/4–1/3 of the starting powder particle size. When sintered with titanium fibers, the porosity can approach 95% with the pore size ranging from one hundred to several hundred micrometers. Porous titanium sintered from hollow titanium spheres can achieve more than 98% porosity, but the pore size is usually greater than 500 μm because of the difficulties of producing small hollow titanium spheres [8]. Figure 28.2a–d shows porous titanium (commercially pure [CP-Ti]) structures sintered from angular titanium powder (a), spherical titanium powder (b), titanium fibers (b), and hollow titanium spheres (d).

28.2.2 Bimodal pore structures

Porous titanium with a bimodal pore structure contains pores with two distinct size ranges, one being macro-pores (500–1000 μm) and the other being micro-pores (<100 μm) in the wall of the large pores. Figure 28.3 shows one such example. Bimodal pore structures may be preferred for implant applications as they are likely to be osteoinductive [11]. The space-holder technique has been commonly used to make porous titanium with a bimodal pore structure [12,13]. In this process, macro-pores are produced by removal of large space holder particles while micro-pores are created

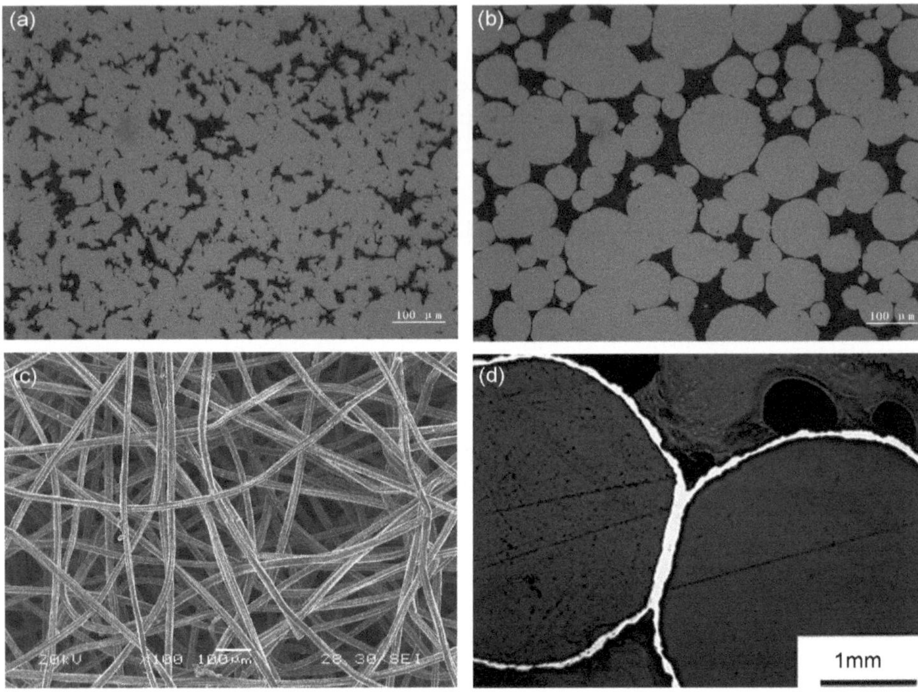

Figure 28.2 Porous titanium (CP-Ti) structures sintered from angular titanium powder (a), spherical titanium powder (b), titanium fibers (c), and hollow titanium spheres (d).

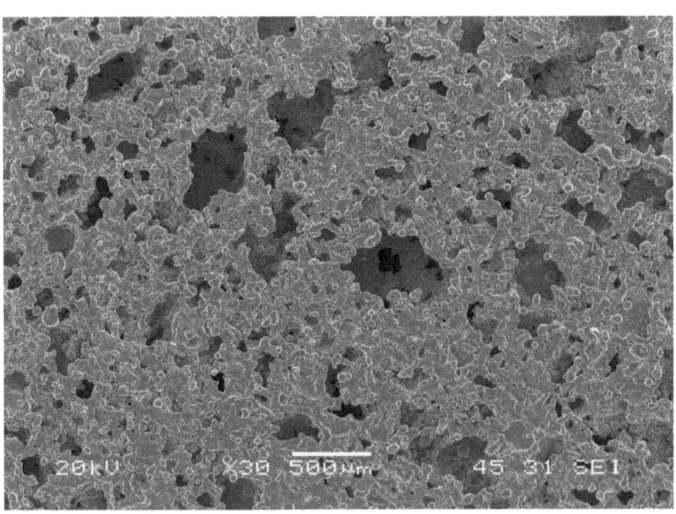

Figure 28.3 Porous titanium (CP-Ti) with a bimodal pore distribution.

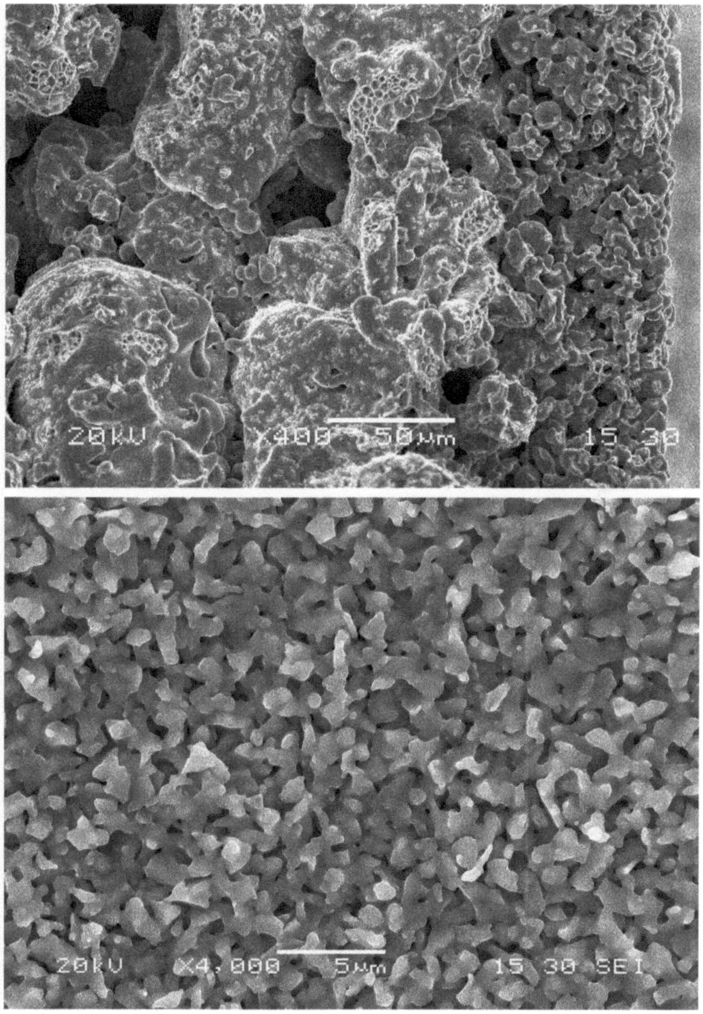

Figure 28.4 Porous titanium (CP-Ti) with a gradient pore structure: (a) longitudinal section morphology and (b) surface topography.

by sintering of titanium powder particles. The resulting porosity can be varied from 40% to 80%. Other techniques, including sintering of titanium powder with a polyurethane replica (the replica technique), sintering of loose granules (200–300 µm) made from HDH Ti powder [14,15], and additive manufacturing, can be used to make bimodal pore structures too.

28.2.3 Gradient pore structures

Porous titanium with a gradient pore structure usually shows pore size changes along its thickness directions (Figure 28.4). Gradient pore structures can be produced by spraying or depositing fine titanium powder onto a tubular or planar porous substrate (often prepared by sintering of coarse titanium powder), followed by sintering [16,17]. Porous titanium with a gradient pore structure is commercially used as precision filters. The thin (<200 μm) fine pore layer acts as an active surface filtration membrane, while the porous substrate provides mechanical strength without significantly reducing the flow rate of the filter element.

28.2.4 Honeycomb structures

Porous titanium with a honeycomb structure consists of millimeter-scaled pores with up to 95% porosity. Both the replica technique and the additive manufacturing method can be used to produce such structures (Figure 28.5). The replica process consists of (1) infiltration of titanium powder slurry into a polyurethane template or coating a polyurethane template with titanium by low temperature arc deposition, (2) removal of the template at 500°C in flowing argon, and (3) subsequent vacuum sintering at

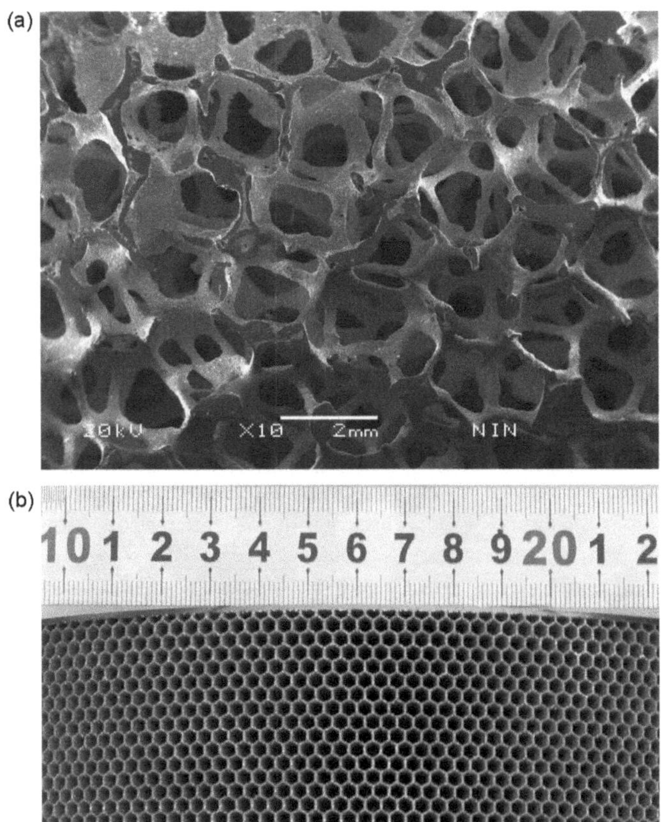

Figure 28.5 Porous titanium (CP-Ti) with a honeycomb structure made by (a) the template method and (b) additive manufacturing.

about 1250°C [18]. Additive manufacturing offers unrivaled freedom in the design and fabrication of porous titanium [19–22]. The current major limitation is pore size, limited to about 100 μm.

28.2.5 Closed-pore structures

As yet, porous titanium with closed pores has not received much attention because of the lack of applications. The gas entrapment process can be used to fabricate such porous titanium materials. In this method, argon gas is first entrapped in a preform by hot isostatic pressing (HIP). The resulting high-pressure argon bubbles are then expanded by exposure to a high temperature, where the densified titanium matrix creeps rapidly and the pressurized pores expand [23]. A variant of this process uses a titanium canister that is kept before expansion, resulting in a titanium sandwich structure (Figure 28.6) [24,25]. The process can be improved by taking advantage of the superplasticity of the matrix during pore expansion that can enable faster foaming and higher porosity (53%) with good strengths [26,27].

An alternative approach is to sinter hollow titanium spheres together (Figure 28.2d) rather than solid titanium powder particles. Hollow titanium spheres can be made using commercially available expanded polystyrene spheres (EPSs) coated with titanium powder-binder slurry by spray coating [8]. Microwave sintering is particularly attractive as the shell thickness of each titanium hollow sphere can be made to be similar to the microwave skin depth [8].

28.3 Properties of porous titanium

28.3.1 Permeability

The permeability of a porous medium measures the capability of a fluid passing through it under a pressure differential. The permeability coefficient, K, for a liquid is given by Darcy's law [28]:

$$K = \frac{Ql\mu}{A \cdot \Delta p} \tag{28.1}$$

where Q is the liquid flow rate in volume per unit time at 25°C, μ is the viscosity, l is the specimen thickness, Δp is the pressure differential, and A is the sample area. Darcy's law for an ideal gas flowing through a porous medium was proposed by Collins [29]:

$$Q_{p_a} = -\frac{KA}{\mu l}\frac{(p_a^2 - p_b^2)}{2} \tag{28.2}$$

where p_a and p_b are the exit and entrance pressures, respectively.

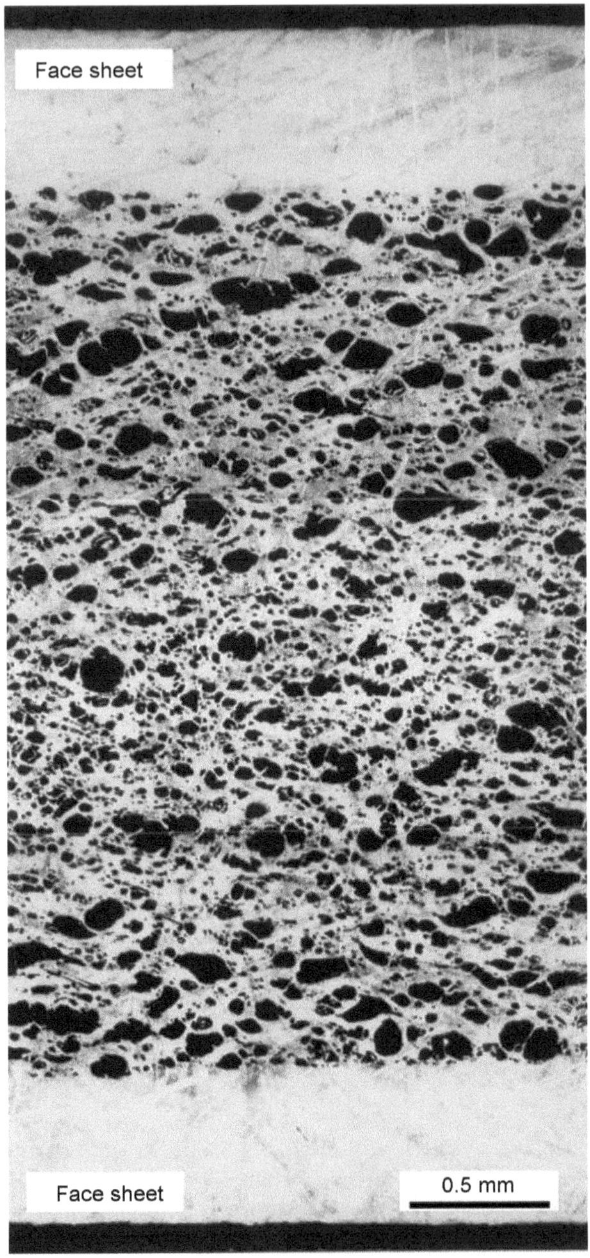

Figure 28.6 Ti-6Al-4V sandwich structure with a porous core made by the gas entrapment technique [25].

The permeability of porous titanium not only depends on the fluid but also on the porous structure. The Kozeny–Carman (KC) equation [30] offers a generic description of the permeability–porosity relationship for porous materials:

$$K = \frac{\phi^3}{c(1-\phi)^2 S^2} \tag{28.3}$$

where K is the permeability coefficient, ϕ is the porosity, c is the KC constant, and S is the specific surface area. When sintered from spherical solid particles of the same diameter d, the KC equation can be rewritten as [28]

$$K = \frac{\phi^3}{180(1-\phi)^2} d^2 \tag{28.4}$$

The pore size is related to the particle size d. Figure 28.7 shows the room temperature permeability coefficient of nitrogen gas through porous titanium (CP-Ti) as a function of the maximum pore size over the porosity range of 33–35%.

28.3.2 Mechanical properties

28.3.2.1 Young's modulus

The elastic modulus of a porous metal depends strongly on its pore structure. This differs from dense metallic materials, whose elastic modulus in general depends only

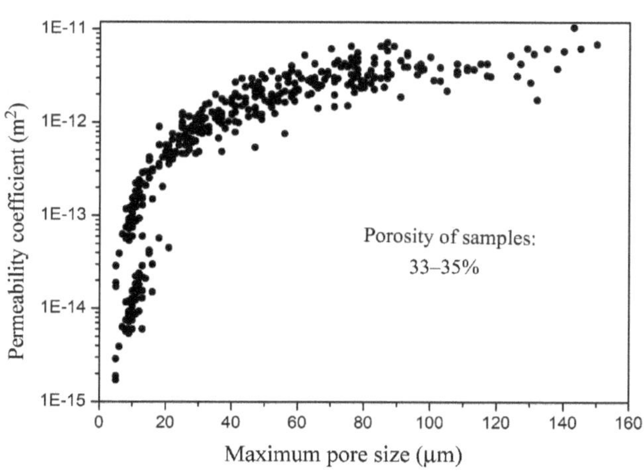

Figure 28.7 Permeability coefficient of nitrogen gas through porous titanium (CP-Ti) as a function of the maximum pore size over the porosity range of 33–35%, measured at room temperature (permeability coefficient is measured according to ISO 4022, maximum pore size according to ISO4003).

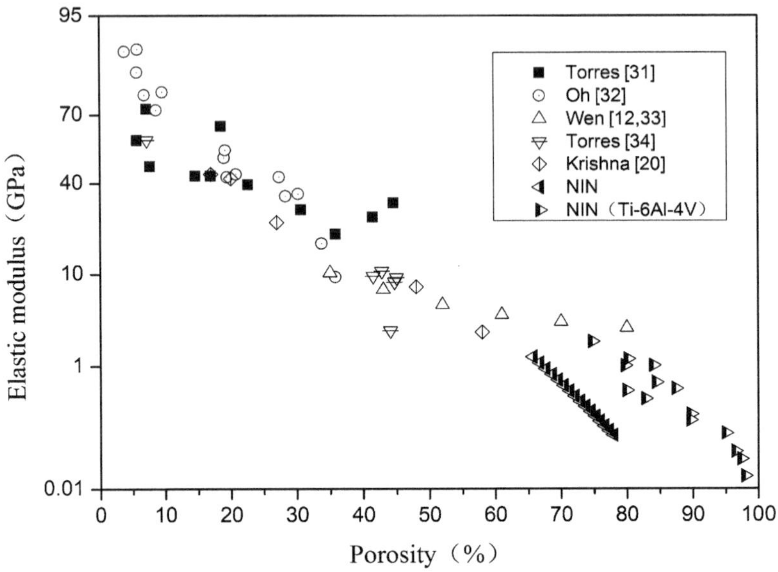

Figure 28.8 Compressive elastic modulus of porous titanium as a function of porosity. (Source: Northwest Institute for Nonferrous Metal Research China.)

marginally on microstructure. Accordingly, the elastic modulus of porous titanium can be readily manipulated to meet the low elastic modulus requirement for medical implant applications. Figure 28.8 shows literature data on Young's modulus versus porosity for porous titanium materials. The Young's modulus decreases with increasing porosity. The pore size and pore morphology also affect the Young's modulus.

28.3.2.2 Compression strength

Porous metals under compression often show a stress–strain curve with a plateau region where the stress is nearly constant over a wide range of strain variations. This behavior enables porous metals to be used for energy-absorbing applications where a large amount of deformation can be absorbed at a relatively low constant stress. Compared with other porous metals, the plateau region for porous titanium is small especially at low porosity (40%) (Figure 28.9) [13]. Some porous titanium materials may behave like brittle materials showing no plateau because of the unfavorable pore structure and/or their high interstitial content (O, N, C). The interstitial impurities should be strictly controlled.

Figure 28.10 summarizes the compressive strength of porous titanium over almost the entire range of porosity. The experimental data generally follow the Gibson–Ashby relationship established for cellular solids [37]:

$$\frac{\sigma_{pl}}{\sigma_{ys}} \approx C_2(1-\phi)^{n_2} \qquad (28.5)$$

where σ_{pl} is the plateau stress, σ_{ys} is the yield strength of the pore-free material, C_2 is a constant related to the cell geometry, and n_2 is an exponential factor, being around 1.5 for ideal stochastic open cellular foams.

28.3.2.3 Fatigue properties

Fatigue properties are required for a variety of applications of porous titanium materials. However, only limited information is available. One study [38] found that the ratio of the fatigue strength of porous Ti-6Al-4V measured at 10^6 cycles to its single-cycle compression yield strength (0.15–0.25) was less than that for solid Ti-6Al-4V (about 0.4). Another study [39] found that the fatigue strength and relative density of porous Ti-6Al-4V can be evaluated using Equation (28.5) assuming n is given by 2.7. This is higher than the n value for aluminum and nickel foams, and almost double the ideal value (n = 1.5) for a stochastic open cellular metal foam.

28.3.3 Corrosion resistance

Porous titanium generally inherits the excellent corrosion resistance of dense titanium in mildly reducing to highly oxidizing environments. However, because of their large specific surface area in contact with the corrosive medium, the corrosion rate of porous titanium is usually higher than that of dense samples. In addition, as corrosion develops deep into the pores, the migration of corrosion products will be distinctly different from the situation with dense titanium materials. This will change the local

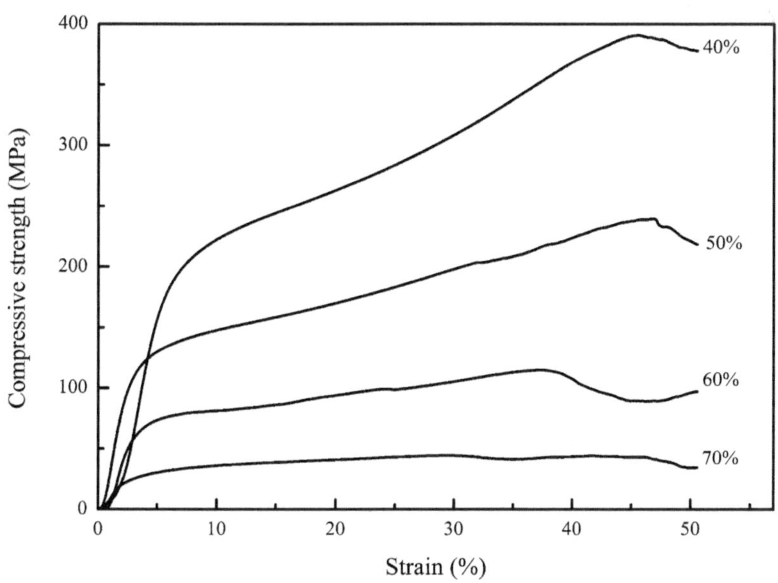

Figure 28.9 Compressive stress–strain curves of porous titanium with porosity being varied from 40% to 70%.

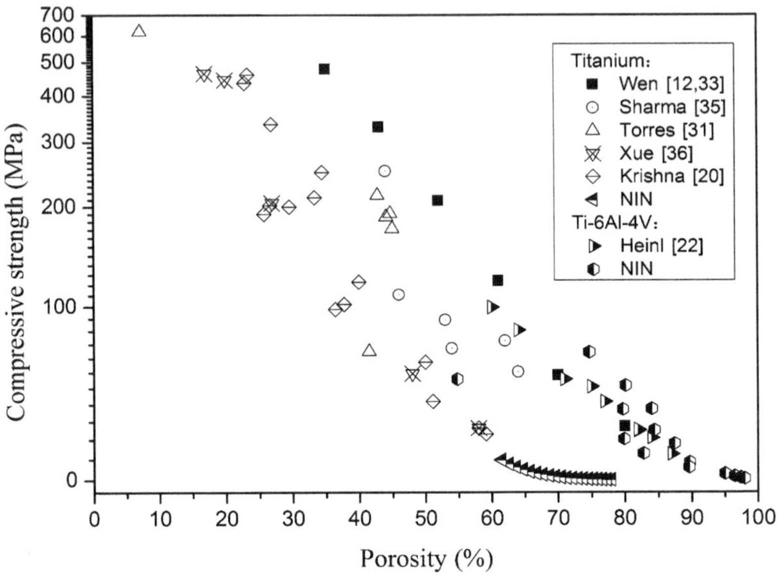

Figure 28.10 Compressive strength of porosity titanium as a function of porosity. (Source: Northwest Institute for Nonferrous Metal Research China.)

corrosive environment. Hence, the use of the overall corrosion rate based on weight losses may not be informative enough to describe the corrosion behavior of porous titanium. A new assessment approach is desired.

Since corrosion changes the porous structure and the properties of the porous titanium [40], it is essential to minimize the corrosion rate. Arensburger [6] found that nitriding porous titanium can increase its resistance to corrosion in 20% HCl solutions. In addition, surface alloying with palladium (Pd) has proved to be effective to substantially improve the corrosion resistance of porous titanium in 20% HCl and 40% H_2SO_4 solutions.

28.4 Commercial applications

Porous titanium has found many important applications. Figure 28.11 outlines the applications of porous titanium classified by the degree of "pore openness" required by specified applications. Discussed below are some selected examples.

28.4.1 Corrosion-resistant filters

Porous titanium has long been used in the chemical processing industry to deal with highly corrosive fluids, such as molten sulfur filtration [41], J acid solution filtration [41], stream filtration in the dairy industry [42], and oilfield injection water filtration

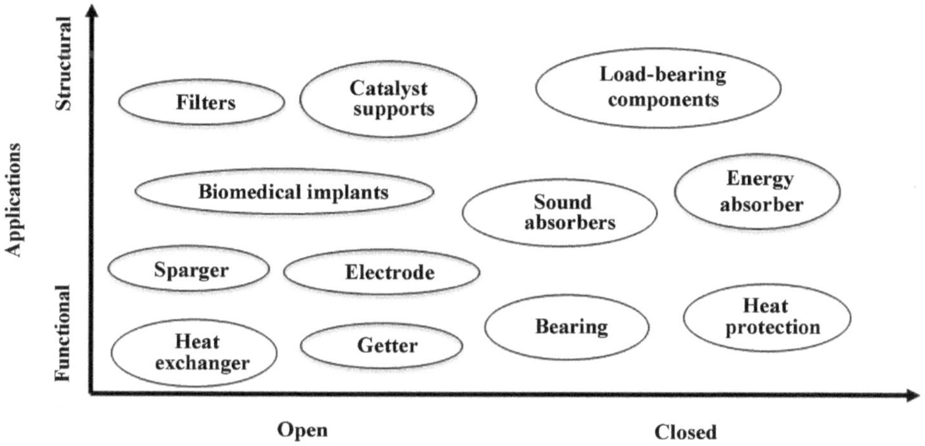

Figure 28.11 Applications of porous titanium classified by the degree of "pore openness" needed.

[43]. The pharmaceutical industry is another area where porous titanium filters have been used because of its inertness to and/or compatibility with most solvents and organic compounds, as well as complex drug ingredients.

Porous titanium filters are now available in a variety of product forms (e.g., sheets, plates, and tubes) from a number of suppliers, including Pall Corporation, Mott Corporation, GKN Sinter Metal, Graver Technology and NIN. Seamless porous titanium tubes with an outside diameter of 110 mm up to 1500 mm long have also been manufactured for several applications. Figure 28.12 shows a few porous titanium products produced by the NIN. The relationship between filter rating and maximum pore size has been studied [44,45]. Their characteristics are listed in Table 28.1 [46].

28.4.2 Gas spargers

Porous metal spargers are widely used in gas–liquid contacting applications in many industrial processes, for example, CO_2 sparging for carbonated beverages and beers, water purification, and chlorine sparging for bleaching pulp in paper manufacturing. These processes require the use of gas spargers by which small gas bubbles can be injected into a liquid or another gas while being resistant to corrosion by the liquid or gas and thermal or mechanical shocks [47]. Porous titanium gas spargers manufactured by the Powder Metallurgy Institute Belarus [48,49] have been used in the water purification systems in Russia, Belarus, and Ukraine for more than 15 years. In addition, it has been recommended that porous titanium be used to sparge high-concentration ozone for disinfecting and deodorizing air or waste treatment (Table 28.2) because of the excellent corrosion resistance to titanium to ozone. In contrast, porous stainless steel gas spargers are better suited to handling low-concentration ozone because of the corrosion caused by ozone.

(a)

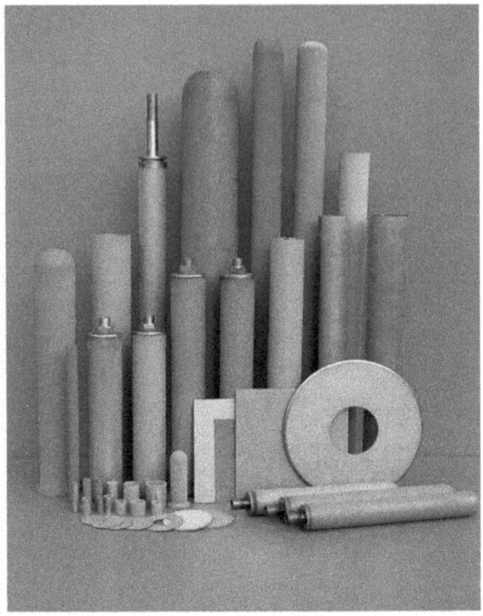

(b)

Figure 28.12 Commercial porous titanium products: (a) seamless tubes (up to 1500-mm long) and discs and (b) sheets (up to 400-mm wide).

Table 28.1 Characteristics of porous titanium filters for liquid filtration, fabricated by cold isostatic pressing (CIP), and vacuum sintering [46]

Filter grade	Filtration effect for specified minimum particle size (μm)		Permeability coefficients (10^{-12} m^2)
	98%	99.9%	
TG003	3	5	0.04
TG006	6	10	0.15
TG010	10	14	0.40
TG020	20	32	1.01
TG035	35	52	2.01
TG060	60	85	3.02
Testing standard	ISO-16889		ISO-4022

Table 28.2 **Sintered ozone spargers and suitable ranges of ozone concentration for sparging [47]**

Materials	Suitable ozone concentration (%)
316L stainless steel	<3
Hastelloy C-276	3–6
Titanium	6–10

Another example is oxygen sparger. During major heart or lung surgeries, it is often necessary to substitute the function of these organs by an extracorporeal device. The lungs are substituted by an oxygenator (artificial lung), which delivers oxygen to the body and removes carbon dioxide from it. Porous titanium has been used as oxygen spargers in such oxygenators because of its excellent biocompatibility. Figure 28.13 shows the oxygenators manufactured by NIN, China.

28.4.3 Electrochemical applications

Porous titanium has been used as current collectors in polymer electrolyte membrane (PEM) water electrolysis cells. The pore structure for these applications has been designed to provide a unique combination of fine pores, which allows water to flow

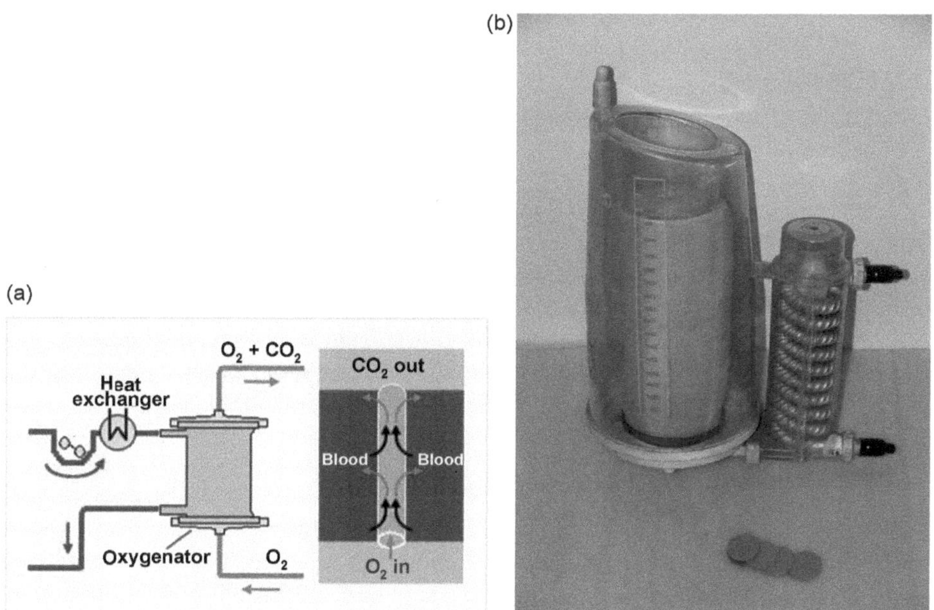

Figure 28.13 (a) Illustration of the basic principles of an oxygenator, and (b) a commercial oxygenator system and the porous titanium discs used in the oxygenator.

Table 28.3 Characteristics of porous titanium used in PEM water electrolysis cells [51]

Characteristics	Range
Thickness of current collectors	0.8–2 mm
Porosity	20–50%
Pore size	5–30 μm
Gas permeability coefficient	10^{-13}–10^{-11} m^2
Specific electric resistance	5–10 mΩ cm

through the electrodes, and large pores, which carries the product gas away from the electrodes [50]. Porous titanium current collectors not only resist the harsh corrosive acidic regime from the proton exchange membrane (pH ~2) but also sustain the high applied overvoltage (~2 V), especially at high current densities. Typical parameters of porous titanium current collectors used in the PEM water electrolysis cells are listed in Table 28.3 [51], and a commercial product electroplated with platinum (Pt) is shown in Figure 28.14.

Porous titanium may also be used as the base material for preparing insoluble anodes in electrochemical processes. For instance, Electrodes International, Inc. [52], has developed a novel insoluble Ti-Pb composite anode manufactured by infiltration of porous titanium with lead. In this innovative design, the lead shields the porous titanium from passivation while the latter stabilizes the lead from spalling, which is a major problem for conventional lead anodes. The Ti–Pb composite anode can be used in the electrowinning of metals such as copper, nickel zinc, and electrolytic manganese dioxide production. NIN has also developed porous titanium–iridium oxide (Ti/IrO_2) electrodes for chlor-alkali industry applications [53].

28.4.4 Gas getters

Titanium is a potent gas absorber of oxygen, nitrogen, carbon dioxide, and hydrogen. In addition, water vapor and methane can be absorbed by titanium too [54], although it may be limited to the surface layers. Porous structures allow the gases to diffuse through the pores for deep and quick absorption. Hence, porous titanium has been used as gas getters in advanced electronic devices. Commercial porous titanium getter products manufactured by SAES are made of Ti–Mo alloys [55].

28.4.5 Dental and orthopedic applications

CP-Ti and a number of titanium alloys are attractive orthopedic and dental implant materials because of their excellent biocompatibility and corrosion resistance. However, most dense titanium implants in use today suffer from the problem of stress-shielding due to the biomechanical mismatch of elastic modulus between the implant and bone [56]. In contrast, both the strength and Young's modulus of porous titanium can be manipulated through adjusting the pore structure for a near-perfect match with

(a)

(b)

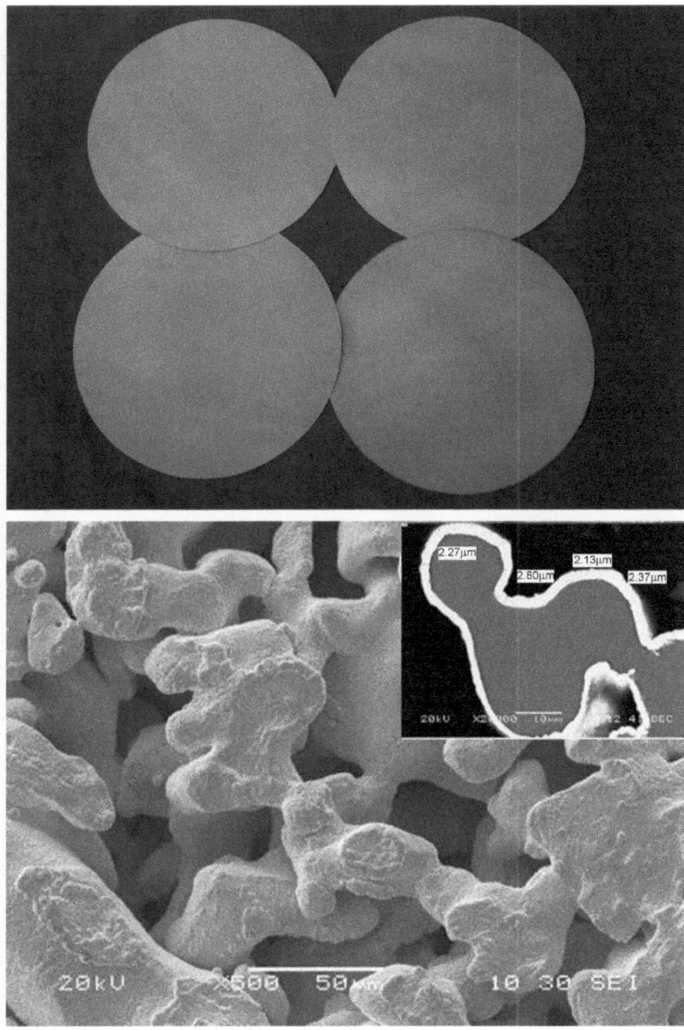

Figure 28.14 (a) Porous titanium plates (ϕ120 mm × 0.3 mm) and (b) microstructure of the porous titanium plates coated with platinum (Pt).

the bone. Also, porous structure can facilitate ingrowth of new-bone tissues and effective transport of body fluids and, after simple chemical and thermal treatments, bone induction can be realized in porous titanium [57]. As a result, the last decade has seen an increasing interest in porous titanium implants [58,59]. Figure 28.15 shows several porous titanium implants developed by NIN. According to Stryker [60], a total of 288 hip arthroplasties were completed using porous titanium acetabular cups in 252 patients from 2008 to 2010. No cup failures occurred after three years of testing. It is anticipated that biomedical applications will become an important market for porous titanium structures.

(a)

(b)

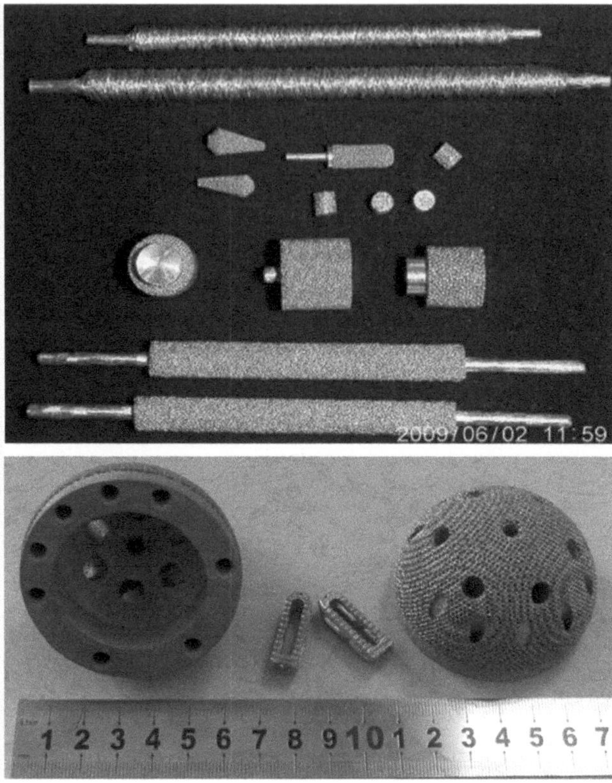

Figure 28.15 Porous titanium implants developed by NIN: (a) dental implants and (b) acetabular cups and lumbar fusion.

28.4.6 Other applications

Porous titanium aluminide can be used as flow restrictors that require small permeability and micro-flow control for high accuracy flow-control [61–63]. The expansion during reaction sintering caused by the Kirkendall Effect enables the realization of tight contact between the porous titanium aluminide core and the dense titanium alloy shell without having to use brazing. Porous titanium can also be used as camera flash cathodes [64].

28.5 Concluding remarks

Porous titanium has evolved into an important class of metallic material and is the material of choice for many industrial applications, including filtration, separation, catalyst supporting, gas absorbing, gas sparging, current collection, and medical implantation. The properties of porous titanium including permeability, elastic modulus, and mechanical strength are determined by the pore structure. Porous titanium can be produced to have uniform, bimodal, gradient, honeycomb, or closed-pore structures with different pore morphologies. The properties of porous titanium can thus be readily manipulated for different applications. Similar to the fabrication of dense titanium materials, the interstitial impurities (O, N, and C) in porous titanium should be strictly controlled for both desired mechanical properties and corrosion resistance. Additive manufacturing or 3D metal printing provides a powerful platform for the design and fabrication of novel porous titanium structures compared to conventional powder metallurgy. Future developments of porous titanium will continue to be driven by application but the new design freedom offered by additive manufacturing is expected to play an increasingly important role.

Acknowledgments

The authors acknowledge the financial support from the Ministry of Science and Technology China under the International Science & Technology Cooperation Program (2011DFA5290) and the National Basic Research Program of China (973 Program, 2011CB610302). In addition, financial support from the National Natural Foundation of China (51134003) is also gratefully acknowledged.

References

[1] M. Qian, Cold compaction and sintering of titanium and its alloys for near-net-shape or preform fabrication, Int. J. Powder Metall. 46 (5) (2010) 29–44.
[2] D.E. Gray, J.O. Ollier, Electric condenser, U.S. Patent 2299228 (Priority date: January 12, 1938) (1942).
[3] W.W. Kittelberger, Production of titanium, U.S. Patent 2,789,943 (Priority date: May 5, 1955) (1957).
[4] V.Y. Kramnik, Y.N. Semenov, E.A. Arutyunov, V.N. Morozov, O.Y. Demchenko, Chemical properties of sintered filters manufactured from spongy titanium wastes, Sov. Powder Metall. Metal Ceram. 3 (4) (1964) 333–336.
[5] A.I. Zhevnovatyi, G.F. Shenberg, Study of the production technology of porous titanium tubes, Sov. Powder Metall. Metal Ceram. 4 (2) (1965) 95–99.
[6] D.S. Arensburger, V.S. Pugin, I.M. Fedorchenko, Corrosion resistance of porous titanium in some aggressive media, Sov. Powder Metall. Metal Ceram. 7 (12) (1968) 977–981.
[7] G.F. Tikhonov, V.K. Sorokin, Preparation of porous titanium by rolling powder, Powder Metall. Metal Ceram. 8 (2) (1969) 167–169.
[8] P. Yu, G. Stephani, S.D. Luo, H. Goehler, M. Qian, Microwave-assisted fabrication of titanium hollow spheres with tailored shell structures for various potential applications, Mater. Lett. 86 (2012) 84–87.
[9] Y.X. Li, A study on porous titanium filter tube by powder metallurgy process, Rare Metal Alloy Process. (1981) 3, 6–10 (in Chinese).
[10] C.F. Song, Y.N. Wang, Properties of porous titanium produced by powder rolling, Rare Metal Alloy Process. (1981) 1, 19–21(in Chinese).
[11] X.D. Zhang, Biomaterials with Intrinsic Osteoinductivity, The 6th World Biomaterials Congress, Hawaii, USA, May 15–20, 2000.

[12] C.E. Wen, M. Mabuchi, Y. Yamada, Processing of biocompatible porous Ti and Mg, Scripta Mater. 45 (2001) 1147–1153.
[13] C. Xiang, Y. Zhang, Z. Li, H. Zhang, Y. Huang, H. Tang, Preparation and compressive behavior of porous titanium prepared by space holder sintering process, Procedia Eng. 27 (2012) 768–774.
[14] K. Yang, H.P. Tang, J. Wang, C.S. Xiang, B.J. Yang, Y.N. Li, J. Liu, A method for preparing a bimodal pore structure of the porous metal material for medical, CN102690967B (Pub. Date Sept. 26, 2012) (2012).
[15] Y. Kun, W. Jian, L. Jie, T. Huiping, Y. Baojun, L. Yaling, Preparation and characterization of porous titanium scaffolds with bimodal porous structure using granulating loose pack sintering method, Rare Metal Mater. Eng. 42 (S2) (2013) 418–422.
[16] P. Neumann, S. Steigert, M.K. Bram, H.P. Bruchkremer, Z. Li, L. Zhao, Filters with a graduated structure and a method for producing the same, U.S. Patent 2004/0050773A1 (Pub. Date Mar. 18, 2004) (2004).
[17] Q. Wang, H. Tang, Q. Zhang, Q. Qiu, J. Wang, Preparation of titanium microfiltration membrane by field-flow fractionation deposition. 2006 Powder Metallurgy World Congress and Exhibition, PM 2006, September 24–28, 2006, PART 2 ed., Trans Tech Publications Ltd, Busan, Republic of Korea, 2007, pp. 985–988.
[18] J.P. Li, C.A.V.K. Blitterswijk, D. Groot, Factors having influence on the rheological properties of Ti6Al4V slurry, J. Mater. Sci. 15 (2004) 951–958.
[19] H.P. Tang, P. Zhao, W.P. Jia, G.Y. Yang, S.L. Lu, A non-weld metal honeycomb structure fabricated by selective electron beam melting, CN201310214682 (Pub. Date Sept. 04, 2013) (2013).
[20] B.V. Krishna, S. Bose, A. Bandyopadhyay, Low stiffness porous Ti structures for load-bearing implants, Acta Biomater. 3 (2007) 997–1006.
[21] D.K. Pattanayak, A. Fukuda, T. Matsushita, M. Takemoto, S. Fujibayashi, K. Sasaki, N. Nishida, T. Nakamura, T. Kokubo, Bioactive Ti metal analogous to human cancellous bone: fabrication by selective laser melting and chemical treatments, Acta Biomater. 7 (2007) 1398–1406.
[22] P. Heinl, L. Müller, C. Körner, R.F. Singer, F.A. Müller, Cellular Ti-6Al-4V structures with interconnected macro porosity for bone implants fabricated by selective electron beam melting, Acta Biomater. 4 (2008) 1536–1544.
[23] M.W. Kearns, P.A. Blenkinsop, A.C. Barber, T.W. Farthing, Manufacture of a novel porous metals, Int. J. Powder Metall. 24 (1988) 59–64.
[24] D.T. Queheillalt, B.W. Choi, D.S. Schwartz, H.N.G. Wadley, Creep expansion of porous Ti-6Al-4V sandwich structures, Metall. Mater. Trans. A 31 (2000) 261–273.
[25] J. Banhart, Manufacture, characterization and application of cellular metals and metal foams, Progr. Mater. Sci. 46 (2001) 559–632.
[26] N.G. Davis, J. Teisen, C. Schun, D.C. Dunand, Solid-state foaming of titanium by superplastic expansion of argon-filled pores, J. Mater. Res. 16 (2001) 1508–1519.
[27] C.D. David, Processing of titanium foams, Adv. Eng. Mater. 6 (6) (2004) 369–376.
[28] J. Bear, Dynamics of Fluid in Porous Media, Elsevier, New York, 1972.
[29] R.E. Collis, Flow of Fluids Through Porous Materials, Reinhold Pub, Corp. 1961.
[30] P.C. Carman, Fluid flow through granular beds, Chem. Eng. Res. Des. 15 (1937) 150–166.
[31] Y. Torres, J.J. Pavón, I. Nieto, J.A. Rodríguez, Conventional powder metallurgy process and characterization of porous titanium for biomedical applications, Metall. Mater. Trans. B42 (2011) 891–900.
[32] I.H. Oh, N. Nomura, N. Masahashi, S. Hanada, Mechanical properties of porous titanium compacts prepared by powder sintering, Scripta Mater. 49 (2003) 1197–1202.

[33] C.E. Wen, Y. Yamada, K. Shimojima, Y. Chino, H. Hosokawa, M. Mabuchi, Novel titanium foam for bone tissue engineering, J. Mater. Res. 17 (2002) 2633–2639.
[34] Y. Torres, J.A. Rodríguez, S. Arias, M. Echeverry, S. Robledo, V. Amigo, J.J. Pavón, Processing, characterization and biological testing of porous titanium obtained by space-holder technique, J. Mater. Sci. 47 (2012) 6565–6576.
[35] M. Sharma, G.K. Gupta, O.P. Modi, B.K. Prasad, A.K. Gupta, Titanium foam through powder metallurgy route using acicular urea particles as space holder, Mater. Lett. 65 (2011) 3199–3201.
[36] W.C. Xue, B.V. Krishna, A. Bandyopadhyay, S. Bose, Processing and biocompatibility evaluation of laser processed porous titanium, Acta Biomater. 3 (2007) 1007–1018.
[37] L.J. Gibson, M.F. Ashby, Cellular Solids: Structure and Properties, second ed., Cambridge University Press, 1999.
[38] N.W. Hrabe, P. Heinl, B. Flinn, C. Kornec, R.K. Bordia, Compression-compression fatigue of selective electron beam melted cellular titanium (Ti-6Al-4V), J. Biomed. Mater. Res. Part B 99B (2011) 313–320.
[39] S.J. Li, L.E. Murr, X.Y. Cheng, Z.B. Zhang, Y.L. Hao, R. Yang, F. Medina, R.B. Wicker, Compression fatigue behavior of Ti-6Al-4V mesh arrays fabricated by electron beam melting, Acta Mater. 60 (2012) 793–802.
[40] K.M. Liu, Corrosion resistant of porous titanium, Rare Metal Mater. Eng. 5 (1989) 55–58 (in Chinese).
[41] Y.X. Li, G.Z. Hu, Applications of powder metallurgy titanium filters, Rare Metal Mater. Eng. 1 (1983) 63–65 (in Chinese).
[42] Information on http:// www.gravertech.com (accessed 30.04.14).
[43] B. Deng, H. Ding, F.L. Geng, Fine filtration technique for effluent from oil field production, Technol. Waste Treatment 32 (2006) 73–75 (in Chinese).
[44] Q. Zhang, Z.D. Zhang, H.R. Wei, Characterization methods of porous material filter rating, Filter Separator 10 (1) (2000) 33–37 (in Chinese).
[45] T. Huiping, Z. Qing, X. Zhengping, G. Yuan, W. Jianyong, The relationship between maximum pore size and filter rating porous metal, Rare Metal Mater. Eng. 36 (S13) (2007) 559–561 (in Chinese).
[46] Chinese Standard GB/T 6887-2007, Sintered metal filter elements (Publication date: Nov. 01, 2007) (2007).
[47] K.L. Rubow, L.L. Stange, Sintered porous metal media in food and beverage processing, Presented at the American Filtration and Separation Society Conference, Tampa, FL, November 14–16, 2001.
[48] V. Savich, A. Taraykovich, S. Bedenko, Improved porous sponge titanium aerators for waste treatment, Powder Metall. 56 (2013) 272–275.
[49] Information on http:// www.pminstitute.by (accessed 06.05.14).
[50] M. Carmo, D.L. Fritz, J. Mergel, D. Stolten, A comprehensive review on PEM water electrolysis, Int. J. Hydrogen Energy 38 (2013) 4901–4934.
[51] S.A. Grigoriev, P. Millet, S.A. Volobuev, V.N. Fateev, Optimization of porous current collectors for PEM water electrolysers, Int. J. Hydrogen Energy 34 (2009) 4968–4973.
[52] A. Sapozhnikova, Insoluble titanium-lead anode for sulfate electrolytes. U.S. Patent 6287433B1 (Date of Patent: Sep. 11, 2001) (2001).
[53] C. Xiang, G. Li, X. Kang, Y. Huang, H. Tang, Preparation and microstructure of porous Ti/IrO_2 electrodes, 12th World Conference on Titanium, Ti 2011, June 19–24, 2011, Science Press, Beijing, China, 2012, pp. 823–826.
[54] V.L. Stout, M.D. Gibbons, Gettering of gas by titanium, J. Appl. Phys. 26 (1955) 1488–1492.

[55] Information on http://www.saeagetter.com (accessed 25.05.14).
[56] M. Niinomi, Mechanical biocompatibilities of titanium alloys for biomedical applications, J. Mech. Behav. Biomed. Mater. I (2008) 30–42.
[57] S. Fujibayashi, M. Neo, H.M. Kim, T. Kokubo, T. Nakamura, Osteoinduction of porous bioactive titanium metal, Biomaterials 25 (2004) 443–450.
[58] P. Singh, P.D. Lee, R.J. Dashwood, T.C. Lindley, Titanium foams for biomedical applications: a review, Mate. Technol. 25 (2010) 127–136.
[59] G. Ryan, A. Pandit, D.P. Apatsidis, Fabrication methods of porous metals for use in orthopaedic applications, Biomaterials 27 (2006) 2651–2670.
[60] Information on http://www.stryker.com (accessed 30.05.14).
[61] H.P. Tang, P. Tan, X.L. Kang, X.T. Kang, S.L. Yu, Q.B. Wang, C. Li, J.L. Zhu, J.C. Liao, Preparation of porous metal core for precise control of micro-flow, CN101112721 (Pub. Date June 03, 2009) (2009).
[62] P. Tan, H.P. Tang, X.T. Kang, Q.B. Wang, C. Li, J.L. Zhu, J.Y. Wang, Y. Ge, A method of connecting the porous core and the dense shell, CN101108421 (Pub. Date Aug. 19, 2009).
[63] P. Tan, H.P. Tang, X.T. Kang, Q.B. Wang, J.L. Zhu, C. Li, Research on TiAl alloy porous metal flow restrictors, Mater. Trans. 50 (2009) 2484–2487.
[64] Q. Zhang, J.Y. Lin, X. Guo, The research of porous titanium cathode, Rare Metal 23 (72–74) (1999) (in Chinese).

Microstructural characterization of as-sintered titanium and titanium alloys

29

Ming Yan
RMIT University, School of Aerospace, Mechanical and Manufacturing Engineering, Centre for Additive Manufacture, Melbourne, Victoria, Australia

29.1 Introduction

Microstructure is one of the three pillars supporting the entire materials science; the well-known triangle relationship between microstructure, processing, and properties highlights the importance of microstructure and the associated microstructural characterization in understanding and developing advanced materials. Microstructural characterization of materials normally consists of three key parts: (a) compositional/chemical analysis, (b) structural (phase) identification, and (c) morphological observation. For example, to predict and understand the mechanical properties of thermomechanically processed Ti-6Al-4V, it is a prerequisite to have the following details [1,2]: (a) thickness of Widmanstatten α-lath, (b) α colony size, (c) β grain size, (d) volume fraction of Widmanstatten α-lath, (e) width of grain boundary α, and (f) mean edge length of α-lath. Owing to the importance of microstructure, it may deserve a whole book to review the progress achieved in the study of the PM Ti and Ti alloys from a microstructural characterization perspective. This chapter focuses on selected topics that are closely relevant to the microstructural characterization of PM Ti and Ti alloys. Since PM Ti is only a small member of the Ti family, in some cases the chapter will have to refer to the results obtained from other processing techniques such as casting or forging where relevant information is insufficient.

29.2 Microstructural features of PM Ti and Ti alloys

It is useful to clarify in the first place that powder metallurgy differs from casting, forging, or other processing techniques in many aspects, and these differences define the microstructural characteristics of PM Ti and Ti alloys. One may note the following for PM Ti and Ti alloys:

- Isothermal sintering is typically carried out in the β phase region [3–8], at a temperature (e.g., at 1300°C) well below the liquidus temperature (~1668°C for CP-Ti) [9–11]. For sintering under pressure such as hot isotactic pressing, hot pressing, or spark plasma sintering, the consolidation temperature can be much lower, for instance, around 900°C [12].

Titanium Powder Metallurgy. http://dx.doi.org/10.1016/B978-0-12-800054-0.00029-0

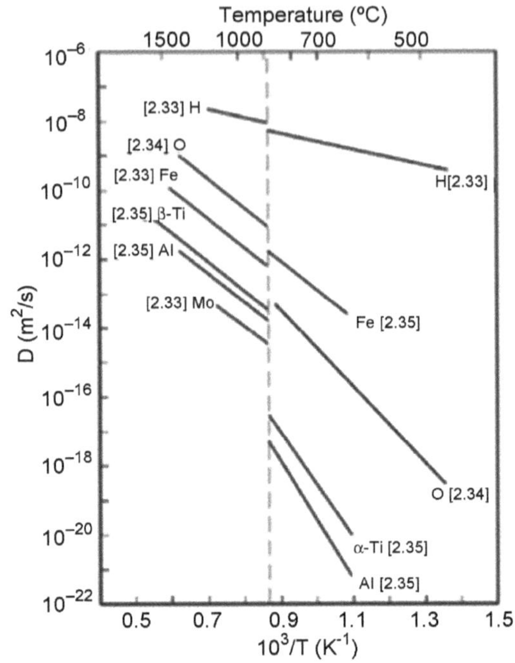

Figure 29.1 Diffusivity of elements in Ti [11]. Note that some elements such as Mo and Sn, Zr and Nb (not shown here) are slow diffusers in Ti.

- Diffusion of elements during sintering (see Figure 29.1) [11] is much slower than that in the liquidus state; and
- Most sintering practices use low heating and cooling rates (e.g., from 5°C/min to 20°C/min) [2–8], which make the sintering process closer to equilibrium conditions than casting.

Hence PM Ti and Ti alloys are expected to show the following microstructural characteristics [2–8]:

- The overall microstructure is close to the equilibrium state because of the low cooling rate.
- There could be some annealing and/or aging effect resulting from the slow cooling process from the isothermal sintering temperature to room temperature. Consequently, aging-induced phases such as isothermal ω (see Section 29.3.4) may form in some β-containing alloys during cooling [13].
- Pores are part of an as-sintered microstructure because of the difficulty to achieve a pore-free microstructure in most cases.

29.3 Common phases in PM Ti and Ti alloys

This section will summarize the common phases observed in PM Ti and Ti alloys, including α, β, ω, martensite phases (α′ and α″), and α_2. Figures 29.2–29.4 [14–16] provide some basic information about α and α_2, which are thermodynamically favorable at room temperature; ω and martensite phases (α′ and α″), which are

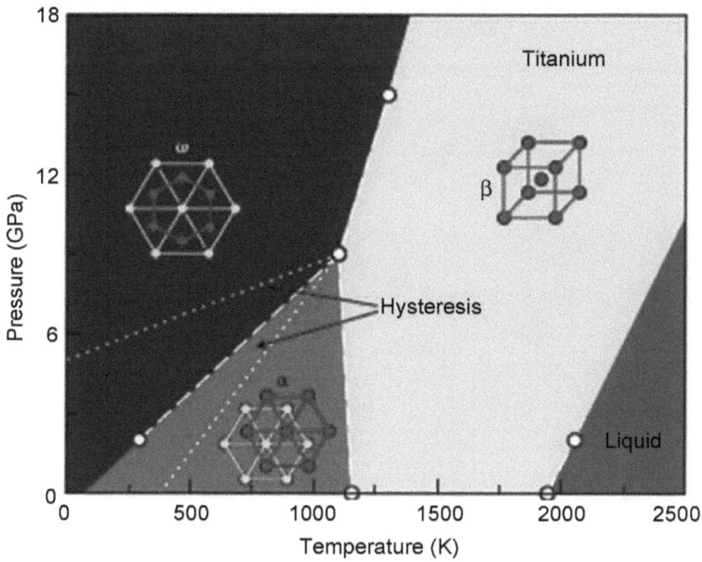

Figure 29.2 Temperature-pressure dependency of α, β, ω, and liquidus phase (L) of CP-Ti. Note that at ambient condition, α phase is the equilibrium phase, while β and ω are metastable phases [14].

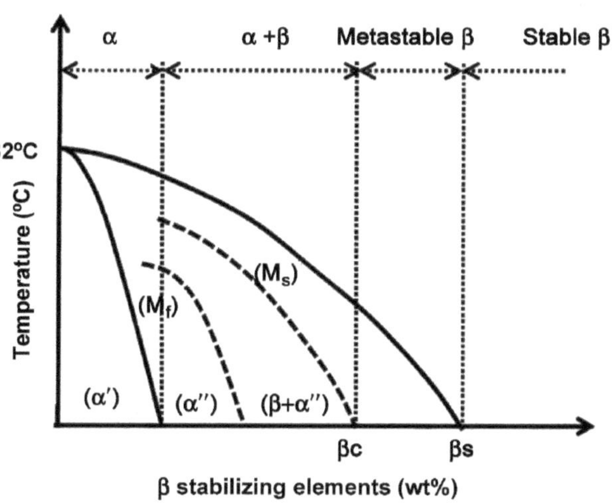

Figure 29.3 Schematic graph to show the compositional difference between α′ and α″ [15]. Note that α′ forms when the β stabilizers are in low concentration. β_c in the figure stands for the critical concentration of β stabilizers to fully retain β phase at the condition of quenching. β_s stands for the concentration that is required to fully stabilize the β phase at room temperature. M_s and M_f mean the start and finish of the martensite phase transformation.

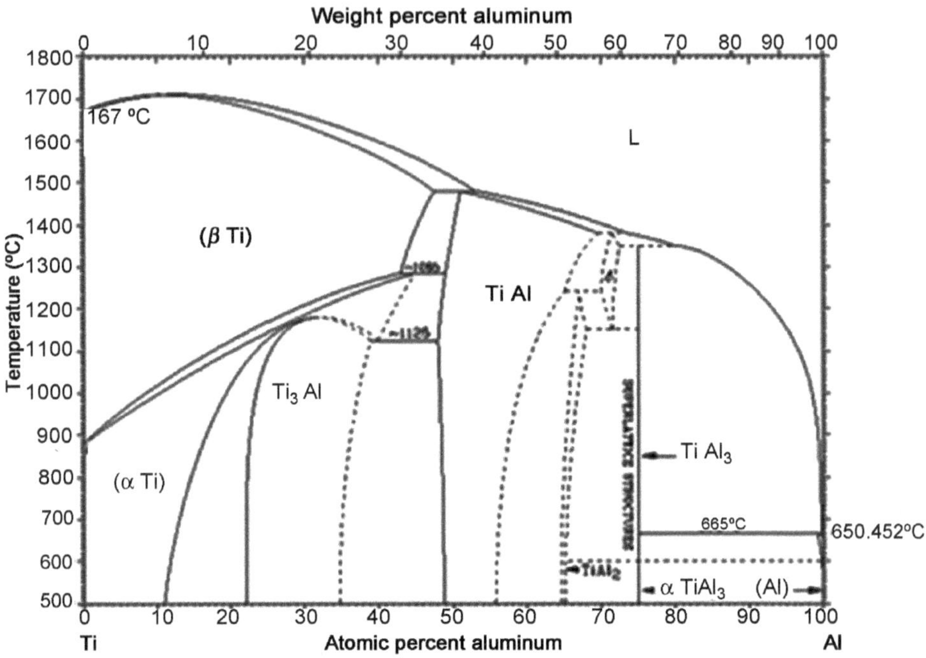

Figure 29.4 Introduction to the α_2 (Ti_3Al) phase using the Ti-Al phase diagram [16].

metastable phases; and β, which can be stable at room temperature provided there are high enough β stabilizers to decrease the β→α transformation temperature to below the room temperature.

29.3.1 The α phase in the PM Ti and Ti alloys

- The α phase has a hexagonal-close-packed (*hcp*) crystal structure with lattice parameters of $a = 0.28$ nm and $c = 0.47$ nm. Both lattice parameters can be modified by alloying elements or interstitial elements. Al, O, and N are strong α stabilizers while some β stabilizers can dissolve into the α phase too.
- The morphology of the α phase, including near equiaxed, lamellar (lath) or acicular, depends on alloy chemistry, processing pathway, and thermal history. Figure 29.5a–c shows the α phase in various types of PM Ti and Ti alloys (α, α+β, and β) [17–19]. The α phase can also exist as nano particles such as the one shown in Figure 29.5d [20].
- Transformation from the ω phase via aging is one of the many pathways to form the α phase; Figure 29.5e shows one such example [21].

29.3.2 The β phase in PM Ti and Ti alloys

1. The β phase is body-centered-cubic (*bcc*) in crystal structure with lattice parameter of $a = 0.33$ nm. It is normally metastable but can be a room-temperature stable phase when there is a sufficient amount of β stabilizers.

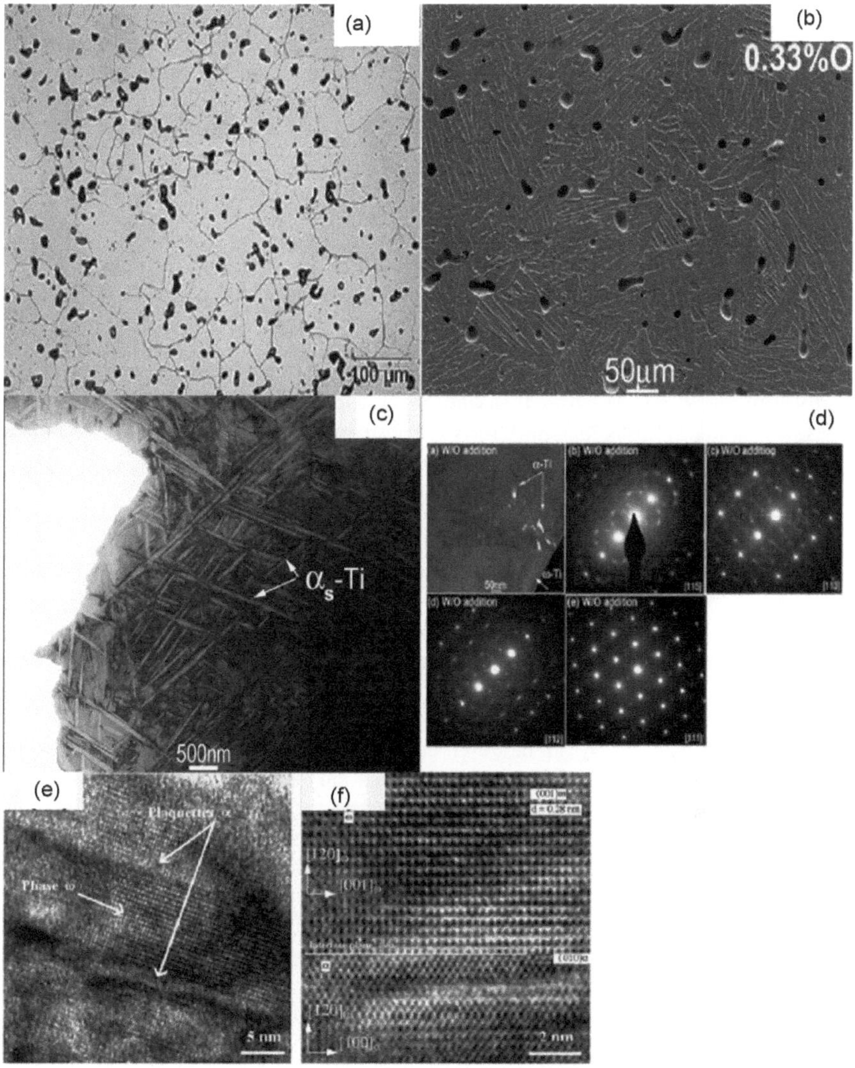

Figure 29.5 α phase with different morphologies in the PM α-Ti (CP-Ti; (a) SEM image) [17], (α+β)-Ti alloy (Ti-6Al-4V; (b) SEM image) [18], and β-Ti alloy (Ti-15Mo; (c) TEM image) [19]. (d) Nanoscale α phase in a PM (α+β)-Ti alloy (Ti-2.25Mo-1.5Fe) with the corresponding electron diffraction results; the ω phase is also present in the microstructure [20]. The high-resolution TEM images in (e) and (f) suggest that the α phase may have nucleated from the ω phase [2].

2. The β stabilizers are normally categorized into two groups [9–11]:

  a. Isomorphic-type β stabilizers like the Mo, V, Nb, and Ta. They form solid solution with Ti [22]; and

  b. Eutectoid-type β stabilizers [23], which can be further grouped into:
  - Fe, Mn, Cr (Subgroup I); and
  - Si, Cu, Ag, Ni, and H (Subgroup II).

3. The rate of the eutectoid reaction, β-Ti(M) (M: β stabilizer) → α-Ti + Ti_xM_y intermetallic, is different for the two subgroups. Figure 29.6a shows the eutectoid microstructure of PM Ti-7Ni alloy [24]. Because of the slow reaction rate, however, the β phase may retain in β-Ti alloys that are stabilized by the first subgroup elements at room temperature [13,25,26]. Figure 29.6b and c show such examples in a typical eutectoid alloy Ti-6Fe-0.5Si [25] and in an as-sintered CP-Ti containing 1280 ppm of Fe [13].
4. Precipitates can be found in the β phase matrix. These precipitates can be, e.g., α phase (e.g., Figure 29.5c) and/or ω phase (e.g., Figure 29.5d).

29.3.3 The ω phase in PM Ti and Ti alloys

The ω phase has a simple hexagonal crystal structure with lattice parameters of $a = 0.460$ nm and $c = 0.282$ nm. It is one of the three basic (solid) phases of Ti (α, β, and ω; see Fig. 29.2) but is less understood compared to the α and β phases, and therefore it is still a subject of interest for research. A concise summary about the phase can be made as follows [27–31]:

- The ω phase can be categorized into athermal ω (formed during quenching, irrelevant to aging/heat treatment) and isothermal ω (formed during aging/heat treatment).
- Athermal ω is believed to be a product of martensite transformation, or due to a displacement/shear mechanism from a crystallographic point of view. Transformation between the athermal ω and β is reversible. Its phase constitution is virtually the same as that of the parent β phase, a common feature of martensitic phase transformation. Figure 29.7 schematically illustrates the formation mechanism of the athermal ω [29].
- Isothermal ω is the one that can be observed in the as-sintered microstructure of some PM Ti and Ti alloys. The time–temperature–transformation (TTT) curve in Figure 29.8 suggests that aging at relatively low temperatures may induce the formation of the isothermal ω in some Ti alloys. The low cooling rate in most of the sintering practices satisfies this requirement (i.e., to provide an aging-like effect). In terms of phase constitution, the isothermal ω is lean in β stabilizers compared to the parent β phase, supported by 3D atom probe tomography (3D-APT) studies (Figure 29.9).
- The orientation relationship between β and ω can be defined as $[1\ 1\ 1]_\beta//[0\ 0\ 0\ 1]_\omega$ and $(1\ 1\ 0)_\beta//(1, 1, -2, 0)_\omega$.
- The ω phase is normally present as fine (~10 nm) particles in Ti and Ti alloys. Figure 29.9 shows a typical transmission electron microscopy (TEM) bright-field image of the ω phase and the corresponding SAED pattern.
- Oxygen tends to suppress the ω formation. Figure 29.10 gives a typical example from β-(Ti-V) alloy.
- Transformation from ω→α is also possible (see the relevant example in Figure 29.5e and 5f), following an orientational relationship of $(2, -1, 0)_\omega//(0\ 0\ 2)_\alpha$, and $[0\ 0\ 1]_\omega//[1\ 0\ 0]_\alpha$.
- TiH_2 PM is receiving increasing attention. Hydrogen is a β stabilizer and can reduce the β→α phase transformation temperature (Figure 29.11), and possibly affect the martensite transformation too [32,33]. This means that the martensite phases such as the α′ and α″ and the athermal ω phase can be suppressed when using TiH_2 as the starting powder material. Figure 29.11 provides such an example: Figure 29.12a (as-cast Ti-15Mo) [34] shows the presence of the ω phase by TEM bright field imaging as well as by electron diffraction while Figure 29.12b (using TiH_2 as the starting powder [19]) shows no clear evidence for the presence of the ω phase.

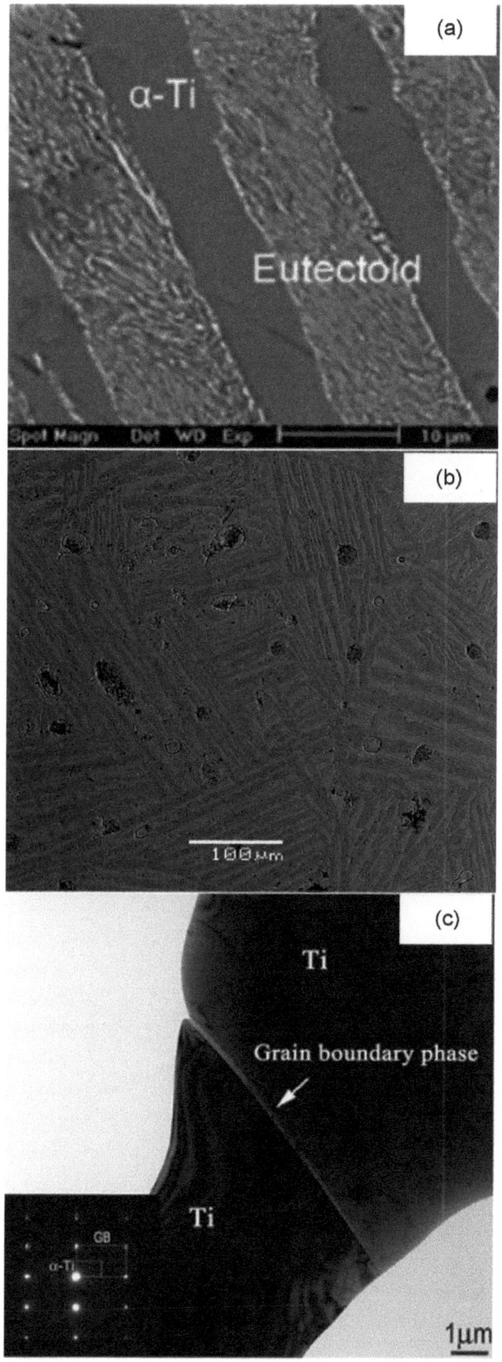

Figure 29.6 Comparison between the two subgroups (I and II) of the eutectoid-type β stabilizers. When alloyed with Ni (the II type), the PM Ti-7Ni alloy shows a eutectoid structure consisting of Ti_2Ni and Ti phases (a) [24]. In contrast, Fe-stabilized (the I type) β-Ti retains as a main phase in Ti-6Fe-0.5Si [25] (b) or a grain boundary phase in an as-sintered CP-Ti [13] (c).

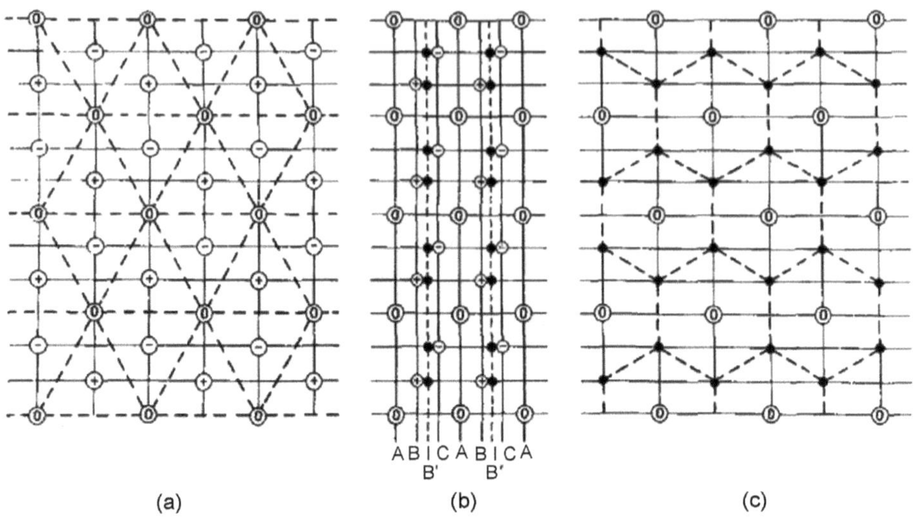

Figure 29.7 Schematic graph to show the formation mechanism of the athermal ω (c) from β phase (a), which is believed to be due to a displacement of atoms (1/3 shift in unit distance along the 1 1 1 direction (b)) [29]. Detailed description of the original figures can be found in Ref. [29].

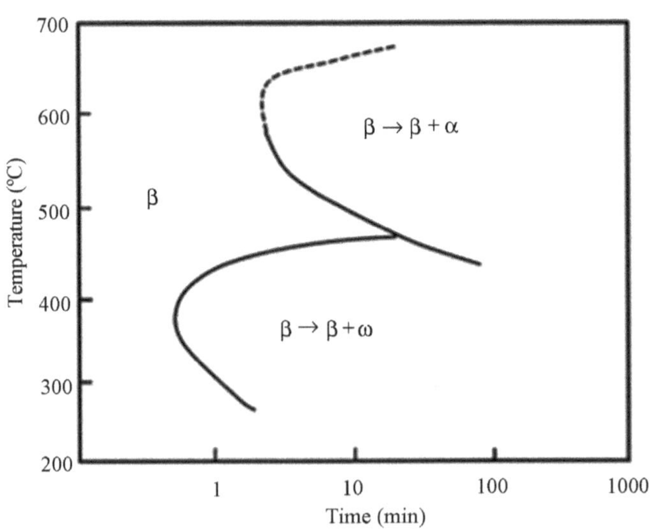

Figure 29.8 Schematic graph of the time–temperature–transformation (TTT) to show the time and temperature requirements to form isothermal ω in the Ti-20V β alloy [27]. Note that the ω phase tends to exist in the lower (aging) temperature region.

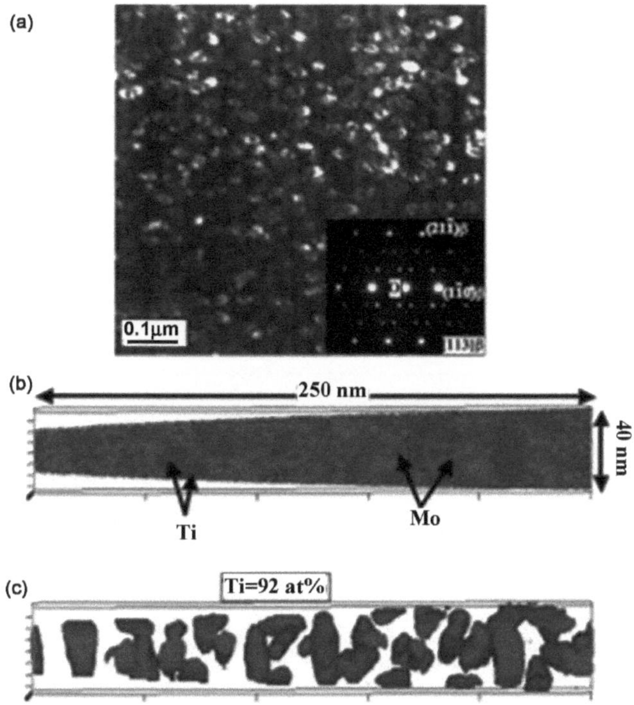

Figure 29.9 Three-dimensional ATP results of the Ti-18Mo β alloy to show that the isothermal ω containing a smaller amount of β stabilizer (Mo in this case) than in the β matrix [30]. (a) TEM bright field image with the corresponding diffraction pattern; (b) reconstruction of the APT results to show the Ti-enriched ω phases and (c) iso-concentration surface processing (Ti = 92 at%) to show the morphology of the ω phases.

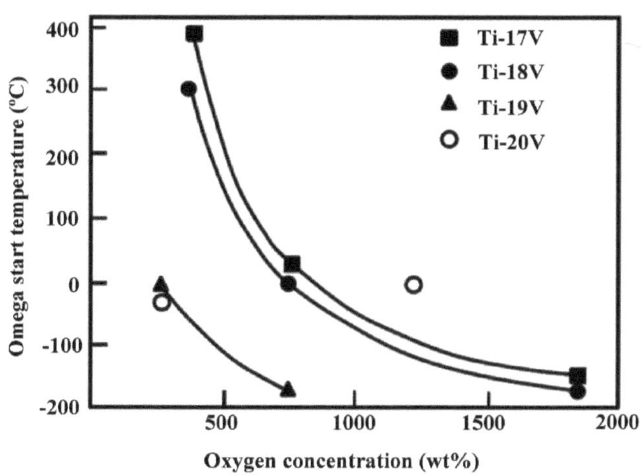

Figure 29.10 Illustrative graph to show that the temperature for the athermal ω phase to start to precipitate from the β matrix in Ti-V β alloys is a function of the oxygen concentration [31].

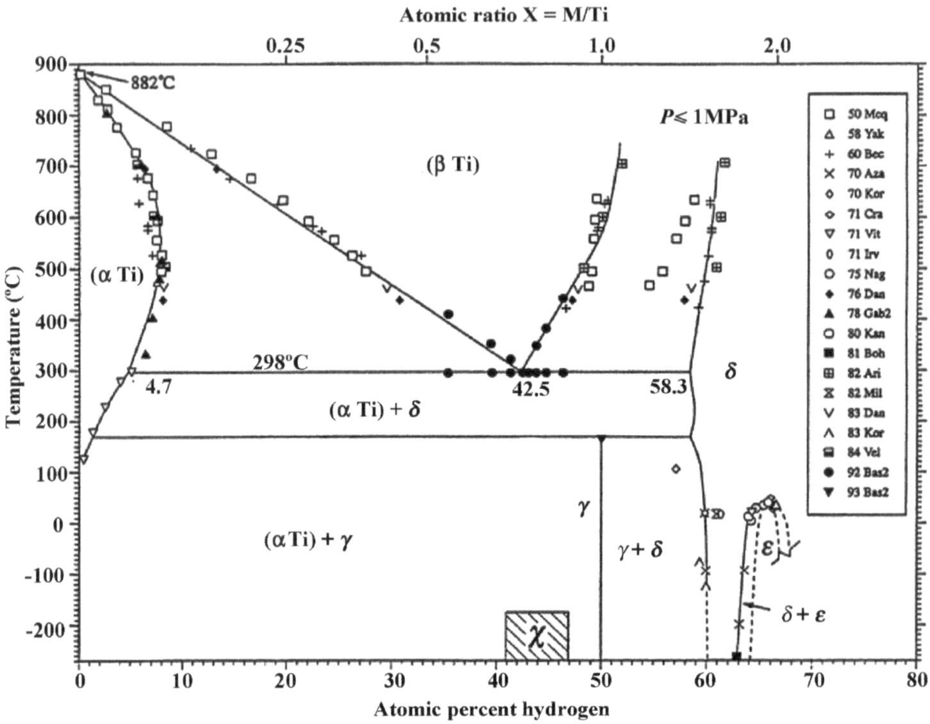

Figure 29.11 The Ti-H phase diagram under pressure less than or equal to 1 MPa [32].

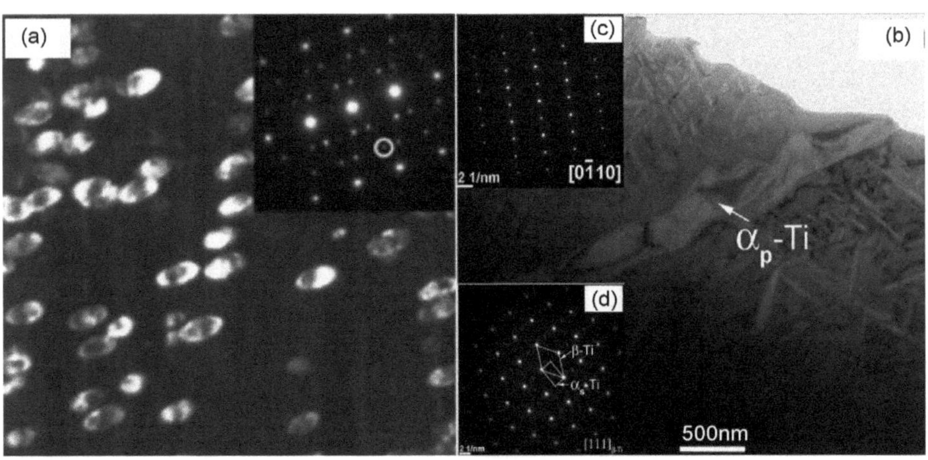

Figure 29.12 TEM results for forged Ti-15Mo that shows the presence of ω nanoparticles by both bright field image and electron diffraction (a) [34]. In contrast, there is no apparent evidence for the existence of the ω phase in the TiH_2-based as-sintered Ti-15Mo [19], although the two alloys have virtually identical composition, see (b) for bright field image, (c) for diffraction of the α phase, and (d) for the β phase.

29.3.4 Martensite phases (α' and α'') in PM Ti and Ti alloys

Owing to the slow cooling rate after isothermal sintering, as-sintered PM Ti materials do not normally contain the metastable, quenching-induced martensite phases (the α' or α'' phase). With the increasing importance of additive manufacturing (cooling rate up to 10^4 K/s) [35,36], the newest PM technology, martensite phases may become regular microstructural features in additively manufactured Ti alloys. Martensite phases in PM Ti and Ti alloys may have the following features [15,37–40]:

- Martensitic transformation can occur in both α- and β-Ti alloys, and α' and α'' are the two most common forms of martensite in Ti and Ti alloys. For a given Ti alloy, α' tends to exist when there is only a low concentration of β stabilizers while α'' normally shows in alloys with a high concentration of β stabilizers (Figure 29.3).
- α' has an *hcp* structure with a = 0.28 nm and c = 0.47 nm. It can have an orientational relationship with the parent β phase as $(0\ 0\ 0\ 1)_{\alpha'}//(1\ 1\ 0)_\beta$ and $[1, 1, -2, 0]_\alpha'//[1\ 1\ 1]_\beta$. The α' phase can show an acicular or lath/lamellar morphology (see Figure 29.13 [39] and Figure 29.14) [40]. It should be noted that α' is very similar to α in terms of both crystal structure and lattice parameters, making it difficult to differentiate between them via X-ray or electron diffraction. The two phases, however, are largely different in phase constitution.
- α'' is an orthorhombic structured phase with lattice parameters of a = 0.30 nm, b = 0.50 nm, and c = 0.47 nm. The $(0\ 0\ 1)_\alpha''//(1\ 1\ 0)_\beta$, and $[1\ 1\ 0]_\alpha''//[1\ 1\ 1]_\beta$ orientational relationship is observable between the parent β phase and the α'' phase. The α'' phase normally tends to show an acicular morphology.
- Martensite phases are expected to decompose into the thermodynamically stable α phase or a combination of ($\alpha + \beta$) upon annealing, following the pathway of (α' and/or α'') $\rightarrow \beta \rightarrow \alpha$ or (α' and/or α'') $\rightarrow \beta \rightarrow (\alpha+\beta)$.

29.3.5 The α_2 phase in PM Ti and Ti alloys

- The α_2 (Ti_3Al) phase is *hcp* structured with lattice parameters of (a = 0.464 nm and c = 0.578 nm).

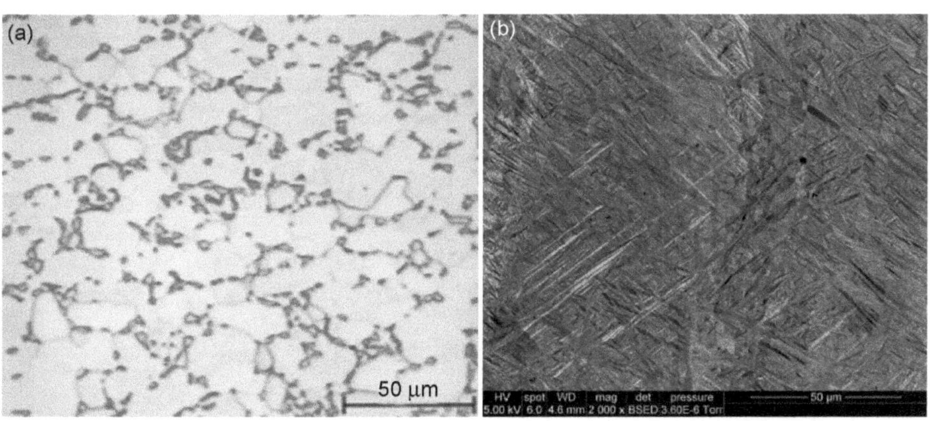

Figure 29.13 Comparison of the microstructure of Ti-6Al-4V: (a) forged [39] and (b) fabricated by selective laser melting (SLM). The former consists of equiaxed $\alpha+\beta$ phases while the latter is composed of martensite phase (α' in this case).

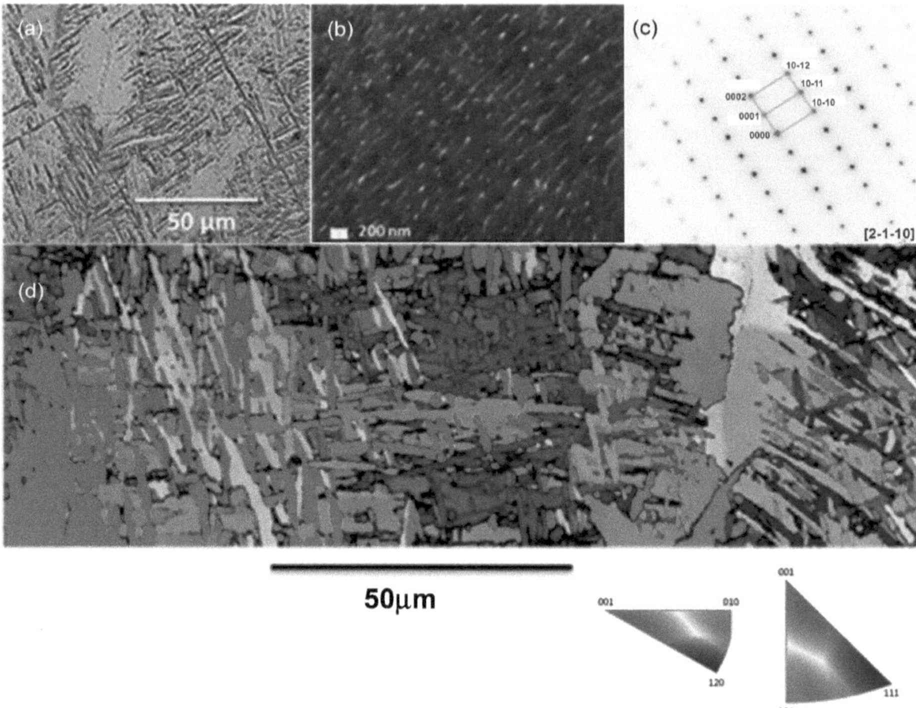

Figure 29.14 Example of the α' phase from an electron-beam-melted (EBMed) Ti-6Al-4V. (a) Optical micrograph showing the (featherless) martensite phase, (b) a high-resolution SEM image to show that β phase coexists with the α′ phase; (c) TEM electron diffraction to confirm the martensite phase; and (d) EBSD image to show the martensite phase and the surrounding microstructure [40].

- The formation of ordered α_2 can readily embrittle a titanium alloy [41].
- Oxygen is a strong α_2 stabilizer which can promote the precipitation of α_2 from the α matrix [17]. The other pathway to form α_2 phase is through annealing [42–46].
- The aging-induced α_2 tends to exist as nanoparticles in PM Ti and Ti alloys. Figure 29.15 shows an example from the PM Ti-15Mo, and Table 29.1 lists the features of the phase collected from some other Ti alloys [18, 42–46].

29.4 Analytical techniques for microstructural characterization of PM Ti and Ti alloys

29.4.1 Comparison of analytical approaches

Table 29.2 summarizes the essential microstructural characterization techniques for various materials including PM Ti materials and their main features [47–50]. Since pores are part of the microstructure of an as-sintered Ti material, techniques such as positron annihilation spectroscopy (PAS), 3D tomography of porosity by either

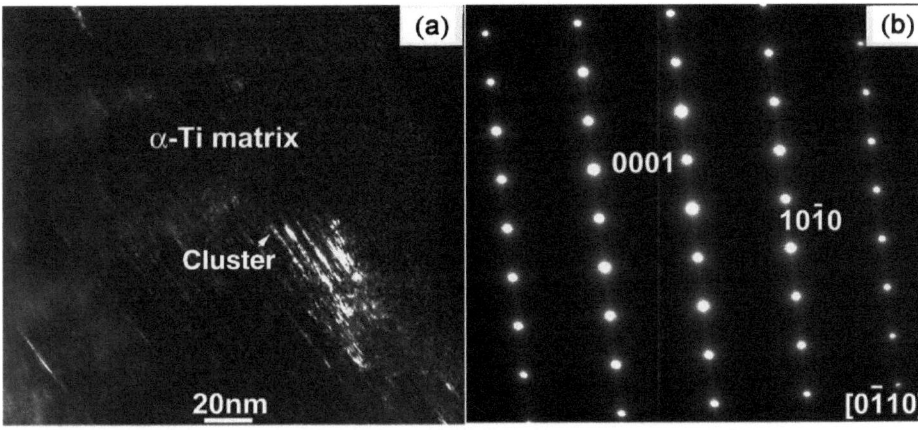

Figure 29.15 TEM results for α_2 phase in PM Ti-6Al-4V. (a) TEM dark-field (DF) image showing the fine clusters inside the α-Ti phase, where the DF operation was taken using the diffuse steaks in (b), and (b) TEM–SAED pattern revealing diffuse scattering along regular diffraction (0 0 0 1) spots of α-Ti [18].

Table 29.1 α_2 characteristics summarized from literature [42–46]

Type	Alloy composition (wt%)	Diameter of α_2 (nm)	Length of α_2 (nm)	References
Near α	Ti-5.6Al-4.8Sn-2Zr-1Mo-0.35Si-0.7Nb	8.2–30.6	19.8–67.1	42
Near α	Ti-6Al-2Sn-4Zr-2Mo	~5	~55	43
Near α	Ti-5.5Al-3.5Sn-3Zn-1Cr-0.25Mo-0.3Si	~5	–	44
α+β	Ti-6Al-2Cr-2Mo-2Sn-2Zr	5–10	20–50	45
α+β	Ti-6Al-2Cr-2Mo-2Sn-2Zr	30	–	46

Table 29.2 Summary of mostly available microstructural characterization approaches [47–50]

Approach	Type	Feature
X-ray photoemission spectroscopy (XPS)	Composition analysis	Surface analysis (signal from ~ <10 nm beneath surface); space resolution ~100 μm
Electron probe microanalyzer (EBM)	Composition analysis	Quantitative analysis; space resolution ~ 1 μm
Energy dispersive X-ray spectroscopy (EDX)	Composition analysis	Semiquantitative (without standard) and quantitative analysis (with standard); space resolution from micrometers (with SEM) to nanometers (with TEM)

(Continued)

Table 29.2 Summary of mostly available microstructural characterization approaches [47–50] *(cont.)*

Approach	Type	Feature
Electron energy loss spectroscopy (EELS)	Composition analysis	Semiquantitative and quantitative analysis; space resolution ~1 nm
Atom probe tomography (APT)	Composition analysis	Semiquantitative; space resolution ~0.1 nm
X-ray diffractometer (XRD)	Structure determination	Space resolution ~10 μm; wide temperature range when using cold or hot stage
Synchrotron radiation (SR)	Structure determination	Space resolution ~10 μm; wide temperature range when using cold or hot stage; time-resolved study
Electron backscatter diffraction (EBSD)	Structure determination at diffraction mode	Space resolution ~10 nm
Transmission electron microscopy (TEM)	Structure determination via electron diffraction	Space resolution 0.1 nm; wide temperature range when using cold or hot stage; time-resolved study
Optical microscopy (OP)	Morphology analysis	Space resolution ~1 μm
EBSD	Morphology analysis at imaging mode	Space resolution ~10 nm
Positron annihilation spectroscopy (PAS)	Morphology analysis	Void/pore analysis; space resolution ~1 nm
Scanning electron microscopy (SEM)	Morphology analysis	Space resolution ~1 nm; wide temperature range when using cold or hot stage
Focused ion beam	Morphology analysis	Space resolution ~1 nm; 3D information
TEM	Morphology analysis via imaging mode	Space resolution ~0.1 nm; wide temperature range when using cold or hot stage
APT	Morphology analysis	Space resolution ~0.1 nm; 3D information

normal optical microscopy, scanning electron microscopy (SEM) or by focused ion beam (FIB) via applying a sequential sectioning method [51,52], micro–computed tomography (micro-CT) are useful to characterize as-sintered Ti materials [53].

29.4.2 Transmission electron microscopy for PM Ti and Ti alloys

TEM characterization involves sample preparation, operations (imaging and diffraction), and interpretation of relatively complex results [54]. As an important characterization technique, TEM has been widely used in the study of PM Ti and Ti alloys. It is a capable analytical tool in many aspects, including the following:

- Phase identification through crystallographic and compositional analysis (see Figure 29.16) [19];
- Morphological observation through imaging operation;

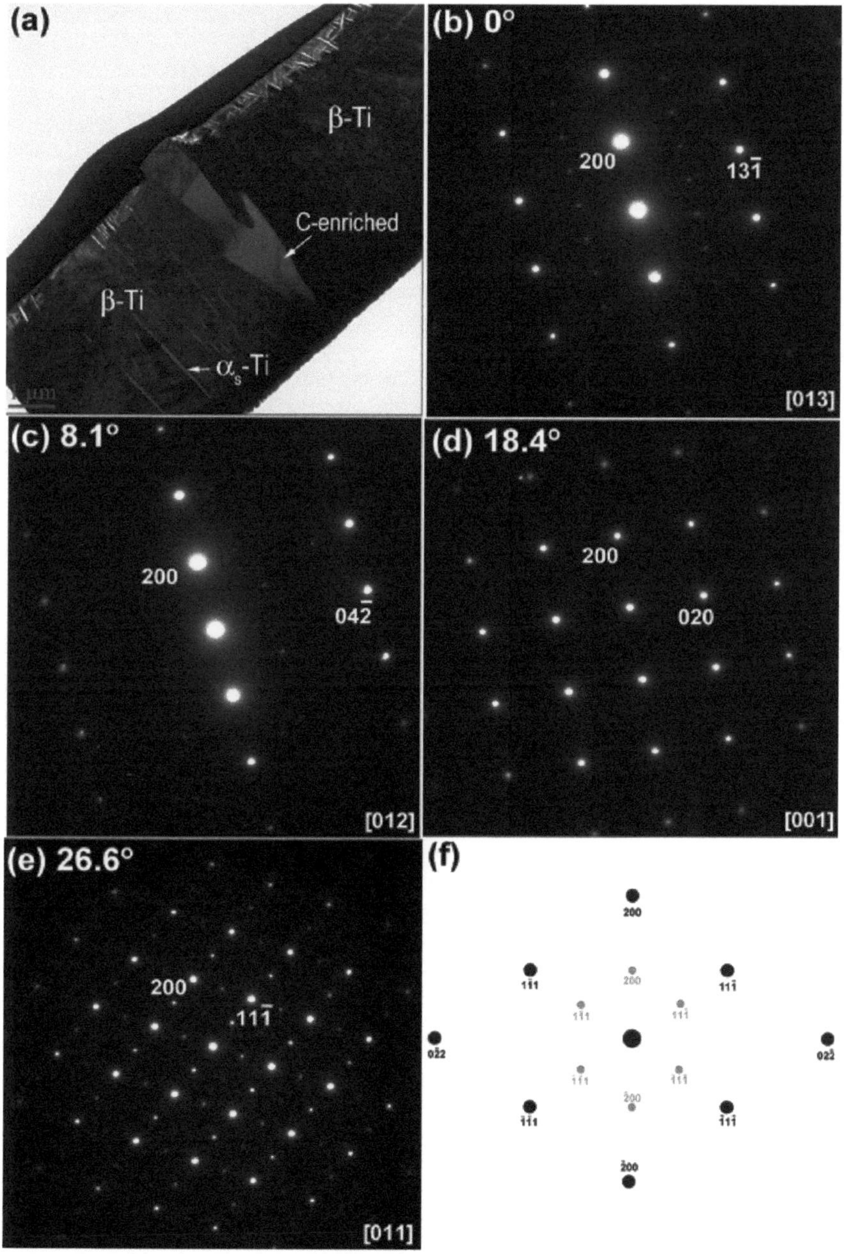

Figure 29.16 A typical example to show that TEM is a capable tool to study key issues such as secondary phases (and then solution limit of carbon in Ti) in PM Ti and Ti alloys (Ti-15Mo in this case). (a) TEM-BF image of the β-Ti matrix, secondary α-Ti (α_s) and carbon-enriched grain boundary phase in an FIBed sample. (b–e) TEM-SAED patterns for the carbon-enriched grain boundary phase, where diffraction maxima can be observed in (b) and (e). (f) Illustration of the other index option for the carbon-enriched GB phase [19].

- Study of phase transformation mechanism through crystallographic analysis; and
- Study of deformation mode.

29.4.3 Focused ion beam in the study of PM Ti and Ti alloys

- When interested phases in a sample are low in volume fraction, it may become difficult to prepare samples by conventional techniques such as precision ion polishing system (PIPS) or electrochemical etching for TEM analysis. Modern FIB is normally equipped with both ion beam (for thinning) and electron beam (for observation, similar to SEM). It can first locate interested features using the electron beam and then thin the sample using the ion beam, making it a potent technique to prepare site-specific TEM samples [55].
- FIB is also an important approach to making high-quality samples for 3D-APT studies [56].
- Aside from sample preparation, with dual-beam FIB, 3D tomography information of interested phases in PM Ti materials (see Figure 29.17) can be revealed using an approach of series sectioning plus imaging and with the assistance of professional computer software packages [19].

29.4.4 Synchrotron radiation in the study of the PM Ti and Ti alloys

High-energy synchrotron radiation (SR) has become increasingly important as a potent and comprehensive characterization approach. The working principle of SR is to project high-flux, high-brilliance photons onto target materials to generate compositional/chemical information (via spectroscopy and spectrochemical analysis), structural information (via diffraction and scattering analysis), and/or image, based on the interaction between the photons and the material to be analyzed, and all these can be completed within a very short period of machine time [57,58]. Many countries in the world are building and/or extending their SR facilities and capabilities, and Table 29.3 lists a few typical SR facilities in the world.

For PM Ti and Ti alloys, SR can be used to provide unique information in the following three aspects:

- Temperature-dependent microstructural evolution (e.g., dynamic decomposition of YH_2 at elevated temperatures, Figure 29.18, in CP-Ti to disclose the decomposition pathway of YH_2 as well as the oxygen-gettering mechanism involved) [59];
- Time-resolved microstructural evolution (e.g., Figure 29.19, which reveals how the reaction proceeds with time to form TiC-TiB_2 composite based on PM Ti) [60]; and
- Structure refinement (e.g., Figure 29.20, which presents a quantitative analysis of the phase constitution in α+β Ti-6Al-2Sn-4Zr-2Mo alloy using the Rietveld method) [61].

29.4.5 3D atom probe tomography in the study of the PM Ti and Ti alloys

Three-dimensional APT uses a position-sensitive detector to record the lateral (i.e., *x* and *y*) position of ions/atoms and a time-of-flight counter to generate the *z* direction

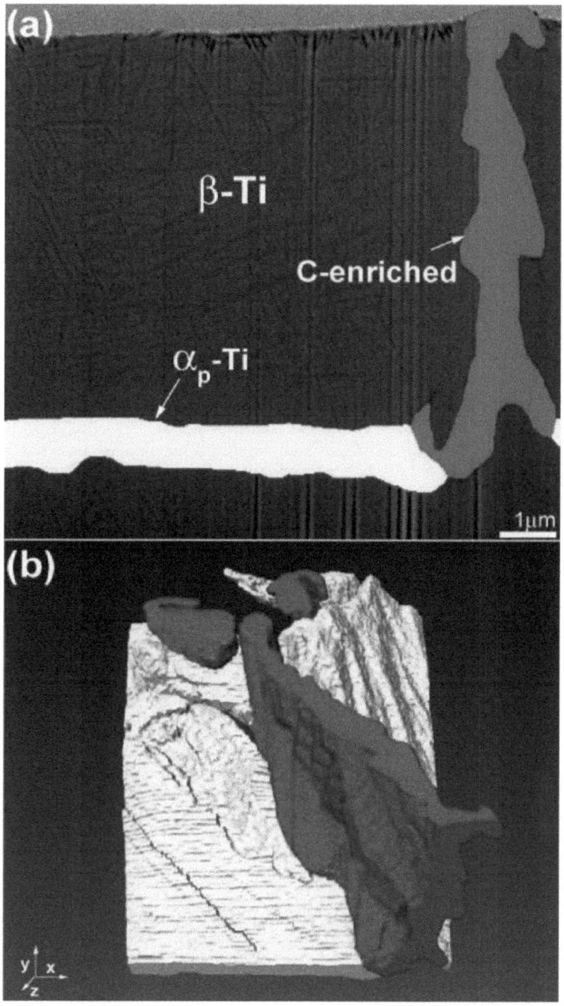

Figure 29.17 Three-dimensional APT example from a PM Ti-15Mo alloy. (a) SEM-SE image to show the carbon-induced phase (dark gray [red color in the web version]), primary α-Ti (light gray [yellow color in the web version]), and the β-Ti matrix. In total, 184 such images/slices were obtained and the movie in Appendix A of the reference shows a serial presentation of these images. (b) A snapshot of the 3D tomography movie (Appendix B of the reference) to show the morphology of the carbon-enriched phase (dark gray [red color in the web version]) and primary α-Ti (light gray [yellow color in the web version]) [19].

information; the 3D information (*x*, *y*, and *z*) of any atom in the sample can therefore be realized [62,63]. 3D-APT is a unique technology that is able to provide compositional and/or morphological information of fine features down to the atomic level; the study aforementioned for the analysis of the ω phase (~10 nm) in PM Ti is a typical example. Figure 29.21 shows another example of using 3D-APT to resolve a key question for PM Ti materials: the compositional characteristics of the ω phase [64].

Table 29.3 Selective examples of the SR facilities in the world

Country	Name	Electron energy (GeV)	Emittance (nm·rad)	Amount of beamlines
Australia	Australian Synchrotron	3	12	9
China	Shanghai Synchrotron Radiation Facility	3.5	3.9	8
France	European Synchrotron Radiation Facility	6	3.9	45
Germany	PETRA III	6	1	14
Japan	SPring-8	8	5.9	57
USA	Advanced Photon Source (Argonne)	7	2.5	34

Note: Reference materials include the books [57,58] as well as information obtained from internet.

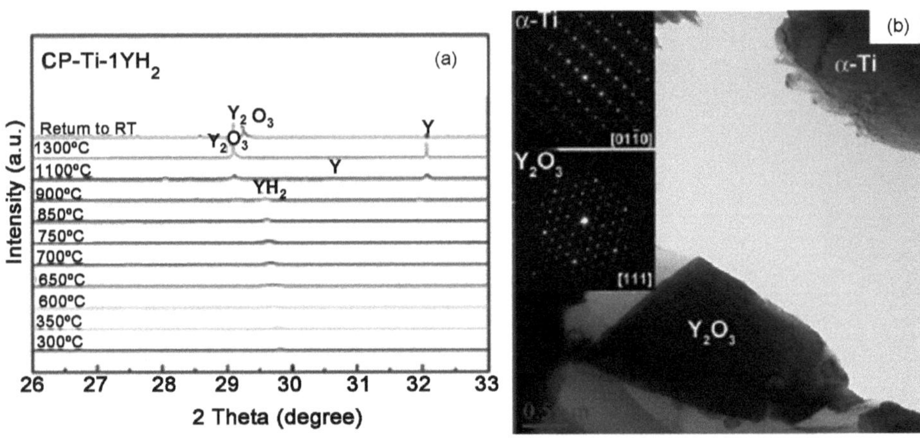

Figure 29.18 SR was used to study the decomposition of YH_2 in a PM CP-Ti [59]: (a) the high-energy SR signals collected from room temperature to sintering temperature (1300 °C) and (b) TEM results to show the room-temperature phase constitution.

29.4.6 Electron backscatter diffraction in the study of the PM Ti and Ti alloys

Electron backscatter diffraction (EBSD) is to correlate Kikuchi bands generated from individual grains of a material to their crystal structure and their orientation. It is an extremely useful analytical approach to analyzing microstructure in the following three aspects: (a) grain size of materials and crystal orientation, (b) texture in microstructure and its relationship with individual grains, and (c) grain boundary misorientation [65]. Figure 29.13d provides an example of using EBSD to study the grain size and orientation of additively manufactured (selective electron beam melting in this case) Ti-6Al-4V [40].

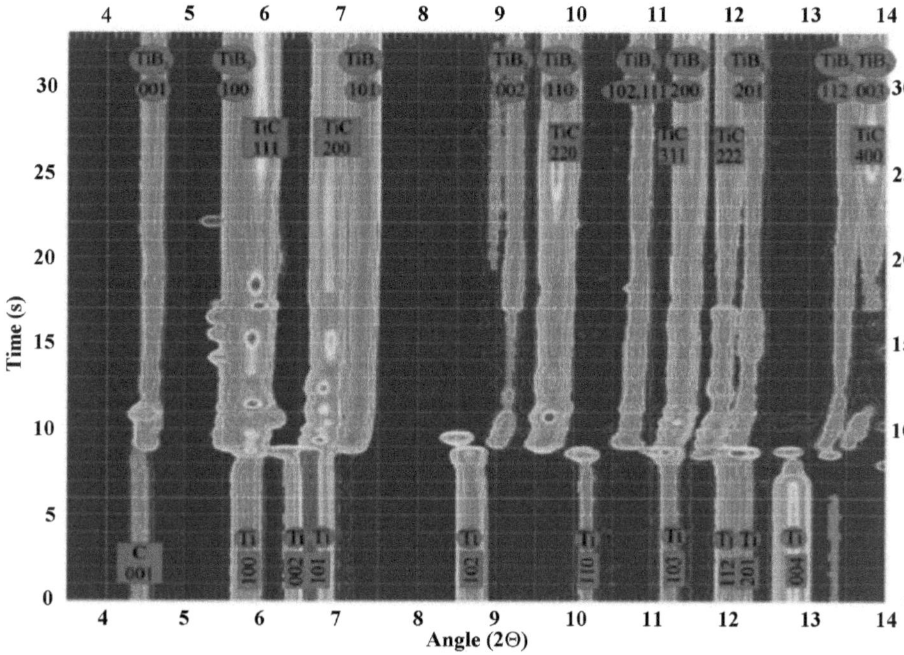

Figure 29.19 Time-resolved high energy SR results to show the reaction sequences in C- and B-doped PM Ti to form the high performance Ti-TiC-TiB_2 composite [60].

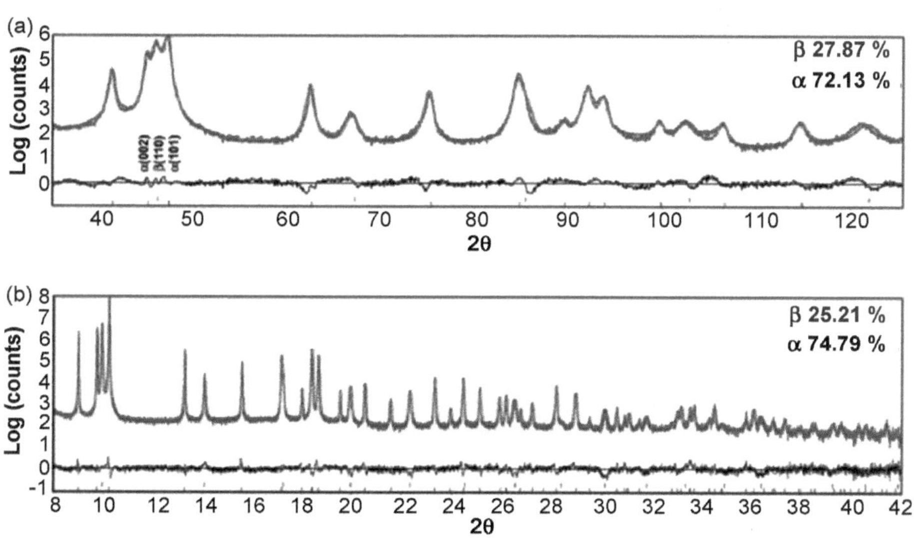

Figure 29.20 Results about the phase constitution in an (α+β)-Ti alloy Ti6264 (Ti-6Al-2Sn-4Zr-2Mo) collected by lab XRD (a) and SR (b), and then analyzed by the Rietveld method [61]. Note that the latter gives sharper peaks and stronger signal intensity than the former, which lead to better accuracy in data analysis.

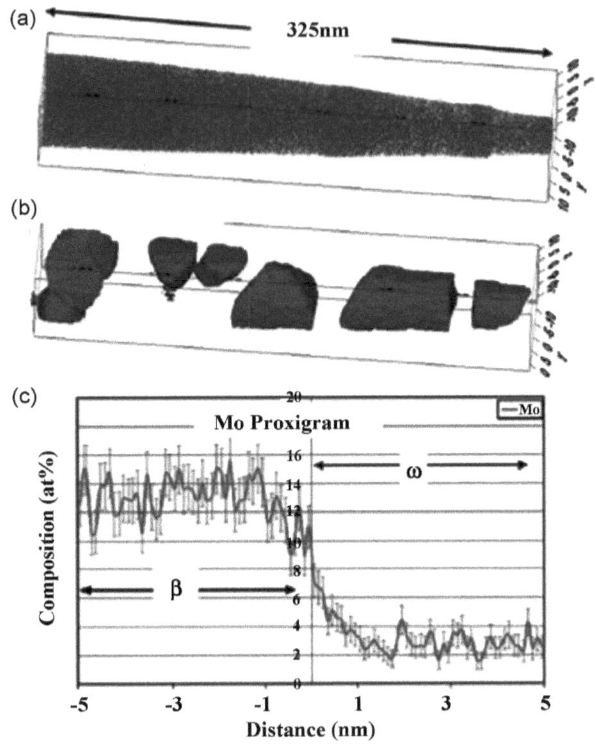

Figure 29.21 Another example to show the 3D-APT in analyzing nanoscale fine cluster (the ω phase in this case) [64]. (a) An atom probe reconstruction of Ti–9Mo aged for 48 h at 475 °C, with Ti atoms in blue and Mo atoms in red showing the well-developed ω precipitates; (b) Ti-enriched regions (β) highlighted by 92% Ti isoconcentration surface; and (c) the Mo proximity histogram using Ti = 92 at% isoconcentration surface showing partitioning of Mo between β and ω phases.

29.5 Concluding remarks

Microstructural characterization is essential in understanding the performance of materials including PM Ti and Ti alloys. It also produces critical knowledge for designing novel PM Ti alloys. Most microstructural characterization techniques discussed in this chapter are well established and readily available. In this regard, a reasonable understanding of the microstructure of PM Ti materials has been developed using many of the microstructural characterization techniques discussed. With the increasing attention to TiH_2 powder metallurgy and additive manufacturing of titanium alloys, the microstructural development that is associated with each process will need to be understood before both technologies are fully embraced by industry. In addition, because the microstructure of the prealloyed powders may have an impact on the final microstructure of the additively manufactured Ti materials, microstructural characterization of the prealloyed powders themselves, particularly the β-Ti type, may need research attention as well.

Acknowledgments

M. Yan acknowledges finical support from a Queensland Smart Future Fellowship (Early Career). Figure input (Figure 29.13b) of Dr. W. Xu of RMIT University is acknowledged.

References

[1] G. Lütjering, Influence of processing on microstructure and mechanical properties of (α+ β) titanium alloys, Mater. Sci. Eng. A 243 (1) (1998) 32–45.
[2] P.C. Collins, B. Welk, T. Searles, J. Tiley, J.C. Russ, H.L. Fraser, Development of methods for the quantification of microstructural features in α+ β-processed α/β titanium alloys, Mater. Sci. Eng. A 508 (1) (2009) 174–182.
[3] F.H. (Sams) Froes, D. Eylon, Powder metallurgy of titanium alloys, Int. Mater. Rev. 35 (1) (1990) 162–184.
[4] Y. Liu, L.F. Chen, H.P. Tang, C.T. Liu, B. Liu, B.Y. Huang, Design of powder metallurgy titanium alloys and composites, Mater. Sci. Eng. A 418 (1) (2006) 25–35.
[5] M. Qian, Cold compaction and sintering of titanium and its alloys for near-net-shape or preform fabrication, Int. J. Powder Metall. 46 (5) (2010) 29–44.
[6] W. Hongtao, Z.Z. Fang, P. Sun, A critical review of mechanical properties of powder metallurgy titanium, Int. J. Powder Metall. 46 (5) (2010) 45–57.
[7] I.M. Robertson, G.B. Schaffer, Comparison of sintering of titanium and titanium hydride powders, Powder Metall. 53 (1) (2010) 12–19.
[8] M. Yan, S.D. Luo, G.B. Schaffer, M. Qian, TEM and XRD characterisation of commercially pure α-Ti made by powder metallurgy and casting, Mater. Lett. 72 (2012) 64–67.
[9] M.J. Donachie, Titanium: A Technical Guide, ASM International, Materials Park, OH 44073-0002 (2000).
[10] L. Christoph, M. Peters, Titanium and Titanium Alloys, Wiley-VCH, Weinheim, 2003.
[11] L. Gerd, J.C. Williams, Titanium, vol. 2, Springer, Berlin, 2003.
[12] Y.F. Yang, H. Imai, K. Kondoh, M. Qian, Comparison of spark plasma sintering of elemental and master alloy powder mixes and prealloyed Ti-6Al-4V powder, Int. J. Powder Metall. 50 (1) (2014) 41–47.
[13] M. Yan, S.D. Luo, G.B. Schaffer, M. Qian, Impurity (Fe, Cl, and P)-induced grain boundary and secondary phases in commercially pure titanium (CP-Ti), Metall. Mater. Trans. A 44 (8) (2013) 3961–3969.
[14] H. Richard G, D.R. Trinkle, J. Bouchet, S.G. Srinivasan, R.C. Albers, J.W. Wilkins, Impurities block the α to ω martensitic transformation in titanium, Nat. Mater. 4 (2) (2005) 129–133.
[15] M.A.H. Gepreel, Recent developments in the study of recrystallization, in: P. Wilson (Ed.), Recent Developments in the Study of Recrystallization, InTech, Croatia, 2013.
[16] J.L. Murray, The Al–Ti (aluminum-titanium) system, Phase Diagrams Binary Titanium Alloys 44 (1987) 12–24.
[17] S.D. Luo, M. Yan, G.B. Schaffer, M. Qian, Sintering of titanium in vacuum by microwave radiation, Metall. Mater. Trans. A 42 (8) (2011) 2466–2474.
[18] M. Yan, M.S. Dargusch, T. Ebel, M. Qian, A transmission electron microscopy and three-dimensional atom probe study of the oxygen-induced fine microstructural features in as-sintered Ti–6Al–4V and their impacts on ductility, Acta Mater. 68 (2014) 196–206.
[19] M. Yan, M. Qian, C. Kong, M.S. Dargusch, Impacts of trace carbon on the microstructure of as-sintered biomedical Ti–15Mo alloy and reassessment of the maximum carbon limit, Acta Biomater. 10 (2) (2014) 1014–1023.

[20] M. Yan, Y. Liu, Y.B. Liu, C. Kong, G.B. Schaffer, M. Qian, Simultaneous gettering of oxygen and chlorine and homogenization of the β phase by rare earth hydride additions to a powder metallurgy Ti–2.25 Mo–1. 5 Fe alloy, Script. Mater. 67 (5) (2012) 491–494.
[21] F. Prima, P. Vermaut, G. Texier, D. Ansel, T. Gloriant, Evidence of α-nanophase heterogeneous nucleation from ω particles in a β-metastable Ti-based alloy by high-resolution electron microscopy, Script. Mater. 54 (4) (2006) 645–648.
[22] Joanne L. Murray, Phase diagrams of binary titanium alloys, ASM Int. (1987) 354.
[23] J.L. Murray, The Fe− Ti (iron-titanium) system, Bull. Alloy Phase Diagr. 2 (3) (1981) 320–334.
[24] S.D. Luo, Y.F. Yang, G.B. Schaffer, M. Qian, The effect of a small addition of boron on the sintering densification, microstructure and mechanical properties of powder metallurgy Ti–7Ni alloy, J. Alloys Compd. 555 (2013) 339–346.
[25] Y.F. Yang, S.D. Luo, G.B. Schaffer, M. Qian, The sintering, sintered microstructure and mechanical properties of Ti-Fe-Si alloys, Metall. Mater. Trans. A 43 (12) (2012) 4896–4906.
[26] B.Y. Chen, K.S. Hwang, K.L. Ng, Effect of cooling process on the α phase formation and mechanical properties of sintered Ti–Fe alloys, Mater. Sci. Eng. A 528 (13) (2011) 4556–4563.
[27] B.S. Hickman, The formation of omega phase in titanium and zirconium alloys: a review, J. Mater. Sci. 4 (6) (1969) 554–563.
[28] D. De Fontaine, N.E. Paton, J.C. Williams, The omega phase transformation in titanium alloys as an example of displacement controlled reactions, Acta Metall. 19 (11) (1971) 1153–1162.
[29] J.C. Williams, D. De Fontaine, N.E. Paton, The ω-phase as an example of an unusual shear transformation, Metall. Trans. 4 (12) (1973) 2701–2708.
[30] A. Devaraj, R.E.A. Williams, S. Nag, R. Srinivasan, H.L. Fraser, R. Banerjee, Three-dimensional morphology and composition of omega precipitates in a binary titanium–molybdenum alloy, Scripta Mater. 61 (7) (2009) 701–704.
[31] S. Nag, R. Banerjee, R. Srinivasan, J.Y. Hwang, M. Harper, H.L. Fraser, ω-Assisted nucleation and growth of α precipitates in the Ti–5Al–5Mo–5V–3Cr–0.5 Fe β titanium alloy, Acta Mater. 57 (7) (2009) 2136–2147.
[32] D. Rengen, I.P. Jones, *In situ* hydride formation in titanium during focused ion milling, J. Electron Microsc. 60 (1) (2011) 1–9.
[33] V. Bhosle, E.G. Baburaj, M. Miranova, K. Salama, Dehydrogenation of nanocrystalline TiH_2 and consequent consolidation to form dense Ti, Metall. Mater. Trans. A 34 (12) (2003) 2793–2799.
[34] A. Devaraj, R.E.A. Williams, S. Nag, R. Srinivasan, H.L. Fraser, R. Banerjee, Investigations of omega precipitation in titanium molybdenum alloys by coupling 3D atom probe tomography and high resolution (S) TEM, Microsc. Microanal. 15 (S2) (2009) 268–269.
[35] L.E. Murr, S.A. Quinones, S.M. Gaytan, M.I. Lopez, A. Rodela, E.Y. Martinez, D.H. Hernandez, E. Martinez, F. Medina, R.B. Wicker, Microstructure and mechanical behavior of Ti–6Al–4V produced by rapid-layer manufacturing, for biomedical applications, J. Mech. Behav. Biomed. Mater. 2 (1) (2009) 20–32.
[36] T. Lore, F. Verhaeghe, T. Craeghs, J.V. Humbeeck, J.-P. Kruth, A study of the microstructural evolution during selective laser melting of Ti–6Al–4V, Acta Mater. 58 (9) (2010) 3303–3312.
[37] M. Hiroaki, S. Watanabe, S. Hanada, α′ Martensite Ti–V–Sn alloys with low Young's modulus and high strength, Mater. Sci. Eng. A 448 (1) (2007) 39–48.

[38] L. Shaoqiang, Z. Chen, Z. Wang, J. Liu, Q. Wang, R. Yang, Microstructure study of a rapid solidification powder metallurgy high temperature titanium alloy, Acta Metall. Sin. (China) 49 (4) (2013) 464–474.

[39] V. Bey, L. Thijs, J.-P. Kruth, J.V. Humbeeck, Heat treatment of Ti6Al4V produced by selective laser melting: microstructure and mechanical properties, J. Alloys Compd. 541 (2012) 177–185.

[40] S.L. Lu, M. Qian, M. Yan, H.P. Tang, D. St. John, Unpublished materials (2014).

[41] J.Y. Lim, C.J. McMahon, D.P. Pope, J.C. Williams, The effect of oxygen on the structure and mechanical behavior of aged Ti-8 Wt pct Al, Metall. Trans. A 7 (1) (1976) 139–144.

[42] S.Z. Zhang, H.Z. Xu, G.P. Li, Y.Y. Liu, R. Yang, Effect of carbon and aging treatment on precipitation of ordered α_2 in Ti–5.6 Al–4. 8 Sn–2Zr–1Mo–0. 35 Si–0. 7 Nd alloy, Mater. Sci. Eng. A 408 (1) (2005) 290–296.

[43] M.G. Mendiratta, A.K. Chakrabarti, J.A. Roberson, Embrittlement of Ti-6Al-2Sn-4Zr-2Mo alloy by α 2-phase precipitation, Metall. Mater. Trans. B 5 (8) (1974) 1949–1951.

[44] A.P. Woodfield, P.J. Postans, M.H. Loretto, R.E. Smallman, The effect of long-term high temperature exposure on the structure and properties of the titanium alloy Ti 5331S, Acta Metall. 36 (3) (1988) 507–515.

[45] X.D. Zhang, J.M.K. Wiezorek, W.A. Baeslack III, D.J. Evans, H.L. Fraser, Precipitation of ordered α_2 phase in Ti–6-22-22 alloy, Acta Mater. 46 (13) (1998) 4485–4495.

[46] X.D. Zhang, D.J. Evans, W.A. Baeslack III, H.L. Fraser, Effect of long term aging on the microstructural stability and mechanical properties of Ti–6Al–2Cr–2Mo–2Sn–2Zr alloy, Mater. Sci. Eng. A 344 (1) (2003) 300–311.

[47] B. David, W.D. Kaplan, Microstructural Characterization of Materials, John Wiley & Sons, The Atrium, Southern Fare, Chichester, West Sussex, United Kingdom, 2008.

[48] E.J. Mittemeijer, Fundamentals of Materials Science: The Microstructure–Property Relationship Using Metals as Model Systems, Springer, Berlin Heidelberg, Germany 2010.

[49] C.T. Forwood, L.M. Clarebrough, Electron Microscopy of Interfaces in Metals and Alloys, CRC Press, Techno House, Redcliffe Way, Bistol BS1 6NX, England, 1991.

[50] G. Wu, Microstructural Characterisation and Its Applications, Chemical Industry Press, Beijing, 2009.

[51] L. Shufeng, B.S. Hisashi Imai, K. Kondoh, Powder metallurgy Ti–TiC metal matrix composites prepared by *in situ* reactive processing of Ti-VGCFs system, Carbon 61 (2013) 216–228.

[52] R. Singh, P.D. Lee, C. Trevor, R.J. Lindley, E.F. Dashwood, T. Imwinkelried, Characterization of the structure and permeability of titanium foams for spinal fusion devices, Acta Biomater. 5 (1) (2009) 477–487.

[53] V.B. Simon, G. Kerckhofs, M. Moesen, G. Pyka, J. Schrooten, J.-P. Kruth, Micro-CT-based improvement of geometrical and mechanical controllability of selective laser melted Ti6Al4V porous structures, Mater. Sci. Eng. A 528 (24) (2011) 7423–7431.

[54] B.W. David, C.B. Carter, Transmission Electron Microscopy: A Textbook for Materials Science, Edit. Plenum Press, New York & London, 1996.

[55] J. Orloff, L. Swanson, M. Utlaut (Eds.), High resolution focused ion beams: FIB and its applications, Kluwer Academic/Plenum Publishers, New York, USA, 2003.

[56] J. Takahashi, K. Kawakami, H. Otsuka, H. Fujii, Atom probe analysis of titanium hydride precipitates, Ultramicroscopy 109 (5) (2009) 568–573.

[57] M. Watanabe, S. Sato, Introduction to Synchrotron Radiation Research, Jiao Tong University Press, Shanghai, 2010.

[58] W. Philip, An Introduction to Synchrotron Radiation: Techniques and Applications, John Wiley & Sons, The Atrium, Southern Fare, Chichester, West Sussex, United Kingdom, 2011.

[59] M. Yan, Y. Liu, G.B. Schaffer, M. Qian, *In situ* synchrotron radiation to understand the pathways for the scavenging of oxygen in commercially pure Ti and Ti–6Al–4V by yttrium hydride, Scripta Mater. 68 (1) (2013) 63–66.

[60] L. Contreras, X. Turrillas, G.B.M. Vaughan, Åke Kvick, M.A. Rodrıguez, Time-resolved XRD study of TiC–TiB_2 composites obtained by SHS, Acta Mater. 52 (16) (2004) 4783–4790.

[61] M.M. Attallah, S. Zabeen, R.J. Cernik, M. Preuss, Comparative determination of the α/β phase fraction in α+ β-titanium alloys using X-ray diffraction and electron microscopy, Mater. Character. 60 (11) (2009) 1248–1256.

[62] K.M. Miller, A. Cerezo, M.G. Hetherington, G.D.W. Smith, Atom Probe Field ion Microscopy, Clarendon Press, Oxford, 1996.

[63] G. Baptiste, M.P. Moody, J.M. Cairney, S.P. Ringer, Atom Probe Microscopy. (160) Springer, New York, NY, USA, 2012.

[64] A. Devaraj, S. Nag, R. Srinivasan, R.E.A. Williams, S. Banerjee, R. Banerjee, H.L. Fraser, Experimental evidence of concurrent compositional and structural instabilities leading to ω precipitation in titanium–molybdenum alloys, Acta Mater. 60 (2) (2012) 596–609.

[65] F.J. Humphreys, Review grain and subgrain characterisation by electron backscatter diffraction, J. Mater. Sci. 36 (16) (2001) 3833–3854.

Future prospects for titanium powder metallurgy markets

30

David Whittaker, Francis H. (Sam) Froes***
*DW Associates 231, Coalway Road, Merryhill Wolverhampton, United Kingdom
**Consultant to the Titanium Industry, Tacoma, WA, USA

30.1 Introduction

In assessing the potential for the development of new powder metallurgy (PM) market opportunities for titanium and titanium alloys, it is a common response to couple this with the need for new, lower-cost powder feedstock materials.

In fact, the answer to the question is not quite as straightforward as this and depends critically on the market sector under consideration.

The various potential application sectors for PM titanium products that are the focus of reported R&D activity separate into two broad categories: (a) The first includes those that are already committed to (wrought or cast route) titanium and titanium alloys as a "material of choice" and are therefore fully aware of the high raw material costs involved. For such sectors, interest in PM titanium is based on the prospects of deriving cost savings from the net-shape capabilities of PM forming technologies or on new product characteristics that a powder-based route might offer. (b) The second comprises those where titanium alloy products are not currently being specified, with the high costs involved being seen as the major impediment. Here, markets would have to be won in products, largely being produced in steels currently, on the basis of "the cost being right." For the application sectors comprising category (a), given a choice between the two broad powder considerations, quality will always take precedence over price. This is not, of course, to say, that the emerging, lower-cost powders could not penetrate such markets, but stringent quality requirements would need to be met. To penetrate markets in category (b), low-cost powder would have to be of paramount importance, although this is not to say that quality requirements are not also of significance. Also, cost implications may narrow the choice of potentially viable PM forming technologies in these sectors.

This chapter will therefore begin by discussing the market sectors that have been penetrated to date by wrought and cast titanium products, in order to identify the sectors that comprise category (a).

Established applications for titanium powders will also be briefly discussed (further details can be found in other chapters of this book).

Potential new PM titanium applications and consequent R&D activities in a range of application sectors that fall within the broad category (a) will then be discussed.

Finally, prospects in a number of application sectors in category (b) will be considered in the context of the raw material types and PM forming technologies that are being addressed in the reported R&D activity.

Titanium Powder Metallurgy. http://dx.doi.org/10.1016/B978-0-12-800054-0.00030-7

30.2 Current markets for titanium

Titanium and titanium alloys offer an outstanding combination of properties and characteristics that drives their selection for a range of applications:

- A low density level, approximately half of that of steel
- The capability for delivering high levels of strength and, consequently, very high-levels of specific strength
- Outstanding corrosion resistance, around four times that of stainless steel, by some measures
- The ability to be readily manufactured or fabricated by standard techniques such as forging, casting, welding, or machining

Markets have been established in high added value sectors, where combinations of these attributes have proven to be imperatives in the given applications.

In certain applications sectors, the somewhat "softer" attributes of the high customer appeal and "high-tech" image of titanium have also made a contribution to market development.

30.2.1 Wrought or cast titanium products

Currently, the global usage of titanium products *in all forms* is in the range 85,000–90,000 tonnes p.a. The largest market share (~45,000 tonnes p.a.) is in chemical processing plant and other industrial markets.

The aerospace sector is the second most significant market (currently around 30,000 tonnes p.a.). However, as will be discussed in a later section of this chapter, increasing aircraft production levels over the next decade and the increasing usage of titanium alloys as the metallic "material of choice" in airframe applications may well tip this balance and make aerospace the leading area of application in the future.

The remaining 10,000–15,000 tonnes p.a. of titanium usage is spread across a range of emerging markets, which includes health care, consumer products, and jewelry, among others.

30.2.2 Established markets for titanium powders

The established methods of titanium and titanium alloy powder production have been considered elsewhere in this book and can be briefly summarized here as follows:

- Crushed titanium sponge fines (from the Hunter process)
- Hydrided–dehydrided (HDHed) (from titanium sponge, scrap, or billet) powder
- Rotating electrode process (REP) or plasma rotating electrode process (PREP) powders (starting from cast + forged titanium bars)
- Gas atomized powders (from cast + forged titanium bars or titanium plus alloying additions)
- Plasma spray atomized (starting from wire)

A range of emerging, potentially lower-cost, processes 2008 published statistics on annual production levels of these various powder types are shown in Table 30.1. This table indicates that total titanium powder production was at a level of around

Table 30.1 Estimated global production levels, in 2008, for titanium powders of various types

Powder type	Production level (tonnes p.a.)
Crushed sponge fines	~5000
HDH powders	500–750
Gas atomized powders	200–300
REP/PREP powders	<25
Plasma spray atomized powders	<20
New powder production technologies	6–10

6000 tonne p.a., that crushed sponge fines accounted for the majority of this production, that HDH and gas atomized powders both made significant contributions to the total, but that markets for the other powder types were very limited.

The majority of the current usage of titanium *powders* is in non-PM applications, such as:

- Nucleants for Al castings
- Alloying additives
- Rocket propellants
- Pyrotechnics
- Reagents for chemical processing
- Thermal spray coatings (in medical implant applications). (N.B. Although thermal spraying may use a particulate feedstock, it cannot be considered to be PM as it does not include a forming or consolidation step followed by sintering; it is appropriately defined as Surface Engineering.)

Established PM applications (i.e., the consolidation of powders to form engineering components) for titanium and titanium alloys are limited to niche applications, such as those listed below, and account for an annual powder usage of no more than a few hundred tonnes:

- Sintered filter elements
- Ti-TiB metal matrix composite automotive engine valves (Toyota)
- Metal injection molded (MIM) titanium parts in medical, aerospace, and consumer goods sectors
- Sputtering targets, used as consumables in thin film physical vapor deposition in electronics (Ti-Mo or Ti-W alloys) or in engineering applications (for the deposition of TiN hard coatings), could be considered to be PM products. These products represent a significant current market for hydride–dehydride (HDH) grades of titanium powder.

Table 30.2 provides estimates of titanium PM production levels in 2008, analyzed by the PM forming technology applied.

Most of the current market penetration, in tonnage terms, is based on either hot isostatic pressing (HIP) of prealloyed powders or conventional uniaxial pressing/sintering of elemental blends. Current MIM titanium markets are at a much lower tonnage level.

Notwithstanding these low current levels of market penetration, titanium is a particularly expensive material and, therefore, the net-shape capabilities of PM process

Table 30.2 Estimated global production levels, in 2008, for titanium PM products, analyzed by forming technology

PM forming technology	Production level (tonnes p.a.)
Hot isostatic pressing (HIP)	250–300
Uniaxial press/sinter	100–150
Metal injection molding (MIM)	10–20

routes make their potential adoption particularly attractive in application sectors, where conventional processing (involving significant finish machining) is very wasteful of material, and this has generated much R&D activity over the past decade or more.

The current level of this activity is evidenced by the fact that an international conference, dedicated solely to titanium PM issues and held in Brisbane in December 2011 [1], attracted 127 registered delegates from 15 countries and comprised 78 papers from authors from a similar range of countries.

30.3 New product opportunities in established market sectors

Much of the reported applied research activity comes from application sectors, which are already committed to the use of titanium products in wrought or cast form.

30.3.1 Aero-engines

Potential gas turbine applications are being driven principally by the need to reduce costs of current titanium component production (wrought and machined), which often has poor final material utilization levels, with "buy-to-fly" ratios of 10:1 or higher. In addition, there is a stated aim of utilizing some of PM's unique capabilities for processing novel compositions and exercising superior control over fine-grained microstructure [2–4]. The target applications range from very large components, such as engine casings, to small brackets and fasteners.

There is also an interest in using appropriate powder-based additive manufacturing (AM) technologies for fabrication of components and repair/refurbishment of worn components in this sector (see chapter on AM).

In order to penetrate this market, knowledge will be required on

- Tensile data
- S-N fatigue curve data
- Fatigue crack growth rate data
- Fracture toughness data

A range of PM forming technologies is being studied in this sector. For the production of large components, near-net-shape (NNS)-HIP has been under active

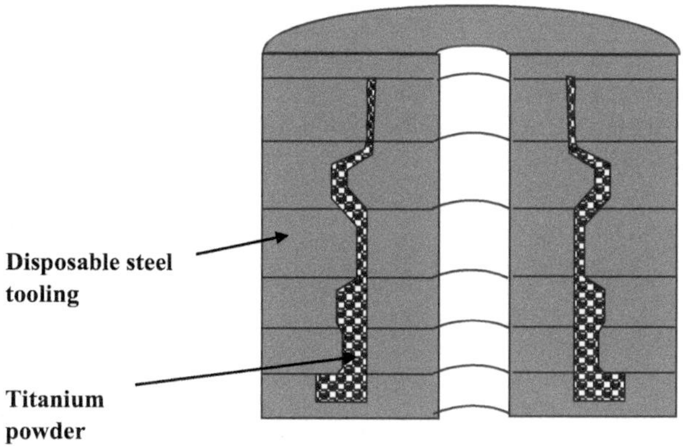

Figure 30.1 Schematic of the near-net-shape HIP technology. (Courtesy Professor Malcolm Ward-Close.)

development for some time (see the chapter on this technology). NNS-HIP steel tooling is based on a concept originally developed by LNT Moscow, in which the elastic and plastic strains induced in the thick-walled, disposable steel tooling and the consolidation characteristics of the titanium powder in the tooling cavities during the HIP cycle are modeled using finite-element methods, so that the final dimensions of the consolidated component can be accurately predicted and controlled. Synertech PM in the USA and Rolls-Royce, in collaboration with the University of Birmingham Interdisciplinary Research Centre, have been particularly active in developing this approach. A schematic of the process is shown in Figure 30.1 [5] and a prototype Ti-6Al-4V component, produced by the technology at the University of Birmingham, is pictured in Figure 30.2 [6].

Figure 30.2 Ti-6Al-4V component produced by near-net-shape HIP. (Courtesy University of Birmingham.)

For NNS-HIP, spherical powders in the size range 45–250 μm are required. Although there has been some use of gas atomized grades in this development work, powders made by PREP are emerging as the preferred option, as they can guarantee freedom from dissolved argon, a source of potential problems in HIP consolidation with inert gas atomized powders.

The use of MIM is being actively considered for the production of small components that are used in large numbers on each engine (N.B. MIM has a requirement for high production volumes to be cost-effective, perhaps even more so than conventional press/sinter PM).

A new UK entrant to MIM production, William Beckett Plastics Ltd., has a stated ambition to move into the manufacture of Ti-6Al-4V aerospace fasteners and has worked with the Mercury Centre of the University of Sheffield to develop and patent a process route. The developed process combines MIM, burnishing, and thread rolling to eliminate all machining requirements and a patented thermomechanical treatment to deliver the required mechanical properties [7].

For MIM, finer spherical powders are required (e.g., $D_{90} = 22$ μm). In reported R&D work, powders produced by each of the following techniques have been used:

- Inert gas atomization
- Plasma spray atomization
- Electrode induction gas atomization (EIGA)
- PREP

AM technologies (both powder bed and deposition methods) are being considered, but often for the addition of "stand-outs" or features to a conventionally produced, base component, in order to reduce machining envelope. Laser deposition is already being used in production for the repair and refurbishment of worn titanium components; for instance, the repair of blade root seals has been discussed in research publications. These deposition technologies can use a feedstock in either a powder or a wire form (see Chapter 24 on AM for further details on this technology).

For AM technologies, spherical powder morphologies with minimal satellites are generally required, typically in the size range 40–150 μm, although recently angular powders have also been used (see the chapter on historical perspective and AM). Relevant traditional production technologies would be the following:

- Inert gas atomization
- Plasma spray atomization
- Plasma spheroidized HDH powders

In all applications, this sector would be looking to use fully prealloyed powders. Ti-6Al-4V is the "workhorse" alloy for the sector. However, a range of other high-strength Ti alloys are used (Rolls-Royce claims to use 10 or more different alloys). Examples of these alloys are Ti-5Al-5V-5Mo-3Cr (Ti-5553) and Ti-10V-2Fe-3Al (Ti-1023). In the gas turbine sector, there is also a significant interest in the use of titanium aluminides.

30.3.1.1 Predictions for market growth

There is significant interest and research in titanium PM evident in all of the leading aero-engine original equipment manufacturers (OEMs).

However, on the basis of publicly available information, it would appear that the interest in the potentially heavier tonnage applications (manufactured by NNS-HIP) currently rests with Rolls-Royce with no equivalent interest/activity having been revealed at GE or Pratt and Whitney.

It might, therefore, be reasonable to set the potential market here at around Rolls-Royce's usage of titanium alloys, that is, rising from the current level of around 2000 tonnes p.a. [2] to perhaps 4000 tonnes p.a. based on the projected increased demand for aircraft (as discussed in the next section on airframes) over the next two decades. (However, as the driving force for the adoption of PM routes is to improve material utilization factors, this might significantly overstate the market potential for powder sales.)

Repair/refurbishment by AM technologies, such as laser deposition, is already established in commercial production. However, other titanium PM technologies are believed to be currently rated at a technology readiness level (TRL) of no more than 3–4.

TRL, which originated in the aerospace sector, is a rating of the maturity of development of a technology, with TRL 1 being an original idea and TRL 9 being technology in full-scale commercial use. At TRL 3–4, a development would still be at the laboratory-scale testing stage.

In such a safety-conscious sector as this, with exacting qualification requirements, it would be surprising therefore if the market were penetrated in less than around a decade.

If the typical buy-to-fly factors are applied as a multiplier on wrought Ti alloy feedstock costs, it is clear that this sector could tolerate the current high costs of commercial Ti alloy powders and still show a raw material cost advantage. However, this is not to say that it would not have an interest in exploiting one of the emerging (potentially lower cost) powder production routes, if the powders were available at the required quality levels.

30.3.2 Airframes

As airframes have evolved from primarily aluminum structures to being predominantly based on polymer composites, titanium alloys have become the main materials of choice for the remaining metallic components. This has arisen because of titanium's closer match with the composite materials in terms of the following:

- Coefficient of thermal expansion (to reduce temperature-related stresses during service)
- Position in the electrochemical series (to reduce galvanic corrosion problems)

From a position of virtually no usage in 1960, titanium components now represent 15–20% by weight of the latest Boeing and Airbus airframes (Figure 30.3).

Many of these component applications are conventionally produced at buy-to-fly ratios above 10:1 and can therefore be considered viable targets for production by

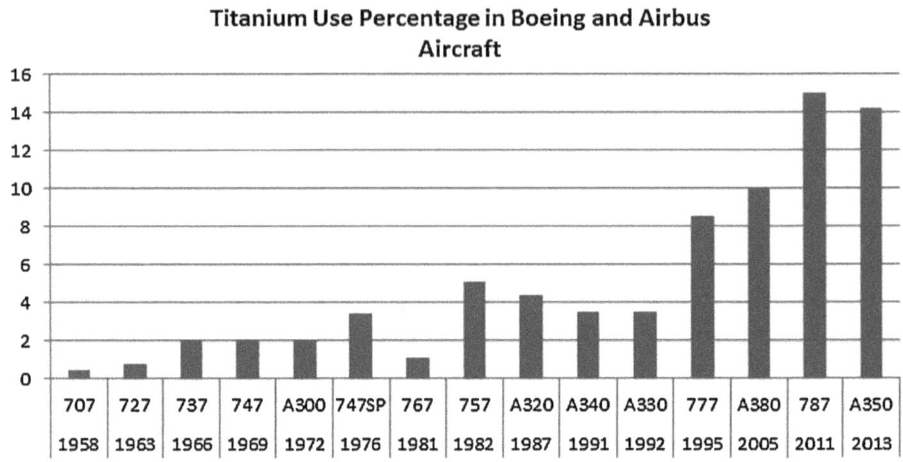

Figure 30.3 Growth in titanium use as a percentage of total gross empty weight on Boeing and Airbus aircraft. Note the decreased use on the 767 was due to a perceived shortage in titanium when this plane was designed. Designers substituted other materials for titanium (such as steel and aluminum). (Image and data courtesy of ASM International.)

a net-shape PM technology. Current development targets span all of the aerospace industry "product criticality" classifications. Components span the whole range from leading edges to small, prismatic parts.

In common with the previous subsection, airframe applications would generally demand knowledge of the following:

- Tensile data
- S-N fatigue curve data
- Fatigue crack growth rate data
- Fracture toughness data

A potentially significant move into the use of cold pressed (and, in some cases, HIPed) blended elemental (BE) powder-route titanium alloys in this sector has been recently signaled by the achievement of Boeing approvals by Dynamet Technology Inc., USA [8]. In a joint collaboration, a Boeing Materials Specification for powder metal titanium alloy, manufactured by Dynamet Technology's PM titanium processing approach, has been developed. As a result, Dynamet has become the sole qualified supplier of Ti-6Al-4V powder metal products to Boeing, opening the door for the production of parts, ranging from fuselage to landing gear components.

Some of these components would be manufactured using Dynamet's patented CHIP technology [9]. This process combines cold isostatic pressing (CIP), sintering, and HIP to produce net-shape parts (Figure 30.4).

However, the company envisages that the performance requirements of some applications will be able to be satisfied by material in the as-sintered condition, without any need for the relatively expensive postsinter HIP process step.

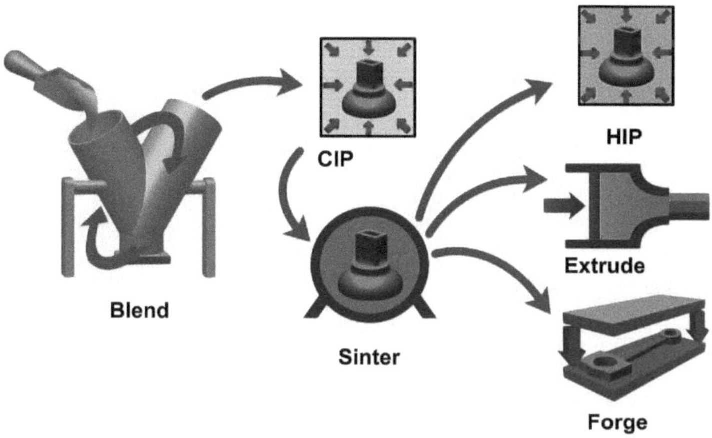

Figure 30.4 Schematic of the Dynamet Technology Inc. CHIP process. (Courtesy Stan Abkowitz.)

In addition to this established or potential use of HIPed or MIM products, the currently high-profile development activity is focused on the building of components by AM technologies.

Unlike the aero-engine sector, there is a major concentration on the building of monolithic AM components by powder-bed technologies, although there are also some applications where the addition of "stand-outs" or features by blown powder deposition technologies is being considered.

In the related space sector, there are AM products already being deployed in satellites and others are in development – some of these components are in titanium alloys.

The powder characteristics for AM are as defined in Section 3.1 and Chapter 24.

As in aero-engine gas turbines, Ti-6Al-4V is the leading alloy of choice, but a range of other high-strength Ti alloys is also deployed.

30.3.2.1 Predictions for market growth

Aircraft production is predicted to be buoyant over the next two decades, with air traffic growing at a compounded annual rate of 4.7%, requiring almost 30,000 new passenger aircraft and freighters at a value of nearly US$4.4 trillion [10].

The interest in development of AM titanium alloys is high in all of the major aircraft producers, although a realistic rating of the current TRL of these developments would be that they are still around TRL 4.

However, aircraft manufacture is an increasingly competitive business, especially in a scenario where these companies are being faced with the need to replace models, which had become major cash cows. There will therefore be a need to take as much cost as possible out of these new models to ensure continuing financial viability. It is envisaged that this aim will contribute to a significant acceleration in the process of taking these AM developments through the TRLs.

Some senior personnel in this sector are predicting that up to 50% of titanium parts may be built by AM within a decade. This would equate to a powder supply opportunity in excess of 10,000 tonnes p.a., which might grow to 20,000 tonnes as aircraft production levels ramp up.

Again, qualification requirements would be stringent, so significant penetration of this market in less than a decade is unlikely.

30.4 Health care

This is a sector that is largely controlled by a small number of multinational suppliers. These companies tend to be quite secretive about their business activities and therefore it is not always straightforward to distinguish between applications that are still at the development stage and those that have already translated into commercial production.

It is known that MIM is being used commercially for the production of dental implants in standard sizes in high volumes. A more recent development in this product type has been revealed in titanium implant tooth anchors, in which two-material MIM is being used to incorporate a porous central portion for tissue ingrowth, as shown in Figure 30.5 [11,12]. Other established medical applications include endoscopy instrument components [13].

A number of innovative medical device applications, based on MIM titanium, have been recently developed by German companies, for example, a component in a sleep apnea device, from OBE GmbH & Co KG (Figure 30.6) [14] and drug delivery ports and heart valves from Element 22 GmbH (formerly Tijet Medizintechnik GmbH) (Figure 30.7) [15].

Figure 30.5 Titanium dental implant with a central porous region for tissue ingrowth. (Courtesy Powder Injection Moulding International.)

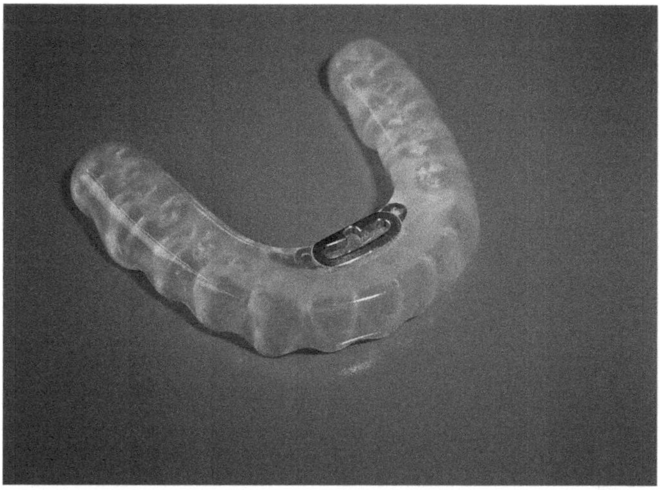

Figure 30.6 Part of a sleep apnea device made by MIM from Ti Grade 4 feedstock. (Courtesy Powder Injection Moulding International.)

It is also known that AM technologies are being used for the manufacture of dental implants in standard sizes, and it believed that AM is already being applied commercially for the production of a range of customized dental and medical implants. In the orthopedic implants sector, titanium alloys compete with Co-Cr-Mo alloys and stainless steels (largely Type 316L) as the "material of choice." The choice for individual operations is often related to the surgeon's personal preference as much as anything. Other drivers for the development of applications in this sector have been reported in

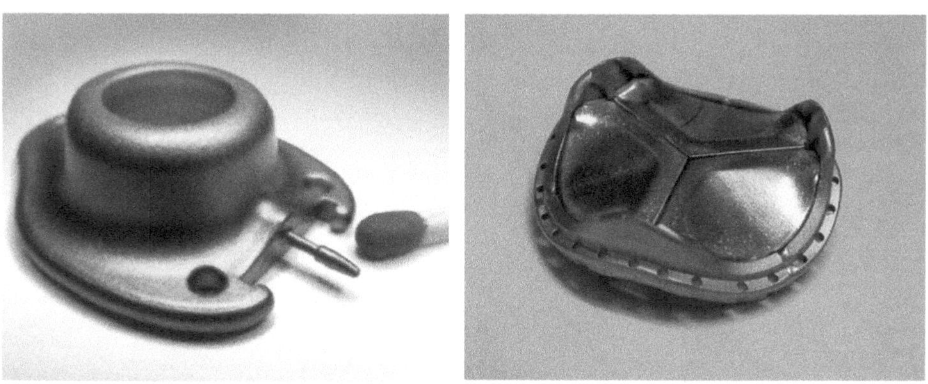

(a) (b)

Figure 30.7 (a) Housing for an implantable port produced by MIM and (b) a heart valve implant assembled from several MIM parts. (Courtesy Powder Injection Moulding International.)

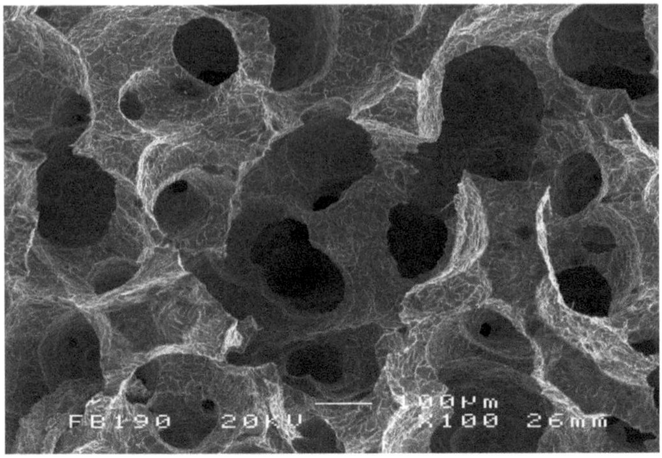

Figure 30.8 Open pore structure created by the HIP + heat treatment process. (Courtesy University of Birmingham and Stanmore Implants Ltd.)

published research work, but here it is often hard to judge as to what has translated into commercial production:

- The provision of a graded structure with porous surface layers to aid bone integration. The most common commercial approach has been to thermally spray a porous surface layer onto a conventionally produced, fully-dense implant. However, a range of other powder-based techniques have been revealed for the production of such a graded structure.
- There is a growing interest in the manufacture of porous scaffolds for tissue engineering. Several technologies have been discussed for this type of application – AM, a range of consolidation techniques of feedstocks with space holder additions (which are subsequently removed by leaching or thermal decomposition) [16], and controlled HIP processing schedules (Figure 30.8) [6]. Metal foam materials might also have a role to play here.

In terms of mechanical properties, many medical implant applications would be looking for materials that give a reasonable match with human bone in terms of compressive strength and stiffness (Young's modulus). Its better match than competing metals in terms of stiffness is one of titanium's competitive advantages in this field.

Other than mechanical properties, biocompatibility is clearly a critical issue for implants. The main processing technologies of interest would appear to be

- MIM and
- AM technologies

Other technologies have also been revealed, including HIP processing for the creation of controlled porous structures and even conventional BE press/sinter PM for the consolidation of feedstocks with shape holder additions.

The powder characteristic required for MIM and AM applications are as defined in Section 3.1.

The main titanium alloy of interest in this sector is generally quoted, again, as being Ti-6Al-4V. However, reported work, particularly on porous scaffolds, has referred to the development of Ti-6Al-7Nb as a more suitable biomaterial than Ti-6Al-4V, because of toxicity concerns with vanadium. There has also been reported attention to the processing of Ni-Ti shape memory alloys for biomedical applications.

30.4.1 Predictions for market growth

Although this sector also has stringent qualification/certification requirements, the fact that several applications have translated to full production already suggests that market penetration may occur more rapidly than in the aerospace sector.

This is not a cost-driven application sector, and it is likely that PM titanium applications will be driven less by cost competition against conventional process routes than by the ability of PM to deliver product attributes not readily derived conventionally, for example, porous structures, individually customized implants.

The sector is one that is already familiar with and committed to the use of titanium alloy products and to the high costs involved.

However, this sector is not believed to be a particularly high tonnage powder usage opportunity, especially in the context that titanium is in competition with two other alloy types as "the material of choice."

30.5 Jewelry

Titanium already has an established presence in jewelry applications (Figure 30.9) based on the following:

- The enormous popularity of body piercing, which requires hypoallergenic material
- The material's light weight and ability to be brightly colored (see Figure 30.9)
- Strong current design trends toward "white" metals
- A perception of titanium by the jewelry-buying public as being "exotic" or "high tech," based on its use in aerospace, motorsport, and sporting goods applications

Figure 30.9 Anodized titanium beads. (Courtesy Jewellery Industry Innovation Centre, School of Jewellery, Birmingham City University, UK.)

Figure 30.10 Additive manufactured titanium parts, currently in development, from a Jack Row design by the Jewellery Industry Innovation Centre, School of Jewellery, Birmingham City University, for Jack Row writing instruments www.jackrow.com

The Engelhard-Clal process has been a long-established PM process in the jewelry sector. This process is used to make gold wedding rings and comprises the manufacture of ring blanks by conventional BE press/sinter PM followed by ring rolling to final dimensions. There is a growing interest in titanium rings and there is no obvious barrier to the process's being adapted for titanium.

MIM of titanium watch bezels and watch bracelet parts is already established in large-scale production, and the manufacture of other jewelry items (e.g., pendants) by MIM has been reported.

The jewelry sector has a significant demand for customized items in small production batches, and the potential for AM for such items is under active research. AM titanium jewelry parts (Figure 30.10) are in current development, building on the sector's familiarity with similar technologies for prototyping.

For jewelry applications, mechanical property requirements are not nonexistent but are relatively modest. Of critical importance in this sector are achievable surface finish, the ability to polish product surfaces, and corrosion resistance.

The need for highly polished finishes implies stringent requirements for levels of non-metallic inclusions in powder materials and freedom from residual porosity in formed and sintered parts.

As in other sectors, MIM would require finer spherical powders (e.g., $D_{90} = 22\ \mu m$). MIM powders for this sector would probably be inert gas atomized.

For AM technologies, traditionally, spherical powder morphologies with minimal satellites would be required, typically in the size range 40–150 μm, with the powder production technologies quoted previously being of relevance.

If the press/sinter PM route were viable, nonspherical powders would be needed for green strength reasons. The obvious current sources of such powders would be the HDH process or sponge fines.

In view of the more modest strength requirements in this sector, there may be more interest in CP-titanium grades than in high-strength alloys. As mentioned previously, quality requirements, in terms of levels of nonmetallic inclusions, may, however, be quite stringent.

30.5.1 *Predictions for market growth*

Once technical viability is proven, there would seem to be less obstacles to early adoption in this sector than most of the others.

However, although the global jewelry sector is large in terms of sales value of products, it is unlikely to be large in terms of tonnage demand for powders.

30.6 Other sectors

There is current development activity on means of direct manufacture of titanium "semi-products" from powders. For instance, Commonwealth Scientific and Industrial Research Organisation, Australia, has discussed its work on sheet rolling (Figure 30.11), extrusion, and spray forming [17] (see Chapters 21 and 22).

Such semiproducts could open up market opportunities in sectors that already use their ingot-route counterparts. These potential applications might include the following:

- Fighting vehicle and personal armor (defense sector)
- Submarine hulls (defense and oil and gas sectors)
- Sour gas well pipes (oil and gas sector)
- Seamless tube products in bicycles frames and tennis racquets (recreational goods sector)
- Sheet products for architectural cladding (built environment sector)

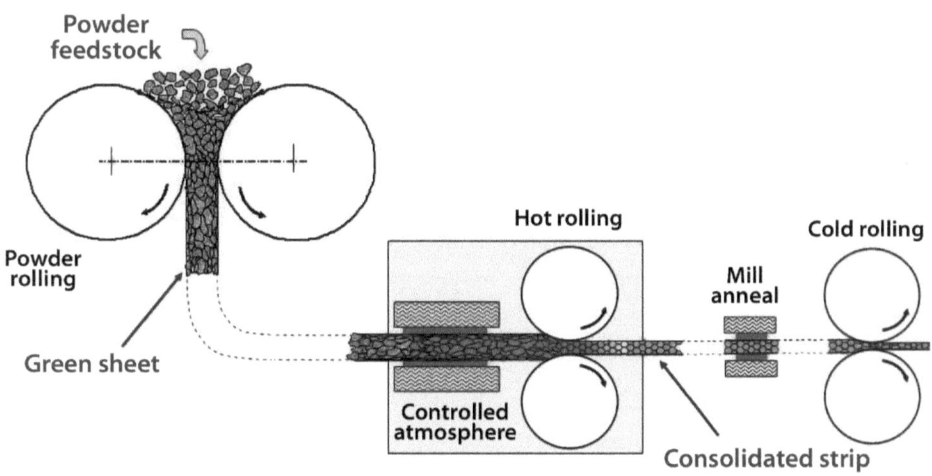

Figure 30.11 Schematic of the process to produce commercial purity titanium sheet at the Commonwealth Scientific and Industrial Research Organisation, Australia. (Courtesy Powder Metallurgy Review.)

- Tubing in power plant turbine steam condensors (energy sector)
- Sheet products in marine applications (marine transport sector)

However, these applications must be regarded as rather speculative at present, until the economics of the powder-based routes are fully defined versus conventional products. There is, therefore, no firm basis for prediction of significant market penetration in the short to medium term.

30.7 Prospects for developing applications in new market sectors – automotive and general engineering

If the appropriate cost targets can be met (discussed later in the market growth prediction statements), recently published studies have identified a wide range of potential titanium applications in automobiles, based on weight saving derived from high specific strength (at ambient and elevated temperatures) or on superior corrosion or oxidation resistance (Figure 30.12) [18]. Systems of interest that could include PM component opportunities are power train (e.g., connecting rods), valve train, exhaust systems, drive shafts, suspension/shock absorber systems, and brake systems.

In reported development work on automotive applications, components such as connecting rods, which are fatigue critical, have been targeted. However, for many applications in this category, components would be specified and designed on the basis of

- Yield strength and
- Ductility

A recent report at the December 2011 Brisbane conference [19] has suggested that mass markets for titanium PM products could be opened up if the following targets could be met:

- around 98% sintered density
- Ultimate tensile strength > 1000 N/mm^2
- Yield strength > 900 N/mm^2
- Elongation $>8\%$

A number of sources [18,19] have reported that an attack on these mass production (but highly cost-sensitive) markets would have to be primarily based on the use of conventional BE press/sinter PM. Of the commercially established powder production processes, HDH is considered to be the only potentially viable option, on the basis of both costs and the irregular particle shape required to promote green strength. However, the hydrided sponge produced by ADMA Products (Chapter 8) is a further option (discussed below).

For such applications, interest is firmly fixed on the BE approach, using HDH-titanium as the base, rather than on prealloyed powders.

A variant on this approach, which had been originally developed by the Ukrainian Academy of Sciences and subsequently exploited by ADMA Products in the United States, is to use TiH_2 powder as the base and to carry out the dehydrogenation stage as part of the sintering process [20,21]. This process can deliver attractive properties in

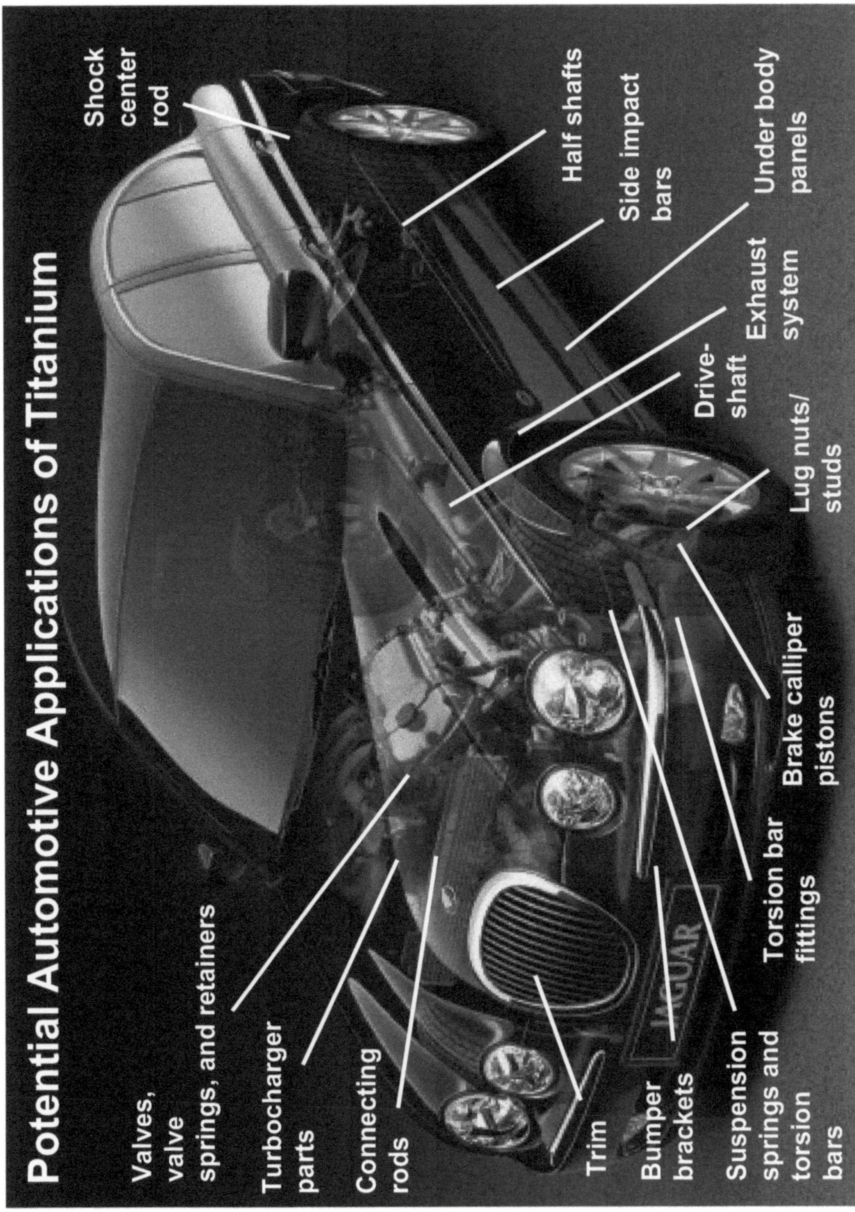

Figure 30.12 Potential automotive applications for titanium. (Source: With permission from Jaguar UK.)

Table 30.3 Room temperature tensile properties of material processed from TiH_2 powder by the ADMA process followed by forging and annealing

PM Ti-6Al-4V	Ultimate tensile strength (MPa)	Yield strength (MPa)	Elongation (%)	Reduction of area (%)
3.5-cm thick	994–1026	911–938	14.0–15.5	34–38
ASTM	897	828	10	25

the as-sintered condition, and these can be enhanced further by postsinter forging and annealing to the levels indicated in Table 30.3. These mechanical properties compare well with those achieved in cast and wrought products. This process has been largely targeted at the automotive sector and a number of automotive parts made by the technique are shown in Figure 30.13.

Figure 30.13 Ti-6Al-4V parts made by a press/sinter approach and titanium hydride. Ti-6Al-4V: (1) connecting rod with big end cap; (2) saddles of inlet and exhaust valves; (3) plate of valve spring; (4) driving pulley of distributing shaft; (5) roller of strap tension gear; (6) screw nut; (7) embedding filter, fuel pump; (8) embedding filter. (Courtesy the Ukrainian Academy of Sciences.)

Alloy systems have been developed that minimize the cost of the alloying elements utilized and maximize sintering activity (to create high sintered density).

Work on the Ti-Fe-Si system, possibly with minor additions of other elements [16,18], has been reported, and the Ti-Cu system [16] has also been identified in this context.

30.7.1 Predictions for market growth

Penetration of the mass-production markets would not be based on selling the relative cost effectiveness of PM processing against competing manufacturing technologies to sectors that have an established commitment to the use of titanium. Rather, the task is to persuade these sectors that the use of titanium in component applications currently using other materials, most often steels, can be viable.

These are very cost-driven sectors, and therefore an important criterion in considering the viability of such a material replacement is always going to be one of economics.

There has, therefore, been much debate in the published literature as to the powder price level, which might be required to open up mass-market opportunities for titanium PM.

Certain sources have suggested a price level of \$10/kg or below, while the author of a paper in Brisbane believes that a target price of \$25/kg would be sufficient to stimulate opportunities [19].

Even if the more optimistic of these projections were true, there would need to be a compelling performance advantage in switching material choice to titanium for the application to become viable. Some of the target applications, discussed earlier, may come in this category.

The possibility of titanium's supplanting steels across the broad range of established PM component applications seems highly unlikely, on the grounds of cost competitiveness even at the target powder price levels.

It is, therefore, difficult to make a prediction of potential market penetration in these sectors. A level of the order of around 10,000 tonnes p.a. might be considered realistic, which compares with an established ferrous PM market approaching 1 million tonnes.

30.8 Concluding discussion

It is incumbent, on anyone attempting an analysis of future market potential, to offer some quantification of market size in the foreseeable future. This, of course, involves a considerable degree of crystal ball gazing.

The following provides some views on the possible penetration of various PM titanium products, categorized on the basis of the material type and forming technology involved, and some first-order estimates of consequent market sizes in a 5-year time frame.

30.8.1 Blended elemental process routes

The potential Dynamet business with Boeing is an example of the use of BE material. However, the statements that either company has placed into the public domain do not provide a good insight into market penetration rates.

Otherwise, the main potential for BE material would seem to lie in the mass markets of automotive and general engineering, where "the price must be right." HDH Ti or TiH_2 may offer sufficient cost effectiveness to open up these markets; otherwise, one or more of the emerging low-cost powder routes would need to deliver on its/their promise.

A "finger in the air" estimate is that there might be an eventual potential market of up to 10,000 tonnes p.a. (22 million pounds) here.

A very rough estimate might be that the market penetration in the 5-year time frame might be around 6800 tonnes p.a. (15 million pounds).

30.8.2 Additive manufacturing

The likely application sectors for AM titanium products would appear to be aerospace (mainly airframe but some aero-engine potential), health care, motorsport, and consumer goods (including jewelry).

In tonnage terms, aerospace might offer the largest opportunity, but qualification issues could inhibit early adoption. However, there is evidence that there is growing pressure on the major OEMs to take as much cost as is feasible out of next-generation aircraft and that this is likely to accelerate the progression of the current AM titanium developments through the TRLs.

Statements have been made in the technical press by senior personnel from the airframe sector that 50% of all titanium parts could be made by AM within 10 years – this would amount to at least 10,000 tonnes (22 million pounds) p.a. of products.

It is believed that the target component applications would have high buy-to-fly ratios in conventional manufacture, so AM could offer cost effectiveness without the availability of lower cost powders.

The other three application sectors cited would be capable of taking products to the market more quickly (i.e., quicker than the 5-year time scale), but tonnages are likely to be more modest than the aerospace sector.

Overall, a market penetration within 5 years of more than 10,000 tonnes (22 million pounds) may be considered to be feasible.

30.8.3 Metal injection molding

MIM Ti already has some established applications in the health care, aerospace, and consumer goods sectors, and it is anticipated that these sectors will be penetrated further in the next 5 years.

There is significant development activity in both the aerospace and health care sectors.

However, MIM parts are characteristically small and light, so the tonnages involved are likely to be quite modest.

Given the estimate of 10–20 tonnes p.a. for current MIM Ti markets, a projection to around 45 tonnes p.a. (100,000 pounds) in 5 years' time seems to be feasible.

On account of their low weight, the material cost contribution to total manufacturing costs is often quite low for MIM parts in general. If this statement also holds true for MIM titanium parts, it is not clear that the availability of a lower-cost spherical powder would have a major impact on market penetration. However, the consensus from the MIM titanium community seems to be that lower cost titanium powder would make a significant difference.

30.8.4 Prealloyed material/processed by HIP (PA/HIP)

There is significant interest in this route for aero-engine parts (with Rolls-Royce in particular), although again qualification issues could delay commercial exploitation.

Rolls-Royce has certainly taken an interest in some of the emerging, potentially lower cost powder sources, but they have applications that could tolerate the costs of PREP powders. The same is believed to hold true for the US-based aero-engine OEMs.

There are some potential airframe applications for this route. Also, there may be potential in other sectors, such as oil and gas and chemical processing plants.

A rough estimate suggests an eventual market of around 2000 tonnes (44 million pounds) p.a. rising to possibly 4000 tonnes (88 million pounds) for this route. However, penetration of these markets within 5 years is likely to be very modest.

30.8.5 HDH materials

Much of the current development activity on press/sinter BE products has centered on the use of HDH titanium powder as the base material, on green strength and cost effectiveness grounds.

Ametek/Reading Alloys are also targeting their plasma spheroidized HDH powder grades at AM, MIM, and HIP applications, although the cost of spheroidization is high at \$50–\$100 per pound (\$110–\$220 per kg).

Given the current size of markets for HDH powders, a projection of around 90 tonnes p.a. (200,000 pounds) of such powders going into PM applications in 5 years' time might be considered realistic.

30.8.6 Spraying applications

The use of titanium-based powders in spray forming applications is likely to be quite modest, with a projected penetration of around 9 tonnes p.a. (20,000 pounds) in the 5-year time frame being suggested.

These projected market penetrations are summarized in Table 30.4.

Table 30.4 Projected market penetrations of various PM titanium product types within a 5-year time frame

Product/technology type	Projected market size in 5 years	
	Tonnes p.a.	Pounds p.a.
Blended elemental (BE) process routes	6,800	15 million
Additive manufacturing (AM)	10,000	22 million
Metal injection molding	45	100,000
Prealloyed/hot isostatically pressed (PA/HIP)	2,000	44 million
Hydride–dehydrided (HDH) materials	90	200,000
Spray formed products	9	20,000

References

[1] M. Qian (Ed.), Powder metallurgy of titanium, powder processing, consolidation and metallurgy of titanium; Proceedings of International Conference, December 4–7, 2011, Brisbane, Australia; Trans Tech Publications Ltd., Switzerland, 2012.
[2] D. Rugg, Presentation at PowdermatriX One Day Meeting, Birmingham, UK, 2004.
[3] W. Voice, M. Hardy, D. Rugg, The Development of Powder Consolidated Components for the Gas Turbine Engine, PM World Congress, Vienna, 2004.
[4] Powder prospects set to take off as production methods change, Metal Powder Rep. 60 (4) (2005) 8–12 (article based on Reference 3).
[5] M. Ward-Close, Presentation at a Special Titanium PM Workshop at EuroPM 2005, Prague.
[6] D. Whittaker, et al., Paper in the Special Titanium PM Workshop at EuroPM 2005, Prague.
[7] Powder Inj. Mould. Int. 7 (2) (2013) 28–29.
[8] News item in Powder Metall. Rev. 2 (4) (2013) 14.
[9] S.M. Abkowitz, D. Abkowitz, H. Fisher, D.H. Main, Affordable PM Titanium – Microstructure, Properties and Products, Paper at PowderMet 2011, MPIF.
[10] Future Journeys, Airbus Global Market Forecast for 2013–2032, available on-line at http://www.airbus.com/company/market/forecast/.
[11] F. Watari, A. Yokoyama, F. Saso, M. Uo, H. Matsuno, T. Kawasaki, Imaging of gradient structure of titanium/apatite functionally graded dental implant, J. Jpn. Inst. Metals 62 (1998) 1095–1101.
[12] Y. Thomas, E. Baril, F. Ilinca, J.F. Hetu, Development of Titanium Dental Implant by MIM: Experiments and Simulation, Advances in Powder Metallurgy and Particulate Materials – 2009, Metal Powder Industries Federation, Princeton, NJ, 2009, pp. 4.81–4.93.
[13] ITB Precisietechniek: MIM titanium drives growth at Dutch MIM producer, Powder Inj. Mould. Int. 5 (4) (2011) 33–40.
[14] R. German, Opportunities in the dental sector for metal and ceramic injection moulding, Powder Inj. Mould. Int. 8 (1) (2014) 31–35.
[15] Element 22 GmbH: pushing the boundaries for titanium MIM in the medical and aerospace sectors, Powder Inj. Mould. Int. 7 (4) (2013) 31–36.
[16] J. Mentz, L. Krone, M. Bram, H.P. Buchkremer, D. Stover, Paper at EuroPM 2005, Prague, October 2005.
[17] J. Barnes, CSIRO, Presentation at Powder Metallurgy of Titanium, Powder Processing, Consolidation and Metallurgy of Titanium; International Conference, December 4–7, 2011, Brisbane, Australia.
[18] B. Kieback, T. Schubert, T. Weissgarber, V. Kruzhanov, V. Arnhold, Presentation in PM Titanium Workshop at Euro PM 2005 Conference, Prague, October, 2005.
[19] M. Qian, Y.F. Yang, M. Yan, S.D. Luo, reference 1, 24–29.
[20] G.I. Abakumov, V.A. Duz, V.S. Moxon, Titanium alloy manufactured by low cost solid state PM processes for military, aerospace and other critical applications, ITA Conference, 2010.
[21] F.H. (Sams) Froes, O. Ivasishin, V.S. Moxon, D.G. Savvakin, K.A. Bondereva, A.M. Demidik, Cost effective synthesis of Ti-6Al-4V alloy components produced via the PM approach, in: F.H. (Sams) Froes et al. (Ed.), Proceedings of the TMS Symposium on High Performance Metallic Materials for Cost Sensitive Applications, held in Seattle, WA, TMS, Warrendale, PA.

A perspective on the future of titanium powder metallurgy

31

Francis H. (Sam) Froes, Ma Qian***
*Consultant to the Titanium Industry, Tacoma, WA, USA
**RMIT University, School of Aerospace, Mechanical and Manufacturing Engineering, Centre for Additive Manufacturing, Melbourne, Victoria, Australia

31.1 Introduction

The chapters in this book have, it is hoped, covered all aspects of the science/technology of titanium PM. In the first chapter, a historical perspective was presented followed by successive chapters covering many technological and scientific aspects of this subject. In the present writeup, projections are made as to the future of titanium PM (see also Ref. [1]), a parallel chapter to Chapter 30 that emphasizes the future market for titanium PM [2]. The focus of this chapter will be on the projections of the most important or promising Ti PM technologies including prealloyed plus hot isostatic pressing (PA/HIP), blended elemental (BE), additive manufacturing (AM) or 3D metal printing, metal injection molding (MIM), and cold spray forming. These technologies have proved to be capable of producing components with both acceptable mechanical property performance and cost with respect to selected applications.

31.2 Prealloyed plus HIP

Early work on this technology used prealloyed (PA) spherical (for good flow and packing characteristics) powder produced by the plasma rotating electrode process (PREP) and involved the ceramic mold process to produce cost-effective, high-integrity near-net shapes (Chapter 1). Because of fears that second-phase particles could get into the compacts from the mold, and concerns regarding the shape-making capabilities, this technology was abandoned and was replaced by a metal can approach. This latter technology has been successfully commercialized and has been used to fabricate large complex engine and rocket parts (Chapter 18), Figure 31.1. Holding back widespread implementation of this fabrication approach has been the high cost of the PA spherical powder (the lowest-cost powder currently is at the $40-per-pound level). With the possibility of lower-cost powder, either from a melt process (Chapter 2) or via one of the developing extraction techniques discussed in other chapters of this book, this technology may take off in the foreseeable future and reach a reasonable percentage of the large complex aerospace shapes. However, it should be emphasized that HIP of complex near-net-shape parts for aerospace applications is a synergetic technology that requires and involves exceptional engineering in design of the HIP tooling and of

Titanium Powder Metallurgy. http://dx.doi.org/10.1016/B978-0-12-800054-0.00031-9

Figure 31.1 Near-net-shape Ti-6Al-4V engine casing fabricated using the prealloyed metal method. (Courtesy Victor Samarov, Synertechpm.)

the entire process including elaborate powder handling technology, well-established out-gassing processes, continuous control of the powder bulk properties, and surface and shape of powders. Some other requirements can be found in Chapter 18.

31.3 Blended elemental

This method of fabricating near-net-shape PM parts has shown major advances since 2012. The Dynamet Technology approach (Chapter 17) has been qualified to fabricate parts for Boeing Commercial, which may lead to a significant number of flying titanium PM components. Another indication that Boeing are serious about titanium PM is the memorandum of understanding that they have recently signed with the Council for Scientific and industrial Research (CSIR), South Africa [3]. Dynamet Technology have also developed metal matrix composites using reinforcement with TiC or TiB_2 (CermeTi), which is being used in the aluminum die casting industry. A further indication of the commercial viability of the Dynamet Technology is the acquisition of the company by RTI International [4]. This acquisition, RTI's second new advanced technology investment in 2014, adds innovative titanium powder research and development, as well as a second distinct powder-based manufacturing technology to RTI Directed Manufacturing, which specializes in 3D printing. The addition of Dynamet Technology enhances RTI's strategic positioning to capitalize on increasing customer demand for lower-cost materials and innovative near-net-shape preforms and components.

As pointed out by Abkowitz (see Chapter 17), the use of inexpensive hydride–dehydride (HDH) titanium powder produced from Kroll sponge is the key to the commercial success of Dynamet's PM process. It allows Dynamet to produce a wide range of affordable PM near-net-shape preformed components. Also central to the success of Dynamet's PM process is that Dynamet has developed critical specifications for its titanium and master alloy powders that control for morphology, particle size, particle distribution, and chemistry.

A variety of inexpensive HDH Ti powder products produced from Kroll sponge exists on the market. They often contain higher oxygen (>0.25 wt%) and residual chlorides (>200 ppm chlorine). Recent developments have shown that both oxygen and chlorine can be effectively scavenged from the HDH Ti powder during heating and isothermal sintering via a small addition (e.g., 0.3 wt%) of rare earth–containing compounds [5–8]. The exceptional potency of rare earth elements scavenging oxygen from titanium has been demonstrated in a recent study, which showed that the introduction of 0.1 wt% yttrium to a titanium alloy (Ti-6Al-2.75Sn-4Zr-0.4Mo-0.45Si) containing just 0.07 wt% oxygen resulted in precipitation of a significant number of fine Y_2O_3 dispersoids in the additively manufactured alloy [9]. Other means of scavenging of oxygen from titanium also exist. This has been discussed in some detail in Chapter 15. As a result of the effective scavenging of oxygen and chlorine during sintering, the tensile elongation increased by up to 90% [5,6]. This offers a potential alternative avenue for the manufacture of competitive PM Ti components from inexpensive HDH Ti powder produced from Kroll sponge.

The development of the hydrogenated titanium sponge approach by ADMA Products in conjunction with the Academy of Science, Kiev (Chapter 8), has led to high-integrity parts that have been fabricated into potential parts for armor use and have been supplied to Boeing Long Beach, where they have been successfully evaluated (see Chapter 25). TiH_2 PM, as an alternative to Ti PM, is expected to receive increasing attention in the future.

So just as with the PA/HIP technique, there should be significant growth of the blended elemental approach, both monolithic parts and metal matrix composite components, for production of titanium PM Parts.

31.4 Additive manufacturing

Additive manufacturing (AM) has reached a level of maturity that allows the production of parts for end users [10] and as a method to repair damaged parts (see Chapter 24). Apart from acceptable property performance, it has been reported that geometrical accuracy can now reach values of ±10 μm and surface roughness can reach Ra values of 10 μm [10]. This means minimal or even no post-AM processing for most applications. As an enabling technology, AM should show rapid growth for the manufacturing of various types of complex titanium components. Predicted below are just four possible pathways to benefit from the outstanding advantage of the technology for the titanium business. Many new opportunities are expected to emerge as times goes on.

Additive manufacturing makes imagination the major limitation. Innovative design will continue to drive the growth of AM of titanium components with novel functionality or properties. The titanium weapons suppressors designed and manufactured by Titanium Industry Development Association (TiDA) New Zealand and its partner Oceania Defence, shown in Figure 31.2 [11], is one such recent successful AM story of titanium. These uniquely designed and additively manufactured titanium weapons suppressors offer excellent performances while being 50% lighter than conventional steel weapons suppressers. Design played a key role in the success. Several such products have gone into batch production. However, there are certain limitations, especially geometric features such as holes (minimum around 100 μm at present) and thin walls or fins, which have to be designed carefully and processed according to certain rules [10].

Development of novel microstructures for outstanding properties through improved fundamental understanding of the AM process of titanium alloys. For instance, Ti-6Al-4V fabricated by selective laser melting (SLM) often consists of columnar prior-β grains filled with acicular α' martensite, and usually displays high yield strength over 1100 MPa but limited and inconsistent tensile elongation up to 8%. As a result, post-SLM heat treatment is often applied for ductility improvement. New developments have shown that through novel design based on both phase transformation and processing optimization α' martensite decomposition can be realized *in situ* during SLM to produce an ultrafine α–β lamellar structure comprising ultrafine (~200–300-nm) α laths and retained β (Figure 31.3). As a result, outstanding tensile mechanical properties (yield strength > 1100 MPa and elongation 11%) are achieved in the SLM as-fabricated Ti-6Al-4V [12]. This shows the capabilities of developing desired microstructures and properties with AM without post-AM heat treatment.

Figure 31.2 SLM at work: titanium weapons suppressors designed and additively manufactured by TiDA and its partner Oceania Defence [11].

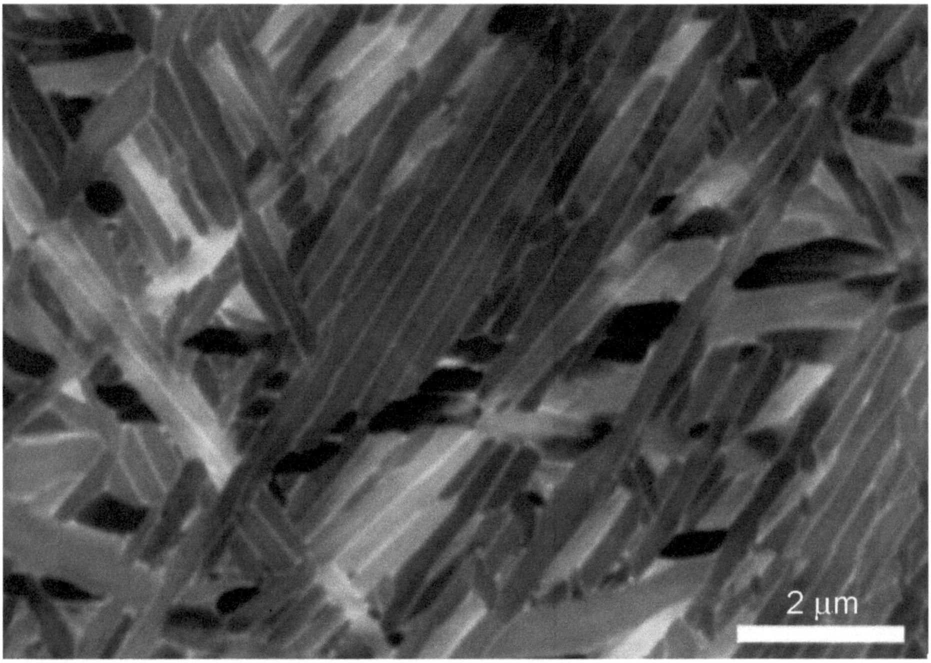

Figure 31.3 Ultrafine α–β lamellar structure comprising 200–300-nm α laths and retained β in SLM as-fabricated Ti-6Al-4V realized via *in situ* α′ martensite decomposition during SLM through novel design based on both phase transformation and processing optimization [12].

A number of advances in AM are possible, some of which are already being pursued as described below.

- Dual- or multiple-chemistry parts by AM. This could include the interior of titanium parts with high levels of mechanical properties and titanium surface regions of high corrosion behavior/oxidation resistance/ease of machining. It could also feature a mixture of different metals such as titanium bonded to steel (with an intermediate layer of for example vanadium or niobium) to reduce the overall weight of a (steel) component, where the steel is the outer layer. This opens a big avenue for innovative design and manufacturing. Some dual-chemistry parts with titanium have found niche applications.
- Metrology needs for metal AM powders. There are several challenges that are currently preventing more widespread adoption of AM technologies. Among these challenges are metrology issues associated with the measurement and characterization of the metal powders used for additive manufacturing systems. The recently released ASTM F3049 *Standard Guide for Characterizing Properties of Metal Powders Used for Additive Manufacturing Processes* marks a milestone in this regard. More developments are expected that will eventually promote the manufacture of cost-effective or lower-cost metal powders for AM.
- The use of two low-cost powders (ADMA Products hydrogenated powder and Metalysis FFC/Cambridge powder) were shown to be readily amenable to AM processing (the Metalysis angular powder albeit after a spheroidization treatment) [13,14], and demonstration auto parts were fabricated from both types of powder (Chapter 2).

The AM technique should see extensive growth both as a method to produce cost-effective high integrity near-net-shapes and as a method of repairing damaged parts. Further growth should occur with use of lower-cost powders and by the development of dual-material concepts.

31.5 Metal injection molding

The MIM technique has proved to be very successful for the fabrication of small parts for the consumer industry (e.g., watch cases), medical implants, and aerospace applications. However the high cost of fine spherical powder (<40 μm) has stunted growth. This problem should be resolved with the introduction of finer spherical powder and suitable nonmelt powders from innovative extraction processes. Effort has been made to develop MIM-TiH_2 as a cost-effective option for the fabrication of MIM-Ti parts. However, noticeable challenges still exist, including the development of effective feedstock compounding methods for good homogeneity, high shrinkage, and dimensional reproducibility (after debinding and dehydriding) and for reduced tool wear too (when compared to the MIM of titanium metal powder). Hence growth of titanium MIM should come from use of lower-cost spherical powder and more widespread acceptance of this technology in industries such as aerospace and auto. MIM has been used for making titanium medical components, even permanent implants, and the two existing ASTM standards target exactly this market. This market is expected to continue to grow.

The scientific aspects of MIM-Ti should focus on the influence of porosity and interstitial content on mechanical properties, as these factors are closely relevant to practical application. In particular, the pickup of carbon appears to be a critical concern for the MIM of some titanium alloys. Research has shown that the solubility limit of carbon in titanium alloys containing a high level (e.g., ≥10 wt%) of molybdenum (Mo), niobium (Nb), and vanadium (V) can be less than 100 ppm by weight. Consequently, MIM of such titanium alloys can lead to a noticeable presence of titanium carbides in the as-sintered microstructure, often distributed along grain boundaries, because of the unavoidable pickup of carbon (300–500 ppm by weight) from the polymeric binder used. The formation of such grain boundary carbides can embrittle the MIM-processed titanium alloys [15,16]. As regards the porosity, experimental data indicate that when the pore size is less than 10 μm, the effect of porosity on the fatigue properties of as-sintered dense titanium materials (>98% of the theoretical density) is negligible [17]. Control of the maximum pore size is more important in this regard.

Growth of titanium MIM should come from use of lower-cost spherical powder and more widespread acceptance of this technology in industries such as aerospace, medical, and auto.

31.6 Cold spray forming

This technique has been successfully applied to the production of some complex small shapes of commercially pure titanium (CP Ti). Dense deposits with acceptable property performance after heat treatment can be produced for select applications, although not as easily as the more commonly used metals (e.g., aluminum). In general, post–cold spray heat treatment is required to improve ductility and/or reduce porosity. Chapter 22 has provided a comprehensive overview of the current status and challenges of the technology. Overall, the cold spray field is still relatively new with few established protocols and/or standards. The noticeable advantages held by the technology for processing Ti and Ti alloys hold promise that development efforts for cold spray forming of titanium will continue.

31.7 Concluding remarks

The future appears bright for the various facets of titanium powder metallurgy technology. This includes the prealloyed/HIP, BE, MIM methods, and especially the AM technique. A latent barrier is that designers typically consider that most powder metallurgy components will be high risk based on historical information as a result of residual porosity and/or contamination. With the dramatically growing market for additive manufacturing and consistent property performance of an increasing number of AM titanium components, the use of metal powders for the manufacture of structural components is expected to become more acceptable than before. This will in turn assist in the further development of titanium powder metallurgy. The bottom line is that a low-cost powder with good flow/packing is needed. On the other hand, research on titanium powder metallurgy has seen increasing momentum over the last two decades, with a number of promising developments or initiatives in the Research Base at present, including those discussed in Chapters 4, 6, 8, 10, 22, and 24. Some breakthroughs may well occur in the next decade or so, leading to increased applications of powder metallurgy titanium components.

Acknowledgments

The authors would like to acknowledge useful discussions and supply of information from the following: Stanley Abkowitz, Susan Abkowitz, Dave Bourell, Bhaskar Dutta, Vlad Duz, Dave Bourell, Rand German, Joe Gromowski, Ian Mellor, Vladimir Moxson, Victor Samarov, Jim Withers, and Fred Yolton.

References

[1] F.H. Froes, Titanium powder metallurgy: developments and opportunities in a sector poised for growth, Powder Metall. Rev. 2 (4) (2013) 29–43.

[2] D. Whittaker, F.H. Froes, Future prospects for titanium powder metallurgy markets, in: M. Qian, F.H. Froes (Eds.), Titanium Powder Metallurgy, Elsevier, Waltham, MA, USA, 2015.

[3] Anon., CSIR Signs Titanium Powder Research MOU with Boeing. <http://www.engineeringnews.co.za/article/csir-signs-titanium-powder-research-mou-with-boeing-2013-06-11>, 2013 (accessed 14.08.14).

[4] Anon., RTI International Metals Increases Advanced Powder Materials Capabilities with Acquisition of Boston-Based Dynamet Technology. <http://investor.rtiintl.com/press-release/rti-international-metals-increases-advanced-powder-materials-capabilities-acquisition->, 2014 (accessed 14.08.14).

[5] M. Yan, Y. Liu, Y.B. Liu, C. Kong, G.B. Schaffer, M. Qian, Simultaneous gettering of oxygen and chlorine and homogenization of the β phase by rare earth hydride additions to a powder metallurgy Ti-2.25 Mo-1.5 Fe alloy, Scripta Mater. 67 (5) (2012) 491–494.

[6] Y.F. Yang, S.D. Luo, G.B. Schaffer, M. Qian, Impurity scavenging, microstructural refinement and mechanical properties of powder metallurgy titanium and titanium alloys by a small addition of cerium silicide, Mater. Sci. Eng. A 573 (2013) 166–174.

[7] M. Yan, Y. Liu, G.B. Schaffer, M. Qian, *In situ* synchrotron radiation to understand the pathways for the scavenging of oxygen in commercially pure Ti and Ti-6Al-4V by yttrium hydride, Scripta Mater. 68 (1) (2013) 63–66.

[8] R.J. Low, M. Qian, G.B. Schaffer, Sintering of titanium with yttrium oxide additions for the scavenging of chlorine impurities, Metall. Mater. Trans. A 43 (13) (2012) 5271–5278.

[9] H.P. Tang, S.L. Lu, W.P. Jia, G.Y. Yang, M. Qian, Selective electron beam melting of titanium and titanium aluminide alloys, Int. J. Powder Metall. 50 (2014) 57–64.

[10] C. Aumund-Kopp, J. Isaza, F. Petzoldt, Fast and functional parts: new ways to use selective laser melting, PM World Congress 2014, Orlando, FL, May 18–22, 2014, Metal Powder Industries Federation.

[11] M. Qian, PM titanium 2013: impurities in PM Ti and Ti alloys, porous titanium and the additive manufacturing of titanium, Powder Metall. Rev. 3 (2) (2014) 53–60.

[12] W. Xu, M. Brandt, S. Sun, J. Elambasseril, Q. Liu, K. Latham, K. Xia, M. Qian, Additive manufacturing of strong and ductile Ti-6Al-4V by selective laser melting via *in situ* martensite decomposition. Acta Mater. 85 (2015) 74–84.

[13] J.C. Withers, V. Shapovalov, R. Storm, R.O. Loutfy, There is low cost titanium componentry today, Key Eng. Mater. 551 (2013) 11–15.

[14] Anon., Metalysis' titanium powder used to 3D print automotive parts. <http://www.materialstoday.com/metals-alloys/products/metalysis-titanium-powder/>, 2013 (accessed 14.08.14).

[15] D. Zhao, K. Chang, T. Ebel, M. Qian, R. Willumeit, M. Yan, F. Pyczak, Titanium carbide precipitation in Ti-22Nb alloy fabricated by metal injection moulding, Powder Metall. 57 (1) (2014) 2–4.

[16] M. Yan, M. Qian, C. Kong, M.S. Dargusch, Impacts of trace carbon on the microstructure of as-sintered biomedical Ti–15Mo alloy and reassessment of the maximum carbon limit, Acta Biomater. 10 (2) (2014) 1014–1023.

[17] T. Furuta, T. Saito, U.S. Patent 5,409,518. Washington, DC, U.S. Patent and Trademark Office, 1995.

Index

D

Kolibri 360 GmbH
Ferdinand-Jühlke-Straße 7
99095 Erfurt, Deutschland
produktsicherheit@kolibri360.de